SIXTH EDITION

DIGITAL FUNDAMENTALS

THOMAS L. FLOYD

Prentice-Hall International, Inc.

Editor: Linda Ludewig
Developmental Editor: Carol Hinklin Robison
Production Editor: Mary Harlan
Design Coordinator: Julia Zonneveld Van Hook
Text Designer: Anne Flanagan
Cover photo: Uniphoto, Inc.
Cover Designer: Brian Deep
Production Manager: Deidra M. Schwartz
Marketing Manager: Debbie Yarnell
Illustrations: Rolin Graphics

This book was set in Times Roman by The Clarinda Company and was printed and bound by Von Hoffmann Press, Inc. The cover was printed by Von Hoffmann Press, Inc.

Printed in the United States of America

10   9   8   7   6   5   4   3   2   1

ISBN: 0-13-573478-9

Prentice-Hall International (UK) Limited, *London*
Prentice-Hall of Australia Pty. Limited, *Sydney*
Prentice-Hall of Canada, Inc., *Toronto*
Prentice-Hall Hispanoamericana, S. A., *Mexico*
Prentice-Hall of India Private Limited, *New Delhi*
Prentice-Hall of Japan, Inc., *Tokyo*
Simon & Schuster Asia Pte. Ltd., *Singapore*
Editora Prentice-Hall do Brasil, Ltda., *Rio de Janeiro*
Prentice-Hall, Inc., *Upper Saddle River, New Jersey*

To Debbie and Cyndi

PREFACE

Digital Fundamentals, Sixth Edition, offers comprehensive coverage in an easy-to-read style. As in previous editions, this edition provides a well balanced coverage of basic concepts, up-to-date technology, practical applications, and troubleshooting. Some topics have been strengthened and improved, and two new chapters on programmable logic devices (PLDs) have been added. This book has been thoroughly reviewed, and every effort has been made to ensure that the coverage is accurate and up to date.

You will probably find more topics than can be covered in one term because of time limitations or program emphasis. This breadth of topics provides flexibility in designing your course to meet the specific goals of your program. For example, some of the mathematical, design-oriented, troubleshooting, or system application topics may not be appropriate for certain programs. Some programs do not cover PLDs or do not provide an introduction to microprocessors in the digital fundamentals course. These topics can be easily omitted or lightly touched on without affecting other coverage.

Features

- Two new chapters on PLDs (Chapter 7 and Chapter 11).
- Digital System Application sections with "Digital Workbench" assignments that include analysis, design, and troubleshooting.
- A revised chapter on memories, including a new section on flash memories.
- A comprehensive glossary at the end of the book.
- New coverage of 5-variable Karnaugh maps.
- A full-color format.
- A related exercise in each worked example.
- An overview and list of objectives at the opening of each chapter.
- An introduction and objectives at the beginning of each section within a chapter.
- Review questions at the end of each chapter section.
- A summary at the end of each chapter.
- A multiple-choice self-test at the end of each chapter.
- Extensive problem sets

The improved ancillary package for this edition includes

- Two lab manuals: *Digital Experiments Emphasizing Systems and Design,* Fourth Edition, by David Buchla, and *Digital Experiments Emphasizing Troubleshooting,* Fourth Edition, by Jerry V. Cox.
- Instructor's Resource Manual.
- Test Item File (hard copy).
- DOS PH Custom Test (Test Item File on disk).
- Transparency masters and 4-color transparencies.
- Electronics Workbench Data Disk.
- Bergwall Video.

Illustration of Chapter Features

Chapter Opener   Each chapter begins with a two-page opener, as shown in Figure P–1. The left page includes a listing of the sections within the chapter and a list of chapter objectives. The right page has a chapter overview, a list of specific devices introduced in the chapter, and a preview of the Digital System Application.

Section Opener   Each section within a chapter begins with a brief introduction that provides a general overview of the material to be covered and a list of section objectives. This is illustrated in Figure P–2.

Section Review   Each section ends with a review consisting of questions or exercises that focus on the main concepts presented in the section. Answers to these section reviews are at the end of the chapter. This is also illustrated in Figure P–2.

Worked Examples and Related Exercises   Frequent examples help to demonstrate and clarify basic concepts or illustrate specific procedures. Each example concludes with a related exercise that reinforces or expands on the example. Some related exercises require a repetition of the example using different parameters or conditions. Others focus on a more limited part of the example or encourage further thought. Answers to all the related

Chapter overview
and a list of devices
introduced in this chapter.

List of chapter objectives.

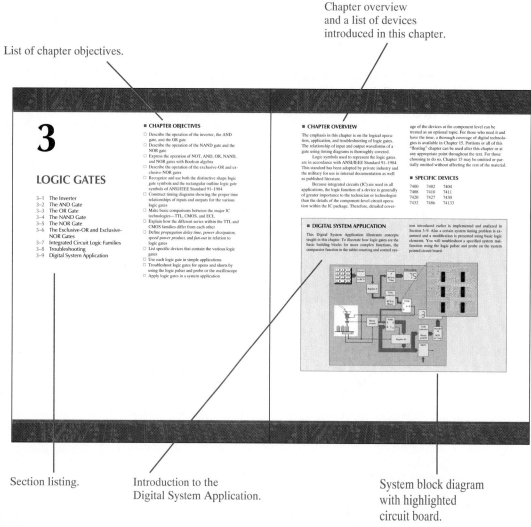

Section listing.

Introduction to the
Digital System Application.

System block diagram
with highlighted
circuit board.

FIGURE P–1
Chapter opener.

FIGURE P–2
Section opener and section review.

Review exercises end each section.

Introductory paragraph and a list of performance-based section objectives begin each section.

exercises are found at the end of the book. A typical worked example and related exercise are shown in Figure P–3.

Troubleshooting Section   Many chapters include special troubleshooting sections that emphasize troubleshooting techniques and the use of test instruments as applied to situations related to chapter topics. These sections are optional and can be omitted without affecting the rest of the material.

FIGURE P–3
An example and related exercise.

Each example is contained within a ruled box.

Each example contains an exercise related to the example.

Digital System Application   The last section of each chapter (except Chapters 14 and 15) presents a practical application of the concepts and devices covered in the chapter. Each application is based on a "real world" system. Many include analysis, design, and troubleshooting elements that are implemented in a series of "Workbench" activities. Some system applications are limited to a single chapter and others extend over two or more chapters. The Digital System Applications and their associated chapters are as follows:

- Tablet counting and control system: Chapters 1, 2, 3, and 4
- Digital control system for a lumber mill: Chapter 5
- Traffic light control system: Chapters 6, 7, 8, 9, and 11
- Security entry system: Chapters 10 and 12
- Satellite antenna positioning system: Chapter 13

Many of the system applications involve realistic representations of printed circuit boards that provide experience in relating schematics to actual circuits and identifying physical devices. Solutions to the Workbench activities are in the Instructor's Resource Manual. Although the Digital System Application is a very effective feature, it is optional and can be omitted or given limited coverage without affecting other material. A portion of a typical Digital System Application section is shown in Figure P–4.

Opener with a list of objectives.

"Digital Workbench" provides a series of analysis, design, or troubleshooting activities. Many applications include a printed circuit board and instrumentation.

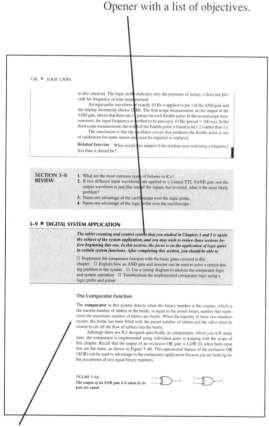

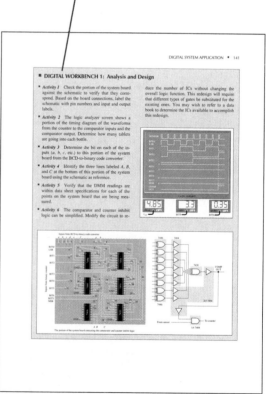

An overall introduction to the system application is provided before the Workbench activities.

FIGURE P–4

Representative pages from a typical Digital System Application section.

Chapter End Matter   At the end of each chapter are a summary, a multiple-choice self test, and a sectionalized problem set (except Chapter 14). The problem sets include basic problems, troubleshooting problems, system application problems, and special design problems when applicable.

Content and Organization

This textbook contains fifteen chapters, beginning with basic digital concepts in Chapter 1 and progressing through number systems, logic gates, Boolean algebra, combinational logic including PLD implementation, sequential logic including PLD implementation, memories, and interfacing, and ending with a basic introduction to microprocessors and microcomputers in Chapter 14.

Chapter 15 covers digital circuit technologies and discusses the operating characteristics and parameters of the major IC families. This chapter can be used in whole or in part at various appropriate points as a "floating" chapter, at the discretion of the instructor. Its placement as the last chapter in the book is intended to facilitate this flexible usage, and a tab edge design is provided for quick and easy reference.

Chapters 7 and 11 provide an introduction to programmable logic devices (PLDs) and can be treated as optional if desired. Either or both of these chapters can be omitted without affecting other topics. Chapter 7 follows the coverage of combinational logic and provides an introduction to PLDs and PLD programming for combinational logic functions. Chapter 11 follows the coverage of sequential logic and continues the PLD coverage from Chapter 7 with an introduction to the implementation of sequential logic using PLDs.

Chapter 14 provides a brief introduction to microprocessor and microcomputer concepts and the development of the major microprocessor families. The Intel 8088 microprocessor is used as a basic model or "launching pad" for teaching basic concepts because of its relative simplicity. The basic elements are common to more recent devices, so the concepts learned can be applied to other, more advanced, microprocessors.

As in previous editions, the ANSI/IEEE std. 91-1984 logic symbols with dependency notation are introduced gradually and conservatively at appropriate points, while the use of the more traditional symbols is retained throughout. Although it is important to become familiar with this standard, its symbols and notation represent, in many cases, a significant departure from many traditional symbols, and thus a gradual and limited introduction is quite appropriate. A full treatment of the ANSI/IEEE standard is provided in the Instructors Resource Manual. Omission of this coverage will not affect the rest of the material.

At the end of the book are the answers to odd-numbered end-of-chapter problems, answers to the related exercises for examples, several representative data sheets, a table of code conversions, a table of powers-of-two, a short coverage of error detection and correction codes, a comprehensive glossary, and the index.

Suggestions for Use

If time limitations or course emphasis restricts the topics that can be covered, as is generally the case, several possibilities for selective coverage exist. Suggestions for light treatment or omission do not imply that a given topic is less important than others, but that, in the context of a specific program, the topic does not require the emphasis that the more fundamental topics do. Since course emphasis, level, and available time vary from one program to another, the omission or lighter treatment of selected topics must be made on an individual basis and, therefore, the following suggestions are intended only as a general guide.

1. Chapters that may be considered for selective coverage:
   Chapter 1   Introductory Digital Concepts
   Chapter 2   Number Systems, Operations, and Codes
   Chapter 4   Boolean Algebra and Logic Simplification
   Chapter 15   Integrated Circuit Technologies

2. Chapters that may be considered for omission:

Chapter 7 Introduction to Programmable Logic Devices
Chapter 11 Sequential Logic Applications of PLDs
Chapter 13 Interfacing
Chapter 14 Introduction to Microprocessors and Microcomputers

3. Troubleshooting sections and/or Digital System Application sections can be omitted. Other specific topics on a section-by-section basis that may be considered for omission are:

2–7 Arithmetic Operations with Signed Numbers
2–9 Octal Numbers
4–10 Karnaugh Map POS Minimization
4–11 Five-Variable Karnaugh Maps
6–3 Ripple Carry versus Look-Ahead Carry Adders
6–10 Parity Generators/Checkers
8–3 Master-Slave Flip-Flops
9–4 Design of Synchronous Counters
9–9 Logic Symbols with Dependency Notation
10–7 Shift Register Counters
10–10 Logic Symbols with Dependency Notation
12–7 Special Types of Memories

Depending on your program, there may be additional topics that could be skimmed over or omitted.

The order in which certain topics appear in the text can be altered at the instructor's discretion. For example, portions of Chapter 2 (Number Systems and Codes) can be covered at a later point in the chapter sequence. As another example, Chapter 7 (Introduction to Programmable Logic Devices) can be delayed until after Chapter 10 and covered in sequence with Chapter 11 (Sequential Logic Applications of PLDs).

Digital System Applications These sections are very useful for motivation and as an introduction to real-world applications of basic concepts and devices. Possible uses are

1. As an integral part of the chapter to illustrate how the concepts and devices can be applied in a practical situation. The Workbench activities can be assigned for homework or as a class miniproject.
2. As extra credit assignments.
3. As in-class activities to promote discussion and interaction and to help answer the "need-to-know" question that many students have.

For the Student

All of the material in this preface is intended to help both you and your instructor make the most effective use of this textbook as a teaching and a learning tool.

Acquiring knowledge and skills in any discipline is hard work, and perhaps even more so in the field of electronics. You must use this book as more than just a reference. You must really dig in by reading, thinking, and doing. Don't expect every concept or procedure to become immediately clear. Most of the topics are not overly difficult but some may take several readings, working many problems, and help from your instructor before you really understand them.

In order to master the material in each section you should first read the material, then go through each example step-by-step, work the related exercises, and complete the review exercises. After completing a chapter, you should first work through the self-test and then, as a minimum, work the problems assigned by your instructor. Check your answers at the end of the chapter or book. Be sure to ask questions in class about anything that you do not understand.

The problems at the end of each chapter provide varying degrees of difficulty. In any technical field, it is important that you work lots of problems. Working through a problem gives you a level of insight and understanding that reading or classroom lectures alone do not provide. Never think that you can fully understand a concept or procedure by simply watching or listening to someone else. In the final analysis, you must do it yourself.

■ ACKNOWLEDGMENTS

I hope you will agree that this sixth edition of *Digital Fundamentals* is the best yet. Many people have contributed valuable ideas and constructive criticism for this new edition. It has been thoroughly reviewed for both content and accuracy.

Again, it is have been a pleasure to work with the people at Prentice Hall. Their enthusiasm and dedication to quality continue to be a source of inspiration to me. My appreciation goes to Dave Garza, Carol Robison, and Mary Harlan for their efforts on this project.

Special thanks to Lois Porter who has, once again, done an amazingly thorough job of copy editing the manuscript, and to Gary Snyder of Bently-Nevada Corporation for his excellent work in checking the manuscript for accuracy. Also, Dave Buchla of Yuba College has provided many valuable suggestions for improvements to the book.

I am grateful to the following reviewers, whose comments and suggestions have been of tremendous help in the development of this edition:

- Rabah Aoufi, DeVry Institute of Technology
- Robert Clark, Front Range Community College
- Paul Dilsner, Seminole Community College
- Ken Dreistadt, Lincoln Technical Institute
- David Hata, Portland Community College
- Robert Jones, DeVry Institute of Technology
- David Lewis, Denver Institute of Technology
- Abul Mohamedulla, Lansing Community College
- Dennis Quatrine, Henry Ford Community College
- Mark Williams, University of Memphis
- Steve Yelton, Cincinnati State Technical and Community College
- Asad Yousuf, Savannah State College

The photographs used in this book were provided by Tektronix and Intel Corporation. The date sheets are courtesy of Texas Instruments.

Last, but not least, a special thanks to my wife, Sheila.

CONTENTS

DIGITAL FUNDAMENTALS

1

INTRODUCTORY DIGITAL CONCEPTS

■ CHAPTER OBJECTIVES

☐ Explain the basic differences between digital systems and analog systems

☐ Show how voltage levels are used to represent digital quantities

☐ Describe various parameters of a pulse waveform such as rise time, fall time, pulse width, frequency, period, and duty cycle

☐ Explain the basic logic operations of NOT, AND, OR, and exclusive-OR

☐ Describe the basic functions of the comparator, adder, code converter, encoder, decoder, multiplexer, demultiplexer, counter, and register

☐ Show how a complete digital system is formed from the basic functions in a practical application

☐ Identify digital integrated circuits according to their complexity and the type of circuit packaging

☐ Identify pin numbers on integrated circuit packages

☐ Recognize digital instruments and understand how they are used in troubleshooting digital circuits and systems

■ CHAPTER OVERVIEW

The concept of a digital computer can be traced to Charles Babbage, who developed a crude mechanical computation device in the 1830s. The first functioning digital computer was built in 1944 at Harvard University, but it was electromechanical, not electronic. Modern digital electronics began in 1946 with an electronic digital computer called ENIAC, which was implemented with vacuum-tube circuits. Even though it took up an entire room, ENIAC didn't have the computing power that your hand-held calculator does.

The term *digital* is derived from the way computers perform operations, by counting digits. For many years, applications of digital electronics were confined to computer systems. Today, digital technology is applied in a wide range of areas in addition to the computer. Such applications as telephone systems, radar, navigation and guidance systems, military systems, medical instrumentation, industrial process control, and consumer electronics use digital techniques. Digital technology has progressed from vacuum-tube circuits to integrated circuits.

■ DIGITAL SYSTEM APPLICATION

The last section in most chapters of this textbook uses a system application to bring together many of the principal topics covered in the chapter. Each system is designed to fit the particular chapter to illustrate how the theory and devices can be used. Throughout the book, five different systems are introduced, some covering two or more chapters.

All of the systems are simplified to make them manageable in the context of the chapter material. Although they are based on actual system requirements, they are designed to accommodate the topical coverage of the chapter and are not intended to necessarily represent the most efficient or ultimate approach in a given application.

Section 1–7 introduces the first system, which is an industrial control system for counting and controlling items for packaging on a conveyor line. It is designed to incorporate all of the logic functions that are introduced in this chapter so that you can see how they are used and how they work together to achieve an overall objective.

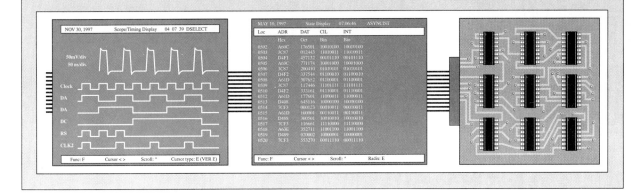

1–1 ■ DIGITAL AND ANALOG QUANTITIES

Electronic circuits can be divided into two broad categories, digital and analog. Digital electronics involves quantities with discrete values, and analog electronics involves quantities with continuous values. Although you will be studying digital fundamentals in this book, you should also know about analog because many applications require both. After completing this section, you should be able to

☐ Define *digital*   ☐ Define *analog*   ☐ Explain the difference between digital and analog quantities   ☐ Give examples of how digital and analog quantities are used in electronics

An **analog*** quantity is one having continuous values. A **digital** quantity is one having a discrete set of values. Most things that can be measured quantitatively appear in nature, in analog form. For example, the air temperature changes over a continuous range of values. During a given day, the temperature does not go from, say, 70° to 71° instantaneously; it takes on all the infinite values in between. If you graphed the temperature on a typical summer day, you would have a smooth, continuous curve similar to Figure 1–1. Other examples of analog quantities are time, pressure, distance, and sound.

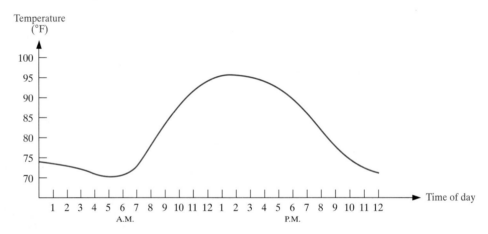

FIGURE 1–1
Graph of an analog quantity (temperature versus time).

Rather than graphing the temperature on a continuous basis, suppose you just take a temperature reading every hour. Now you have sampled values representing the temperature at discrete points in time over a 24-hour period, as indicated in Figure 1–2. You have effectively converted an analog quantity to a form that can now be digitized by representing each sampled value by a digital code. It is important to realize that Figure 1–2 is not the digital representation of the analog quantity.

Digital has certain advantages over analog in electronics applications. For one thing, digital data can be processed and transmitted more efficiently and reliably than analog data. Also, digital data has a great advantage when storage is necessary. For example, music when converted to digital form can be stored more compactly and reproduced with greater accuracy and clarity than is possible when it is in analog form.

---

*Boldface terms in the text are defined in the glossary.

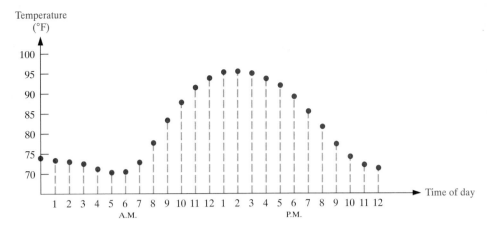

FIGURE 1–2

Sampled-value representation of the analog quantity in Figure 1–1. Each value represented by a dot can be digitized by representing it as a digital code.

An Analog Electronic System

A public address system, used to amplify sound so that it can be heard by a large audience, is one example of analog electronics. The basic diagram in Figure 1–3 illustrates that sound waves, which are analog in nature, are picked by a microphone and converted to a small analog voltage called the audio signal. This voltage varies continuously as the volume and frequency of the sound changes and is applied to the input of a linear amplifier. The output of the amplifier, which is an increased reproduction of input voltage, goes to the speaker(s). The speaker changes the amplified audio signal back to sound waves having a much greater volume than the original sound waves picked up by the microphone.

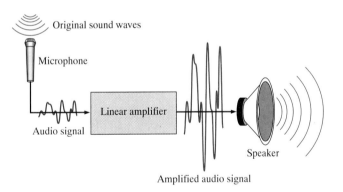

FIGURE 1–3

A basic public address system.

A System Using Digital and Analog Methods

The compact disk (CD) player provides an example of a system in which both digital and analog circuits are used. The simplified diagram in Figure 1–4 illustrates the basic principle. Music in digital form is stored on the compact disk. A laser diode optical system picks up the digital data from the rotating disk and transfers it to the **digital-to-analog converter (DAC).** The DAC changes the digital data into an analog signal that is an electrical reproduction of the original music. This signal is amplified and sent to the speaker for you to enjoy. When the music was originally recorded on the CD, a process, essentially the reverse of the one described here, using an **analog-to-digital converter (ADC)** was used.

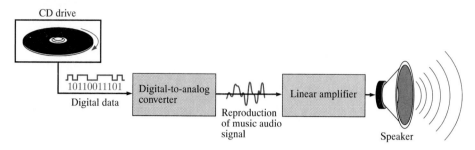

FIGURE 1–4
Basic principle of a CD player. Only one channel is shown.

**SECTION 1–1
REVIEW**

Answers to section reviews are found at the end of the chapter.
1. Define *digital.*
2. Define *analog.*
3. Explain the difference between a digital quantity and an analog quantity.
4. Give an example of a system that is analog and one that is a combination of both digital and analog. Name a system that is entirely digital.

1–2 ■ BINARY DIGITS, LOGIC LEVELS, AND DIGITAL WAVEFORMS

Digital electronics involves circuits and systems in which there are only two possible states. These states are represented by two different voltage levels: A HIGH and a LOW. The two states can also be represented by current levels, open and closed switches, or lamps turned on and off. In digital systems, combinations of the two states, called codes, are used to represent numbers, symbols, alphabetic characters, and other types of information. The two-state number system is called binary, and its two digits are 0 and 1. A binary digit is called a bit. After completing this section, you should be able to

□ Define *binary*   □ Define *bit*   □ Name the bits in a binary system   □ Explain how voltage levels are used to represent bits   □ Explain how voltage levels are interpreted by a digital circuit   □ Describe the general characteristics of a pulse   □ Determine the amplitude, rise time, fall time, and width of a pulse   □ Identify and describe the characteristics of a digital waveform   □ Determine the amplitude, period, frequency, and duty cycle of a digital waveform   □ Explain what a timing diagram is and state its purpose   □ Explain serial and parallel data transfer and state the advantage and disadvantage of each

Binary Digits

The two digits in the **binary** system, 1 and 0, are called **bits,** which is a contraction of *binary digit.* In digital circuits, two different voltage levels are used to represent the two bits. A 1 is represented by the higher voltage, which we will refer to as a HIGH and a 0 is represented by the lower voltage level, which we will refer to as a LOW. This is called **positive logic** and will be used throughout the book.

$$\text{HIGH} = 1 \quad \text{and} \quad \text{LOW} = 0$$

A much less common system in which a 1 is represented by a LOW and a 0 is represented by a HIGH is called *negative logic.*

Groups of bits, called *codes,* are used to represent numbers, letters, symbols, instructions, and anything else required in a given application.

Logic Levels

The voltages used to represent a 1 and a 0 are called *logic levels.* Ideally, one voltage level represents a HIGH and one voltage level represents a LOW. In a practical digital circuit, however, a HIGH can be any voltage between a specified minimum value and a specified maximum value. Likewise, a LOW can be any voltage between a specified minimum and a specified maximum.

Figure 1–5 illustrates the general range of LOWs and HIGHs for a digital circuit. The variable $V_{H(max)}$ represents the maximum HIGH voltage value, and $V_{H(min)}$ represents the minimum HIGH voltage value. The maximum LOW voltage value is represented by $V_{L(max)}$, and the minimum LOW voltage value is represented by $V_{L(min)}$. The range of voltages between $V_{L(max)}$ and $V_{H(min)}$ is a range of uncertainty. A voltage in the range of uncertainty can appear as either a HIGH or a LOW to a given circuit; one can never be sure. Therefore, the values in the uncertain range are unused. For example, the HIGH values for a certain type of digital circuit may range from 2 V to 5 V and the LOW values may range from 0 V to 0.8 V. So, if a voltage of 3.5 V is applied, the circuit will accept it as a HIGH or binary 1. If a voltage of 0.5 V is applied, the circuit will accept it as a LOW or binary 0.

FIGURE 1–5
Logic level ranges of voltage for a digital circuit.

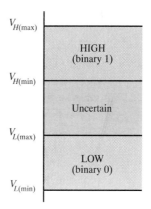

Digital Waveforms

Digital waveforms consist of voltage levels that are changing back and forth between the HIGH and LOW states. Figure 1–6(a) shows that a single positive-going **pulse** is generated when the voltage (or current) goes from its normally LOW level to its HIGH level and then back to its LOW level. The negative-going pulse in Figure 1–6(b) is generated when the voltage goes from its normally HIGH level to its LOW level and back to its HIGH level. A digital waveform is made up of a series of pulses.

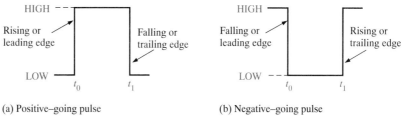

FIGURE 1–6
Ideal pulses.

The Pulse   As indicated in Figure 1–6, the pulse has two edges: a **leading edge** which occurs first at time t_0 and a **trailing edge** which occurs last at time t_1. For a positive-going pulse, the leading edge is a rising edge, and the trailing edge is a falling edge. The pulses in Figure 1–6 are ideal because the rising and falling edges change in zero time (instantaneously). In practice, these transitions never occur instantaneously, although for most digital work you can assume ideal pulses.

   Figure 1–7 shows a nonideal pulse. The time required for the pulse to go from its LOW level to its HIGH level is called the **rise time** (t_r), and the time required for the transition from the HIGH level to the LOW level is called the **fall time** (t_f). In practice, it is common to measure rise time from 10% of the pulse **amplitude** (height from baseline) to 90% of the pulse amplitude, and to measure the fall time from 90% to 10% of the pulse amplitude, as indicated in Figure 1–7. The bottom 10% and the top 10% of the pulse are not included in the rise and fall times because of the nonlinearities in the waveform in these areas.

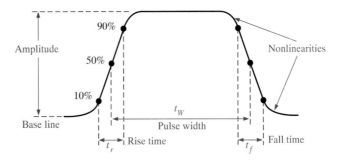

FIGURE 1–7
Nonideal pulse characteristics.

   The **pulse width** (t_W) is a measure of the duration of the pulse and is often defined as the time interval between the 50% points on the rising and falling edges, as indicated in Figure 1–7.

Overshoot and Ringing   Two commonly observed but undesirable pulse characteristics are overshoot and ringing. Positive overshoot and negative overshoot are caused by a capacitive effect in the circuit or measuring instrument that results in the voltage exceeding the normal HIGH and LOW levels for a short time on the rising and falling edges, as indicated in Figure 1–8(a). Ringing on the rising and falling edges of a pulse is actually an oscillation caused by capacitance and inductance in the circuit, as indicated in Figure 1–8(b). The ringing dies out after a short time.

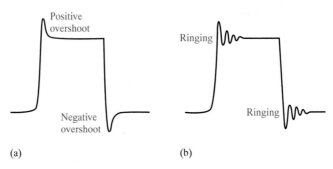

FIGURE 1–8
Overshoot and ringing.

Waveform Characteristics   Most waveforms encountered in digital systems are composed of series of pulses, sometimes called *pulse trains,* and can be classified as either periodic or nonperiodic. A **periodic** pulse waveform is one that repeats itself at a fixed interval, called a **period** (T). The frequency (f) is the rate at which it repeats itself and is measured in hertz (Hz). A nonperiodic pulse waveform, of course, does not repeat itself at fixed intervals and may be composed of pulses of differing pulse widths and/or differing time intervals between the pulses. An example of each type is shown in Figure 1–9.

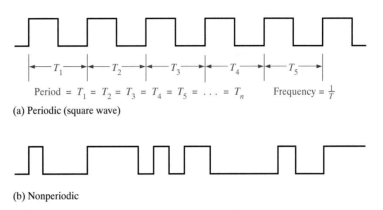

Period $= T_1 = T_2 = T_3 = T_4 = T_5 = \ldots = T_n$ Frequency $= \frac{1}{T}$

(a) Periodic (square wave)

(b) Nonperiodic

FIGURE 1–9
Examples of digital waveforms.

The **frequency** (f) of a pulse waveform is the reciprocal of the period. The relationship between frequency and period is expressed as follows:

$$f = \frac{1}{T} \tag{1–1}$$

$$T = \frac{1}{f} \tag{1–2}$$

An important characteristic of a periodic digital waveform is its duty cycle. The **duty cycle** is defined as the ratio of the pulse width (t_W) to the period (T) expressed as a percentage.

$$\text{Duty cycle} = \left(\frac{t_W}{T}\right)100\% \tag{1–3}$$

EXAMPLE 1–1

A portion of a periodic digital waveform is shown in Figure 1–10. The measurements are in milliseconds. Determine the following:
(a) period   **(b)** frequency   **(c)** duty cycle

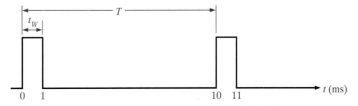

FIGURE 1–10

Solution

(a) The period is measured from the edge of one pulse to the corresponding edge of the next pulse. In this case T is measured from leading edge to leading edge, as indicated. T equals 10 ms.

(b) $f = \dfrac{1}{T} = \dfrac{1}{10 \text{ ms}} = 100$ Hz

(c) Duty cycle $= \left(\dfrac{t_W}{T}\right)100\% = \left(\dfrac{1 \text{ ms}}{10 \text{ ms}}\right)100\% = 10\%$

Related Exercise   A periodic digital waveform has a pulse width of 25 μs and a period of 150 μs. Determine the frequency and the duty cycle.*

A Digital Waveform Carries Binary Information

Binary information that is handled by digital systems appears as waveforms that represent sequences of bits. When the waveform is HIGH, a binary 1 is present; when the waveform is LOW, a binary 0 is present. Each bit in a sequence occupies a defined time interval called a **bit time.**

The Clock   In many digital systems, all waveforms are synchronized with a basic timing waveform called the **clock.** The clock is a periodic waveform in which each interval between pulses (the period) equals one bit time.

An example of a clock waveform is shown in Figure 1–11. Notice that, in this case, each change in level of waveform A occurs at the leading edge of the clock waveform. In other cases, level changes occur at the trailing edge of the clock. During each bit time of the clock, waveform A is either HIGH or LOW. These HIGHs and LOWs represent a sequence of bits as indicated. A group of several bits can be used as a piece of binary information, such as a number or a letter.

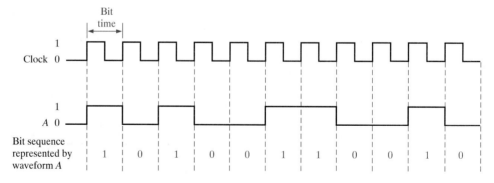

FIGURE 1–11
Example of a clock waveform synchronized with a waveform representation of a sequence of bits.

Timing Diagrams

A **timing diagram** is a graph of digital waveforms showing the proper time relationship of all the waveforms and how each waveform changes in relation to the others. Figure 1–11 is an example of a simple timing diagram that shows how the clock waveform and waveform A are related.

A timing diagram can consist of any number of related waveforms. By looking at a timing diagram, you can determine the states (HIGH or LOW) of all the waveforms at any specified point in time and the exact time that a waveform changes state relative to the other waveforms. Figure 1–12 is an example of a timing diagram made up of four waveforms.

---

*Answers to all related exercises are found at the end of the book.

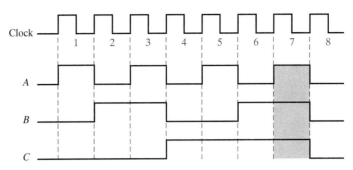

FIGURE 1–12
Example of a timing diagram.

From this timing diagram you can see, for example, that all three waveforms (*A*, *B*, and *C*) are HIGH only during bit time 7 and they all change back LOW at the end of bit time 7.

Data Transfer

Data refers to groups of bits that convey some type of information. Binary data, which are represented by digital waveforms, must be transferred from one circuit to another within a digital system or from one system to another in order to accomplish a given purpose. For example, numbers stored in binary form in the memory of a computer must be transferred to the central processing unit in order to be added. The sum of the addition must then be transferred to a monitor for display and/or transferred back to the memory. In digital systems as in Figure 1–13, binary data are transferred in two ways: serial and parallel.

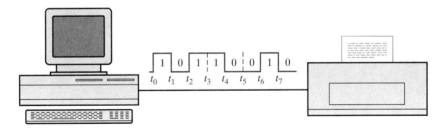

(a) Serial transfer of binary data from computer to printer. Interval t_0 to t_1 is first.

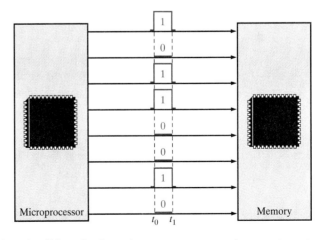

(b) Parallel transfer of binary data from microprocessor to memory in a computer system. t_0 is the beginning time.

FIGURE 1–13
Illustration of serial and parallel transfer of binary data.

When bits are transferred in **serial** form from one point to another, they are sent one bit at a time along a single conductor, as illustrated in Figure 1–13(a) for the case of a computer-to-printer transfer. During the time interval from t_0 to t_1, the first bit is transferred. During the time interval from t_1 to t_2, the second bit is transferred, and so on. To transfer eight bits in series, it takes eight time intervals.

When bits are transferred in **parallel** form, all the bits in a group are sent out on separate lines at the same time. There is one line for each bit, as shown in Figure 1–13(b) for the case of eight bits being transferred from the microprocessor to the memory in a computer. To transfer eight bits in parallel, it takes one time interval compared to eight time intervals for the serial transfer.

To summarize, the advantage of serial transfer of binary data is that only one line is required. In parallel transfer, a number of lines equal to the number of bits to be transferred at one time is required. The disadvantage of serial transfer is that it takes longer to transfer a given number of bits than with parallel transfer. For example, if one bit can be transferred in $1\,\mu s$, then it takes $8\,\mu s$ to serially transfer eight bits but only $1\,\mu s$ to parallel transfer eight bits. The disadvantage of parallel transfer is that it takes more lines.

EXAMPLE 1–2

(a) Determine the total time required to serially transfer the eight bits contained in waveform A of Figure 1–14, and indicate the sequence of bits. The left-most bit is the first to be transferred. The 100 kHz clock is used as reference.

(b) What is the total time to transfer the same eight bits in parallel?

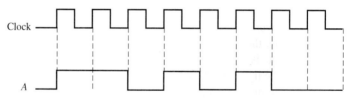

FIGURE 1–14

Solution

(a) Since the frequency of the clock is 100 kHz, the period is

$$T = \frac{1}{f} = \frac{1}{100 \text{ kHz}} = 10\,\mu s$$

It takes $10\,\mu s$ to transfer each bit in the waveform. The total transfer time is

$$8 \times 10\,\mu s = 80\,\mu s$$

To determine the sequence of bits, examine the waveform in Figure 1–14 during each bit time. If waveform A is HIGH during the bit time, a 1 is transferred. If waveform A is LOW during the bit time, a 0 is transferred. The bit sequence is illustrated in Figure 1–15. The left-most bit is the first to be transferred.

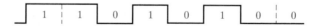

FIGURE 1–15

(b) A parallel transfer would take $10\,\mu s$ for all eight bits.

Related Exercise If binary data are transferred at the rate of 10 million bits per second (10 Mbits/s), how long will it take to parallel transfer 16 bits on 16 lines? How long will it take to serially transfer 16 bits?

1–3 ■ BASIC LOGIC OPERATIONS

In its basic form, logic is the realm of human reasoning that tells you a certain proposition (declarative statement) is true if certain conditions are true. Propositions can be classified as true or false. Many situations, problems, and processes that you encounter in your daily lives can be expressed in the form of propositional, or logic, functions. Since such functions are true/false or yes/no statements, digital circuits with their two-state characteristics are applicable. After completing this section, you should be able to

□ List four basic logic operations □ Define the NOT operation □ Define the AND operation □ Define the OR operation □ Define the exclusive-OR operation

Several propositions, when combined, form propositional, or logic, functions. For example, the propositional statement "The light is on" will be true if "The bulb is not burned out" is true and if "The switch is on" is true. Therefore, this logical statement can be made: The light is on if and only if the bulb is not burned out *and* the switch is on. In this example the first statement is true only if the last two statements are true. The first statement ("The light is on") is then the basic proposition, and the other two statements are the conditions on which the proposition depends.

In the 1850s, the Irish logician and mathematician George Boole developed a mathematical system for formulating logic statements with symbols so that problems can be written and solved in a manner similar to ordinary algebra. Boolean algebra, as it is known today, finds application in the design and analysis of digital systems and will be covered in detail in Chapter 4.

The term **logic** is applied to digital circuits used to implement logic functions. Several kinds of digital **circuits** are the basic elements that form the building blocks for such complex digital systems as the computer. We will now look at these elements and discuss their functions in a very general way. Later chapters will cover these circuits in detail.

Four basic logic operations are indicated by standard distinctive shape symbols in Figure 1–16. Other standard symbols for these logic operations will be introduced in Chapter 3. The lines connected to each symbol are the **inputs** and **outputs.** The inputs are on the left of each symbol and the output is on the right. A circuit that performs a specified logic operation (AND, OR, exclusive-OR) is called a logic **gate.** AND and OR gates can have any number of inputs, as indicated by the dashes in the figure.

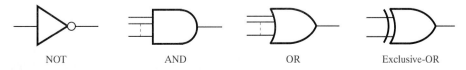

NOT   AND   OR   Exclusive-OR

FIGURE 1–16
The basic logic operations.

In logic operations, the true/false conditions mentioned earlier are represented by a HIGH (true) and a LOW (false). Each of the four basic logic operations produces a unique response to a given set of conditions.

NOT

The **NOT** operation changes one logic level to the opposite logic level, as indicated in Figure 1–17. When the input is HIGH, the output is LOW. When the input is LOW, the output is HIGH. In either case, the output is *not* the same as the input. The NOT operation is implemented by a logic circuit known as an **inverter.**

FIGURE 1–17
The NOT operation.

AND

The **AND** operation produces a HIGH output if and only if all the inputs are HIGH, as indicated in Figure 1–18 for the case of two inputs. When one input is HIGH *and* the other input is HIGH, the output is HIGH. When any or all inputs are LOW, the output is LOW. The AND operation is implemented by a logic circuit known as an *AND gate.*

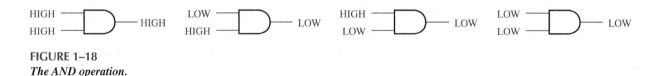

FIGURE 1–18
The AND operation.

OR

The **OR** operation produces a HIGH output when any of the inputs is HIGH, as indicated in Figure 1–19 for the case of two inputs. When one input is HIGH *or* the other input is HIGH *or* both inputs are HIGH, the output is HIGH. When both inputs are LOW, the output is LOW. The OR operation is implemented by a logic circuit known as an *OR gate.*

FIGURE 1–19
The OR operation.

Exclusive-OR

The **exclusive-OR** operation produces a HIGH output when one and only one of two inputs is HIGH, as indicated in Figure 1–20. If both inputs are HIGH *or* if both inputs are LOW, the output is LOW. The exclusive-OR operation is implemented by an *exclusive-OR gate,* which is a combination of the AND, OR, and NOT operations.

FIGURE 1–20
The exclusive-OR operation.

1–4 ■ BASIC LOGIC FUNCTIONS

The inverter and the basic gates can be combined to form more complex logic circuits that perform many useful operations and that are used to build complete digital systems. Some of the common logic functions are comparison, arithmetic, code conversion, encoding, decoding, data selection, storage, and counting. This section provides a general overview of these important functions so that you can begin to see how they form the building blocks of digital systems such as computers. Each of the basic logic functions will be covered in detail in later chapters. After completing this section, you should be able to

☐ Identify eight basic types of logic functions   ☐ Describe a basic magnitude comparator   ☐ List the four arithmetic functions   ☐ Describe a basic adder   ☐ Describe a basic encoder   ☐ Describe a basic decoder   ☐ Define multiplexing and demultiplexing   ☐ State how data storage is accomplished   ☐ Describe the function of a basic counter

The Comparison Function

Magnitude comparison is performed by a logic circuit called a **comparator,** covered in Chapter 6. A comparator compares two quantities and indicates whether or not they are equal. For example, suppose you have two numbers and wish to know if they are equal or not equal and, if not equal, which is greater. The comparison function is represented in Figure 1–21. One number is applied in binary form to input A and the other number in binary form is applied to input B. The outputs indicate the relationship of the two numbers by producing a HIGH level on the proper output line. Suppose that a binary representation of the number 2 is applied to input A and a binary representation of the number 5 is applied to input B. (We will discuss the binary representation of numbers and symbols in the next chapter.) A HIGH level will appear on the $A < B$ (A is less than B) output, indicating the relationship between the two numbers (2 is less than 5). The wide arrows represent a group of parallel lines on which the bits are transferred.

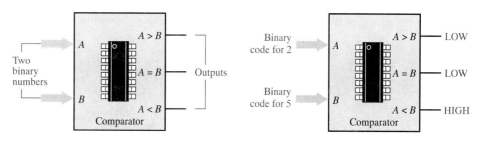

(a) Basic magnitude comparator

(b) Example: A less than B ($2 < 5$)

FIGURE 1–21
The comparison function.

The Arithmetic Functions

Addition   Addition is performed by a logic circuit called an **adder,** covered in Chapter 6. An adder adds two binary numbers (on inputs A and B with a carry input C_{in}) and generates a sum (Σ) and a carry output (C_{out}), as shown in Figure 1–22(a). Figure 1–22(b) illustrates the addition of 3 and 9. You know that the sum is 12; the adder indicates this result by producing 2 on the sum output and 1 on the carry output. Assume the carry input in this example to be 0.

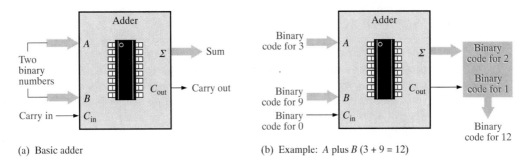

(a) Basic adder

(b) Example: A plus B ($3 + 9 = 12$)

FIGURE 1–22
The addition function.

Subtraction   Subtraction is also performed by a digital circuit. A **subtracter** requires three inputs: the two numbers that are to be subtracted and a borrow input. The two outputs are the difference and the borrow output. When, for instance, 5 is subtracted from 8 with no borrow input, the difference is 3 with no borrow output. You will see in Chapter 2 how subtraction can actually be performed by an adder because subtraction is simply a special case of addition.

Multiplication   Multiplication is performed by a digital circuit called a *multiplier.* Since numbers are always multiplied two at a time, two inputs are required. The output of the multiplier is the product. Since multiplication is simply a series of additions with shifts in the positions of the partial products, it can be performed by using an adder in conjunction with other circuits.

Division   Division can be performed with a series of subtractions, comparisons, and shifts, and thus it can also be done using an adder in conjunction with other circuits. Two inputs to the divider are required, and the outputs generated are the quotient and the remainder.

The Code Conversion Function

A **code** is a set of bits arranged in a unique pattern and used to represent specified information. A code converter changes a form of coded information into another coded form. Examples are conversion between binary and other codes such as the binary coded decimal (BCD) and the Gray code. Various types of codes are covered in Chapter 2, and code converters are covered in Chapter 6.

The Encoding Function

The encoding function is performed by a circuit called an **encoder,** covered in Chapter 6. The encoder converts information, such as a decimal number or an alphabetic character, into some coded form. For example, a certain type of encoder converts each of the decimal digits, 0 through 9, to a binary code. A HIGH level on the input corresponding to a specific decimal digit produces the proper binary code on the output lines.

Figure 1–23 is a simple illustration of an encoder used to convert (encode) a calculator keystroke into a binary code that can be processed by the calculator circuits.

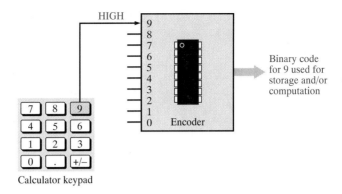

FIGURE 1–23
An encoder used to encode a calculator keystroke into a binary code for storage or for calculation.

The Decoding Function

The decoding function is performed by a circuit called a **decoder,** covered in Chapter 6. The decoder converts coded information, such as a binary number, into a noncoded form, such as a decimal form. For example, a particular type of decoder converts a 4-bit binary code into the appropriate decimal digit.

Figure 1–24 is a simple illustration of one type of decoder that is used to activate a 7-segment display. Each of the seven segments of the display is connected to an output line from the decoder. When a special binary code appears on the decoder inputs, the appropriate output lines are activated and light the proper segments to display the decimal digit corresponding to the binary code.

FIGURE 1–24
A decoder used to convert a special binary code into a 7-segment decimal readout.

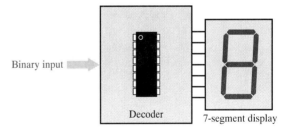

The Data Selection Function

Two types of circuits that select data are the multiplexer and the demultiplexer. The **multiplexer,** or mux for short, is a circuit that switches digital data from several input lines onto a single output line in a specified time sequence. Functionally, a multiplexer can be represented by an electronic switch operation that sequentially connects each of the input lines to the output line. The **demultiplexer** (demux) is a circuit that switches digital data from one input line to several output lines in a specified time sequence. Essentially, the demux is a mux in reverse.

Multiplexing and demultiplexing are used when data from several sources are to be transmitted over one line to a distant location and redistributed to several destinations. Figure 1–25 illustrates this type of application where digital data from three computers are sent out along a single line to three other computers at another location.

In Figure 1–25, binary data from computer A are connected to the output line during time interval Δt_1 and transmitted to the demultiplexer that connects them to computer D. Then, during interval Δt_2, the multiplexer switches to the input from computer B and the demultiplexer switches the output to computer E. During interval Δt_3, the multiplexer switches to the input from computer C and the demultiplexer switches the output to computer F.

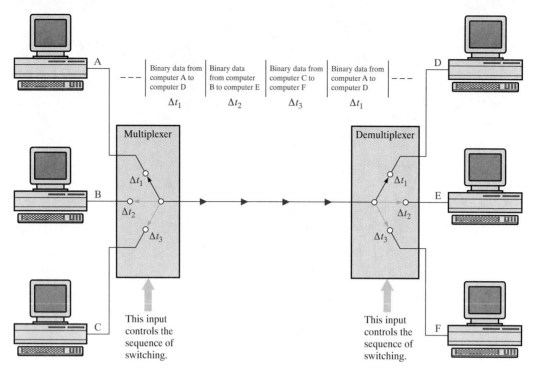

FIGURE 1–25
Illustration of a basic multiplexing/demultiplexing application.

To summarize, during the first time interval, computer A sends data to computer D. During the second time interval, computer B sends data to computer E. During the third time interval, computer C sends data to computer F. After this, the first two computers again communicate and the sequence repeats. Because the time is divided up among several sets of systems where each has its turn to send and receive data, this process is called *time division multiplexing* (TDM).

The Storage Function

Storage is a function that is required in most digital systems, and its purpose is to retain binary data for a period of time. Some storage devices are used for short-term storage and some for long-term storage. A storage device "memorizes" a bit or groups of bits and retains it as long as necessary. Common types of storage devices are flip-flops, registers, semiconductor memories, magnetic disks, magnetic tape, and optical disks.

Flip-flops   The **flip-flop** is a bistable (two stable states) logic circuit that can store only one bit at a time, either a 1 or a 0. The output of a flip-flop indicates which bit it is storing. A HIGH output indicates that a 1 is stored and a LOW output indicates that a 0 is stored. Flip-flops are covered in Chapter 8.

Registers   A **register** is formed by combining several flip-flops so that groups of bits can be stored. For example, an 8-bit register is constructed from eight flip-flops. In addition to storing bits, registers can be used to shift the bits from one position to another within the register or out of the register to another circuit; therefore, these devices are known as *shift registers,* covered in Chapter 10.

The two basic types of shift registers are serial and parallel. The bits are stored in a serial shift register one at a time, as illustrated in Figure 1–26. A good analogy to this is loading passengers onto a bus single file through the door. They also exit the bus single file.

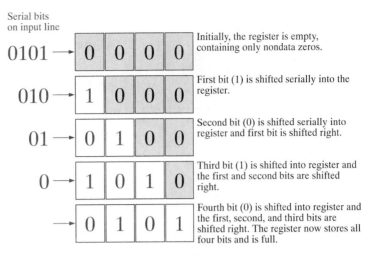

FIGURE 1–26
Example of the operation of a serial shift register.

FIGURE 1–27
Example of the operation of a parallel shift register.

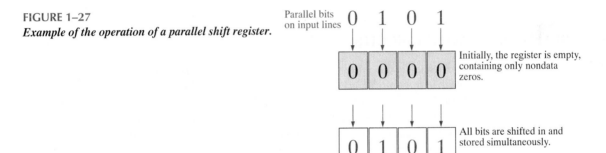

The bits are stored in a parallel register simultaneously from parallel lines, as shown in Figure 1–27. For this case, a good analogy is loading passengers on a roller coaster where they enter the cars in parallel.

Semiconductor Memories Semiconductor memories are devices typically used for storing large numbers of bits. In one type of memory, called the *read-only m*emory or ROM, the binary data are permanently or semipermanently stored and cannot be readily changed. In the *random-a*ccess *m*emory or RAM, the binary data are temporarily stored and can be easily changed. Memories are covered in Chapter 12.

Magnetic Memories Magnetic disk memories are used for mass storage of binary data. Examples are the so-called floppy disks used in computers and the computer's internal hard disk. Magneto-optic disks use laser beams to store and retrieve data. Magnetic tape is also used in memory applications.

The Counting Function

The counting function is very important in digital systems. There are many types of digital **counters,** but their basic purpose is to count events represented by changing levels or pulses or to generate a particular code sequence. To count, the counter must "remember" the present number so that it can go to the next proper number in sequence. Therefore, storage capability is an important characteristic of all counters, and flip-flops are generally used to implement them. Figure 1–28 illustrates the basic idea of counter operation. Counters are covered in Chapter 9.

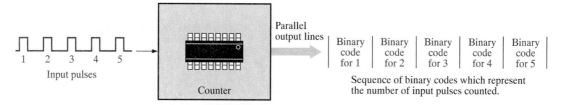

FIGURE 1–28
Illustration of basic counter operation.

1–5 ■ DIGITAL INTEGRATED CIRCUITS

All the logic elements and functions that have been discussed—and many more—are available in integrated circuit (IC) form. Modern digital systems use ICs almost exclusively in their designs because of their small size, high reliability, low cost, and low power consumption. It is important to be able to recognize the IC packages and to know how the pin connections are numbered, as well as to be familiar with the way in which circuit complexities and circuit technologies determine the various IC classifications. After completing this section, you should be able to

☐ Recognize the difference between through-hole devices and surface-mount devices ☐ Identify dual-in-line packages (DIP) ☐ Identify small-outline integrated circuit packages (SOIC) ☐ Identify plastic leaded chip carrier packages (PLCC) ☐ Identify leadless ceramic chip carrier packages (LCCC) ☐ Identify flat packs ☐ Determine pin numbers on various types of IC packages ☐ Explain the circuit complexity classifications of integrated circuits

A monolithic **integrated circuit (IC)** is an electronic circuit that is constructed entirely on a single small chip of silicon. All the components that make up the circuit—transistors, diodes, resistors, and capacitors—are an integral part of that single chip.

Figure 1–29 shows a cutaway view of one type of IC package, with the circuit chip shown within the package. Points on the chip are connected to the package pins to allow input and output connections to the outside world.

FIGURE 1–29
Cutaway view of one type of IC package showing the chip mounted inside, with connections to input and output pins.

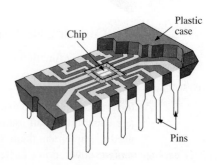

IC Packages

Integrated circuit packages are classified according to the way they are mounted on printed circuit (PC) boards as either through-hole mounted or surface mounted. The through-hole type packages have pins (leads) that are inserted through holes in the PC board and can be soldered to conductors on the opposite side. The most common type of through-hole package is the **dual-in-line package (DIP)** shown in Figure 1–30(a).

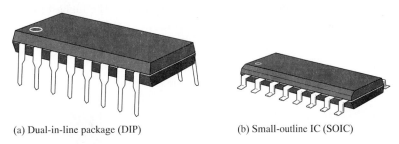

(a) Dual-in-line package (DIP)   (b) Small-outline IC (SOIC)

FIGURE 1–30
Examples of through-hole and surface-mounted devices. The DIP is larger than the SOIC with the same number of leads. This particular DIP is approximately 0.785 in. long, and the SOIC is approximately 0.385 in. long.

Another type of IC package uses **surface-mount technology (SMT).** Surface mounting is a newer, space-saving alternative to through-hole mounting. The holes through the PC board are unnecessary for SMT. The pins of surface-mounted packages are soldered directly to conductors on one side of the board, leaving the other side free for additional circuits. Also, for a circuit with the same number of pins, a surface-mounted package is much smaller than a dual-in-line package because the pins are placed closer together. An example of a surface-mounted package is the small-outline integrated circuit (SOIC) shown in Figure 1–30(b).

Four common types of SMT packages are the **SOIC** (small-outline IC), the **PLCC** (plastic leaded chip carrier), the **LCCC** (leadless ceramic chip carrier), and the **flat pack.** These types of packages are available in various sizes depending on the number of leads (more leads are required for more complex circuits). Examples of each type are shown in Figure 1–31. As you can see, the leads of the SOIC are formed into a "gull-wing" shape. The leads of the PLCC are turned under the package in a J-type shape. Instead of leads, the LCCC has metal contacts molded into its ceramic body. The leads of the flat pack extend straight out from the body.

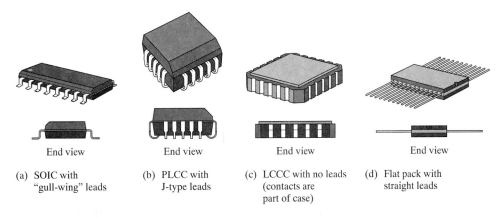

| End view | End view | End view | End view |

(a) SOIC with "gull-wing" leads   (b) PLCC with J-type leads   (c) LCCC with no leads (contacts are part of case)   (d) Flat pack with straight leads

FIGURE 1–31
Examples of SMT package configurations.

Pin Numbering

All IC packages have a standard format for numbering the pins (leads). The dual-in-line packages (DIPs), the small-outline IC packages (SOICs), and the flat packs have the numbering arrangement illustrated in Figure 1–32(a) for a 16-pin package. Looking at the top of the package, pin 1 is indicated by an identifier that can be either a small dot, a notch, or a beveled edge. The dot is always next to pin 1. Also, with the notch oriented upward, pin 1 is always the top left pin, as indicated. Starting with pin 1, the pin numbers increase as you go down, then across and up. The highest pin number is always to the right of the notch or opposite the dot.

FIGURE 1–32

Pin numbering for standard IC packages. Top views are shown.

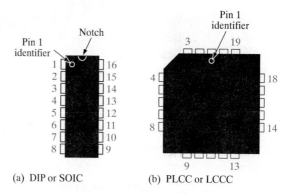

(a) DIP or SOIC   (b) PLCC or LCCC

The PLCC and LCCC packages have leads arranged on all four sides. Pin 1 is indicated by a dot or other index mark and is located at the center of one set of leads. The pin numbers increase going counterclockwise as viewed from the top of the package. The highest pin number is always to the right of pin 1. Figure 1–32(b) illustrates this format for a 20-pin PLCC package.

Integrated Circuit Complexity Classifications

Integrated circuits are classified according to their complexity. They are listed here from the least complex to the most complex. The complexity figures stated here for SSI, MSI, LSI, VLSI, and ULSI are generally accepted, but definitions may vary from one source to another.

- **Small-scale integration (SSI)** describes circuits that have up to twelve equivalent gate circuits on a single chip, and they include basic gates and flip-flops.
- **Medium-scale integration (MSI)** describes circuits that have from 12 to 99 equivalent gates on a chip. They include logic functions such as encoders, decoders, counters, registers, multiplexers, arithmetic circuits, small memories, and others.
- **Large-scale integration (LSI)** is a classification of ICs with complexities of 100 to 9999 equivalent gates per chip, including memories.
- **Very large-scale integration (VLSI)** describes integrated circuits with complexities of 10,000 to 99,999 equivalent gates per chip.
- **Ultra large-scale integration (ULSI)** describes very large memories, larger **microprocessors,** and larger single-chip computers. Complexities of 100,000 and greater are classified as ULSI.

Integrated Circuit Technologies

The types of transistors with which all integrated circuits are implemented are either bipolar junction transistors or MOSFETs (metal-oxide semiconductor field-effect transistors). Two types of digital circuit technology that use bipolar junction transistors are TTL

The Volts/Division Control Notice that in Figure 1–35 there are two VOLTS/DIV switches, one for channel 1 (CH1) and one for channel 2 (CH2). The VOLTS/DIV selector switch sets the number of volts to be represented by each major division on the vertical scale. For example, Figure 1–36 shows a digital waveform on the scope screen with the VOLTS/DIV switch set at 1 V. This means that each of the major vertical divisions is 1 V. The pulses are three divisions high, and, since each division is 1 V, the amplitude of the pulses is 3 V (3 divisions $\times$ 1 V/division = 3 V).

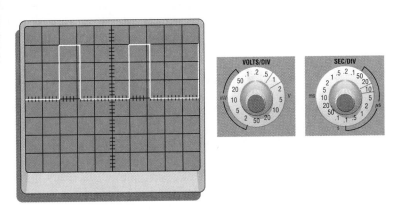

FIGURE 1–36

The Seconds/Division Control The SEC/DIV selector switch sets the number of seconds, milliseconds, or microseconds to be represented by each major division on the horizontal scale. It actually controls how fast the electron beam sweeps horizontally across the screen. In Figure 1–36, the SEC/DIV switch is set at 10 μs. This means that each major horizontal division is 10 μs. Since a full cycle of the waveform covers 4 divisions, the period of the waveform is 40 μs; that is, (4 divisions $\times$ 10 μs/division) $\times$ 40 μs. From this, $f = 1/40\,\mu s = 25$ kHz.

So, as you can see, it is easy to measure the period and then calculate the frequency by counting the number of major divisions covered by one cycle and then multiplying by the SEC/DIV setting.

Other Oscilloscope Controls

The following descriptions refer to the oscilloscope in Figure 1–35 but apply also to most general-purpose scopes.

Power Switch The power switch turns the power to the scope on and off. A light indicates when the power is on.

Intensity The intensity control knob varies the brightness of the trace on the screen. Caution should be used so that the intensity is not left too high for an extended period of time, especially when the beam forms a motionless dot on the screen. Damage to the screen can result from excessive intensity.

Focus This control focuses the beam so that it converges to a tiny point at the screen. An out-of-focus condition results in a fuzzy trace.

Horizontal Position These control knobs (coarse and fine) adjust the neutral horizontal position of the beam. They are used to reposition horizontally a waveform display for more convenient viewing or measurement.

Vertical Position The two vertical position controls move each trace up or down for easier measurement or observation.

AC-GND-DC Switch This switch, located below the VOLTS/DIV control, allows the input signal to be ac coupled, dc coupled, or grounded. The ac coupling eliminates any dc component on the input signal. The dc coupling permits dc values to be displayed. The ground position allows a 0 V reference to be established on the screen.

Signal Inputs The signals to be displayed are connected into the channel 1 (CH1) and/or channel 2 (Ch2) input connectors. These connections are normally done with a special probe that minimizes the loading effect of the scope's input resistance on the circuit being measured. Oscilloscope voltage probes are generally either ×1 (nonattenuating) or ×10 (attenuates by 10). When a ×10 probe is used, the VOLTS/DIV setting must be multiplied by 10. All applications in this book assume ×1 voltage probes.

Mode Switches These switches provide for displaying either or both channel inputs, inverting channel 2 signal, adding two waveforms, and selecting between alternate and chopped mode of sweep.

Trigger Controls The trigger controls allow the beam to be triggered from various selected sources. The triggering of the beam causes it to begin its sweep across the screen. It can be triggered from an internally generated signal derived from an input signal, or from the line voltage, or from an externally applied trigger signal. The modes of triggering are auto, normal, single-sweep, and TV. In the auto mode, sweep occurs in the absence of an adequate trigger signal. In the normal mode, a trigger signal must be present for the sweep to occur. The TV mode provides triggering on the TV field or TV line signals. The slope switch allows the triggering to occur on either the positive-going slope or the negative-going slope of the trigger waveform. The level control selects the voltage level on the trigger signal at which the triggering occurs.

Basically, the trigger controls provide for synchronization of the horizontal sweep waveform and the input signal waveform. As a result, the display of the input signal is stable on the screen, rather than appearing to drift across the screen.

The Logic Analyzer

A typical logic analyzer is shown in Figure 1–37. This instrument can detect and display digital data in several formats.

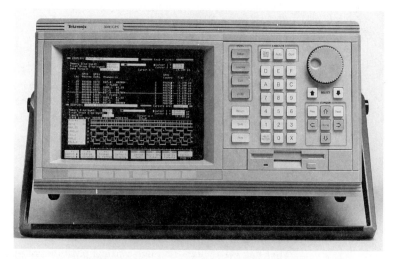

FIGURE 1–37
A typical logic analyzer (courtesy of Tektronix, Inc.).

Oscilloscope Format   The logic analyzer can be used to display single or dual waveforms on the screen, as indicated in Figure 1–38(a), so that characteristics of individual pulses or waveform parameters can be measured.

Timing Diagram Format   The logic analyzer can display typically up to sixteen waveforms in proper time relationship, as indicated in Figure 1–38(b), so that you can analyze sets of waveforms and determine how they change in time with respect to each other.

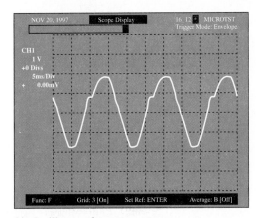

(a) Oscilloscope format

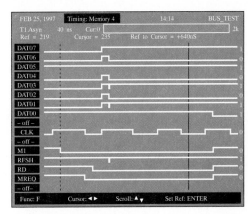

(b) Timing diagram format

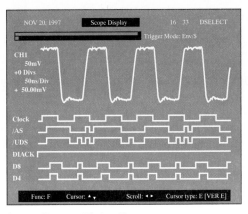

(c) Oscilloscope/Timing diagram

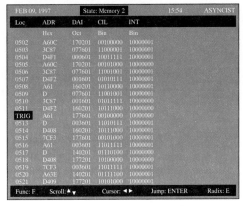

(d) State table format

FIGURE 1–38
Illustration of typical logic analyzer display formats.

Oscilloscope/Timing Diagram Combination   In this format, as indicated in Figure 1–38(c), both individual waveforms and a complete timing diagram can be displayed simultaneously. This allows you to examine the details of a certain waveform while having a timing diagram available.

State Table Format   The logic analyzer can display binary data in tabular form, as illustrated in Figure 1–38(d). For example, various memory locations in a microprocessor-based system can be examined to determine the contents. The data can be displayed in a variety of number systems and codes such as binary, hexadecimal, octal, binary coded decimal (BCD), and ASCII. These number systems and codes are the topics of the next chapter.

The Logic Probe, Pulser, and Current Probe

The logic probe is a convenient, hand-held tool that provides a means of troubleshooting a digital circuit by sensing various conditions at a point in a circuit, as illustrated in Figure 1–39. The probe can detect high-level voltage, low-level voltage, single pulses, repetitive

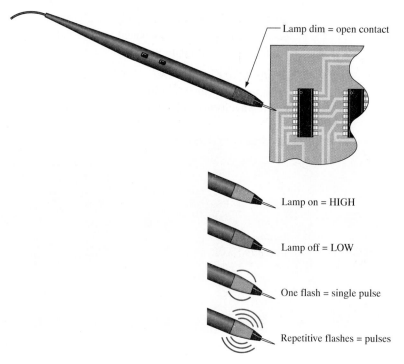

FIGURE 1–39
Illustration of how a logic probe is used to detect various voltage conditions at a given point in a circuit.

pulses, and opens on a circuit board. The probe lamp indicates the condition that exists at a certain point, as indicated in the figure.

The logic pulser is a pulse source that produces a repetitive pulse waveform that can be used to force a condition in a circuit. You can apply pulses at one point in a circuit with the pulser and check another point for resulting pulses with a logic probe. Also the pulser can be used in conjunction with the current probe as indicated in Figure 1–40. The current probe senses when there is pulsating current in a line and is particularly useful for locating shorts on a circuit board.

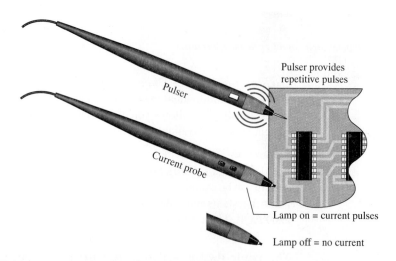

FIGURE 1–40
Illustration of how a logic pulser and a current probe can be used to pulse a given point and check for resulting current in another part of the circuit.

The DC Power Supply

The dc power supply is an indispensable instrument on any test bench. The power supply converts ac power from the standard wall outlet into regulated dc voltage. All digital circuits require dc voltage to operate. For example, TTL circuits require approximately +5 V. The power supply is used when a new circuit is breadboarded or when a circuit board is pulled from a system for testing and is no longer operating from the internal system power supply. A typical test bench power supply is shown in Figure 1–41(a).

The Function Generator

The function generator is a versatile signal source that provides pulse waveforms, as well as sine wave and triangular waveforms. Many function generators have logic-compatible outputs to provide proper level waveforms as inputs to digital circuits in order to check the operation. One type of function generator is shown in Figure 1–41(b).

The Digital Multimeter

No test bench is complete without a digital multimeter (DMM). This instrument is used for measuring dc and ac voltage, dc and ac current, and resistance. Figure 1–41(c) shows typical test bench and portable DMMs.

(a) DC power supply

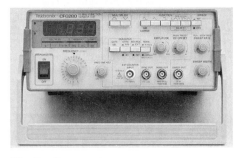

(b) Function generator

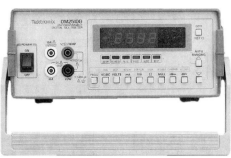

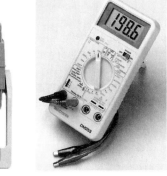

(c) Two types of digital multimeter (DMM)

FIGURE 1–41
Typical test instruments (courtesy of Tektronix, Inc.).

SECTION 1–6 REVIEW

1. How many major horizontal divisions are there on an oscilloscope screen?
2. How many major vertical divisions are there on an oscilloscope screen?
3. What type of instrument can display up to sixteen waveforms on its screen?
4. What is indicated when the lamp on a logic probe is flashing repetitively?

1–7 ■ DIGITAL SYSTEM APPLICATION

In this section, an interesting but simplified system application of the logic elements and functions that were discussed in previous sections is presented. It is important that an electronic technician or technologist understand how various digital functions can operate together as a total system to perform a specified task. It is also important to begin to think in terms of system-level operation because, in practice, a large part of your work will involve systems rather than individual functions. Of course, to understand systems, you must first understand the basic elements and functions that make up a system.

This section introduces you to the system concept. The example here will show you how logic functions can work together to perform a higher-level task and will get you started thinking at the system level. The specific system used here to illustrate the system concept is not necessarily the approach that would be used in practice, although it could be. In modern industrial applications like the one discussed here, instruments known as programmable controllers are often used. After completing this section, you should be able to

□ Begin to think in terms of system-level operation □ Explain the purpose of each logic function in the example system □ Discuss how the various functions work together to accomplish a desired task

The System

A pharmaceutical company uses the system shown in the block diagram of Figure 1–42 for automatically counting and bottling tablets. The tablets are fed into a large funnel-like hopper. The narrow neck of the funnel allows only one tablet at a time to fall into a bottle on the conveyor belt below.

The digital system controls the number of tablets going into each bottle and displays a continually updated total near the conveyor line as well as at a remote location in another part of the plant. As you can see, this system utilizes all the basic logic functions that were introduced in Section 1–4.

The general operation is as follows. An optical sensor at the bottom of the funnel neck detects each tablet that passes and produces an electrical pulse. This pulse goes to the counter and advances it by one count; thus, at any time during the filling of a bottle, the counter holds the binary representation of the number of tablets in the bottle. The binary count is transferred on parallel lines to the B input of the comparator (Comp). A preset binary number equal to the number of tablets that are to go into each bottle is placed on the A input of the comparator. The preset number comes from the keypad and the associated circuits. When the desired number of tablets is entered on the keypad, it is encoded and then stored by register A until a change in the quantity of tablets per bottle is required.

Suppose, for example, that each bottle is to hold fifty tablets. When the number in the counter reaches fifty, the $A = B$ output of the comparator goes HIGH, indicating that the bottle is full.

The HIGH output of the comparator immediately closes the valve in the neck of the funnel to stop the flow of tablets, and at the same time it activates the conveyor to move the next bottle into place under the funnel. When the next bottle is positioned properly under

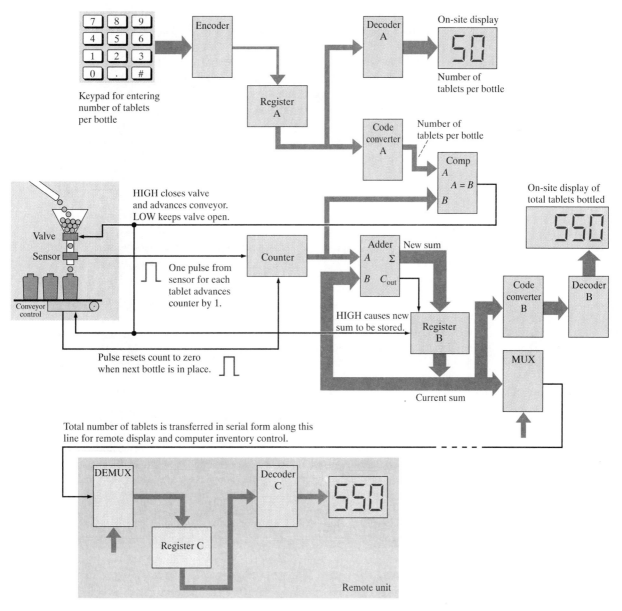

FIGURE 1–42
Simplified basic block diagram for a tablet counting and bottling control system.

the neck of the funnel, the conveyor control circuit produces a pulse that resets the counter to zero. The $A = B$ output of the comparator goes back LOW, opening the funnel valve to restart the flow of tablets.

In the display portion of the system, the number in the counter is transferred in parallel to the A input of the adder. The B input of the adder comes from register B that holds the total number of tablets bottled, up through the last bottle filled. For example, if ten bottles have been filled and each bottle holds fifty tablets, register B contains the binary representation for 500. Then, when the next bottle has been filled, the number 50 appears on the A input of the adder, and the number 500 is on the B input. The adder produces a new sum of 550, which is stored in register B, replacing the previous sum of 500.

The content of register B is transferred in parallel to the decoder, which changes it from binary form to decimal form for display on a readout near the conveyor line. The content of the register is also transferred to a multiplexer (mux) so that it can be transmitted along a single line to a remote location some distance away. (It is more economical to run a single line than to run several parallel lines when significant distances are involved.) At the remote location, the serial data are demultiplexed and sent to register C. From there the data are then decoded for display on the remote readout.

Again, you should be aware that this system does not necessarily represent the ultimate or most efficient way to implement this hypothetical process. Although there are certainly other approaches, this particular approach has been selected in order to illustrate all of the logic functions that were introduced in Section 1–4 and that will be covered in detail in future chapters. It shows you one application of the various functional devices at the system level and how they can be connected to accomplish a specific objective. You will see this system again in the next chapter.

SECTION 1–7 REVIEW

1. Explain the purpose of the comparator (comp) in the system in Figure 1–42.
2. What actions take place when the $A = B$ output of the comparator goes HIGH?
3. What is the content of each register at any given time?

■ SUMMARY

- An analog quantity has continuous values.
- A digital quantity has a discrete set of values.
- A binary digit is called a bit.
- A pulse is characterized by rise time, fall time, pulse width, and amplitude.
- The frequency of a periodic waveform is the reciprocal of the period.
- A timing diagram is an arrangement of two or more waveforms showing their relationship with respect to time.
- Four basic logic operations are NOT, AND, OR, and exclusive-OR.
- The basic logic functions are comparison, arithmetic, code conversion, decoding, encoding, data selection, storage, and counting.
- The two broad physical categories of IC packages are through-hole mounted and surface mounted.
- The categories of ICs in terms of circuit complexity are SSI (small-scale integration), MSI (medium-scale integration), LSI, VLSI, and ULSI (large-scale, very large-scale, and ultra large-scale integration).
- Common instruments used in testing and troubleshooting digital circuits are the oscilloscope, logic analyzer, logic probe, pulser, current probe, dc power supply, function generator, and digital multimeter.

■ SELF-TEST

Answers are found at the end of the book.

1. A quantity having continuous values is
   (a) a digital quantity   (b) an analog quantity
   (c) a binary quantity   (d) a natural quantity
2. The term *bit* means
   (a) a small amount of data   (b) a 1 or a 0
   (c) binary digit   (d) both answers (b) and (c)
3. The time interval on the leading edge of a pulse between 10% and 90% of the amplitude is the
   (a) rise time   (b) fall time   (c) pulse width   (d) period
4. A pulse in a certain waveform occurs ever 10 ms. The frequency is
   (a) 1 kHz   (b) 1 Hz   (c) 100 Hz   (d) 10 Hz
5. In a certain digital waveform, the period is twice the pulse width. The duty cycle is
   (a) 100%   (b) 200%   (c) 50%
6. An inverter
   (a) performs the NOT operation   (b) changes a HIGH to a LOW
   (c) changes a LOW to a HIGH   (d) does all of the above

7. The output of an AND gate is HIGH when
    (a) any input is HIGH   (b) all inputs are HIGH
    (c) no inputs are HIGH   (d) both answers (a) and (b)

8. The output of an OR gate is HIGH when
    (a) any input is HIGH   (b) all inputs are HIGH
    (c) no inputs are HIGH   (d) both answers (a) and (b)

9. The output of an exclusive-OR gate is HIGH when
    (a) one and only one input is HIGH   (b) both inputs are HIGH
    (c) no inputs are HIGH   (d) both answers (a) and (b)

10. An example of a data storage device is
    (a) the logic gate   (b) the flip-flop   (c) the comparator
    (d) the register   (e) both answers (b) and (d)

11. An IC package containing four AND gates is an example of
    (a) MSI   (b) SMT   (c) SOIC   (d) SSI

12. An LSI device has a circuit complexity of
    (a) 12 to 99 equivalent gates   (b) 100 to 9999 equivalent gates
    (c) 2000 to 5000 equivalent gates   (d) 10,000 to 99,999 equivalent gates

■ PROBLEMS

Answers to selected odd-numbered problems are found at the end of the book.

SECTION 1–1   Digital and Analog Quantities

1. Name two advantages of digital data as compared to analog data.

2. Name an analog quantity other than temperature and sound.

SECTION 1–2   Binary Digits, Logic Levels, and Digital Waveforms

3. Define the sequence of bits (1s and 0s) represented by each of the following sequences of levels:
    (a) HIGH, HIGH, LOW, HIGH, LOW, LOW, LOW, HIGH
    (b) LOW, LOW, LOW, HIGH, LOW, HIGH, LOW, HIGH, LOW

4. List the sequence of levels (HIGH and LOW) that represent each of the following bit sequences:
    (a) 1 0 1 1 1 0 1   (b) 1 1 1 0 1 0 0 1

5. For the pulse shown in Figure 1–43, graphically determine the following:
    (a) rise time   (b) fall time   (c) pulse width   (d) amplitude

6. Determine the period of the digital waveform in Figure 1–44.

7. What is the frequency of the waveform in Figure 1–44?

8. Is the pulse waveform in Figure 1–44 periodic or nonperiodic?

9. Determine the duty cycle of the waveform in Figure 1–44.

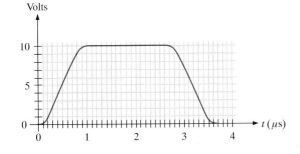

FIGURE 1–43

FIGURE 1–44

10. Determine the bit sequence represented by the waveform in Figure 1–45. A bit time is 1 μs in this case.

11. What is the total serial transfer time for the eight bits in Figure 1–45? What is the total parallel transfer time?

FIGURE 1–45

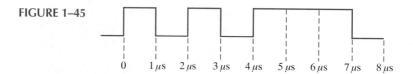

SECTION 1–3 Basic Logic Operations

12. A logic circuit requires HIGHs on all its inputs to make the output HIGH. What type of logic circuit is it?

13. A basic 2-input logic circuit has a HIGH on one input and a LOW on the other input, and the output is LOW. Identify the circuit.

14. A basic 2-input logic circuit has a HIGH on one input and a LOW on the other input, and the output is HIGH. What type of logic circuit is it?

SECTION 1–4 Basic Logic Functions

15. Name the function of each block in Figure 1–46 based on your observation of the inputs and outputs.

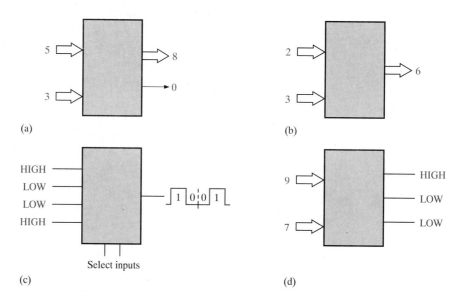

FIGURE 1–46

16. A pulse waveform with a frequency of 10 kHz is applied to the input of a counter. During 100 ms, how many pulses are counted?

17. Consider a register that can store eight bits. Assume that it has been reset so that it contains zeros in all positions. If you transfer four alternating bits (0101) serially into the register, beginning with a 1 and shifting to the right, what will the total content of the register be as soon as the fourth bit is stored?

SECTION 1–5 Digital Integrated Circuits

18. An integrated circuit chip has a complexity of 200 equivalent gates. How is it classified?

19. Explain the main difference between the DIP and SMT packages.

20. Label the pin numbers on the packages in Figure 1–47. Top views are shown.

FIGURE 1–47

(a) (b)

SECTION 1–6 Testing and Troubleshooting Instruments

21. A pulse is displayed on the screen of an oscilloscope, and you measure the base line as 1 V and the top of the pulse as 8 V. What is the amplitude?

22. A logic probe is applied to a contact point on an IC that is operating in a system. The lamp on the probe flashes repeatedly. What does this indicate?

SECTION 1–7 Digital System Application

23. Define the term *system*.

24. In the system depicted in Figure 1–42, why are the multiplexer and demultiplexer necessary?

25. What action can be taken to change the number of tablets per bottle in the system of Figure 1–42?

■ **ANSWERS TO SECTION REVIEWS**

SECTION 1–1

1. Digital means discrete.   **2.** Analog means continuous.

3. A digital quantity has a discrete set of values and an analog quantity has continuous values.

4. A PA system is analog. A CD player is analog and digital. A computer is all digital.

SECTION 1–2

1. Binary means having two states or values.   **2.** A bit is a binary digit.

3. The bits are 1 and 0.

4. Rise time: from 10% to 90% of amplitude. Fall time: from 90% to 10% of amplitude.

5. Frequency is the reciprocal of the period.

6. A clock waveform is a basic timing waveform from which other waveforms are derived.

7. A timing diagram shows time relationships of waveforms.

8. Parallel transfer is faster than serial transfer.

SECTION 1–3

1. When the input is LOW   **2.** When all inputs are HIGH

3. When any or all inputs are HIGH   **4.** When the inputs are different

5. An inverter is a NOT circuit.

6. A logic gate is a circuit that performs a logic operation (AND, OR, exclusive-OR).

SECTION 1–4

1. A comparator compares the magnitudes of two input numbers.

2. Add, subtract, multiply, and divide

3. Encoding is changing a familiar form such as decimal to a coded form such as binary.
4. Decoding is changing a code to a familiar form such as binary to decimal.
5. Multiplexing puts data from many sources onto one line. Demultiplexing takes data from one line and distributes it to many destinations.
6. Flip-flops, registers, semiconductor memories, magnetic disks
7. A counter counts events with a sequence of binary states.

SECTION 1–5

1. An IC is a circuit with all components integrated on a single silicon chip.
2. DIP—dual-in-line package; SMT—surface-mount technology; SOIC—small outline integrated circuit; SSI—small-scale integration; MSI—medium-scale integration; LSI—large-scale integration; VLSI—very large-scale integration; ULSI—ultra large-scale integration
3. **(a)** MSI   **(b)** LSI   **(c)** SSI   **(d)** VLSI   **(e)** ULSI

SECTION 1–6

1. Ten   2. Eight   3. Logic analyzer
4. A repetitive pulse waveform is indicated.

SECTION 1–7

1. The comparator determines when the tablet count reaches the preset number of tablets per bottle.
2. The dispenser valve is closed, the next bottle is moved into place by the conveyor, and the new sum is stored in register B.
3. Register A stores the preset number of tablets per bottle. Register B stores the total number of tablets bottled.

2

NUMBER SYSTEMS, OPERATIONS, AND CODES

■ CHAPTER OBJECTIVES

☐ Count in the binary number system

☐ Convert from decimal to binary and from binary to decimal

☐ Add, subtract, multiply, and divide binary numbers

☐ Determine the 1's and 2's complements of a binary number

☐ Express signed binary numbers in sign-magnitude, 1's complement, and 2's complement forms

☐ Carry out arithmetic operations with signed binary numbers

☐ Convert between the binary and hexadecimal number systems

☐ Add numbers in hexadecimal form

☐ Convert between the binary and octal number systems

☐ Express decimal numbers in binary coded decimal (BCD) form

☐ Add BCD numbers

☐ Convert between the binary system and the Gray code

☐ Express quantities in excess-3 code

☐ Interpret the American Standard Code for Information Interchange (ASCII)

☐ Use binary numbers and codes in a system application

■ CHAPTER OVERVIEW

The binary number system and digital codes are funda-mental to digital electronics. In this chapter, the binary number system and its relationship to other number systems such as decimal, hexadecimal, and octal is the principal focus. Arithmetic operations with binary numbers are covered to provide a basis for understand-ing how computers and many other types of digital systems work. Also, digital codes such as binary coded decimal (BCD), the Gray code, the excess-3 code, and the ASCII are covered. The parity method for detecting errors in codes is introduced.

■ DIGITAL SYSTEM APPLICATION

This Digital System Application illustrates the concepts taught in this chapter. To demonstrate how binary numbers and codes are used in an application, the tablet counting and control system that was introduced in Chapter 1 is used as an example. In the digital system application (Sec-tion 2–12) you will examine the binary states (numbers and codes) at various points in the system and see how the system sequences through a counting cycle. You will ana-lyze the system to determine the binary states for certain specified conditions.

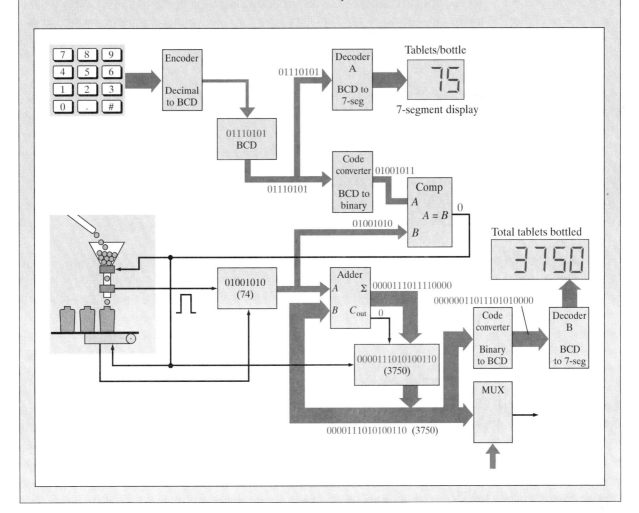

2–1 ■ DECIMAL NUMBERS

You are familiar with the decimal number system because you use decimal numbers every day. Although decimal numbers are commonplace, their weighted structure is often not understood. In this section, the structure of decimal numbers is reviewed. This review will help you more easily understand the structure of the binary number system, which is important in digital electronics. After completing this section, you should be able to

☐ Explain why the decimal number system is a weighted system   ☐ Define *powers-of-ten*   ☐ Determine the weight of each digit in a decimal number

In the **decimal** number system each of the ten digits, 0 through 9, represents a certain quantity. The ten symbols (**digits**) do not limit you to expressing only ten different quantities because you use the various digits in appropriate positions within a number to indicate the magnitude of the quantity. You can express quantities up through nine before running out of digits; if you wish to express a quantity greater than nine, you use two or more digits, and the position of each digit within the number tells you the magnitude it represents. If, for example, you wish to express the quantity twenty-three, you use (by their respective positions in the number) the digit 2 to represent the quantity twenty and the digit 3 to represent the quantity three, as illustrated below.

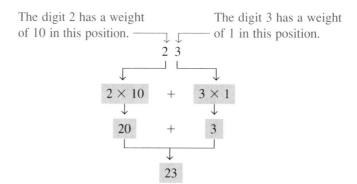

The position of each digit in a decimal number indicates the magnitude of the quantity represented and can be assigned a **weight.** The weights for whole numbers are positive powers-of-ten that increase from right to left, beginning with $10^0 = 1$.

$$\ldots\ 10^5\ 10^4\ 10^3\ 10^2\ 10^1\ 10^0$$

For fractional numbers, the weights are negative powers of ten that decrease from left to right beginning with 10^{-1}.

$$10^2\ 10^1\ 10^0.10^{-1}\ 10^{-2}\ 10^{-3}\ \ldots$$
$$\text{↑}\text{—Decimal point}$$

The value of a decimal number is the sum of the digits after each digit has been multiplied by its weight, as Examples 2–1 and 2–2 illustrate.

EXAMPLE 2–1

Express the decimal number 47 as a sum of the values of each digit.

Solution The digit 4 has a weight of 10 (10^1), as indicated by its position. The digit 7 has a weight of 1 (10^0), as indicated by its position.

$$47 = (4 \times 10^1) + (7 \times 10^0)$$
$$= (4 \times 10) + (7 \times 1) = 40 + 7$$

Related Exercise Determine the value of each digit in 939.

EXAMPLE 2–2

Express the decimal number 568.25 as a sum of the values of each digit.

Solution The whole number digit 5 has a weight of 100 (10^2), the digit 6 has a weight of 10 (10^1), the digit 8 has a weight of 1 (10^0), the fractional digit 2 has a weight of 0.1 (10^{-1}), and the fractional digit 5 has a weight of 0.01 (10^{-2}).

$$568.25 = (5 \times 10^2) + (6 \times 10^1) + (8 \times 10^0) + (2 \times 10^{-1}) + (5 \times 10^{-2})$$
$$= (5 \times 100) + (6 \times 10) + (8 \times 1) + (2 \times 0.1) + (5 \times 0.01)$$
$$= 500 + 60 + 8 + 0.2 + 0.05$$

Related Exercise Determine the value of each digit in 67.924.

CALCULATOR TIP

Powers-of-ten can be found on the calculator using the y^x key with the following steps:

1. Enter *y*, which is the number ten: [1] [0]

2. Press y^x

3. Enter *x*, which is the power (3 for example): [3]

4. The display shows [1000]

SECTION 2–1 REVIEW

1. What weight does the digit 7 have in each of the following numbers?
   (a) 1370   **(b)** 6725   **(c)** 7051   **(d)** 58.72
2. Express each of the following decimal numbers as a sum of the products obtained by multiplying each digit by its appropriate weight:
   (a) 51   **(b)** 137   **(c)** 1492   **(d)** 106.58

2–2 ▪ BINARY NUMBERS

The binary number system is simply another way to count. The binary system is less complicated than the decimal system because it has only two digits. It may seem more difficult at first because it is unfamiliar to you.

The decimal system with its ten digits is a base-ten system; the binary system with its two digits is a base-two system. The two binary digits (bits) are 1 and 0. The position of a 1 or 0 in a binary number indicates its weight, or value within the number, just as the position of a decimal digit determines the value of that digit. The weights in a binary number are based on powers-of-two. After completing this section, you should be able to

□ Count in binary   □ Determine the largest decimal number that can be represented by a given number of bits   □ Convert a binary number to a decimal number

Counting in Binary

To learn to count in the **binary** system, first look at how you count in the decimal system. You start at zero and count up to nine before you run out of digits. You then start another digit position (to the left) and continue counting 10 through 99. At this point you have exhausted all two-digit combinations, so a third digit position is needed to count from 100 through 999.

A comparable situation occurs when you count in binary, except that you have only two digits, called **bits.** Begin counting: 0, 1. At this point you have used both digits, so include another digit position and continue: 10, 11. You have now exhausted all combinations of two digits, so a third position is required. With three digit positions you can continue to count: 100, 101, 110, and 111. Now you need a fourth digit position to continue, and so on. A binary count of zero through fifteen is shown in Table 2–1. Notice the patterns with which the 1s and 0s alternate in each column.

TABLE 2–1

Decimal Number	Binary Number			
0	0	0	0	0
1	0	0	0	1
2	0	0	1	0
3	0	0	1	1
4	0	1	0	0
5	0	1	0	1
6	0	1	1	0
7	0	1	1	1
8	1	0	0	0
9	1	0	0	1
10	1	0	1	0
11	1	0	1	1
12	1	1	0	0
13	1	1	0	1
14	1	1	1	0
15	1	1	1	1

As you have seen in Table 2–1, four bits are required to count from zero to 15. In general, with n bits you can count up to a number equal to $2^n - 1$.

$$\text{Largest decimal number} = 2^n - 1$$

For example, with five bits ($n = 5$) you can count from zero to thirty-one:

$$2^5 - 1 = 32 - 1 = 31$$

With six bits ($n = 6$) you can count from zero to sixty-three:

$$2^6 - 1 = 64 - 1 = 63$$

A table of powers-of-two is given in Appendix C.

CALCULATOR TIP

Powers-of-two can be found on the calculator using the $\boxed{y^x}$ key with the following steps:

1. Enter y, the base number 2: $\boxed{2}$
2. Press $\boxed{y^x}$
3. Enter x, the power (4 for example): $\boxed{4}$
4. The display shows $\boxed{\qquad 16}$

The Weighting Structure of Binary Numbers

A binary number is a weighted number. The right-most bit is the **least significant bit (LSB)** in a binary whole number and has a weight of $2^0 = 1$. The weights increase from right to left by a power of two for each bit. The left-most bit is the **most significant bit (MSB)**; its weight depends on the size of the binary number.

Fractional numbers can also be represented in binary by placing bits to the right of the binary point, just as fractional decimal digits are placed to the right of the decimal point. The left-most bit is the MSB in a binary fractional number and has a weight of $2^{-1} = 0.5$. The fractional weights decrease from left to right by a negative power of two for each bit.

The weight structure of a binary number is

$$2^{n-1} \ldots 2^3 \ 2^2 \ 2^1 \ 2^0.2^{-1} \ 2^{-2} \ldots 2^{-n}$$
$$\underset{\text{Binary point}}{\uparrow}$$

where n is the number of bits from the binary point. Thus, all the bits to the left of the binary point have weights that are positive powers of two, as previously discussed for whole numbers. All bits to the right of the binary point have weights that are negative powers of two, or fractional weights.

The powers of two and their equivalent decimal weights for an 8-bit binary whole number and a 6-bit binary fractional number are shown in Table 2–2. Notice that the weight doubles for each positive power of two and that the weight is halved for each negative power of two. You can easily extend the table by doubling the weight of the most significant positive power of two and halving the weight of the least significant negative power of two; for eample, $2^9 = 512$ and $2^{-7} = 0.0078125$.

TABLE 2–2
Binary weights

Positive Powers of Two (whole numbers)									Negative Powers of Two (fractional number)					
2^8	2^7	2^6	2^5	2^4	2^3	2^2	2^1	2^0	2^{-1}	2^{-2}	2^{-3}	2^{-4}	2^{-5}	2^{-6}
256	128	64	32	16	8	4	2	1	1/2	1/4	1/8	1/16	1/32	1/64
									0.5	0.25	0.125	0.0625	0.03125	0.015625

Evaluating Binary Numbers

The decimal value of any binary number can be determined by adding the weights of all bits that are 1 and discarding the weights of all bits that are 0. The following two examples will illustrate this.

EXAMPLE 2–3

Determine the decimal value of the binary whole number 1101101.

Solution Determine the weight of each bit that is a 1 and then find the sum of the weights.

$$\text{Weight:} \quad 2^6 \ 2^5 \ 2^4 \ 2^3 \ 2^2 \ 2^1 \ 2^0$$
$$\text{Binary number:} \quad 1 \ \ 1 \ \ 0 \ \ 1 \ \ 1 \ \ 0 \ \ 1$$

$$1101101 = 2^6 + 2^5 + 2^3 + 2^2 + 2^0$$
$$= 64 + 32 + 8 + 4 + 1 = 109$$

Related Exercise Evaluate the binary number 10010001.

EXAMPLE 2–4

Determine the decimal value of the fractional binary number 0.1011.

Solution First, determine the weight of each bit that is a 1, and then sum the weights.

$$\text{Weight:} \quad 2^{-1} \ 2^{-2} \ 2^{-3} \ 2^{-4}$$
$$\text{Binary number:} \quad 0 \ . \ 1 \ \ \ 0 \ \ \ 1 \ \ \ 1$$

$$0.1011 = 2^{-1} + 2^{-3} + 2^{-4}$$
$$= 0.5 + 0.125 + 0.0625 = 0.6875$$

Related Exercise Evaluate the binary number 10.111.

SECTION 2–2 REVIEW

1. What is the largest decimal number that can be represented in binary with eight bits?
2. Determine the weight of the 1 in the binary number 10000.
3. Evaluate the binary number 10111101.011.

2–3 ■ DECIMAL-TO-BINARY CONVERSION

In Section 2–2 you learned how to determine the equivalent decimal value of a binary number. Now you will learn two ways of converting from a decimal number to a binary number. After completing this section, you should be able to

☐ Convert a decimal number to binary using the sum-of-weights method ☐ Convert a decimal whole number to binary using the repeated division-by-2 method ☐ Convert a decimal fraction to binary using the repeated multiplication-by-2 method

Sum-of-Weights Method

One way to find the binary number that is equivalent to a given decimal number is to determine the set of binary weights whose sum is equal to the decimal number. An easy way to remember binary weights is that the lowest weight is 1 (2^0) and that by doubling any weight, you get the next higher weight; thus, a list of seven binary weights would be 64, 32, 16, 8, 4, 2, 1 as you learned in the last section. The decimal number 9, for example, can be expressed as the sum of binary weights as follows:

$$9 = 8 + 1 \quad \text{or} \quad 9 = 2^3 + 2^0$$

Placing 1s in the appropriate weight positions, 2^3 and 2^0, and 0s in the 2^2 and 2^1 positions determines the binary number for decimal 9:

$$
\begin{array}{cccc}
2^3 & 2^2 & 2^1 & 2^0 \\
1 & 0 & 0 & 1
\end{array}
\quad \text{Binary number for nine}
$$

EXAMPLE 2–5

Convert the following decimal numbers to binary form:
(a) 12   (b) 25   (c) 58   (d) 82

Solution
(a) $12 = 8 + 4 = 2^3 + 2^2$ ⟶ 1100
(b) $25 = 16 + 8 + 1 = 2^4 + 2^3 + 2^0$ ⟶ 11001
(c) $58 = 32 + 16 + 8 + 2 = 2^5 + 2^4 + 2^3 + 2^1$ ⟶ 111010
(d) $82 = 64 + 16 + 2 = 2^6 + 2^4 + 2^1$ ⟶ 1010010

Related Exercise   Convert the decimal number 125 to binary.

Repeated Division-by-2 Method

A systematic method of converting whole numbers from decimal to binary is the *repeated division-by-2* process. For example, to convert the decimal number 12 to binary, begin by dividing 12 by 2. Then divide each resulting quotient by 2 until there is a 0 whole-number quotient. The **remainders** generated by each division form the binary number. The first remainder to be produced is the least significant bit (LSB) in the binary number, and the last remainder to be produced is the most significant bit (MSB). This procedure is shown in the following steps for converting the decimal number 12 to binary.

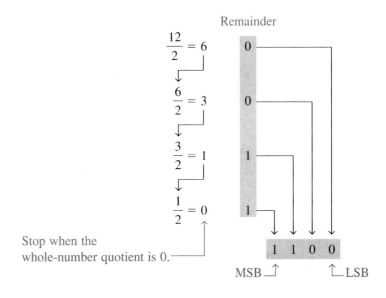

Remainder

$$\frac{12}{2} = 6 \qquad 0$$

$$\frac{6}{2} = 3 \qquad 0$$

$$\frac{3}{2} = 1 \qquad 1$$

$$\frac{1}{2} = 0 \qquad 1$$

Stop when the whole-number quotient is 0.

1 1 0 0

MSB ⌐   ⌐ LSB

CALCULATOR TIP

When using your calculator to convert from decimal to binary, if the result of a division is a whole number with no fractional part, the remainder is always 0. For example,

$\boxed{1}\boxed{2}\boxed{\div}\boxed{2}\boxed{=}$ $\boxed{\qquad\qquad 6}$ *remainder is 0*

If the result of a division is a whole number with a fractional part or just a fractional number, the remainder is always 1. For example,

$\boxed{3}\boxed{\div}\boxed{2}\boxed{=}$ $\boxed{\qquad\qquad 1.5}$ *remainder is 1*

$\boxed{1}\boxed{\div}\boxed{2}\boxed{=}$ $\boxed{\qquad\qquad 0.5}$ *remainder is 1*

EXAMPLE 2–6

Convert the following decimal numbers to binary:
(a) 19 **(b)** 45

Solution

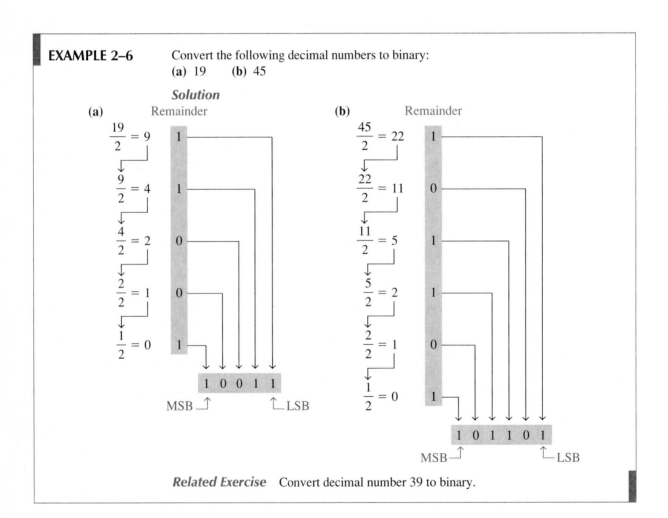

Related Exercise Convert decimal number 39 to binary.

Converting Decimal Fractions to Binary

Examples 2–5 and 2–6 demonstrated whole-number conversions. Now let's look at fractional conversions. An easy way to remember fractional binary weights is that the most significant weight is 0.5 (2^{-1}) and that by halving any weight, you get the next lower weight; thus a list of four fractional binary weights would be 0.5, 0.25, 0.125, 0.0625.

Sum-of-Weights The sum-of-weights method can be applied to fractional decimal numbers, as shown in the following example:

$$0.625 = 0.5 + 0.125 = 2^{-1} + 2^{-3} = 0.101$$

There is a 1 in the 2^{-1} position, a 0 in the 2^{-2} position, and a 1 in the 2^{-3} position.

Repeated Multiplication by 2 As you have seen, decimal whole numbers can be converted to binary by repeated division by 2. Decimal fractions can be converted to binary by repeated multiplication by 2. For example, to convert the decimal fraction 0.3125 to binary, begin by multiplying 0.3125 by 2 and then multiplying each resulting fractional part of the product by 2 until the fractional product is zero or until the desired number of decimal places is reached. The carried digits, or **carries,** generated by the multiplications produce the binary number. The first carry produced is the MSB, and the last carry is the LSB. This procedure is illustrated as follows:

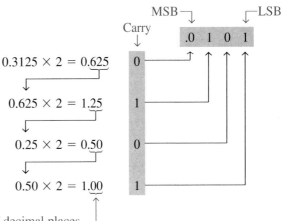

$$0.3125 \times 2 = 0.625 \qquad 0$$
$$0.625 \times 2 = 1.25 \qquad 1$$
$$0.25 \times 2 = 0.50 \qquad 0$$
$$0.50 \times 2 = 1.00 \qquad 1$$

Continue to the desired number of decimal places
or stop when fractional part is all zeros.

**SECTION 2–3
REVIEW**

1. Convert each decimal number to binary by using the sum-of-weights method:
   (a) 23 **(b)** 57 **(c)** 45.5
2. Convert each decimal number to binary by using the repeated division-by-2 method (repeated multiplication-by-2 for fractions):
   (a) 14 **(b)** 21 **(c)** 0.375

2–4 ■ BINARY ARITHMETIC

Binary arithmetic is essential in all digital computers and in many other types of digital systems. To understand digital systems, you must know the basics of binary addition, subtraction, multiplication, and division. This section provides an introduction that will be expanded in later sections. After completing this section, you should be able to

☐ Add binary numbers ☐ Subtract binary numbers ☐ Multiply binary numbers
☐ Divide binary numbers

Binary Addition

The four basic rules for adding binary digits are as follows:

$0 + 0 = 0$	Sum of 0 with a carry of 0
$0 + 1 = 1$	Sum of 1 with a carry of 0
$1 + 0 = 1$	Sum of 1 with a carry of 0
$1 + 1 = 10$	Sum of 0 with a carry of 1

Notice that the first three rules result in a single bit and in the fourth rule the addition of two 1s yields a binary two (10). When binary numbers are added, the last condition creates a sum of 0 in a given column and a carry of 1 over to the next column to the left, as illustrated in the following addition of $11 + 1$:

$$
\begin{array}{ccc}
\text{Carry} & \text{Carry} & \\
1 & 1 & \\
0 & 1 & 1 \\
+\,0 & 0 & 1 \\
\hline
1 & 0 & 0
\end{array}
$$

In the right column, $1 + 1 = 0$ with a carry of 1 to the next column to the left. In the middle column, $1 + 1 + 0 = 0$ with a carry of 1 to the next column to the left. In the left column, $1 + 0 + 0 = 1$.

When there is a carry of 1, you have a situation in which three bits are being added (a bit in each of the two numbers and a carry bit). This situation is illustrated as follows:

Carry bits		
$1 + 0 + 0 = 01$	Sum of 1 with a carry of 0
$1 + 1 + 0 = 10$	Sum of 0 with a carry of 1
$1 + 0 + 1 = 10$	Sum of 0 with a carry of 1
$1 + 1 + 1 = 11$	Sum of 1 with a carry of 1

Example 2–7 illustrates binary addition.

EXAMPLE 2–7

Add the following binary numbers:
(a) $11 + 11$ (b) $100 + 10$ (c) $111 + 11$ (d) $110 + 100$

Solution The equivalent decimal addition is also shown for reference.

$$
\textbf{(a)}\quad
\begin{array}{r}
11 \\
+\,11 \\
\hline
110
\end{array}
\quad
\begin{array}{r}
3 \\
+\,3 \\
\hline
6
\end{array}
\qquad
\textbf{(b)}\quad
\begin{array}{r}
100 \\
+\,10 \\
\hline
110
\end{array}
\quad
\begin{array}{r}
4 \\
+\,2 \\
\hline
6
\end{array}
\qquad
\textbf{(c)}\quad
\begin{array}{r}
111 \\
+\,11 \\
\hline
1010
\end{array}
\quad
\begin{array}{r}
7 \\
+\,3 \\
\hline
10
\end{array}
\qquad
\textbf{(d)}\quad
\begin{array}{r}
110 \\
+\,100 \\
\hline
1010
\end{array}
\quad
\begin{array}{r}
6 \\
+\,4 \\
\hline
10
\end{array}
$$

Related Exercise Add 1111 and 1100.

Binary Subtraction

The four basic rules for subtracting binary digits are as follows:

$0 - 0 = 0$	
$1 - 1 = 0$	
$1 - 0 = 1$	
$10 - 1 = 1$	$0 - 1$ with a borrow of 1

When subtracting numbers, you sometimes have to borrow from the next column to the left. A borrow is required in binary only when you try to subtract a 1 from a 0. In this case, when a 1 is borrowed from the next column to the left, a 10 is created in the column being subtracted, and the last of the four basic rules just listed must be applied. Examples 2–8 and 2–9 illustrate binary subtraction, with the equivalent decimal subtraction also shown.

EXAMPLE 2–8

Perform the following binary subtractions:
(a) $11 - 01$ **(b)** $11 - 10$

Solution   **(a)**
$$\begin{array}{r} 11 \\ - 01 \\ \hline 10 \end{array} \quad \begin{array}{r} 3 \\ - 1 \\ \hline 2 \end{array} \qquad \textbf{(b)} \quad \begin{array}{r} 11 \\ - 10 \\ \hline 01 \end{array} \quad \begin{array}{r} 3 \\ - 2 \\ \hline 1 \end{array}$$

No borrows were required in this example. The binary number 01 is the same as 1.

Related Exercise   Subtract 100 from 111.

EXAMPLE 2–9

Subtract 011 from 101.

Solution
$$\begin{array}{r} 101 \\ - 011 \\ \hline 010 \end{array} \quad \begin{array}{r} 5 \\ - 3 \\ \hline 2 \end{array}$$

Let's examine exactly what was done to subtract the two binary numbers. Begin with the right column.

Left column:
When a 1 is borrowed,
a 0 is left, so $0 - 0 = 0$.

Middle column:
Borrow 1 from next column
to the left, making a 10 in this
column, then $10 - 1 = 1$.

$$\begin{array}{r} \overset{0}{1}{}^{1}01 \\ - \ 0 \ 11 \\ \hline 0 \ 10 \end{array}$$

Right column:
$1 - 1 = 0$

Related Exercise   Subtract 101 from 110.

Binary Multiplication

The four basic rules for multiplying binary digits are as follows:

$$0 \times 0 = 0$$
$$0 \times 1 = 0$$
$$1 \times 0 = 0$$
$$1 \times 1 = 1$$

Multiplication is performed with binary numbers in the same manner as with decimal numbers. It involves forming partial products, shifting each successive partial product left one place, and then adding all the partial products. Example 2–10 will illustrate the procedure, with the equivalent decimal multiplication shown for reference.

EXAMPLE 2–10

Perform the following binary multiplications:

(a) 1×11 (b) 11×11 (c) 101×111 (d) 1001×1011

Solution

(a)
```
   11      3
 × 1     × 1
   11      3
```

(b)
```
                      11       3
                    × 11     × 3
         Partial ⎰   11        9
         products ⎱ + 11
                    1001
```

(c)
```
                           111      7
                         × 101    × 5
          Partial ⎰       111      35
          products ⎱      000
                        + 111
                         100011
```

(d)
```
                           1011      11
                         × 1001    ×  9
          Partial ⎰       1011      99
          products ⎰      0000
                    ⎱     0000
                        + 1011
                         1100011
```

Related Exercise   Multiply 1101×1010.

Binary Division

Division in binary follows the same procedure as division in decimal, as Example 2–11 illustrates. The equivalent decimal divisions are also given.

EXAMPLE 2–11

Perform the following binary divisions:

(a) $110 \div 11$ (b) $110 \div 10$

Solution

```
            10      2
(a) 11)110   3)6
        11       6
       000       0
```

```
            11       3
(b) 10)110    2)6
        10        6
        10        0
        10
        00
```

Related Exercise   Divide 1100 by 100.

SECTION 2–4 REVIEW

1. Perform the following binary additions:
   (a) $1101 + 1010$ (b) $10111 + 01101$
2. Perform the following binary subtractions:
   (a) $1101 - 0100$ (b) $1001 - 0111$
3. Perform the indicated binary operations:
   (a) 110×111 (b) $1100 \div 011$

2–5 ■ 1'S AND 2'S COMPLEMENTS OF BINARY NUMBERS

The 1's complement and the 2's complement of a binary number are important because they permit the representation of negative numbers. The method of 2's complement arithmetic is commonly used in computers to handle negative numbers. After completing this section, you should be able to

☐ Convert a binary number to its 1's complement ☐ Convert a binary number to its 2's complement using either of two methods

Obtaining the 1's Complement of a Binary Number

The 1's **complement** of a binary number is found by simply changing all 1s to 0s and all 0s to 1s, as illustrated below:

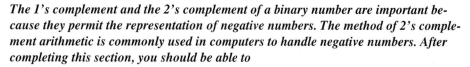

```
1 0 1 1 0 0 1 0      Binary number
↓ ↓ ↓ ↓ ↓ ↓ ↓ ↓
0 1 0 0 1 1 0 1      1's complement
```

Obtaining the 2's Complement of a Binary Number

The 2's complement of a binary number is found by adding 1 to the LSB of the 1's complement.

2's complement = (1's complement) + 1

Example 2–12 shows how to find the 2's complement.

EXAMPLE 2–12

Find the 2's complement of 10110010.

Solution

```
  10110010      Binary number
  01001101      1's complement
+        1      Add 1
  01001110      2's complement
```

Related Exercise   Determine the 2's complement of 11001011.

An alternative method of obtaining the 2's complement of a binary number is as follows:

1. Start at the right with the LSB and write the bits as they are up to and including the first 1.
2. Take the 1's complements of the remaining bits.

Example 2–13 illustrates these steps.

EXAMPLE 2–13

Find the 2's complement of 10111000.

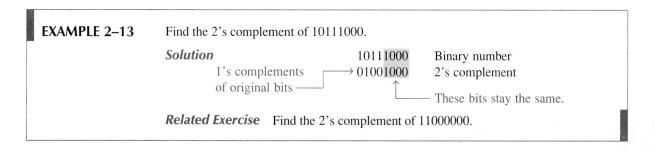

Solution   10111000   Binary number
1's complements ⟶ 01001000   2's complement
of original bits
                                                              These bits stay the same.

Related Exercise   Find the 2's complement of 11000000.

To convert from a 1's or 2's complement back to the true (uncomplemented) binary form, use the two procedures described previously. To go from the 1's complement back to true binary, reverse all the bits. To go from the 2's complement form back to true binary, take the 1's complement and add 1 to the least significant bit.

SECTION 2–5 REVIEW

1. Determine the 1's complement of each binary number:
   (a) 00011010   (b) 11110111   (c) 10001101
2. Determine the 2's complement of each binary number:
   (a) 00010110   (b) 11111100   (c) 10010001

2–6 ■ SIGNED NUMBERS

Digital systems, such as the computer, must be able to handle both positive and negative numbers. A signed binary number consists of both sign and magnitude information. The sign indicates whether a number is positive or negative and the magnitude is the value of the number. There are three ways in which signed numbers can be represented in binary form: sign-magnitude, 1's complement, and 2's complement. After completing this section, you should be able to

☐ Express positive and negative numbers in sign-magnitude   ☐ Express positive and negative numbers in 1's complement   ☐ Express positive and negative numbers in 2's complement   ☐ Evaluate signed binary numbers

The Sign Bit

The left-most bit in a signed binary number is the **sign bit,** which tells you whether the number is positive or negative. A 0 is for positive, and a 1 is for negative.

Sign-Magnitude System

When a signed binary number is represented in sign-magnitude, the left-most bit is the sign bit and the remaining bits are the magnitude bits. The magnitude bits are in true (uncomplemented) binary for both positive and negative numbers. For example, the decimal number +25 is expressed as an 8-bit signed binary number using the sign-magnitude system as

$$00011001$$

Sign bit ⬏   ⬑ Magnitude bits

The decimal number −25 is expressed as

$$10011001$$

Notice that the only difference between +25 and −25 is the sign bit because the magnitude bits are in true binary for both positive and negative numbers.

In the sign-magnitude system, a negative number has the same magnitude bits as the corresponding positive number but the sign bit is a 1.

1's Complement System

Positive numbers in the 1's complement system are represented the same way as the positive sign-magnitude numbers. Negative numbers, however, are the 1's complements of the corresponding positive numbers. For example, the decimal number -25 is expressed as the 1's complement of $+25$ (00011001) as

$$11100110$$

In the 1's complement system, a negative number is the 1's complement of the corresponding positive number.

2's Complement System

Positive numbers in the 2's complement system are represented the same way as in the sign magnitude and 1's complement systems. Negative numbers are the 2's complements of the corresponding positive numbers. Again let's take -25 and express it as the 2's complement of $+25$ (00011001):

$$11100111$$

In the 2's complement system, a negative number is the 2's complement of the corresponding positive number.

In computers, the 2's complement system is the most widely used for handling signed numbers for reasons that are discussed later.

EXAMPLE 2–14

Express the decimal number -39 as an 8-bit number in sign-magnitude, 1's complement, and 2's complement systems.

Solution　First, write the 8-bit binary number for $+39$:

$$00100111$$

In the sign-magnitude system, -39 is produced by changing the sign bit to a 1 and leaving the magnitude bits as they are. The number is

$$10100111$$

In the 1's complement system, -39 is produced by taking the 1's complement of $+39$ (00100111):

$$11011000$$

In the 2's complement system, -39 is produced by taking the 2's complement of $+39$ (00100111) as follows:

$$
\begin{array}{ll}
11011000 & \text{1's complement} \\
+\quad\quad\ 1 & \\
\hline
11011001 & \text{2's complement}
\end{array}
$$

Related Exercise　Express $+19$ and -19 in sign-magnitude, 1's complement, and 2's complement.

Evaluation of Signed Numbers

Sign-magnitude   Positive and negative numbers in the sign-magnitude system are evaluated by summing the weights in all the magnitude bit positions where there are 1s and ignoring those positions where there are zeros. The sign is determined by examination of the sign bit. The evaluation is illustrated by Example 2–15.

EXAMPLE 2–15

Determine the decimal value of this signed binary number expressed in sign-magnitude: 10010101.

Solution   The seven magnitude bits and their powers-of-two weights are as follows:

$$2^6 \ 2^5 \ 2^4 \ 2^3 \ 2^2 \ 2^1 \ 2^0$$
$$0 \ \ 0 \ \ 1 \ \ 0 \ \ 1 \ \ 0 \ \ 1$$

Summing the weights where there are 1s,

$$16 + 4 + 1 = 21$$

The sign bit is 1; therefore, the number is -21.

Related Exercise   Determine the decimal value of the sign-magnitude number 01110111.

1's Complement   Positive numbers in the 1's complement system are evaluated by summing the weights in all bit positions where there are 1s and ignoring those positions where there are zeros. Negative numbers are evaluated by assigning a negative value to the weight of the sign bit, summing all the weights where there are 1s, and adding 1 to the result. This is illustrated by Example 2–16.

EXAMPLE 2–16

Determine the decimal values of the signed binary numbers expressed in 1's complement: **(a)** 00010111   **(b)** 11101000.

Solution
(a) The bits and their powers-of-two weights for the positive number are as follows:

$$-2^7 \ 2^6 \ 2^5 \ 2^4 \ 2^3 \ 2^2 \ 2^1 \ 2^0$$
$$0 \ \ 0 \ \ 0 \ \ 1 \ \ 0 \ \ 1 \ \ 1 \ \ 1$$

Summing the weights where there are 1s,

$$16 + 4 + 2 + 1 = +23$$

(b) The bits and their powers-of-two weights for the negative number are as follows. Notice that the negative sign bit has a weight of -2^7 or -128.

$$-2^7 \ 2^6 \ 2^5 \ 2^4 \ 2^3 \ 2^2 \ 2^1 \ 2^0$$
$$1 \ \ 1 \ \ 1 \ \ 0 \ \ 1 \ \ 0 \ \ 0 \ \ 0$$

Summing the weights where there are 1s,

$$-128 + 64 + 32 + 8 = -24$$

Adding 1 to the result, the final number is

$$-24 + 1 = -23$$

Related Exercise   Determine the decimal value of the 1's complement number 11101011.

2's Complement Positive and negative numbers in the 2's complement system are evaluated by summing the weights in all bit positions where there are 1s and ignoring those positions where there are zeros. The weight of the sign bit in a negative number is given a negative value. This is illustrated by Example 2–17.

EXAMPLE 2–17 Determine the decimal values of the signed binary numbers expressed in 2's complement: **(a)** 01010110   **(b)** 10101010.

Solution
(a) The bits and their powers-of-two weights for the positive number are as follows:

$$-2^7 \ 2^6 \ 2^5 \ 2^4 \ 2^3 \ 2^2 \ 2^1 \ 2^0$$
$$0 \ \ 1 \ \ 0 \ \ 1 \ \ 0 \ \ 1 \ \ 1 \ \ 0$$

Summing the weights where there are 1's,

$$64 + 16 + 4 + 2 = +86$$

(b) The bits and their powers-of-two weights for the negative number are as follows. Notice that the negative sign bit has a weight of $-2^7 = -128$.

$$-2^7 \ 2^6 \ 2^5 \ 2^4 \ 2^3 \ 2^2 \ 2^1 \ 2^0$$
$$1 \ \ 0 \ \ 1 \ \ 0 \ \ 1 \ \ 0 \ \ 1 \ \ 0$$

Summing the weights where there are 1's,

$$-128 + 32 + 8 + 2 = -86$$

Related Exercise Determine the decimal value of the 2's complement number 111010111.

From these examples, you can see one of the reasons why the 2's complement system is preferred for representing signed numbers: It simply requires a summation of weights regardless of whether the number is positive or negative. The sign-magnitude system requires two steps—sum the weights of the magnitude bits and examine the sign bit to determine if the number is positive or negative. The 1's complement system requires adding 1 to the summation of weights for negative numbers but not for positive numbers. Also, the 1's complement system is generally not used because two representations of zero (00000000 or 11111111) are possible.

The 2's complement system is preferred and is used in most computers because it makes arithmetic operations easier, as you will see in Section 2–7.

Range of Signed Numbers That Can Be Represented

We have used eight bit numbers for illustration because the 8-bit grouping is standard in most computers and has been given the special name **byte.** Using eight bits, 256 different numbers can be represented. Combining two bytes to get sixteen bits, 65,536 different numbers can be represented. Combining four bytes to get 32 bits, 4.295×10^9 different numbers can be represented, and so on. The formula for finding the number of different combinations of *n* bits is

$$\text{Total combinations} = 2^n$$

For 2's complement signed numbers, the range of values for *n*-bit numbers is

$$-(2^{n-1}) \quad \text{to} \quad +(2^{n-1} - 1)$$

where in each case there is one sign bit and $n - 1$ magnitude bits. For example, with four bits you can represent numbers in 2's complement ranging from $-(2^3) = -8$ to $2^3 - 1 = +7$. Similarly, with eight bits, you can go from -128 to $+127$, with sixteen bits you can go from $-32,768$ to $+32,767$, and so on.

SECTION 2–6 REVIEW	**1.** Express the decimal number $+9$ as an 8-bit binary number in the sign-magnitude system. **2.** Express the decimal number -33 as an 8-bit binary number in the 1's complement system. **3.** Express the decimal number -46 as an 8-bit binary number in the 2's complement system.

2–7 ■ ARITHMETIC OPERATIONS WITH SIGNED NUMBERS

In the last section, you learned how signed numbers are represented in three different systems. In this section, you will learn how signed numbers are added, subtracted, multiplied, and divided. Because the 2's complement system for representing signed numbers is the most widely used in computers and microprocessor-based systems, the coverage in this section is limited to 2's complement arithmetic. The processes covered can be extended to the other systems if necessary. After completing this section, you should be able to

☐ Add signed binary numbers ☐ Explain how computers add strings of numbers
☐ Define *overflow* ☐ Subtract signed binary numbers ☐ Multiply signed binary numbers using the direct addition method ☐ Multiply signed binary numbers using the partial products method ☐ Divide signed binary numbers

Addition

The two numbers in an addition are the **addend** and the **augend.** The result is the **sum.** There are four cases that can occur when two signed binary numbers are added:

1. Both numbers positive
2. Positive number with magnitude larger than negative number
3. Negative number with magnitude larger than positive number
4. Both numbers negative

Let's take one case at a time using 8-bit signed numbers as examples. The equivalent decimal numbers are shown for reference.

Both numbers positive:

$$
\begin{array}{rr}
00000111 & 7 \\
+\ 00000100 & +\ 4 \\
\hline
00001011 & 11
\end{array}
$$

The sum is positive and is therefore in true (uncomplemented) binary.

Positive number with magnitude larger than negative number:

$$
\begin{array}{rr}
00001111 & 15 \\
+\ 11111010 & +\ -6 \\
\hline
\text{Discard carry} \longrightarrow \mathbf{1}\ 00001001 & 9
\end{array}
$$

The final carry bit is discarded. The sum is positive and therefore in true (uncomplemented) binary.

Negative number with magnitude larger than positive number:

$$\begin{array}{rr} 00010000 & 16 \\ +\ 11101000 & +\ -24 \\ \hline 11111000 & -8 \end{array}$$

The sum is negative and therefore in 2's complement form.

Both numbers negative:

$$\begin{array}{rr} 11111011 & -5 \\ +\ 11110111 & +\ -9 \\ \hline \text{Discard carry} \longrightarrow \boxed{1}\ 11110010 & -14 \end{array}$$

The final carry bit is discarded. The sum is negative and therefore in 2's complement form.

In a computer, the negative numbers are stored in 2's complement form so, as you can see, the addition process is very simple: *Add the two numbers and discard any final carry bit.*

Overflow Condition   When two numbers are added and the number of bits required to represent the sum exceeds the number of bits in the two numbers, an **overflow** results as indicated by an incorrect sign bit. An overflow can occur only when both numbers are positive or both numbers are negative. The following 8-bit example will illustrate this condition.

$$\begin{array}{rr} 01111101 & 125 \\ +\ 00111010 & +\ \ 58 \\ \hline 10110111 & 183 \end{array}$$

Sign incorrect ⟶
Magnitude incorrect ⟶

In this example the sum of 183 requires eight magnitude bits. Since there are seven magnitude bits in the numbers (one bit is the sign), there is a carry into the sign bit which produces the overflow indication.

Numbers Are Added Two at a Time   Now let's look at the addition of a string of numbers, added two at a time. This can be accomplished by adding the first two numbers, then adding the third number to the sum of the first two, then adding the fourth number to this result, and so on. This is how computers add strings of numbers. The addition of numbers taken two at a time is illustrated in Example 2–18.

EXAMPLE 2–18

Add the signed numbers: 01000100, 00011011, 00001110, and 00010010.

Solution   The equivalent decimal additions are given for reference.

$$\begin{array}{rll} 68 & 01000100 & \\ +\ 27 & +\ 00011011 & \text{Add 1st two numbers} \\ \hline 95 & 01011111 & \text{1st sum} \\ +\ 14 & +\ 00001110 & \text{Add 3rd number} \\ \hline 109 & 01101101 & \text{2nd sum} \\ +\ 18 & +\ 00010010 & \text{Add 4th number} \\ \hline 127 & 01111111 & \text{Final sum} \end{array}$$

Related Exercise   Add 00110011, 10111111, and 01100011. These are signed numbers.

Subtraction

Subtraction is a special case of addition. For example, subtracting +6 (the **subtrahend**) from +9 (the **minuend**) is equivalent to adding −6 to +9. Basically, *the subtraction operation changes the sign of the subtrahend and adds it to the minuend.* The result of a subtraction is called the **difference.**

> **The sign of a positive or negative binary number is changed by taking its 2's complement.**

For example, taking the 2's complement of the positive number 00000100 (+4), you get 11111100, which is −4 as the following sum-of-weights evaluation shows:

$$-128 + 64 + 32 + 16 + 8 + 4 = -4$$

As another example, taking the 2's complement of the negative number 11101101 (−19), you get 00010011, which is +19 as the following evaluation shows:

$$16 + 2 + 1 = 19$$

Since subtraction is simply an addition with the sign of the subtrahend changed, the process is stated as follows:

> **To subtract two signed numbers, take the 2's complement of the subtrahend and add, discarding any final carry bit.**

Example 2–19 illustrates the subtraction process.

EXAMPLE 2–19

Perform each of the following subtractions of the signed numbers:
(a) 00001000 − 00000011   **(b)** 00001100 − 11110111
(c) 11100111 − 00010011   **(d)** 10001000 − 11100010

Solution   Like in other examples, the equivalent decimal subtractions are given for reference.
(a) In this case, $8 − 3 = 8 + (−3) = 5$.

$$\begin{array}{ll} 00001000 & \text{Minuend } (+8) \\ +\ 11111101 & \text{2's complement of subtrahend } (-3) \\ \hline \mathbf{1}\ 00000101 & \text{Difference } (+5) \end{array}$$

Discard carry → **1** 00000101

(b) In this case, $12 − (−9) = 12 + 9 = 21$.

$$\begin{array}{ll} 00001100 & \text{Minuend } (+12) \\ +\ 00001001 & \text{2's complement of subtrahend } (+9) \\ \hline 00010101 & \text{Difference } (+21) \end{array}$$

(c) In this case, $−25 − (+19) = −25 + (−19) = −44$.

$$\begin{array}{ll} 11100111 & \text{Minuend } (-25) \\ +\ 11101101 & \text{2's complement of subtrahend } (-19) \\ \hline \mathbf{1}\ 11010100 & \text{Difference } (-44) \end{array}$$

Discard carry → **1** 11010100

(d) In this case, $−120 − (−30) = −120 + 30 = −90$

$$\begin{array}{ll} 10001000 & \text{Minuend } (-120) \\ +\ 00011110 & \text{2's complement of subtrahend } (+30) \\ \hline 10100110 & \text{Difference } (-90) \end{array}$$

Related Exercise   Subtract 01000111 from 01011000.

Multiplication

The numbers in a multiplication are the **multiplicand,** the **multiplier,** and the **product.** These are illustrated in the following decimal multiplication:

$$
\begin{array}{rl}
8 & \text{Multiplicand} \\
\times\,3 & \text{Multiplier} \\
\hline
24 & \text{Product}
\end{array}
$$

The multiplication operation in many computers is accomplished using addition. As you have already seen, subtraction is done with an adder; now let's see how multiplication is done.

Direct addition and partial products are two basic methods for performing multiplication using addition. In the direct addition method, you add the multiplicand a number of times equal to the multiplier. In the previous decimal example (3×8), three multiplicands are added: $8 + 8 + 8 = 24$. The disadvantage of this approach is that it becomes very lengthy if the multiplier is a large number. If, for example, you multiply 75×350, 350 must be added to itself 75 times. Incidentally, this is why the term *times* is used to mean multiply.

The partial products method is perhaps the more common one because it is the way you multiply long hand. The multiplicand is multiplied by each multiplier digit beginning with the least significant digit. The result of the multiplication of the multiplicand by a multiplier digit is called a *partial product.* Each successive partial product is moved one place to the left and when all the partial products have been produced, they are added to get the final product. Here is a decimal example.

$$
\begin{array}{rl}
239 & \text{Multiplicand} \\
\times\,123 & \text{Multiplier} \\
\hline
717 & \text{1st partial product } (3 \times 239) \\
478 & \text{2nd partial product } (2 \times 239) \\
+\,239 & \text{3rd partial product } (1 \times 239) \\
\hline
29{,}397 & \text{Final product}
\end{array}
$$

The sign of the product of a multiplication depends on the signs of the multiplicand and the multiplier according to the following two rules:

- **If the signs are the same, the product is positive.**
- **If the signs are different, the product is negative.**

When two binary numbers are multiplied, both numbers must be in true (uncomplemented) form. The direct addition method is illustrated in Example 2–20 adding two binary numbers at a time.

EXAMPLE 2–20

Multiply the signed binary numbers: 01001101 (multiplicand) and 00000100 (multiplier).

Solution Since both numbers are positive, they are in true form, and the product will be positive. The decimal value of the multiplier is 4, so the multiplicand is added to itself four times as follows:

$$
\begin{array}{rl}
01001101 & \text{1st time} \\
+\,01001101 & \text{2nd time} \\
\hline
10011010 & \text{Partial sum} \\
+\,01001101 & \text{3rd time} \\
\hline
11100111 & \text{Partial sum} \\
+\,01001101 & \text{4th time} \\
\hline
100110100 & \text{Product}
\end{array}
$$

Since the sign bit of the multiplicand is 0, it has no effect on the outcome. All of the bits in the product are magnitude bits.

Related Exercise   Multiply 01100001 by 00000110.

Now let's look at the partial products method of binary multiplication. The basic steps in the process are as follows:

Step 1. Determine if the signs of the multiplicand and multiplier are the same or different. This determines what the sign of the product will be.

Step 2. Change any negative number to true (uncomplemented) form. Since most computers store negative numbers in 2's complement, then a 2's complement operation is required to get the negative number into true form.

Step 3. Starting with the least significant multiplier bit, generate the partial products. When the multiplier bit is 1, the partial product is the same as the multiplicand. When the multiplier bit is 0, the partial product is zero. Shift each successive partial product one bit to the left.

Step 4. Add each successive partial product to the sum of the previous partial products to get the final product.

Step 5. If the sign bit that was determined in step 1 is negative, take the 2's complement of the product. If positive, leave the product in true form. Attach the sign bit to the product.

Example 2–21 illustrates these steps.

EXAMPLE 2–21

Multiply the signed binary numbers: 01010011 (multiplicand) and 11000101 (multiplier).

Solution

Step 1. The sign bit of the multiplicand is 0 and the sign bit of the multiplier is 1. The sign bit of the product will be 1 (negative).

Step 2. Take the 2's complement of the multiplier to put it in true form:

$$11000101 \longrightarrow 00111011$$

Steps 3 and 4. The multiplication proceeds as follows. Notice that only the magnitude bits are used in these steps.

1010011	Multiplicand
× 0111011	Multiplier
1010011	1st partial product
+ 1010011	2nd partial product
11111001	Sum of 1st and 2nd
+ 0000000	3rd partial product
011111001	Sum
+ 1010011	4th partial product
1110010001	Sum
+ 1010011	5th partial product
100011000001	Sum
+ 1010011	6th partial product
1001100100001	Sum
+ 0000000	7th partial product
1001100100001	Final product

Step 5. Since the sign of the product is a 1 as determined in step 1, take the 2's complement of the product.

$$1001100100001 \longrightarrow 0110011011111$$

Attach the sign bit.

$$\longrightarrow 1 \; 0110011011111$$

Related Exercise Verify the multiplication is correct by converting to decimal numbers and performing the multiplication.

Division

The numbers in a division are the **dividend,** the **divisor,** and the **quotient.** These are illustrated in the following standard division format.

$$\frac{\text{dividend}}{\text{divisor}} = \text{quotient}$$

The division operation in computers is accomplished using subtraction. Since subtraction is done with an adder, division can also be accomplished with an adder.

The result of a division is called the *quotient;* the quotient is the number of times that the divisor will go into the dividend. This means that the divisor can be subtracted from the dividend a number of times equal to the quotient, as illustrated by dividing 21 by 7:

$$
\begin{array}{rl}
21 & \text{Dividend} \\
- \; 7 & \text{1st subtraction of divisor} \\
\hline
14 & \text{1st partial remainder} \\
- \; 7 & \text{2nd subtraction of divisor} \\
\hline
7 & \text{2nd partial remainder} \\
- \; 7 & \text{3rd subtraction of divisor} \\
\hline
0 & \text{Zero remainder}
\end{array}
$$

In this simple example, the divisor was subtracted from the dividend three times before a remainder of zero was obtained. Therefore, the quotient is 3.

The sign of the quotient depends on the signs of the dividend and the divisor according to the following two rules:

- **If the signs are the same, the quotient is positive.**
- **If the signs are different, the quotient is negative.**

When two binary numbers are divided, both numbers must be in true (uncomplemented) form. The basic steps in a division process are as follows:

Step 1. Determine if the signs of the dividend and divisor are the same or different. This determines what the sign of the quotient will be. Initialize the quotient to zero.

Step 2. Subtract the divisor from the dividend using 2's complement addition to get the first partial remainder and add 1 to the quotient. If this partial remainder is positive, go to step 3. If the partial remainder is zero or negative, the division is complete.

Step 3. Subtract the divisor from the partial remainder and add 1 to the quotient. If the result is positive, repeat for the next partial remainder. If the result is zero or negative, the division is complete.

Continue to subtract the divisor from the dividend and the partial remainders until there is a zero or a negative result. Count the number of times that the divisor is subtracted and you have the quotient. Example 2–22 illustrates these steps using 8-bit signed binary numbers.

EXAMPLE 2–22   Divide 01100100 by 00011001.

Solution

Step 1.   The signs of both numbers are positive, so the quotient will be positive. The quotient is initially zero: 00000000.

Step 2.   Subtract the divisor from the dividend using 2's complement addition (remember that final carries are discarded):

01100100	Dividend
+ 11100111	2's complement of divisor
01001011	Positive 1st partial remainder

Add 1 to quotient: 00000000 + 00000001 = 00000001.

Step 3.   Subtract the divisor from the 1st partial remainder using 2's complement addition:

01001011	1st partial remainder
+ 11100111	2's complement of divisor
00110010	Positive 2nd partial remainder

Add 1 to quotient: 00000001 + 00000001 = 00000010.

Step 4.   Subtract the divisor from the 2nd partial remainder using 2's complement addition:

00110010	2nd partial remainder
+ 11100111	2's complement of divisor
00011001	Positive 3rd partial remainder

Add 1 to quotient: 00000010 + 00000001 = 00000011.

Step 5.   Subtract the divisor from the 3rd partial remainder using 2's complement addition:

00011001	3rd partial remainder
+ 11100111	2's complement of divisor
00000000	Zero remainder

Add 1 to quotient: 00000011 + 00000001 = 00000100 (final quotient). The process is complete.

Related Exercise   Verify that the process is correct by converting to decimal numbers and performing the division.

SECTION 2–7 REVIEW

1. List the four cases when numbers are added.
2. Add 00100001 and 10111100.
3. Subtract 00110010 from 01110111.
4. What is the sign of the product when two negative numbers are multiplied?
5. Multiply 01111111 by 00000101.
6. What is the sign of the quotient when a positive number is divided by a negative number?
7. Divide 00110000 by 00001100.

2–8 ▪ HEXADECIMAL NUMBERS

The hexadecimal number system has sixteen digits and is used primarily as a "short-hand" way of displaying binary numbers because it is very easy to convert between binary and hexadecimal. As you are probably aware, long binary numbers are difficult to read and write because it is easy to drop or transpose a bit. Since computers and microprocessors understand only 1s and 0s, it is necessary to use these digits when you program in "machine language." Imagine writing a sixteen bit instruction for a microprocessor system in 1s and 0s. It is much more efficient to use hexadecimal or octal; octal numbers are covered in the next section. Hexadecimal is frequently used in computer and microprocessor applications. After completing this section, you should be able to

☐ List the hexadecimal digits   ☐ Count in hexadecimal   ☐ Convert from binary to hexadecimal   ☐ Convert from hexadecimal to binary   ☐ Convert from hexadecimal to decimal   ☐ Convert from decimal to hexadecimal   ☐ Add hexadecimal numbers   ☐ Subtract hexadecimal numbers

The **hexadecimal** system has a base of sixteen; that is, it is composed of 16 digits and alphabetic **characters.** Most digital systems process binary data in groups that are multiples of four bits, making the hexadecimal number very convenient because each hexadecimal digit represents a 4-bit binary number (as listed in Table 2–3).

Ten **numeric** digits and six alphabetic characters make up the hexadecimal number system. The use of letters A, B, C, D, E, and F to represent numbers may seem strange at first, but keep in mind that any number system is only a set of sequential symbols. If you understand what quantities these symbols represent, then the form of the symbols themselves is less important once you get accustomed to using them. We will use the subscript 16 to designate hexadecimal numbers to avoid confusion with decimal numbers.

TABLE 2–3

Decimal	Binary	Hexadecimal
0	0000	0
1	0001	1
2	0010	2
3	0011	3
4	0100	4
5	0101	5
6	0110	6
7	0111	7
8	1000	8
9	1001	9
10	1010	A
11	1011	B
12	1100	C
13	1101	D
14	1110	E
15	1111	F

Counting in Hexadecimal

How do you count in hexadecimal once you get to F? Simply start over with another column and continue as follows:

10, 11, 12, 13, 14, 15, 16, 17, 18, 19, 1A, 1B, 1C, 1D, 1E, 1F, 20, 21, 22, 23, 24, 25, 26, 27, 28, 29, 2A, 2B, 2C, 2D, 2E, 2F, 30, 31, . . .

With two hexadecimal digits, you can count up to FF_{16}, which is decimal 255. To count beyond this, three hexadecimal digits are needed. For instance, 100_{16} is decimal 256, 101_{16} is decimal 257, and so forth. The maximum 3-digit hexadecimal number is FFF_{16}, or decimal 4095. The maximum 4-digit hexadecimal number is $FFFF_{16}$, which is decimal 65,535.

Binary-to-Hexadecimal Conversion

Converting a binary number to hexadecimal is a very straightforward procedure. Simply break the binary number into 4-bit groups, starting at the right-most bit and replace each 4-bit group with the equivalent hexadecimal symbol as illustrated in Example 2–23.

EXAMPLE 2–23

Convert the following binary numbers to hexadecimal:
(a) 1100101001010111   (b) 111111000101101001

Solution

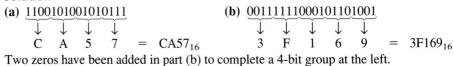

Two zeros have been added in part (b) to complete a 4-bit group at the left.

Related Exercise   Convert the binary number 1001111011110011100 to hexadecimal.

Hexadecimal-to-Binary Conversion

To convert from a hexadecimal number to a binary number, reverse the process and replace each hexadecimal symbol with the appropriate four bits as illustrated in Example 2–24.

EXAMPLE 2–24

Determine the binary numbers for the following hexadecimal numbers:
(a) $10A4_{16}$ (b) $CF8E_{16}$ (c) 9742_{16}

Solution

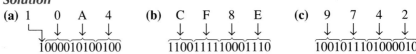

In part (a), the MSB is understood to have three zeros preceding it, thus forming a 4-bit group.

Related Exercise   Convert the hexadecimal number 6BD3 to binary.

It should be clear that it is much easier to deal with a hexadecimal number than the equivalent binary number. Since conversion is so easy, the hexadecimal system is widely used for representing binary numbers in programming, printouts, and displays.

Hexadecimal-to-Decimal Conversion

One way to find the decimal equivalent of a hexadecimal number is to first convert the hexadecimal number to binary and then convert from binary to decimal. Example 2–25 illustrates this procedure.

EXAMPLE 2–25

Convert the following hexadecimal numbers to decimal:
(a) $1C_{16}$ **(b)** $A85_{16}$

Solution   Remember, convert the hexadecimal number to binary first, then to decimal.

(a)   1   C
     ↓   ↓
     $\overline{00011100}$ $= 2^4 + 2^3 + 2^2 = 16 + 8 + 4 = 28_{10}$

(b)   A   8   5
     ↓   ↓   ↓
     $\overline{101010000101}$ $= 2^{11} + 2^9 + 2^7 + 2^2 + 2^0 = 2048 + 512 + 128 + 4 + 1 = 2693_{10}$

Related Exercise   Convert the hexadecimal number 6BD to decimal.

Another way to convert a hexadecimal number to its decimal equivalent is to multiply the decimal value of each hexadecimal digit by its weight and then take the sum of these products. The weights of a hexadecimal number are increasing powers of 16 (from right to left). For a 4-digit hexadecimal number, the weights are

$$16^3 \quad 16^2 \quad 16^1 \quad 16^0$$
$$4096 \quad 256 \quad 16 \quad 1$$

Example 2–26 shows this conversion method.

EXAMPLE 2–26

Convert the following hexadecimal numbers to decimal:
(a) $E5_{16}$ **(b)** $B2F8_{16}$

Solution   Recall from Table 2–3 that letters A through F represent decimal numbers 10 through 15, respectively.

(a) $E5_{16} = (E \times 16) + (5 \times 1) = (14 \times 16) + (5 \times 1) = 224 + 5 = 229_{10}$
(b) $B2F8_{16} = (B \times 4096) + (2 \times 256) + (F \times 16) + (8 \times 1)$
            $= (11 \times 4096) + (2 \times 256) + (15 \times 16) + (8 \times 1)$
            $= \quad 45,056 \quad + \quad 512 \quad + \quad 240 \quad + \quad 8 \quad = 45,816_{10}$

Related Exercise   Convert $60A_{16}$ to decimal.

CALCULATOR TIP

Powers-of-sixteen can be found on the calculator using the key $\boxed{y^x}$ with the following steps:

1. Enter the base number 16: $\boxed{1}$ $\boxed{6}$

2. Press $\boxed{y^x}$

3. Enter the power (3 for example): $\boxed{3}$

4. The display shows $\boxed{\qquad\qquad 4096\quad}$

Decimal-to-Hexadecimal Conversion

Repeated division of a decimal number by 16 will produce the equivalent hexadecimal number, formed by the remainders of the divisions. The first remainder produced is the least significant digit (LSD). Each successive division by 16 yields a remainder that becomes a digit in the equivalent hexadecimal number. This procedure is similar to repeated division by 2 for decimal-to-binary conversion that was covered in Section 2–3. Example 2–27 illustrates the procedure. Note that when a quotient has a fractional part, the fractional part is multiplied by the divisor to get the remainder.

EXAMPLE 2–27

Convert the decimal number 650 to hexadecimal by repeated division by 16.

Solution

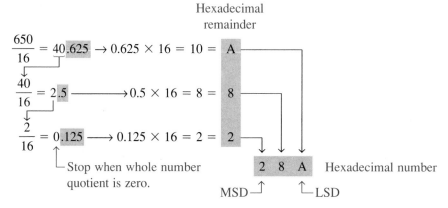

Related Exercise   Convert decimal 2591 to hexadecimal.

CALCULATOR TIP

When using your calculator to convert from decimal to hexadecimal, if the result of a division is a whole number with no fractional part, the remainder is always 0. For example,

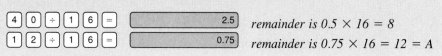

If the result of a division is a whole number with a fractional part or just a fractional number, the remainder equals the fractional part times 16. For example,

Hexadecimal Addition

Addition can be done directly with hexadecimal numbers by remembering that the hexadecimal digits 0 through 9 are equivalent to decimal digits 0 through 9 and that hexadecimal digits A through F are equivalent to decimal numbers 10 through 15. When adding two hexadecimal numbers, use the following rules. (Decimal numbers are indicated by a subscript 10.)

1. In any given column of an addition problem, think of the two hexadecimal digits in terms of their decimal value. For instance, $5_{16} = 5_{10}$ and $C_{16} = 12_{10}$.

2. If the sum of these two digits is 15_{10} or less, bring down the corresponding hexadecimal digit.
3. If the sum of these two digits is greater than 15_{10}, bring down the amount of the sum that exceeds 16_{10}, and carry a 1 to the next column.

EXAMPLE 2–28

Add the following hexadecimal numbers:
(a) $23_{16} + 16_{16}$ (b) $58_{16} + 22_{16}$ (c) $2B_{16} + 84_{16}$ (d) $DF_{16} + AC_{16}$

Solution

(a) 23_{16} right column: $3_{16} + 6_{16} = 3_{10} + 6_{10} = 9_{10} = 9_{16}$
    $\underline{+\ 16_{16}}$ left column: $2_{16} + 1_{16} = 2_{10} + 1_{10} = 3_{10} = 3_{16}$
    39_{16}

(b) 58_{16} right column: $8_{16} + 2_{16} = 8_{10} + 2_{10} = 10_{10} = A_{16}$
    $\underline{+\ 22_{16}}$ left column: $5_{16} + 2_{16} = 5_{10} + 2_{10} = 7_{10} = 7_{16}$
    $7A_{16}$

(c) $2B_{16}$ right column: $B_{16} + 4_{16} = 11_{10} + 4_{10} = 15_{10} = F_{16}$
    $\underline{+\ 84_{16}}$ left column: $2_{16} + 8_{16} = 2_{10} + 8_{10} = 10_{10} = A_{16}$
    AF_{16}

(d) DF_{16} right column: $F_{16} + C_{16} = 15_{10} + 12_{10} = 27_{10}$
    $\underline{+\ AC_{16}}$ $27_{10} - 16_{10} = 11_{10} = B_{16}$ with a 1 carry
    $18B_{16}$ left column: $D_{16} + A_{16} + 1_{16} = 13_{10} + 10_{10} + 1_{10} = 24_{10}$
                        $24_{10} - 16_{10} = 8_{10} = 8_{16}$ with a 1 carry

Related Exercise   Add $4C_{16}$ and $3A_{16}$.

Hexadecimal Subtraction Using 2's Complement Method

Since a hexadecimal number can be used to represent a binary number, it can also be used to represent the 2's complement of a binary number. For instance, the hexadecimal representation of 11001001_2 is $C9_{16}$. The 2's complement of this binary number is 00110111, which is written in hexadecimal as 37_{16}.

As you have learned, the 2's complement allows you to subtract by adding binary numbers. You can also use this method for hexadecimal subtraction, as in Example 2–29.

EXAMPLE 2–29

Subtract the following hexadecimal numbers:
(a) $84_{16} - 2A_{16}$ (b) $C3_{16} - 0B_{16}$

Solution

(a) $2A_{16} = 00101010$
    2's complement of $2A_{16} = 11010110 = D6_{16}$

            84_{16}
        $\underline{+\ D6_{16}}$ Add
        $\cancel{1}5A_{16}$ Drop carry, as in 2's complement addition

The difference is $5A_{16}$.

(b) $0B_{16} = 00001011$

2's complement of $0B_{16} = 11110101 = F5_{16}$

$$
\begin{array}{l}
C3_{16} \\
+ \ F5_{16} \qquad \text{Add}\\
\hline
\cancel{1}B8_{16} \qquad \text{Drop carry}
\end{array}
$$

The difference is $B8_{16}$.

Related Exercise Subtract 173_{16} from BCD_{16}.

SECTION 2–8 REVIEW

1. Convert the binary numbers to hexadecimal:
   (a) 10110011   (b) 110011101000
2. Convert the hexadecimal numbers to binary:
   (a) 57_{16} (b) $3A5_{16}$ (c) $F80B_{16}$
3. Convert $9B30_{16}$ to decimal.
4. Convert the decimal number 573 to hexadecimal.
5. Add the hexadecimal numbers directly:
   (a) $18_{16} + 34_{16}$ (b) $3F_{16} + 2A_{16}$
6. Subtract the hexadecimal numbers:
   (a) $75_{16} - 21_{16}$ (b) $94_{16} - 5C_{16}$

2–9 ■ OCTAL NUMBERS

Like the hexadecimal system, the octal system provides a convenient way to express binary numbers and codes. It is used less frequently than hexadecimal in conjunction with computers and microprocessors to express binary quantities for input and output purposes. After completing this section, you should be able to

□ Write the digits of the octal number system   □ Convert from octal to decimal
□ Convert from decimal to octal   □ Convert from octal to binary   □ Convert from binary to octal

The **octal** number system is composed of eight digits, which are

$$0, 1, 2, 3, 4, 5, 6, 7$$

To count above 7, begin another column and start over:

$$10, 11, 12, 13, 14, 15, 16, 17, 20, 21$$

and so on. Counting in octal is similar to counting in decimal, except that the digits 8 and 9 are not used. To distinguish octal numbers from decimal numbers or hexadecimal numbers, we will use the subscript 8 to indicate an octal number. For instance, 15_8 in octal is equivalent to 13_{10} in decimal and D in hexadecimal.

Octal-to-Decimal Conversion

Since the octal number system has a base of eight, each successive digit position is an increasing power of eight, beginning in the right-most column with 8^0. The evaluation of an octal number in terms of its decimal equivalent is accomplished by multiplying each digit by its weight and summing the products, as illustrated here for 2374_8:

$$\text{Weight: } 8^3 \ 8^2 \ 8^1 \ 8^0$$
$$\text{Octal number: } 2 \ 3 \ 7 \ 4$$

$$\begin{aligned}
2374_8 &= (2 \times 8^3) + (3 \times 8^2) + (7 \times 8^1) + (4 \times 8^0) \\
&= (2 \times 512) + (3 \times 64) + (7 \times 8) + (4 \times 1) \\
&= \quad 1024 \quad + \quad 192 \quad + \quad 56 \quad + \quad 4 \quad = 1276_{10}
\end{aligned}$$

Decimal-to-Octal Conversion

A method of converting a decimal number to an octal number is the repeated division-by-8 method, which is similar to the method used in the conversion of decimal numbers to binary or to hexadecimal. To show how it works, we convert the decimal number 359 to octal. Each successive division by 8 yields a remainder that becomes a digit in the equivalent octal number. The first remainder generated is the least significant digit (LSD).

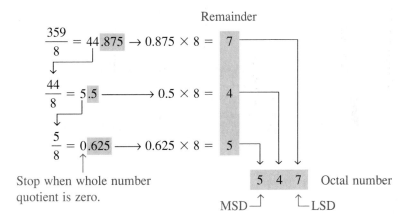

CALCULATOR TIP

When using your calculator to convert from decimal to octal, if the result of a division is a whole number with no fractional part, the remainder is always 0. For example,

[4][0][÷][8][=] [_____5] *remainder is 0*

If the result of a division is a whole number with a fractional part or just a fractional number, the remainder equals the fractional part times 8. For example,

[9][÷][8][=] [_____1.125] *remainder is 0.125 × 8 = 1*
[5][÷][8][=] [_____0.625] *remainder is 0.625 × 8 = 5*

Octal-to-Binary Conversion

Because each octal digit can be represented by a 3-bit binary number, it is very easy to convert from octal to binary. Each octal digit is represented by three bits as shown in Table 2–4.

TABLE 2–4
Octal/binary conversion

Octal Digit	0	1	2	3	4	5	6	7
Binary	000	001	010	011	100	101	110	111

To convert an octal number to a binary number, simply replace each octal digit with the appropriate three bits. This procedure is illustrated in Example 2–30.

EXAMPLE 2–30

Convert each of the following octal numbers to binary:
(a) 13_8 (b) 25_8 (c) 140_8 (d) 7526_8

Solution

         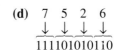

Related Exercise   Convert each of the binary numbers to decimal and verify that each value agrees with the decimal value of the corresponding octal number.

Binary-to-Octal Conversion

Conversion of a binary number to an octal number is the reverse of the octal-to-binary conversion. The procedure is as follows: Start with the right-most group of three bits and, moving from right to left, convert each 3-bit group to the equivalent octal digit. If there are not three bits available for the left-most group, add either one or two zeros to make a complete group. These leading zeros do not affect the value of the binary number.

EXAMPLE 2–31

Convert each of the following binary numbers to octal:
(a) 110101   (b) 101111001   (c) 100110011010   (d) 11010000100

Solution

(a) 110101
    ↓ ↓
    6 5 = 65_8

(b) 101111001
    ↓ ↓ ↓
    5 7 1 = 571_8

(c) 100110011010
    ↓ ↓ ↓ ↓
    4 6 3 2 = 4632_8

(d) 011010000100
    ↓ ↓ ↓ ↓
    3 2 0 4 = 3204_8

Related Exercise   Convert the binary number 1010101000111110010 to octal.

SECTION 2–9 REVIEW

1. Convert the octal numbers to decimal:
   (a) 73_8 (b) 125_8
2. Convert the decimal numbers to octal:
   (a) 98_{10} (b) 163_{10}
3. Convert the octal numbers to binary:
   (a) 46_8 (b) 723_8 (c) 5624_8
4. Convert the binary numbers to octal:
   (a) 110101111   (b) 1001100010   (c) 10111111001

2–10 ■ BINARY CODED DECIMAL (BCD)

Binary coded decimal (BCD) is a way to express each of the decimal digits with a binary code. Since there are only ten code groups in the BCD system, it is very easy to convert between decimal and BCD. Because we like to read and write in decimal, the BCD code provides an excellent interface to binary systems. Examples of such interfaces are keypad inputs and digital readouts. After completing this section, you should be able to

☐ Convert each decimal digit to BCD   ☐ Express decimal numbers in BCD   ☐ Convert from BCD to decimal   ☐ Add BCD numbers

The 8421 Code

The 8421 code is a type of **binary coded decimal (BCD)** code. Binary coded decimal means that each decimal digit, 0 through 9, is represented by a binary **code** of four bits. The designation 8421 indicates the binary weights of the four bits (2^3, 2^2, 2^1, 2^0). The ease of conversion between 8421 code numbers and the familiar decimal numbers is the main advantage of this code. All you have to remember are the ten binary combinations that represent the ten decimal digits as shown in Table 2–5. The 8421 code is the predominant BCD code, and when we refer to BCD, we always mean the 8421 code unless otherwise stated.

TABLE 2–5
Decimal/BCD conversion

Decimal Digit	0	1	2	3	4	5	6	7	8	9
BCD	0000	0001	0010	0011	0100	0101	0110	0111	1000	1001

Invalid Codes   You should realize that, with four bits, sixteen numbers (0000 through 1111) can be represented but that, in the 8421 code, only ten of these are used. The six code combinations that are not used—1010, 1011, 1100, 1101, 1110, and 1111—are invalid in the 8421 BCD code.

To express any decimal number in BCD, simply replace each decimal digit with the appropriate 4-bit code, as shown by Example 2–32.

EXAMPLE 2–32

Convert each of the following decimal numbers to BCD:
(a) 35   (b) 98   (c) 170   (d) 2469

Solution

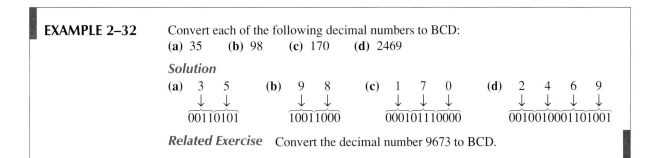

Related Exercise   Convert the decimal number 9673 to BCD.

It is equally easy to determine a decimal number from a BCD number. Start at the right-most bit and break the code into groups of four bits. Then write the decimal digit represented by each 4-bit group. Example 2–33 illustrates.

EXAMPLE 2–33

Convert each of the following BCD codes to decimal:
(a) 10000110   (b) 001101010001   (c) 1001010001110000

Solution
(a) 10000110   (b) 001101010001   (c) 1001010001110000
      ↓   ↓   ↓   ↓   ↓   ↓   ↓   ↓   ↓
      8   6   3   5   1   9   4   7   0

Related Exercise Convert the BCD code 10000010001001110110 to decimal.

BCD Addition

BCD is a numerical code and can be used in arithmetic operations. Addition is the most important operation because the other three operations (subtraction, multiplication, and division) can be accomplished by the use of addition. Here is how to add two BCD numbers:

Step 1. Add the two BCD numbers, using the rules for binary addition in Section 2–4.

Step 2. If a 4-bit sum is equal to or less than 9, it is a valid BCD number.

Step 3. If a 4-bit sum is greater than 9, or if a carry out of the 4-bit group is generated, it is an invalid result. Add 6 (0110) to the 4-bit sum in order to skip the six invalid states and return the code to 8421. If a carry results when 6 is added, simply add the carry to the next 4-bit group.

Example 2–34 illustrates BCD additions in which the sum in each 4-bit column is equal to or less than 9, and the 4-bit sums are therefore valid BCD numbers. Example 2–35 illustrates the procedure in the case of invalid sums (greater than 9 or a carry).

EXAMPLE 2–34

Add the following BCD numbers:
(a) 0011 + 0100   (b) 00100011 + 00010101
(c) 10000110 + 00010011   (d) 010001010000 + 010000010111

Solution The decimal number addition is shown for comparison.

(a)   0011   3   (b)   0010 0011   23
    + 0100   + 4   + 0001 0101   + 15
    ------   ---   -----------   ----
      0111   7   0011 1000   38

(c)   1000 0110   86   (d)   0100 0101 0000   450
    + 0001 0011   + 13   + 0100 0001 0111   + 417
    -----------   ---   -----------------   -----
      1001 1001   99   1000 0110 0111   867

Note that in each case the sum in any 4-bit column does not exceed 9, and the results are valid BCD numbers.

Related Exercise Add the BCD numbers: 1001000001000011 + 0000100100100101.

EXAMPLE 2–35

Add the following BCD numbers:
(a) 1001 + 0100   (b) 1001 + 1001
(c) 00010110 + 00010101   (d) 01100111 + 01010011

Solution   The decimal number additions are shown for comparison.

(a)

```
            1001              9
          + 0100            + 4
            1101   Invalid BCD number (>9)    13
          + 0110   Add 6
   0001     0011   Valid BCD number
    ↓        ↓
    1        3
```

(b)

```
            1001              9
          + 1001            + 9
       1    0010   Invalid because of carry    18
          + 0110   Add 6
   0001     1000   Valid BCD number
    ↓        ↓
    1        8
```

(c)

```
   0001     0110                              16
 + 0001     0110                            + 15
   0010     1011   Right group is invalid (>9),    31
                   left group is valid.
          + 0110   Add 6 to invalid code. Add
                   carry, 0001, to next group.
   0011     0001   Valid BCD number
    ↓        ↓
    3        1
```

(d)

```
        0110     0111                          67
      + 0101     0011                        + 53
        1011     1010   Both groups are invalid (>9)   120
      + 0110   + 0110   Add 6 to both groups
0001    0010     0000   Valid BCD number
 ↓       ↓        ↓
 1       2        0
```

Related Exercise   Add the BCD numbers: 01001000 + 00110100.

SECTION 2–10 REVIEW

1. What is the binary weight of each 1 in the following BCD numbers?
   (a) 0010   **(b)** 1000   **(c)** 0001   **(d)** 0100
2. Convert the following decimal numbers to BCD:
   (a) 6   **(b)** 15   **(c)** 273   **(d)** 849
3. What decimal numbers are represented by each BCD code?
   (a) 10001001   **(b)** 001001111000   **(c)** 000101010111
4. In BCD addition, when is a 4-bit sum invalid?

2–11 ■ DIGITAL CODES

There are many specialized codes used in digital systems. You have just learned about the BCD code; now let's look at a few others. Some codes are strictly numeric, like BCD, and others are alphanumeric; that is, they are used to represent numbers, letters, symbols, and instructions. The three codes introduced in this section are the Gray code, the excess-3 code, and the ASCII code. Also, the detection of errors in codes using a parity bit is covered. Other error detection and correction codes are covered in Appendix B. After completing this section, you should be able to

☐ Explain the advantage of the Gray code ☐ Convert between Gray code and binary
☐ Convert from decimal to excess-3 code ☐ Use the ASCII code ☐ Identify errors in codes based on the parity method

The Gray Code

The **Gray code** is unweighted and is not an arithmetic code; that is, there are no specific weights assigned to the bit positions. The important feature of the Gray code is that *it exhibits only a single bit change from one code number to the next.* This property is important in many applications, such as shaft position encoders, where error susceptibility increases with the number of bit changes between adjacent numbers in a sequence.

Table 2–6 is a listing of the 4-bit Gray code for decimal numbers 0 through 15. Binary numbers are shown in the table for reference. Like binary numbers, *the Gray code can have any number of bits.* Notice the single-bit change between successive Gray code numbers. For instance, in going from decimal 3 to decimal 4, the Gray code changes from 0010 to 0110, while the binary code changes from 0011 to 0100, a change of three bits. The only bit change is in the third bit from the right in the Gray code; the others remain the same.

TABLE 2–6
Four-bit Gray code

Decimal	Binary	Gray Code	Decimal	Binary	Gray Code
0	0000	0000	8	1000	1100
1	0001	0001	9	1001	1101
2	0010	0011	10	1010	1111
3	0011	0010	11	1011	1110
4	0100	0110	12	1100	1010
5	0101	0111	13	1101	1011
6	0110	0101	14	1110	1001
7	0111	0100	15	1111	1000

Binary-to-Gray Code Conversion Conversion between binary code and Gray code is sometimes useful. The following rules explain how to convert from a binary number to a Gray code number:

1. The most significant bit (left-most) in the Gray code is the same as the corresponding MSB in the binary number.
2. Going from left to right, add each adjacent pair of binary code bits to get the next Gray code bit. Discard carries.

For example, the conversion of the binary number 10110 to Gray code is as follows:

Step 1. The left-most Gray code digit is the same as the left-most binary code bit.

$$1 \quad 0 \quad 1 \quad 1 \quad 0 \quad \text{Binary}$$
$$\downarrow$$
$$1 \quad\quad\quad\quad\quad\quad \text{Gray}$$

Step 2. Add the left-most binary code bit to the adjacent one.

$$\begin{array}{ccccc} \boxed{1 \; + \; 0} & 1 & 1 & 0 & \text{Binary} \\ \downarrow & & & & \\ 1 \qquad 1 & & & & \text{Gray} \end{array}$$

Step 3. Add the next adjacent pair.

$$\begin{array}{ccccc} 1 & \boxed{0 \; + \; 1} & 1 & 0 & \text{Binary} \\ & \downarrow & & & \\ 1 \quad 1 \quad 1 & & & \text{Gray} \end{array}$$

Step 4. Add the next adjacent pair and discard the carry.

$$\begin{array}{ccccc} 1 & 0 & \boxed{1 \; + \; 1} & 0 & \text{Binary} \\ & & \downarrow & & \\ 1 \quad 1 \quad 1 \quad 0 & & \text{Gray} \end{array}$$

Step 5. Add the last adjacent pair.

$$\begin{array}{ccccc} 1 & 0 & 1 & \boxed{1 \; + \; 0} & \text{Binary} \\ & & & \downarrow & \\ 1 \quad 1 \quad 1 \quad 0 \quad 1 & \text{Gray} \end{array}$$

The conversion is now complete; the Gray code is 11101.

Gray-to-Binary Conversion To convert from Gray code to binary, use a similar method; however, but there are some differences. The following rules apply:

1. The most significant bit (left-most) in the binary code is the same as the corresponding bit in the Gray code.
2. Add each binary code bit generated to the Gray code bit in the next adjacent position. Discard carries.

For example, the conversion of the Gray code number 11011 to binary is as follows:

Step 1. The left-most bits are the same.

$$\begin{array}{ccccc} 1 & 1 & 0 & 1 & 1 \quad \text{Gray} \\ \downarrow & & & & \\ 1 & & & & \text{Binary} \end{array}$$

Step 2. Add the last binary code bit just generated to the Gray code bit in the next position. Discard the carry.

$$\begin{array}{ccccc} 1 & 1 & 0 & 1 & 1 \quad \text{Gray} \\ & {}_{+}\!\downarrow & & & \\ 1 & 0 & & & \text{Binary} \end{array}$$

Step 3. Add the last binary code bit generated to the next Gray code bit.

$$\begin{array}{ccccc} 1 & 1 & 0 & 1 & 1 \quad \text{Gray} \\ & & {}_{+}\!\downarrow & & \\ 1 & 0 & 0 & & \text{Binary} \end{array}$$

Step 4. Add the last binary code bit generated to the next Gray code bit.

$$\begin{array}{ccccc} 1 & 1 & 0 & 1 & 1 \quad \text{Gray} \\ & & & {}_{+}\!\downarrow & \\ 1 & 0 & 0 & 1 & \text{Binary} \end{array}$$

Step 5. Add the last binary code bit generated to the next Gray code bit. Discard the carry.

$$\begin{array}{ccccc} 1 & 1 & 0 & 1 & 1 \\ & & & & \\ 1 & 0 & 0 & 1 & 0 \end{array} \quad \begin{array}{l} \text{Gray} \\ \\ \text{Binary} \end{array}$$

This completes the conversion. The final binary number is 10010.

EXAMPLE 2–36

(a) Convert the binary number 11000110 to Gray code.
(b) Convert the Gray code 10101111 to binary.

Solution
(a) Binary to Gray code:

$$\begin{array}{cccccccc} 1 & + & 1 & + & 0 & + & 0 & + & 0 & + & 1 & + & 1 & + & 0 \\ \downarrow & & \downarrow & & \downarrow & & \downarrow & & \downarrow & & \downarrow & & \downarrow & & \downarrow \\ 1 & & 0 & & 1 & & 0 & & 0 & & 1 & & 0 & & 1 \end{array}$$

(b) Gray code to binary:

$$\begin{array}{cccccccc} 1 & 0 & 1 & 0 & 1 & 1 & 1 & 1 \\ \downarrow & \nearrow \downarrow & \nearrow \downarrow & \nearrow \downarrow & \nearrow \downarrow & \nearrow \downarrow & \nearrow \downarrow & \nearrow \downarrow \\ 1 & 1 & 0 & 0 & 1 & 0 & 1 & 0 \end{array}$$

Related Exercise   (a) Convert binary 101101 to Gray code. (b) Convert Gray code 100111 to binary.

Application of the Gray Code to a Shaft Position Encoder   A simplified diagram of a 3-bit shaft encoder mechanism is shown in Figure 2–1. Basically, there are three concentric conductive rings that are segmented into eight sectors. The more sectors there are, the more accurately the position can be represented, but we are using only eight for purposes of illustration. Each sector of each ring is fixed at either a high-level or a low-level voltage

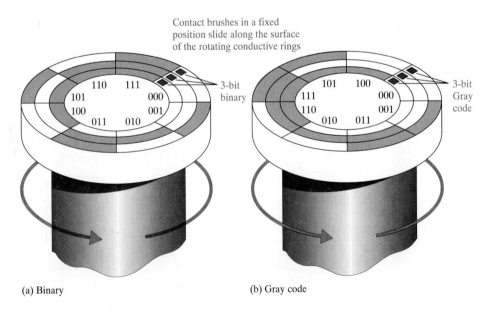

(a) Binary

(b) Gray code

FIGURE 2–1
A simplified illustration of how the Gray code solves the error problem in shaft position encoders.

to represent 1s and 0s. A 1 is indicated by a color sector and a 0 by a white sector. As the rings rotate with the shaft, they make contact with a brush arrangement that is in a fixed position and to which output lines are connected. As the shaft rotates counterclockwise through 360°, the eight sectors move past the three brushes producing a 3-bit binary output that indicates the shaft position.

In Figure 2–1(a), the sectors are arranged in a straight binary pattern, so that the brushes go from 000 to 001 to 010 to 011, and so on. When the brushes are on color sectors, they output a 1 and when on white sectors, they output a 0. If one brush is slightly ahead of the others during the transition from one sector to the next, an erroneous output can occur. Consider what happens when the brushes are on the 111 sector and about to enter the 000 sector. If the MSB brush is slightly ahead, the position would be incorrectly indicated by a transitional 011 instead of a 111 or a 000. In this type of application, it is virtually impossible to maintain precise mechanical alignment of all the brushes; therefore, some error will always occur at many of the transitions between sectors.

The Gray code is used to eliminate the error problem which is inherent in the binary code. As shown in Figure 2–1(b), the Gray code assures that only one bit will change between adjacent sectors. This means that even though the brushes may not be in precise alignment, there will never be a transitional error. For example, let's again consider what happens when the brushes are on the 111 sector and about to move into the next sector, 101. The only two possible outputs during the transition are 111 and 101, no matter how the brushes are aligned. A similar situation occurs at the transitions between each of the other sectors.

The Excess-3 Code

Excess-3 is a digital code related to BCD that is derived by adding 3 to each decimal digit and then converting the result of that addition to 4-bit binary. Since no definite weights can be assigned to the four digit position, excess-3 is an unweighted code that has advantages in certain arithmetic operations. The excess-3 code for decimal 2 is

$$\begin{array}{r} 2 \\ +\ 3 \\ \hline 5 \end{array} \rightarrow 0101$$

The excess-3 code for decimal 9 is

$$\begin{array}{r} 9 \\ +\ 3 \\ \hline 12 \end{array} \rightarrow 1100$$

The excess-3 code for each decimal digit is found by the same procedure. The entire code is shown in Table 2–7.

TABLE 2–7 *Excess-3 code*		
Decimal	**BCD**	**Excess-3**
0	0000	0011
1	0001	0100
2	0010	0101
3	0011	0110
4	0100	0111
5	0101	1000
6	0110	1001
7	0111	1010
8	1000	1011
9	1001	1100

Notice that ten of the possible 16 code combinations are used in the excess-3 code. The six invalid combinations are 0000, 0001, 0010, 1101, 1110, and 1111.

EXAMPLE 2–37

Convert each of the following decimal numbers to excess-3 code:
(a) 13 **(b)** 430

Solution First, add 3 to each digit in the decimal number, and then convert each resulting 4-bit sum to its equivalent binary code.

(a)
$$
\begin{array}{cc}
1 & 3 \\
+\,3 & +\,3 \\
\hline
4 & 6 \\
\downarrow & \downarrow \\
0100 & 0110
\end{array}
\quad \text{Excess-3}
$$

(b)
$$
\begin{array}{ccc}
4 & 3 & 0 \\
+\,3 & +\,3 & +\,3 \\
\hline
7 & 6 & 3 \\
\downarrow & \downarrow & \downarrow \\
0111 & 0110 & 0011
\end{array}
\quad \text{Excess-3}
$$

Related Exercise Convert decimal 928 to excess-3.

Self-Complementing Property The key feature of the excess-3 code is that it is *self-complementing*. This means that the 1's complement of an excess-3 number is the excess-3 code for the 9's complement of the corresponding decimal number. The 9's complement of a decimal number is found by subtracting each digit in the number from 9. For example, the 9's complement of 4 is 5. The excess-3 code for decimal 4 is 0111. The 1's complement of this is 1000, which is the excess-3 code for the decimal 5 (and 5 is the 9's complement of 4).

The usefulness of the 9's complement and thus excess-3 stems from the fact that subtraction of a smaller decimal number from a larger one can be accomplished by *adding* the 9's complement (1's complement of the excess-3 code) of the subtrahend (in this case the smaller number) to the minuend and then adding the carry to the result. When subtracting a larger number from a smaller one, there is no carry, and the result is in 9's complement form and negative. This procedure has a distinct advantage over BCD in certain types of arithmetic logic.

Alphanumeric Codes

In order to communicate, you need not only numbers, but also letters and other symbols. In the strictest sense, **alphanumeric** codes are codes that represent numbers and alphabetic characters (letters). Most such codes, however, also represent other characters such as symbols and various instructions necessary for conveying information.

At a minimum, an alphanumeric code must represent 10 decimal digits and 26 letters of the alphabet, for a total of 36 items. This number requires six bits in each code combination because five bits are insufficient ($2^5 = 32$). There are 64 total combinations of six bits, so there are 28 unused code combinations. Obviously, in many applications, symbols other than just numbers and letters are necessary to communicate completely. You need spaces, periods, colons, semicolons, question marks, etc. You also need instructions to tell the receiving system what to do with the information. So with codes that are six bits long, you can handle decimal numbers, the alphabet, and 28 other symbols. This should give you an idea of the requirements for a basic alphanumeric code. The ASCII is the most common alphanumeric code and is covered next.

ASCII

The **American Standard Code for Information Interchange (ASCII)**—pronounced "askee"—is a universally accepted alphanumeric code used in most computers and other electronic equipment. Most computer keyboards are standardized with the ASCII. When you enter a letter, a number, or control command, the corresponding ASCII code goes into the computer.

ASCII has 128 characters and symbols represented by a 7-bit binary code. Actually, ASCII can be considered an 8-bit code with the MSB always 0. This 8-bit code is 00 through 7F in hexadecimal. The first thirty-two ASCII characters are nongraphic commands that are never printed or displayed and used only for control purposes. Examples of the control characters are "null", "line feed", "start of text", and "escape". The other characters are graphic symbols that can be printed or displayed and include the letters of the alphabet (lowercase and uppercase), the ten decimal digits, punctuation signs, and other commonly used symbols.

Table 2–8 is a listing of the ASCII code showing the decimal, hexadecimal, and binary representations for each character and symbol. The left section of the table lists the names of the 32 control characters (00 through 1F hexadecimal). The graphic symbols are listed in the rest of the table (20 through 7F hexadecimal).

EXAMPLE 2–38

Determine the binary ASCII codes that are entered from the computer's keyboard when the following BASIC program statement is typed in. Also express each code in hexadecimal.

```
20 PRINT "A=";X
```

Solution   The ASCII code for each symbol is found in Table 2–8.

Symbol	Binary	Hexadecimal
2	0110010	32_{16}
0	0110000	30_{16}
Space	0100000	20_{16}
P	1010000	50_{16}
R	1010010	52_{16}
I	1001001	49_{16}
N	1001110	$4E_{16}$
T	1010100	54_{16}
Space	0100000	20_{16}
"	0100010	22_{16}
A	1000001	41_{16}
=	0111101	$3D_{16}$
"	0100010	22_{16}
;	0111011	$3B_{16}$
X	1011000	58_{16}

Related Exercise   Determine the sequence of ASCII codes required for the following program statement and express them in hexadecimal:

```
"80 INPUT Y"
```

The ASCII Control Characters   The first thirty-two codes in the ASCII table (Table 2–8) represent the control characters. These are used to allow devices such as a computer and printer to communicate with each other when passing information and data. Table 2–9 lists the control characters and the control key function that allows them to be entered di-

TABLE 2-8
American Standard Code for Information Interchange (ASCII)

Control Characters				Graphic Symbols											
Name	Dec	Binary	Hex	Symbol	Dec	Binary	Hex	Symbol	Dec	Binary	Hex	Symbol	Dec	Binary	Hex
NUL	0	0000000	00	space	32	0100000	20	@	64	1000000	40	`	96	1100000	60
SOH	1	0000001	01	!	33	0100001	21	A	65	1000001	41	a	97	1100001	61
STX	2	0000010	02	"	34	0100010	22	B	66	1000010	42	b	98	1100010	62
ETX	3	0000011	03	#	35	0100011	23	C	67	1000011	43	c	99	1100011	63
EOT	4	0000100	04	$	36	0100100	24	D	68	1000100	44	d	100	1100100	64
ENQ	5	0000101	05	%	37	0100101	25	E	69	1000101	45	e	101	1100101	65
ACK	6	0000110	06	&	38	0100110	26	F	70	1000110	46	f	102	1100110	66
BEL	7	0000111	07	`	39	0100111	27	G	71	1000111	47	g	103	1100111	67
BS	8	0001000	08	(	40	0101000	28	H	72	1001000	48	h	104	1101000	68
HT	9	0001001	09	)	41	0101001	29	I	73	1001001	49	i	105	1101001	69
LF	10	0001010	0A	*	42	0101010	2A	J	74	1001010	4A	j	106	1101010	6A
VT	11	0001011	0B	+	43	0101011	2B	K	75	1001011	4B	k	107	1101011	6B
FF	12	0001100	0C	,	44	0101100	2C	L	76	1001100	4C	l	108	1101100	6C
CR	13	0001101	0D	-	45	0101101	2D	M	77	1001101	4D	m	109	1101101	6D
SO	14	0001110	0E	.	46	0101110	2E	N	78	1001110	4E	n	110	1101110	6E
SI	15	0001111	0F	/	47	0101111	2F	O	79	1001111	4F	o	111	1101111	6F
DLE	16	0010000	10	0	48	0110000	30	P	80	1010000	50	p	112	1110000	70
DC1	17	0010001	11	1	49	0110001	31	Q	81	1010001	51	q	113	1110001	71
DC2	18	0010010	12	2	50	0110010	32	R	82	1010010	52	r	114	1110010	72
DC3	19	0010011	13	3	51	0110011	33	S	83	1010011	53	s	115	1110011	73
DC4	20	0010100	14	4	52	0110100	34	T	84	1010100	54	t	116	1110100	74
NAK	21	0010101	15	5	53	0110101	35	U	85	1010101	55	u	117	1110101	75
SYN	22	0010110	16	6	54	0110110	36	V	86	1010110	56	v	118	1110110	76
ETB	23	0010111	17	7	55	0110111	37	W	87	1010111	57	w	119	1110111	77
CAN	24	0011000	18	8	56	0111000	38	X	88	1011000	58	x	120	1111000	78
EM	25	0011001	19	9	57	0111001	39	Y	89	1011001	59	y	121	1111001	79
SUB	26	0011010	1A	:	58	0111010	3A	Z	90	1011010	5A	z	122	1111010	7A
ESC	27	0011011	1B	;	59	0111011	3B	[	91	1011011	5B	{	123	1111011	7B
FS	28	0011100	1C	<	60	0111100	3C	\	92	1011100	5C	|	124	1111100	7C
GS	29	0011101	1D	=	61	0111101	3D	]	93	1011101	5D	}	125	1111101	7D
RS	30	0011110	1E	>	62	0111110	3E	^	94	1011110	5E	~	126	1111110	7E
US	31	0011111	1F	?	63	0111111	3F	_	95	1011111	5F	Del	127	1111111	7F

TABLE 2–9
ASCII control characters

Name	Decimal	Hex	Key	Description
NUL	0	00	CTRL @	null character
SOH	1	01	CTRL A	start of header
STX	2	02	CTRL B	start of text
ETX	3	03	CTRL C	end of text
EOT	4	04	CTRL D	end of transmission
ENQ	5	05	CTRL E	enquire
ACK	6	06	CTRL F	acknowledge
BEL	7	07	CTRL G	bell
BS	8	08	CTRL H	backspace
HT	9	09	CTRL I	horizontal tab
LF	10	0A	CTRL J	line feed
VT	11	0B	CTRL K	vertical tab
FF	12	0C	CTRL L	form feed (new page)
CR	13	0D	CTRL M	carriage return
SO	14	0E	CTRL N	shift out
SI	15	0F	CTRL O	shift in
DLE	16	10	CTRL P	data link escape
DC1	17	11	CTRL Q	device control 1
DC2	18	12	CTRL R	device control 2
DC3	19	13	CTRL S	device control 3
DC4	20	14	CTRL T	device control 4
NAK	21	15	CTRL U	negative acknowledge
SYN	22	16	CTRL V	synchronize
ETB	23	17	CTRL W	end of transmission block
CAN	24	18	CTRL X	cancel
EM	25	19	CTRL Y	end of medium
SUB	26	1A	CTRL Z	substitute
ESC	27	1B	CTRL [	escape
FS	28	1C	CTRL /	file separator
GS	29	1D	CTRL]	group separator
RS	30	1E	CTRL ^	record separator
US	31	1F	CTRL _	unit separator

rectly from an ASCII keyboard by pressing the control key (CTRL) and the corresponding symbol. A brief description of each control character is also given.

Extended ASCII Characters

In addition to the 128 standard ASCII characters, there are an additional 128 characters that were adopted by IBM for use in their PCs. Because of the popularity of the PC, these particular extended ASCII characters are also used in applications other than PCs and have become essentially an unofficial standard.

The extended ASCII characters are represented by an 8-bit code series from hexadecimal 80 to hexadecimal FF. The extended ASCII contains characters in the following general categories:

1. Foreign (non-English) alphabetic characters
2. Foreign currency symbols
3. Greek letters
4. Mathematical symbols
5. Drawing characters
6. Bar graphing characters
7. Shading characters

TABLE 2–10

Extended ASCII characters

Symbol	Dec	Hex	Symbol	Dec	Hex	Symbol	Dec	Hex	Symbol	Dec	Hex
Ç	128	80	á	160	A0	└	192	C0	α	224	E0
ü	129	81	í	161	A1	┴	193	C1	β	225	E1
é	130	82	ó	162	A2	┬	194	C2	Γ	226	E2
â	131	83	ú	163	A3	├	195	C3	π	227	E3
ä	132	84	ñ	164	A4	─	196	C4	Σ	228	E4
à	133	85	Ñ	165	A5	┼	197	C5	σ	229	E5
å	134	86	ª	166	A6	╞	198	C6	μ	230	E6
ç	135	87	º	167	A7	╟	199	C7	τ	231	E7
ê	136	88	¿	168	A8	╚	200	C8	Φ	232	E8
ë	137	89	⌐	169	A9	╒	201	C9	Θ	233	E9
è	138	8A	¬	170	AA	╩	202	CA·	Ω	234	EA·
ï	139	8B	½	171	AB	╦	203	CB	δ	235	EB
î	140	8C	¼	172	AC	╠	204	CC	∞	236	EC
ì	141	8D	¡	173	AD	═	205	CD	φ	237	ED
Ä	142	8E	«	174	AE	╬	206	CE	ε	238	EE
Å	143	8F	»	175	AF	╧	207	CF	∩	239	EF
É	144	90	░	176	B0	╨	208	D0	≡	240	F0
æ	145	91	▒	177	B1	╤	209	D1	±	241	F1
Æ	146	92	▓	178	B2	╥	210	D2	≥	242	F2
ô	147	93	│	179	B3	╙	211	D3	≤	243	F3
ö	148	94	┤	180	B4	╘	212	D4	⌠	244	F4
ò	149	95	╡	181	B5	╒	213	D5		245	F5
û	150	96	╢	182	B6	╓	214	D6	÷	246	F6
ù	151	97	╖	183	B7	╫	215	D7	≈	247	F7
ÿ	152	98	╕	184	B8	╪	216	D8	°	248	F8
Ö	153	99	╣	185	B9	┘	217	D9	•	249	F9
Ü	154	9A	║	186	BA	┌	218	DA	·	250	FA
¢	155	9B	╗	187	BB	█	219	DB	√	251	FB
£	156	9C	╝	188	BC	▄	220	DC	η	252	FC
¥	157	9D	╜	189	BD	▌	221	DD	²	253	FD
Pτ	158	9E	╛	190	BE	▐	222	DE	■	254	FE
ƒ	159	9F	┐	191	BF	▀	223	DF		255	FF

Table 2–10 is a list of the extended ASCII character set with their decimal and hexadecimal representations.

Parity Method for Error Detection

Many systems use a parity bit as a means for bit **error detection.** Any group of bits contains either an even or an odd number of 1s. A **parity** bit is attached to a group of bits to make the total number of 1s in a group always even or always odd. An even parity bit makes the total number of 1s even, and an odd parity bit makes the total odd.

A given system operates with even or odd parity, but not both. For instance, if a system operates with even parity, a check is made on each group of bits received to make sure the total number of 1s in that group is even. If there is an odd number of 1s, an error has occurred.

As an illustration of how parity bits are attached to a code, Table 2–11 lists the parity bits for each BCD number for both even and odd parity. The parity bit for each BCD number is in the *P* column.

TABLE 2–11
The BCD code with parity bits

Even Parity		Odd Parity	
P	BCD	P	BCD
0	0000	1	0000
1	0001	0	0001
1	0010	0	0010
0	0011	1	0011
1	0100	0	0100
0	0101	1	0101
0	0110	1	0110
1	0111	0	0111
1	1000	0	1000
0	1001	1	1001

The parity bit can be attached to the code at either the beginning or the end, depending on system design. Notice that the total number of 1s, including the parity bit, is always even for even parity and always odd for odd parity.

Detecting an Error   A parity bit provides for the detection of a single bit error (or any odd number of errors, which is very unlikely) but cannot check for two errors in one group. For instance, let's assume that we wish to transmit the BCD code 0101. (Parity can be used with any number of bits; we are using four for illustration.) The total code transmitted, including the even parity bit, is

$$\overset{\downarrow}{0}0101$$

Even parity bit
BCD code

Now let's assume that an error occurs in the third bit from the left (the 1 becomes a 0) as follows:

$$\overset{\downarrow}{0}0001$$

Even parity bit
Bit error

When this code is received, the parity check circuitry determines that there is only a single 1 (odd number), when there should be an even number of 1s. Because an even number of 1s does not appear in the code when it is received, an error is indicated.

An odd parity bit also provides in a similar manner for the detection of a single error in a given group of bits.

See Appendix B for a coverage of more error detection codes and the Hamming error detection and correction code.

EXAMPLE 2–39

Assign the proper even parity bit to the following code groups:
(a) 1010   (b) 111000   (c) 101101
(d) 100011100101   (e) 101101011111

Solution   Make the parity bit either 1 or 0 to make the total number of 1s even. The parity bit will be the left-most bit (shaded).

(a) `0` 1010   **(b)** `1` 111000   **(c)** `0` 101101

(d) `0` 100011100101   **(e)** `1` 101101011111

Related Exercise Add an even parity bit to the 7-bit ASCII code for the letter K.

EXAMPLE 2–40

An odd parity system receives the following code groups: 10110, 11010, 110011, 110101110100, and 1100010101010. Determine which groups, if any, are in error.

Solution Since odd parity is required, any group with an even number of 1s in incorrect. The following groups are in error: 110011 and 1100010101010.

Related Exercise The following ASCII character is received by an odd parity system: 00110111. Is it correct?

SECTION 2–11 REVIEW

1. Convert the following binary numbers to the Gray code:
   (a) 1100   **(b)** 1010   **(c)** 11010
2. Convert the following Gray codes to binary:
   (a) 1000   **(b)** 1010   **(c)** 11101
3. Convert the following decimal numbers to excess-3 code:
   (a) 3   **(b)** 87   **(c)** 349
4. What is the ASCII representation for each of the following characters? Express each as a bit pattern and in hexadecimal notation.
   (a) K   **(b)** r   **(c)** $ **(d)** +
5. Add an even parity bit to each ASCII code found in question 4.

2–12 ■ DIGITAL SYSTEM APPLICATION

The system that was introduced in Chapter 1 provides a simple way to illustrate how binary numbers and codes can be used in digital systems. Recall that the system controls the number of tablets for each bottle on an assembly line in a pharmaceutical plant. In this section, you will see an example of the binary numbers and codes that exist at various points in this system during a part of its sequence. You will also determine numbers and codes throughout the system under certain specified conditions. After completing this section, you should be able to

☐ Explain the purpose of each functional block in the system in terms of the numbers and codes ☐ Analyze the system under specified conditions in terms of the numbers and codes at various points

System Operation

The block diagrams in Figures 2–2, 2–3, and 2–4 show the state of the system at certain points in its sequence. The binary numbers and BCD codes indicate the inputs and outputs of each block and the contents of the storage devices. The wide arrows represent parallel lines that connect the various blocks in the system and provide paths for parallel transfer of data. For example, eight bits of a given binary number are simultaneously present on eight parallel lines. Parallel lines such as these are commonly called *buses* in digital systems terminology.

The block diagram of the system is shown in a simplified form, leaving out many details that are not essential at this time. Also, the remote unit that was shown in Chapter 1 is omitted to keep things as simple as possible.

Initializing the System   At the beginning, the system must be initialized so that it starts in the proper binary states. First, the counter and the registers must be reset so the count can start from zero. Next, the number of tablets (up to 99) that will be put into each bottle is entered on the keypad, converted from decimal to BCD by the encoder, and stored in Register A.

In Figure 2–2, the decimal number 75 is entered on the keypad, indicating that 75 tablets will go into each bottle. Each keystroke is encoded into BCD and sent, four bits at a time, to be stored in Register A. The 8-bit BCD code is placed on the register's output lines and is applied to the decoder which converts it from BCD to a 7-segment code to light the display. The same BCD code is simultaneously applied to the code converter which converts it to binary. The resulting binary number for 75 appears at the

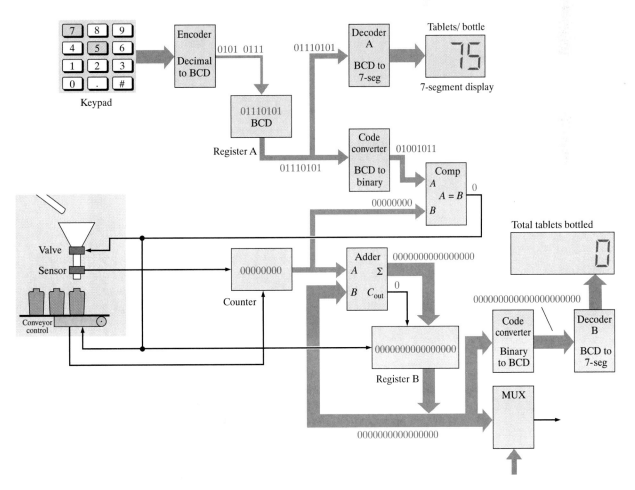

FIGURE 2–2
The system is in its initial state.

A input to the comparator (comp). Now the system is initialized and ready to start counting tablets.

The System at Work Now assume that the system has counted a total of 50 bottles (3750 tablets), and it is in the process of counting tablets for the fifty-first bottle. In fact, it has counted 74 tablets as indicated by the contents of the counter in Figure 2–3. Notice that Register B is holding the previous binary sum (3750) which is converted to BCD and then to 7-segment format for the display. Register B is updated each time another bottle is filled but not during the filling of a bottle.

The output of the comparator is LOW (0) because the *A* and *B* inputs are not equal at this time. Input *A* has a binary 75 and input *B* has a binary 74. The output of the adder always indicates the most current sum, in this case 3824, but this number is not stored in Register B until the bottle is full.

In the next state, as shown in Figure 2–4, the counter has reached a binary count of 75. As a result, the *A* = *B* output of the comparator goes HIGH (1) and causes several things to happen simultaneously. It stops the flow of tablets, it allows the new binary sum on the output of the adder (3825) to be stored in Register B, and it initiates an advance of the conveyor to get the next bottle into position. The binary number in Register B, which is now the current sum, is applied to the code converter and converted to BCD and then to 7-segment format to update the display to 3825. As soon as the next bottle is in place, the counter is reset to zero and the cycle starts over.

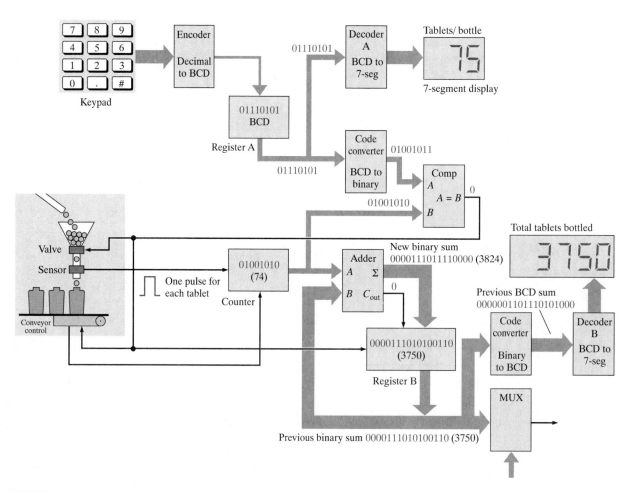

FIGURE 2–3
The system has counted 50 bottles of tablets and is working on the next bottle.

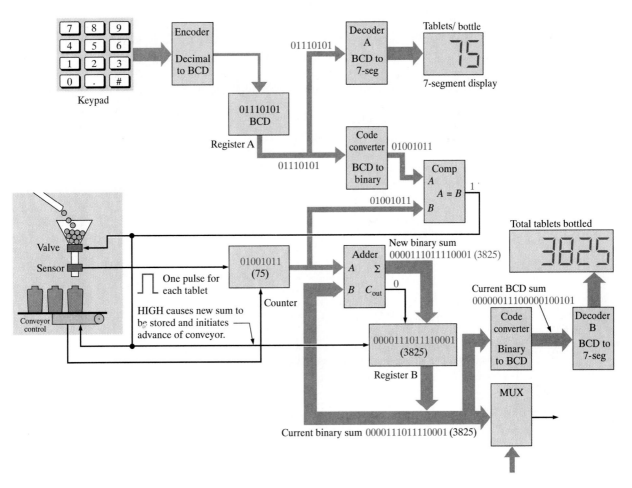

FIGURE 2–4
The system has just counted its fifty-first bottle of tablets.

System Analysis

- *Activity 1   Initialization*
  - (a) The system is set up for 60 tablets per bottle. Specify the contents of Register A and the data on input *A* of the comparator.
  - (b) What is the maximum binary number to which the counter will go when the system is initialized for 60 tablets per bottle?
- *Activity 2   Maximum Count*
  - (a) Determine the maximum number of full bottles that can be counted by the system if it is set for 75 tablets per bottle. The number of bits shown at each point is the maximum number of bits.
  - (b) Specify the binary number in Register B at the time the last bottle has been filled.
  - (c) Specify the BCD number at the output of the binary to BCD code converter at this time.
- *Activity 3   Sequencing*
  - (a) If the system is set for 30 tablets per bottle and at a given point, 523 bottles have been counted, determine the BCD number in Register A and the binary number in Register B.

    (b) If the counter is at the binary number 00010110, determine the binary output of the adder for the condition in part (a).

    (c) How many total tablets have actually been counted in part (b)?

    (d) What is the current number on the total display and what will the next number be?

■ *Activity 4*   *Hexadecimal*

    (a) Express each binary number in Figure 2–2 in hexadecimal.

    (b) Express each binary number in Figure 2–3 in hexadecimal.

    (c) Express each binary number in Figure 2–4 in hexadecimal.

    (d) If the counter is at $4C_{16}$ and then it counts nine more tablets, what will be its hexadecimal count?

    (e) If the display showing the total tablets in Figure 2–3 were a hexadecimal display, what would it show?

SECTION 2–12 REVIEW

1. In the system discussed in this section, when does the comparator output go to a HIGH level?
2. Under what condition does the display showing number of tablets per bottle change?
3. Under what condition does the display showing total tablets change?

■ **SUMMARY**

- A binary number is a weighted number in which the weight of each whole number digit is a positive power of two and the weight of each fractional digit is a negative power of two. The whole number weights increase from right to left—from least significant digit to most significant.
- A binary number can be converted to a decimal number by summing the decimal values of the weights of all the 1s in the binary number.
- A decimal whole number can be converted to binary by using the sum-of-weights or the repeated division-by-2 method.
- A decimal fraction can be converted to binary by using the sum-of-weights or the repeated multiplication-by-2 method.
- The basic rules for binary addition are as follows:

$$0 + 0 = 0$$
$$0 + 1 = 1$$
$$1 + 0 = 1$$
$$1 + 1 = 10$$

- The basic rules for binary subtraction are as follows:

$$0 - 0 = 0$$
$$1 - 1 = 0$$
$$1 - 0 = 1$$
$$10 - 1 = 1$$

- The 1's complement of a binary number is derived by changing 1s to 0s and 0s to 1s.
- The 2's complement of a binary number can be derived by adding 1 to the 1's complement.
- Binary subtraction can be accomplished with addition by using the 1's or 2's complement method.
- A positive binary number is represented by a 0 sign bit.
- A negative binary number is represented by a 1 sign bit.
- For arithmetic operations, negative binary numbers are represented in 1's complement or 2's complement form.

■ In an addition operation, an overflow is possible when both numbers are positive or when both numbers are negative. An incorrect sign bit in the sum indicates the occurrence of an overflow.
■ The hexadecimal number system consists of 16 digits and characters, 0 through 9 and A through F.
■ One hexadecimal digit represents a four-bit binary number, and its primary usefulness is in simplifying bit patterns and making them easier to read.
■ A decimal number can be converted to hexadecimal by the repeated division-by-16 method.
■ The octal number system consists of eight digits, 0 through 7.
■ A decimal number can be converted to octal by using the repeated division-by-8 method.
■ Octal-to-binary conversion is accomplished by simply replacing each octal digit with its 3-bit binary equivalent. The process is reversed for binary-to-octal conversion.
■ A decimal number is converted to BCD by replacing each decimal digit with the appropriate 4-bit binary code.
■ The ASCII is a seven-bit alphanumeric code that is widely used in computer systems for input and output of information.

■ SELF-TEST

1. $2 \times 10^1 + 8 \times 10^0$ is equal to
   (a) 10   (b) 280   (c) 2.8   (d) 28

2. The binary number 1101 is equal to the decimal number
   (a) 13   (b) 49   (c) 11   (d) 3

3. The binary number 11011101 is equal to the decimal number
   (a) 121   (b) 221   (c) 441   (d) 256

4. The decimal number 17 is equal to the binary number
   (a) 10010   (b) 11000   (c) 10001   (d) 01001

5. The decimal number 175 is equal to the binary number
   (a) 11001111   (b) 10101110   (c) 10101111   (d) 11101111

6. The sum of 11010 + 01111 equals
   (a) 101001   (b) 101010   (c) 110101   (d) 101000

7. The difference of 110 − 010 equals
   (a) 001   (b) 010   (c) 101   (d) 100

8. The 1's complement of 10111001 is
   (a) 01000111   (b) 01000110   (c) 11000110   (d) 10101010

9. The 2's complement of 11001000 is
   (a) 00110111   (b) 00110001   (c) 01001000   (d) 00111000

10. The decimal number $+122$ is expressed in the 2's complement system as
    (a) 01111010   (b) 11111010   (c) 01000101   (d) 10000101

11. The decimal number -34 is expressed in the 2's complement system as
    (a) 01011110   (b) 10100010   (c) 11011110   (d) 01011101

12. In the 2's complement system, the binary number 10010011 is equal to the decimal number
    (a) -19 (b) $+109$ (c) $+91$ (d) -109

13. The binary number 101100111001010100001 can be written in octal as
    (a) 5471230_8 (b) 5471241_8 (c) 2634521_8 (d) 23162501_8

14. The binary number 100011010100001101111 can be written in hexadecimal as
    (a) $AD467_{16}$ (b) $8C46F_{16}$ (c) $8D46F_{16}$ (d) $AE46F_{16}$

15. The binary number for $F7A9_{16}$ is
    (a) 1111011110101001   (b) 1110111110101001
    (c) 1111111010110001   (d) 1111011010101001

16. The BCD number for decimal 473 is
    (a) 111011010   (b) 110001110011   (c) 010001110011   (d) 010011110011

17. The word STOP in ASCII is
    (a) 1010011101010010011111010000   (b) 1010010100110010011101010000
    (c) 1001010110110110011101010001   (d) 1010011101010010011101100100

■ **PROBLEMS**

SECTION 2–1   Decimal Numbers

1. What is the weight of the digit 6 in each of the following decimal numbers?
   (a) 1386   (b) 54,692   (c) 671,920

2. Express each of the following decimal numbers as a power of ten:
   (a) 10   (b) 100   (c) 10,000   (d) 1,000,000

3. Give the value of each digit in the following decimal numbers:
   (a) 471   (b) 9356   (c) 125,000

4. How high can you count with four decimal digits?

SECTION 2–2   Binary Numbers

5. Convert the following binary numbers to decimal:
   (a) 11   (b) 100   (c) 111   (d) 1000
   (e) 1001   (f) 1100   (g) 1011   (h) 1111

6. Convert the following binary numbers to decimal:
   (a) 1110   (b) 1010   (c) 11100   (d) 10000
   (e) 10101   (f) 11101   (g) 10111   (h) 11111

7. Convert each binary number to decimal:
   (a) 110011.11   (b) 101010.01   (c) 1000001.111
   (d) 1111000.101   (e) 1011100.10101   (f) 1110001.0001
   (g) 1011010.1010   (h) 1111111.11111

8. What is the highest decimal number that can be represented by each of the following numbers of binary digits (bits)?
   (a) two   (b) three   (c) four   (d) five   (e) six
   (f) seven   (g) eight   (h) nine   (i) ten   (j) eleven

9. How many bits are required to represent the following decimal numbers?
   (a) 17   (b) 35   (c) 49   (d) 68
   (e) 81   (f) 114   (g) 132   (h) 205

10. Generate the binary sequence for each decimal sequence:
    (a) 0 through 7   (b) 8 through 15   (c) 16 through 31
    (d) 32 through 63   (e) 64 through 75

SECTION 2–3   Decimal-to-Binary Conversion

11. Convert each decimal number to binary by using the sum-of-weights method:
    (a) 10   (b) 17   (c) 24   (d) 48
    (e) 61   (f) 93   (g) 125   (h) 186

12. Convert each decimal fraction to binary using the sum-of-weights method:
    (a) 0.32   (b) 0.246   (c) 0.0981

13. Convert each decimal number to binary using repeated division by 2:
    (a) 15   (b) 21   (c) 28   (d) 34
    (e) 40   (f) 59   (g) 65   (h) 73

14. Convert each decimal fraction to binary using repeated multiplication by 2:
    (a) 0.98   (b) 0.347   (c) 0.9028

SECTION 2–4   Binary Arithmetic

15. Add the binary numbers:
    (a) 11 + 01   (b) 10 + 10   (c) 101 + 11
    (d) 111 + 110   (e) 1001 + 101   (f) 1101 + 1011

16. Use direct subtraction on the following binary numbers:
    (a) 11 − 1   (b) 101 − 100   (c) 110 − 101
    (d) 1110 − 11   (e) 1100 − 1001   (f) 11010 − 10111

17. Perform the following binary multiplications:

(a) 11×11 (b) 100×10 (c) 111×101

(d) 1001×110 (e) 1101×1101 (f) 1110×1101

18. Divide the binary numbers as indicated:

(a) $100 \div 10$ (b) $1001 \div 11$ (c) $1100 \div 100$

SECTION 2–5   1's and 2's Complements of Binary Numbers

19. Determine the 1's complement of each binary number:

(a) 101   (b) 110   (c) 1010

(d) 11010111   (e) 1110101   (f) 00001

20. Determine the 2's complement of each binary number using either method:

(a) 10   (b) 111   (c) 1001   (d) 1101

(e) 11100   (f) 10011   (g) 10110000   (h) 00111101

SECTION 2–6   Signed Numbers

21. Express each decimal number in binary as an 8-bit sign-magnitude number:

(a) $+29$ (b) -85 (c) $+100$ (d) -123

22. Express each decimal number as an 8-bit number in the 1's complement system:

(a) -34 (b) $+57$ (c) -99 (d) $+115$

23. Express each decimal number as an 8-bit number in the 2's complement system:

(a) $+12$ (b) -68 (c) $+101$ (d) -125

24. Determine the decimal value of each signed binary number in the sign-magnitude system:

(a) 10011001   (b) 01110100   (c) 10111111

25. Determine the decimal value of each signed binary number in the 1's complement system:

(a) 10011001   (b) 01110100   (c) 10111111

26. Determine the decimal value of each signed binary number in the 2's complement system:

(a) 10011001   (b) 01110100   (c) 10111111

SECTION 2–7   Arithmetic Operations with Signed Numbers

27. Convert each pair of decimal numbers to binary and add using the 2's complement system:

(a) 33 and 15   (b) 56 and -27 (c) -46 and 25   (d) -110 and -84

28. Perform each addition in the 2's complement system:

(a) $00010110 + 00110011$ (b) $01110000 + 10101111$

29. Perform each addition in the 2's complement system:

(a) $10001100 + 00111001$ (b) $11011001 + 11100111$

30. Perform each subtraction in the 2's complement system:

(a) $00110011 - 00010000$ (b) $01100101 - 11101000$

31. Multiply 01101010 by 11110001 in the 2's complement system.

32. Divide 01000100 by 00011001 in the 2's complement system.

SECTION 2–8   Hexadecimal Numbers

33. Convert each hexadecimal number to binary:

(a) 38_{16} (b) 59_{16} (c) $A14_{16}$ (d) $5C8_{16}$

(e) 4100_{16} (f) $FB17_{16}$ (g) $8A9D_{16}$

34. Convert each binary number to hexadecimal:

(a) 1110   (b) 10   (c) 10111

(d) 10100110   (e) 1111110000   (f) 100110000010

35. Convert each hexadecimal number to decimal:

(a) 23_{16} (b) 92_{16} (c) $1A_{16}$ (d) $8D_{16}$

(e) $F3_{16}$ (f) EB_{16} (g) $5C2_{16}$ (h) 700_{16}

36. Convert each decimal number to hexadecimal:

(a) 8   (b) 14   (c) 33   (d) 52

(e) 284   (f) 2890   (g) 4019   (h) 6500

37. Perform the following additions:

(a) $37_{16} + 29_{16}$ (b) $A0_{16} + 6B_{16}$ (c) $FF_{16} + BB_{16}$

38. Perform the following subtractions:

(a) $51_{16} - 40_{16}$ (b) $C8_{16} - 3A_{16}$ (c) $FD_{16} - 88_{16}$

SECTION 2–9   Octal Numbers

39. Convert each octal number to decimal:

(a) 12_8 (b) 27_8 (c) 56_8 (d) 64_8 (e) 103_8

(f) 557_8 (g) 163_8 (h) 1024_8 (i) 7765_8

40. Convert each decimal number to octal by repeated division by 8:

(a) 15   (b) 27   (c) 46   (d) 70

(e) 100   (f) 142   (g) 219   (h) 435

41. Convert each octal number to binary:

(a) 13_8 (b) 57_8 (c) 101_8 (d) 321_8 (e) 540_8

(f) 4653_8 (g) 13271_8 (h) 45600_8 (i) 100213_8

42. Convert each binary number to octal:

(a) 111   (b) 10   (c) 110111

(d) 101010   (e) 1100   (f) 1011110

(g) 101100011001   (h) 10110000011   (i) 111111101111000

SECTION 2–10   Binary Coded Decimal (BCD)

43. Convert each of the following decimal numbers to 8421 BCD:

(a) 10   (b) 13   (c) 18   (d) 21   (e) 25   (f) 36

(g) 44   (h) 57   (i) 69   (j) 98   (k) 125   (l) 156

44. Convert each of the decimal numbers in Problem 43 to straight binary, and compare the number of bits required with that required for BCD.

45. Convert the following decimal numbers to BCD:

(a) 104   (b) 128   (c) 132   (d) 150   (e) 186

(f) 210   (g) 359   (h) 547   (i) 1051

46. Convert each of the BCD numbers to decimal:

(a) 0001   (b) 0110   (c) 1001

(d) 00011000   (e) 11001   (f) 00110010

(g) 1000101   (h) 10011000   (i) 100001110000

47. Convert each of the BCD numbers to decimal:

(a) 10000000   (b) 1000110111

(c) 1101000110   (d) 10000100001

(e) 11101010100   (f) 100000000000

(g) 100101111000   (h) 1011010000011

(i) 1001000000011000   (j) 0110011001100111

48. Add the following BCD numbers:

(a) 0010 + 0001   (b) 0101 + 0011

(c) 0111 + 0010   (d) 1000 + 0001

(e) 00011000 + 00010001   (f) 01100100 + 00110011

(g) 01000000 + 01000111   (h) 10000101 + 00010011

49. Add the following BCD numbers:

(a) 1000 + 0110   (b) 0111 + 0101

(c) 1001 + 1000   (d) 1001 + 0111

(e) 00100101 + 00100111   (f) 01010001 + 01011000

(g) 10011000 + 10010111   (h) 010101100001 + 011100001000

50. Convert each pair of decimal numbers to BCD, and add as indicated:

    (a) 4 + 3   **(b)** 5 + 2   **(c)** 6 + 4   **(d)** 17 + 12

    (e) 28 + 23   **(f)** 65 + 58   **(g)** 113 + 101   **(h)** 295 + 157

SECTION 2–11   Digital Codes

51. In a certain application a 4-bit binary sequence cycles from 1111 to 0000 periodically. There are four bit changes, and because of circuit delays, these changes may not occur at the same instant. For example, if the LSB changes first, the number will appear as 1110 during the transition from 1111 to 0000 and may be misinterpreted by the system. Illustrate how the Gray code avoids this problem.

52. Convert each binary number to Gray code:

    (a) 11011   **(b)** 1001010   **(c)** 1111011101110

53. Convert each Gray code to binary:

    (a) 1010   **(b)** 00010   **(c)** 11000010001

54. Convert each of the following decimal numbers to excess-3 code:

    (a) 1   **(b)** 3   **(c)** 6   **(d)** 10   **(e)** 18

    (f) 29   **(g)** 56   **(h)** 75   **(i)** 107

55. Convert each excess-3 code number to decimal:

    (a) 0011   **(b)** 1001   **(c)** 0111

    (d) 01000110   **(e)** 01111100   **(f)** 10000101

56. Decode the following ASCII coded message:

    1001000 1100101 1101100 1101100 1101111 0101110
    0100000 1001000 1101111 1110111 0100000 1100001
    1110010 1100101 0100000 1111001 1101111 1110101
    0111111

57. Write the message in Problem 56 in hexadecimal.

58. Convert the following computer program statement to ASCII:

```
30 INPUT A, B
```

59. Determine which of the following even parity codes are in error:

    (a) 100110010   **(b)** 011101010   **(c)** 10111111010001010

60. Determine which of the following odd parity codes are in error:

    (a) 11110110   **(b)** 00110001   **(c)** 01010101010101010

61. Attach the proper even parity bit to each of the following bytes of data.

    (a) 10100100   **(b)** 00001001   **(c)** 11111110

SECTION 2–12   Digital System Application

62. In the tablet counting and control system, what limits the maximum number of tablets per bottle that can be preset into the system?

63. Assume that the system is set up for 25 tablets per bottle. Determine the binary numbers at each of the following locations in the system after 15 tablets have gone into the 1000th bottle?

    (a) In register A   **(b)** Input A to comparator   **(c)** In the counter

    (d) In register B   **(e)** Output of adder

64. Determine the maximum number of total tablets that the system can count. Explain what changes would be required to increase this number.

■ ANSWERS TO SECTION REVIEWS

SECTION 2–1

1. **(a)** 1370: 10   **(b)** 6725: 100   **(c)** 7051: 1000   **(d)** 58.72: 0.1

2. **(a)** $51 = 5 \times 10 + 1 \times 1$ **(b)** $137 = 1 \times 100 + 3 \times 10 + 7 \times 1$

    (c) $1492 = 1 \times 1000 + 4 \times 100 + 9 \times 10 + 2 \times 1$

    (d) $106.58 = 1 \times 100 + 0 \times 10 + 6 \times 1 + 5 \times 0.1 + 8 \times 0.01$

SECTION 2–2

1. $2^8 - 1 = 255$ **2.** Weight is 16.   **3.** $10111101.011 = 189.375$

SECTION 2–3

1. **(a)** $23 = 10111$ **(b)** $57 = 111001$ **(c)** $45.5 = 101101.1$
2. **(a)** $14 = 1110$ **(b)** $21 = 10101$ **(c)** $0.375 = 0.011$

SECTION 2–4

1. **(a)** $1101 + 1010 = 10111$ **(b)** $10111 + 01101 = 100100$
2. **(a)** $1101 - 0100 = 1001$ **(b)** $1001 - 0111 = 0010$
3. **(a)** $110 \times 111 = 101010$ **(b)** $1100 \div 011 = 100$

SECTION 2–5

1. **(a)** 1's comp of $00011010 = 11100101$ **(b)** 1's comp of $11110111 = 00001000$
    (c) 1's comp of $10001101 = 01110010$
2. **(a)** 2's comp of $00010110 = 11101010$ **(b)** 2's comp of $11111100 = 00000100$
    (c) 2's comp of $10010001 = 01101111$

SECTION 2–6

1. S&M: $+9 = 00001001$ **2.** 1's comp: $-33 = 11011110$
3. 2's comp: $-46 = 11010010$

SECTION 2–7

1. Cases of addition: positive number is larger, negative number is larger, both are positive, both are negative
2. $00100001 + 10111100 = 11011101$
3. $01110111 - 00110010 = 01000101$
4. Sign of product is positive.
5. $00000101 \times 01111111 = 01001111011$
6. Sign of quotient is negative.
7. $00110000 \div 00001100 = 00000100$

SECTION 2–8

1. **(a)** $10110011 = B3_{16}$ **(b)** $110011101000 = CE8_{16}$
2. **(a)** $57_{16} = 01010111$ **(b)** $3A5_{16} = 001110100101$
    (c) $F8OB_{16} = 1111100000001011$
3. $9B30_{16} = 39,728_{10}$ **4.** $573_{10} = 23D_{16}$
5. **(a)** $18_{16} + 34_{16} = 4C_{16}$ **(b)** $3F_{16} + 2A_{16} = 69_{16}$
6. **(a)** $75_{16} - 21_{16} = 54_{16}$ **(b)** $94_{16} - 5C_{16} = 38_{16}$

SECTION 2–9

1. **(a)** $73_8 = 59_{10}$ **(b)** $125_8 = 85_{10}$ **2.** **(a)** $98_{10} = 142_8$ **(b)** $163_{10} = 243_8$
3. **(a)** $46_8 = 100110$ **(b)** $723_8 = 111010011$ **(c)** $5624_8 = 101110010100$
4. **(a)** $110101111 = 657_8$ **(b)** $1001100010 = 1142_8$ **(c)** $10111111001 = 2771_8$

SECTION 2–10

1. **(a)** 0010: 2   **(b)** 1000: 8   **(c)** 0001: 1   **(d)** 0100: 4
2. **(a)** $6_{10} = 0110$ **(b)** $15_{10} = 00010101$ **(c)** $273_{10} = 001001110011$
    (d) $849_{10} = 100001001001$

3. (a) $10001001 = 89_{10}$ **(b)** $001001111000 = 278_{10}$ **(c)** $000101010111 = 157_{10}$

4. A 4-bit sum is invalid when it is greater than 9_{10}.

SECTION 2–11

1. (a) $1100_2 = 1010$ Gray **(b)** $1010_2 = 1111$ Gray **(c)** $11010_2 = 10111$ Gray

2. (a) 1000 Gray $= 1111_2$ **(b)** 1010 Gray $= 1100_2$ **(c)** 11101 Gray $= 10110_2$

3. (a) $3_{10} = 0110$ xs3 **(b)** $87_{10} = 10111010$ xs3 **(c)** $349_{10} = 011001111100$ xs3

4. (a) K: $1001011 \rightarrow 4B_{16}$ **(b)** r: $1110010 \rightarrow 72_{16}$

   (c) \$: $0100100 \rightarrow 24_{16}$ **(d)** +: $0101011 \rightarrow 2B_{16}$

5. (a) *0*1001011 **(b)** *0*1110010 **(c)** *0*0100100 **(d)** *0*0101011

SECTION 2–12

1. The comparator output goes HIGH when the actual tablets counted equals the preset number.

2. The number of tablets/bottle display changes when a new number is entered on the keypad.

3. The total tablets display changes each time another bottle is filled.

3

LOGIC GATES

■ CHAPTER OBJECTIVES

☐ Describe the operation of the inverter, the AND gate, and the OR gate

☐ Describe the operation of the NAND gate and the NOR gate

☐ Express the operation of NOT, AND, OR, NAND, and NOR gates with Boolean algebra

☐ Describe the operation of the exclusive-OR and exclusive-NOR gates

☐ Recognize and use both the distinctive shape logic gate symbols and the rectangular outline logic gate symbols of ANSI/IEEE Standard 91–1984

☐ Construct timing diagrams showing the proper time relationships of inputs and outputs for the various logic gates

☐ Make basic comparisons between the major IC technologies—TTL, CMOS, and ECL

☐ Explain how the different series within the TTL and CMOS families differ from each other

☐ Define *propagation delay time, power dissipation, speed-power product,* and *fan-out* in relation to logic gates

☐ List specific devices that contain the various logic gates

☐ Use each logic gate in simple applications

☐ Troubleshoot logic gates for opens and shorts by using the logic pulser and probe or the oscilloscope

☐ Apply logic gates in a system application

■ CHAPTER OVERVIEW

The emphasis in this chapter is on the logical operation, application, and troubleshooting of logic gates. The relationship of input and output waveforms of a gate using timing diagrams is thoroughly covered.

Logic symbols used to represent the logic gates are in accordance with ANSI/IEEE Standard 91–1984. This standard has been adopted by private industry and the military for use in internal documentation as well as published literature.

Because integrated circuits (IC) are used in all applications, the logic function of a device is generally of greater importance to the technician or technologist than the details of the component-level circuit operation within the IC package. Therefore, detailed coverage of the devices at the component level can be treated as an optional topic. For those who need it and have the time, a thorough coverage of digital technologies is available in Chapter 15. Portions or all of this "floating" chapter can be used after this chapter or at any appropriate point throughout the text. For those choosing to do so, Chapter 15 may be omitted or partially omitted without affecting the rest of the material.

■ SPECIFIC DEVICES

7400	7402	7404
7408	7410	7411
7420	7427	7430
7432	7486	74133

■ DIGITAL SYSTEM APPLICATION

This Digital System Application illustrates concepts taught in this chapter. To illustrate how logic gates are the basic building blocks for more complex functions, the comparator function in the tablet counting and control system introduced earlier is implemented and analyzed in Section 3–9. Also a certain system timing problem is examined and a modification is presented using basic logic elements. You will troubleshoot a specified system malfunction using the logic pulser and probe on the system printed circuit board.

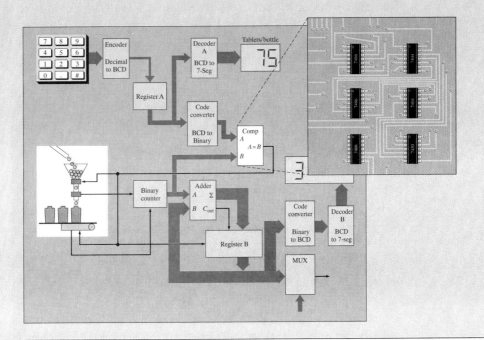

3–1 ■ THE INVERTER

The inverter (NOT circuit) performs the operation called inversion or complementation. The inverter changes one logic level to the opposite level. In terms of bits, it changes a 1 to a 0 and a 0 to a 1. After completing this section, you should be able to

☐ Identify negation and polarity indicators   ☐ Identify an inverter by either its distinctive shape symbol or its rectangular outline symbol   ☐ Produce the truth table for an inverter   ☐ Describe the logical operation of an inverter

Standard logic symbols for the **inverter** are shown in Figure 3–1. Part (a) shows the distinctive shape symbols, and part (b) shows the rectangular outline symbols. In this text, distinctive shape symbols are used; however, the rectangular outline symbols are found in many industry publications, and you should become familiar with them as well. (Logic symbols are in accordance with **ANSI/IEEE** Standard 91–1984.)

The Negation and Polarity Indicators

The negation indicator is a "bubble" (○) that indicates **inversion** or **complementation** when it appears on the input or output of a logic element, as shown in Figure 3–1(a). Generally, inputs are on the left of a logic symbol and the output is on the right. When appearing on the input, the bubble means that a 0 is the active or asserted input state. When appearing on the output, the bubble means that a 0 is the active or asserted output state. The absence of a bubble on the input or output means that a 1 is the active or asserted state.

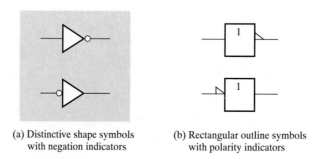

(a) Distinctive shape symbols with negation indicators

(b) Rectangular outline symbols with polarity indicators

FIGURE 3–1
Standard logic symbols for the inverter (ANSI/IEEE Std. 91–1984).

The polarity or level indicator is a "triangle" (◿) that indicates inversion when it appears on the input or output of a logic element, as shown in Figure 3–1(b). When appearing on the input, it means that a LOW level is the active or asserted input state. When appearing on the output, it means that a LOW level is the active or asserted output state.

Either indicator (bubble or triangle) can be used both on distinctive shape symbols and on rectangular outlines. Figure 3–1(a) indicates the principal inverter symbols used in this text. Note that a change in the placement of the negation or polarity indicator does not imply a change in the way an inverter operates.

Inverter Truth Table

When a HIGH level is applied to an inverter input, a LOW level will appear on its output. When a LOW level is applied to its input, a HIGH will appear on its output. This operation is summarized in Table 3–1, which shows the output for each possible input in terms of levels and corresponding bits. A table such as this is called a **truth table.**

TABLE 3–1
Inverter truth table

Input	Output
LOW (0)	HIGH (1)
HIGH (1)	LOW (0)

Inverter Operation

Figure 3–2 shows the output of an inverter for a pulse input, where t_1 and t_2 indicate the corresponding points on the input and output pulse waveforms.

When the input is LOW, the output is HIGH; when the input is HIGH, the output is LOW, thereby producing an inverted output pulse.

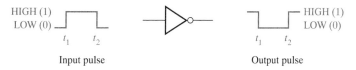

Input pulse   Output pulse

FIGURE 3–2
Inverter with pulse input.

Timing Diagrams

Recall from Chapter 1 that a **timing diagram** is basically a graph that accurately displays the relationship of two or more waveforms with respect to time. For example, the time relationship of the output pulse to the input pulse in Figure 3–2 can be shown with a simple timing diagram by aligning the two pulses so that the occurrences of the edges appear in the proper time relationship. The rising edge of the input pulse and the falling edge of the output pulse occur at the same time (ideally). Similarly, the falling edge of the input pulse and the rising edge of the output pulse occur at the same time (ideally). This timing relationship is shown in Figure 3–3. Timing diagrams are useful for illustrating the relationship of digital waveforms with multiple pulses, as Example 3–1 illustrates.

FIGURE 3–3
Timing diagram for the case in Figure 3–2.

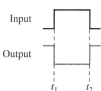

EXAMPLE 3–1   A waveform is applied to an inverter in Figure 3–4. Determine the output waveform corresponding to the input and sketch the timing diagram. According to the placement of the bubble, what is the active output state?

FIGURE 3–4

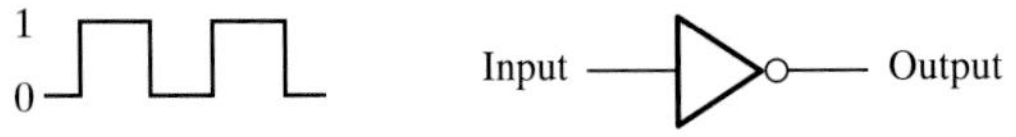

Solution   The output waveform is exactly opposite to the input (inverted), as shown in Figure 3–5, which is the basic timing diagram. The active output state is 0.

FIGURE 3–5

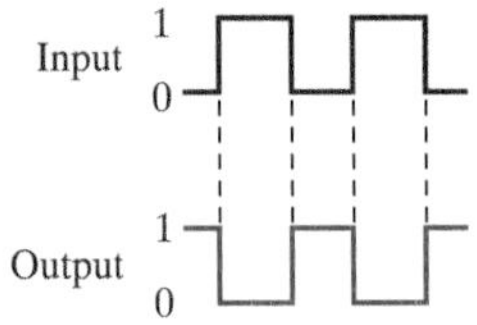

Related Exercise   If the inverter is shown with the negative indicator (bubble) on the input instead of the output, how is the timing diagram affected?

Logic Expression for the Inverter

In **Boolean algebra,** which is the mathematics of logic circuits and will be covered thoroughly in Chapter 4, a variable is designated by a letter. The **complement** of a variable is designated by a bar over the letter. A variable can take on a value of either 1 or 0. If a given variable is 1, its complement is 0 and vice versa.

The operation of an inverter (NOT circuit) can be expressed as follows: If the input variable is called A and the output variable is called X, then

$$X = \overline{A}$$

This expression states that the output is the complement of the input, so if $A = 0$, then $X = 1$, and if $A = 1$, then $X = 0$. Figure 3–6 illustrates this. The complemented variable $\overline{A}$ can be read as "*A* bar" or "*A* not."

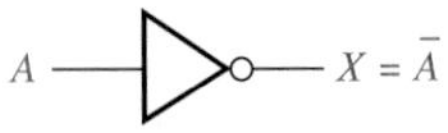

FIGURE 3–6
The inverter complements an input variable.

An Inverter Application

Figure 3–7 shows a circuit for producing the 1's complement of an 8-bit binary number. The bits of the binary number are applied to the inverter inputs and the 1's complement of the number appears on the outputs.

FIGURE 3–7
Example of a 1's complement circuit using inverters.

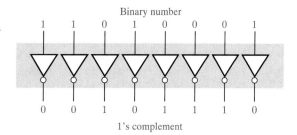

Binary number
1 1 0 1 0 0 0 1

0 0 1 0 1 1 1 0
1's complement

**SECTION 3–1
REVIEW**

1. When a 1 is on the input of an inverter, what is the output?
2. An active HIGH pulse (HIGH level when asserted, LOW level when not) is required on an inverter input.
   (a) Draw the appropriate logic symbol, using the distinctive shape and the negation indicator, for the inverter in this application.
   (b) Describe the output when a positive-going pulse is applied to the input of an inverter.

3–2 ■ THE AND GATE

The AND gate is one of the basic gates from which all logic functions are constructed. An AND gate can have two or more inputs and performs what is known as logical multiplication. After completing this section, you should be able to

☐ Identify an AND gate by its distinctive shape symbol or by its rectangular outline symbol   ☐ Describe the logical operation of an AND gate   ☐ Generate the truth table for an AND gate with any number of inputs   ☐ Produce a timing diagram for an AND gate with any specified input waveforms   ☐ Write the logic expression for an AND gate with any number of inputs   ☐ Discuss examples of AND gate applications

The term **gate** is used to describe a circuit that performs a basic logic operation. The AND gate is composed of two or more inputs and a single output, as indicated by the standard logic symbols shown in Figure 3–8. Inputs are on the left, and the output is on the right in each symbol. Gates with two inputs are shown; however, an AND gate can have any number of inputs greater than one. Although examples of both distinctive shape symbols and rectangular outline symbols are shown, the distinctive shape symbol, shown in part (a), is used predominantly in this book.

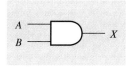 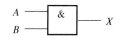

(a) Distinctive shape

(b) Rectangular outline with AND (&) qualifying symbol

FIGURE 3–8
Standard logic symbols for the AND gate showing two inputs (ANSI/IEEE Std. 91–1984).

Logical Operation of the AND Gate

The **AND gate** produces a HIGH output *only* when *all* of the inputs are HIGH. When any of the inputs is LOW, the output is LOW. Therefore, the basic purpose of an AND gate is to determine when certain conditions are simultaneously true, as indicated by HIGH levels on all of its inputs, and to produce a HIGH on its output to indicate that all these conditions are true. The inputs of the 2-input AND gate in Figure 3–8 are labeled *A* and *B*, and the output is labeled *X*. The gate operation can be stated as follows:

> **In a 2-input AND gate operation, output *X* is HIGH if inputs *A* and *B* are HIGH; *X* is LOW if either *A* or *B* is LOW, or if both *A* and *B* are LOW.**

Figure 3–9 illustrates a 2-input AND gate with all four possibilities of input combinations and the resulting output for each.

FIGURE 3–9
All possible logic levels for a 2-input AND gate.

AND Gate Truth Table

The logical operation of a gate can be expressed with a truth table that lists all input combinations with the corresponding outputs, as illustrated in Table 3–2 for a 2-input AND gate. The truth table can be expanded to any number of inputs. Although the terms HIGH and LOW tend to give a "physical" sense to input and output states, the truth table is shown with 1s and 0s, since a HIGH is equivalent to a 1 and a LOW is equivalent to a 0 in positive logic. For any AND gate, regardless of the number of inputs, the output is HIGH *only* when *all* inputs are HIGH.

TABLE 3–2
Truth table for a 2-input AND gate

Inputs		Output
A	*B*	*X*
0	0	0
0	1	0
1	0	0
1	1	1

1 = HIGH, 0 = LOW

The total number of possible combinations of binary inputs to a gate is determined by the following formula:

$$N = 2^n \qquad (3\text{–}1)$$

where N is the number of possible input combinations and n is the number of input variables. To illustrate,

For two input variables: $N = 2^2 = 4$ combinations
For three input variables: $N = 2^3 = 8$ combinations
For four input variables: $N = 2^4 = 16$ combinations

You can determine the number of input bit combinations for gates with any number of inputs by using Equation (3–1).

EXAMPLE 3–2

(a) Develop the truth table for a 3-input AND gate.
(b) Determine the total number of possible input combinations for a 5-input AND gate.

Solution

(a) There are eight possible input combinations ($2^3 = 8$) for a 3-input AND gate. The input side of the truth table (Table 3–3) shows all eight combinations of three bits. The output side is all 0s except when all three input bits are 1s.

(b) $N = 2^5 = 32$. There are 32 possible combinations of input bits for a 5-input AND gate.

TABLE 3–3

Inputs			Output
A	B	C	X
0	0	0	0
0	0	1	0
0	1	0	0
0	1	1	0
1	0	0	0
1	0	1	0
1	1	0	0
1	1	1	1

Related Exercise Develop the truth table for a 4-input AND gate.

Pulsed Operation

In a majority of applications, the inputs to a gate are not stationary levels but are voltage waveforms that change frequently between HIGH and LOW logic levels. Now let's look at the operation of AND gates with pulse waveform inputs, keeping in mind that an AND gate obeys the truth table operation regardless of whether its inputs are constant levels or levels that change back and forth.

Let's examine the pulsed operation of the AND gate by looking at the inputs with respect to each other in order to determine the output level at any given time. In Figure 3–10, the inputs are both HIGH (1) during the time interval t_1, making the output HIGH (1) during

FIGURE 3–10
Example of pulsed AND gate operation with a timing diagram showing input and output relationships.

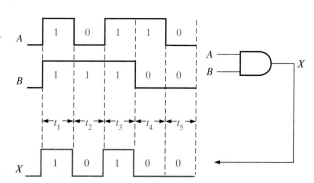

this interval. During time interval t_2, input A is LOW (0) and input B is HIGH (1), so the output is LOW (0). During time interval t_3, both inputs are HIGH (1) again, and therefore the output is HIGH (1). During time interval t_4, input A is high (1) and input B is LOW (0), resulting in a LOW (0) output. Finally, during time interval t_5, input A is LOW (0), input B is LOW (0), and the output is therefore LOW (0). As you know, a diagram of input and output waveforms showing time relationships is called a *timing diagram.*

EXAMPLE 3–3

If two waveforms, A and B, are applied to the AND gate inputs as in Figure 3–11, what is the resulting output waveform?

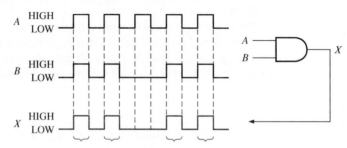

A and B are both HIGH during these four time intervals.
Therefore X is HIGH.

FIGURE 3–11

Solution   The output waveform X is HIGH only when both A and B are HIGH as shown in the timing diagram.

Related Exercise   Determine the output waveform and draw a timing diagram if the second and fourth pulses in waveform A of Figure 3–11 are omitted.

Remember, when analyzing the pulsed operation of logic gates, it is important to pay careful attention to the time relationships of all the inputs to each other and to the output.

EXAMPLE 3–4

For the two input waveforms, A and B, in Figure 3–12, sketch the output waveform, showing its proper relation to the inputs.

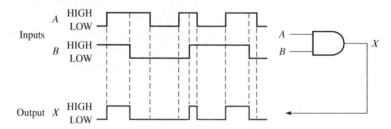

FIGURE 3–12

Solution   The output waveform is HIGH only when both of the inputs are HIGH as shown in the timing diagram.

Related Exercise   Show the output waveform if the B input to the AND gate in Figure 3–12 is always HIGH.

EXAMPLE 3–5   For the 3-input AND gate in Figure 3–13, determine the output waveform in relation to the inputs.

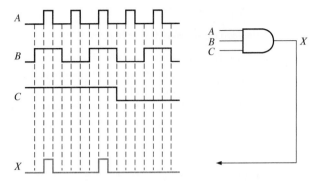

FIGURE 3–13

Solution   The output waveform X of the 3-input AND gate is HIGH only when all three inputs A, B, and C are HIGH.

Related Exercise   What is the output waveform of the AND gate in Figure 3–13 if the C input is always HIGH?

Logic Expressions for the AND Gate

The logical AND function of two variables is represented mathematically either by placing a dot between the two variables, as $A \cdot B$, or by simply writing the adjacent letters without the dot, as AB. We will normally use the latter notation because it is easier to write.

Multiplication in Boolean algebra follows the same basic rules governing binary multiplication, which were discussed in Chapter 2 and are as follows:

$$0 \cdot 0 = 0$$
$$0 \cdot 1 = 0$$
$$1 \cdot 0 = 0$$
$$1 \cdot 1 = 1$$

Boolean multiplication is the same as the AND function.

The operation of a 2-input AND gate can be expressed in equation form as follows: If one input variable is A, the other input variable is B, and the output variable is X, then the Boolean expression is

$$X = AB$$

Figure 3–14(a) shows the gate with the input and output variables indicated.

To extend the AND expression to more than two input variables, simply use a new letter for each input variable. The function of a 3-input AND gate, for example, can be expressed as $X = ABC$, where A, B, and C are the input variables. The expression for a 4-input AND gate can be $X = ABCD$, and so on. Parts (b) and (c) of Figure 3–14 show AND gates with three and four input variables, respectively.

$X = AB$ $X = ABC$ $X = ABCD$

(a)   (b)   (c)

FIGURE 3–14
Boolean expressions for AND gates.

You can evaluate an AND gate operation by using the Boolean expressions for the output. For example, each variable on the inputs can be either a 1 or a 0, so for the 2-input AND gate, make substitutions in the equation for the output, $X = AB$, as shown in Table 3–4. This evaluation shows that the output X of an AND gate is a 1 (HIGH) only when both inputs are 1s (HIGHs). A similar analysis can be made for any number of input variables.

TABLE 3–4

A	B	$AB = X$
0	0	$0 \cdot 0 = 0$
0	1	$0 \cdot 1 = 0$
1	0	$1 \cdot 0 = 0$
1	1	$1 \cdot 1 = 1$

Applications

The AND Gate as an Enable/Inhibit Device   A common application of the AND gate is to **enable** (that is, to allow) the passage of a signal (pulse waveform) from one point to another at certain times and to inhibit (prevent) the passage at other times.

A simple example of this particular use of the AND gate is shown in Figure 3–15, where the AND gate controls the passage of a signal (waveform A) to a digital counter. The purpose of this circuit is to measure the frequency of waveform A. The Enable pulse has a width of precisely 1 s. When the Enable pulse is HIGH, waveform A passes through the gate to the counter, and when the Enable pulse is LOW, the signal is prevented from passing through (inhibited).

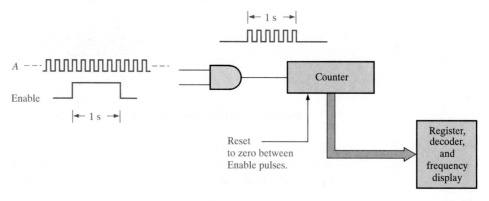

FIGURE 3–15
An AND gate performing an Enable/Inhibit function for a frequency counter.

During the 1 second (1 s) interval of the Enable pulse, a certain number of pulses in waveform A pass through the AND gate to the counter. The number of pulses passing through during 1 s is equal to the frequency of waveform A. For example, if 1000 pulses pass through the gate in the 1 s interval of the Enable pulse, there are 1000 pulses/s, or a frequency of 1000 Hz.

The counter counts the number of pulses per second and produces a binary output that goes to a decoding and display circuit to produce a readout of the frequency. The Enable pulse repeats at certain intervals and a new updated count is made so that if the frequency changes, the new value will be displayed. Between Enable pulses, the counter is reset so that it starts at zero each time an Enable pulse occurs. The current frequency count is stored in a register so that the display is unaffected by the resetting of the counter.

A Seat Belt Alarm System   In Figure 3–16, an AND gate is used in a simple automobile seat belt alarm system to detect when the ignition switch is on *and* the seat belt is unbuckled. If the ignition switch is on, a HIGH is produced on input *A* of the AND gate. If the seat belt is not properly buckled, a HIGH is produced on input *B* of the AND gate. Also, when the ignition switch is turned on, a timer is started that produces a HIGH on input *C* for 30 s.

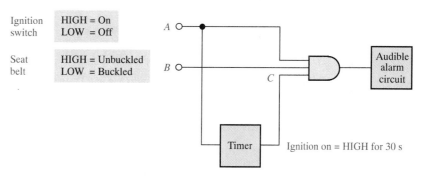

FIGURE 3–16
A simple seat belt alarm circuit using an AND gate.

If all three conditions exist—that is, if the ignition is on *and* the seat belt is unbuckled *and* the timer is running—the output of the AND gate is HIGH, and an audible alarm is energized to remind the driver.

SECTION 3–2 REVIEW	1. When is the output of an AND gate HIGH?
	2. When is the output of an AND gate LOW?
	3. Describe the truth table for a 5-input AND gate.

3–3 ■ THE OR GATE

The OR gate is another of the basic gates from which all logic functions are constructed. An OR gate can have two or more inputs and performs what is known as logical addition. After completing this section, you should be able to

☐ Identify an OR gate by its distinctive shape symbol or by its rectangular outline symbol   ☐ Describe the logical operation of an OR gate   ☐ Generate the truth table for an OR gate with any number of inputs ☐ Produce a timing diagram for an OR gate with any specified input waveforms   ☐ Write the logic expression for an OR gate with any number of inputs   ☐ Discuss examples of OR gate applications

An **OR gate** has two or more inputs and one output, as indicated by the standard logic symbols in Figure 3–17, where OR gates with two inputs are illustrated. An OR gate can have any number of inputs greater than one. Although both distinctive shape and rectangular outline symbols are shown for familiarization, the distinctive shape OR gate symbol will be used in this text.

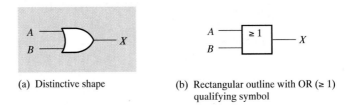

(a) Distinctive shape

(b) Rectangular outline with OR (≥ 1) qualifying symbol

FIGURE 3–17
Standard logic symbols for the OR gate showing two inputs (ANSI/IEEE Std. 91–1984).

Logical Operation of the OR Gate

The OR gate produces a HIGH on the output when *any* of the inputs is HIGH. The output is LOW only when all of the inputs are LOW. Therefore, an OR gate determines when one or more of its inputs are HIGH and produces a HIGH on its output to indicate this condition. The inputs of the 2-input OR gate in Figure 3–17 are labeled A and B, and the output is labeled X. The operation of the gate can be stated as follows:

> **In a 2-input OR gate operation, output X is HIGH if either input A or input B is HIGH, or if both A and B are HIGH; X is LOW if both A and B are LOW.**

The HIGH level is the active output level for the OR gate. Figure 3–18 illustrates the logic operation for a 2-input OR gate for all four possible input combinations.

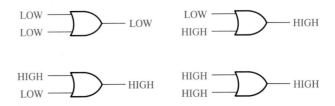

FIGURE 3–18
All possible logic levels for a 2-input OR gate.

OR Gate Truth Table

The logical operation of the 2-input OR gate is described in Table 3–5. This truth table can be expanded for any number of inputs; but regardless of the number of inputs, the output is HIGH when one or more of the inputs are HIGH.

TABLE 3–5
Truth table for a 2-input OR gate

Inputs		Output
A	B	X
0	0	0
0	1	1
1	0	1
1	1	1

1 ≡ HIGH, 0 ≡ LOW.

Pulsed Operation

Now let's look at the operation of an OR gate with pulsed inputs, keeping in mind its logical operation. Again, the important thing in the analysis of gate operation with pulsed waveforms is the time relationship of all the waveforms involved. For example, in Figure 3–19, inputs A and B are both HIGH (1) during time interval t_1, making the output HIGH (1). During time interval t_2, input A is LOW (0), but because input B is HIGH (1), the output is HIGH (1). Both inputs are LOW (0) during time interval t_3, so there is a LOW (0) output during this time. During time interval t_4, the output is HIGH (1) because input A is HIGH (1).

FIGURE 3–19

Example of pulsed OR gate operation with a timing diagram showing input and output time relationships.

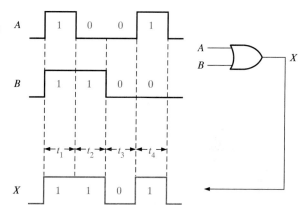

In this illustration, we have simply applied the truth table operation of the OR gate to each of the time intervals during which the levels are nonchanging. Examples 3–6 through 3–8 further illustrate OR gate operation with waveforms on the inputs.

EXAMPLE 3–6

If the two input waveforms, A and B, in Figure 3–20 are applied to the OR gate, what is the resulting output waveform?

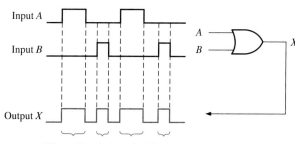

When either input or both inputs are HIGH, the output is HIGH.

FIGURE 3–20

Solution The output waveform X of a 2-input OR gate is HIGH when either or both inputs are HIGH as shown in the timing diagram. In this case, both inputs are never HIGH at the same time.

Related Exercise Determine the output waveform and draw the timing diagram if input A is changed such that it is HIGH from the beginning of the existing first pulse to the end of the existing second pulse.

EXAMPLE 3–7

For the two input waveforms, A and B, in Figure 3–21, sketch the output waveform, showing its proper relation to the inputs.

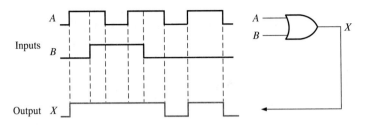

FIGURE 3–21

Solution   When either or both inputs are HIGH, the output is HIGH as shown by the output waveform X in the timing diagram.

Related Exercise   Determine the output waveform and draw the timing diagram if the middle pulse of input A is omitted.

EXAMPLE 3–8

For the 3-input OR gate in Figure 3–22, determine the output waveform in proper time relation to the inputs.

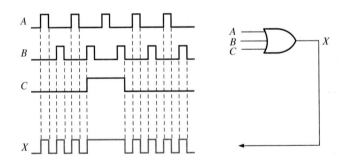

FIGURE 3–22

Solution   The output is HIGH when one or more of the inputs are HIGH as indicated by the output waveform X in the timing diagram.

Related Exercise   Determine the output waveform and draw the timing diagram if input C is always LOW.

Logic Expressions for the OR Gate

The logical OR function of two variables is represented mathematically by a $+$ between the two variables, for example, $A + B$.

Addition in Boolean algebra involves variables whose values are either binary 1 or binary 0. The basic rules for Boolean addition are as follows:

$$0 + 0 = 0$$
$$0 + 1 = 1$$
$$1 + 0 = 1$$
$$1 + 1 = 1$$

Boolean addition is the same as the OR function.

Notice that Boolean addition differs from binary addition in the case where two 1s are added.

The operation of a 2-input OR gate can be expressed as follows: If one input variable is A, if the other input variable is B, and if the output variable is X, then the Boolean expression is

$$X = A + B$$

Figure 3–23(a) shows the gate logic symbol with input and output variables labeled.

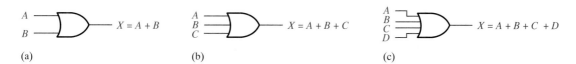

(a)   (b)   (c)

FIGURE 3–23
Boolean expressions for OR gates.

To extend the OR expression to more than two input variables, a new letter is used for each additional variable. For instance, the function of a 3-input OR gate can be expressed as $X = A + B + C$. The expression for a 4-input OR gate can be written as $X = A + B + C + D$, and so on. Parts (b) and (c) of Figure 3–23 show OR gates with three and four input variables, respectively.

OR gate operation can be evaluated by using the Boolean expressions for the output X by substituting all possible combinations of 1 and 0 values for the input variables, as shown in Table 3–6 for a 2-input OR gate. This evaluation shows that the output X of an OR gate is a 1 (HIGH) when any one or more of the inputs are 1 (HIGH). A similar analysis can be extended to OR gates with any number of input variables.

TABLE 3–6

A	B	$A + B = X$
0	0	$0 + 0 = 0$
0	1	$0 + 1 = 1$
1	0	$1 + 0 = 1$
1	1	$1 + 1 = 1$

An Application

Intrusion Detection   A simplified portion of an intrusion detection and alarm system is shown in Figure 3–24. This system could be used for one room in a home—a room with two windows and a door. The sensors are magnetic switches that produce a HIGH output when open and a LOW output when closed. As long as the windows and the door are secured, the switches are closed and all three of the OR gate inputs are LOW. When one of the windows or the door is opened, a HIGH is produced on that input to the OR gate and the gate output goes HIGH. It then activates an alarm circuit to warn of the intrusion.

FIGURE 3–24
A simplified intrusion detection system using an OR gate.

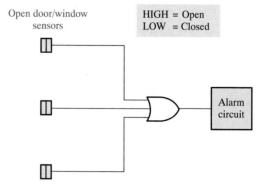

Open door/window sensors

HIGH = Open
LOW = Closed

Alarm circuit

SECTION 3–3 REVIEW	1. When is the output of an OR gate HIGH? 2. When is the output of an OR gate LOW? 3. Describe the truth table for a three-input OR gate.

3–4 ■ THE NAND GATE

The NAND gate is a popular logic element because it can be used as a universal gate; that is, NAND gates can be used in combination to perform the AND, OR, and inverter operations. The universal property of the NAND gate will be examined thoroughly in Chapter 5. After completing this section, you should be able to

☐ Identify a NAND gate by its distinctive shape symbol or by its rectangular outline symbol ☐ Describe the logical operation of a NAND gate ☐ Develop the truth table for a NAND gate with any number of inputs ☐ Produce a timing diagram for a NAND gate with any specified input waveforms ☐ Write the logic expression for a NAND gate with any number of inputs ☐ Describe NAND gate operation in terms of its negative-OR equivalent ☐ Discuss examples of NAND gate applications

The term *NAND* is a contraction of NOT-AND and implies an AND function with a complemented (inverted) output. The standard logic symbol for a 2-input NAND gate and its equivalency to an AND gate followed by an inverter are shown in Figure 3–25(a). A rectangular outline symbol is shown in part (b).

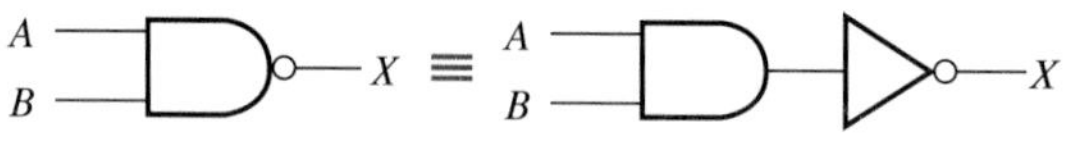

(a) Distinctive shape, 2-input NAND gate and its NOT/AND equivalent

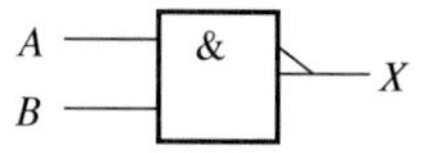

(b) Rectangular outline, 2-input NAND gate with polarity indicator

FIGURE 3–25
Standard NAND gate logic symbols (ANSI/IEEE Std. 91–1984).

Logical Operation of the NAND Gate

The **NAND gate** produces a LOW output only when all the inputs are HIGH. When any of the inputs is LOW, the output will be HIGH. For the specific case of a 2-input NAND gate, as shown in Figure 3–25 with the inputs labeled A and B and the output labeled X, the operation can be stated as follows:

In a 2-input NAND gate operation, output *X* is LOW if inputs *A* and *B* are HIGH; *X* is HIGH if either *A* or *B* is LOW, or if both *A* and *B* are LOW.

Note that this operation is opposite that of the AND in terms of the output level. In a NAND gate, the LOW level (0) is the active output level, as indicated by the bubble on the output. Figure 3–26 illustrates the logical operation of a 2-input NAND gate for all four input combinations, and Table 3–7 is the truth table summarizing the logical operation of the 2-input NAND gate.

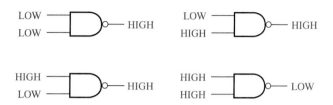

FIGURE 3–26
Logical operation of a 2-input NAND gate.

TABLE 3–7
Truth table for a 2-input NAND gate

Inputs		Output
A	B	X
0	0	1
0	1	1
1	0	1
1	1	0

1 ≡ HIGH, 0 ≡ LOW.

Pulsed Operation

Now let's look at the pulsed operation of the NAND gate. Remember from the truth table that the only time a LOW output occurs is when all of the inputs are HIGH. Examples 3–9 and 3–10 illustrate pulsed operation.

EXAMPLE 3–9

If the two waveforms *A* and *B* shown in Figure 3–27 are applied to the NAND gate inputs, determine the resulting output waveform.

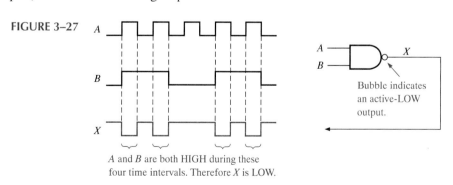

FIGURE 3–27

A and *B* are both HIGH during these four time intervals. Therefore *X* is LOW.

Bubble indicates an active-LOW output.

Solution   Output waveform *X* is LOW only during the four time intervals when both inputs *A* and *B* are HIGH as shown in the timing diagram.

Related Exercise   Determine the output waveform and draw the timing diagram if input *B* is inverted.

EXAMPLE 3–10

Sketch the output waveform for the 3-input NAND gate in Figure 3–28, showing its proper time relationship to the inputs.

FIGURE 3–28

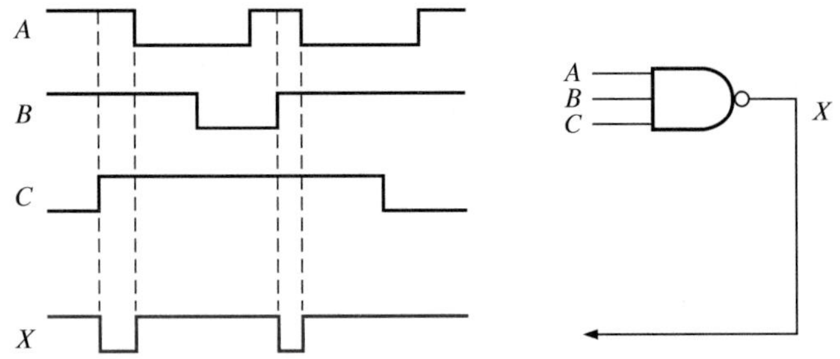

Solution The output waveform X is LOW only when all three inputs are HIGH as shown in the timing diagram.

Related Exercise Determine the output waveform and draw the timing diagram if input A is inverted.

Negative-OR Equivalent Operation of the NAND Gate Inherent in the NAND gate's operation is the fact that one or more LOW inputs produce a HIGH output. Table 3–7 shows that output X is HIGH (1) when any of the inputs, A and B, are LOW (0). From this viewpoint, the NAND gate can be used for an OR operation that requires one or more LOW inputs to produce a HIGH output. This mode of operation is referred to as **negative-OR**. The term *negative* in this context means that the inputs are defined to be in the active or asserted state when LOW.

> **In the operation of a 2-input NAND gate functioning as a negative-OR gate, output X is HIGH if either input A or input B is LOW, or if both A and B are LOW.**

When the NAND gate is looking for one or more LOWs on its inputs rather than for all HIGHs, it is acting as a negative-OR gate and is represented by the standard logic symbol in Figure 3–29. Although the two symbols in Figure 3–29 represent the same physical gate, they serve to define its role or mode of operation in a particular application, as illustrated by Examples 3–11 through 3–13.

FIGURE 3–29
Standard symbols representing the two equivalent operations of the NAND gate.

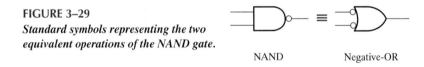

NAND   Negative-OR

EXAMPLE 3–11

A manufacturing plant uses two tanks to store a certain liquid chemical that is required in a manufacturing process. Each tank has a sensor that detects when the chemical level drops to 25% of full. The sensors produce a 5 V level when the tanks are more than one-quarter full. When the volume of chemical in a tank drops to one-quarter full, the sensor puts out a 0 V level.

It is required that a single green light-emitting diode (LED) on an indicator panel show when both tanks are more than one quarter full. Show how a NAND gate can be used to implement this function.

Solution   Figure 3–30 shows a NAND gate with its two inputs connected to the tank level sensors and its output connected to the indicator panel.

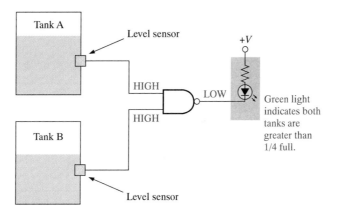

FIGURE 3–30

As long as *both* sensor outputs are HIGH (5 V), indicating that both tanks are more than one-quarter full, the NAND gate output is LOW (0 V). The green LED circuit is arranged so that a LOW voltage turns it on.

Related Exercise   How can the circuit of Figure 3–30 be modified to monitor the levels in three tanks rather than two?

EXAMPLE 3–12

The supervisor of the manufacturing process described in Example 3–11 has decided that he would prefer to have a red LED display come on when at least one of the tanks falls to the quarter-full level rather than have the green LED display indicate when both are above one quarter. Show how this requirement can be implemented.

Solution   Figure 3–31 shows the NAND gate operating as a negative-OR gate to detect the occurrence of at least one LOW on its inputs. A sensor puts out a LOW voltage if the volume in its tank goes to one-quarter full or less. When this happens, the gate output goes HIGH. The red LED circuit in the panel is arranged so that a HIGH voltage turns it on.

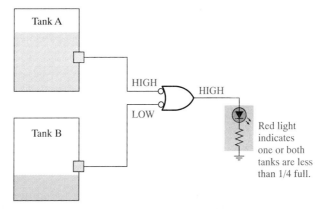

FIGURE 3–31

Notice that, in this example and in Example 3–11, the same 2-input NAND gate is used, but a different gate symbol is used in the schematic, illustrating the functional difference of the NAND and negative-OR gates.

Related Exercise How can the circuit in Figure 3–31 be modified to monitor four tanks rather than two?

EXAMPLE 3–13

For the 4-input NAND gate in Figure 3–32, operating as a negative-OR, determine the output with respect to the inputs.

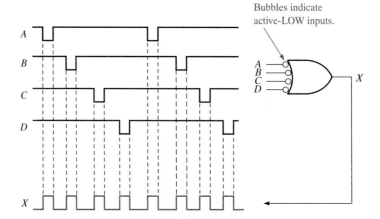

FIGURE 3–32

Solution The output waveform X is HIGH any time an input is LOW as shown in the timing diagram.

Related Exercise Determine the output waveform if input A is inverted before it is applied to the gate.

Logic Expressions for the NAND Gate

The Boolean expression for the output of a 2-input NAND gate is

$$X = \overline{AB}$$

This expression says that the two input variables, A and B, are first ANDed and then complemented, as indicated by the bar over the AND expression. This is a logical description in equation form of the operation of a NAND gate with two inputs. If you evaluate this expression for all possible values of the two input variables, the results are in Table 3–8.

TABLE 3–8

A	B	$\overline{AB} = X$
0	0	$\overline{0 \cdot 0} = \overline{0} = 1$
0	1	$\overline{0 \cdot 1} = \overline{0} = 1$
1	0	$\overline{1 \cdot 0} = \overline{0} = 1$
1	1	$\overline{1 \cdot 1} = \overline{1} = 0$

Thus, once an expression is determined for a given logic function, that function can be evaluated for all possible values of the variables. The evaluation tells you exactly what the output of the logic circuit is for each of the input conditions, and it therefore gives you a complete description of the circuit's logical operation. The NAND expression can be extended to more than two input variables by including additional letters to represent the other variables.

SECTION 3–4 REVIEW	1. When is the output of a NAND gate LOW? 2. When is the output of a NAND gate HIGH? 3. Describe the functional differences between a NAND gate and a negative-OR gate. Do they both have the same truth table? 4. Write the output expression for a NAND gate with inputs A, B, and C.

3–5 ■ THE NOR GATE

The NOR gate, like the NAND gate, is a useful logic element because it can also be used as a universal gate; that is, NOR gates can be used in combination to perform the AND, OR, and inverter operations. The universal property of the NOR gate will be examined thoroughly in Chapter 5. After completing this section, you should be able to

□ Identify a NOR gate by its distinctive shape symbol or by its rectangular outline symbol □ Describe the logical operation of a NOR gate □ Develop the truth table for a NOR gate with any number of inputs □ Produce a timing diagram for a NOR gate with any specified input waveforms □ Write the logic expression for a NOR gate with any number of inputs □ Describe NOR gate operation in terms of its negative-AND equivalent □ Discuss examples of NOR gate applications

The term *NOR* is a contraction of NOT-OR and implies an OR function with an inverted output. A standard logic symbol for a 2-input NOR gate and its equivalent OR gate followed by an inverter are shown in Figure 3–33(a). A rectangular outline symbol is shown in part (b).

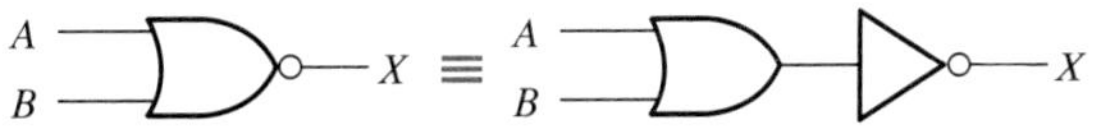

(a) Distinctive shape, 2-input NOR gate and its NOT/OR equivalent

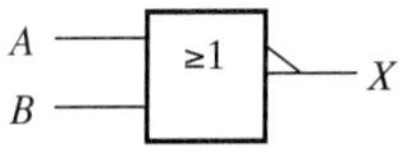

(b) Rectangular outline, 2-input NOR gate with polarity indicator

FIGURE 3–33
Standard NOR gate logic symbols (ANSI/IEEE Std. 91–1984).

Logical Operation of the NOR Gate

The **NOR gate** produces a LOW output when *any* of its inputs is HIGH. Only when all of its inputs are LOW is the output HIGH. For the specific case of a 2-input NOR gate, as shown in Figure 3–33 with the inputs labeled A and B and the output labeled X. The operation can be stated as follows:

In a 2-input NOR gate operation, output X is LOW if either input A or input B is HIGH, or if both A and B are HIGH; X is HIGH if both A and B are LOW.

This operation results in an output level opposite that of the OR gate. In a NOR gate, the LOW output is the active output level as indicated by the bubble on the output. Figure 3–34 illustrates the logical operation of a 2-input NOR gate for all four possible input combinations, and Table 3–9 is the truth table for the 2-input NOR gate.

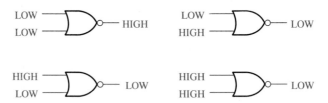

FIGURE 3–34
Logical operation of a 2-input NOR gate.

TABLE 3–9
Truth table for a 2-input NOR gate

Inputs		Output
A	B	X
0	0	1
0	1	0
1	0	0
1	1	0

1 ≡ HIGH, 0 ≡ LOW.

Pulsed Operation

The next two examples illustrate the logic operation of the NOR gate with pulsed inputs. Again, as with the other types of gates, we will simply follow the truth table operation to determine the output waveforms in the proper time relationship to the inputs.

EXAMPLE 3–14

If the two waveforms shown in Figure 3–35 are applied to the NOR gate, what is the resulting output waveform?

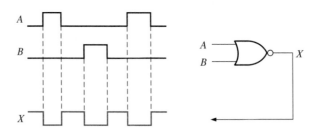

FIGURE 3–35

Solution Whenever any input of a NOR gate is HIGH, the output is LOW as shown by the output waveform X in the timing diagram.

Related Exercise Invert input B and determine the output waveform in relation to the inputs.

EXAMPLE 3–15

Sketch the output waveform for the 3-input NOR gate in Figure 3–36, showing the proper time relation to the inputs.

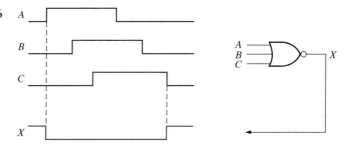

FIGURE 3–36

Solution The output X is LOW when any input is HIGH as shown by the output waveform X in the timing diagram.

Related Exercise With the B and C inputs inverted, determine the output and draw the timing diagram.

Negative-AND Equivalent Operation of the NOR Gate The NOR gate, like the NAND, has another aspect of its operation that is inherent in the way it logically functions. Table 3–9 shows that a HIGH is produced on the gate output only if all of the inputs are LOW. From this viewpoint, the NOR gate can be used for an AND operation that requires all LOW inputs to produce a HIGH output. This mode of operation is called **negative-AND.** The term *negative* in this context means that the inputs are defined to be in the active or asserted state when LOW.

> **In the operation of a 2-input NOR gate functioning as a negative-AND gate, output X is HIGH if both inputs A and B are LOW.**

When the NOR gate is looking for all LOWs on its inputs rather than for one or more HIGHs, it is acting as a negative-AND gate and is represented by the standard symbol in Figure 3–37. It is important to remember that the two symbols in Figure 3–37 represent the same physical gate and serve only to distinguish between the two modes of its logical operation. The following three examples illustrate this.

FIGURE 3–37
Standard symbols representing the two equivalent operations of the NOR gate.

NOR ≡ Negative-AND

EXAMPLE 3–16

A device is needed to indicate when two LOW levels occur simultaneously on its inputs and to produce a HIGH output as an indication. Specify the device.

Solution A negative-AND gate is required to produce a HIGH output when both inputs are LOW, as shown in Figure 3–38.

FIGURE 3–38

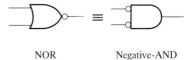

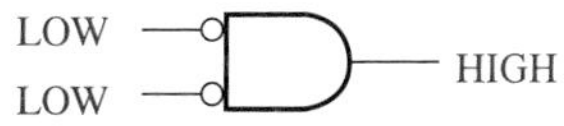

Related Exercise A device is needed to indicate when one or two HIGH levels occur on its inputs and to produce a LOW output as an indication. Specify the device.

EXAMPLE 3–17

In an aircraft, as part of its functional monitoring system, a circuit is required to indicate the status of the landing gear prior to landing. A green LED display turns on if all three gears are properly extended when the "gear down" switch has been activated in preparation for landing. A red LED display turns on if any of the gears fail to extend properly prior to landing. When a landing gear is extended, its sensor produces a LOW voltage. When a landing gear is retracted, its sensor produces a HIGH voltage. Implement a circuit to meet this requirement.

Solution   Power is applied to the circuit only when the "gear down" switch is activated. Use a NOR gate for each of the two requirements as shown in Figure 3–39. One NOR gate operates as a negative-AND to detect a LOW from each of the three landing gear sensors. When all three of the gate inputs are LOW, the three landing gear are properly extended and the resulting HIGH output from the negative-AND gate turns on the green LED display. The other NOR gate operates as a NOR to detect if one or more of the landing gear remain retracted when the "gear down" switch is activated. When one or more of the landing gear remain retracted, the resulting HIGH from the sensor is detected by the NOR gate, which produces a LOW output to turn on the red LED warning display.

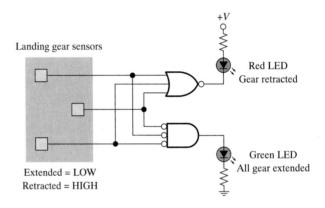

FIGURE 3–39

Related Exercise   What type of gate should be used to detect if all three landing gear are retracted after takeoff, assuming a LOW output is required to activate an LED display?

EXAMPLE 3–18

For the 4-input NOR gate operating as a negative-AND in Figure 3–40, determine the output relative to the inputs.

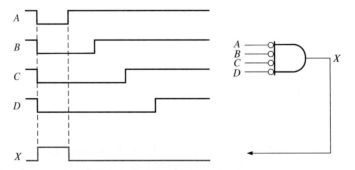

FIGURE 3–40

Solution Any time all of the inputs are LOW, the output is HIGH as shown by output waveform X in the timing diagram.

Related Exercise Determine the output with input D inverted and draw the timing diagram.

Logic Expressions for the NOR Gate

The Boolean expression for the output of a 2-input NOR gate can be written as

$$X = \overline{A + B}$$

This equation says that the two input variables are first ORed and then complemented, as indicated by the bar over the OR expression. Evaluating this expression, you get the results in Table 3–10. The NOR expression can be extended to more than two input variables by including additional letters to represent the other variables.

TABLE 3–10

A	B	$\overline{A + B} = X$
0	0	$\overline{0 + 0} = \overline{0} = 1$
0	1	$\overline{0 + 1} = \overline{1} = 0$
1	0	$\overline{1 + 0} = \overline{1} = 0$
1	1	$\overline{1 + 1} = \overline{1} = 0$

SECTION 3–5 REVIEW

1. When is the output of a NOR gate HIGH?
2. When is the output of a NOR gate LOW?
3. Describe the functional difference between a NOR gate and a negative-AND gate. Do they both have the same truth table?
4. Write the output expression for a 3-input NOR with input variables A, B, and C.

3–6 ▪ THE EXCLUSIVE-OR AND EXCLUSIVE-NOR GATES

The exclusive-OR and exclusive-NOR gates are actually formed by a combination of other gates already discussed, as you will see in Chapter 5. However, because of their fundamental importance in many applications, these gates are treated as basic logic elements with their own unique symbols. After completing this section, you should be able to

□ Identify the exclusive-OR and exclusive-NOR gates by their distinctive shape symbols or by their rectangular outline symbols □ Describe the logical operation of exclusive-OR and exclusive-NOR gates □ Show the truth tables for exclusive-OR and exclusive-NOR gates □ Produce a timing diagram for an exclusive-OR or exclusive-NOR gate with any specified input waveforms □ Discuss examples of exclusive-OR and exclusive-NOR gate applications

The Exclusive-OR Gate

Standard symbols for the exclusive-OR (XOR for short) gate are shown in Figure 3–41. The XOR gate has only two inputs. Unlike the other gates we have discussed, it never has more than two inputs.

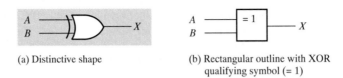

(a) Distinctive shape

(b) Rectangular outline with XOR qualifying symbol (= 1)

FIGURE 3–41
Standard logic symbols for the exclusive-OR gate.

The output of an **exclusive-OR gate** is HIGH *only* when the two inputs are at opposite logic levels. This operation can be stated as follows with reference to inputs A and B and output X:

In an exclusive-OR gate operation, output X is HIGH if input A is LOW and input B is HIGH, or if input A is HIGH and input B is LOW; X is LOW if A and B are both HIGH or both LOW.

The four possible input combinations and the resulting outputs for the XOR gate are illustrated in Figure 3–42. The HIGH level is the active output level and occurs only when the inputs are at opposite levels. The logical operation of the XOR gate is summarized in the truth table shown in Table 3–11.

FIGURE 3–42
All possible logic levels for an exclusive-OR gate.

LOW
LOW ── LOW

LOW
HIGH ── HIGH

HIGH
LOW ── HIGH

HIGH
HIGH ── LOW

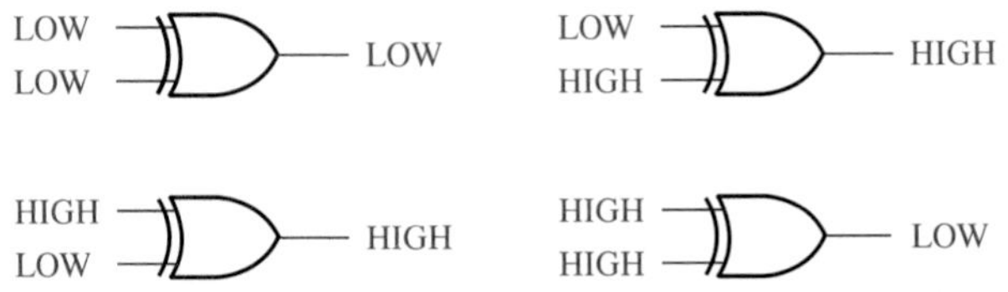

TABLE 3–11
Truth table for an exclusive-OR gate

Inputs		Output
A	B	X
0	0	0
0	1	1
1	0	1
1	1	0

EXAMPLE 3–19

A certain system contains two identical circuits operating in parallel. As long as both are operating properly, the outputs of both circuits are always the same. If one of the circuits fails, the outputs will be at opposite levels at some time. Devise a way to detect that a failure has occurred in one of the circuits.

Solution The outputs of the circuits are connected to the inputs of an XOR gate as shown in Figure 3–43. A failure in either one of the circuits causes the XOR inputs to be at opposite levels. This condition produces a HIGH on the output of the XOR gate, indicating a failure in one of the circuits.

FIGURE 3–43

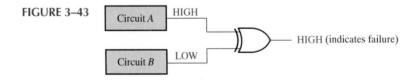

Circuit A ── HIGH

Circuit B ── LOW

── HIGH (indicates failure)

Related Exercise Will the exclusive-OR gate always detect simultaneous failures in both circuits of Figure 3–43? If not, under what condition?

The Exclusive-NOR Gate

Standard symbols for the **exclusive-NOR** (XNOR) **gate** are shown in Figure 3–44. Like the XOR gate, the XNOR has only two inputs. The bubble on the output of the XNOR symbol indicates that its output is opposite that of the XOR gate. When the two input logic levels are opposite, the output of the exclusive-NOR gate is LOW. The operation can be stated as follows (A and B are inputs, X is output):

> **In an exclusive-NOR gate operation, output X is LOW if input A is LOW and input B is HIGH, or if A is HIGH and B is LOW; X is HIGH if A and B are both HIGH or both LOW.**

The four possible input combinations and the resulting outputs for the XNOR gate are shown in Figure 3–45. The logical operation is summarized in Table 3–12. Notice that the output is HIGH when the same level is on both inputs.

FIGURE 3–44
Standard logic symbols for the exclusive-NOR gate.

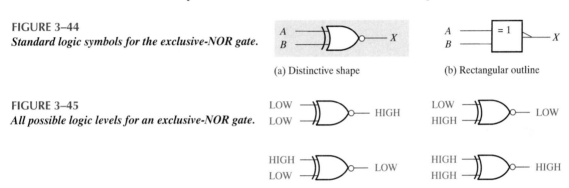

(a) Distinctive shape (b) Rectangular outline

FIGURE 3–45
All possible logic levels for an exclusive-NOR gate.

TABLE 3–12
Truth table for an exclusive-NOR gate

| Inputs | | Output |
A	B	X
0	0	1
0	1	0
1	0	0
1	1	1

Pulsed Operation

As we have done with the other gates, let's examine the operation of the XOR and XNOR gates under pulsed input conditions. As before, we apply the truth table operation during each distinct time interval of the pulsed inputs, as illustrated in Figure 3–46 for an XOR gate. You can see that the input waveforms A and B are at opposite levels during time in-

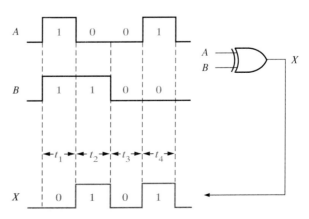

FIGURE 3–46
Example of pulsed exclusive-OR gate operation.

tervals t_2 and t_4. Therefore, the output X is HIGH during these two times. Since both inputs are the same level, either both HIGH or both LOW, during time intervals t_1 and t_3, the output is LOW during those times as shown in the timing diagram.

EXAMPLE 3–20

Determine the output waveforms for the XOR gate and for the XNOR gate, given the input waveforms, A and B, in Figure 3–47.

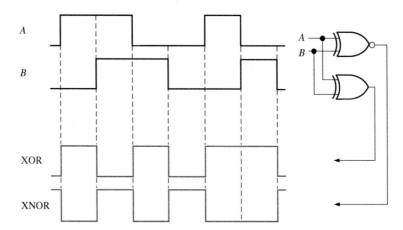

FIGURE 3–47

Solution   The output waveforms are shown in Figure 3–47. Notice that the XOR output is HIGH only when both inputs are at opposite levels. Notice that the XNOR output is HIGH only when both inputs are the same.

Related Exercise   Determine the output waveforms if the two inputs, A and B, are inverted.

An Application

The Exclusive-OR Gate as a Two-Bit Adder   Recall from Chapter 2 that the basic rules for binary addition are as follows: $0 + 0 = 0, 0 + 1 = 1, 1 + 0 = 1$, and $1 + 1 = 10$. An examination of the truth table for the XOR gate will show you that its output is the binary sum of the two input bits. In the case where the inputs are both 1s, the output is the sum 0, but you lose the carry of 1. In Chapter 6 you will see how XOR gates are combined to make complete adding circuits. Figure 3–48 illustrates the XOR gate used as a basic adder.

FIGURE 3–48
The XOR gate used to add two bits.

Input bits		Output (sum)
A	B	Σ
0	0	0
0	1	1
1	0	1
1	1	0 (without 1 carry)

1. When is the output of an XOR gate HIGH?
2. When is the output of an XNOR gate HIGH?
3. How can you use an XOR gate to detect when two bits are different?

3–7 ▪ INTEGRATED CIRCUIT LOGIC FAMILIES

In the previous sections, you learned about the logical operation of the various gates. In this section, we introduce the two most widely used types of digital integrated circuit families, TTL and CMOS, and look at several specific devices in these IC families. A third type of integrated circuit family, ECL, is also introduced.

This coverage is important because it gives you basic information on the specific devices you will be using in the lab and in industry. A more detailed and thorough coverage of the circuitry, specifications, and parameters of IC families is provided in Chapter 15. After completing this section, you should be able to

☐ Briefly describe three logic families: TTL, CMOS, and ECL ☐ Compare some of the basic performance parameters of the three families ☐ Define *propagation delay, power dissipation, fan-out,* and *speed-power product* ☐ Interpret basic data sheet information ☐ Name several specific devices

TTL

The term **TTL** stands for *t*ransistor-*t*ransistor *l*ogic, which refers to the use of **bipolar** junction transistors in the circuit technology used to construct the gates at the chip level.

TTL consists of a series of logic circuits: standard TTL, low-power TTL. **Schottky** TTL, low-power Schottky TTL, advanced low-power Schottky TTL, and advanced Schottky TTL. The differences in these various types of TTL are in their performance characteristics, such as propagation delay times, power dissipation, and fan-out, which are explained later. The TTL family has a number prefix of 54 or 74, followed by a letter or letters that specify the series, as shown in Table 3–13. The prefix 54 indicates an operating temperature range of −55°C to 125°C (generally for military use). The prefix 74 indicates a temperature range of 0°C to 70°C (for commercial use). We will use the prefix 74 throughout the book. The term *quad* used in the table means four individual gates per package.

TABLE 3–13
TTL series designations

TTL Series	Prefix Designation	Example of Device
Standard TTL	54 or 74 (no letter)	7400 (quad NAND gates)
Low-power TTL	54L or 74L	74L00 (quad NAND gates)
Schottky TTL	54S or 74S	74S00 (quad NAND gates)
Low-power Schottky TTL	54LS or 74LS	74LS00 (quad NAND gates)
Advanced low-power Schottky TTL	54ALS or 74ALS	74ALS00 (quad NAND gates)
Advanced Schottky TTL	54AS or 74AS	74AS00 (quad NAND gates)

CMOS

The term **CMOS** stands for *c*omplementary *m*etal-*o*xide *s*emiconductor. Whereas TTL uses bipolar transistors in its circuit technology, CMOS uses field-effect transistors. Logic functions are the same, however, whether the device is implemented with TTL or CMOS technologies. The circuit technologies make a difference, not in logic function, but only in performance characteristics.

Several series of CMOS logic circuits are available, but they fall basically into two process technology categories: metal-gate CMOS and silicon-gate CMOS. The older, metal-gate technology is the 4000 series. The newer, silicon-gate technology consists of the 74C, the 74HC, and the 74HCT. All of the 74 series CMOS devices are both pin compatible and function compatible with the TTL series. That is, a TTL IC and a CMOS IC of the same number have the inputs, outputs, supply voltage, and ground on the same pins, as well as the same logic gates. In addition, the 74HCT series is voltage-level compatible with TTL and requires no special interfacing as do the 74C and 74HC series. Other differences in the various types of 74 series CMOS are in their performance characteristics.

ECL

The term **ECL** stands for *emitter coupled logic* which is a bipolar circuit technology. ECL has the fastest switching speed of any logic family but its power consumption is much higher. The variety of devices that are available in ECL is very limited compared to TTL and CMOS; however, many complex functions and special-purpose circuits are available.

Performance Characteristics

Four important characteristics for the performance of logic circuits are propagation delay time, power dissipation, fan-out, and speed-power product.

Propagation Delay Time   The **propagation delay time** limits the switching speed (frequency) at which logic circuits can operate. The terms *low speed* and *high speed,* when applied to logic circuits, refer to the propagation delays; the shorter the propagation delay, the higher the speed of the circuit.

The propagation delay time of a gate is basically the time interval between the application of an input pulse and the occurrence of the resulting output pulse. There are two different propagation delay times associated with a logic gate:

1. t_{PHL}: The time between a specified reference point on the input pulse and a corresponding reference point on the output pulse, with the output changing from the HIGH level to the LOW level.
2. t_{PLH}: The time between a specified reference point on the input pulse and a corresponding reference point on the output pulse, with the output changing from the LOW level to the HIGH level.

EXAMPLE 3–21   Show the propagation delay times of the inverter in Figure 3–49(a).

FIGURE 3–49

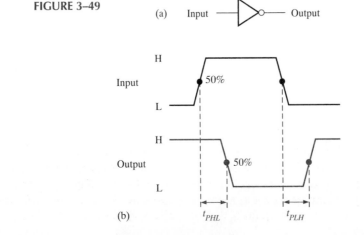

(a)   Input ——▷○—— Output

(b)

Solution   The propagation delay times, t_{PHL} and t_{PLH}, are indicated in part (b) of the figure. In this case, the delays are measured between the 50% points of the corresponding edges of the input and output pulses. The values of t_{PHL} and t_{PLH} are not necessarily equal but in many cases they are the same.

Related Exercise   One type of logic gate has a specified maximum t_{PLH} and t_{PHL} of 10 ns. For another type of gate the value is 4 ns. Which gate can operate at the highest frequency?

Power Dissipation   The **power dissipation** of a logic gate equals the dc supply voltage V_{CC} times the average supply current I_{CC}. Normally, the value of I_{CC} for a LOW gate output is higher than for a HIGH output. The manufacturer's data sheet usually specifies these values as I_{CCL} and I_{CCH}. The average I_{CC} is then determined, based on a 50% duty cycle operation of the gate (LOW half the time and HIGH half the time).

Fan-out   The **fan-out** of a gate is the maximum number of inputs of the same series IC family that the gate can drive while maintaining its output levels within specified limits. That is, the fan-out specifies the maximum load that a given gate is capable of handling. For example, a standard TTL gate has a fan-out of 10 **unit loads.** This means that it can drive no more than 10 inputs of other standard TTL gates and still operate reliably. If the fan-out is exceeded, specified operation is not guaranteed. Figure 3–50 shows a gate driving 10 other gates.

FIGURE 3–50
The standard TTL NAND gate output fans out to a maximum of ten standard TTL gate inputs.

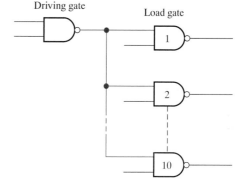

Speed-Power Product   The **speed-power product** is sometimes specified by the manufacturer as a measure of the performance of a logic circuit. It is the product of the propagation delay time and the power dissipation at a specified frequency. The smaller the speed-power product is, the better the performance. The speed-power product is expressed as energy in joules, symbolized by J. For example, the speed-power product (SPP) of a 74HC CMOS gate at a frequency of 100 kHz is

$$SPP = (8 \text{ ns})(0.17 \text{ mW}) = 1.36 \text{ pJ}$$

Table 3–14 provides a comparison of some performance characteristics of various CMOS, TTL, and ECL logic families.

Specific Devices

A wide variety of SSI (small-scale integration) logic gate configurations are available in both the TTL and the CMOS families. When specific devices in the SSI and MSI categories are referred to in this book, standard TTL will generally be used for illustration, although other types will occasionally be used. Keep in mind that most of the specific devices are also available in the other TTL series as well as in CMOS.

TABLE 3–14

Comparison of typical performance characteristics of CMOS, TTL, and ECL logic gates

Technology	CMOS* (silicon-gate)	CMOS* (metal-gate)	TTL Std.	TTL LS	TTL S	TTL ALS	TTL AS	ECL
Device series	74HC	4000B	74	74LS	74S	74ALS	74AS	10KH
Power dissipation: Static @ 100 kHz	2.5 nW 0.17 mW	1 μW 0.1 mW	10 mW 10 mW	2 mW 2 mW	19 mW 19 mW	1 mW 1 mW	8.5 mW 8.5 mW	25 mW 25 mW
Propagation delay time	8 ns	50 ns	10 ns	10 ns	3 ns	4 ns	1.5 ns	1 ns
Fan-out (same series)			10	20	20	20	40	

*Propagation delay is dependent on the dc supply voltage, V_{CC}. Power dissipation and fanout are functions of frequency.

A selection of typical logic gate ICs is now presented. These devices are commonly housed in the dual in-line package (DIP) or a surface-mount (SMT) package. For simplicity, V_{CC} and ground connections to each gate are normally not shown in a logic diagram. On most 14-pin packages V_{CC} is pin 14 and ground is pin 7. On most 16-pin packages V_{CC} is pin 16 and ground is pin 8. Figure 3–51 shows typical 14-pin packages for pin numbering and size comparison. Although not shown to scale, you can see that the SOIC package is significantly smaller than the DIP. Notice that the dimensions are given in inches.

In Figures 3–52 through 3–57, each device is represented by a distinctive shape logic diagram, with the pin numbers indicated in parentheses. Additionally, each device is shown as a rectangular outline logic symbol. The two representations are equivalent. Although the standard 74 series designation is used for illustration, most of these devices are also available in most of the other TTL and CMOS series: 74LS, 74S, 74ALS, 74AS, 74HC, and 74HCT.

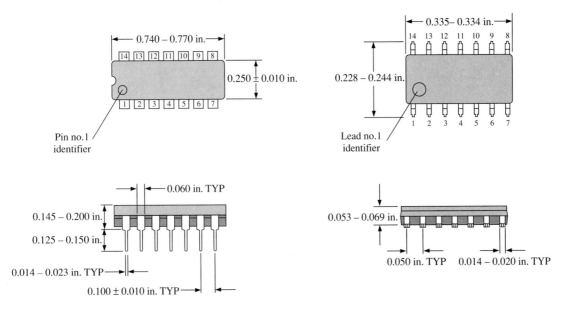

(a) 14-pin dual in-line package (DIP) for feedthrough mounting

(b) 14-pin small outline package (SOIC) for surface mounting

FIGURE 3–51

Typical dual in-line and small outline packages showing pin numbers and basic dimensions.

Hex Inverter The 7404 hex inverter is a standard TTL device consisting of six inverters in a 14-pin package, as shown in Figure 3–52.

AND Gates Several configurations of AND gates are available in IC form. The 7408 has four 2-input AND gates (it is called a quad 2-input AND); the 7411 has three 3-input AND gates (a triple 3-input AND); and the 7421 has two 4-input AND gates (a dual 4-input AND). These gates are shown in Figure 3–53.

FIGURE 3–52
7404 hex inverter (pin numbers in parentheses).

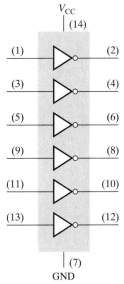

(a) Distinctive shape logic diagram

(b) Rectangular outline logic symbol with polarity indicators. The inverter qualifying symbol (1) appears in the top block and applies to all blocks below.

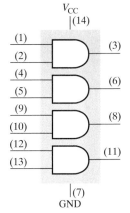

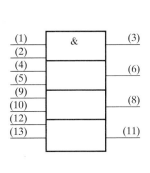

(a) 7408 quad 2-input AND

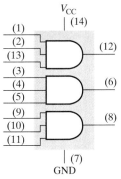

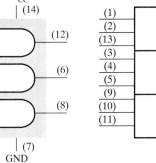

(b) 7411 triple 3-input AND

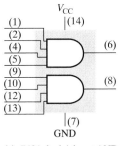

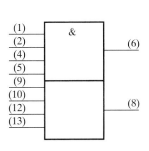

(c) 7421 dual 4-input AND

FIGURE 3–53
AND gates.

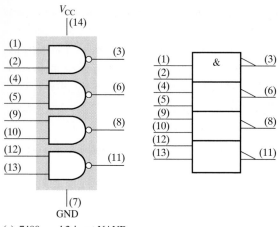

(a) 7400 quad 2-input NAND

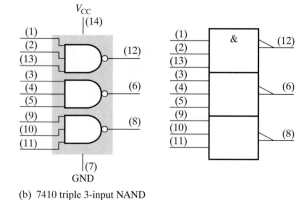

(b) 7410 triple 3-input NAND

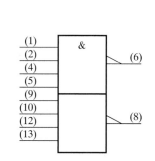

(c) 7420 dual 4-input NAND

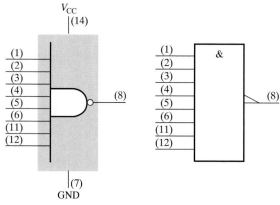

(d) 7430 single 8-input NAND

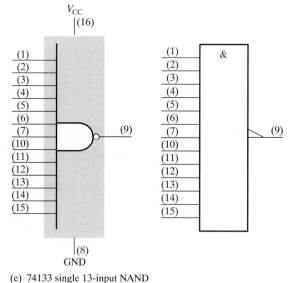

(e) 74133 single 13-input NAND

FIGURE 3–54
NAND gates.

NAND Gates A variety of NAND gates are available, including the 7400 with four 2-input gates, the 7410 with three 3-input gates, the 7420 with two 4-input gates, the 7430 with one 8-input gate, and the 74133 with one 13-input gate. These gates are shown in Figure 3–54. Notice that the 74133 requires 16 pins.

OR Gates The 7432 has four 2-input OR gates, as shown in Figure 3–55.

NOR Gates Examples of NOR gate configurations are shown in Figure 3–56. The 7402 has four 2-input gates, and the 7427 has three 3-input gates.

Exclusive-OR Gates The 7486 has four exclusive-OR gates in the package, as shown in Figure 3–57.

FIGURE 3–55
The 7432 quad 2-input OR gates.

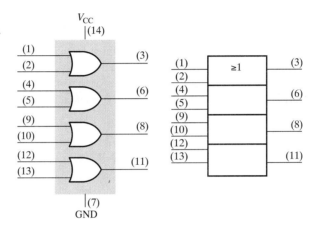

FIGURE 3–56
NOR gates.

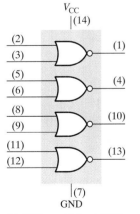
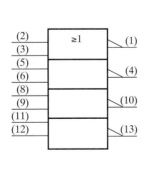

(a) 7402 quad 2-input NOR

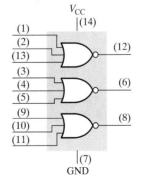
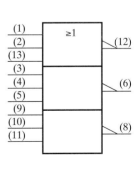

(b) 7427 triple 3-input NOR

FIGURE 3–57
The 7486 quad exclusive-OR gates.

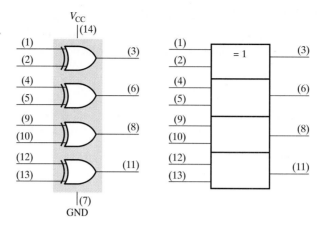

Data Sheet

A typical data sheet is divided into three main sections: recommended operating conditions, electrical characteristics, and switching characteristics. As an example, Figure 3–58 shows the arrangement of a data sheet for the 5400/7400 quad 2-input NAND gate. Additional sample data sheets are given in Appendix A.

Explanation of Data Sheet Parameters

The following list describes the parameters of the data sheet in Figure 3–58:

1. V_{CC}: The dc voltage source that supplies power to the device. Below the specified minimum, reliable operation cannot be guaranteed. Above the specified maximum, damage to the device may occur.

2. I_{OH}: The output current that the gate provides (sources) to a load when the output is at the HIGH level. By convention, the current out of a terminal is assigned a negative value. Figure 3–59(a) illustrates this parameter.

3. I_{OL}: The output current that the gate accepts from a load (sinks) when the output is at the LOW level. By convention, the current into a terminal is assigned a positive value. Figure 3–59(b) illustrates this parameter.

4. V_{IH}: The value of input voltage that can be accepted as a HIGH level by the gate.

5. V_{IL}: The value of input voltage that can be accepted as a LOW level by the gate.

6. V_{OH}: The value of HIGH level voltage that the gate produces on its output.

7. V_{OL}: The value of LOW level voltage that the gate produces on its output.

8. I_{IH}: The value of input current to a gate for a HIGH level input voltage. Figure 3–59(c) illustrates this parameter.

9. I_{IL}: The value of input current from a gate for a LOW level input voltage. Figure 3–59(d) illustrates this parameter.

10. I_{OS}: The output current when the gate output is shorted to ground and with input conditions that would normally establish a HIGH level output. Figure 3–59(e) illustrates this parameter.

11. I_{CCH}: The total current from the V_{CC} supply when all gate outputs are at the HIGH level.

12. I_{CCL}: The total current from the V_{CC} supply when all gate outputs are at the LOW level.

13. t_{PLH}: Propagation delay time from input to output for a LOW to HIGH output transition.

14. t_{PHL}: Propagation delay time from input to output for a HIGH to LOW output transition.

Parameter	5400			7400			Units
	Minimum	Typical	Maximum	Minimum	Typical	Maximum	
Supply voltage (V_{CC})	4.5	5.0	5.5	4.75	5.0	5.25	V
Operating free-air temperature range	−55	25	125	0	25	70	C°
HIGH level output current (I_{OH})			−400			−400	μA
LOW level output current (I_{OL})			16			16	mA

(a) Recommended operating conditions

Parameter		Limits			Units	Test Conditions[1]
		Minimum	Typical[2]	Maximum		
HIGH level input voltage (V_{IH})		2.0			V	
LOW level input voltage (V_{IL})				0.8	V	
HIGH level output voltage (V_{OH})		2.4	3.4		V	V_{CC} = min., I_{OH} = 0.4 mA, V_{IN} = 0.8 V
LOW level output voltage (V_{OL})			0.2	0.4	V	V_{CC} = min., I_{OL} = 16 mA, V_{IN} = 2.0 V
HIGH level input current (I_{IH})				40	μA	V_{CC} = max., V_{IN} = 2.4 V
LOW level input current (I_{IL})				−1.6	mA	V_{CC} = max., V_{IN} = 0.4 V
Short-circuit output current[3] (I_{OS})	5400	−20		−55	mA	V_{CC} = max.
	7400	−18		−55	mA	
Total supply current with outputs HIGH (I_{CCH})			4.0	8.0	mA	V_{CC} = max.
Total supply current with outputs LOW (I_{CCL})			12	22	mA	V_{CC} = max.

(b) Electrical characteristics over operating temperature range (unless otherwise noted)

Parameter	Limits			Units	Test Conditions
	Minimum	Typical	Maximum		
Propagation delay time, LOW-to-HIGH output(t_{PLH})		11	22	ns	V_{CC} = 5.0 V, C_{LOAD} = 15 pF, R_{LOAD} = 400 Ω
Propagation delay time, HIGH-to-LOW output(t_{PHL})		7.0	15	ns	

(c) Switching characteristics (T_A = 25°C)

NOTES:

[1] For conditions shown as min. or max., use the appropriate value specified under recommended operating conditions for the applicable device type.

[2] Typical limits are at V_{CC} = 5.0 V, 25°C

[3] Not more than one output should be shorted at a time. Duration of short not to exceed 1 s.

FIGURE 3–58

Data sheet for the 5400/7400 quad 2-input NAND gate.

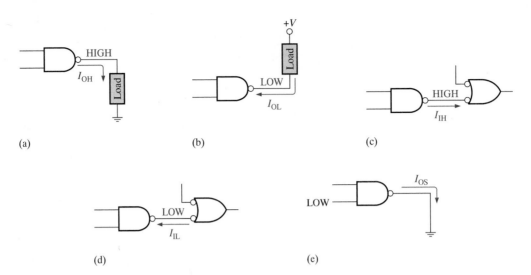

FIGURE 3–59
Illustration of some data sheet parameters.

SECTION 3–7 REVIEW

1. A positive pulse is applied to an inverter input. The time from the leading edge of the input to the leading edge of the output is 10 ns. The time from the trailing edge of the input to the trailing edge of the output is 8 ns. What are the values of t_{PLH} and t_{PHL}?
2. Define I_{CCL} and I_{CCH}.
3. Define *fan-out*.
4. Name the type and series of IC technology that exhibits each of the following performance characteristics:
   (a) fastest switching time (lowest propagation delay)
   (b) lowest power dissipation
   (c) highest fan-out
5. Calculate the speed-power product for an advanced Schottky (74AS) gate at a frequency of 100 kHz.

3–8 ■ TROUBLESHOOTING

Troubleshooting is the process of recognizing, isolating, and correcting a fault or failure in a circuit or system. To be an effective troubleshooter, you must understand how the circuit or system is supposed to work and be able to recognize incorrect performance. For example, to determine whether or not a certain logic gate is faulty, you must know what the output should be for given inputs. At this point it may be helpful to review Section 1–6, on pulsers, probes, and oscilloscopes. After completing this section, you should be able to

☐ Test for internally open inputs and outputs in IC gates using the logic pulser and probe   ☐ Recognize the effects of a shorted IC input or output   ☐ Test for external faults on a PC board   ☐ Troubleshoot a simple frequency counter using an oscilloscope in addition to the logic pulser and probe

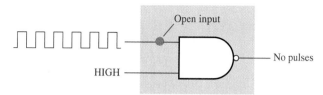

(a) Pulsing the open input will produce no pulses on the output.

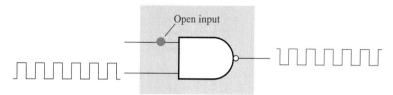

(b) Pulsing the good input will produce output pulses for TTL NAND and AND gates because an open input acts as a HIGH. It is uncertain for CMOS.

FIGURE 3–60
The effect of an open input on a NAND gate.

Internal Failures of IC Logic Gates

Opens and shorts are the most common types of internal gate failures. These can occur on the inputs or on the output of a gate inside the IC package.

Effects of an Internally Open Input   An internal open is the result of an open component on the chip or a break in the tiny wire connecting the IC chip to the package pin. An open input prevents a pulser signal on that input from getting to the output of the gate, as illustrated in Figure 3–60(a) for the case of a 2-input NAND gate. An open TTL input acts effectively as a HIGH level so pulses applied to the good input get through to the NAND gate output as shown in Figure 3–60(b).

Conditions for Testing Gates   When testing a NAND gate or an AND gate, always make sure that the inputs that are not being pulsed are HIGH to enable the gate. When checking a NOR gate or an OR gate, always make sure that the inputs that are not being pulsed are LOW. When checking an XOR or XNOR gate, the level of the nonpulsed input does not matter because the pulses on the other input will force the inputs to alternate between the same level and opposite levels.

Troubleshooting an Open Input   Troubleshooting this type of failure is most easily accomplished with a logic pulser and probe, as demonstrated in Figure 3–61 for the case of a 2-input NAND gate package.

The first step in troubleshooting an IC that is suspected of being faulty is to make sure that the dc supply voltage (V_{CC}) and ground are at the appropriate pins of the IC. Next, using a logic pulser to apply continuous pulses to one of the inputs to the gate, make sure that the other inputs are HIGH (in the case of a NAND gate). In Figure 3–61(a), start by pulsing pin 13, which has been determined to be one of the inputs to the suspected gate. If pulse activity is indicated on the output (pin 11 in this case) by a flashing probe, then the pin 13 input is not open. By the way, this also proves that the output is not open. Next, pulse the other gate input (pin 12). The probe lamp is off indicating that there are no pulses on the output at pin 11 and that the output is LOW, as shown in Figure 3–61(b). Notice that the input not being pulsed must be HIGH for the case of a NAND gate or AND gate. If this were a NOR gate, the input not being pulsed would have to be LOW.

Effects of an Internally Open Output   An internally open gate output prevents a signal on any of the inputs from getting to the output. Therefore, no matter what the input conditions are, the output is unaffected. The level at the output pin of the IC will

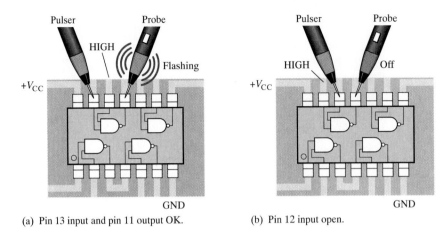

(a) Pin 13 input and pin 11 output OK. (b) Pin 12 input open.

FIGURE 3–61
Troubleshooting a NAND gate for an open input.

depend upon what it is externally connected to. It could be either HIGH, LOW, or floating (not fixed to any reference). In any case, a logic probe at the output pin will not be flashing.

Troubleshooting an Open Output Figure 3–62 illustrates troubleshooting an open NOR gate output with a pulser and probe. In part (a), one of the inputs of the suspected gate (pin 11 in this case) is pulsed, and the probe on the output (pin 13) indicates no pulse activity. In part (b), the other input (pin 12) is pulsed and again there is no indication of pulses on the output. Under the condition that the input that is not being pulsed is at a LOW level, this test shows that the output is internally open.

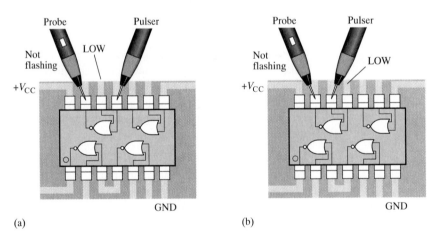

(a) (b)

FIGURE 3–62
Troubleshooting a NOR gate for an open output.

Shorted Input or Output Although not as common as an open, an internal short to the dc supply voltage, the ground, another input, or an output can occur. When an input or output is shorted to the supply voltage, it will be stuck in the HIGH state. If an input or output is shorted to the ground, it will be stuck in the LOW state (0 V). If two inputs or an input and an output are shorted together, they will always be at the same level.

We will reserve further discussion of troubleshooting shorts until Chapter 5, where the use of a current tracer to locate a short will be demonstrated in situations in which several gates are interconnected.

External Opens and Shorts

Many failures involving digital ICs are due to faults that are external to the IC package. These include bad solder connections, solder splashes, wire clippings, improperly etched printed circuit (PC) boards, and cracks or breaks in wires or printed circuit interconnections. These open or shorted conditions have the same effect on the logic gate as the internal faults, and troubleshooting is done in basically the same ways. A visual inspection of any circuit that is suspected of being faulty is the first thing a technician should do.

EXAMPLE 3–22

You are checking a 7410 triple 3-input NAND gate IC that is one of many ICs located on a printed circuit board. You have checked pins 1 and 2 with your logic probe, and they are both HIGH. Now your logic pulser is placed on pin 13, and your logic probe is placed first on pin 12 and then on the connecting PC board trace as indicated in Figure 3–63. Based on the responses of the probe, what is the most likely problem?

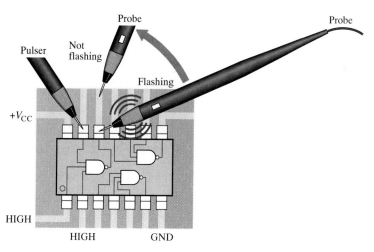

FIGURE 3–63

Solution The flashing indicator on the probe shows that there is pulse activity on the gate output at pin 12 but no activity on the PC board trace. The gate is working properly, but the signal is not getting from pin 12 of the IC to the PC board trace.

Most likely there is a bad solder connection between pin 12 of the IC and the PC board, which is creating an open. You should resolder that point and check it again.

Related Exercise If the probe does not flash at either point in Figure 3–63, what fault(s) does this indicate?

In most cases, you will be troubleshooting ICs that are mounted on printed circuit boards or prototype assemblies and interconnected with other ICs. As you progress through this book, you will learn how different types of digital ICs are used together to perform system functions. At this point, however, we are concentrating on individual IC gates. This limitation does not prevent us from looking at the system concept at a very basic and simplified level, as we have already done several times.

To continue the emphasis on systems, Examples 3–23 and 3–24 deal with troubleshooting the frequency counter that was introduced in Section 3–2.

EXAMPLE 3–23   After trying to operate the frequency counter shown in Figure 3–64, you find that it constantly reads out all 0s on its display, regardless of the input frequency. Determine the cause of this malfunction.

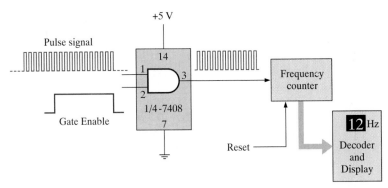

(a) This is how the counter should be working with a 12 Hz input signal.

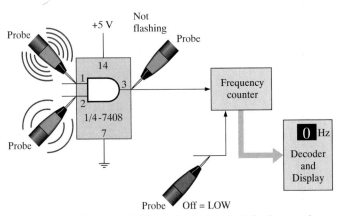

(b) The display is showing a frequency of 0 Hz and the logic probe indications are shown.

FIGURE 3–64

Solution   Here are three possible causes:

1. A constant active or asserted level on the counter reset input, which keeps the counter at zero.
2. No pulse signal on the input to the counter because of an internal open or short in the counter. This problem would keep the counter from advancing after being reset to zero.
3. No pulse signal on the input to the counter because of an open AND gate output or the absence of input signals, again keeping the counter from advancing from zero.

   Figure 3–64(a) gives an example of how the frequency counter should be working with a 12 Hz pulse waveform on the input to the AND gate. Part (b) shows that the display is improperly indicating 0 Hz.
   The first step is to make sure that V_{CC} and ground are connected to all the right places; assume that they are found to be OK. Next, check for pulses on both inputs to the AND gate. The probe indicates that there is pulse activity on both of these inputs. A probe check of the counter reset shows a LOW level which is known to be the unasserted level (assume this information was found on the data sheet for the counter) and, therefore, this is not the problem. The next probe check on pin 3 of the 7408 shows that there is no pulse activity on the output of the AND gate, indicating

that the gate output is internally open. Replace the 7408 IC and check the operation again.

Related Exercise   If pin 2 of the 7408 AND gate is open, what indication should you see on the display?

EXAMPLE 3–24   The frequency counter shown in Figure 3–65 appears to measure the frequency of input signals incorrectly. It is found that when a signal with a precisely known frequency is applied to pin 1 of the AND gate, the display indicates a higher frequency. Determine what is wrong.

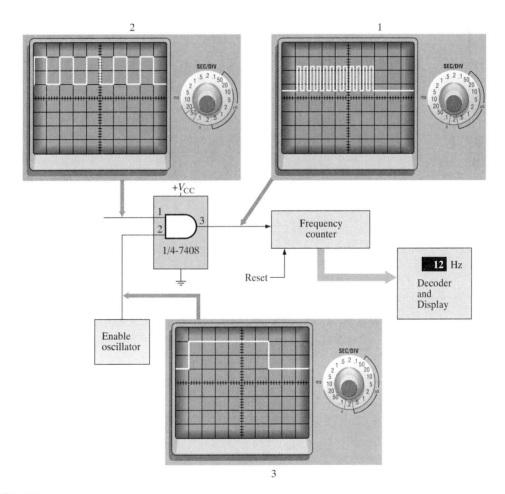

FIGURE 3–65

Solution   Recall from Section 3–2 that the input pulses are allowed to pass through the AND gate for exactly 1 s. The number of pulses counted in 1 s is equal to the frequency in hertz (cycles per second). Therefore, the 1 s interval, which is produced by the Enable pulse on pin 2 of the AND gate, is very critical to an accurate frequency measurement. The Enable pulses are produced internally by a precision oscillator circuit. The pulse must be exactly 1 s in width and in this case it occurs every 3 s to update the count. Just prior to each Enable pulse, the counter is reset to zero so that it starts a new count each time.

Since the counter appears to be counting more pulses than it should to produce a frequency readout that is too high, the Enable pulse is the primary suspect. Exact time-interval measurements must be made, so an oscilloscope is used instead of a logic probe in this situation. The logic probe indicates only the presence of pulses; it does not provide for frequency or time measurement.

An input pulse waveform of exactly 10 Hz is applied to pin 1 of the AND gate and the display incorrectly shows 12 Hz. The first scope measurement, on the output of the AND gate, shows that there are 12 pulses for each Enable pulse. In the second scope measurement, the input frequency is verified to be precisely 10 Hz (period = 100 ms). In the third scope measurement, the width of the Enable pulse is found to be 1.2 s rather than 1 s.

The conclusion is that the oscillator circuit that produces the Enable pulse is out of calibration for some reason and must be repaired or replaced.

Related Exercise   What would you suspect if the readout were indicating a frequency less than it should be?

SECTION 3–8 REVIEW

1. What are the most common types of failures in ICs?
2. If two different input waveforms are applied to a 2-input TTL NAND gate and the output waveform is just like one of the inputs, but inverted, what is the most likely problem?
3. Name one advantage of the oscilloscope over the logic probe.
4. Name one advantage of the logic probe over the oscilloscope.

3–9 ■ DIGITAL SYSTEM APPLICATION

The tablet counting and control system that you studied in Chapters 1 and 2 is again the subject of the system application, and you may wish to review those sections before beginning this one. In this section, the focus is on the application of logic gates in certain system functions. After completing this section, you should be able to

☐ Implement the comparator function with the basic gates covered in this chapter   ☐ Explain how an AND gate and inverter can be used to solve a certain timing problem in the system   ☐ Use a timing diagram to analyze the comparator logic and system operation   ☐ Troubleshoot the implemented comparator logic using a logic probe and pulser

The Comparator Function

The **comparator** in this system detects when the binary number in the counter, which is the current number of tablets in the bottle, is equal to the preset binary number that represents the maximum number of tablets per bottle. When the equality of these two numbers occurs, the bottle has been filled with the preset number of tablets and the valve must be closed to cut off the flow of tablets into the bottle.

Although there are ICs designed specifically as comparators, which you will study later, the comparator is implemented using individual gates in keeping with the scope of this chapter. Recall that the output of an exclusive-OR gate is LOW (0) when both input bits are the same, as shown in Figure 3–66. This operational feature of the exclusive-OR (XOR) can be used to advantage in the comparator application because you are looking for the occurrence of two equal binary numbers.

FIGURE 3–66
The output of an XOR gate is 0 when its inputs are equal.

In this system application, the number of tablets entered on the keypad appears as a fixed 8-bit binary number on the output of the BCD-to-binary code converter. The binary counter also contains an 8-bit number that represents the number of tablets actually in the bottle at any given time and this number increases as each tablet is counted. Both of these 8-bit binary numbers are applied to the inputs of the comparator as shown in Figure 3–67.

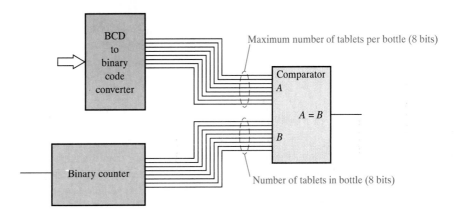

FIGURE 3–67
Block diagram of the portion of the system containing the comparator, counter, and code converter.

Implementing the Comparator

One XOR gate can accept two bits on its inputs and indicate by its output level if the two bits are equal or not equal. To compare the two 8-bit numbers, eight XOR gates are required. Each gate compares the corresponding bits of the two numbers; that is, one gate compares the LSBs of the two numbers, the next gate compares the next two bits, and so on. Finally, the eighth gate compares the MSBs of the two numbers. An 8-input negative-AND (NOR) gate is then used to detect when all the XOR outputs are LOW, indicating equal bits. The basic logic is shown in Figure 3–68(a).

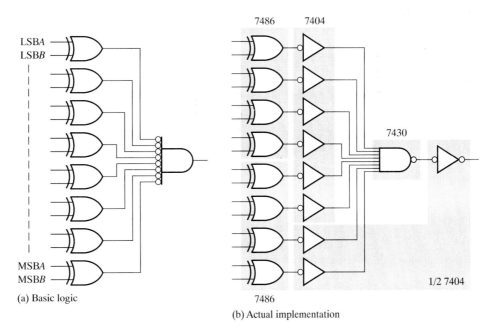

FIGURE 3–68
Implementation of the comparator logic.

An 8-input NOR gate is not available in an IC package to implement the required negative-AND; therefore, you must use an 8-input NAND gate. Since an XOR gate produces a LOW (0) when its inputs are the same (both 1 or both 0), an inverter is required on the output of each XOR gate to produce a HIGH for the NAND gate.

Figure 3–68(b) shows the actual implementation using two 7486 quad XOR gate packages, two 7404 hex inverter packages, and one 7430 8-input NAND gate package. This actual implementation is logically the same as the basic implementation shown in part (a). When the two 8-bit input numbers are equal, the output of the NAND gate is LOW. The NAND gate output must be inverted to provide the required HIGH to control the valve and other circuits. There is a way to implement the comparator with fewer devices, but this is reserved as a workbench activity.

A System Problem

The HIGH output from the comparator causes the valve in the tablet "hopper" to close and shut off the flow, causes the conveyor to move the next bottle into position, and causes the adder output to be stored in register B (see Figure 2–3).

A timing problem results when the output of the comparator goes HIGH. When the last tablet passes the sensor and the counter advances to the binary count equal to the preset number of tablets, there is a possibility that another tablet has already passed through the valve and will cause the counter to advance to its next number, thus making the comparator output go back LOW.

Solution To correct this timing problem, the inverter and AND gate shown in Figure 3–69 are used to inhibit the sensor pulse to the counter as soon as the binary numbers are equal. As long as the comparator output is LOW, the HIGH on the inverter output enables the AND gate and allows the sensor pulses to pass through the gate to the counter. When the comparator output goes HIGH, the inverter output goes LOW and prevents another sensor pulse from getting through the AND gate, thus keeping the counter from advancing if one more tablet passes the sensor.

FIGURE 3–69
Partial system diagram showing the counter inhibit logic.

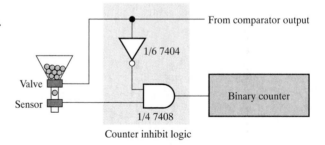

From comparator output

1/6 7404

Valve

Sensor

1/4 7408

Binary counter

Counter inhibit logic

■ THE DIGITAL WORKBENCH

Analysis and Design

The system is assumed to be operating properly. The comparator logic and the counter inhibit logic are implemented on the portion of the system board shown on Digital Workbench 1. The comparator logic consists of two 7486 quad XOR packages, two 7404 hex inverter packages, and one 7430 8-input NAND gate package. The counter inhibit circuit uses one of the inverters left over from the 7404 and one of the AND gates from a 7408 quad AND gate package. This design will be simplified in the Workbench Activity 6.

The numbers on the system board indicate corresponding points that are connected by conductive traces on the back side of the board by feedthroughs at each pad. Using these numbers, you can determine which points are connected together even though the connecting traces are not visible in the diagram.

Perform the activities listed on Digital Workbench 1.

■ DIGITAL WORKBENCH 1: Analysis and Design

■ *Activity 1*   Check the portion of the system board against the schematic to verify that they correspond. Based on the board connections, label the schematic with pin numbers and input and output labels.

■ *Activity 2*   The logic analyzer screen shows a portion of the timing diagram of the waveforms from the counter to the comparator inputs and the comparator output. Determine how many tablets are going into each bottle.

■ *Activity 3*   Determine the bit on each of the inputs (*a, b, c,* etc.) to this portion of the system board from the BCD-to-binary code converter.

■ *Activity 4*   Identify the three lines labeled *A, B,* and *C* at the bottom of this portion of the system board using the schematic as reference.

■ *Activity 5*   Verify that the DMM readings are within data sheet specifications for each of the points on the system board that are being measured.

■ *Activity 6*   The comparator and counter inhibit logic can be simplified. Modify the circuit to re-

duce the number of ICs without changing the overall logic function. This redesign will require that different types of gates be substituted for the existing ones. You may wish to refer to a data book to determine the ICs available to accomplish this redesign.

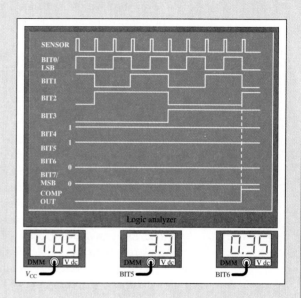

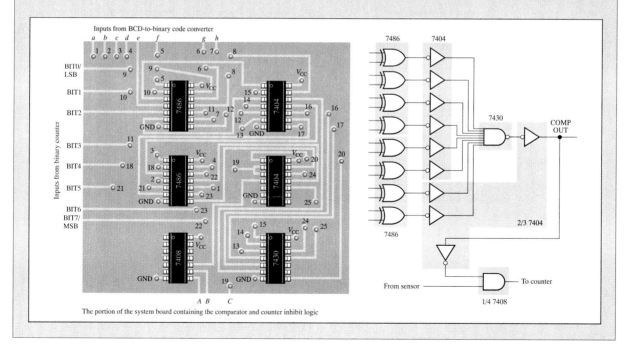

The portion of the system board containing the comparator and counter inhibit logic

Troubleshooting

Observation   A system problem has developed in which the tablets continue to flow after the bottle has been filled and the bottles do not advance on the conveyor.

Conclusion   There appears to be two problems: (1) the valve is not closing and (2) the conveyor is not advancing. However, it is unlikely that both of these elements will fail at the same time, so you must look for a common cause and assume that the valve and the conveyor are not faulty. What controls both the valve and the conveyor advance? When the output of the comparator goes HIGH, it should cause the valve to close and the conveyor to move the next bottle into position. Something may be keeping the comparator output from going HIGH.

Verify the Conclusion   If the output of the comparator stays LOW, the conclusion is correct. If it goes HIGH, perhaps both the valve and the conveyor have indeed failed and the conclusion is wrong. Assume that the comparator output is found to be always LOW and continue.

Priority Number One   The first priority after you have determined that the fault appears to be in the system circuit board and not in the valve or the conveyor mechanism is to minimize the system down-time (down-time means lost money). First, check to make sure the tablet sensor is producing pulses to the counter and that the system power supply is working. Assuming that they are, turn off the power to the system and remove the system circuit board. Insert a spare system board that is known to be good (assume there is one always available). Turn the power back on and initialize the system. Assume that it now runs properly.

   The next step is to find the problem with the faulty board and repair it so that there will be a good spare available when another problem occurs.

Possible Faults   The next step is to list failures in the comparator and counter inhibit logic and failures external to the logic circuits that could keep the comparator output LOW when it should be HIGH. Refer to the schematic in Figure 3–70 on p. 144.

1. Binary counter not working
2. Inverter 9 output or input open or output shorted
3. Inverter 10 input open or output shorted
4. NAND gate output open
5. An input open or output shorted on one of the inverters 1 through 8
6. An output open on one of the XOR gates
7. An input on one of the XOR gates is open or shorted in such a way as to prevent a comparison from occurring
8. The sensor input to the AND gate open
9. AND gate output open

Isolate the Faulty Element   With the faulty system board on the workbench, power and ground are connected and a pulse is applied to the sensor input. A DMM can be used to make sure that V_{CC} and ground are getting to each IC, and a probe or scope can be used to check the counter. Assuming that power and ground are OK and that the counter appears to be working, check the comparator and counter inhibit logic circuits using a logic pulser and probe as illustrated by various measurements on Digital Workbench 2. Although the probability is high that only one fault exists, several possibilites are presented to provide practice in arriving at a conclusion based on a set of measurements. Keeping in mind that all possible faults are not included, perform the activities listed on Digital Workbench 2.

■ DIGITAL WORKBENCH 2: Troubleshooting

■ *Activity 1* A pulser and logic probe are applied to the system board as shown. The probe light is off, indicating no pulse activity. Determine whether or not this test indicates a fault. If a fault is indicated, specify the faulty device or devices. If you can't isolate the fault (assuming one exists) to a single IC, explain why you can't and state what further tests are needed.

■ *Activity 2* The table indicates seven *independent* sets of pulser/probe measurements on the portion of the system board that contains the comparator and counter inhibit logic. For each set of measurements, determine whether or not a fault is indicated. If a fault is indicated, specify what the fault is and how to repair it. If a set of measurements is inconclusive, explain why.

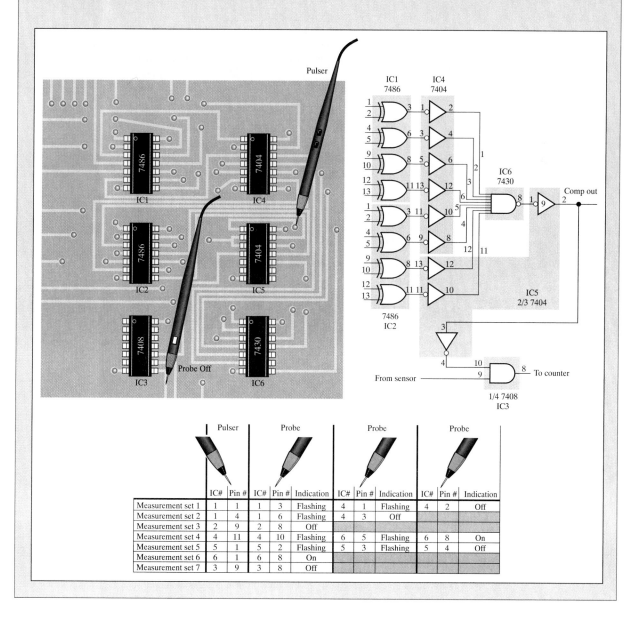

	Pulser		Probe			Probe			Probe	
	IC#	Pin #	IC#	Pin #	Indication	IC#	Pin #	Indication	IC#	Pin #	Indication
Measurement set 1	1	1	1	3	Flashing	4	1	Flashing	4	2	Off
Measurement set 2	1	4	1	6	Flashing	4	3	Off			
Measurement set 3	2	9	2	8	Off						
Measurement set 4	4	11	4	10	Flashing	6	5	Flashing	6	8	On
Measurement set 5	5	1	5	2	Flashing	5	3	Flashing	5	4	Off
Measurement set 6	6	1	6	8	On						
Measurement set 7	3	9	3	8	Off						

FIGURE 3–70

Logic diagram of comparator and counter inhibit circuits.

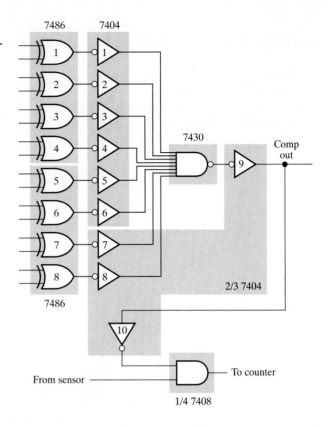

SECTION 3–9 REVIEW

1. In Digital Workbench 1, why do BIT4, BIT5, BIT6, and BIT7 never change?

2. In Digital Workbench 1, what is the maximum voltage that should be displayed by the DMM on the left? By the DMM on the right?

3. Why can't the NAND gate in the comparator logic be fully tested with a pulser and a logic probe?

■ SUMMARY

- The inverter output is the complement of the input.
- The AND gate output is HIGH if and only if all the inputs are HIGH.
- The OR gate output is HIGH if any of the inputs is HIGH.
- The NAND gate output is LOW if and only if all the inputs are HIGH.
- The NAND can be viewed as a negative-OR whose output is HIGH when any input is LOW.
- The NOR gate output is LOW if any of the inputs is HIGH.
- The NOR can be viewed as a negative-AND whose output is HIGH if and only if all the inputs are LOW.
- The exclusive-OR gate output is HIGH when the inputs are not the same.
- The exclusive-NOR gate output is LOW when the inputs are not the same.
- TTL and ECL are made with bipolar junction transistors.
- CMOS is made with MOS field-effect transistors.
- As a rule, ECL has the fastest switching speed, TTL is next, and CMOS is the slowest.
- As a rule, CMOS has the lowest power consumption, TTL is next, and ECL has the highest power consumption.
- The meanings of prefix letters in the 54/74 family of TTL and CMOS devices are as follows:
  No letter: standard TTL
  L: Low power TTL
  S: Schottky TLL
  LS: Low power Schottky TTL

ALS: Advanced low-power Schottky TTL
AS: Advanced Schottky TTL
C: Standard CMOS
HC: High-speed CMOS
HCT: High-speed CMOS with TTL compatibility
■ Distinctive shape symbols and truth tables for various logic gates (limited to 2 inputs) are shown in Figure 3–71.

FIGURE 3–71

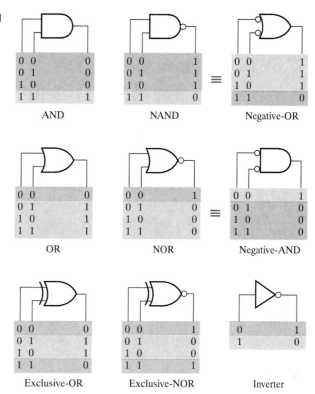

Note: Active states are shown in color.

■ **SELF-TEST**

1. When the input to an inverter is HIGH (1), the output is
   (a) HIGH or 1   (b) LOW or 1   (c) HIGH or 0   (d) LOW or 0

2. An inverter performs an operation known as
   (a) complementation   (b) assertion
   (c) inversion   (d) both aswers (a) and (c)

3. The output of an AND gate with inputs A, B, and C is a 1 (HIGH) when
   (a) $A = 1, B = 1, C = 1$ (b) $A = 1, B = 0, C = 1$ (c) $A = 0, B = 0, C = 0$

4. The output of an OR gate with inputs A, B, and C is a 1 (HIGH) when
   (a) $A = 1, B = 1, C = 1$ (b) $A = 0, B = 0, C = 1$ (c) $A = 0, B = 0, C = 0$
   (d) answers (a), (b), and (c)   (e) only answers (a) and (b)

5. A pulse is applied to each input of a 2-input NAND gate. One pulse goes HIGH at $t = 0$ and goes back LOW at $t = 1$ ms. The other pulse goes HIGH at $t = 0.8$ ms and goes back LOW at $t = 3$ ms. The output pulse can be described as follows:
   (a) It goes LOW at $t = 0$ and back HIGH at $t = 3$ ms.
   (b) It goes LOW at $t = 0.8$ ms and back HIGH at $t = 3$ ms.
   (c) It goes LOW at $t = 0.8$ ms and back HIGH at $t = 1$ ms.
   (d) It goes HIGH at $t = 0.8$ ms and back LOW at $t = 1$ ms.

6. A pulse is applied to each input of a 2-input NOR gate. One pulse goes HIGH at $t = 0$ and goes back LOW at $t = 1$ ms. The other pulse goes HIGH at $t = 0.8$ ms and goes back LOW at $t = 3$ ms. The output pulse can be described as follows:

(a) It goes LOW at $t = 0$ and back HIGH at $t = 3$ ms.

(b) It goes LOW at $t = 0.8$ ms and back HIGH at $t = 3$ ms.

(c) It goes LOW at $t = 0.8$ ms and back HIGH at $t = 1$ ms.

(d) It goes HIGH at $t = 0.8$ ms and back LOW at $t = 1$ ms.

7. A pulse is applied to each input of an exclusive-OR gate. One pulse goes HIGH at $t = 0$ and goes back LOW at $t = 1$ ms. The other pulse goes HIGH at $t = 0.8$ ms and goes back LOW at $t = 3$ ms. The output pulse can be described as follows:

(a) It goes HIGH at $t = 0$ and back LOW at $t = 3$ ms.

(b) It goes HIGH at $t = 0$ and back LOW at $t = 0.8$ ms.

(c) It goes HIGH at $t = 1$ ms and back LOW at $t = 3$ ms.

(d) both answers (b) and (c)

8. A positive-going pulse is applied to an inverter. The time interval from the leading edge of the input to the leading edge of the output is 7 ns. This parameter is

(a) speed-power product   (b) propagation delay, t_{PHL}

(c) propagation delay, t_{PLH} (d) pulse width

9. The TTL family with the fastest switching speed is

(a) Standard   (b) ALS   (c) S   (d) AS

10. If power consumption were the major criterion in the design of a digital system, the logic family that you would probably use is

(a) Standard TTL   (b) ALS TTL   (c) ECL   (d) CMOS

11. To measure the period of a pulse waveform, you must use

(a) a DMM   (b) a logic probe   (c) an oscilloscope   (d) a pulser

12. Once you measure the period of a pulse waveform, the frequency is found by

(a) using another setting   (b) measuring the duty cycle

(c) finding the reciprocal of the period   (d) using another type of instrument

■ PROBLEMS

SECTION 1–1   The Inverter

1. The input waveform shown in Figure 3–72 is applied to an inverter. Sketch the timing diagram of the output waveform in proper relation to the input.

2. A network of cascaded inverters is shown in Figure 3–73. If a HIGH is applied to point A, determine the logic levels at points B through F.

FIGURE 3–72

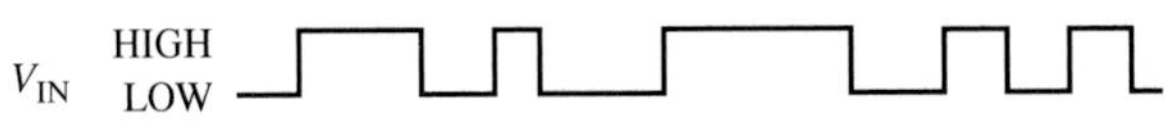

FIGURE 3–73

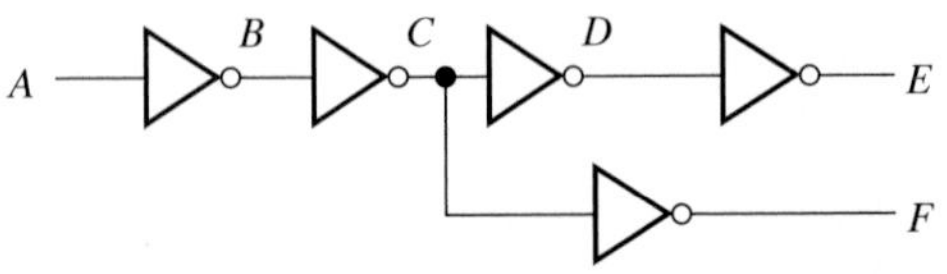

SECTION 3–2   The AND Gate

3. Determine the output, X, for a 2-input AND gate with the input waveforms shown in Figure 3–74. Show the proper relationship of output to inputs with a timing diagram.

4. Repeat Problem 3 for the waveforms in Figure 3–75.

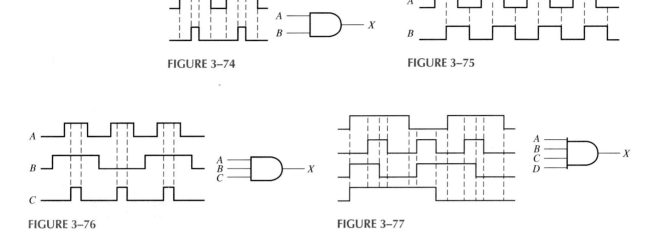

FIGURE 3–74

FIGURE 3–75

FIGURE 3–76

FIGURE 3–77

5. The input waveforms applied to a 3-input AND gate are as indicated in Figure 3–76. Show the output waveform in proper relation to the inputs with a timing diagram.

6. The input waveforms applied to a 4-input gate are as indicated in Figure 3–77. Show the output waveform in proper relation to the inputs with a timing diagram.

SECTION 3–3   The OR Gate

7. Determine the output for a 2-input OR gate when the input waveforms are as in Figure 3–75 and draw a timing diagram.

8. Repeat Problem 5 for a 3-input OR gate.

9. Repeat Problem 6 for a 4-input OR gate.

10. For the five input waveforms in Figure 3–78, determine the output for a 5-input AND gate and the output for a 5-input OR gate. Draw the timing diagram.

FIGURE 3–78

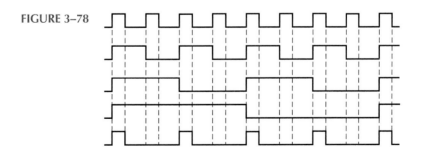

SECTION 3–4   The NAND Gate

11. For the set of input waveforms in Figure 3–79, determine the output for the gate shown and draw the timing diagram.

FIGURE 3–79

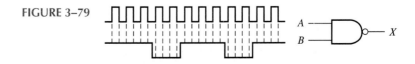

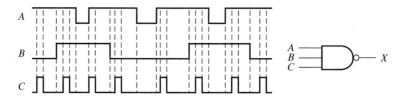

FIGURE 3–80

12. Determine the gate output for the input waveforms in Figure 3–80 and draw the timing diagram.
13. Determine the output waveform in Figure 3–81.
14. As you have learned, the two logic symbols shown in Figure 3–82 represent equivalent operations. The difference between the two is strictly from a functional viewpoint. For the NAND symbol, look for two HIGHs on the inputs to give a LOW output. For the negative-OR, look for at least one LOW on the inputs to give a HIGH on the output. Using these two functional viewpoints, show that each gate will produce the same output for the given inputs.

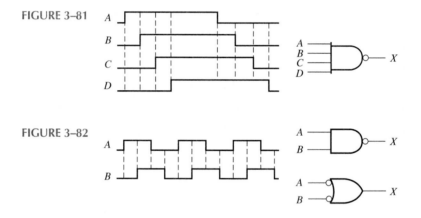

FIGURE 3–81

FIGURE 3–82

SECTION 3–5 The NOR Gate

15. Repeat Problem 11 for a 2-input NOR gate.
16. Determine the output waveform in Figure 3–83 and draw the timing diagram.
17. Repeat Problem 13 for a 4-input NOR gate.
18. The NAND and the negative-OR symbols represent equivalent operations, but they are functionally different. For the NOR symbol, look for at least one HIGH on the inputs to give a LOW on the output. For the negative-AND, look for two LOWs on the inputs to give a HIGH output. Using these two functional points of view, show that both gates in Figure 3–84 will produce the same output for the given inputs.

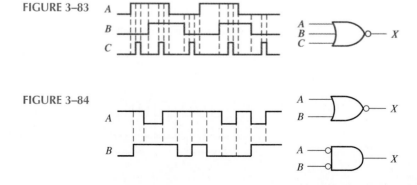

FIGURE 3–83

FIGURE 3–84

SECTION 3-6   The Exclusive-OR and Exclusive-NOR Gates

19. How does an exclusive-OR gate differ from an OR gate in its logical operation?

20. Repeat Problem 11 for an exclusive-OR gate.

21. Repeat Problem 11 for an exclusive-NOR gate.

22. Determine the output of an exclusive-OR gate for the inputs shown in Figure 3–75 and draw a timing diagram.

SECTION 3-7   Integrated Circuit Logic Families

23. In the comparison of certain logic devices, it is noted that the power dissipation for one particular type increases as the frequency increases. Is the device TTL or CMOS?

24. Using Table 3–14, determine which logic series offers the best performance considering both switching speed and power dissipation at 100 kHz. Note: Find the speed-power product of each and compare the results.

25. Determine t_{PLH} and t_{PHL} from the oscilloscope display in Figure 3–85.

FIGURE 3-85

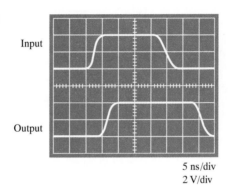

Input

Output

5 ns/div
2 V/div

26. Gate A has $t_{PLH} = t_{PHL} = 6$ ns. Gate B has $t_{PLH} = t_{PHL} = 10$ ns. Which gate can be operated at a higher frequency?

27. If a logic gate operates on a dc supply voltage of +5 V and draws an average current of 4 mA, what is its power dissipation?

28. The variable I_{CCH} represents the dc supply current from V_{CC} when all outputs of an IC are HIGH. The variable I_{CCL} represents the dc supply current when all outputs are LOW. For a 74LS00 IC, determine the typical power dissipation when all four gate outputs are HIGH. (See data sheet in Appendix A.)

 ### SECTION 3-8   Troubleshooting

29. Examine the conditions indicated in Figure 3–86, and identify the faulty gates.

FIGURE 3-86

(a)   (b)   (c)

(d)   (e)   (f)

30. Determine the faulty gates in Figure 3–87 by analyzing the timing diagrams.

31. Using a logic probe and pulser, you make the observations indicated in Figure 3–88. For each observation determine the most likely gate failure.

32. The seat belt alarm circuit in Figure 3–16 has malfunctioned. You find that when the ignition switch is turned on and the seat belt is unbuckled, the alarm comes on and will not go off. What is the most likely problem? How do you troubleshoot it?

FIGURE 3–87

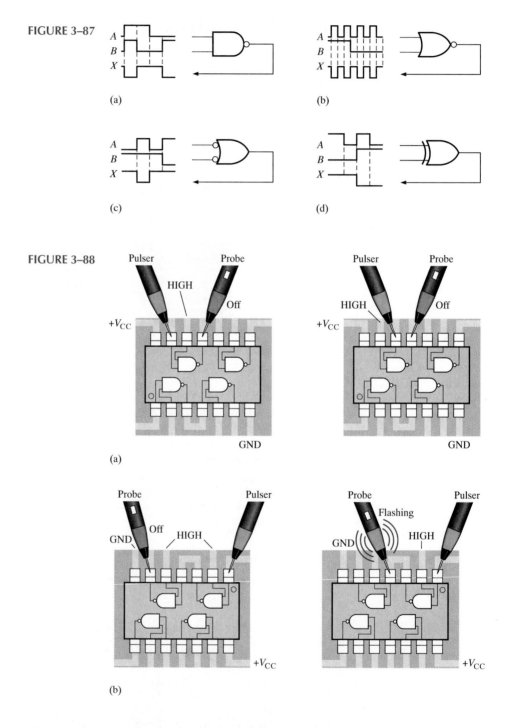

(a)

(b)

(c)

(d)

FIGURE 3–88

(a)

(b)

33. Every time the ignition switch is turned on in the circuit of Figure 3–16, the alarm comes on for thirty seconds, even when the seat belt is buckled. What is the most probable cause of this malfunction?

SECTION 3–9   Digital System Application

34. In relation to the system problem discussed in the section, the counter inhibit logic actually solves only one part of the timing problem. It keeps the counter from advancing an extra count which causes the comparator output to go back LOW. What is the remaining problem?

35. Assume that the tablet counting and control system is set for 75 tablets per bottle. Draw the timing diagram that you should see on the logic analyzer for the ten sensor pulses up to and including the final count. Use the logic analyzer display in Digital Workbench 1 as a guide.

36. Determine the logic levels (expressed as 1s and 0s) that should appear on the inputs *a* through *h* of the system board in Digital Workbench 1 for each of the following tablets-per-bottle settings:

   (a) 30 tablets/bottle   **(b)** 75 tablets/bottle   **(c)** 99 tablets/bottle

37. Assume that the tablet counting and control system is set for 50 tablets per bottle but it is putting only 18 tablets in each bottle. List the possible system malfunctions that would cause this problem in terms of inputs to the system board in Digital Workbench 1.

38. With the board pulled from the system and on your workbench, suggest a procedure for testing the comparator logic.

Special Design Problems

39. Sensors are used to monitor the pressure and the temperature of a chemical solution stored in a vat. The circuitry for each sensor produces a HIGH voltage when a specified maximum value is exceeded. An alarm requiring a LOW voltage input must be activated when either the pressure or the temperature is excessive. Design a circuit for this application?

40. In a certain automated manufacturing process, electrical components are automatically inserted in a PC board. Before the insertion tool is activated, the PC board must be properly positioned, and the component to be inserted must be in the chamber. Each of these prerequisite conditions is indicated by a HIGH voltage. The insertion tool requires a LOW voltage to activate it. Design a circuit to implement this process.

41. Modify the frequency counter in Figure 3–15 to operate with an Enable pulse that is active-LOW rather than HIGH during the 1 s interval.

42. Assume that the Enable signal in Figure 3–15 has the waveform shown in Figure 3–89. Assume that waveform *B* is also available. Devise a circuit that will produce an active-HIGH reset pulse to the counter only during the time that the Enable signal is LOW.

43. Design a circuit to fit in the colored block of Figure 3–90 that will cause the headlights of an automobile to be turned off automatically 15 s after the ignition switch is turned off, if the light switch is left on. Assume that a LOW is required to turn the lights off.

FIGURE 3–89

Enable

B

FIGURE 3–90

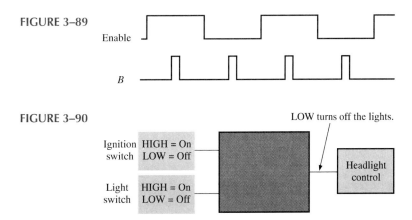

44. Modify the logic circuit for the intrusion alarm in Figure 3–24 so that two additional rooms, each with two windows and one door, can be protected.

45. In the tablet counting and control system, the tablet feeder and conveyor mechanisms have been changed to a different model. In the new model, the valve requires a LOW level to close and the conveyor requires a LOW to advance. Register B still requires a HIGH level to store the new total count. Modify the comparator logic to accommodate this requirement.

■ **ANSWERS TO SECTION REVIEWS**

SECTION 3–1

1. When the inverter input is 1, the output is 0.

2. **(a)** **(b)** A negative-going pulse is on the output (HIGH to LOW and back HIGH).

SECTION 3–2

1. An AND gate output is HIGH when all inputs are HIGH.

2. An AND gate output is LOW when one or more inputs are LOW.

3. Five-input AND: $X = 1$ when $ABCDE = 11111$ and $X = 0$ for all other combinations of $ABCDE$.

SECTION 3–3

1. An OR gate output is HIGH when one or more inputs are HIGH.

2. An OR gate output is LOW when all inputs are LOW.

3. Three-input OR: $X = 0$ when $ABC = 000$, and $X = 1$ for all other combinations of ABC.

SECTION 3–4

1. A NAND output is LOW when all inputs are HIGH.

2. An NAND output is HIGH when one or more inputs are LOW.

3. NAND: active-LOW output for all HIGH inputs; negative-OR: active-HIGH output for one or more LOW inputs; they have the same truth tables.

4. $X = \overline{ABC}$

SECTION 3–5

1. A NOR output is HIGH when all inputs are LOW.

2. A NOR output is LOW when one or more inputs are HIGH.

3. NOR: active-LOW output for one or more HIGH inputs; negative-AND: active-HIGH output for all LOW inputs; they have the same truth tables.

4. $X = \overline{A + B + C}$

SECTION 3–6

1. An XOR output is HIGH when the inputs are at opposite levels.

2. An XNOR output is HIGH when the inputs are at the same levels.

3. Apply the bits to the XOR inputs; when the output is HIGH, the bits are different.

SECTION 3–7

1. $t_{PLH} = 8$ ns; $t_{PHL} = 10$ ns

2. I_{CCL}: supply current with gate output LOW; I_{CCH}: supply current with gate output HIGH

3. Fan-out is the maximum number of loads that a logic circuit can drive.

4. **(a)** Fastest: 10KH ECL **(b)** Lowest power: CMOS, 74HC (static), 4000 (100 kHz)
   (c) Highest fan-out: TTL AS

5. SPP = (1.5 ns)(8.5 mW) = 12.8 pJ

SECTION 3–8

1. Opens and shorts are the most common failures.

2. An open input which effectively makes input HIGH

3. Waveforms can be observed with a scope but not with a logic probe.

4. A probe is easier to use.

SECTION 3–9

1. BIT4, BIT5, BIT6, and BIT7 never change because the count does not exceed 60: 00111100.

2. $V_{\text{left}} = 5.5$ V; $V_{\text{right}} = 0.4$ V

3. All eight NAND inputs can't be forced HIGH at the same time.

4

BOOLEAN ALGEBRA AND LOGIC SIMPLIFICATION

■ CHAPTER OBJECTIVES

☐ Apply the basic laws and rules of Boolean algebra

☐ Apply DeMorgan's theorems to Boolean expressions

☐ Describe gate networks with Boolean expressions

☐ Evaluate Boolean expressions

☐ Simplify expressions by using the laws and rules of Boolean algebra

☐ Convert any Boolean expression into a sum-of-products (SOP) form

☐ Convert any Boolean expression into a product-of-sums (POS) form

☐ Use a Karnaugh map to simplify Boolean expressions

☐ Use a Karnaugh map to simplify truth table functions

☐ Utilize "don't care" conditions to simplify logic functions

☐ Apply Boolean algebra and the Karnaugh map method to a system application

In 1854, George Boole published a work titled *An Investigation of the Laws of Thought, on Which Are Founded the Mathematical Theories of Logic and Probabilities.* It was in this publication that a "logical algebra," known today as Boolean algebra, was formulated. Boolean algebra is a convenient and systematic way of expressing and analyzing the operation of logic circuits. Claude Shannon was the first to apply Boole's work to the analysis and design of logic circuits. In 1938, Shannon wrote a thesis at MIT titled *A Symbolic Analysis of Relay and Switching Circuits.*

This chapter covers the laws, rules, and theorems of Boolean algebra and their application to digital circuits. You will learn how to define a given circuit with a Boolean expression and then evaluate its operation. You will also learn how to simplify logic circuits using the methods of Boolean algebra and Karnaugh maps.

■ DIGITAL SYSTEM APPLICATION

This Digital System Application illustrates concepts taught in this chapter. The 7-segment display logic in the tablet counting and control system is a good way to illustrate the application of Boolean algebra and the Karnaugh map method to obtain the simplest possible implementation in the design of logic circuits. Therefore, in Section 4–12, the focus is on the BCD-to-7-segment logic that drives the two system displays indicated below.

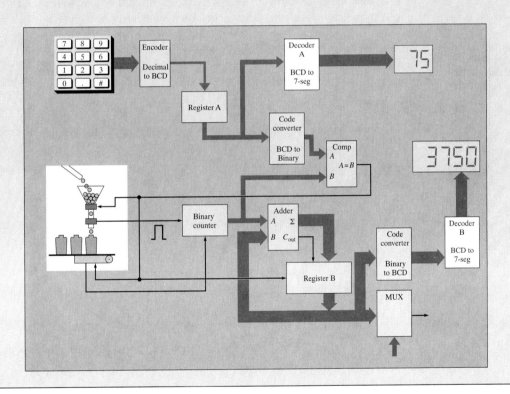

4–1 ■ BOOLEAN OPERATIONS AND EXPRESSIONS

Boolean algebra is the mathematics of digital systems. A basic knowledge of Boolean algebra is indispensable to the study and analysis of logic circuits. In the last chapter, Boolean operations and expressions in terms of their relationship to NOT, AND, OR, NAND, and NOR gates were introduced. This section reviews that material and provides additional definitions and information. After completing this section, you should be able to

☐ Define *variable* ☐ Define *literal* ☐ Identify a sum term ☐ Evaluate a sum term ☐ Identify a product term ☐ Evaluate a product term ☐ Explain Boolean addition ☐ Explain Boolean multiplication

Variable, complement, and *literal* are terms used in Boolean algebra. A **variable** is a symbol (usually an italic uppercase letter) used to represent a logical quantity. Any single variable can have a 1 or a 0 value. The **complement** is the inverse of a variable and is indicated by a bar over the variable (overbar). For example, the complement of the variable A is $\overline{A}$. If $A = 1$, then $\overline{A} = 0$. If $A = 0$, then $\overline{A} = 1$. The complement of the variable A is read as "A not" or "A bar." Sometimes a prime symbol rather than an overbar is used to denote the complement of a variable; for example, B′ indicates the complement of B. In this book, only the overbar is used. A **literal** is a variable or the complement of a variable.

Boolean Addition

Recall from Chapter 3 that **Boolean addition** is equivalent to the OR operation and the basic rules are as follows:

$$0 + 0 = 0 \quad 1 + 0 = 1$$
$$0 + 1 = 1 \quad 1 + 1 = 1$$

In Boolean algebra, a **sum term** is a sum of literals. In logic circuits, a sum term is produced by an OR operation with no AND operations involved. Some examples of sum terms are $A + B$, $A + \overline{B}$, $A + B + \overline{C}$, and $\overline{A} + B + C + \overline{D}$.

A sum term is equal to 1 when one or more of the literals in the term are 1. A sum term is equal to 0 if and only if each of the literals is 0.

EXAMPLE 4–1

Determine the values of A, B, C, and D which make the sum term $A + \overline{B} + C + \overline{D}$ equal to 0.

Solution For the sum term to be 0, each of the literals in the term must be 0. Therefore, $A = 0$, $B = 1$ so that $\overline{B} = 0$, $C = 0$, and $D = 1$ so that $\overline{D} = 0$.

$$A + \overline{B} + C + \overline{D} = 0 + \overline{1} + 0 + \overline{1} = 0 + 0 + 0 + 0 = 0$$

Related Exercise Determine the values of A and B which make the sum term $\overline{A} + B$ equal to 0.

Boolean Multiplication

Also recall from Chapter 3 that **Boolean multiplication** is equivalent to the AND operation and the basic rules are as follows:

$$0 \cdot 0 = 0 \quad 0 \cdot 1 = 0$$
$$1 \cdot 0 = 0 \quad 1 \cdot 1 = 1$$

In Boolean algebra, a **product term** is the product of literals. In logic circuits, a product term is produced by an AND operation with no OR operations involved. Some examples of product terms are AB, $A\overline{B}$, ABC, and $A\overline{B}C\overline{D}$.

A product term is equal to 1 if and only if each of the literals in the term is 1. A product term is equal to 0 when one or more of the literals are 0.

EXAMPLE 4–2

Determine the values of A, B, C, and D which make the product term $A\overline{B}C\overline{D}$ equal to 1.

Solution   For the product term to be 1, each of the literals in the term must be 1. Therefore, $A = 1$, $B = 0$ so that $\overline{B} = 1$, $C = 1$, and $D = 0$ so that $\overline{D} = 1$.

$$A\overline{B}C\overline{D} = 1 \cdot \overline{0} \cdot 1 \cdot \overline{0} = 1 \cdot 1 \cdot 1 \cdot 1 = 1$$

Related Exercise   Determine the values of A and B which make the product term $\overline{A}\,\overline{B}$ equal to 1.

SECTION 4–1 REVIEW

1. If $A = 0$, what does $\overline{A}$ equal?
2. Determine the values of A, B, and C that make the sum term $\overline{A} + \overline{B} + C$ equal to 0.
3. Determine the values of A, B, and C that make the product term $A\overline{B}C$ equal to 1.

4–2 ■ LAWS AND RULES OF BOOLEAN ALGEBRA

As in other areas of mathematics, there are certain well-developed rules and laws that must be followed in order to properly apply Boolean algebra. The most important of these are presented in this section. After completing this section, you should be able to

□ Apply the commutative laws of addition and multiplication   □ Apply the associative laws of addition and multiplication   □ Apply the distributive law   □ Apply twelve basic rules of Boolean algebra

Laws of Boolean Algebra

The basic laws of Boolean algebra—the **commutative laws,** the **associative laws,** and the **distributive law**—are the same as in ordinary algebra. Each of the laws is illustrated with two or three variables, but the number of variables is not limited to this.

Commutative Laws   The commutative law of addition for two variables is written algebraically as

$$A + B = B + A \tag{4–1}$$

This law states that the order in which the variables are ORed makes no difference. Remember, in Boolean algebra terminology as applied to logic circuits, addition and the OR operation are the same. Figure 4–1 illustrates the commutative law as applied to the OR gate and shows that it doesn't matter to which input each variable is applied. (The symbol ≡ means "equivalent to.")

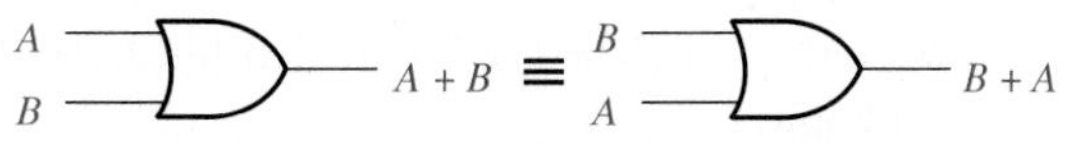

FIGURE 4–1
Application of commutative law of addition.

The commutative law of multiplication for two variables is

$$AB = BA \qquad\qquad (4\text{--}2)$$

This law states that the order in which the variables are ANDed makes no difference. Figure 4–2 illustrates this law as applied to the AND gate.

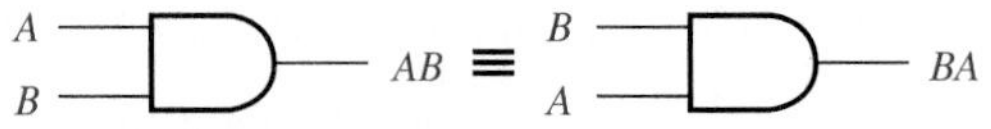

FIGURE 4–2
Application of commutative law of multiplication.

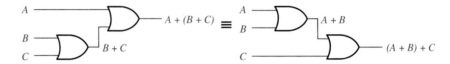

Associative Laws   The associative law of addition is written algebraically as follows for three variables:

$$A + (B + C) = (A + B) + C \qquad\qquad (4\text{--}3)$$

This law states that in the ORing of more than two variables, the result is the same regardless of the grouping of the variables. Figure 4–3 illustrates this law as applied to OR gates.

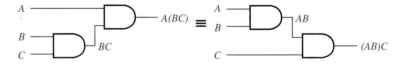

FIGURE 4–3
Application of associative law of addition.

The associative law of multiplication is written as follows for three variables:

$$A(BC) = (AB)C \qquad\qquad (4\text{--}4)$$

This law states that it makes no difference in what order the variables are grouped when ANDing more than two variables. Figure 4–4 illustrates this law as applied to AND gates.

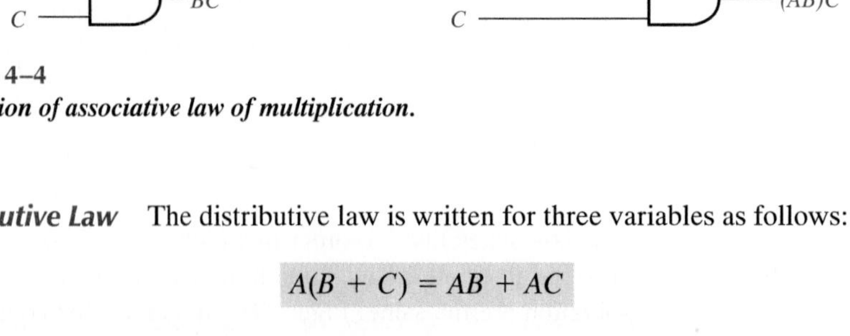

FIGURE 4–4
Application of associative law of multiplication.

Distributive Law   The distributive law is written for three variables as follows:

$$A(B + C) = AB + AC \qquad\qquad (4\text{--}5)$$

This law states that ORing two or more variables and ANDing the result with a single variable is equivalent to ANDing the single variable with each of the two or more variables and then ORing the products. The distributive law also expresses the process of *factoring* in which the common variable A is factored out of the product terms, for example, $AB + AC = A (B + C)$. Figure 4–5 illustrates the distributive law in terms of gate implementation.

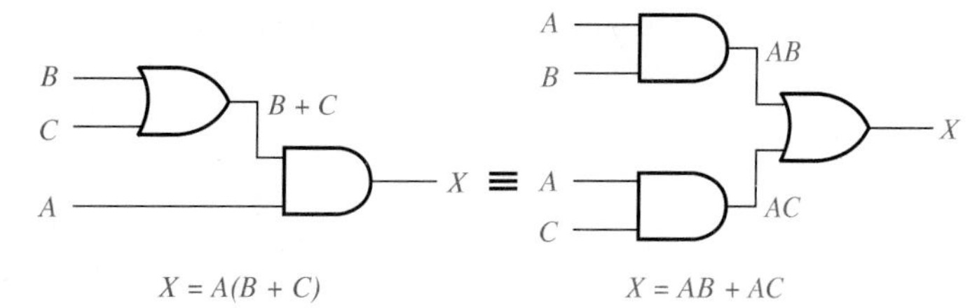

$$X = A(B + C) \qquad\qquad X = AB + AC$$

FIGURE 4–5
Application of distributive law.

Rules of Boolean Algebra

Table 4–1 lists 12 basic rules that are useful in manipulating and simplifying **Boolean expressions.** Rules 1 through 9 will be viewed in terms of their application to logic gates. Rules 10 through 12 will be derived in terms of the simpler rules and the laws previously discussed.

TABLE 4–1
Basic rules of Boolean algebra

1. $A + 0 = A$	**7.** $A \cdot A = A$
2. $A + 1 = 1$	**8.** $A \cdot \overline{A} = 0$
3. $A \cdot 0 = 0$	**9.** $\overline{\overline{A}} = A$
4. $A \cdot 1 = A$	**10.** $A + AB = A$
5. $A + A = A$	**11.** $A + \overline{A}B = A + B$
6. $A + \overline{A} = 1$	**12.** $(A + B)(A + C) = A + BC$

A, B, or C can represent a single variable or a combination of variables.

1. $A + 0 = A$ A variable ORed with 0 is always equal to the variable. If A is 1, the output is 1, which is equal to A. If A is 0, the output is 0, which is also equal to A. This rule is illustrated in Figure 4–6, where the lower input is fixed at 0.

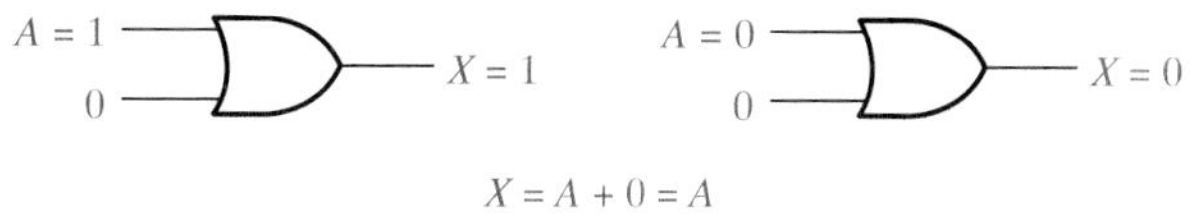

FIGURE 4–6
Rule 1: $A + 0 = A$.

$$X = A + 0 = A$$

2. $A + 1 = 1$ A variable ORed with 1 is always equal to 1. A 1 on an input to an OR gate produces a 1 on the output, regardless of the value of the variable on the other input. This rule is illustrated in Figure 4–7, where the lower input is fixed at 1.

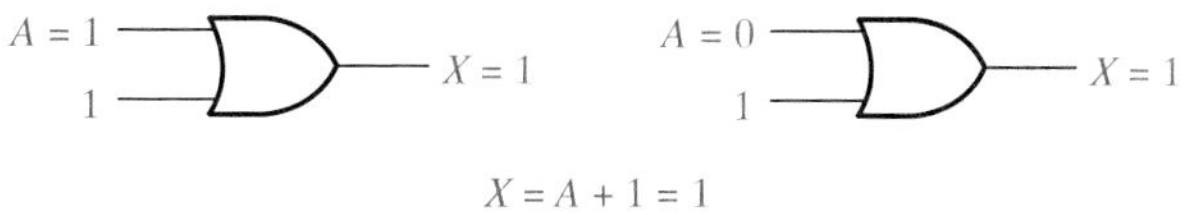

FIGURE 4–7
Rule 2: $A + 1 = 1$.

$$X = A + 1 = 1$$

3. $A \cdot 0 = 0$ A variable ANDed with 0 is always equal to 0. Any time one input to an AND gate is 0, the output is 0, regardless of the value of the variable on the other input. This rule is illustrated in Figure 4–8, where the lower input is fixed at 0.

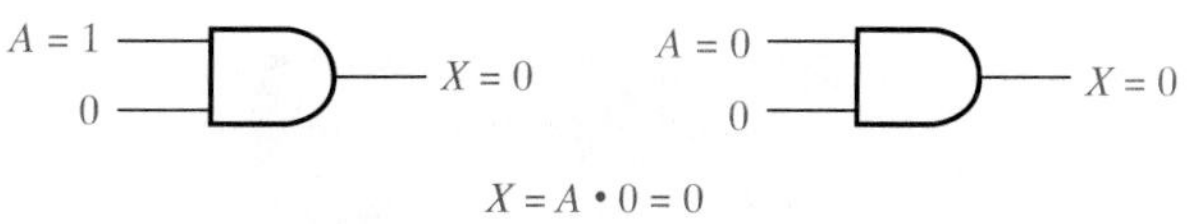

FIGURE 4–8
Rule 3: $A \cdot 0 = 0$.

$$X = A \cdot 0 = 0$$

4. $A \cdot 1 = A$ A variable ANDed with 1 is always equal to the variable. If the variable A is 0, the output of the AND gate is 0. If the variable A is 1, the output of the AND gate is 1 because both inputs are now 1s. This rule is shown in Figure 4–9, where the lower input is fixed at 1.

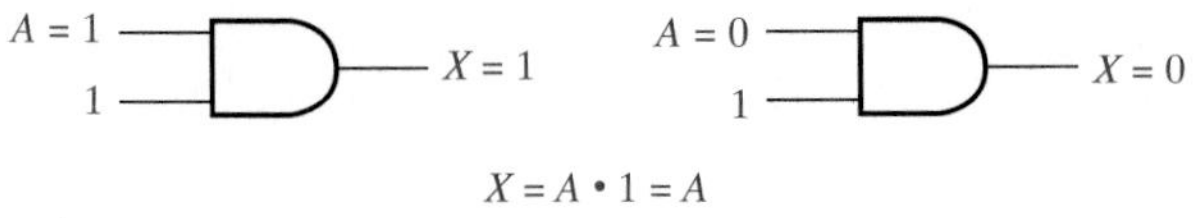

FIGURE 4–9
Rule 4: $A \cdot 1 = A$.

$$X = A \cdot 1 = A$$

5. $A + A = A$ A variable ORed with itself is always equal to the variable. If A is 0, then $0 + 0 = 0$, and if A is 1, then $1 + 1 = 1$. This is shown in Figure 4–10, where both inputs are the same variable.

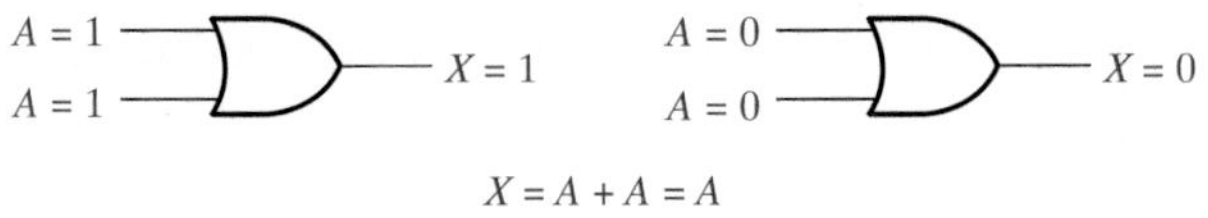

FIGURE 4–10
Rule 5: $A + A = A$.

$$X = A + A = A$$

6. $A + \overline{A} = 1$ A variable ORed with its complement is always equal to 1. If A is 0, then $0 + \overline{0} = 0 + 1 = 1$. If A is 1, then $1 + \overline{1} = 1 + 0 = 1$. See Figure 4–11, where one input is the complement of the other.

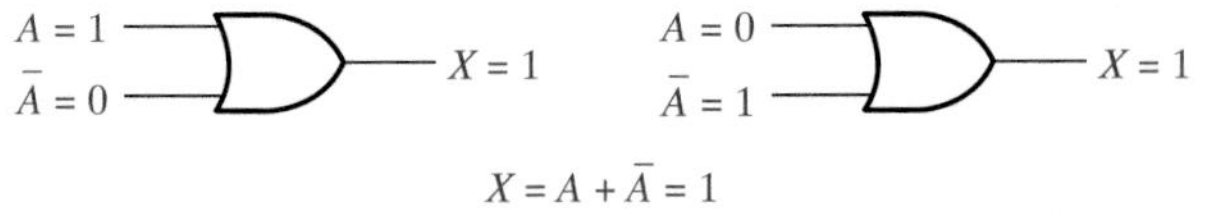

FIGURE 4–11
Rule 6: $A + \overline{A} = 1$.

$$X = A + \overline{A} = 1$$

7. $A \cdot A = A$ A variable ANDed with itself is always equal to the variable. If $A = 0$, then $0 \cdot 0 = 0$, and if $A = 1$, then $1 \cdot 1 = 1$. Figure 4–12 illustrates this rule.

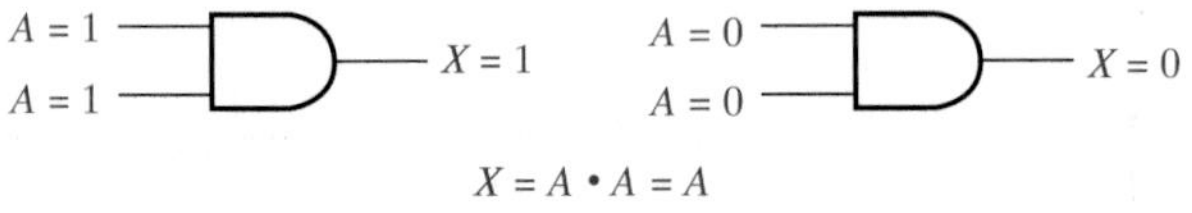

FIGURE 4–12
Rule 7: $A \cdot A = A$.

$$X = A \cdot A = A$$

8. $A \cdot \overline{A} = 0$ A variable ANDed with its complement is always equal to 0. Either A or $\overline{A}$ will always be 0; and when a 0 is applied to the input of an AND gate, the output will be 0 also. Figure 4–13 illustrates this rule.

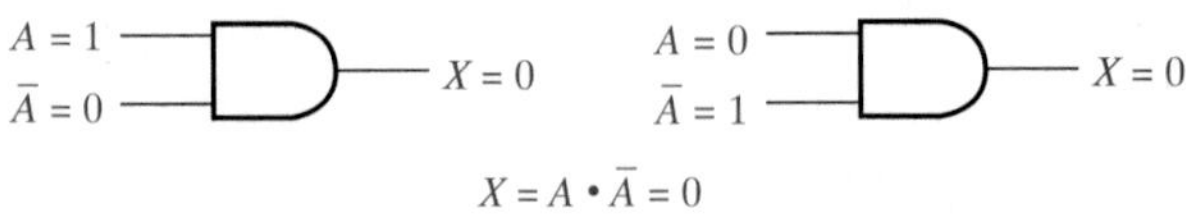

FIGURE 4–13
Rule 8: $A \cdot \overline{A} = 0$.

$$X = A \cdot \overline{A} = 0$$

9. $\overline{\overline{A}} = A$ The double complement of a variable is always equal to the variable. If you start with the variable A and complement (invert) it once, you get $\overline{A}$. If you then take $\overline{A}$ and complement (invert) it, you get A, which is the original variable. This rule is shown in Figure 4–14 using inverters.

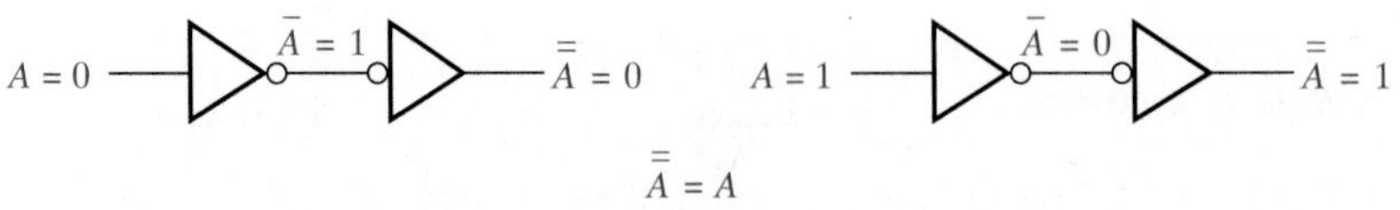

FIGURE 4–14
Rule 9: $\overline{\overline{A}} = A$.

$$\overline{\overline{A}} = A$$

10. $A + AB = A$ This rule can be proved by applying the distributive law, rule 2, and rule 4 as follows:

$$A + AB = A(1 + B) \quad \text{Factoring (distributive law)}$$
$$= A \cdot 1 \quad \text{Rule 2: } (1 + B) = 1$$
$$= A \quad \text{Rule 4: } A \cdot 1 = A$$

The proof is shown in the truth table (Table 4–2).

TABLE 4–2

A	B	AB	A + AB
0	0	0	0
0	1	0	0
1	0	0	1
1	1	1	1

equal

11. $A + \overline{A}B = A + B$ This rule can be proved as follows:

$$A + \overline{A}B = (A + AB) + \overline{A}B \quad \text{Rule 10: } A = A + AB$$
$$= (AA + AB) + \overline{A}B \quad \text{Rule 7: } A = AA$$
$$= AA + AB + A\overline{A} + \overline{A}B \quad \text{Rule 8: Adding } A\overline{A} = 0$$
$$= (A + \overline{A})(A + B) \quad \text{Factoring}$$
$$= 1 \cdot (A + B) \quad \text{Rule 6: } A + \overline{A} = 1$$
$$= A + B \quad \text{Rule 4: drop the 1}$$

The proof is shown in the truth table (Table 4–3).

TABLE 4–3

A	B	$\overline{A}B$	$A + \overline{A}B$	A + B
0	0	0	0	0
0	1	1	1	1
1	0	0	1	1
1	1	0	1	1

equal

12. $(A + B)(A + C) = A + BC$ This rule can be proved as follows:

$$(A + B)(A + C) = AA + AC + AB + BC \quad \text{Distributive law}$$
$$= A + AC + AB + BC \quad \text{Rule 7: } AA = A$$
$$= A(1 + C) + AB + BC \quad \text{Factoring (distributive law)}$$
$$= A \cdot 1 + AB + BC \quad \text{Rule 2: } 1 + C = 1$$
$$= A(1 + B) + BC \quad \text{Factoring (distributive law)}$$
$$= A \cdot 1 + BC \quad \text{Rule 2: } 1 + B = 1$$
$$= A + BC \quad \text{Rule 4: } A \cdot 1 = A$$

The proof is shown in the truth table (Table 4–4).

TABLE 4–4

A	B	C	A + B	A + C	(A + B)(A + C)	BC	A + BC
0	0	0	0	0	0	0	0
0	0	1	0	1	0	0	0
0	1	0	1	0	0	0	0
0	1	1	1	1	1	1	1
1	0	0	1	1	1	0	1
1	0	1	1	1	1	0	1
1	1	0	1	1	1	0	1
1	1	1	1	1	1	1	1

↑——————— equal ———————↑

SECTION 4–2 REVIEW

1. Apply the associative law of addition to the expression $A + (B + C + D)$.
2. Apply the distributive law to the expression $A(B + C + D)$.

4–3 ■ DEMORGAN'S THEOREMS

DeMorgan, a mathematician who was acquainted with Boole, proposed two theorems that are an important part of Boolean algebra. In practical terms, DeMorgan's theorems provide mathematical verification of the equivalency of the NAND and negative-OR gates and the equivalency of the NOR and negative-AND gates, which were discussed in Chapter 3. After completing this section, you should be able to

☐ State DeMorgan's theorems ☐ Relate DeMorgan's theorems to the equivalency of the NAND and negative-OR gates and to the equivalency of the NOR and negative-AND gates ☐ Apply DeMorgan's theorems to the simplification of Boolean expressions

One of DeMorgan's theorems is stated as follows:

The complement of a product of variables is equal to the sum of the complements of the variables.

Stated another way,

The complement of two or more variables ANDed is equivalent to the OR of the complements of the individual variables.

The formula for expressing this theorem for two variables is

$$\overline{XY} = \overline{X} + \overline{Y} \tag{4–6}$$

DeMorgan's second theorem is stated as follows:

The complement of a sum of variables is equal to the product of the complements of the variables.

Stated another way,

The complement of two or more variables ORed is equivalent to the AND of the complements of the individual variables.

The formula for expressing this theorem for two variables is

$$\overline{X + Y} = \overline{X}\,\overline{Y} \tag{4–7}$$

Figure 4–15 shows the gate equivalencies and truth tables for Equations (4–6) and (4–7).

X	Y	$\overline{XY}$	$\overline{X} + \overline{Y}$
0	0	1	1
0	1	1	1
1	0	1	1
1	1	0	0

X	Y	$\overline{X + Y}$	$\overline{X}\,\overline{Y}$
0	0	1	1
0	1	0	0
1	0	0	0
1	1	0	0

FIGURE 4–15

Gate equivalencies and corresponding truth tables illustrating DeMorgan's theorems. Notice the equality of the two output columns in each table. This shows that the equivalent gates perform the same function.

As stated, DeMorgan's theorems also apply to expressions in which there are more than two variables. The following examples illustrate the application of DeMorgan's theorems to 3-variable and 4-variable expressions.

EXAMPLE 4–3

Apply DeMorgan's theorems to the expressions $\overline{XYZ}$ and $\overline{X + Y + Z}$.

Solution
$$\overline{XYZ} = \overline{X} + \overline{Y} + \overline{Z}$$
$$\overline{X + Y + Z} = \overline{X}\,\overline{Y}\,\overline{Z}$$

Related Exercise   Apply DeMorgan's theorem to the expression $\overline{\overline{X} + \overline{Y} + \overline{Z}}$.

EXAMPLE 4–4

Apply DeMorgan's theorems to the expressions $\overline{WXYZ}$ and $\overline{W + X + Y + Z}$.

Solution
$$\overline{WXYZ} = \overline{W} + \overline{X} + \overline{Y} + \overline{Z}$$
$$\overline{W + X + Y + Z} = \overline{W}\,\overline{X}\,\overline{Y}\,\overline{Z}$$

Related Exercise   Apply DeMorgan's theorem to the expression $\overline{\overline{W}\,\overline{X}\,\overline{Y}\,\overline{Z}}$.

Each variable in DeMorgan's theorems as stated in Equations (4–6) and (4–7) can also represent a combination of other variables. For example, X can be equal to the term $AB + C$, and Y can be equal to the term $A + BC$. So if you apply DeMorgan's theorem for two variables as stated by $\overline{XY} = \overline{X} + \overline{Y}$ to the expression $\overline{(AB + C)(A + BC)}$, you get the following result:

$$\overline{(AB + C)(A + BC)} = \overline{(AB + C)} + \overline{(A + BC)}$$

Notice that in the preceding result you have two terms, $\overline{AB + C}$ and $\overline{A + BC}$, to each of which you can again apply DeMorgan's theorem $\overline{X + Y} = \overline{X}\,\overline{Y}$ individually, as follows:

$$\overline{(AB + C)} + \overline{(A + BC)} = (\overline{AB})\overline{C} + \overline{A}(\overline{BC})$$

Notice that you still have two terms in the expression to which DeMorgan's theorem can again be applied. These terms are $\overline{AB}$ and $\overline{BC}$. A final application of DeMorgan's theorem gives the following result:

$$(\overline{AB})\overline{C} + \overline{A}(\overline{BC}) = (\overline{A} + \overline{B})\overline{C} + \overline{A}(\overline{B} + \overline{C})$$

Although this result can be simplified further by the use of Boolean rules and laws, De-Morgan's theorems cannot be used any more.

Applying DeMorgan's Theorems

The following procedure illustrates the application of DeMorgan's theorems and Boolean algebra to the specific expression (brackets are the same as parentheses):

$$\overline{\overline{A + B\overline{\overline{C}}} + D(\overline{E + \overline{\overline{F}}})}$$

Step 1. Identify the terms to which you can apply DeMorgan's theorems, and think of each term as a single variable. Let $\overline{A + B\overline{\overline{C}}} = X$ and $D(\overline{E + \overline{\overline{F}}}) = Y$

Step 2. Since $\overline{X + Y} = \overline{X}\,\overline{Y}$,

$$\overline{[\overline{A + B\overline{\overline{C}}}] + [D(\overline{E + \overline{\overline{F}}})]} = [\overline{\overline{A + B\overline{\overline{C}}}}][\overline{D(\overline{E + \overline{\overline{F}}})}]$$

Step 3. Use rule 9 ($\overline{\overline{A}} = A$) to cancel the double bars over the left term (this is not part of DeMorgan's theorem):

$$[\overline{\overline{A + B\overline{\overline{C}}}}][\overline{D(\overline{E + \overline{\overline{F}}})}] = [A + B\overline{C}][\overline{D(\overline{E + \overline{\overline{F}}})}]$$

Step 4. In the right term, let $Z = \overline{E + \overline{\overline{F}}}$.

$$[A + B\overline{C}][\overline{D(\overline{E + \overline{\overline{F}}})}] = [A + B\overline{C}][\overline{DZ}]$$

Step 5. Since $\overline{DZ} = \overline{D} + \overline{Z}$,

$$[A + B\overline{C}][\overline{D(\overline{E + \overline{\overline{F}}})}] = (A + B\overline{C})(\overline{D} + (\overline{\overline{E + \overline{\overline{F}}}}))$$

Step 6. Use rule 9 ($\overline{\overline{A}} = A$) to cancel the double bars over the $E + \overline{F}$ part of the term:

$$(A + B\overline{C})(\overline{D} + \overline{\overline{E + \overline{F}}}) = (A + B\overline{C})(\overline{D} + E + \overline{F})$$

The following three examples will further illustrate how to use DeMorgan's theorems.

EXAMPLE 4–5

Apply DeMorgan's theorems to each of the following expressions:

(a) $\overline{(A + B + C)D}$ **(b)** $\overline{ABC + DEF}$ **(c)** $\overline{A\overline{B} + \overline{C}D + EF}$

Solution

(a) Let $A + B + C = X$ and $D = Y$. The expression $\overline{(A + B + C)D}$ is of the form $\overline{XY} = \overline{X} + \overline{Y}$ and can be rewritten as

$$\overline{(A + B + C)D} = \overline{A + B + C} + \overline{D}$$

Next, apply DeMorgan's theorem to the term $\overline{A + B + C}$:

$$\overline{A + B + C} + \overline{D} = \overline{A}\,\overline{B}\,\overline{C} + \overline{D}$$

(b) Let $ABC = X$ and $DEF = Y$. The expression $\overline{ABC + DEF}$ is of the form $\overline{X + Y} = \overline{X}\,\overline{Y}$ and can be rewritten as

$$\overline{ABC + DEF} = (\overline{ABC})(\overline{DEF})$$

Next, apply DeMorgan's theorem to each of the terms $\overline{ABC}$ and $\overline{DEF}$:

$$(\overline{ABC})(\overline{DEF}) = (\overline{A} + \overline{B} + \overline{C})(\overline{D} + \overline{E} + \overline{F})$$

(c) Let $A\overline{B} = X$, $\overline{C}D = Y$, and $EF = Z$. The expression $\overline{A\overline{B} + \overline{C}D + EF}$ is of the form $\overline{X + Y + Z} = \overline{X}\,\overline{Y}\,\overline{Z}$ and can be rewritten as

$$\overline{A\overline{B} + \overline{C}D + EF} = (\overline{A\overline{B}})(\overline{\overline{C}D})(\overline{EF})$$

Next, apply DeMorgan's theorem to each of the terms $\overline{A\overline{B}}$, $\overline{\overline{C}D}$, and $\overline{EF}$:

$$(\overline{A\overline{B}})(\overline{\overline{C}D})(\overline{EF}) = (\overline{A} + B)(C + \overline{D})(\overline{E} + \overline{F})$$

Related Exercise Apply DeMorgan's theorems to the expression $\overline{ABC + D + E}$.

EXAMPLE 4–6

Apply DeMorgan's theorems to each expression:

(a) $\overline{(\overline{A} + B) + \overline{C}}$ (b) $\overline{(\overline{A} + B) + CD}$ (c) $\overline{(A + B)\overline{C}\,\overline{D} + E + \overline{F}}$

Solution

(a) $\overline{(\overline{A} + B) + \overline{C}} = (\overline{\overline{A} + B})\overline{\overline{C}} = (A + B)C$

(b) $\overline{(\overline{A} + B) + CD} = (\overline{\overline{A} + B})\overline{CD} = (\overline{\overline{A}}\,\overline{B})(\overline{C} + \overline{D}) = A\overline{B}(\overline{C} + \overline{D})$

(c) $\overline{(A + B)\overline{C}\,\overline{D} + E + \overline{F}} = [\overline{(A + B)\overline{C}\,\overline{D}}]\,(\overline{E + \overline{F}}) = (\overline{A}\,\overline{B} + C + D)\overline{E}F$

Related Exercise Apply DeMorgan's theorems to the expression $\overline{\overline{AB}(C + \overline{D}) + E}$.

EXAMPLE 4–7

The Boolean expression for an exclusive-OR gate is $A\overline{B} + \overline{A}B$. With this as a starting point, develop an expression for the exclusive-NOR gate, using DeMorgan's theorems and any other rules or laws that are applicable.

Solution Start by complementing the exclusive-OR expression and then applying DeMorgan's theorems as follows:

$$\overline{A\overline{B} + \overline{A}B} = (\overline{A\overline{B}})(\overline{\overline{A}B}) = (\overline{A} + B)(\overline{\overline{A}} + \overline{B}) = (\overline{A} + B)(A + \overline{B})$$

Next, apply the distributive law and rule 8 ($A \cdot \overline{A} = 0$):

$$(\overline{A} + B)(A + \overline{B}) = \overline{A}A + \overline{A}\,\overline{B} + AB + B\overline{B} = \overline{A}\,\overline{B} + AB$$

The final expression for the XNOR is $\overline{A}\,\overline{B} + AB$. Note that this expression equals 1 any time both variables are 0s or both variables are 1s.

Related Exercise Starting with the expression for a 4-input NAND gate, use DeMorgan's theorems to develop an expression for a 4-input negative-OR gate.

SECTION 4–3 REVIEW

1. Apply DeMorgan's theorems to the following expressions:

(a) $\overline{ABC + (\overline{D} + E)}$ (b) $\overline{(A + B)\overline{C}}$ (c) $\overline{A + B + \overline{C} + \overline{DE}}$

4–4 ■ BOOLEAN ANALYSIS OF LOGIC CIRCUITS

Boolean algebra provides a concise way to express the operation of a logic circuit formed by a combination of logic gates so that the output can be readily determined for various combinations of input values. After completing this section, you should be able to

□ Determine the Boolean expression for a combination of gates   □ Evaluate the logic operation of a circuit from the Boolean expression   □ Construct a truth table

The Boolean Expression for a Logic Circuit

To derive the Boolean expression of a given logic circuit, begin at the left-most inputs and work toward the final output, writing the expression for each gate. For the example circuit in Figure 4–16, the Boolean expression is determined as follows:

1. The expression for the left-most AND gate with inputs C and D is CD.
2. The output of the left-most AND gate is one of the inputs to the OR gate and B is the other input. Therefore, the expression for the OR gate is $B + CD$.
3. The output of the OR gate is one of the inputs to the right-most AND gate and A is the other input. Therefore, the expression for this AND gate is $A(B + CD)$, which is the final output expression for the entire circuit.

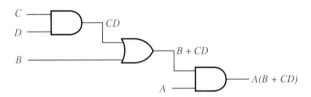

FIGURE 4–16
A logic circuit showing the development of the Boolean expression for the output.

Constructing the Truth Table for a Logic Circuit

Once the Boolean expression for a given logic circuit has been determined, a truth table that shows the output of the logic circuit for all possible values of the input variables can be developed. The procedure requires that we evaluate the Boolean expression for all possible combinations of values for the input variables. In the case of the circuit in Figure 4–16, there are four input variables (A, B, C, and D) and therefore sixteen ($2^4 = 16$) combinations of values are possible.

Evaluating the Expression To evaluate the expression $A(B + CD)$, first let's find the values of the variables that make the expression equal to 1, using the rules for Boolean addition and multiplication. In this case, the expression equals 1 only if $A = 1$ and $B + CD = 1$ because

$$A(B + CD) = 1 \cdot 1 = 1$$

Now let's determine when the $B + CD$ term equals 1. The term $B + CD = 1$ if either $B = 1$ or $CD = 1$ or if both B and CD equal 1 because

$$B + CD = 1 + 0 = 1$$
$$B + CD = 0 + 1 = 1$$
$$B + CD = 1 + 1 = 1$$

The term $CD = 1$ only if $C = 1$ and $D = 1$.

To summarize, the expression $A(B + CD) = 1$ when $A = 1$ and $B = 1$ regardless of the values of C and D or when $A = 1$ and $C = 1$ and $D = 1$, regardless of the value of B. The expression $A(B + CD) = 0$ for all other value combinations of the variables.

Putting the Results in Truth Table Format The first step is to list the sixteen input variable combinations of 1s and 0s in a binary sequence as shown in Table 4–5. Next, place a 1 in the output column for each combination of input variables that was determined in the evaluation. Finally, place a 0 in the output column for all other combinations of input variables. These results are shown in the truth table in Table 4–5.

TABLE 4–5
Truth table for the logic circuit in Figure 4–16

A	B	C	D	A(B + CD)
0	0	0	0	0
0	0	0	1	0
0	0	1	0	0
0	0	1	1	0
0	1	0	0	0
0	1	0	1	0
0	1	1	0	0
0	1	1	1	0
1	0	0	0	0
1	0	0	1	0
1	0	1	0	0
1	0	1	1	1
1	1	0	0	1
1	1	0	1	1
1	1	1	0	1
1	1	1	1	1

(Inputs: A, B, C, D; Output: A(B + CD))

SECTION 4–4 REVIEW

1. Replace the AND gates with OR gates and the OR gate with an AND gate in Figure 4–16 and determine the Boolean expression for the output.
2. Construct a truth table for the circuit in Question 1.

4–5 ■ SIMPLIFICATION USING BOOLEAN ALGEBRA

Many times in the application of Boolean algebra, you have to reduce a particular expression to its simplest form or change its form to a more convenient one to implement the expression most efficiently. The approach taken in this section is to use the basic laws, rules, and theorems of Boolean algebra to manipulate and simplify an expression. This method depends on a thorough knowledge of Boolean algebra and considerable practice in its application, not to mention a little ingenuity and cleverness. After completing this section, you should be able to

☐ Apply the laws, rules, and theorems of Boolean algebra to simplify general expressions

A simplified Boolean expression uses the fewest gates possible to implement a given expression. Four examples follow to illustrate Boolean simplification step by step.

EXAMPLE 4–8

Using Boolean algebra techniques, simplify this expression:

$$AB + A(B + C) + B(B + C)$$

Solution The following is not necessarily the only approach.

Step 1. Apply the distributive law to the second and third terms in the expression, as follows:

$$AB + AB + AC + BB + BC$$

Step 2. Apply rule 7 ($BB = B$) to the fourth term:

$$AB + AB + AC + B + BC$$

Step 3. Apply rule 5 ($AB + AB = AB$) to the first two terms:

$$AB + AC + B + BC$$

Step 4. Apply rule 10 ($B + BC = B$) to the last two terms:

$$AB + AC + B$$

Step 5. Apply rule 10 ($AB + B = B$) to the first and third terms:

$$B + AC$$

At this point the expression is simplified as much as possible. Once you gain experience in applying Boolean algebra, you can combine many individual steps.

Related Exercise Simplify the Boolean expression $A\overline{B} + A(\overline{B + C}) + B(\overline{B + C})$.

Figure 4–17 shows that the simplification process in Example 4–8 has significantly reduced the number of logic gates required to implement the expression. Part (a) shows that five gates are required to implement the expression in its original form; only two gates are needed for the simplified expression, shown in part (b). It is important to realize that these two gate networks are equivalent. That is, for any combination of levels on the A, B, and C inputs, you get the same output from either circuit.

FIGURE 4–17
Gate networks for Example 4–8.

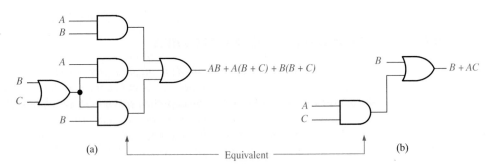

(a)   Equivalent   (b)

EXAMPLE 4–9

Simplify the Boolean expression:

$$[A\overline{B}(C + BD) + \overline{A}\ \overline{B}]C$$

Note that brackets and parentheses mean the same thing: the term inside is multiplied (ANDed) with the term outside.

Solution
Step 1. Apply the distributive law to the terms within the brackets:

$$(A\overline{B}C + A\overline{B}BD + \overline{A}\ \overline{B})C$$

Step 2. Apply rule 8 ($\overline{B}B = 0$) to the second term within the parentheses:

$$(A\overline{B}C + A \cdot 0 \cdot D + \overline{A}\ \overline{B})C$$

Step 3. Apply rule 3 ($A \cdot 0 \cdot D = 0$) to the second term within the parentheses:

$$(A\overline{B}C + 0 + \overline{A}\ \overline{B})C$$

Step 4. Apply rule 1 (drop the 0) within the parentheses:

$$(A\bar{B}C + \bar{A}\,\bar{B})C$$

Step 5. Apply the distributive law:

$$A\bar{B}CC + \bar{A}\,\bar{B}C$$

Step 6. Apply rule 7 ($CC = C$) to the first term:

$$A\bar{B}C + \bar{A}\,\bar{B}C$$

Step 7. Factor out $\bar{B}C$:

$$\bar{B}C(A + \bar{A})$$

Step 8. Apply rule 6 ($A + \bar{A} = 1$):

$$\bar{B}C \cdot 1$$

Step 9. Apply rule 4 (drop the 1):

$$\bar{B}C$$

Related Exercise   Simplify the Boolean expression $[AB(C + \bar{B}D) + \bar{A}B]CD$.

EXAMPLE 4–10   Simplify the Boolean expression:

$$\bar{A}BC + A\bar{B}\,\bar{C} + \bar{A}\,\bar{B}\,\bar{C} + A\bar{B}C + ABC$$

Solution
Step 1. Factor BC out of the first and last terms:

$$BC(\bar{A} + A) + A\bar{B}\,\bar{C} + \bar{A}\,\bar{B}\,\bar{C} + A\bar{B}C$$

Step 2. Apply rule 6 ($\bar{A} + A = 1$) to the term in parentheses, and factor $A\bar{B}$ from the second and last terms:

$$BC \cdot 1 + A\bar{B}(\bar{C} + C) + \bar{A}\,\bar{B}\,\bar{C}$$

Step 3. Apply rule 4 (drop the 1) to the first term and rule 6 ($\bar{C} + C = 1$) to the term in parentheses:

$$BC + A\bar{B} \cdot 1 + \bar{A}\,\bar{B}\,\bar{C}$$

Step 4. Apply rule 4 (drop the 1) to the second term:

$$BC + A\bar{B} + \bar{A}\,\bar{B}\,\bar{C}$$

Step 5. Factor $\bar{B}$ from the second and third terms:

$$BC + \bar{B}(A + \bar{A}\,\bar{C})$$

Step 6. Apply rule 11 ($A + \bar{A}\,\bar{C} = A + \bar{C}$) to the term in parentheses:

$$BC + \bar{B}(A + \bar{C})$$

Step 7. Use the distributive and commutative laws to get the following expression:

$$BC + A\bar{B} + \bar{B}\,\bar{C}$$

Related Exercise   Simplify the Boolean expression $AB\bar{C} + \bar{A}\,\bar{B}C + \bar{A}BC + \bar{A}\,\bar{B}\,\bar{C}$.

EXAMPLE 4–11   Simplify the following Boolean expression:

$$\overline{AB + AC} + \overline{A}\,\overline{B}C$$

Solution

Step 1. Apply DeMorgan's theorem to the first term:

$$(\overline{AB})(\overline{AC}) + \overline{A}\,\overline{B}C$$

Step 2. Apply DeMorgan's theorem to each term in parentheses:

$$(\overline{A} + \overline{B})(\overline{A} + \overline{C}) + \overline{A}\,\overline{B}C$$

Step 3. Apply the distributive law to the two terms in parentheses:

$$\overline{A}\,\overline{A} + \overline{A}\,\overline{C} + \overline{A}\,\overline{B} + \overline{B}\,\overline{C} + \overline{A}\,\overline{B}C$$

Step 4. Apply rule 7 $(\overline{A}\,\overline{A} = \overline{A})$ to the first term, and apply rule 10 $[\overline{A}\,\overline{B} + \overline{A}\,\overline{B}C = \overline{A}\,\overline{B}(1 + C) = \overline{A}\,\overline{B}]$ to the third and last terms:

$$\overline{A} + \overline{A}\,\overline{C} + \overline{A}\,\overline{B} + \overline{B}\,\overline{C}$$

Step 5. Apply rule 10 $[\overline{A} + \overline{A}\,\overline{C} = \overline{A}(1 + \overline{C}) = \overline{A}]$ to the first and second terms:

$$\overline{A} + \overline{A}\,\overline{B} + \overline{B}\,\overline{C}$$

Step 6. Apply rule 10 $[\overline{A} + \overline{A}\,\overline{B} = \overline{A}(1 + \overline{B}) = \overline{A}]$ to the first and second terms:

$$\overline{A} + \overline{B}\,\overline{C}$$

Related Exercise   Simplify the Boolean expression $\overline{AB} + AC + \overline{A}\,\overline{B}\,\overline{C}$.

SECTION 4–5 REVIEW

1. Simplify the following Boolean expressions if possible:
   (a) $A + AB + A\overline{B}C$ **(b)** $(\overline{A} + B)C + ABC$ **(c)** $A\overline{B}C(BD + CDE) + A\overline{C}$
2. Implement each expression in Question 1 as originally stated with the appropriate logic gates. Then implement the simplified expression, and compare the number of gates.

4–6 ■ STANDARD FORMS OF BOOLEAN EXPRESSIONS

All Boolean expressions, regardless of their form, can be converted into either of two standard forms: the sum-of-products form or the product-of-sums form. Standardization makes the evaluation, simplification, and implementation of Boolean expressions much more systematic and easier. After completing this section, you should be able to

□ Identify a sum-of-products expression   □ Determine the domain of a Boolean expression   □ Convert any sum-of-products expression to a standard form   □ Evaluate a standard sum-of-products expression in terms of binary values   □ Identify a product-of-sums expression   □ Convert any product-of-sums expression to a standard form   □ Evaluate a standard product-of-sums expression in terms of binary values   □ Convert from one standard form to the other

The Sum-of-Products (SOP) Form

A product term was defined in Section 4–1 as a term consisting of the product (Boolean multiplication) of literals (variables or their complements). When two or more product terms are summed by Boolean addition, the resulting expression is a **sum-of-products** (SOP). Some examples are

$$AB + ABC$$

$$ABC + CDE + \overline{B}C\overline{D}$$

$$\overline{A}B + \overline{A}B\overline{C} + AC$$

Also, an SOP expression can contain a single-variable term, as in $A + \overline{A}\,\overline{B}C + BC\overline{D}$. Refer to the simplification examples in the last section, and you will see that each of the final expressions was either a single product term or in SOP form. In an SOP expression, a single overbar cannot extend over more than one variable, although more than one variable in a term can have an overbar. For example, an SOP expression can have the term $\overline{A}\,\overline{B}\,\overline{C}$ but not $\overline{ABC}$.

Domain of a Boolean Expression   The **domain** of a general Boolean expression is the set of variables contained in the expression in either complemented or uncomplemented form. For example, the domain of the expression $\overline{A}B + A\overline{B}C$ is the set of variables A, B, C and the domain of the expression $AB\overline{C} + C\overline{D}E + \overline{B}C\overline{D}$ is the set of variables A, B, C, D, E.

Implementation of an SOP Expression   Implementing an SOP expression simply requires ORing the outputs of two or more AND gates. A product term is produced by an AND operation, and the addition of two or more product terms is produced by an OR operation. Therefore, an SOP expression can be implemented by AND-OR logic in which the outputs of a number (equal to the number of product terms in the expression) of AND gates feed into the inputs of an OR gate, as shown in Figure 4–18 for the expression $AB + BCD + AC$. The output X of the OR gate equals the SOP expression.

FIGURE 4–18
Implementation of the SOP expression $AB + BCD + AC$.

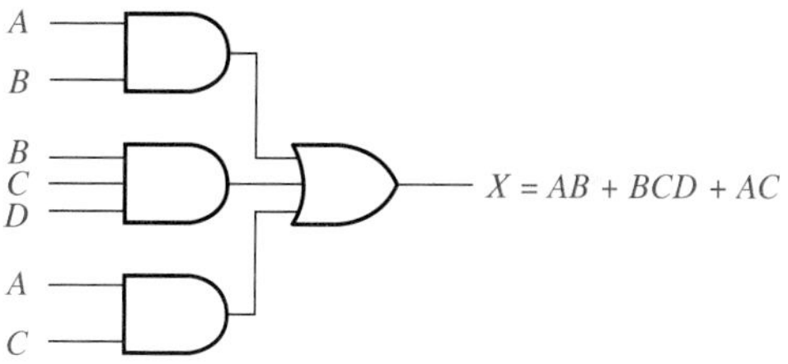

$$X = AB + BCD + AC$$

Conversion of a General Expression to SOP Form

Any logic expression can be changed into SOP form by applying Boolean algebra techniques. For example, the expression $A(B + CD)$ can be converted to SOP form by applying the distributive law:

$$A(B + CD) = AB + ACD$$

EXAMPLE 4–12   Convert each of the following Boolean expressions to SOP form:

(a) $AB + B(CD + EF)$ (b) $(A + B)(B + C + D)$ (c) $\overline{(\overline{A + B}) + C}$

Solution
(a) $AB + B(CD + EF) = AB + BCD + BEF$
(b) $(A + B)(B + C + D) = AB + AC + AD + BB + BC + BD$
(c) $\overline{(\overline{A + B}) + C} = (\overline{\overline{A + B}})\overline{C} = (A + B)\overline{C} = A\overline{C} + B\overline{C}$

Related Exercise   Convert $\overline{A}B\overline{C} + (A + \overline{B})(B + \overline{C} + A\overline{B})$ to SOP form.

The Standard SOP Form

So far, you have seen SOP expressions in which some of the product terms do not contain all of the variables in the domain of the expression. For example, the expression $\overline{A}B\overline{C} + A\overline{B}D + \overline{A}BC\overline{D}$ has a domain made up of the variables A, B, C, and D. Notice that the complete set of variables in the domain is not represented in the first two terms of the expression; that is, D or $\overline{D}$ is missing from the first term and C or $\overline{C}$ is missing from the second term.

A standard SOP expression is one in which *all* the variables in the domain appear in each product term in the expression. For example, $A\overline{B}CD + \overline{A}\,BC\overline{D} + AB\overline{C}\,\overline{D}$ is a standard SOP expression. Standard SOP expressions are important in constructing truth tables, covered in Section 4–7, and in the Karnaugh map simplification method, which is covered in the Section 4–8. Any nonstandard SOP expression (referred to simply as SOP) can be converted to the standard form using Boolean algebra.

Converting Product Terms to Standard SOP   Each product term in an SOP expression that does not contain all the variables in the domain can be expanded to standard form to include all variables in the domain and their complements. As stated in the following steps, a nonstandard SOP expression is converted into standard form using Boolean algebra rule 6 $(A + \overline{A} = 1)$ from Table 4–1 that states: A variable added to its complement equals 1.

Step 1.   Multiply each nonstandard product term by a term made up of the sum of a missing variable and its complement. This results in two product terms. As you know, you can multiply anything by 1 without changing its value.

Step 2.   Repeat Step 1 until all resulting product terms contain all variables in the domain in either complemented or uncomplemented form. In converting a product term to standard form, the number of product terms is doubled for each missing variable, as Example 4–13 shows.

EXAMPLE 4–13   Convert the following Boolean expression into standard SOP form:

$$A\overline{B}C + \overline{A}\,\overline{B} + AB\overline{C}D$$

Solution   The domain of this SOP expression is A, B, C, D. Take one term at a time. The first term, $A\overline{B}C$, is missing variable D or $\overline{D}$, so multiply the first term by $D + \overline{D}$ as follows:

$$A\overline{B}C = A\overline{B}C(D + \overline{D}) = A\overline{B}CD + A\overline{B}C\overline{D}$$

In this case, two standard product terms are the result.

The second term, $\overline{A}\,\overline{B}$, is missing variables C or $\overline{C}$ and D or $\overline{D}$, so first multiply the second term by $C + \overline{C}$ as follows:

$$\overline{A}\,\overline{B} = \overline{A}\,\overline{B}(C + \overline{C}) = \overline{A}\,\overline{B}C + \overline{A}\,\overline{B}\,\overline{C}$$

The two resulting terms are missing variable D or $\overline{D}$, so multiply both terms by $D + \overline{D}$ as follows:

$$\overline{A}\,\overline{B} = \overline{A}\,\overline{B}C + \overline{A}\,\overline{B}\,\overline{C} = \overline{A}\,\overline{B}C(D + \overline{D}) + \overline{A}\,\overline{B}\,\overline{C}(D + \overline{D})$$
$$= \overline{A}\,\overline{B}CD + \overline{A}\,\overline{B}C\overline{D} + \overline{A}\,\overline{B}\,\overline{C}D + \overline{A}\,\overline{B}\,\overline{C}\,\overline{D}$$

In this case, four standard product terms are the result.

The third term, $AB\overline{C}D$, is already in standard form. The standard SOP form of the original expression is as follows:

$$A\overline{B}C + \overline{A}\,\overline{B} + AB\overline{C}D = A\overline{B}CD + A\overline{B}C\overline{D} + \overline{A}\,\overline{B}CD + \overline{A}\,\overline{B}C\overline{D} + \overline{A}\,\overline{B}\,\overline{C}D + \overline{A}\,\overline{B}\,\overline{C}\,\overline{D} + AB\overline{C}D$$

Related Exercise   Convert the expression $W\overline{X}Y + \overline{X}Y\overline{Z} + WX\overline{Y}$ to standard SOP form.

Binary Representation of a Standard Product Term   A standard product term is equal to 1 for only one combination of variable values. For example, the product term $A\overline{B}C\overline{D}$ is equal to 1 when $A = 1$, $B = 0$, $C = 1$, and $D = 0$, as shown below, and is 0 for all other combinations of values for the variables.

$$A\overline{B}C\overline{D} = 1 \cdot \overline{0} \cdot 1 \cdot \overline{0} = 1 \cdot 1 \cdot 1 \cdot 1 = 1$$

In this case, the product term has a binary value of 1010 (decimal ten).

Remember, a product term is implemented with an AND gate whose output is 1 if and only if each of its inputs is 1. Inverters are used to produce the complements of the variables as required.

An SOP expression is equal to 1 if and only if one or more of the product terms in the expression is equal to 1.

EXAMPLE 4–14

Determine the binary values for which the following standard SOP expression is equal to 1:

$$ABCD + A\overline{B}\,\overline{C}D + \overline{A}\,\overline{B}\,\overline{C}\,\overline{D}$$

Solution   The term $ABCD$ is equal to 1 when $A = 1$, $B = 1$, $C = 1$, and $D = 1$.

$$ABCD = 1 \cdot 1 \cdot 1 \cdot 1 = 1$$

The term $A\overline{B}\,\overline{C}D$ is equal to 1 when $A = 1$, $B = 0$, $C = 0$, and $D = 1$.

$$A\overline{B}\,\overline{C}D = 1 \cdot \overline{0} \cdot \overline{0} \cdot 1 = 1 \cdot 1 \cdot 1 \cdot 1 = 1$$

The term $\overline{A}\,\overline{B}\,\overline{C}\,\overline{D}$ is equal to 1 when $A = 0$, $B = 0$, $C = 0$, and $D = 0$.

$$\overline{A}\,\overline{B}\,\overline{C}\,\overline{D} = \overline{0} \cdot \overline{0} \cdot \overline{0} \cdot \overline{0} = 1 \cdot 1 \cdot 1 \cdot 1 = 1$$

The SOP expression equals 1 when any or all of the three product terms is 1.

Related Exercise   Determine the binary values for which the following SOP expression is equal to 1:

$$\overline{X}YZ + X\overline{Y}Z + XY\overline{Z} + \overline{X}Y\overline{Z} + XYZ$$

Is this a standard SOP expression?

The Product-of-Sums (POS) Form

A sum term was defined in Section 4–1 as a term consisting of the sum (Boolean addition) of literals (variables or their complements). When two or more sum terms are multiplied, the resulting expression is a **product-of-sums** (POS). Some examples are

$$(\overline{A} + B)(A + \overline{B} + C)$$
$$(\overline{A} + \overline{B} + \overline{C})(C + \overline{D} + E)(\overline{B} + C + D)$$
$$(A + B)(A + \overline{B} + C)(\overline{A} + C)$$

A POS expression can contain a single-variable term, as in $\overline{A}(A + \overline{B} + C)(\overline{B} + \overline{C} + D)$. In a POS expression, a single overbar cannot extend over more than one variable, although more than one variable in a term can have an overbar. For example, a POS expression can have the term $\overline{A} + \overline{B} + \overline{C}$ but not $\overline{A + B + C}$.

Implementation of a POS Expression   Implementing a POS expression simply requires ANDing the outputs of two or more OR gates. A sum term is produced by an OR operation, and the product of two or more sum terms is produced by an AND operation. Therefore, a POS expression can be implemented by logic in which the outputs of a number (equal

FIGURE 4–19

Implementation of the POS expression
(A + B)(B + C + D)(A + C).

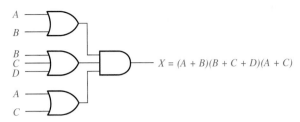

$$X = (A + B)(B + C + D)(A + C)$$

to the number of sum terms in the expression) of OR gates feed into the inputs of an AND gate, as shown in Figure 4–19 for the expression $(A + B)(B + C + D)(A + C)$. The output X of the AND gate equals the POS expression.

The Standard POS Form

So far, you have seen POS expressions in which some of the sum terms do not contain all of the variables in the domain of the expression. For example, the expression

$$(A + \overline{B} + C)(A + B + \overline{D})(A + \overline{B} + \overline{C} + D)$$

has a domain made up of the variables A, B, C, and D. Notice that the complete set of variables in the domain is not represented in the first two terms of the expression; that is, D or $\overline{D}$ is missing from the first term and C or $\overline{C}$ is missing from the second term.

A standard POS expression is one in which *all* the variables in the domain appear in each sum term in the expression. For example,

$$(\overline{A} + \overline{B} + \overline{C} + \overline{D})(A + \overline{B} + C + D)(A + B + \overline{C} + D)$$

is a standard POS expression. Any nonstandard POS expression (referred to simply as POS) can be converted to the standard form using Boolean algebra.

Converting a Sum Term to Standard POS Each sum term in an POS expression that does not contain all the variables in the domain can be expanded to standard form to include all variables in the domain and their complements. As illustrated in the following steps, a nonstandard POS expression is converted into standard form using Boolean algebra rule 8 ($A \cdot \overline{A} = 0$) from Table 4–1 that states: A variable multiplied by its complement equals 0.

Step 1. Add to each nonstandard product term a term made up of the product of a missing variable and its complement. This results in two sum terms. As you know, you can add 0 to anything without changing its value.

Step 2. Apply rule 12 from Table 4–1: $A + BC = (A + B)(A + C)$

Step 3. Repeat Step 1 until all resulting sum terms contain all variables in the domain in either complemented or uncomplemented form.

EXAMPLE 4–15

Convert the following Boolean expression into standard POS form:

$$(A + \overline{B} + C)(\overline{B} + C + \overline{D})(A + \overline{B} + \overline{C} + D)$$

Solution The domain of this POS expression is A, B, C, D. Take one term at a time. The first term, $A + \overline{B} + C$, is missing variable D or $\overline{D}$, so add $D\overline{D}$ and apply rule 12 as follows:

$$A + \overline{B} + C = A + \overline{B} + C + D\overline{D} = (A + \overline{B} + C + D)(A + \overline{B} + C + \overline{D})$$

The second term, $\overline{B} + C + \overline{D}$, is missing variable A or $\overline{A}$, so add $A\overline{A}$ and apply rule 12 as follows:

$$\overline{B} + C + \overline{D} = \overline{B} + C + \overline{D} + A\overline{A} = (A + \overline{B} + C + \overline{D})(\overline{A} + \overline{B} + C + \overline{D})$$

The third term, $A + \overline{B} + \overline{C} + D$, is already in standard form. The standard POS form of the original expression is as follows:

$$(A + \overline{B} + C)(\overline{B} + C + \overline{D})(A + \overline{B} + \overline{C} + D) =$$
$$(A + \overline{B} + C + D)(A + \overline{B} + C + \overline{D})(A + \overline{B} + \overline{C} + \overline{D})(\overline{A} + \overline{B} + C + \overline{D})(A + \overline{B} + \overline{C} + D)$$

Related Exercise Convert the expression $(A + \overline{B})(B + C)$ to standard POS form.

Binary Representation of a Standard Sum Term A standard sum term is equal to 0 for only one combination of variable values. For example, the sum term $A + \overline{B} + C + \overline{D}$ is 0 when $A = 0, B = 1, C = 0,$ and $D = 1$, as shown below, and is 1 for all other combinations of values for the variables.

$$A + \overline{B} + C + \overline{D} = 0 + \overline{1} + 0 + \overline{1} = 0 + 0 + 0 + 0 = 0$$

In this case, the sum term has a binary value of 0101 (decimal 5). Remember, a sum term is implemented with an OR gate whose output is 0 if and only if each of its inputs is 0. Inverters are used to produce the complements of the variables as required.

A POS expression is equal to 0 if and only if one or more of the sum terms in the expression is equal to 0.

EXAMPLE 4–16

Determine the binary values of the variables for which the following standard POS expression is equal to 0:

$$(A + B + C + D)(A + \overline{B} + \overline{C} + D)(\overline{A} + \overline{B} + \overline{C} + \overline{D})$$

Solution The term $A + B + C + D$ is equal to 0 when $A = 0, B = 0, C = 0,$ and $D = 0$.

$$A + B + C + D = 0 + 0 + 0 + 0 = 0$$

The term $A + \overline{B} + \overline{C} + D$ is equal to 0 when $A = 0, B = 1, C = 1,$ and $D = 0$.

$$A + \overline{B} + \overline{C} + D = 0 + \overline{1} + \overline{1} + 0 = 0 + 0 + 0 + 0 = 0$$

The term $\overline{A} + \overline{B} + \overline{C} + \overline{D}$ is equal to 0 when $A = 1, B = 1, C = 1,$ and $D = 1$.

$$\overline{A} + \overline{B} + \overline{C} + \overline{D} = \overline{1} + \overline{1} + \overline{1} + \overline{1} = 0 + 0 + 0 + 0 = 0$$

The POS expression equals 0 when any of the three sum terms equals 0.

Related Exercise Determine the binary values for which the following POS expression is equal to 0:

$$(X + \overline{Y} + Z)(\overline{X} + Y + Z)(X + Y + \overline{Z})(\overline{X} + \overline{Y} + \overline{Z})(X + \overline{Y} + \overline{Z})$$

Is this a standard POS expression?

Converting Standard SOP to Standard POS

The binary values of the product terms in a given standard SOP expression are not present in the equivalent standard POS expression. Also, the binary values that are not represented in the SOP expression are present in the equivalent POS expression. Therefore, to convert from standard SOP to standard POS, the following steps are taken:

Step 1. Evaluate each product term in the SOP expression. That is, determine the binary numbers which represent the product terms.

Step 2. Determine all of the binary numbers not included in the evaluation in Step 1.

Step 3. Write the equivalent sum term for each binary number from Step 2 and express in POS form.

Using a similar procedure, you can go from POS to SOP.

EXAMPLE 4–17

Convert the following SOP expression to an equivalent POS expression:

$$\overline{A}\,\overline{B}\,\overline{C} + \overline{A}B\overline{C} + \overline{A}BC + A\overline{B}C + ABC$$

Solution The evaluation is as follows:

$$000 + 010 + 011 + 101 + 111$$

Since there are three variables in the domain of this expression, there are a total of eight (2^3) possible combinations. The SOP expression contains five of these combinations, so the POS must contain the other three which are 001, 100, and 110. Remember, these are the binary values that make the sum term 0. The equivalent POS expression is

$$(A + B + \overline{C})(\overline{A} + B + C)(\overline{A} + \overline{B} + C)$$

Related Exercise Verify that the SOP and POS expressions in this example are equivalent by substituting binary values into each.

SECTION 4–6 REVIEW

1. Identify each of the following expressions as SOP, standard SOP, POS, or standard POS:
   (a) $AB + \overline{A}BD + \overline{A}C\overline{D}$ (b) $(A + \overline{B} + C)(A + B + \overline{C})$
   (c) $\overline{A}BC + AB\overline{C}$ (d) $A(A + \overline{C})(A + B)$
2. Convert each SOP expression in Question 1 to standard form.
3. Convert each POS expression in Question 1 to standard form.

4–7 ■ BOOLEAN EXPRESSIONS AND TRUTH TABLES

All standard Boolean expressions can be easily converted into truth table format using binary values for each term in the expression. The truth table is a common way of presenting, in a concise format, the logical operation of a circuit. Also, standard SOP or POS expressions can be determined from a truth table. You will find truth tables in logic data sheets and other literature related to the operation of digital circuits and systems. After completing this section, you should be able to

☐ Convert a standard SOP expression into truth table format   ☐ Convert a standard POS expression into truth table format   ☐ Derive a standard expression from a truth table   ☐ Properly interpret truth table data

Converting SOP Expressions to Truth Table Format

Recall from Section 4–6 that an SOP expression is equal to 1 only if at least one of the product terms is equal to 1. A truth table is simply a list of the possible combinations of input variable values and the corresponding output values (1 or 0). For an expression with a domain of two variables, there are four different combinations of those variables ($2^2 = 4$). For an expression with a domain of three variables, there are eight different combinations of those variables ($2^3 = 8$). For an expression with a domain of four variables, there are sixteen different combinations of those variables ($2^4 = 16$), and so on.

The first step in constructing a truth table is to list all possible combinations of binary values of the variables in the expression. Next, convert the SOP expression to standard form if it is not already. Finally, place a 1 in the output column (X) for each binary value that makes the standard SOP expression a 1 and place a 0 for all the remaining binary values. This procedure is illustrated in Example 4–18.

EXAMPLE 4–18

Develop a truth table for the standard SOP expression $\overline{A}\,\overline{B}C + A\overline{B}\,\overline{C} + ABC$.

Solution There are three variables in the domain, so there are eight possible combinations of binary values of the variables as listed in the left three columns of Table 4–6. The binary values that make the product terms in the expression equal to 1 are $\overline{A}\,\overline{B}C$: 001; $A\overline{B}\,\overline{C}$: 100; and ABC: 111. For each of these binary values, a 1 is placed in the output column as shown in the table. For each of the remaining binary combinations, a 0 is placed in the output column.

TABLE 4–6

Inputs			Output
A	B	C	X
0	0	0	0
0	0	1	1
0	1	0	0
0	1	1	0
1	0	0	1
1	0	1	0
1	1	0	0
1	1	1	1

Related Exercise Create a truth table for the standard SOP expression $\overline{A}B\overline{C} + A\overline{B}C$.

Converting POS Expressions to Truth Table Format

Recall that a POS expression is equal to 0 only if at least one of the sum terms is equal to 0. To construct a truth table from a POS expression, list all the possible combinations of binary values of the variables just as was done for the SOP expression. Next, convert the POS expression to standard form if it is not already. Finally, place a 0 in the output column (X) for each binary value that makes the expression a 0 and place a 1 for all the remaining binary values. This procedure is illustrated in Example 4–19.

EXAMPLE 4–19

Determine the truth table for the following standard POS expression:

$$(A + B + C)(A + \overline{B} + C)(A + \overline{B} + \overline{C})(\overline{A} + B + \overline{C})(\overline{A} + \overline{B} + C)$$

Solution There are three variables in the domain and the eight possible binary values are listed in the left three columns of Table 4–7. The binary values that make the sum terms in the expression equal to 0 are $A + B + C$: 000; $A + \overline{B} + C$: 010; $A + \overline{B} + \overline{C}$: 011; $\overline{A} + B + \overline{C}$: 101; and $\overline{A} + \overline{B} + C$: 110. For each of these binary values, a 0 is placed in the output column as shown in the table. For each of the remaining binary combinations, a 1 is placed in the output column.

TABLE 4–7

	Inputs		Output
A	B	C	X
0	0	0	0
0	0	1	1
0	1	0	0
0	1	1	0
1	0	0	1
1	0	1	0
1	1	0	0
1	1	1	1

Notice that the truth table in this example is the same as the one in Example 4–18. This means that the SOP expression in the previous example and the POS expression in this example are equivalent.

Related Exercise Develop a truth table for the following standard POS expression:

$$(A + \overline{B} + C)(A + B + \overline{C})(\overline{A} + \overline{B} + \overline{C})$$

Determining Standard Expressions from a Truth Table

To determine the standard SOP expression represented by a truth table, list the binary values of the input variables for which the output is 1. Convert each binary value to the corresponding product term by replacing each 1 with the corresponding variable and each 0 with the corresponding variable complement. For example, the binary value 1010 is converted to a product term as follows:

$$1010 \longrightarrow A\overline{B}C\overline{D}$$

To determine the standard POS expression represented by a truth table, list the binary values for which the output is 0. Convert each binary value to the corresponding sum term by replacing each 1 with the corresponding variable complement and each 0 with the corresponding variable. For example, the binary value 1001 is converted to a sum term as follows:

$$1001 \longrightarrow \overline{A} + B + C + \overline{D}$$

EXAMPLE 4–20 From the truth table in Table 4–8, determine the standard SOP expression and the equivalent standard POS expression.

TABLE 4–8

	Inputs		Output
A	B	C	X
0	0	0	0
0	0	1	0
0	1	0	0
0	1	1	1
1	0	0	1
1	0	1	0
1	1	0	1
1	1	1	1

Solution There are four 1s in the output column and the corresponding binary values are 011, 100, 110, and 111. These binary values are converted to product terms as follows:

$$011 \longrightarrow \overline{A}BC$$
$$100 \longrightarrow A\overline{B}\,\overline{C}$$
$$110 \longrightarrow AB\overline{C}$$
$$111 \longrightarrow ABC$$

The resulting standard SOP expression for the output X is

$$X = \overline{A}BC + A\overline{B}\,\overline{C} + AB\overline{C} + ABC$$

For the POS expression, the output is 0 for binary values 000, 001, 010, and 101. These binary values are converted to sum terms as follows:

$$000 \longrightarrow A + B + C$$
$$001 \longrightarrow A + B + \overline{C}$$
$$010 \longrightarrow A + \overline{B} + C$$
$$101 \longrightarrow \overline{A} + B + \overline{C}$$

The resulting standard POS expression for the output X is

$$X = (A + B + C)(A + B + \overline{C})(A + \overline{B} + C)(\overline{A} + B + \overline{C})$$

Related Exercise By substitution of binary values, show that the SOP and the POS expressions derived in this example are equivalent; that is, for any binary value they should either both be 1 or both be 0, depending on the binary value.

SECTION 4–7 REVIEW

1. If a certain Boolean expression has a domain of five variables, how many binary values will be in its truth table?
2. In a certain truth table, the output is a 1 for the binary value 0110. Convert this binary value to the corresponding product term using variables $W, X, Y,$ and Z.
3. In a certain truth table, the output is a 0 for the binary value 1100. Convert this binary value to the corresponding sum term using variables $W, X, Y,$ and Z.

4–8 ■ THE KARNAUGH MAP

The Karnaugh map provides a systematic method for simplifying Boolean expressions and, if properly used, will produce the simplest SOP or POS expression possible. As you have seen, the effectiveness of algebraic simplification depends on your familiarity with all the laws, rules, and theorems of Boolean algebra and on your ability to apply them. The Karnaugh map, on the other hand, basically provides a "cookbook" method for simplification. After completing this section, you should be able to

□ Construct a Karnaugh map for three or four variables   □ Determine the binary value of each cell in a Karnaugh map   □ Determine the standard product term represented by each cell in a Karnaugh map   □ Explain cell adjacency and identify adjacent cells

A **Karnaugh map** is similar to a truth table because it presents all of the possible values of input variables and the resulting output for each value. Instead of being organized into columns and rows like a truth table, the Karnaugh map is an array of **cells** in which each cell represents a binary value of the input variables. The cells are arranged in a way so that simplification of a given expression is simply a matter of properly grouping the cells. Karnaugh maps can be used for expressions with two, three, four, and five variables, but we will discuss only 3-variable and 4-variable situations to illustrate the principles. Section 4–11 deals with five variables using a 32-cell Karnaugh map. Another method, which is beyond the scope of this book, called the Quine-McClusky method can be used for higher numbers of variables.

The number of cells in a Karnaugh map is equal to the total number of possible input variable combinations as is the number of rows in a truth table. For three variables, the number of cells is $2^3 = 8$. For four variables, the number of cells is $2^4 = 16$.

The 3-Variable Karnaugh Map

The 3-variable Karnaugh map is an array of eight cells, as shown in Figure 4–20(a). In this case, *A, B,* and *C* are used for the variables although other letters could be used. Binary values of *A* and *B* are along the left side (notice the sequence) and the values of *C* are across the top.

The value of a given cell is the binary values of *A* and *B* at the left in the same row combined with the value of *C* at the top in the same column. For example, the cell in the upper left corner has a binary value of 000 and the cell in the lower right corner has a binary value of 101. Figure 4–20(b) shows the standard product terms that are represented by each cell in the Karnaugh map.

FIGURE 4–20
A 3-variable Karnaugh map.

AB \ *C*	0	1
00		
01		
11		
10		

(a)

AB \ *C*	0	1
00	$\overline{A}\overline{B}\overline{C}$	$\overline{A}\overline{B}C$
01	$\overline{A}B\overline{C}$	$\overline{A}BC$
11	$AB\overline{C}$	ABC
10	$A\overline{B}\overline{C}$	$A\overline{B}C$

(b)

The 4-Variable Karnaugh Map

The 4-variable Karnaugh map is an array of sixteen cells, as shown in Figure 4–21(a). Binary values of *A* and *B* are along the left side and the values of *C* and *D* are across the top.

The value of a given cell is the binary values of *A* and *B* at the left in the same row combined with the binary values of *C* and *D* at the top in the same column. For example, the cell in the upper right corner has a binary value of 0010 and the cell in the lower right corner has a binary value of 1010. Figure 4–21(b) shows the standard product terms that are represented by each cell in the 4-variable Karnaugh map.

FIGURE 4–21
A 4-variable Karnaugh map.

AB \ *CD*	00	01	11	10
00				
01				
11				
10				

(a)

AB \ *CD*	00	01	11	10
00	$\overline{A}\overline{B}\overline{C}\overline{D}$	$\overline{A}\overline{B}\overline{C}D$	$\overline{A}\overline{B}CD$	$\overline{A}\overline{B}C\overline{D}$
01	$\overline{A}B\overline{C}\overline{D}$	$\overline{A}B\overline{C}D$	$\overline{A}BCD$	$\overline{A}BC\overline{D}$
11	$AB\overline{C}\overline{D}$	$AB\overline{C}D$	$ABCD$	$ABC\overline{D}$
10	$A\overline{B}\overline{C}\overline{D}$	$A\overline{B}\overline{C}D$	$A\overline{B}CD$	$A\overline{B}C\overline{D}$

(b)

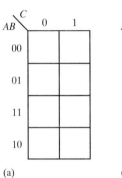

Cell Adjacency

The cells in a Karnaugh map are arranged so that there is only a single-variable change between adjacent cells. **Adjacency** is defined by a single-variable change. Cells that differ by only one variable are adjacent. For example, in the 3-variable map the 010 cell is adjacent to the 000 cell, the 011 cell, and the 110 cell. Cells with values that differ by more than one variable are not adjacent. For example, the 010 cell is not adjacent to the 001 cell, the 111 cell, the 100 cell, or the 101 cell.

Physically, each cell is adjacent to the cells that are immediately next to it on any of its four sides. A cell is not adjacent to the cells that diagonally touch any of its corners. Also, the cells in the top row are adjacent to the corresponding cells in the bottom row and the cells in the outer left column are adjacent to the corresponding cells in the outer right column. This is called "wrap-around" adjacency because you can think of the map as wrapping around from top to bottom to form a cylinder or from left to right to form a cylinder. Figure 4–22 illustrates the cell adjacencies with a 4-variable map, although the same rules for adjacency apply to Karnaugh maps with any number of cells.

FIGURE 4–22

Cell adjacencies on a Karnaugh map. Arrows point between adjacent cells.

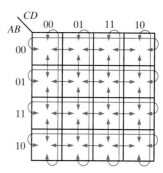

SECTION 4–8 REVIEW

1. In a 3-variable Karnaugh map, what is the binary value for the cell in each of the following locations:
    (a) upper left corner   (b) lower right corner
    (c) lower left corner   (d) upper right corner
2. What is the standard product term for each cell in Question 1 for variables X, Y, and Z?
3. Repeat Question 1 for a 4-variable map.
4. Repeat Question 2 for a 4-variable map using variables W, X, Y, and Z.

4–9 ■ KARNAUGH MAP SOP MINIMIZATION

As stated in the last section, the Karnaugh map is used for simplifying Boolean expressions to their minimum form. A minimized SOP expression contains the fewest possible terms with the fewest possible variables per term. Generally, a minimum SOP expression can be implemented with fewer logic gates than a standard expression and this is the basic purpose in the simplification process. After completing this section, you should be able to

□ Map a standard SOP expression on a Karnaugh map   □ Combine the 1s on the map into maximum groups   □ Determine the minimum product term for each group on the map   □ Combine the minimum product terms to form a minimum SOP expression   □ Convert a truth table into a Karnaugh map for simplification of the represented expression   □ Use "don't care" conditions on a Karnaugh map

Mapping a Standard SOP Expression

For an SOP expression in standard form, a 1 is placed on the Karnaugh map for each product term in the expression. Each 1 is placed in a cell corresponding to the value of a product term. For example, for the product term $A\overline{B}C$, a 1 goes in the 101 cell on a 3-variable map.

When an SOP expression is completely mapped, there will be a number of 1s on the Karnaugh map equal to the number of product terms in the standard SOP expression. The cells that do not have a 1 are the cells for which the expression is 0. Usually, when working with SOP expressions, the 0s are left off the map. The following steps and the illustration in Figure 4–23 show the mapping process.

Step 1. Determine the binary value of each product term in the standard SOP expression. After some practice, the evaluation of terms is usually done mentally.

Step 2. As each product term is evaluated, place a 1 on the Karnaugh map in the cell having the same value as the product term.

FIGURE 4–23
Example of mapping a standard SOP expression.

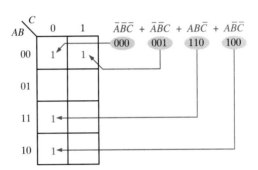

The following two examples will further illustrate the mapping process.

EXAMPLE 4–21

Map the following standard SOP expression on a Karnaugh map:

$$\overline{A}\,\overline{B}C + \overline{A}B\overline{C} + AB\overline{C} + ABC$$

Solution The expression is evaluated as shown below and a 1 is placed on the 3-variable Karnaugh map in Figure 4–24 for each standard product term in the expression.

$$\begin{array}{cccc} \overline{A}\,\overline{B}C & + & \overline{A}B\overline{C} & + & AB\overline{C} & + & ABC \\ 001 & & 010 & & 110 & & 111 \end{array}$$

FIGURE 4–24

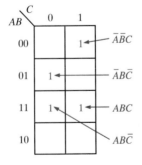

Related Exercise Map the standard SOP expression $A\overline{B}C + AB\,\overline{C} + \overline{A}BC$ on a Karnaugh map.

EXAMPLE 4–22   Map the following standard SOP expression on a Karnaugh map:

$$\overline{A}\,\overline{B}CD + \overline{A}B\overline{C}\,\overline{D} + AB\overline{C}D + ABCD + AB\overline{C}\,\overline{D} + \overline{A}\,\overline{B}\,\overline{C}D + A\overline{B}C\overline{D}$$

Solution   The expression is evaluated as shown below and a 1 is placed on the 4-variable Karnaugh map in Figure 4–25 for each standard product term in the expression.

$$\overline{A}\,\overline{B}CD + \overline{A}B\overline{C}\,\overline{D} + AB\overline{C}D + ABCD + AB\overline{C}\,\overline{D} + \overline{A}\,\overline{B}\,\overline{C}D + A\overline{B}C\overline{D}$$
$$\quad 0011 \qquad 0100 \qquad 1101 \qquad 1111 \qquad 1100 \qquad 0001 \qquad 1010$$

FIGURE 4–25

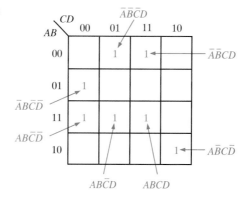

Related Exercise   Map the standard SOP expression $\overline{A}BC\overline{D} + ABC\overline{D} + AB\overline{C}\,\overline{D} + ABCD$ on a Karnaugh map.

Mapping a Nonstandard SOP Expression

A Boolean expression must first be in standard form before using the Karnaugh map. If an expression is not in standard form, then it must be converted to standard form by the procedure covered in Section 4–6 or by numerical expansion. Since an expression should be evaluated before mapping anyway, numerical expansion is probably the most efficient approach.

Numerical Expansion of a Nonstandard Product Term   Recall that a nonstandard product term has one or more missing variables. For example, assume that one of the product terms in a certain 3-variable SOP expression is $A\overline{B}$. This term can be expanded numerically to standard form as follows. First, write the binary value of the two variables and attach a 0 for the missing variable $\overline{C}$: 100. Next, write the binary value of the two variables and attach a 1 for the missing variable C: 101. The two resulting binary numbers are the values of the standard SOP terms $A\overline{B}\,\overline{C}$ and $A\overline{B}C$.

As another example, assume that one of the product terms in a 3-variable expression is B (remember that a single variable counts as a product term in an SOP expression). This term can be expanded numerically to standard form as follows: Write the binary value of the variable; then attach all possible values for the missing variables A and C as follows:

$$B$$
$$010$$
$$011$$
$$110$$
$$111$$

The four resulting binary numbers are the values of the standard SOP terms $\overline{AB}\overline{C}$, $\overline{A}BC$, $AB\overline{C}$, and ABC.

EXAMPLE 4–23

Map the following SOP expression on a Karnaugh map: $\overline{A} + A\overline{B} + AB\overline{C}$.

Solution   The SOP expression is obviously not in standard form because each product term does not have three variables. The first term is missing two variables, the second term is missing one variable, and the third term is standard. First expand the terms numerically as follows:

$$\overline{A} \quad + A\overline{B} \quad + AB\overline{C}$$

$$
\begin{array}{lll}
000 & 100 & 110 \\
001 & 101 & \\
010 & & \\
011 & &
\end{array}
$$

Each of the resulting binary values is mapped by placing a 1 in the appropriate cell of the 3-variable Karnaugh map in Figure 4–26.

FIGURE 4–26

AB \ C	0	1
00	1	1
01	1	1
11	1	
10	1	1

Related Exercise   Map the SOP expression $BC + \overline{A}\,C$ on a Karnaugh map.

EXAMPLE 4–24

Map the following SOP expression on a Karnaugh map:

$$\overline{B}\,\overline{C} + A\overline{B} + AB\overline{C} + AB\overline{C}\overline{D} + \overline{A}\,\overline{B}\,\overline{C}D + A\overline{B}CD$$

Solution   The SOP expression is obviously not in standard form because each product term does not have four variables. The first and second terms are both missing two variables, the third term is missing one variable, and the rest of the terms are standard. First expand the terms numerically as follows:

$$\overline{B}\,\overline{C} + A\overline{B} \quad + AB\overline{C} + AB\overline{C}\overline{D} + \overline{A}\,\overline{B}\,\overline{C}D + A\overline{B}CD$$

$$
\begin{array}{llllll}
0000 & 1000 & 1100 & 1010 & 0001 & 1011 \\
0001 & 1001 & 1101 & & & \\
1000 & 1010 & & & & \\
1001 & 1011 & & & &
\end{array}
$$

Each of the resulting binary values is mapped by placing a 1 in the appropriate cell of the 4-variable Karnaugh map in Figure 4–27. Notice that some of the values in the expanded expression are redundant.

FIGURE 4–27

AB \ CD	00	01	11	10
00	1	1		
01				
11	1	1		
10	1	1	1	1

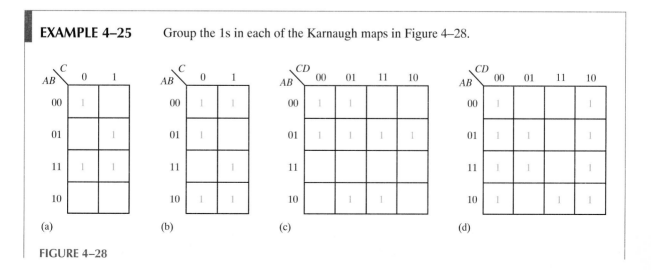

Related Exercise Map the expression $A + \overline{C}D + AC\overline{D} + \overline{A}BC\overline{D}$ on a Karnaugh map.

Karnaugh Map Simplification of SOP Expressions

The process that results in an expression containing the fewest possible terms with the fewest possible variables is called **minimization.** After an SOP expression has been mapped, there are three steps in the process of obtaining a minimum SOP expression: grouping the 1s, determining the product term for each group, and summing the resulting product terms.

Grouping the 1s You can group 1s on the Karnaugh map according to the following rules by enclosing those adjacent cells containing 1s. The goal is to maximize the size of the groups and to minimize the number of groups.

1. A group must contain either 1, 2, 4, 8, or 16 cells. In the case of a 3-variable map, eight cells is the maximum group.
2. Each cell in a group must be adjacent to one or more cells in that same group, but all cells in the group do not have to be adjacent to each other.
3. Always include the largest possible number of 1s in a group in accordance with rule 1.
4. Each 1 on the map must be included in at least one group. The 1s already in a group can be included in another group as long as the overlapping groups include noncommon 1s.

EXAMPLE 4–25 Group the 1s in each of the Karnaugh maps in Figure 4–28.

AB \ C	0	1
00	1	
01		1
11	1	1
10		

(a)

AB \ C	0	1
00	1	1
01	1	
11		1
10	1	1

(b)

AB \ CD	00	01	11	10
00	1	1		
01	1	1	1	1
11				
10		1	1	

(c)

AB \ CD	00	01	11	10
00	1			1
01	1	1		1
11	1	1		1
10	1		1	1

(d)

FIGURE 4–28

Solution The groupings are shown in Figure 4–29. In some cases, there may be more than one way to group the 1s to form maximum groupings.

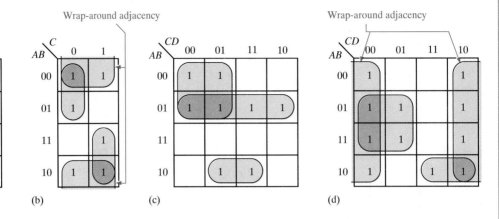

FIGURE 4–29

Related Exercise Determine if there are other ways to group the 1s in Figure 4–29 to obtain a minimum number of maximum groupings.

Determining the Minimum SOP Expression from the Map When all the 1s representing the standard product terms in an expression are properly mapped and grouped, the process of determining the resulting minimum SOP expression begins. The following rules are applied to find the minimum product terms and the minimum SOP expression:

1. Group the cells that have 1s. Each group of cells containing 1s creates one product term composed of all variables that occur in only one form (either uncomplemented or complemented) within the group. Variables that occur both uncomplemented and complemented within the group are eliminated. These are called *contradictory variables.*
2. Determine the minimum product terms for each group.
   (a) For a 3-variable map:
      (1) A 1-cell group yields a 3-variable product term
      (2) A 2-cell group yields a 2-variable product term
      (3) A 4-cell group yields a 1-variable term
      (4) An 8-cell group yields a value of 1 for the expression
   (b) For a 4-variable map:
      (1) A 1-cell group yields a 4-variable product term
      (2) A 2-cell group yields a 3-variable product term
      (3) A 4-cell group yields a 2-variable product term
      (4) An 8-cell group yields a 1-variable term
      (5) A 16-cell group yields a value of 1 for the expression
3. When all the minimum product terms are derived from the Karnaugh map, they are summed to form the minimum SOP expression.

EXAMPLE 4–26

Determine the product terms for the Karnaugh map in Figure 4–30 and write the resulting minimum SOP expression.

Solution In Figure 4–30, the product term for the 8-cell group is B because the cells within that group contain both A and $\bar{A}$, C and $\bar{C}$, and D and $\bar{D}$, so these variables are

FIGURE 4–30

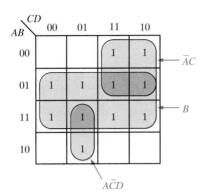

eliminated. The 4-cell group contains B, $\overline{B}$, D, and $\overline{D}$, leaving the product term $\overline{A}C$. The 2-cell group contains B and $\overline{B}$, leaving $A\overline{C}D$ as the product term. Notice how overlapping is used to maximize the size of the groups. The resulting minimum SOP expression is the sum of these product terms:

$$B + \overline{A}C + A\overline{C}D$$

Related Exercise For the Karnaugh map in Figure 4–30, add a 1 in the lower right cell (1010) and determine the resulting SOP expression.

EXAMPLE 4–27 Determine the product terms for each of the Karnaugh maps in Figure 4–31 and write the resulting minimum SOP expression.

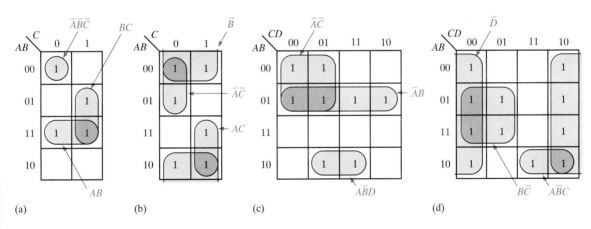

(a) (b) (c) (d)

FIGURE 4–31

Solution The resulting minimum product terms for each group are shown in Figure 4–31. The minimum SOP expressions for each of the Karnaugh maps in the figure are

(a) $AB + BC + \overline{A}\,\overline{B}\,\overline{C}$ **(b)** $\overline{B} + \overline{A}\,\overline{C} + AC$
(c) $\overline{A}B + \overline{A}\,\overline{C} + A\overline{B}D$ **(d)** $\overline{D} + A\overline{B}C + B\overline{C}$

Related Exercise For the Karnaugh map in Figure 4–31(d) add a 1 in the 0111 cell and determine the resulting SOP expression.

EXAMPLE 4–28

Use a Karnaugh map to minimize the following standard SOP expression:

$$A\overline{B}C + \overline{A}BC + \overline{A}\,\overline{B}C + \overline{A}\,\overline{B}\,\overline{C} + A\overline{B}\,\overline{C}$$

Solution The binary values of the expression are

$$101 + 011 + 001 + 000 + 100$$

The standard SOP expression is mapped and the cells are grouped as shown in Figure 4–32.

FIGURE 4–32

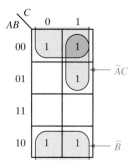

Notice the "wrap around" 4-cell group that includes the top row and the bottom row of 1s. The remaining 1 is absorbed in an overlapping group of two cells. The group of four 1s produces a single variable term, $\overline{B}$. This is determined by observing that within the group, $\overline{B}$ is the only variable that does not change from cell to cell. The group of two 1s produces a 2-variable term $\overline{A}C$. This is determined by observing that within the group, $\overline{A}$ and C do not change from one cell to the next. The product terms for each group are shown and the resulting minimum SOP expression is

$$\overline{B} + \overline{A}C$$

Keep in mind that this minimum expression is equivalent to the original standard expression.

Related Exercise Use a Karnaugh map to simplify the following standard SOP expression:

$$X\overline{Y}Z + XY\overline{Z} + \overline{X}YZ + \overline{X}Y\overline{Z} + X\overline{Y}\,\overline{Z} + XYZ$$

EXAMPLE 4–29

Use a Karnaugh map to minimize the following SOP expression:

$$\overline{B}\,\overline{C}\,\overline{D} + \overline{A}B\overline{C}\,\overline{D} + AB\overline{C}\,\overline{D} + \overline{A}\,\overline{B}CD + \overline{A}BCD + \overline{A}\,\overline{B}C\overline{D} + \overline{A}BC\overline{D} + ABC\overline{D} + AB\overline{C}D$$

Solution The first term $\overline{B}\,\overline{C}\,\overline{D}$ must be expanded into $A\overline{B}\,\overline{C}\,\overline{D}$ and $\overline{A}\,\overline{B}\,\overline{C}\,\overline{D}$ to get a standard SOP expression, which is then mapped; and the cells are grouped as shown in Figure 4–33.

Notice that both groups exhibit "wrap around" adjacency. The group of eight is formed because the cells in the outer columns are adjacent. The group of four is formed to pick up the remaining two 1s because the top and bottom cells are adjacent. The product terms for each group are shown and the resulting minimum SOP expression is

$$\overline{D} + \overline{B}C$$

Keep in mind that this minimum expression is equivalent to the original standard expression.

FIGURE 4–33

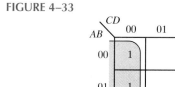

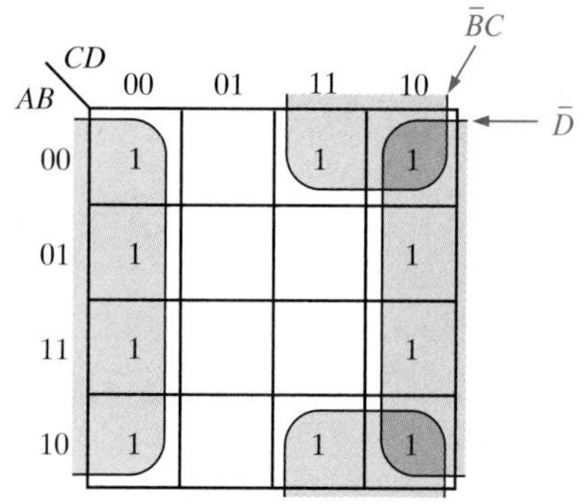

Related Exercise   Use a Karnaugh map to simplify the following SOP expression:

$$\overline{W}\,\overline{X}\,\overline{Y}\,\overline{Z} + W\overline{X}YZ + W\overline{X}\,\overline{Y}Z + \overline{W}YZ + W\overline{X}\,\overline{Y}\,\overline{Z}$$

Mapping Directly from a Truth Table

You have seen how to map a Boolean expression; now you will learn how to go directly from a truth table to a Karnaugh map. Recall that a truth table gives the output of a Boolean expression for all possible input variable combinations. An example of a Boolean expression and its truth table representation is shown in Figure 4–34. Notice in the truth table that the output X is 1 for four different input variable combinations. The 1s in the output column of the truth table are mapped directly onto a Karnaugh map into the cells corresponding to the values of the associated input variable combinations, as shown in Figure 4–34. In the figure you can see that the Boolean expression, the truth table, and the Karnaugh map are simply different ways to represent a logic function.

FIGURE 4–34
Example of mapping directly from a truth table to a Karnaugh map.

$$X = \overline{A}\,\overline{B}\,\overline{C} + A\overline{B}\,\overline{C} + AB\overline{C} + ABC$$

Inputs			Output
A	B	C	X
0	0	0	1
0	0	1	0
0	1	0	0
0	1	1	0
1	0	0	1
1	0	1	0
1	1	0	1
1	1	1	1

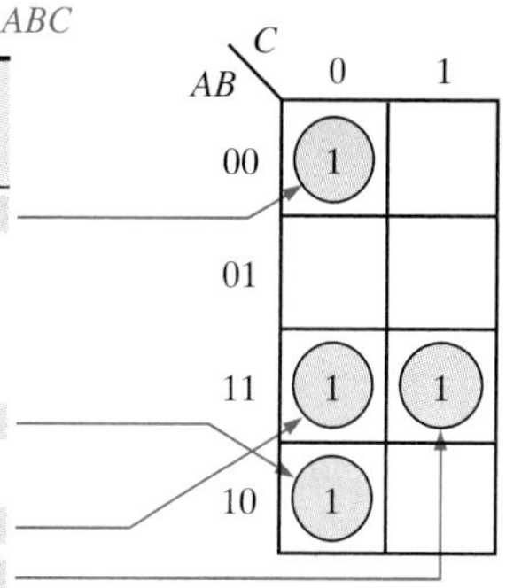

AB \ C	0	1
00	1	
01		
11	1	1
10	1	

"Don't Care" Conditions

Sometimes a situation arises in which some input variable combinations are not allowed. For example, recall that in the BCD code covered in Chapter 2, there are six invalid combinations: 1010, 1011, 1100, 1101, 1110, and 1111. Since these unallowed states will never occur in an application involving the BCD code, they can be treated as **"don't care"** terms with respect to their effect on the output. That is, for these "don't care" terms either a 1 or a 0 may be assigned to the output; it really does not matter since they will never occur.

The "don't care" terms can be used to advantage on the Karnaugh map. Figure 4–35 shows that for each "don't care" term, an X is placed in the cell. When grouping the 1s, Xs can be treated as 1s to make a larger grouping or as 0s if they cannot be used to advantage. The larger a group, the simpler the resulting term will be.

The truth table in Figure 4–35(a) describes a logic function that has a 1 output only when the BCD code for 7, 8, or 9 is present on the inputs. Taking advantage of the "don't

Inputs	Output
$ABCD$	Y
0 0 0 0	0
0 0 0 1	0
0 0 1 0	0
0 0 1 1	0
0 1 0 0	0
0 1 0 1	0
0 1 1 0	0
0 1 1 1	1
1 0 0 0	1
1 0 0 1	1
1 0 1 0	X
1 0 1 1	X
1 1 0 0	X
1 1 0 1	X
1 1 1 0	X
1 1 1 1	X

Don't cares

(a) Truth table

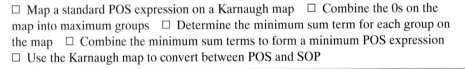

(b) Without "don't cares" $Y = \overline{A}\,\overline{B}C + \overline{A}BCD$
  With "don't cares" $Y = A + BCD$

FIGURE 4–35
Example of the use of "don't care" conditions to simplify an expression.

cares" and using them as 1s, the resulting expression for the function is $A + BCD$, as indicated. If the "don't cares" are not used as 1s, the resulting expression is $A\overline{B}\,\overline{C} + \overline{A}BCD$. So you can see the advantage of using "don't care" terms to get the simplest expression.

SECTION 4–9 REVIEW

1. Lay out Karnaugh maps for three and four variables.
2. Group the 1s and write the simplified SOP expression for the Karnaugh map in Figure 4–24.
3. Write the original standard SOP expressions for each of the Karnaugh maps in Figure 4–31.

4–10 ■ KARNAUGH MAP POS MINIMIZATION

In the last section, you studied the minimization of an SOP expression using a Karnaugh map. In this section, we will focus on POS expressions. The approaches are much the same except that with POS expressions, 0s representing the standard sum terms are placed on the Karnaugh map instead of 1s. This section can be omitted without impact on future material. After completing this section, you should be able to

☐ Map a standard POS expression on a Karnaugh map   ☐ Combine the 0s on the map into maximum groups   ☐ Determine the minimum sum term for each group on the map   ☐ Combine the minimum sum terms to form a minimum POS expression   ☐ Use the Karnaugh map to convert between POS and SOP

Mapping a Standard POS Expression

For a POS expression in standard form, a 0 is placed on the Karnaugh map for each sum term in the expression. Each 0 is placed in a cell corresponding to the value of a sum term. For example, for the sum term $A + \overline{B} + C$, a 0 goes in the 010 cell on a 3-variable map.

When a POS expression is completely mapped, there will be a number of 0s on the Karnaugh map equal to the number of sum terms in the standard POS expression. The cells that do not have a 0 are the cells for which the expression is 1. Usually, when working with POS expressions, the 1s are left off. The following steps and the illustration in Figure 4–36 show the mapping process.

Step 1. Determine the binary value of each sum term in the standard POS expression. This is the binary value that makes the term equal to 0.

Step 2. As each sum term is evaluated, place a 0 on the Karnaugh map in the corresponding cell.

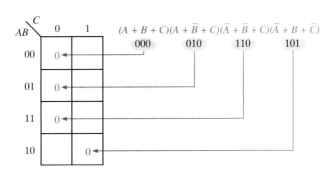

FIGURE 4–36
Example of mapping a standard POS expression.

Example 4–30 will further illustrate the mapping process.

EXAMPLE 4–30

Map the following standard POS expression on a Karnaugh map:

$$(\bar{A} + \bar{B} + C + D)(\bar{A} + B + \bar{C} + \bar{D})(A + B + \bar{C} + D)(\bar{A} + \bar{B} + \bar{C} + \bar{D})(A + B + \bar{C} + \bar{D})$$

Solution The expression is evaluated as shown below and a 0 is placed on the 4-variable Karnaugh map in Figure 4–37 for each standard sum term in the expression.

$$(\bar{A} + \bar{B} + C + D)(\bar{A} + B + \bar{C} + \bar{D})(A + B + \bar{C} + D)(\bar{A} + \bar{B} + \bar{C} + \bar{D})(A + B + \bar{C} + \bar{D})$$
$$\quad 1100 \qquad\qquad 1011 \qquad\qquad 0010 \qquad\qquad 1111 \qquad\qquad 0011$$

FIGURE 4–37

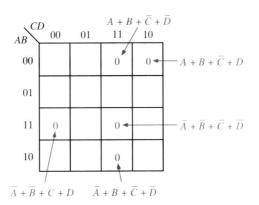

Related Exercise Map the following standard POS expression on a Karnaugh map:

$$(A + \bar{B} + \bar{C} + D)(A + B + C + \bar{D})(A + B + C + D)(\bar{A} + B + \bar{C} + D)$$

Karnaugh Map Simplification of POS Expressions

The process for minimizing a POS expression is basically the same as for an SOP expression except that you group 0s to produce minimum sum terms instead of grouping 1s to produce minimum product terms. The rules for grouping the 0s are the same as those for grouping the 1s that you learned in Section 4–9.

EXAMPLE 4–31   Use a Karnaugh map to minimize the following standard POS expression:

$$(A + B + C)(A + B + \overline{C})(A + \overline{B} + C)(A + \overline{B} + \overline{C})(\overline{A} + \overline{B} + C)$$

Solution   The combinations of binary values of the expression are

$$(0 + 0 + 0)(0 + 0 + 1)(0 + 1 + 0)(0 + 1 + 1)(1 + 1 + 0)$$

The standard POS expression is mapped and the cells are grouped as shown in Figure 4–38.

FIGURE 4–38

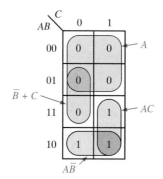

Notice how the 0 in the 110 cell is included into a 2-cell group by utilizing the 0 in the 4-cell group. The sum terms for each group are shown in the figure and the resulting minimum POS expression is

$$A(\overline{B} + C)$$

Keep in mind that this minimum POS expression is equivalent to the original standard POS expression.

Grouping the 1s as shown yields an SOP expression that is equivalent to grouping the 0s:

$$AC + A\overline{B} = A(\overline{B} + C)$$

Related Exercise   Use a Karnaugh map to simplify the following standard POS expression:

$$(X + \overline{Y} + Z)(X + \overline{Y} + \overline{Z})(\overline{X} + \overline{Y} + Z)(\overline{X} + Y + Z)$$

EXAMPLE 4–32   Use a Karnaugh map to minimize the following POS expression:

$$(B + C + D)(A + B + \overline{C} + D)(\overline{A} + B + C + \overline{D})(A + \overline{B} + C + D)(\overline{A} + \overline{B} + C + D)$$

Solution The first term must be expanded into $\overline{A} + B + C + D$ and $A + B + C + D$ to get a standard POS expression, which is then mapped; and the cells are grouped as shown in Figure 4–39. The sum terms for each group are shown and the resulting minimum POS expression is

$$(C + D)(A + B + D)(\overline{A} + B + C)$$

Keep in mind that this minimum POS expression is equivalent to the original standard POS expression.

FIGURE 4–39

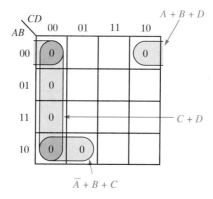

Related Exercise Use a Karnaugh map to simplify the following POS expression:

$$(W + \overline{X} + Y + \overline{Z})(W + X + Y + Z)(W + \overline{X} + \overline{Y} + Z)(\overline{W} + \overline{X} + Z)$$

Converting Between POS and SOP Using the Karnaugh Map

When a POS expression is mapped, it can easily be converted to the equivalent SOP form directly from the Karnaugh map. Also, given a mapped SOP expression, an equivalent POS expression can be derived directly from the map. This provides a good way to compare both minimum forms of an expression to determine if one of them can be implemented with fewer gates than the other.

For a POS expression, all the cells that do not contain 0s contain 1s, from which the SOP expression is derived. Likewise, for an SOP expression, all the cells that do not contain 1s contain 0s, from which the POS expression is derived. Example 4–33 illustrates this conversion.

EXAMPLE 4–33

Using a Karnaugh map, convert the following standard POS expression into a minimum POS expression, a standard SOP expression, and a minimum SOP expression.

$$(\overline{A} + \overline{B} + C + D)(A + \overline{B} + C + D)(A + B + C + \overline{D})$$
$$(A + B + \overline{C} + D)(\overline{A} + B + C + \overline{D})(A + B + \overline{C} + D)$$

Solution The 0s for the standard POS expression are mapped and grouped to obtain the minimum POS expression in Figure 4–40(a). In Figure 4–40(b), 1s are added to the cells that do not contain 0s. From each cell containing a 1, a standard product term is obtained as indicated. These product terms form the standard SOP expression. In Figure 4–40(c), the 1s are grouped and a minimum SOP expression is obtained.

FIGURE 4–40

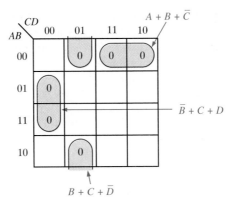

(a) Minimum POS: $(A + B + \bar{C})(\bar{B} + C + D)(B + C + \bar{D})$

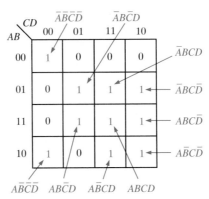

(b) Standard SOP:
$\bar{A}\bar{B}\bar{C}\bar{D} + \bar{A}B\bar{C}D + \bar{A}BCD + \bar{A}BC\bar{D} + ABC\bar{D} + A\bar{B}C\bar{D} + A\bar{B}\bar{C}\bar{D} + AB\bar{C}D + A\bar{B}CD + ABCD$

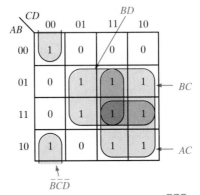

(c) Minimum SOP: $AC + BC + BD + \bar{B}\bar{C}\bar{D}$

Related Exercise Use a Karnaugh map to convert the following expression to minimum SOP form:

$$(W + \bar{X} + Y + \bar{Z})(\bar{W} + X + \bar{Y} + \bar{Z})(\bar{W} + \bar{X} + \bar{Y} + Z)(\bar{W} + \bar{X} + \bar{Z})$$

SECTION 4–10 REVIEW

1. What is the difference in mapping a POS expression and an SOP expression?
2. What is the standard sum term for a 0 in cell 1011?
3. What is the standard product term for a 1 in cell 0010?

4–11 ■ FIVE-VARIABLE KARNAUGH MAPS

Boolean functions with five variables can be simplified using a 32-cell Karnaugh map. Actually, two 4-variable maps (16 cells each) are used to construct a 5-variable map. You already know the cell adjacencies within each of the 4-variable maps and how to form groups of cells containing 1s to simplify an SOP expression. All you need to learn for five variables is the cell adjacencies between the two 4-variable maps and how to group those adjacent 1s. This section can be omitted without affecting future material. After completing this section, you should be able to

☐ Determine cell adjacencies in a 5-variable map   ☐ Form maximum cell groupings in a 5-variable map   ☐ Minimize 5-variable Boolean expressions using the Karnaugh map

A Karnaugh map for five variables *(ABCDE)* can be constructed using two 4-variable maps with which you are already familiar. Each map contains 16 cells with all combinations of variables *B, C, D,* and *E.* One map is for $A = 0$ and the other is for $A = 1$, as shown in Figure 4–41.

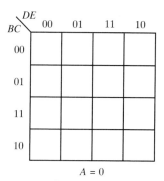
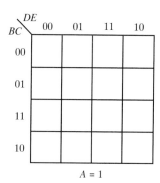

FIGURE 4–41
A 5-variable Karnaugh map.

Cell Adjacencies

You already know how to determine adjacent cells within the 4-variable map. The best way to visualize cell adjacencies between the two 16-cell maps is to imagine that the $A = 0$ map is placed on top of the $A = 1$ map. Each cell in the $A = 0$ map is adjacent to the cell directly below it in the $A = 1$ map.

To illustrate, an example with four groups is shown in Figure 4–42 with the maps in a 3-dimensional arrangement. The 1s in the yellow cells form an 8-bit group (four in the $A = 0$ map combined with four in the $A = 1$ map). The 1s in the orange cells form a 4-bit group. The 1s in the light red cells form a 4-bit group only in the $A = 0$ map. The 1 in the gray cell in the $A = 1$ map is grouped with the 1 in the lower right light red cell in the $A = 0$ map to form a 2-bit group.

The Simplified Boolean Expression

The original SOP Boolean expression that is plotted on the Karnaugh map in Figure 4–42 contains seventeen 5-variable terms because there are seventeen 1s on the map. As you know, only the variables that do not change within a group remain in the expression for that group. The simplified expression taken from the map is developed as follows:

- The term for the yellow group is $D\overline{E}$.
- The term for the orange group is $\overline{B}CE$.

FIGURE 4–42
Illustration of groupings of 1s in adjacent cells of a 5-variable map.

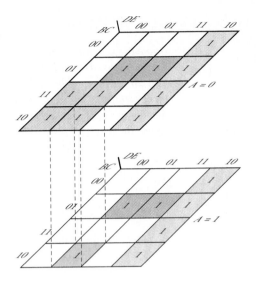

- The term for the light red group is $\overline{A}B\overline{D}$.
- The term for the gray cell grouped with the red cell is $B\overline{C}\,\overline{D}E$.

Combining these terms into the simplified SOP expression yields

$$X = D\overline{E} + \overline{B}CE + \overline{A}B\overline{D} + B\overline{C}\,\overline{D}E$$

EXAMPLE 4–34   Use a Karnaugh map to minimize the following standard SOP 5-variable expression:

$$X = \overline{A}\,\overline{B}\,\overline{C}\,\overline{D}\,\overline{E} + \overline{A}\,\overline{B}CD\,\overline{E} + \overline{A}BC\overline{D}\,\overline{E} + \overline{A}B\overline{C}\,\overline{D}\,\overline{E} + \overline{A}\,\overline{B}\,\overline{C}\,DE + \overline{A}BCDE$$
$$+ \overline{A}BCDE + A\overline{B}\,\overline{C}\,\overline{D}\,\overline{E} + A\overline{B}\,\overline{C}\,DE + ABC\overline{D}E + ABCDE + A\overline{B}CDE$$

Solution   The SOP expression is mapped in Figure 4–43 and the groupings and their corresponding terms are indicated. Combining the terms yields the following minimized SOP expression:

$$X = \overline{A}\,\overline{D}\,\overline{E} + \overline{B}\,\overline{C}\,\overline{D} + BCE + ACDE$$

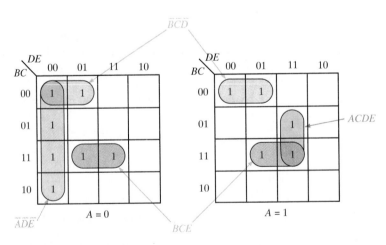

FIGURE 4–43

Related Exercise   Minimize the following expression:

$$Y = \overline{A}\,\overline{B}\,\overline{C}\,\overline{D}\,\overline{E} + \overline{A}\,\overline{B}CD\,\overline{E} + \overline{A}BC\overline{D}\,\overline{E} + \overline{A}B\overline{C}\,\overline{D}\,\overline{E} + A\overline{B}\,\overline{C}\,\overline{D}\,\overline{E} + \overline{A}BC\overline{D}\,\overline{E} + ABC\overline{D}\,\overline{E} + AB\overline{C}\,\overline{D}\,\overline{E}$$
$$+ \overline{A}\,\overline{B}\,\overline{C}D\overline{E} + \overline{A}\,\overline{B}CD\overline{E} + \overline{A}BCD\overline{E} + \overline{A}BC\overline{D}\overline{E} + A\overline{B}\,\overline{C}D\overline{E} + A\overline{B}CD\overline{E} + ABCD\overline{E} + AB\overline{C}D\overline{E}$$

4–12 ■ DIGITAL SYSTEM APPLICATION

Seven-segment displays are used in everything from automobile instruments to Z-meters. The tablet counting and control system that has been the focus of the system applications in the previous chapters, has two 7-segment displays. These displays are used with logic circuits that decode a binary coded decimal (BCD) number and activate the appropriate digits on the display. In this section, we will focus on a minimum-gate design using decoder logic to illustrate applications of Boolean expressions and the Karnaugh map. After completing this section, you should be able to

□ Appreciate the approach to the design of logic circuits   □ Explain the operation of a 7-segment display   □ Translate a specified requirement into a Boolean expression and/or truth table and then into a minimum-gate logic circuit using a Karnaugh map

The 7-Segment Display

Figure 4–44 shows a common display format composed of seven elements or segments. Energizing certain combinations of these segments can cause each of the ten decimal digits to be displayed. Figure 4–45 illustrates this method of digital display for each of the ten digits by using a red segment to represent one that is energized. To produce a 1, segments *b* and *c* are energized; to produce a 2, segments *a, b, g, e,* and *d* are used; and so on.

FIGURE 4–44
Seven-segment display format showing arrangement of segments.

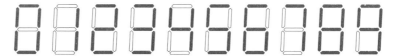

FIGURE 4–45
Display of decimal digits with a 7-segment device.

LED Displays   One common type of 7-segment display consists of light-emitting diodes **(LED)** arranged as shown in Figure 4–46. Each segment is an LED that emits light when there is current through it. In Figure 4–46(a) the common-anode arrangement requires the driving circuit to provide a low-level voltage in order to activate a given segment. When a LOW is applied to a segment input, the LED is turned on, and there is current through it. In Figure 4–46(b) the common-cathode arrangement requires the driver to provide a high-level voltage to activate a segment. When a HIGH is applied to a segment input, the LED is turned on and there is current through it.

LCD Displays   Another common type of 7-segment display is the liquid crystal display **(LCD).** LCDs operate by polarizing light so that a nonactivated segment reflects incident light and thus appears invisible against its background. An activated segment does not re-

FIGURE 4–46
Arrangements of 7-segment LED displays.

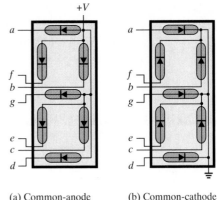

(a) Common-anode   (b) Common-cathode

flect incident light and thus appears dark. LCDs consume much less power than LEDs but cannot be seen in the dark, while LEDs can.

LCDs operate from a low-frequency signal voltage (30 Hz to 60 Hz) applied between the segment and a common element called the *backplane* (bp). The basic operation is as follows. Figure 4–47 shows a square wave used as the source signal. Each segment in the display is driven by an exclusive-OR gate with one input connected to an output of the 7-segment decoder/driver and the other input connected to the signal source. When the decoder/driver output is HIGH (1), the exclusive-OR output is a square wave that is 180° out-of-phase with the source signal, as shown in Figure 4–47(a). You can verify this by reviewing the truth table operation of the exclusive-OR. The resulting voltage between the LCD segment and the backplane is also a square wave because when $V_{seg} = 1$, $V_{bp} = 0$, and vice versa. The voltage difference turns the segment on.

When the decoder/driver output is LOW (0), the exclusive-OR output is a square wave that is in-phase with the source signal, as shown in Figure 4–47(b). The resulting voltage difference between the segment and the backplane is 0 because $V_{seg} = V_{bp}$. This condition turns the segment off.

FIGURE 4–47
Basic operation of an LCD.

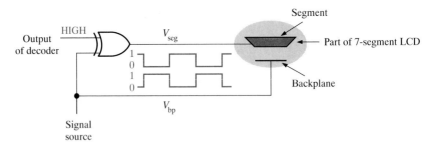

(a) Segment activated (on)

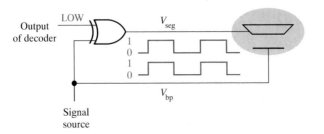

(b) Segment not activated (off)

For driving LCDs, TTL is not recommended, because its low-level voltage is typi-cally a few tenths of a volt, thus creating a dc component across the LCD, which degrades its performance. Therefore, CMOS is used in LCD applications.

Segment Decoding Logic

Each segment is used for various decimal digits, but no one segment is used for all ten dig-its. Therefore each segment must be activated by its own decoding circuit that detects the occurrence of any of the numbers in which the segment is used. From Figures 4–44 and 4–45, the segments that are required to be activated for each digit are determined and listed in Table 4–9.

TABLE 4–9
Active segments for each decimal digit

Digit	Segments Activated
0	a, b, c, d, e, f
1	b, c
2	a, b, d, e, g
3	a, b, c, d, g
4	b, c, f, g
5	a, c, d, f, g
6	a, c, d, e, f, g
7	a, b, c
8	a, b, c, d, e, f, g
9	a, b, c, d, f, g

Truth Table for the Segment Logic   The segment decoding logic requires four binary coded decimal (BCD) inputs and seven outputs, one for each segment in the display, as indi-cated in the block diagram of Figure 4–48. The multiple-output truth table, shown in Table 4–10, is actually seven truth tables in one and could be separated into a separate table for each segment. A 1 in the segment output columns of the table indicates an activated segment.

Since the BCD code does not include the binary values 1010, 1011, 1100, 1101, 1110, and 1111, these combinations will never appear on the inputs and can therefore be treated as "don't care" (X) conditions, as indicated in the truth table. To conform with the practice of most IC manufacturers, A represents the least significant bit and D represents the most significant bit in this application.

FIGURE 4–48
Block diagram of 7-segment logic and display.

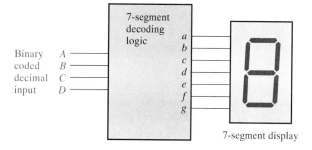

Boolean Expressions for the Segment Logic   From the truth table, a standard SOP or POS expression can be written for each segment. For example, the standard SOP expres-sion for segment a is

$$a = \overline{D}\,\overline{C}\,\overline{B}\,\overline{A} + \overline{D}\,\overline{C}B\overline{A} + \overline{D}\,\overline{C}BA + \overline{D}C\overline{B}A + \overline{D}CB\overline{A} + \overline{D}CBA + D\overline{C}\,\overline{B}\,\overline{A} + D\overline{C}\,\overline{B}A$$

TABLE 4–10

Truth table for 7-segment logic

Decimal Digit	Inputs				Segment Outputs						
	D	*C*	*B*	*A*	*a*	*b*	*c*	*d*	*e*	*f*	*g*
0	0	0	0	0	1	1	1	1	1	1	0
1	0	0	0	1	0	1	1	0	0	0	0
2	0	0	1	0	1	1	0	1	1	0	1
3	0	0	1	1	1	1	1	1	0	0	1
4	0	1	0	0	0	1	1	0	0	1	1
5	0	1	0	1	1	0	1	1	0	1	1
6	0	1	1	0	1	0	1	1	1	1	1
7	0	1	1	1	1	1	1	0	0	0	0
8	1	0	0	0	1	1	1	1	1	1	1
9	1	0	0	1	1	1	1	1	0	1	1
10	1	0	1	0	X	X	X	X	X	X	X
11	1	0	1	1	X	X	X	X	X	X	X
12	1	1	0	0	X	X	X	X	X	X	X
13	1	1	0	1	X	X	X	X	X	X	X
14	1	1	1	0	X	X	X	X	X	X	X
15	1	1	1	1	X	X	X	X	X	X	X

Output = 1 means segment is activated (on)

Output = 0 means segment is not activated (off)

Output = X means "don't care"

and the standard SOP expression for segment *e* is

$$e = \overline{D}\,\overline{C}\,\overline{B}\,\overline{A} + \overline{D}\,\overline{C}B\overline{A} + \overline{D}CB\overline{A} + D\overline{C}\,\overline{B}\,\overline{A}$$

Expressions for the other segments can be similarly developed. As you can see, the expression for segment *a* has eight product terms and the expression for segment *e* has four product terms representing each of the BCD inputs that activate that segment. This means that the standard SOP implementation of segment-*a* logic requires an AND-OR circuit consisting of eight 4-input AND gates and one 8-input OR gate. The implementation of segment-*e* logic requires four 4-input AND gates and one 4-input OR gate. In both cases, four inverters are required to produce the complement of each variable.

Karnaugh Map Minimization of the Segment Logic   We will begin by obtaining a minimum SOP expression for segment *a*. A Karnaugh map for segment *a* is shown in Figure 4–49 and the following steps are carried out:

Step 1.   The 1s are mapped directly from Table 4–10.
Step 2.   All of the "don't cares" (X) are placed on the map.
Step 3.   The 1s are grouped as shown. "Don't cares" and overlapping of cells are utilized to form the largest groups possible.
Step 4.   Write the minimum product term for each group and sum the terms to form the minimum SOP expression.

Keep in mind that "don't cares" do not have to be included in a group, but in this case all of them are used. Also, notice that the 1s in the corner cells are grouped with a "don't care" using the "wrap around" adjacency of the corner cells.

Standard SOP expression:
$$\overline{D}\,\overline{C}\,\overline{B}\,\overline{A} + \overline{D}\,\overline{C}B\overline{A} + \overline{D}\,C\overline{B}A + \overline{D}\,CB\overline{A} + D\,\overline{C}\,\overline{B}\,\overline{A} + D\,\overline{C}BA + D\,C\overline{B}\,\overline{A} + D\,C\overline{B}A$$

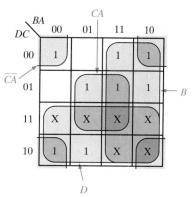

Minimum SOP expression: $D + B + CA + \overline{C}\,\overline{A}$

FIGURE 4–49
Karnaugh map minimization of the segment-a logic expression.

Minimum Implementation of Segment-a Logic The minimum SOP expression taken from the Karnaugh map in Figure 4–49 for the segment-*a* logic is

$$D + B + CA + \overline{C}\,\overline{A}$$

This expression can be implemented with two 2-input AND gates, one 4-input OR gate and two inverters as shown in Figure 4–50. Compare this to the standard SOP implementation for segment-*a* logic discussed earlier; you'll see that the number of gates and inverters has been reduced from thirteen to five and, as a result, the number of interconnections has been significantly reduced.

The minimum logic for each of the remaining six segments (*b, c, d, e, f,* and *g)* can be obtained with a similar approach.

FIGURE 4–50
The minimum logic implementation for segment a of the 7-segment display.

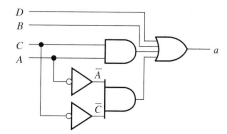

Further Multiple-Output Logic Simplification Once the minimum logic for each individual segment has been developed, there is a possibility of additional simplification by elimination of any duplicate product terms or complemented single-variable terms. For example, if the minimum SOP for segment *a* has a product term that is also found in, say, the minimum SOP for segment *b*, then that duplicate term can be implemented with only one AND gate with its output commonly connected to the inputs of the OR gates for the two segments. Likewise, the complement of a variable needs only to be implemented with one inverter and then the output of that inverter can be connected where it is needed provided that its fan-out is not exceeded.

▪ **Activity 1** Determine the minimum logic for segment *b*.
▪ **Activity 2** Determine the minimum logic for segment *c*.
▪ **Activity 3** Determine the minimum logic for segment *d*.
▪ **Activity 4** Determine the minimum logic for segment *e*.
▪ **Activity 5** Determine the minimum logic for segment *f*.
▪ **Activity 6** Determine the minimum logic for segment *g*.
▪ **Activity 7** Determine if there are any duplicate product terms in the 7-segment logic expressions and, if so, eliminate all but one of the duplicate AND gates from the overall implementation.
▪ **Activity 8** Draw a logic diagram for the complete 7-segment decoding logic by combining all of the individual segment logic circuits and eliminating any duplicate gates or inverters.
▪ **Activity 9** Specify by type number the specific integrated circuits required to implement the logic diagram in Activity 8 using only those that have been introduced up to this point. Utilize all available gates in a package where possible to keep the number of ICs to a minimum.

SECTION 4–12 REVIEW

1. What is the advantage of using "don't care" conditions in a given application?
2. How many duplicate terms did you find in the segment logic in Activity 7?
3. How many gates and inverters were you able to eliminate as a result of duplication in Activity 8?

▪ **SUMMARY**

▪ Gate symbols and Boolean expressions for the outputs of an inverter and two-input gates are shown in Figure 4–51.

▪ Commutative laws: $A + B = B + A$
$AB = BA$

▪ Associative laws: $A + (B + C) = (A + B) + C$
$A(BC) = (AB)C$

▪ Distributive law: $A(B + C) = AB + AC$

▪ Boolean rules:
1. $A + 0 = A$
2. $A + 1 = 1$
3. $A \cdot 0 = 0$
4. $A \cdot 1 = A$
5. $A + A = A$
6. $A + \overline{A} = 1$
7. $A \cdot A = A$
8. $A \cdot \overline{A} = 0$
9. $\overline{\overline{A}} = A$
10. $A + AB = A$
11. $A + \overline{A}B = A + B$
12. $(A + B)(A + C) = A + BC$

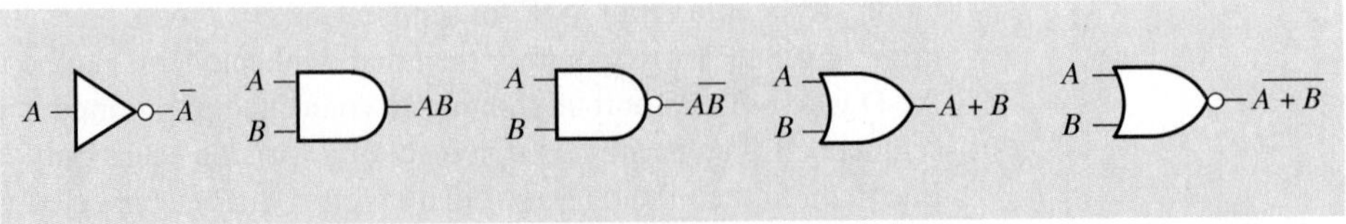

FIGURE 4–51

▪ DeMorgan's theorems:

1. The complement of a product is equal to the sum of the complements of the terms in the product:

$$\overline{XY} = \overline{X} + \overline{Y}$$

2. The complement of a sum is equal to the product of the complements of the terms in the sum:

$$\overline{X + Y} = \overline{X}\,\overline{Y}$$

▪ Karnaugh maps for 3 and 4 variables are shown in Figure 4–52. A 5-variable map is formed from two 4-variable maps.

FIGURE 4–52

3-variable   4-variable

■ **SELF-TEST**

1. The complement of a variable is always

(a) 0   (b) 1   (c) equal to the variable   (d) the inverse of the variable

2. The Boolean expression $A + \overline{B} + C$ is

(a) a sum term   (b) a literal term   (c) a product term   (d) a complemented term

3. The Boolean expression $A\overline{B}C\overline{D}$ is

(a) a sum term   (b) a product term   (c) a literal term   (d) always 1

4. The domain of the expression $A\overline{B}CD + A\overline{B} + \overline{C}D + B$ is

(a) A and D (b) B only   (c) A, B, C, and D (d) none of these

5. According to the commutative law of addition,

(a) $AB = BA$ (b) $A = A + A$

(c) $A + (B + C) = (A + B) + C$ (d) $A + B = B + A$

6. According to the associative law of multiplication,

(a) $B = BB$ (b) $A(BC) = (AB)C$ (c) $A + B = B + A$ (d) $B + B(B + 0)$

7. According to the distributive law,

(a) $A(B + C) = AB + AC$ (b) $A(BC) = ABC$ (c) $A(A + 1) = A$ (d) $A + AB = A$

8. Which one of the following is *not* a valid rule of Boolean algebra?

(a) $A + 1 = 1$ (b) $A = \overline{A}$ (c) $AA = A$ (d) $A + 0 = A$

9. Which of the following rules states that if one input of an AND gate is always 1, the output is equal to the other input?

(a) $A + 1 = 1$ (b) $A + A = A$ (c) $A \cdot A = A$ (d) $A \cdot 1 = A$

10. According to DeMorgan's theorems, the following equality(s) are correct:

(a) $\overline{AB} = \overline{A} + \overline{B}$ (b) $\overline{XYZ} = \overline{X} + \overline{Y} + \overline{Z}$

(c) $\overline{A + B + C} = \overline{A}\,\overline{B}\,\overline{C}$ (d) all of these

11. The Boolean expression $X = AB + CD$ represents

(a) two ORs ANDed together   (b) a 4-input AND gate

(c) two ANDs ORed together   (d) an exclusive-OR

12. An example of a sum-of-products expression is
   (a) $A + B(C + D)$ **(b)** $\overline{A}B + A\overline{C} + AB\overline{C}$
   (c) $(\overline{A} + B + C)(A + \overline{B} + C)$ **(d)** both answers (a) and (b)

13. An example of a product-of-sums expression is
   (a) $A(B + C) + A\overline{C}$ **(b)** $(A + B)(\overline{A} + B + \overline{C})$
   (c) $\overline{A} + \overline{B} + BC$ **(d)** both answers (a) and (b)

14. An example of a standard SOP expression is
   (a) $\overline{A}B + A\overline{B}C + AB\overline{D}$ **(b)** $A\overline{B}C + A\overline{C}D$
   (c) $A\overline{B} + \overline{A}B + AB$ **(d)** $AB\overline{C}D + \overline{A}B + \overline{A}$

15. A 3-variable Karnaugh map has
   (a) eight cells   **(b)** three cells   **(c)** sixteen cells   **(d)** four cells

16. In a 4-variable Karnaugh map, a 2-variable product term is produced by
   (a) a 2-cell group of 1s   **(b)** an 8-cell group of 1s
   (c) a 4-cell group of 1s   **(d)** a 4-cell group of 0s

17. On a Karnaugh map, grouping the 0s produces
   (a) a product-of-sums expression   **(b)** a sum-of-products expression
   (c) a "don't care" condition   **(d)** AND-OR logic

18. A 5-variable Karnaugh map has
   (a) sixteen cells   **(b)** thirty-two cells   **(c)** sixty-four cells

■ PROBLEMS

SECTION 4–1   Boolean Operations and Expressions

1. Using Boolean notation, write an expression that is a 1 whenever one or more of its variables (A, B, C, and D) are 1s.

2. Write an expression that is a 1 only if all of its variables (A, B, C, D, and E) are 1s.

3. Write an expression that is a 1 when one or more of its variables (A, B, and C) are 0s.

4. Evaluate the following operations:
   (a) $0 + 0 + 1$ **(b)** $1 + 1 + 1$ **(c)** $1 \cdot 0 \cdot 0$
   (d) $1 \cdot 1 \cdot 1$ **(e)** $1 \cdot 0 \cdot 1$ **(f)** $1 \cdot 1 + 0 \cdot 1 \cdot 1$

5. Find the values of the variables that make each product term 1 and each sum term 0.
   (a) AB **(b)** $A\overline{B}C$ **(c)** $A + B$ **(d)** $\overline{A} + B + \overline{C}$
   (e) $\overline{A} + \overline{B} + C$ **(f)** $\overline{A} + B$ **(g)** $A\overline{B}\,\overline{C}$

6. Find the value of X for all possible values of the variables.
   (a) $X = (A + B)C + B$ **(b)** $X = \overline{(A + B)}C$ **(c)** $X = A\overline{B}C + AB$
   (d) $X = (A + B)(\overline{A} + B)$ **(e)** $X = (A + BC)(\overline{B} + \overline{C})$

SECTION 4–2   Laws and Rules of Boolean Algebra

7. Identify the law of Boolean algebra upon which each of the following equalities is based:
   (a) $A\overline{B} + CD + A\overline{C}D + B = B + A\overline{B} + A\overline{C}D + CD$
   (b) $AB\overline{C}D + \overline{ABC} = D\overline{C}BA + \overline{CBA}$
   (c) $AB(CD + E\overline{F} + GH) = ABCD + ABE\overline{F} + ABGH$

8. Identify the Boolean rule(s) on which each of the following equalities is based:
   (a) $\overline{\overline{AB + CD}} + \overline{EF} = AB + CD + \overline{EF}$ **(b)** $AAB + AB\overline{C} + AB\overline{B} = AB\overline{C}$
   (c) $A(BC + BC) + AC = A(BC) + AC$ **(d)** $AB(C + \overline{C}) + AC = AB + AC$
   (e) $A\overline{B} + A\overline{B}C = A\overline{B}$ **(f)** $ABC + \overline{AB} + \overline{AB}CD = ABC + \overline{AB} + D$

SECTION 4–3   DeMorgan's Theorems

9. Apply DeMorgan's theorems to each expression:
   (a) $\overline{A + \overline{B}}$ **(b)** $\overline{\overline{AB}}$ **(c)** $\overline{A + B + C}$ **(d)** $\overline{ABC}$
   (e) $\overline{A(B + C)}$ **(f)** $\overline{AB + CD}$ **(g)** $\overline{AB + CD}$ **(h)** $\overline{(A + \overline{B})(\overline{C} + D)}$

10. Apply DeMorgan's theorems to each expression:

(a) $\overline{A\overline{B}(C + \overline{D})}$

(b) $\overline{AB(CD + EF)}$

(c) $\overline{(A + \overline{B} + C + \overline{D}) + ABC\overline{D}}$

(d) $\overline{(\overline{A} + B + C + D)(A\overline{B}\ \overline{C}D)}$

(e) $\overline{\overline{AB}(CD + \overline{E}F)(\overline{AB} + \overline{CD})}$

11. Apply DeMorgan's theorems to the following:

(a) $\overline{\overline{(ABC)(EFG)} + \overline{(HIJ)(\overline{KLM})}}$

(b) $\overline{(A + B\overline{C} + CD) + B\overline{C}}$

(c) $\overline{\overline{(A + B)(\overline{C + D})(E + F)(\overline{G + H})}}$

SECTION 4–4 Boolean Analysis of Logic Circuits

12. Write the Boolean expression for each of the logic gates in Figure 4–53.

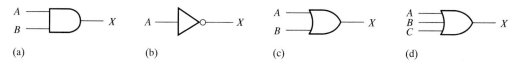

(a)   (b)   (c)   (d)

FIGURE 4–53

13. Write the Boolean expression for each of the logic circuits in Figure 4–54.

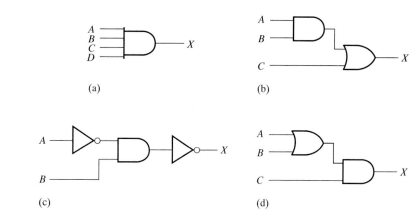

(a)   (b)

(c)   (d)

FIGURE 4–54

14. Draw the logic circuit represented by each of the following expressions:

(a) $A + B + C$ (b) ABC (c) $AB + C$ (d) $AB + CD$

15. Draw the logic circuit represented by each expression:

(a) $A\overline{B} + \overline{A}B$ (b) $AB + \overline{A}\ \overline{B} + \overline{A}BC$

(c) $\overline{A}B(C + \overline{D})$ (d) $A + B[C + D(B + \overline{C})]$

16. Construct a truth table for each of the following Boolean expressions:

(a) $A + B$ (b) AB (c) $AB + BC$

(d) $(A + B)C$ (e) $(A + B)(\overline{B} + C)$

SECTION 4–5 Simplification Using Boolean Algebra

17. Using Boolean algebra techniques, simplify the following expressions as much as possible:

(a) $A(A + B)$ (b) $A(\overline{A} + AB)$ (c) $BC + \overline{B}C$

(d) $A(A + \overline{A}B)$ (e) $A\overline{B}C + \overline{A}BC + \overline{A}\ \overline{B}C$

18. Using Boolean algebra, simplify the following expressions:

(a) $(A + \bar{B})(A + C)$ (b) $\overline{AB} + \overline{AB}\bar{C} + \overline{AB}CD + \overline{AB}\bar{C}\,\overline{DE}$

(c) $AB + \overline{ABC} + A$ (d) $(A + \bar{A})(AB + AB\bar{C})$

(e) $AB + (\bar{A} + \bar{B})C + AB$

19. Using Boolean algebra, simplify each expression:

(a) $BD + B(D + E) + \bar{D}(D + F)$ (b) $\bar{A}\,\bar{B}C + (A + B + \bar{C}) + \bar{A}\,\bar{B}\,\bar{C}\,D$

(c) $(B + BC)(B + \bar{B}C)(B + D)$ (d) $ABCD + AB(\overline{CD}) + (\overline{AB})CD$

(e) $ABC[AB + \bar{C}(BC + AC)]$

20. Determine which of the logic circuits in Figure 4–55 are equivalent.

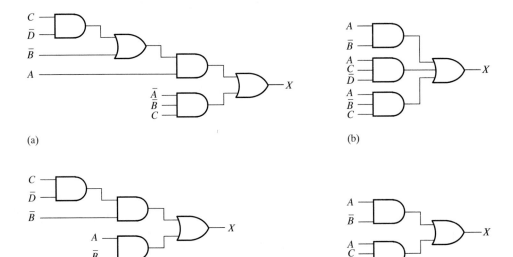

(a)   (b)

(c)   (d)

FIGURE 4–55

SECTION 4–6   Standard Forms of Boolean Expressions

21. Convert the following expressions to sum-of-product (SOP) forms:

(a) $(A + B)(C + \bar{B})$ (b) $(A + \bar{B}C)C$ (c) $(A + C)(AB + AC)$

22. Convert the following expressions to sum-of-product (SOP) forms:

(a) $AB + CD(A\bar{B} + CD)$ (b) $AB(\bar{B}\,\bar{C} + BD)$ (c) $A + B[AC + (B + \bar{C})D]$

23. Define the domain of each SOP expression in Problem 21 and convert the expression to standard SOP form.

24. Convert each SOP expression in Problem 22 to standard SOP form.

25. Determine the binary value of each term in the standard SOP expressions from Problem 23.

26. Determine the binary value of each term in the standard SOP expressions from Problem 24.

27. Convert each standard SOP expression in Problem 23 to standard POS form.

28. Convert each standard SOP expression in Problem 24 to standard POS form.

SECTION 4–7   Boolean Expressions and Truth Tables

29. Develop a truth table for each of the following standard SOP expressions:

(a) $A\bar{B}C + \bar{A}B\bar{C} + ABC$ (b) $\overline{XYZ} + \bar{X}\,\bar{Y}Z + XY\bar{Z} + X\bar{Y}Z + \bar{X}YZ$

30. Develop a truth table for each of the following standard SOP expressions:

(a) $\bar{A}B\bar{C}D + \bar{A}BC\bar{D} + A\bar{B}\,\bar{C}D + \overline{ABCD}$ (b) $WXYZ + WX\bar{Y}\bar{Z} + \bar{W}XYZ + W\bar{X}YZ + WX\bar{Y}Z$

31. Develop a truth table for each of the SOP expressions:

(a) $\bar{A}B + AB\bar{C} + \bar{A}C + A\bar{B}C$ (b) $\bar{X} + Y\bar{Z} + WZ + X\bar{Y}Z$

32. Develop a truth table for each of the standard POS expressions:

   (a) $(\overline{A} + \overline{B} + \overline{C})(A + B + C)(A + \overline{B} + C)$

   (b) $(\overline{A} + B + \overline{C} + D)(A + \overline{B} + C + \overline{D})(A + \overline{B} + \overline{C} + D)(\overline{A} + B + C + \overline{D})$

33. Develop a truth table for each of the standard POS expressions:

   (a) $(A + B)(A + C)(A + B + C)$ (b) $(A + \overline{B})(A + \overline{B} + \overline{C})(B + C + \overline{D})(\overline{A} + B + \overline{C} + D)$

34. For each truth table in Figure 4–56, derive a standard SOP and a standard POS expression.

ABC	X
0 0 0	0
0 0 1	1
0 1 0	0
0 1 1	0
1 0 0	1
1 0 1	1
1 1 0	0
1 1 1	1

(a)

ABC	X
0 0 0	0
0 0 1	0
0 1 0	0
0 1 1	0
1 0 0	0
1 0 1	1
1 1 0	1
1 1 1	1

(b)

ABCD	X
0 0 0 0	1
0 0 0 1	1
0 0 1 0	0
0 0 1 1	1
0 1 0 0	0
0 1 0 1	1
0 1 1 0	1
0 1 1 1	0
1 0 0 0	0
1 0 0 1	1
1 0 1 0	0
1 0 1 1	0
1 1 0 0	1
1 1 0 1	0
1 1 1 0	0
1 1 1 1	0

(c)

ABCD	X
0 0 0 0	0
0 0 0 1	0
0 0 1 0	1
0 0 1 1	0
0 1 0 0	1
0 1 0 1	1
0 1 1 0	0
0 1 1 1	1
1 0 0 0	0
1 0 0 1	0
1 0 1 0	0
1 0 1 1	1
1 1 0 0	1
1 1 0 1	0
1 1 1 0	0
1 1 1 1	1

(d)

FIGURE 4–56

SECTION 4–8   The Karnaugh Map

35. Sketch a 3-variable Karnaugh map and label each cell according to its binary value.

36. Sketch a 4-variable Karnaugh map and label each cell according to its binary value.

37. Write the standard product term for each cell in a 3-variable Karnaugh map.

SECTION 4–9   Karnaugh Map SOP Minimization

38. Use a Karnaugh map to find the minimum SOP form for each expression:

   (a) $\overline{A}\,\overline{B}\,\overline{C} + \overline{A}\,\overline{B}C + A\overline{B}C$ (b) $AC(\overline{B} + C)$

   (c) $\overline{A}(BC + B\overline{C}) + A(BC + B\overline{C})$ (d) $\overline{A}\,\overline{B}\,\overline{C} + A\overline{B}\,\overline{C} + \overline{A}B\overline{C} + AB\overline{C}$

39. Use a Karnaugh map to simplify each expression to a minimum SOP form:

   (a) $\overline{A}\,\overline{B}\,\overline{C} + A\overline{B}C + \overline{A}B\overline{C} + AB\overline{C}$ (b) $AC[\overline{B} + B(B + \overline{C})]$

   (c) $DE\overline{F} + \overline{D}E\overline{F} + \overline{D}\,\overline{E}\,\overline{F}$

40. Expand each expression to a standard SOP form:

   (a) $AB + A\overline{B}C + ABC$ (b) $A + BC$

   (c) $A\overline{B}\,\overline{C}D + AC\overline{D} + B\overline{C}D + \overline{A}BC\overline{D}$ (d) $A\overline{B} + A\overline{B}\,\overline{C}D + CD + B\overline{C}D + ABCD$

41. Minimize each expression in Problem 40 with a Karnaugh map.

42. Use a Karnaugh map to reduce each expression to a minimum SOP form.

   (a) $A + B\overline{C} + CD$

   (b) $\overline{A}\,\overline{B}\,\overline{C}\,\overline{D} + \overline{A}\,\overline{B}\,\overline{C}D + ABCD + ABC\overline{D}$

   (c) $\overline{A}B(\overline{C}\,\overline{D} + \overline{C}D) + AB(\overline{C}\,\overline{D} + \overline{C}D) + A\overline{B}\,\overline{C}D$

   (d) $(\overline{A}\,\overline{B} + A\overline{B})(CD + C\overline{D})$

   (e) $\overline{A}\,\overline{B} + A\overline{B} + \overline{C}\,\overline{D} + C\overline{D}$

43. Reduce the function specified in the truth table in Figure 4–57 to its minimum SOP form by using a Karnaugh map.

44. Use the Karnaugh map method to implement in the minimum SOP form the logic function specified in the truth table in Figure 4–58.

45. Solve Problem 44 for a situation in which the last six binary combinations are not allowed.

FIGURE 4–57

Inputs	Output
A B C	X
0 0 0	1
0 0 1	1
0 1 0	0
0 1 1	1
1 0 0	1
1 0 1	1
1 1 0	0
1 1 1	1

FIGURE 4–58

Inputs	Output
A B C D	X
0 0 0 0	0
0 0 0 1	1
0 0 1 0	1
0 0 1 1	0
0 1 0 0	0
0 1 0 1	0
0 1 1 0	1
0 1 1 1	1
1 0 0 0	1
1 0 0 1	0
1 0 1 0	1
1 0 1 1	0
1 1 0 0	1
1 1 0 1	1
1 1 1 0	0
1 1 1 1	1

SECTION 4–10   Karnaugh Map POS Minimization

46. Use a Karnaugh map to find the minimum POS for each expression:

(a) $(A + B + C)(\overline{A} + \overline{B} + \overline{C})(A + \overline{B} + C)$

(b) $(X + \overline{Y})(\overline{X} + Z)(X + \overline{Y} + \overline{Z})(\overline{X} + \overline{Y} + Z)$

(c) $A(B + \overline{C})(\overline{A} + C)(A + \overline{B} + C)(\overline{A} + B + \overline{C})$

47. Use a Karnaugh map to simplify each expression to minimum POS form:

(a) $(A + \overline{B} + C + \overline{D})(\overline{A} + B + \overline{C} + D)(\overline{A} + \overline{B} + \overline{C} + \overline{D})$

(b) $(X + \overline{Y})(W + \overline{Z})(\overline{X} + \overline{Y} + \overline{Z})(W + X + Y + Z)$

48. For the function specified in the truth table of Figure 4–57, determine the minimum POS expression using a Karnaugh map.

49. Determine the minimum POS expression for the function in the truth table of Figure 4–58.

50. Convert each of the following POS expressions to minimum SOP expressions using a Karnaugh map.

(a) $(A + \overline{B})(A + \overline{C})(\overline{A} + \overline{B} + C)$ (b) $(\overline{A} + B)(\overline{A} + \overline{B} + \overline{C})(B + \overline{C} + D)(A + \overline{B} + C + \overline{D})$

SECTION 4–11   Five-Variable Karnaugh Maps

51. Minimize the following SOP expression using a Karnaugh Map:

$$X = \overline{A}B\overline{C}D\overline{E} + \overline{A}\,\overline{B}\,\overline{C}DE + A\overline{B}\,\overline{C}DE + ABC\,\overline{D}\,\overline{E} + \overline{A}BCDE + \overline{A}BC\overline{D}E$$
$$+ \overline{A}\,\overline{B}\,\overline{C}\,\overline{D}\,\overline{E} + \overline{A}\,\overline{B}CDE + AB\overline{C}D\overline{E} + AB\overline{C}DE$$

52. Apply the Karnaugh map method to minimize the following SOP expression:

$$A = \overline{V}WXYZ + V\overline{W}XYZ + VW\overline{X}YZ + VWX\overline{Y}Z + VWXY\overline{Z} + \overline{V}\,\overline{W}\,\overline{X}\,\overline{Y}\,Z$$
$$+ \overline{V}\,\overline{W}\,\overline{X}Y\overline{Z} + \overline{V}\,\overline{W}X\overline{Y}\,\overline{Z} + VW\overline{X}\,\overline{Y}\,\overline{Z}$$

■ System Applications

SECTION 4–12 Digital System Application

53. If you are required to choose a type of digital display for low light conditions, will you select LED or LCD 7-segment displays? Why?

54. Explain why the codes 1010, 1011, 1100, 1101, 1110, and 1111 fall into the "don't care" category in 7-segment display applications.

55. For segment *b,* how many fewer gates and inverters does it take to implement the minimum SOP expression than the standard SOP expression?

56. Repeat Problem 55 for the logic for segments *c* through *g.*

Special Design Problems

57. The logic for segment *a* in Figure 4–50 produces a HIGH output to activate the segment and so do the circuits for each of the other segments. If a type of 7-segment display is used that requires a LOW to activate each segment, modify the segment logic accordingly.

58. Redesign the logic for segment *a* using a minimum POS approach. Which is simpler, minimum POS or the minimum SOP?

59. Repeat Problem 58 for segments *b* through *g.*

60. Summarize the results of your redesign effort in Problems 58 and 59 and recommend the best design based on fewer ICs. Specify the types of ICs.

■ ANSWERS TO SECTION REVIEWS

SECTION 4–1

1. $\overline{A} = \overline{0} = 1$ **2.** $A = 1, B = 1, C = 0; \overline{A} + \overline{B} + C = \overline{1} + \overline{1} + 0 = 0 + 0 + 0 = 0$

3. $A = 1, B = 0, C = 1; A\overline{B}C = 1 \cdot \overline{0} \cdot 1 = 1 \cdot 1 \cdot 1 = 1$

SECTION 4–2

1. $A + (B + C + D) = (A + B + C) + D$ **2.** $A(B + C + D) = AB + AC + AD$

SECTION 4–3

1. **(a)** $\overline{ABC} + \overline{(\overline{D} + E)} = \overline{A} + \overline{B} + \overline{C} + D\overline{E}$ **(b)** $\overline{(A + B)C} = \overline{A}\,\overline{B} + \overline{C}$

(c) $\overline{A + B + C} + \overline{\overline{D}E} = \overline{A}\,\overline{B}\,\overline{C} + D + \overline{E}$

SECTION 4–4

1. $(C + D)B + A$

2. Abbreviated truth table: The expression is a 1 when A is 1 or when B and C are 1s or when B and D are 1s. The expression is 0 for all other variable combinations.

SECTION 4–5

1. **(a)** $A + AB + A\overline{B}C = A$ **(b)** $(\overline{A} + B)C + ABC = C(\overline{A} + B)$

(c) $A\overline{B}C(BD + CDE) + A\overline{C} = A(\overline{C} + \overline{B}DE)$

2. **(a)** *Original:* 2 AND gates, 1 OR gate, 1 inverter; *Simplified:* No gates (straight connection)

(b) *Original:* 2 OR gates, 2 AND gates, 1 inverter; *Simplified:* 1 OR gate, 1 AND gate, 1 inverter

(c) *Original:* 5 AND gates, 2 OR gates, 2 inverters; *Simplified:* 2 AND gates, 1 OR gate, 2 inverters

SECTION 4–6

1. **(a)** SOP **(b)** standard POS **(c)** standard SOP **(d)** POS

2. **(a)** $AB\overline{C}\,\overline{D} + AB\overline{C}D + ABC\overline{D} + ABCD + \overline{A}B\overline{C}D + \overline{A}BCD + \overline{A}\,\overline{B}C\overline{D} + \overline{A}BC\overline{D}$

(c) Already standard

3. **(b)** Already standard

(d) $(A + \overline{B} + \overline{C})(A + \overline{B} + C)(A + B + \overline{C})(A + B + C)$

SECTION 4–7

1. $2^5 = 32$ **2.** $0110 \longrightarrow \overline{W}XY\overline{Z}$ **3.** $1100 \longrightarrow \overline{W} + \overline{X} + Y + Z$

SECTION 4–8

1. (a) upper left cell: 000   **(b)** lower right cell: 101   **(c)** lower left cell: 100
    (d) upper right cell: 001
2. (a) upper left cell: $\overline{X}\,\overline{Y}\,\overline{Z}$ **(b)** lower right cell: $X\overline{Y}Z$ **(c)** lower left cell: $X\overline{Y}\,\overline{Z}$
    (d) upper right cell: $\overline{X}\,\overline{Y}Z$
3. (a) upper left cell: 0000   **(b)** lower right cell: 1010   **(c)** lower left cell: 1000
    (d) upper right cell: 0010
4. (a) upper left cell: $\overline{W}\,\overline{X}\,\overline{Y}\,\overline{Z}$ **(b)** lower right cell: $W\overline{X}Y\overline{Z}$ **(c)** lower left cell: $W\overline{X}\,\overline{Y}\,\overline{Z}$
    (d) upper right cell: $\overline{W}\,\overline{X}Y\overline{Z}$

SECTION 4–9

1. 8-cell map for 3 variables; 16-cell map for 4 variables
2. $AB + B\overline{C} + \overline{A}\,\overline{B}C$
3. (a) $\overline{A}\,\overline{B}\,\overline{C} + \overline{A}BC + ABC + AB\overline{C}$
    (b) $\overline{A}\,\overline{B}\,\overline{C} + \overline{A}\,\overline{B}C + \overline{A}B\overline{C} + \overline{A}BC + A\overline{B}\,\overline{C} + A\overline{B}C$
    (c) $\overline{A}\,\overline{B}\,\overline{C}\,\overline{D} + \overline{A}\,\overline{B}\,C D + \overline{A}B\overline{C}\,\overline{D} + \overline{A}BCD + \overline{A}BC\overline{D} + A\overline{B}\,\overline{C}D + A\overline{B}CD + A\overline{B}CD$
    (d) $\overline{A}\,\overline{B}\,\overline{C}\,\overline{D} + \overline{A}B\overline{C}\,\overline{D} + AB\overline{C}\,\overline{D} + A\overline{B}\,\overline{C}\,\overline{D} + \overline{A}B\overline{C}D + AB\overline{C}D + A\overline{B}\overline{C}D + \overline{A}\,\overline{B}C\overline{D} + \overline{A}BC\overline{D} + ABC\overline{D} + A\overline{B}C\overline{D}$

SECTION 4–10

1. In mapping a POS expression, 0s are placed in cells whose value makes the standard sum term zero; and in mapping an SOP expression 1s are placed in cells having the same values as the product terms.
2. 0 in the 1011 cell: $\overline{A} + B + \overline{C} + \overline{D}$ **3.** 1 in the 0010 cell: $\overline{A}\,\overline{B}C\overline{D}$

SECTION 4–11

1. There are 32 combinations of 5 variables ($2^5 = 32$).
2. $X = 1$ because the function is 1 for all possible combinations of 5 variables.

SECTION 4–12

1. "Don't cares" simplify a logic expression.
2. You should have found seven duplicate terms in the segment logic.
3. Eight AND gates and eleven inverters were eliminated.

5

COMBINATIONAL LOGIC

■ **CHAPTER OBJECTIVES**

☐ Analyze basic combinational logic circuits, such as AND-OR, AND-OR-Invert, exclusive-OR, exclusive-NOR, and other general combinational networks

☐ Use AND-OR and AND-OR-Invert circuits to implement sum-of-products (SOP) and product-of-sums (POS) expressions

☐ Write the Boolean output expression for any combinational logic circuit

☐ Develop a truth table from the output expression for a combinational logic circuit

☐ Use the Karnaugh map to expand an output expression containing suppressed variable terms into a full SOP form

☐ Design a combinational logic circuit for a given Boolean output expression

☐ Design a combinational logic circuit for a given truth table

☐ Simplify a combinational logic circuit to its minimum form

☐ Use NAND gates to implement any combinational logic function

☐ Use NOR gates to implement any combinational logic function

☐ Troubleshoot faulty logic networks

☐ Troubleshoot logic circuits by using signal tracing and waveform analysis

CHAPTER OVERVIEW

In Chapters 3 and 4, logic gates were discussed on an individual basis and in simple combinations. You were introduced to SOP and POS implementations, which are basic forms of combinational logic. When logic gates are connected together to produce a specified output for certain specified combinations of input variables, with no storage involved, the resulting circuit is in the category of **combinational logic.** In combinational logic, the output level is at all times dependent on the combination of input levels. This chapter expands on the material introduced in earlier chapters with a coverage of the analysis, design, and troubleshooting of various combinational logic circuits.

SPECIFIC DEVICES

74HC58	74LS00	74LS04
74LS10	74LS51	74LS54

DIGITAL SYSTEM APPLICATION

This Digital System Application illustrates the concepts taught in this chapter by demonstrating how combinational logic can be used for a specific purpose in a practical application. In Section 5–7, a digital circuit is used to control a lumber-processing operation in a furniture factory. The digital circuit operates from four switch inputs to control four motors that run the system. Based on the switch settings, the logic circuit determines which motors should be on and which should be off and shuts the system down if the switches are set improperly.

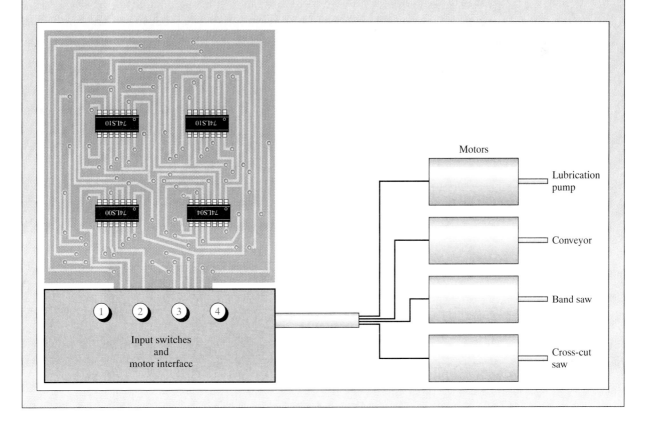

5–1 ■ SPECIAL COMBINATIONAL LOGIC CIRCUITS

In Chapter 4, you learned that SOP expressions are implemented with an AND gate for each product term and one OR gate for summing all of the product terms. This SOP implementation is called AND-OR logic and is the basic form for realizing standard Boolean functions. In this section, the AND-OR and the AND-OR-Invert are examined; and the exclusive-OR and exclusive-NOR gates, which are actually a form of AND-OR logic, are covered. After completing this section, you should be able to

☐ Analyze and apply AND-OR circuits ☐ Analyze and apply AND-OR-Invert circuits ☐ Analyze and apply exclusive-OR gates ☐ Analyze and apply exclusive-NOR gates

AND-OR Logic

Figure 5–1(a) shows an AND-OR circuit consisting of two 2-input AND gates and one 2-input OR gate; Figure 5–1(b) is the ANSI standard rectangular outline symbol. The Boolean expressions for the AND gate outputs and the resulting SOP expression for the output Y are shown on the diagram. In general, an AND-OR circuit can have any number of AND gates each with any number of inputs.

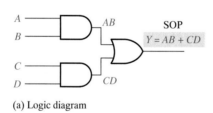

(a) Logic diagram

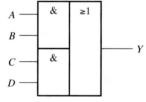

(b) ANSI standard rectangular outline symbol

FIGURE 5–1
An example of AND-OR logic.

TABLE 5–1
Truth table for the AND-OR logic in Figure 5–1

Inputs						Output
A	B	C	D	AB	CD	Y
0	0	0	0	0	0	0
0	0	0	1	0	0	0
0	0	1	0	0	0	0
0	0	1	1	0	1	1
0	1	0	0	0	0	0
0	1	0	1	0	0	0
0	1	1	0	0	0	0
0	1	1	1	0	1	1
1	0	0	0	0	0	0
1	0	0	1	0	0	0
1	0	1	0	0	0	0
1	0	1	1	0	1	1
1	1	0	0	1	0	1
1	1	0	1	1	0	1
1	1	1	0	1	0	1
1	1	1	1	1	1	1

The truth table for the 4-input AND-OR logic circuit is shown in Table 5–1. The intermediate AND gate outputs (the AB and CD columns) are also shown in the table.

An AND-OR circuit directly implements an SOP expression, assuming the complements (if any) of the variables are available. The logical operation of the AND-OR circuit in Figure 5–1 is stated as follows:

> **In a 4-input AND-OR logic circuit, the output Y is HIGH (1) only if both input A and input B are HIGH (1), or both input C and input D are HIGH (1), or all inputs are HIGH (1).**

EXAMPLE 5–1

In a certain chemical-processing plant, three different liquid chemicals are used in a manufacturing process. The three chemicals are stored in three different tanks. A level sensor in each tank produces a HIGH voltage when the level of chemical in the tank drops below a specified point.

Design a circuit that monitors the chemical level in each tank and indicates when the level in any two of the tanks drops below the specified point.

Solution   The AND-OR circuit in Figure 5–2 has inputs from the sensors on tanks A, B, and C as shown. The AND gate G_1 checks the levels in tanks A and B, gate G_2 checks tanks A and C, and gate G_3 checks tanks B and C. When the chemical level in any two of the tanks gets too low, one of the AND gates will have HIGHs on both of its inputs causing its output to be HIGH, and so the final output Y from the OR gate is HIGH. This HIGH output is then used to activate an indicator such as a lamp or audible alarm, as shown in the figure.

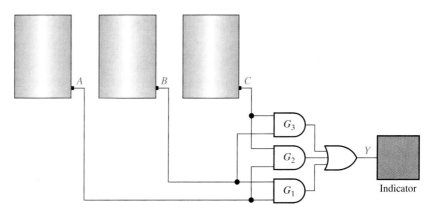

FIGURE 5–2

Related Exercise   Write the Boolean SOP expression for the AND-OR logic in Figure 5–2.

A Specific AND-OR Integrated Circuit   The 74HC58 is an example of AND-OR logic in an IC. This is a CMOS device that includes two separate AND-OR circuits in a single package. One circuit has two 2-input AND gates and the other circuit has two 3-input AND gates, as shown in Figure 5–3(a). Pin numbers for dual-in-line or SOIC packages are shown. The standard rectangular outline symbol is given in part (b).

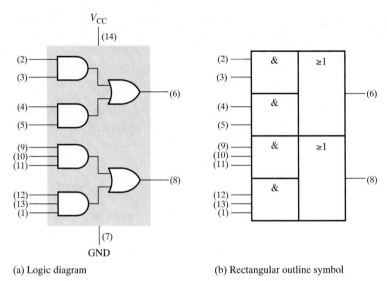

(a) Logic diagram   (b) Rectangular outline symbol

FIGURE 5–3
The 74HC58 dual AND-OR.

AND-OR-Invert Logic

When the output of an AND-OR circuit is complemented (inverted), it results in an AND-OR-Invert circuit. Recall that AND-OR logic directly implements SOP expressions. POS expressions can be implemented with AND-OR-Invert logic. This is illustrated as follows starting with a POS expression and developing the corresponding AND-OR-Invert expression.

$$Y = (\overline{A} + \overline{B})(\overline{C} + \overline{D}) = (\overline{AB})(\overline{CD}) = \overline{(\overline{AB})(\overline{CD})} = \overline{\overline{AB}} + \overline{\overline{CD}} = \overline{AB + CD}$$

The logic diagram in Figure 5–4(a) shows an AND-OR-Invert circuit and the development of the POS output expression. The ANSI standard rectangular outline symbol is shown in part (b). In general, an AND-OR Invert circuit can have any number of AND gates each with any number of inputs.

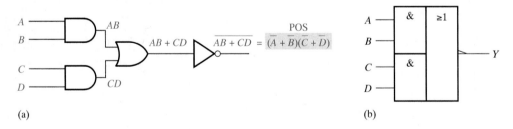

(a)   (b)

FIGURE 5–4
An AND-OR-Invert circuit produces a POS output.

The logical operation of the AND-OR-Invert circuit in Figure 5–4 is stated as follows:

In a 4-input AND-OR-Invert logic circuit, the output Y is LOW (0) only if both input A and input B are HIGH (1), or both input C and input D are HIGH (1), or all inputs are HIGH (1).

A truth table can be developed from the AND-OR truth table in Table 5–1 by simply changing all 1s to 0s and all 0s to 1s in the output column.

EXAMPLE 5–2

The sensors in the chemical tanks of Example 5–1 are being replaced by a new model that produces a LOW voltage instead of a HIGH voltage when the level of the chemical in the tank drops below a critical point.

Modify the circuit in Figure 5–2 to operate with the different input levels and still produce a HIGH output to activate the indicator when the level in any two of the tanks drops below the critical point. Show the logic diagram.

Solution   The AND-OR-Invert circuit in Figure 5–5 has inputs from the sensors on tanks A, B, and C as shown. The AND gate G_1 checks the levels in tanks A and B, gate G_2 checks tanks A and C, and gate G_3 checks tanks B and C. When the chemical level in any two of the tanks gets too low, each AND gate will have a LOW on at least one input causing its output to be LOW and, thus, the final output Y from the inverter is HIGH. This HIGH output is then used to activate an indicator.

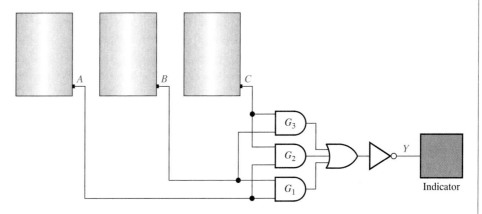

FIGURE 5–5

Related Exercise   Write the Boolean expression for the AND-OR-Invert logic in Figure 5–5 and show that the output is HIGH (1) when any two of the inputs A, B, and C are LOW (0).

Specific AND-OR-Invert Integrated Circuits   The 74LS51 and the 74LS54 are examples of AND-OR-Invert logic. Both are TTL low-power Schottky devices but are available in other logic families as well. The 74LS51 includes two separate AND-OR-Invert circuits in a single package. One circuit has two 2-input AND gates and the other circuit has two 3-input AND gates as shown in Figure 5–6(a). The 74LS54 is a single AND-OR-Invert circuit that has two 2-input AND gates and two 3-input AND gates as shown in Figure 5–6(b). Notice that the inversion is indicated by a bubble on the output of the OR gates in each case, showing that the OR-Invert part of the circuit is effectively a NOR gate.

Exclusive-OR Logic

The exclusive-OR gate and the 7486 IC were introduced in Chapter 3. Although, because of its importance, this circuit is considered a type of logic gate with its own unique symbol, it is actually a combination of two AND gates, one OR gate, and two inverters as shown in Figure 5–7(a). The two standard logic symbols are shown in parts (b) and (c).

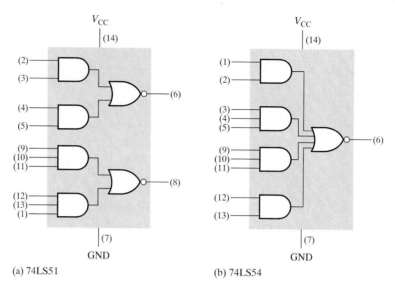

FIGURE 5–6
The 74LS51 dual 2-wide AND-OR-Invert and the 74LS54 single 4-wide AND-OR-Invert. The term 2-wide means there are two AND gates per device and 4-wide means there are four AND gates per device.

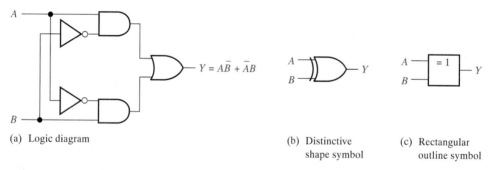

(a) Logic diagram

(b) Distinctive
shape symbol

(c) Rectangular
outline symbol

FIGURE 5–7
Exclusive-OR logic diagram and symbols.

The output expression for the circuit in Figure 5-7 is

$$Y = A\overline{B} + \overline{A}B$$

Evaluation of this expression results in the truth table in Table 5–2. Notice that the output is HIGH only when the two inputs are at opposite levels. A special exclusive-OR operator $\oplus$ is often used, so the expression $Y = A\overline{B} + \overline{A}B$ can be stated as "Y is equal to A exclusive-OR B" and can be written as

$$Y = A \oplus B$$

TABLE 5–2
Truth Table for an Exclusive-OR

A	B	Y
0	0	0
0	1	1
1	0	1
1	1	0

Exclusive-NOR Logic

As you know, the complement of the exclusive-OR function is the exclusive-NOR, which is derived as follows:

$$Y = \overline{A\overline{B} + \overline{A}B} = (\overline{A\overline{B}})(\overline{\overline{A}B}) = (\overline{A} + B)(A + \overline{B}) = \overline{A}\,\overline{B} + AB$$

Notice that the output Y is HIGH only when the two inputs, A and B, are at the same level.

The exclusive-NOR can be implemented by simply inverting the output of an exclusive-OR, as shown in Figure 5–8(a), or by directly implementing the expression $\overline{A}\,\overline{B} + AB$, as shown in part (b).

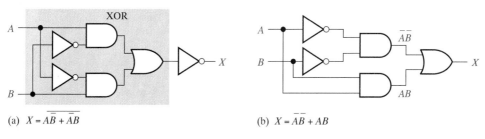

(a) $X = A\overline{B} + \overline{A}B$ (b) $X = \overline{A}\,\overline{B} + AB$

FIGURE 5–8

Two equivalent ways of implementing the exclusive-NOR.

SECTION 5–1 REVIEW

1. Determine the output (1 or 0) of a 4-variable AND-OR-Invert circuit for each of the following input conditions:
   (a) $A = 1, B = 0, C = 1, D = 0$ **(b)** $A = 1, B = 1, C = 0, D = 1$
   (c) $A = 0, B = 1, C = 1, D = 1$
2. Determine the output (1 or 0) of an exclusive-OR gate for each of the following input conditions:
   (a) $A = 1, B = 0$ **(b)** $A = 1, B = 1$
   (c) $A = 0, B = 1$ **(d)** $A = 0, B = 0$
3. Develop the truth table for a certain 3-input logic circuit with the output expression
   $X = \overline{A}\overline{B}C + \overline{A}BC + \overline{A}\,\overline{B}\,\overline{C} + A\overline{B}\overline{C} + ABC$.
4. Draw the logic diagram for an exclusive-NOR circuit.

5–2 ■ IMPLEMENTING COMBINATIONAL LOGIC

In this section, examples are used to illustrate how to implement a logic circuit from a Boolean expression or a truth table. Minimization of a logic circuit using the methods covered in Chapter 4 is also discussed. After completing this section, you should be able to

☐ Implement a logic circuit from a Boolean expression   ☐ Implement a logic circuit from a truth table   ☐ Minimize a logic circuit

---

From a Boolean Expression to a Logic Circuit

Let's examine the following Boolean expression:

$$X = AB + CDE$$

A brief inspection reveals that this expression is composed of two terms, AB and CDE, with a total of five variables. The first term is formed by ANDing A with B, and the second term is formed by ANDing C, D, and E. The two terms are then ORed to form the output X. These operations are indicated in the structure of the expression as follows:

$$X = \underset{\uparrow}{\boxed{AB}} + \underset{}{\boxed{CDE}} \quad \text{AND}$$
$$\text{OR}$$

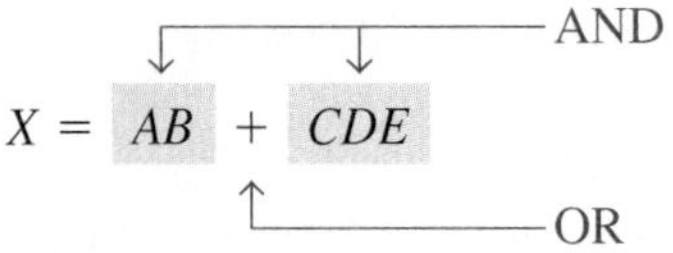

Note that in this particular expression, the AND operations forming the two individual terms, AB and CDE, must be performed before the terms can be ORed.

To implement this Boolean expression, a 2-input AND gate is required to form the term AB, and a 3-input AND gate is needed to form the term CDE. A 2-input OR gate is then required to combine the two AND terms. The resulting logic circuit is shown in Figure 5–9.

FIGURE 5–9
Logic circuit for X = AB + CDE.

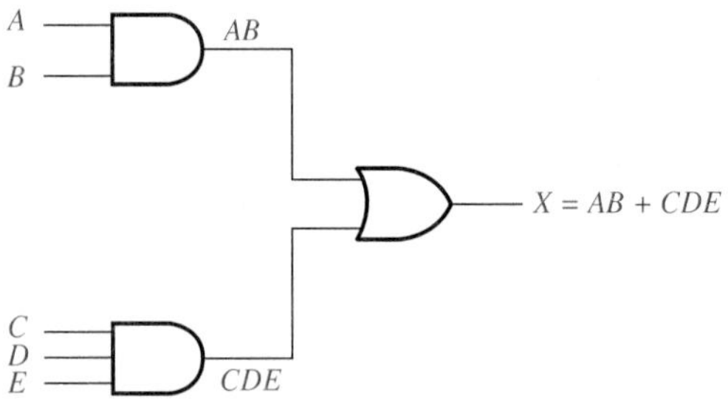

As another example, let's implement the following expression:

$$X = AB(C\overline{D} + EF)$$

A breakdown of this expression shows that the terms A, B, and $C\overline{D} + EF$ are ANDed. The term $C\overline{D} + EF$ is formed by first ANDing C and $\overline{D}$ and ANDing E and F, and then ORing these two terms. This structure is indicated in relation to the expression as follows:

$$X = AB(C\overline{D} + EF)$$

- AND
- NOT
- OR
- AND

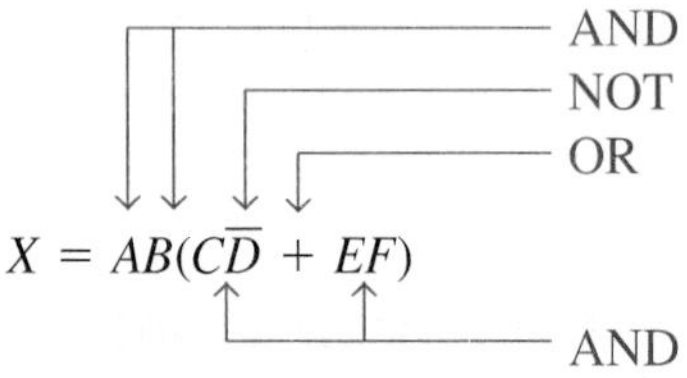

Before the expression can be formed, you must have the term $C\overline{D} + EF$; but before you can get this term, you must have the terms $C\overline{D}$ and EF; but before you can get the term $C\overline{D}$, you must have $\overline{D}$. So, as you can see, there is a chain of logic operations that must be done in the proper order.

The logic gates required to implement $X = AB(C\overline{D} + EF)$ are as follows:

1. One inverter to form $\overline{D}$
2. Two 2-input AND gates to form $C\overline{D}$ and EF
3. One 2-input OR gate to form $C\overline{D} + EF$
4. One 3-input AND gate to form X

The logic circuit for this expression is shown in Figure 5–10(a). Notice that there is a maximum of three gates and an inverter between an input and output in this circuit (from input D to output). Often the total propagation delay time through a logic circuit is a major consideration. Propagation delays are additive, so the more gates or inverters between input and output, the greater the propagation delay time.

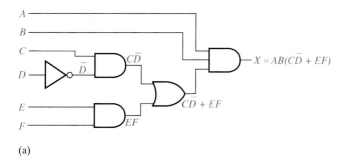

(a)

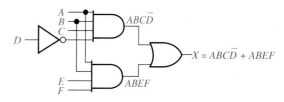

(b) Sum-of-products implementation of the circuit in part (a)

FIGURE 5–10
Logic circuits for X = AB(C̄D + EF) = ABC̄D + ABEF.

Unless an intermediate term, such as $\overline{C}D + EF$ in Figure 5–10(a), is required as an output for some other purpose, it is usually best to reduce a circuit to its SOP form. The expression is converted to SOP as follows, and the resulting circuit is shown in Figure 5–10(b).

$$AB(\overline{C}D + EF) = AB\overline{C}D + ABEF$$

From a Truth Table to a Logic Circuit

If you begin with a truth table instead of an expression, you can write the SOP expression from the truth table and then implement the logic circuit. Table 5–3 specifies a logic function.

TABLE 5–3

Inputs			Output
A	B	C	X
0	0	0	0
0	0	1	0
0	1	0	0
0	1	1	1 $\longrightarrow \overline{A}BC$
1	0	0	1 $\longrightarrow A\overline{B}\,\overline{C}$
1	0	1	0
1	1	0	0
1	1	1	0

The Boolean SOP expression obtained from the truth table by ORing the product terms for which $X = 1$ (the shaded rows) is

$$X = \overline{A}BC + A\overline{B}\,\overline{C}$$

The first term in the expression is formed by ANDing the three variables $\overline{A}$, B, and C. The second term is formed by ANDing the three variables A, $\overline{B}$, and $\overline{C}$.

The logic gates required to implement this expression are as follows: three inverters to form the $\overline{A}$, $\overline{B}$, and $\overline{C}$ variables; two 3-input AND gates to form the terms $\overline{A}BC$ and $A\overline{B}\,\overline{C}$; and one 2-input OR gate to form the final output function, $\overline{A}BC + A\overline{B}\,\overline{C}$.

FIGURE 5–11
Logic circuit for $X = \overline{A}BC + A\overline{B}\,\overline{C}$.

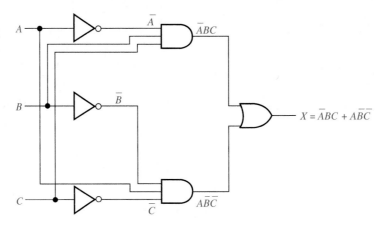

The implementation of this logic function is illustrated in Figure 5-11.

EXAMPLE 5–3

Design a logic circuit to implement the operation specified in the truth table of Table 5–4.

TABLE 5–4

Inputs			Output
A	B	C	X
0	0	0	0
0	0	1	0
0	1	0	0
0	1	1	1
1	0	0	0
1	0	1	1
1	1	0	1
1	1	1	0

Solution   Notice that $X = 1$ for only three of the input conditions. Therefore, the logic expression is

$$X = \overline{A}BC + A\overline{B}C + AB\overline{C}$$

The logic gates required are three inverters, three 3-input AND gates and one 3-input OR gate. The logic circuit is shown in Figure 5–12.

FIGURE 5–12

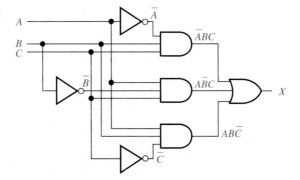

Related Exercise   Determine if the logic circuit of Figure 5–12 can be simplified.

EXAMPLE 5–4

Develop a logic circuit with four input variables that will only produce a 1 output when three input variables are 1s.

Solution Out of sixteen possible combinations of four variables, the combinations in which there are three 1s are listed in Table 5–5, along with the corresponding product term for each.

TABLE 5–5

A	B	C	D	
0	1	1	1	$\longrightarrow \overline{A}BCD$
1	0	1	1	$\longrightarrow A\overline{B}CD$
1	1	0	1	$\longrightarrow AB\overline{C}D$
1	1	1	0	$\longrightarrow ABC\overline{D}$

The product terms are ORed to get the following expression:

$$X = \overline{A}BCD + A\overline{B}CD + AB\overline{C}D + ABC\overline{D}$$

This expression is implemented in Figure 5–13 with AND-OR logic.

FIGURE 5–13

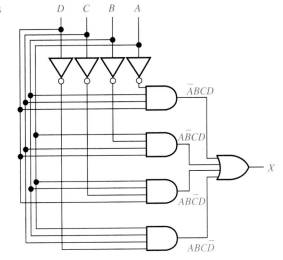

Related Exercise Determine if the logic circuit of Figure 5–13 can be simplified.

EXAMPLE 5–5

Reduce the combinational logic circuit in Figure 5–14 to a minimum form.

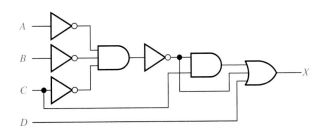

FIGURE 5–14

Solution   The expression for the output of the circuit is

$$X = (\overline{\overline{A}\,\overline{B}\,\overline{C}})C + \overline{\overline{A}\,\overline{B}\,\overline{C}} + D$$

Applying DeMorgan's theorem and Boolean algebra,

$$X = (\overline{\overline{A}} + \overline{\overline{B}} + \overline{\overline{C}})C + \overline{\overline{A}} + \overline{\overline{B}} + \overline{\overline{C}} + D$$
$$= AC + BC + CC + A + B + C + D$$
$$= AC + BC + C + A + B + \cancel{C} + D$$
$$= C(A + B + 1) + A + B + D$$
$$X = A + B + C + D$$

The simplified circuit is a 4-input OR gate as shown in Figure 5–15.

FIGURE 5–15

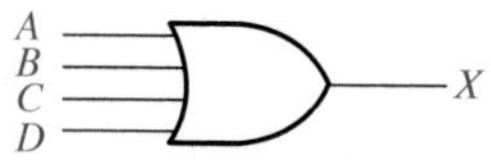

A
B
C
D

— X

Related Exercise   Verify the minimized expression $(A + B + C + D)$ using a Karnaugh map.

EXAMPLE 5–6

Minimize the combinational logic circuit in Figure 5–16. Inverters are not shown.

FIGURE 5–16

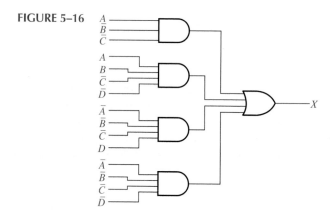

Solution   The output expression is

$$X = A\overline{B}\,\overline{C} + AB\overline{C}\,\overline{D} + \overline{A}\,B\,\overline{C}D + \overline{A}\,B\,\overline{C}\,\overline{D}$$

Expanding the first term to include the missing variables D and $\overline{D}$,

$$X = A\overline{B}\,\overline{C}(D + \overline{D}) + AB\overline{C}\,\overline{D} + \overline{A}\,B\,\overline{C}D + \overline{A}\,B\,\overline{C}\,\overline{D}$$
$$= A\overline{B}\,\overline{C}D + A\overline{B}\,\overline{C}\,\overline{D} + AB\overline{C}\,\overline{D} + \overline{A}\,B\,\overline{C}D + \overline{A}\,B\,\overline{C}\,\overline{D}$$

This expanded SOP expression is mapped and simplified on the Karnaugh map in Figure 5–17(a). The simplified implementation is shown in part (b). Inverters are not shown.

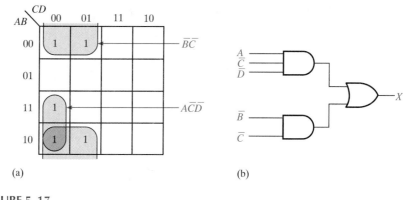

FIGURE 5–17

Related Exercise Develop the POS equivalent of the circuit in Figure 5–17(b).

SECTION 5–2 REVIEW

1. Implement the following Boolean expressions as they are stated:
   (a) $X = ABC + AB + AC$ (b) $X = AB(C + DE)$
2. Develop a logic circuit that will produce a 1 on its output only when all three inputs are 1s or when all three inputs are 0s.
3. Reduce the circuits in Question 1 to minimum SOP form.

5–3 ■ THE UNIVERSAL PROPERTY OF NAND AND NOR GATES

Up to this point, combinational circuits implemented with AND gates, OR gates, and inverters have been studied. In this section, the universal property of the NAND gate and the NOR gate is discussed. The universality of the NAND gate means that it can be used as an inverter and that combinations of NAND gates can be used to implement the AND, OR, and NOR operations. Similarly, the NOR gate can be used to implement the inverter, AND, OR, and NAND operations. After completing this section, you should be able to

☐ Use NAND gates to implement the inverter, the AND gate, the OR gate, and the NOR gate   ☐ Use NOR gates to implement the inverter, the AND gate, the OR gate, and the NAND gate

The NAND Gate as a Universal Logic Element

The NAND gate is a **universal gate** because it can be used to produce the NOT operation, the AND operation, the OR operation, and the NOR operation. An inverter can be made from a NAND gate by connecting all of the inputs together and creating, in effect, a single input, as shown in Figure 5–18(a) for a 2-input gate. An AND operation can be generated by the use of NAND gates alone, as shown in Figure 5–18(b). An OR operation can be produced with NAND gates, as illustrated in part (c). Finally, a NOR operation is produced, as shown in part (d).

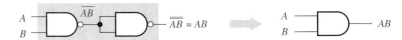

(a) A NAND gate used as an inverter

(b) Two NAND gates used as an AND gate

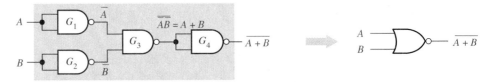

(c) Three NAND gates used as an OR gate

(d) Four NAND gates used as a NOR gate

FIGURE 5–18
Universal application of NAND gates.

In Figure 5–18(b) a NAND gate is used to invert (complement) a NAND output to form the AND operation, as indicated in the following equation:

$$X = \overline{\overline{AB}} = AB$$

In Figure 5–18(c) NAND gates G_1 and G_2 are used to invert the two input variables before they are applied to NAND gate G_3. The final OR output is derived as follows by application of DeMorgan's theorem:

$$X = \overline{\overline{A}\,\overline{B}} = A + B$$

In Figure 5–18(d) NAND gate G_4 is used as an inverter to produce the NOR operation $\overline{A + B}$.

The NOR Gate as a Universal Logic Element

Like the NAND gate, the NOR gate can be used to produce the NOT, AND, OR, and NAND operations. A NOT circuit, or inverter, can be made from a NOR gate by connecting all of the inputs together to effectively create a single input, as shown in Figure 5–19(a) with a 2-input example. Also, an OR gate can be produced from NOR gates, as illustrated in Figure 5–19(b). An AND gate can be constructed by the use of NOR gates, as shown in Figure 5–19(c). In this case the NOR gates G_1 and G_2 are used as inverters, and the final output is derived by the use of DeMorgan's theorem as follows:

$$X = \overline{\overline{A} + \overline{B}} = AB$$

Figure 5–19(d) shows how NOR gates are used to form a NAND operation.

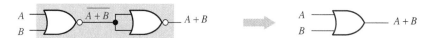

(a) A NOR gate used as an inverter

(b) Two NOR gates used as an OR gate

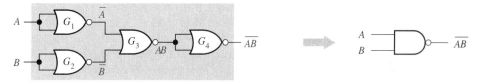

(c) Three NOR gates used as an AND gate

(d) Four NOR gates used as a NAND gate

FIGURE 5–19
Universal application of NOR gates.

5–4 ■ COMBINATIONAL LOGIC USING NAND AND NOR GATES

In this section, you will see how NAND and NOR gates can be used to implement a logic function. Recall from Chapter 3 that the NAND gate also exhibits an equivalent operation called the negative-OR and that the NOR gate exhibits an equivalent operation called the negative-AND. You will see how the use of the appropriate symbols to represent the equivalent operations makes "reading" a logic diagram easier. After completing this section, you should be able to

□ Use NAND gates to implement a logic function   □ Use NOR gates to implement a logic function   □ Use the appropriate dual symbol in a logic diagram

NAND Logic

As you have learned, a NAND gate can function as either a NAND or a negative-OR because, by DeMorgan's theorem,

$$\overline{AB} = \overline{A} + \overline{B}$$

NAND ———↑   ↑——— negative-OR

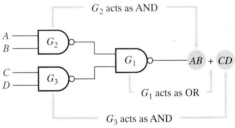

FIGURE 5–20
NAND logic for X = AB + CD.

Consider the NAND logic in Figure 5–20. The output expression is developed in the following steps:

$$X = \overline{(\overline{AB})(\overline{CD})}$$
$$= (\overline{\overline{A} + \overline{B}})(\overline{\overline{C} + \overline{D}})$$
$$= (\overline{\overline{A} + \overline{B}}) + (\overline{\overline{C} + \overline{D}})$$
$$= \overline{\overline{A}}\,\overline{\overline{B}} + \overline{\overline{C}}\,\overline{\overline{D}}$$
$$= AB + CD$$

Notice that in the last step of the development of the output expression, $AB + CD$ is in the form of two AND terms ORed together. This form of the expression shows that gates G_2 and G_3 in Figure 5–20 act as AND gates and that gate G_1 acts as an OR gate, as illustrated in Figure 5–21(a). This circuit is redrawn in part (b) with NAND symbols for gates G_2 and G_3 and a negative-OR symbol for gate G_1.

Notice in Figure 5–21(b) the bubble-to-bubble connections between the outputs of gates G_2 and G_3 and the inputs of gate G_1. *Since a bubble represents an inversion, two connected bubbles represent a double inversion and therefore cancel each other.* This inversion cancellation can be seen in the previous development of the output expression $AB + CD$ and is indicated by the absence of barred terms in the output expression. Thus, the circuit in Figure 5–21(b) is *effectively* an AND-OR circuit as shown in Figure 5–21(c).

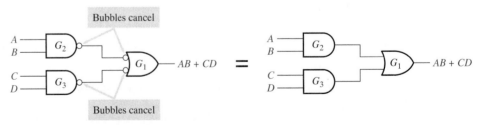

(a) Original NAND logic diagram showing effective gate operation relative to the output expression

(b) Equivalent NAND/Negative-OR logic diagram

(c) AND-OR equivalent

FIGURE 5–21
Development of the AND-OR equivalent of the circuit in Figure 5–20.

NAND Logic Diagrams

All logic diagrams using NAND gates should be drawn with each gate represented by either a NAND symbol or the equivalent negative-OR symbol to reflect the operation of the gate within the logic circuit. The NAND symbol and the negative-OR symbol are called **dual symbols.**

If we begin by representing the output gate with a negative-OR symbol, then we will use the NAND symbol for the level of gates right before that and will alternate the symbols for successive levels of gates as we move away from the output. Always use the gate symbols in such a way that every connection between a gate output and a gate input is either bubble-to-bubble or nonbubble-to-nonbubble. Never connect a bubble output to a nonbubble input or vice versa.

Figure 5–22 illustrates the procedure of using the appropriate dual symbols for a NAND circuit with several gate levels. Although using all NAND symbols as in Figure 5–22(a) is correct and there is nothing wrong with it, the diagram in part (b) is much easier to "read" and is the preferred method. The shape of the gate indicates the way its inputs will appear in the output expression and thus shows how the gate functions within the logic circuit. For a NAND symbol the inputs appear ANDed in the output expression, and for a negative-OR symbol the inputs appear ORed in the output expression, as Figure 5–22(b) illustrates. You can see that the dual-symbol diagram in part (b) makes it much easier to determine the output expression directly from the logic diagram. This is because *each gate symbol indicates the relationship of its input variables as they appear in the output expression.*

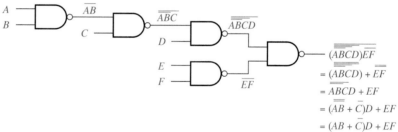

(a) Several Boolean steps are required to arrive at final output expression.

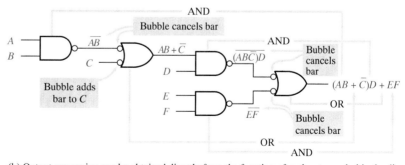

(b) Output expression can be obtained directly from the function of each gate symbol in the diagram.

FIGURE 5–22

Example of the use of the appropriate dual symbols in a NAND logic diagram.

EXAMPLE 5–7   Redraw the logic diagram and develop the output expression for the circuit in Figure 5–23, using the appropriate dual symbols.

FIGURE 5–23

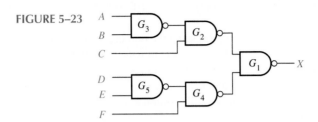

Solution   Redraw the network in Figure 5–23 with the use of equivalent negative-OR symbols as shown in Figure 5–24. Writing the expression for X directly from the indicated logic operation of each gate gives $X = (\bar{A} + \bar{B})C + (\bar{D} + \bar{E})F$.

FIGURE 5–24

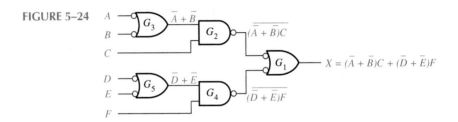

Related Exercise   Derive the output expression from Figure 5–23 and show it is equivalent to the expression in the solution.

EXAMPLE 5–8   Implement each expression with NAND logic: (a) $ABC + DE$ (b) $ABC + \bar{D} + \bar{E}$

Solution   See Figure 5–25.

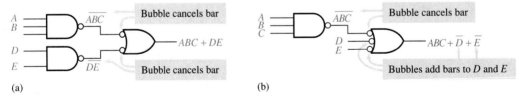

(a)   (b)

FIGURE 5–25

Related Exercise   Convert the NAND circuits in Figure 5–25(a) and (b) to equivalent AND-OR logic.

NOR Logic

A NOR gate can function as either a NOR or a negative-AND, as shown by DeMorgan's theorem:

$$\overline{A + B} = \bar{A}\,\bar{B}$$

NOR ⎯⎯⎯↑   ↑⎿⎯⎯ negative-AND

FIGURE 5–26
NOR logic for X = (A + B)(C + D).

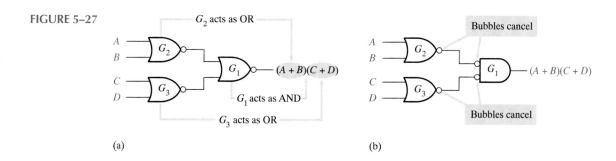

Consider the NOR logic in Figure 5–26. The output expression is developed as follows:

$$X = \overline{\overline{A + B} + \overline{C + D}} = (\overline{\overline{A + B}})\,(\overline{\overline{C + D}}) = (A + B)(C + D)$$

Notice that the expression $(A + B)(C + D)$ consists of two OR terms ANDed together. This shows that gates G_2 and G_3 act as OR gates and gate G_1 acts as an AND gate, as illustrated in Figure 5–27(a). This circuit is redrawn in part (b) with a negative-AND symbol for gate G_1.

FIGURE 5–27

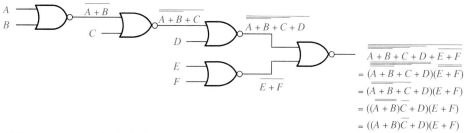

(a)

(b)

As with NAND logic, the purpose for using the dual symbols is to make the logic diagram easier to read and analyze, as illustrated in the NOR logic network in Figure 5–28. When the circuit in part (a) is redrawn with dual symbols in part (b), notice that all output-

FIGURE 5–28
Example of the use of the appropriate dual symbols in a NOR logic diagram.

$$A + B + C + D + E + F$$
$$= (\overline{A + B + C + D})(\overline{E + F})$$
$$= (A + B + C + D)(E + F)$$
$$= ((A + B)\overline{C} + D)(E + F)$$
$$= ((A + B)\overline{C} + D)(E + F)$$

(a) Final output expression is obtained after several Boolean steps.

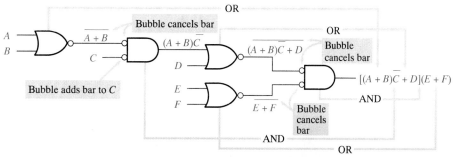

(b) Output expression can be obtained directly from the function of each gate symbol in the diagram.

to-input connections between gates are bubble-to-bubble or nonbubble-to-nonbubble. Again, you can see that the shape of each gate symbol indicates the type of term (AND or OR) that it produces in the output expression, thus making the output expression easier to determine and the logic diagram easier to analyze.

EXAMPLE 5–9

Using appropriate dual symbols, redraw the logic diagram and develop the output expression for the circuit in Figure 5–29.

FIGURE 5–29

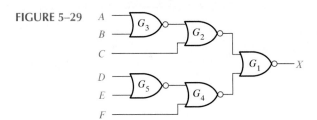

Solution Redraw the network with the equivalent negative-AND symbols as shown in Figure 5–30. Writing the expression for X directly from the indicated logic operation of each gate,

$$X = (\overline{A}\,\overline{B} + C)(\overline{D}\,\overline{E} + F)$$

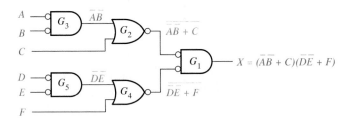

FIGURE 5–30

Related Exercise Prove that the output of the NOR circuit in Figure 5–29 is the same as for the circuit in Figure 5–30.

SECTION 5–4 REVIEW

1. Implement the expression $X = \overline{(\overline{A} + \overline{B} + \overline{C})DE}$ by using NAND logic.
2. Implement the expression $X = \overline{A}\,\overline{B}\,\overline{C} + (D + E)$ with NOR logic.

5–5 ■ OPERATION WITH PULSE WAVEFORMS

Several examples of general combinational logic circuits with pulse waveform inputs are examined in this section. Keep in mind that the logical operation of each gate is the same for pulse inputs as for constant-level inputs. The output of a logic circuit at any given time depends on the inputs at that particular time, so the relationship of the time-varying inputs is of primary importance. After completing this section, you should be able to

☐ Analyze any combinational logic circuit with pulse waveform inputs ☐ Develop a timing diagram for any given combinational logic circuit with specified inputs

The following is a review of the logical operation of individual gates for use in analyzing combinational circuits with pulse waveform inputs:

1. The output of an AND gate is HIGH only when all inputs are HIGH at the same time.
2. The output of an OR gate is HIGH any time at least one of its inputs is HIGH. The output is LOW only when all inputs are LOW at the same time.
3. The output of a NAND gate is LOW only when all inputs are HIGH at the same time.
4. The output of a NOR gate is LOW any time at least one of its inputs is HIGH. The output is HIGH only when all inputs are LOW at the same time.

The logical operation of any gate is the same regardless of whether its inputs are pulsed or constant levels. The nature of the inputs (pulsed or constant levels) does not alter the truth table of a circuit. The following general examples illustrate the analysis of combinational logic circuits with pulsed inputs.

EXAMPLE 5–10

Determine the final output waveform X for the circuit in Figure 5–31, with inputs A, B, and C as shown.

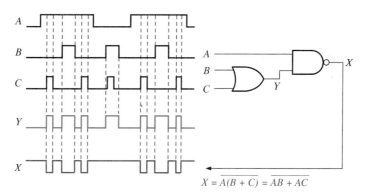

$$X = \overline{A(B + C)} = \overline{AB} + \overline{AC}$$

FIGURE 5–31

Solution The output expression, $\overline{AB} + \overline{AC}$, indicates that the output X is LOW when both A and B are HIGH or when both A and C are HIGH or when all inputs are HIGH. The output waveform X is shown in the timing diagram of Figure 5–31. The intermediate waveform Y at the output of the OR gate is shown in blue.

Related Exercise Determine the output waveform if input A is a constant HIGH level.

EXAMPLE 5–11

Draw the timing diagram for the circuit in Figure 5–32 showing the outputs of G_1, G_2, and G_3 if the input waveforms, A and B, are as indicated.

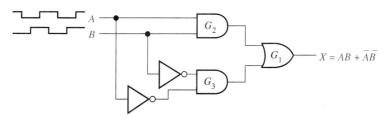

$$X = AB + \overline{A}\,\overline{B}$$

FIGURE 5–32

Solution   When both inputs are HIGH or when both inputs are LOW, the output X is HIGH as shown in Figure 5–33. Notice that this is an exclusive-NOR circuit. The intermediate outputs of gates G_2 and G_3 are also shown in Figure 5–33.

FIGURE 5–33

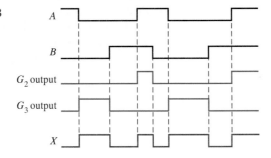

Related Exercise   Determine the output X in Figure 5–32 if input B is inverted.

EXAMPLE 5–12   Determine the output waveform X for the logic circuit in Figure 5–34(a) by first finding the intermediate waveform at each of points Y_1, Y_2, Y_3, and Y_4. The input waveforms are shown in Figure 5–34(b).

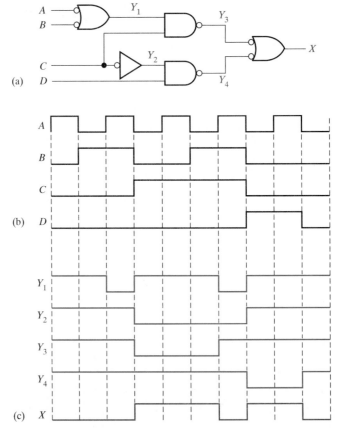

FIGURE 5–34

Solution All the intermediate waveforms and the final output waveform are shown in the timing diagram of Figure 5–34(c).

Related Exercise Determine the waveforms Y_1, Y_2, Y_3, Y_4 and X if input A is inverted.

EXAMPLE 5–13

Determine the output waveform X for the circuit in Example 5–12, Figure 5–34(a), directly from the output expression.

Solution The output expression for the circuit is developed in Figure 5–35. The SOP form indicates that the output is HIGH when A is LOW and C is HIGH or when B is LOW and C is HIGH or when C is LOW and D is HIGH.

The result is shown in Figure 5–36 and is the same as the one obtained by the intermediate-waveform method in Example 5–12. The corresponding product terms for each waveform condition that results in a HIGH output are indicated.

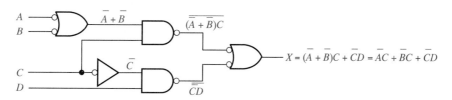

FIGURE 5–35

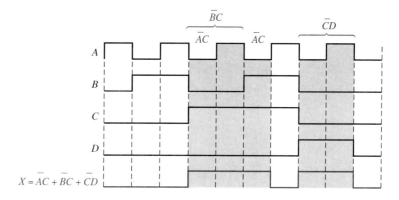

FIGURE 5–36

Related Exercise Repeat this example if all the input waveforms are inverted.

SECTION 5–5 REVIEW

1. One pulse with $t_W = 50\,\mu s$ is applied to one of the inputs of an exclusive-OR circuit. A second positive pulse with $t_W = 10\,\mu s$ is applied to the other input beginning $15\,\mu s$ after the leading edge of the first pulse. Sketch the output in relation to the inputs.
2. The pulse waveforms A and B in Figure 5–31 are applied to the exclusive-NOR circuit in Figure 5–32. Sketch a complete timing diagram.

5–6 ■ TROUBLESHOOTING

The preceding sections have given you some insight into the operation of combinational logic networks and the relationships of inputs and outputs. This type of understanding is essential when you troubleshoot digital circuits because you must know what logic levels or waveforms to look for throughout the circuit for a given set of input conditions.

In this section, the logic pulser, the probe, and the current tracer are used to troubleshoot a logic circuit when a gate output is connected to several gate inputs. Also, an example of signal tracing and waveform analysis methods is presented using an oscilloscope for locating a fault in a combinational logic circuit. After completing this section, you should be able to

☐ Define a circuit node ☐ Use a logic probe to find a faulty circuit node ☐ Use a pulser and probe to find an open gate output ☐ Use a pulser and current tracer to find a shorted gate input or output ☐ Use an oscilloscope for signal tracing in a combinational logic circuit

In a combinational logic circuit, the output of one gate may be connected to two or more gate inputs as shown in Figure 5–37. The interconnecting paths share a common electrical point known as a **node.**

FIGURE 5–37
Illustration of a node in a logic circuit.

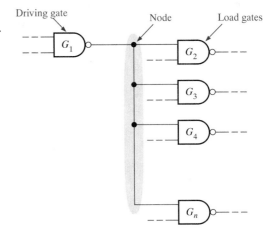

Gate G_1 in Figure 5–37 is driving the node, and the other gates represent loads connected to the node. A driving gate can drive a number of load gate inputs up to its specified fan-out. Several types of failures are possible in this situation. Some of these failure modes are difficult to isolate to a single bad gate because all the gates connected to the node are affected. Common types of failures are the following:

1. *Open output in driving gate.* This failure will cause a loss of signal to all load gates.
2. *Open input in a load gate.* This failure will not affect the operation of any of the other gates connected to the node, but it will result in loss of signal output from the faulty gate.
3. *Shorted output in driving gate.* This failure can cause the node to be stuck in the LOW state.
4. *Shorted input in a load gate.* This failure can also cause the node to be stuck in the LOW state.

Some approaches to troubleshooting each of the faults just listed are presented next.

Open Output in Driving Gate

In this situation there is no pulse activity on the node. With circuit power on, an open node will normally result in a "floating" level and may be indicated by a dim lamp on the logic probe, as illustrated in Figure 5–38.

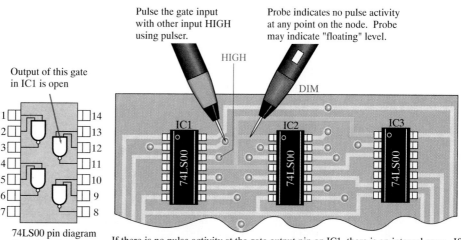

FIGURE 5–38
Open output in driving gate.

Open Input in a Load Gate

If the check for an open driver output is negative, then a check for an open input in a load gate should be performed. Apply the logic pulser tip to the node with all nonpulsed inputs HIGH. Then check the output of each gate for pulse activity with the logic probe, as illustrated in Figure 5–39. If one of the inputs that is normally connected to the node is open, no pulses will be detected on that gate's output.

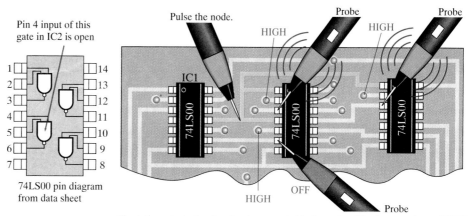

FIGURE 5–39
Open input in a load gate.

Shorted Output in Driving Gate

This fault can cause the node to be stuck LOW, as previously mentioned. A quick check with a pulser and a logic probe will indicate this, as shown in Figure 5–40(a). A short to ground in the driving gate's output or in any load gate input will cause this symptom, and further checks must therefore be made to isolate the short to a particular gate.

If the driving gate's output is internally shorted to ground, then, essentially, no current activity will be present on any of the connections to the node. Thus, a current tracer will indicate no activity with circuit power on, as illustrated in Figure 5–40(b).

To further verify a shorted output, a pulser and a current tracer can be used with the circuit power off, as shown in Figure 5–40(c). When current pulses are applied to the node with the pulser, all of the current will be into the shorted output, and none through the circuit paths into the load gate inputs.

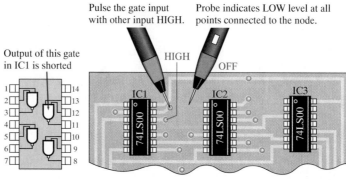

(a) Node is "stuck" LOW.

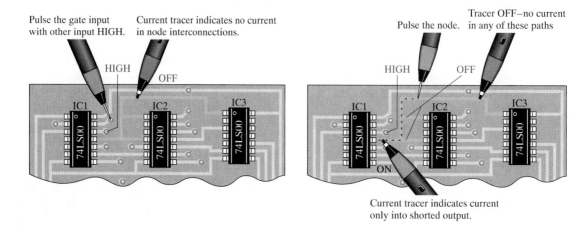

(b) No current in any node interconnection

(c) There is current from point of pulser application directly to short. There is no current in other paths connected to node.

FIGURE 5–40
Shorted output in driving gate.

Shorted Input in a Load Gate

If one of the load gate inputs is internally shorted to ground, the node will be stuck in the LOW state. Again, as in the case of a shorted output, the logic pulser and current tracer can be used to isolate the faulty gate.

When the node is pulsed with circuit power off, essentially all the current will be into the shorted input, and tracing its path with the current tracer will lead to the shorted input, as illustrated in Figure 5–41.

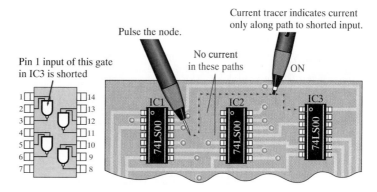

FIGURE 5–41
Shorted input in a load gate.

Signal Tracing and Waveform Analysis

Although the methods of isolating an open or a short at a node point are very useful from time to time, a more general troubleshooting technique called **signal tracing** is of great value to the technician or technologist in just about every troubleshooting situation. Waveform measurement is accomplished with an oscilloscope or a logic analyzer.

Basically, the signal tracing method requires that you observe the waveforms and their time relationships at all accessible points in the logic circuit. You can begin at the inputs and, from an analysis of the waveform timing diagram for each point, determine where an incorrect waveform first occurs. With this procedure you can usually isolate the fault to a specific gate. A procedure beginning at the output and working back toward the inputs can also be used.

The general procedure for signal tracing starting at the inputs is outlined as follows:

- Within a system, define the section of logic that is suspected of being faulty.
- Start at the inputs to the section of logic under examination. We assume, for this discussion, that the input waveforms coming from other sections of the system have been found to be correct.
- For each gate, beginning at the input and working toward the output of the logic network, observe the output waveform of the gate and compare it with the input waveforms by using the oscilloscope or the logic analyzer.
- Determine if the output waveform is correct, using your knowledge of the logical operation of the gate.
- If the output is incorrect, the gate under test may be faulty. Pull the IC containing the gate that is suspected of being faulty, and test it out-of-circuit. If the gate is found to be faulty, replace the IC. If it works correctly, the fault is in the external circuitry or in another IC to which the tested one is connected.
- If the output is correct, go to the next gate. Continue checking each gate until an incorrect waveform is observed.

Figure 5–42 is an example that illustrates the general procedure for a specific logic circuit in the following steps:

Step 1.   Observe the output of gate G_1 relative to the inputs. If it is correct, check the inverter next. If the output is not correct, the gate or its connections are bad; or, if the output is LOW, the input to gate G_2 may be shorted.

Step 2.   Observe the output of the inverter relative to the input. If it is correct, check gate G_2 next. If the output is not correct, the inverter or its connections are bad; or, if the output is LOW, the input to gate G_3 may be shorted.

Step 3.   Observe the output of gate G_2 relative to the inputs. If it is correct, check gate G_3 next. If the output is not correct, the gate or its connections are bad; or, if the output is LOW, the input to gate G_4 may be shorted.

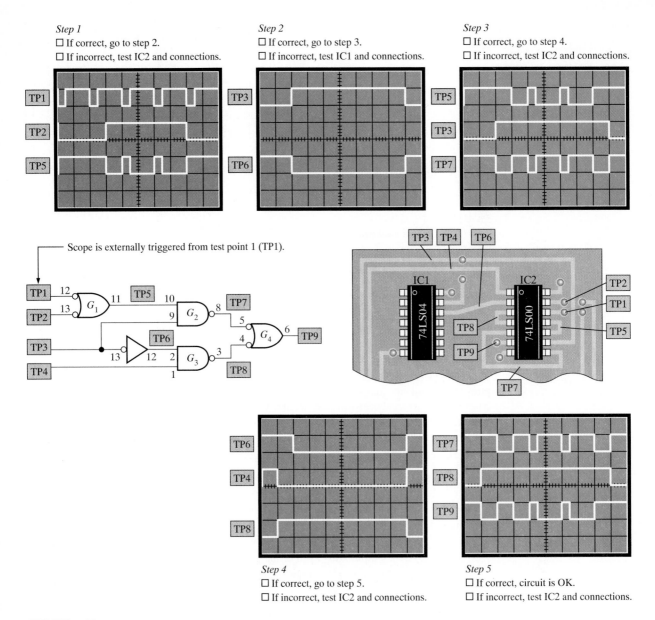

FIGURE 5–42
Example of signal tracing and waveform analysis in a portion of a digital system.

Step 4. Observe the output of gate G_3 relative to the inputs. If it is correct, check gate G_4 next. If the output is not correct, the gate or its connections are bad; or, if the output is LOW, the input to gate G_4 may be shorted.

Step 5. Observe the output of gate G_4 relative to the inputs. If it is correct, the circuit is OK. If the output is not correct, the gate or its connections are bad.

EXAMPLE 5–14 Determine the fault in the TTL logic network of Figure 5–43(a) by using waveform analysis. The color waveforms are correct, and you have observed the waveforms shown in black in Figure 5–43(b).

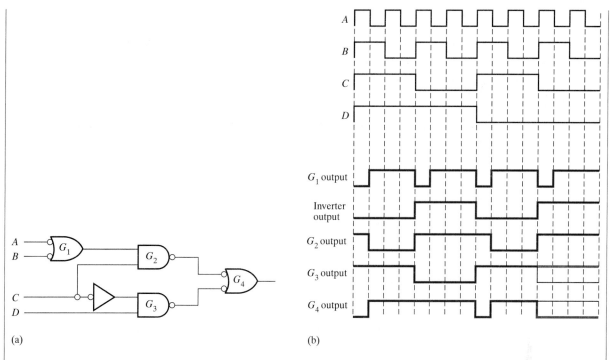

(a)

(b)

FIGURE 5–43

Solution
1. Determine what the correct waveform should be for each gate. The correct waveforms are shown in color, superimposed on the actual measured waveforms, in Figure 5–43(b).
2. Compare waveforms gate by gate until you find a measured waveform that does not match the correct waveform.

In this example, everything tested is correct until gate G_3. The output of this gate is not correct as the differences in the waveforms indicate. An analysis of the waveforms indicates that if the D input to gate G_3 is open and acting as a HIGH, you will get the output waveform measured (shown in color). Notice that the output of G_4 is correct for the inputs measured, although the input from G_3 is incorrect.

Replace the IC containing G_3, and check the circuit's operation again.

Related Exercise For the inputs in Figure 5–43(b), determine the output waveform for the logic circuit (output of G_1) if the inverter has an open output.

SECTION 5–6 REVIEW
1. List four common internal failures in logic gates.
2. One input of a NOR gate is externally shorted to $+V_{CC}$. How does this condition affect the gate operation?
3. Determine the output of gate G_4 in Figure 5–43(a), with inputs as shown in part (b), for the following faults:
   (a) one input to G_1 shorted to ground
   (b) the inverter input shorted to ground
   (c) an open output in G_3

5–7 ■ DIGITAL SYSTEM APPLICATION

In this system application, a digital control system for a portion of a lumber-processing operation in a furniture factory is introduced. This system controls four motors that run a conveyor, its lubrication pump, and two saws. After completing this section, you should be able to

□ Describe the system operation   □ Design the control logic for a portion of the system   □ Troubleshoot the system board   □ Develop a test procedure for the control logic

Basic System Operation

The system uses four manual on/off switches, control logic, and motor drive interface to control the conveyor lubrication pump motor, the conveyor motor, the cross-cut saw motor, and the band saw motor, as indicated by the block diagram in Figure 5–44. This system application focuses on the control logic portion of the system.

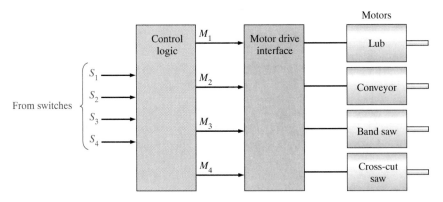

FIGURE 5–44
Basic block diagram for the control system.

Operational Requirements   Switch input S_1 controls the lubrication pump motor (output M_1). Switch input S_2 controls the conveyor motor (output M_2). Switch input S_3 controls the band saw motor (output M_3). Switch input S_4 controls the cross-cut saw motor (output M_4).

The motor that provides for lubrication of the conveyor must be on ($M_1 = 1$) when the conveyor is on. The motor that drives the conveyor is on ($M_2 = 1$) only when both switch 1 and switch 2 are on ($S_1 = 1$ and $S_2 = 1$). The band saw motor is on ($M_3 = 1$) when switch 3 is on ($S_3 = 1$), and the cross-cut saw motor is on ($M_4 = 1$) when switch 4 is on ($S_4 = 1$). The band saw and cross-cut saw motors require no lubrication, but they must never be on at the same time. If switches 3 and 4 are both turned on at the same time, the system must be completely shut down, including the conveyor and lubrication motors. Also, the cross-cut saw and the conveyor cannot be on at the same time. The control logic controls the motors to prevent any of the unallowed conditions from occurring should the switches be improperly operated.

Truth Table for the Control Logic

Table 5–6 is a truth table that shows the states of the switches and motors based on the conditions previously described. A 1 indicates the ON state and a 0 indicates the OFF state. Since there are four switches, there are sixteen possible ON/OFF combinations of the

TABLE 5–6
Truth table for the control logic

Inputs				Outputs				
S_1	S_2	S_3	S_4	M_1 Lub	M_2 Conv	M_3 Band	M_4 Cross	Comments
0	0	0	0	0	0	0	0	Shut down
0	0	0	1	0	0	0	1	OK
0	0	1	0	0	0	1	0	OK
0	0	1	1	0	0	0	0	Unallowed (2 saws on). Shut down
0	1	0	0	0	0	0	0	Unallowed (No lub with conveyor). Shut down
0	1	0	1	0	0	0	0	Unallowed (conveyor and cross cut). Shut down
0	1	1	0	0	0	0	0	Unallowed (No lub with conveyor). Shut down
0	1	1	1	0	0	0	0	Unallowed (2 saws and conveyor). Shut down
1	0	0	0	1	0	0	0	OK
1	0	0	1	1	0	0	1	OK
1	0	1	0	1	0	1	0	OK
1	0	1	1	0	0	0	0	Unallowed (2 saws). Shut down
1	1	0	0	1	1	0	0	OK
1	1	0	1	0	0	0	0	Unallowed (conveyor and cross cut). Shut down
1	1	1	0	1	1	1	0	OK
1	1	1	1	0	0	0	0	Unallowed (all on). Shut down

switches and, as a result, sixteen possible ON/OFF combinations for the motors. The states of the switches are the input variables and the states of the motors are the output variables. The unallowed conditions in this case are not treated as "don't cares." They are switch input conditions that should not occur but if they do, the system must be shut down with all motors off.

Design of the Control Logic

There are four separate logic circuits, one for each of the motors. Let's begin by designing the logic circuit for the lubrication pump motor (output M_1). The first step is to transfer the data from the truth table to a Karnaugh map and develop an SOP expression.

The switch variables S_1, S_2, S_3, and S_4 are map variables and the states of M_1 are plotted and grouped as shown in Figure 5–45(a). The 0s on the map are for switch conditions when the motor is off and the 1s are for switch conditions when the motor is on. The resulting SOP expression for the lubrication pump motor logic results in the NAND implementation shown in part (b).

FIGURE 5–45
Karnaugh map simplification and implementation for the lubrication motor logic.

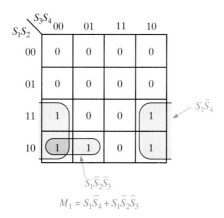

$$M_1 = S_1\bar{S}_4 + S_1\bar{S}_2\bar{S}_3$$

(a)

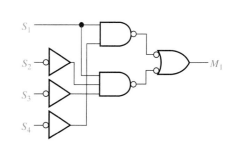

(b)

■ THE DIGITAL WORKBENCH

The lubrication motor logic has been designed and now its your turn to complete the control logic design for the remaining three motors. When you have completed the design, it should agree with the logic on the system board shown on the workbench. All of the connections to the V_{CC} and ground pads are on the back side of the board. Circuit interconnections on the back side of the board feed through to the pads on the component side. When you trace out the circuit, common pads are generally oriented vertically or horizontally and not at an angle with respect to each other so that it should be relatively easy to establish the unseen interconnections. Figure 5–46 on p. 246 shows the pin diagrams for the ICs on the board.

■ DIGITAL WORKBENCH 1: Design

- *Activity 1* Using Table 5–6 and the Karnaugh map method, design the logic for controlling the conveyor motor using NAND gates and inverters.

- *Activity 2* Design the logic for controlling the band saw motor.

- *Activity 3* Design the logic for controlling the cross-cut saw motor.

- *Activity 4* Combine the logic for each of the four motor controls into a complete logic diagram. Specify the specific devices to be used.

- *Activity 5* Trace out the printed circuit board and draw the resulting logic diagram with the aid of the IC logic pin diagrams in Figure 5–46. The logic that you designed should agree with the logic on the circuit board (with the exception of some pin numbers).

- *Activity 6* Label the inputs and outputs on the circuit board using the variables in the truth table.

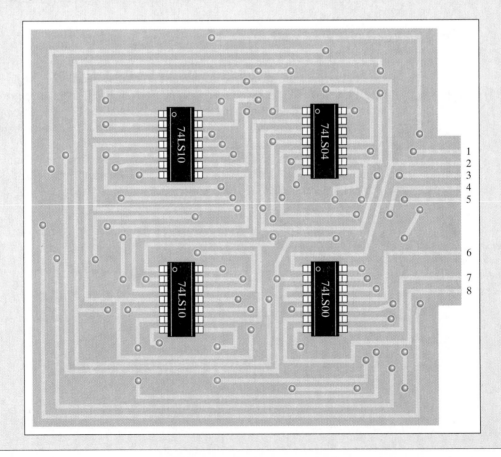

▪ DIGITAL WORK BENCH 2: Testing and Troubleshooting

▪ *Activity 1* To test the control logic board, a device (binary counter) that produces a sequence of binary numbers for 0 through 15 is connected to the board to simulate the switch inputs according to Table 5–6. A logic analyzer is used to observe the output waveforms. For each set of output waveforms, determine if the board is operating properly and, if not, specify the possible fault or faults.

▪ *Activity 2* Devise a manual method that can be used to check out the control logic.

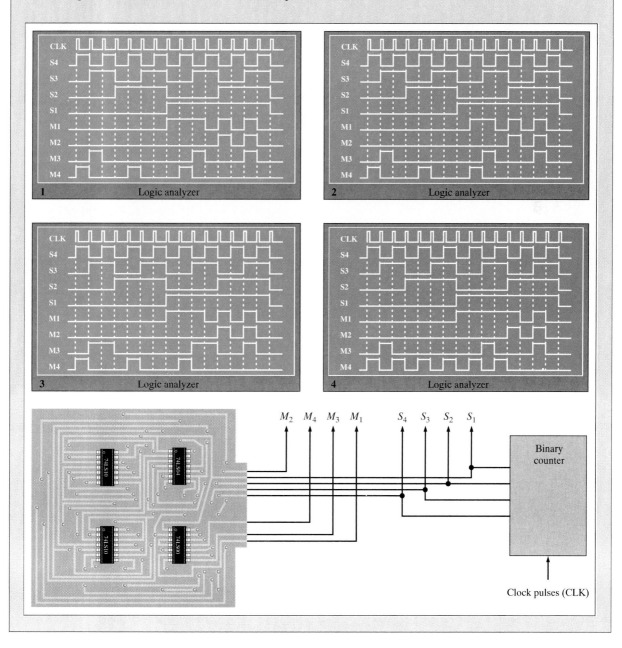

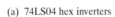

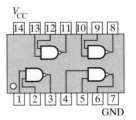

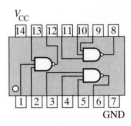

(a) 74LS04 hex inverters

(b) 74LS00 quad 2-input NAND

(c) 74LS10 triple 3-input NAND

FIGURE 5–46
Logic pin diagrams.

SECTION 5–7 REVIEW	1. State the purpose of the control logic in this application. 2. What happens if switches 1 and 3 are on at the same time with switch 2 and switch 4 off? 3. What happens if switches 2 and 4 are on at the same time with switch 1 and switch 3 off?

■ SUMMARY

- AND-OR logic produces an output expression in SOP form.
- AND-OR-Invert logic produces a complemented SOP form, which is actually a POS form.
- The operational symbol for exclusive-OR is $\oplus$. An exclusive-OR expression can be stated in two equivalent ways:

$$A\overline{B} + \overline{A}B = A \oplus B$$

- To perform analysis of a logic circuit, start with the logic circuit, and develop the Boolean output expression or the truth table or both.
- Implementation of a logic circuit is the process in which you start with the Boolean output expression or the truth table and develop a logic circuit that produces that output function.
- All NAND or NOR logic diagrams should be drawn using appropriate dual symbols so that bubble outputs are connected to bubble inputs and nonbubble outputs are connected to nonbubble inputs.
- When two negation indicators (bubbles) are connected, they effectively cancel each other.

■ SELF-TEST

1. The output expression for an AND-OR circuit having one AND gate with inputs A, B, C, and D and one AND gate with inputs E and F is
   (a) $ABCDEF$ (b) $A + B + C + D + E + F$
   (c) $(A + B + C + D)(E + F)$ (d) $ABCD + EF$

2. A logic circuit with an output $X = A\overline{B}C + A\overline{C}$ consists of
   (a) two AND gates and one OR gate
   (b) two AND gates, one OR gate, and two inverters
   (c) two OR gates, one AND gate, and two inverters
   (d) two AND gates, one OR gate, and one inverter

3. To implement the expression $\overline{A}BCD + A\overline{B}CD + AB\overline{C}\,\overline{D}$, it takes one OR gate and
   (a) one AND gate   (b) three AND gates
   (c) three AND gates and four inverters   (d) three AND gates and three inverters

4. The expression $\overline{A}BCD + ABC\overline{D} + A\overline{B}\,\overline{C}D$

   (a) cannot be simplified

   (b) can be simplified to $\overline{A}BC + A\overline{B}$

   (c) can be simplified to $ABC\overline{D} + \overline{A}B\overline{C}$

   (d) is none of the above answers

5. The output expression for an AND-OR-Invert circuit having one AND gate with inputs A, B, C, and D and one AND gate with inputs E and F is

   (a) $ABCD + EF$

   (b) $\overline{A} + \overline{B} + \overline{C} + \overline{D} + \overline{E} + \overline{F}$

   (c) $\overline{(A + B + C + D)(E + F)}$

   (d) $(\overline{A} + \overline{B} + \overline{C} + \overline{D})(\overline{E} + \overline{F})$

6. An exclusive-OR function is expressed as

   (a) $\overline{A}\,\overline{B} + AB$

   (b) $\overline{A}B + A\overline{B}$

   (c) $(\overline{A} + B)(A + \overline{B})$

   (d) $(\overline{A} + \overline{B}) + (A + B)$

7. The AND operation can be produced with

   (a) two NAND gates   (b) three NAND gates

   (c) one NOR gate   (d) three NOR gates

8. The OR operation can be produced with

   (a) two NOR gates   (b) three NAND gates

   (c) four NAND gates   (d) both answers (a) and (b)

9. When using dual symbols in a logic diagram,

   (a) bubble outputs are connected to bubble inputs

   (b) the NAND symbols produce the AND operations

   (c) the negative-OR symbols produce the OR operations

   (d) all of these answers are true

   (e) none of these answers is true

10. All Boolean expressions can be implemented with

    (a) NAND gates only

    (b) NOR gates only

    (c) combinations of NAND and NOR gates

    (d) combinations of AND gates, OR gates, and inverters

    (e) any of these

■ **PROBLEMS**

SECTION 5–1   Special Combinational Logic Circuits

1. Sketch the distinctive-shape logic diagram for a 3-wide, 4-input AND-OR-Invert circuit. Also draw the standard rectangular-outline symbol.

2. Write the output expression for each circuit in Figure 5–47.

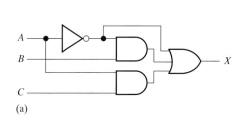

(a)

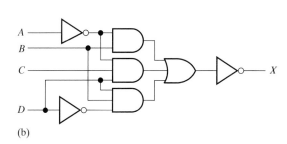

(b)

FIGURE 5–47

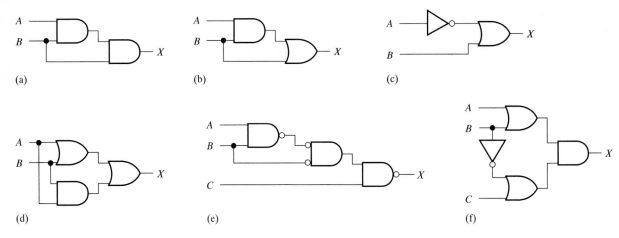

FIGURE 5–48

3. Write the output expression for each circuit as it appears in Figure 5–48.

4. Write the output expression for each circuit as it appears in Figure 5–49 and then change each circuit to an equivalent AND-OR configuration.

5. Develop the truth table for each circuit in Figure 5–48.

6. Develop the truth table for each circuit in Figure 5–49.

7. Show that an exclusive-NOR circuit produces a POS output.

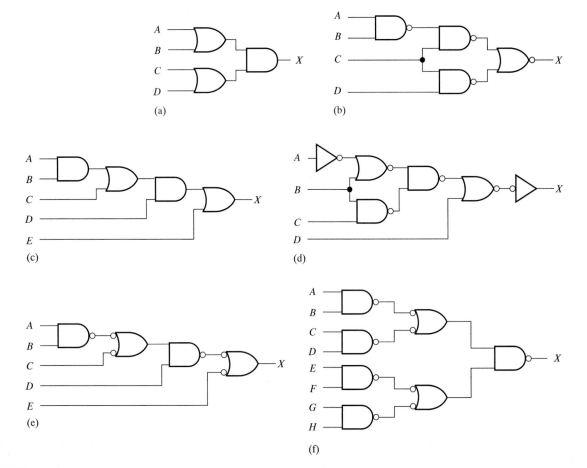

FIGURE 5–49

SECTION 5–2 Implementing Combinational Logic

8. Use AND gates, OR gates, or combinations of both to implement the following logic expressions as stated:

    (a) $X = AB$ (b) $X = A + B$ (c) $X = AB + C$

    (d) $X = ABC + D$ (e) $X = A + B + C$ (f) $X = ABCD$

    (g) $X = A(CD + B)$ (h) $X = AB(C + DEF) + CE(A + B + F)$

9. Use AND gates, OR gates, and inverters as needed to implement the following logic expressions as stated:

    (a) $X = AB + \overline{B}C$ (b) $X = A(B + \overline{C})$

    (c) $X = A\overline{B} + AB$ (d) $X = \overline{ABC} + B(EF + \overline{G})$

    (e) $X = A[BC(A + B + C + D)]$ (f) $X = B(C\overline{DE} + \overline{E}FG)(\overline{AB} + C)$

10. Use NAND gates, NOR gates, or combinations of both to implement the following logic expressions as stated:

    (a) $X = \overline{A}B + CD + (\overline{A + B})(ACD + \overline{BE})$ (b) $X = ABC\,\overline{D} + D\overline{E}F + \overline{AF}$

    (c) $X = \overline{A}[B + \overline{C}(D + E)]$

11. Implement a logic circuit for the truth table in Table 5–7.

12. Implement a logic circuit for the truth table in Table 5–8.

TABLE 5–7

Inputs			Output
A	B	C	X
0	0	0	1
0	0	1	0
0	1	0	1
0	1	1	0
1	0	0	1
1	0	1	0
1	1	0	1
1	1	1	1

TABLE 5–8

Inputs				Output
A	B	C	D	X
0	0	0	0	0
0	0	0	1	0
0	0	1	0	1
0	0	1	1	1
0	1	0	0	1
0	1	0	1	0
0	1	1	0	0
0	1	1	1	0
1	0	0	0	1
1	0	0	1	1
1	0	1	0	1
1	0	1	1	1
1	1	0	0	0
1	1	0	1	0
1	1	1	0	0
1	1	1	1	1

13. Simplify the circuit in Figure 5–50 as much as possible, and verify that the simplified circuit is equivalent to the original by showing that the truth tables are identical.

14. Repeat Problem 13 for the circuit in Figure 5–51.

15. Minimize the gates required to implement the functions in each part of Problem 9 in SOP form.

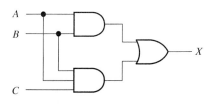

FIGURE 5–50

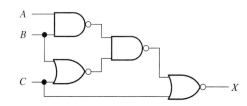

FIGURE 5–51

16. Minimize the gates required to implement the functions in each part of Problem 10 in SOP form.

17. Minimize the gates required to implement the function of the circuit in each part of Figure 5–49 in SOP form.

SECTION 5–3   The Universal Property of NAND and NOR Gates

18. Implement the logic circuits in Figure 5–47 using only NAND gates.

19. Implement the logic circuits in Figure 5–51 using only NAND gates.

20. Repeat Problem 18 using only NOR gates.

21. Repeat Problem 19 using only NOR gates.

SECTION 5–4   Combinational Logic Using NAND and NOR Gates

22. Show how the following expressions can be implemented as stated using only NOR gates:

(a) $X = ABC$ (b) $X = \overline{ABC}$ (c) $X = A + B$

(d) $X = A + B + \overline{C}$ (e) $X = \overline{AB} + \overline{CD}$ (f) $X = (A + B)(C + D)$

(g) $X = AB[C(\overline{DE} + \overline{AB}) + \overline{BCE}]$

23. Repeat Problem 22 using only NAND gates.

24. Implement each function in Problem 8 by using only NAND gates.

25. Implement each function in Problem 9 by using only NAND gates.

SECTION 5–5   Operation with Pulse Waveforms

26. Given the logic circuit and the input waveforms in Figure 5–52, sketch the output waveform.

FIGURE 5–52

27. For the logic circuit in Figure 5–53, sketch the output waveform in proper relationship to the inputs.

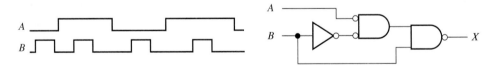

FIGURE 5–53

28. For the input waveforms in Figure 5–54, what logic circuit will generate the output waveform shown?

FIGURE 5–54

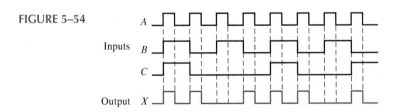

29. Repeat Problem 28 for the waveforms in Figure 5–55.

FIGURE 5–55

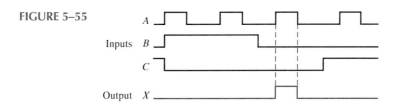

30. For the circuit in Figure 5–56, sketch the waveforms at the numbered points in the proper relationship to each other.

FIGURE 5–56

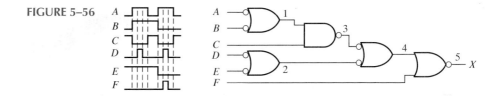

31. Assuming a propagation delay through each gate of 10 nanoseconds (ns), determine if the *desired* output waveform X in Figure 5–57 (a pulse with a minimum $t_W = 25$ ns positioned as shown) will be generated properly with the given inputs.

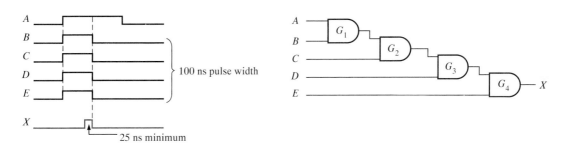

FIGURE 5–57

![SECTION 5–6 Troubleshooting]

SECTION 5–6 Troubleshooting

32. For the logic circuit and the input waveforms in Figure 5–58, the indicated output waveform is observed. Determine if this is the correct output waveform.

FIGURE 5–58

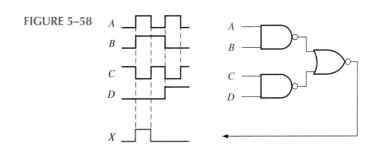

33. The output waveform in Figure 5–59 is incorrect for the inputs that are applied to the circuit. Assuming that one gate in the circuit has failed, with its output either an apparent constant HIGH or a constant LOW, determine the faulty gate and the type of failure (output open or shorted).

FIGURE 5–59

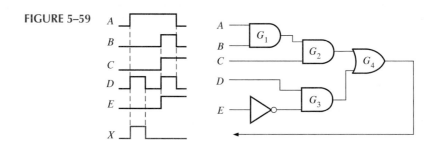

34. Repeat Problem 33 for the circuit in Figure 5–60, with input and output waveforms as shown.

35. The node in Figure 5–61 has been found to be stuck in the LOW state. A pulser and current tracer yield the series of indications shown in the diagram. From these observations, what do you conclude?

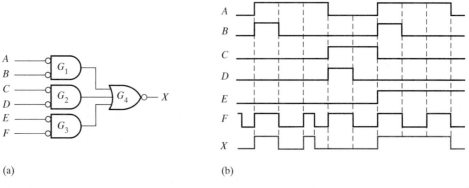

FIGURE 5–60

FIGURE 5–61

36. Figure 5–62(a) is a logic network under test. Figure 5–62(b) shows the waveforms as observed on a logic analyzer. The output waveform is incorrect for the inputs that are applied to the cir-

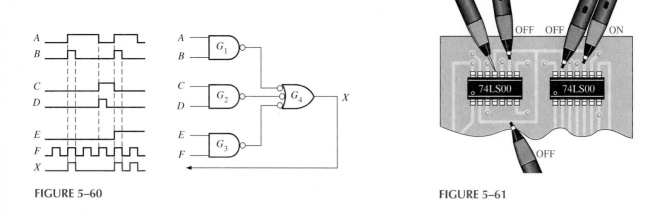

(a)

(b)

FIGURE 5–62

cuit. Assuming that one gate in the network has failed, with its output either an apparent constant HIGH or a constant LOW, determine the faulty gate and the type of failure.

37. The logic circuit in Figure 5–63 has the input waveforms shown.

    (a) Determine the correct output waveform in relation to the inputs.

    (b) Determine the output waveform if the output of gate G_3 is open.

    (c) Determine the output waveform if the upper input to gate G_5 is shorted to ground.

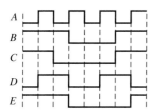
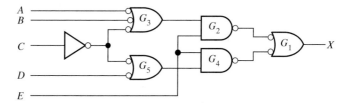

FIGURE 5–63

38. The logic circuit in Figure 5–64 has only one intermediate test point available besides the output, as indicated. For the inputs shown, you observe the indicated waveform at the test point. Is this waveform correct? If not, what are the possible faults that would cause it to appear as it does?

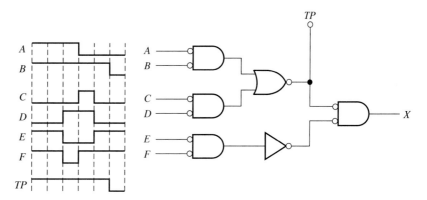

FIGURE 5–64

■ System Applications

SECTION 5–7   Digital System Application

39. Implement the lubrication motor logic in Figure 5–45 on p. 243 with NOR gates and inverters.

40. Repeat Activities 1 through 3 in Digital Workbench 1 using NOR gates and inverters.

41. Modify the motor control logic for the condition $S_1S_2S_3S_4 = 0101$ to shut down the conveyor but keep the cross-cut saw running.

Special Design Problems

42. Design a logic circuit to produce a HIGH output if and only if the input, represented by a 4-bit binary number, is greater than twelve or less than three. First develop the truth table and then draw the logic diagram.

43. Develop the logic circuit necessary to meet the following requirements:
A lamp in a room is to be operated from two switches, one at the back door and one at the front door. The lamp is to be on if the front switch is on and the back switch is off, or if the front switch is off and the back switch is on. The lamp is to be off if both switches are off or if both switches are on. Let a HIGH output represent the on condition and a LOW output represent the off condition.

44. Design a circuit to add to the control logic developed in Digital Workbench 1 that will indicate a shut-down condition with a red light.

45. Develop the NAND logic for a hexadecimal keypad encoder that will convert each key closure to binary.

■ **ANSWERS TO SECTION REVIEWS**

SECTION 5–1

1. (a) $\overline{AB + CD} = \overline{1 \cdot 0 + 1 \cdot 0} = 1$ (b) $\overline{AB + CD} = \overline{1 \cdot 1 + 0 \cdot 1} = 0$
   (c) $\overline{AB + CD} = \overline{0 \cdot 1 + 1 \cdot 1} = 0$

2. (a) $A\overline{B} + \overline{A}B = 1 \cdot \overline{0} + \overline{1} \cdot 0 = 1$ (b) $A\overline{B} + \overline{A}B = 1 \cdot \overline{1} + \overline{1} \cdot 1 = 0$
   (c) $A\overline{B} + \overline{A}B = 0 \cdot \overline{1} + \overline{0} \cdot 1 = 1$ (d) $A\overline{B} + \overline{A}B = 0 \cdot \overline{0} + \overline{0} \cdot 0 = 0$

3. $X = 1$ when $ABC = 000, 011, 101, 110,$ and 111; $X = 0$ when $ABC = 001, 010,$ and 100

4. $X = AB + \overline{A}\,\overline{B}$; the circuit consists of two AND gates, one OR gate, and two inverters. See Figure 5–8(b) for diagram.

SECTION 5–2

1. (a) $X = ABC + AB + AC$: three AND gates, one OR gate
   (b) $X = AB(C + DE)$: three AND gates, one OR gate

2. $X = ABC + \overline{A}\,\overline{B}\,\overline{C}$; two AND gates, one OR gate, and three inverters

3. (a) $X = AB(C + 1) + AC = AB + AC$ (b) $X = AB(C + DE) = ABC + ABDE$

SECTION 5–3

1. (a) $X = \overline{A} + B$: a 2-input NAND gate with A and $\overline{B}$ on its inputs
   (b) $X = A\overline{B}$: a 2-input NAND with A and $\overline{B}$ on its inputs, followed by one NAND used as an inverter

2. (a) $X = \overline{A} + B$: a 2-input NOR with inputs $\overline{A}$ and B, followed by one NOR used as an inverter
   (b) $X = A\overline{B}$: a 2-input NOR with $\overline{A}$ and B on its inputs

SECTION 5–4

1. $X = \overline{(\overline{A + B} + \overline{C})DE}$: a 3-input NAND with inputs, A, B, and C, with its output connected to a second 3-input NAND with two other inputs, D and E

2. $X = \overline{\overline{A}\,\overline{B}\,\overline{C} + \overline{(D + E)}}$: a 3-input NOR with inputs A, B, and C, with its output connected to a second 3-input NOR with two other inputs, D and E

SECTION 5–5

1. The exclusive-OR output is a $15\,\mu s$ pulse followed by a $25\,\mu s$ pulse, with a separation of $10\,\mu s$ between the pulses.

2. The output of the exclusive-NOR is HIGH when both inputs are HIGH or when both inputs are LOW.

SECTION 5–6

1. Common gate failures are input or output open; input or output shorted to ground.

2. Input shorted to V_{CC} causes output to be stuck LOW.

3. (a) G_4 output is HIGH until rising edge of seventh pulse, then it goes LOW.
   (b) G_4 output is the same as input D.
   (c) G_4 output is the same as G_2 output shown in Figure 5–43(b).

SECTION 5–7

1. The control logic decodes switch inputs and produces motor control outputs.

2. Condition is allowed. The bandsaw runs.

3. Unallowed condition because the conveyor and the cross-cut saw can't be on at same time (shutdown) and also because the conveyor is on with no lubrication.

6

FUNCTIONS OF COMBINATIONAL LOGIC

■ CHAPTER OBJECTIVES

- ☐ Distinguish between half-adders and full-adders
- ☐ Use full-adders to implement multibit parallel binary adders
- ☐ Explain the differences between ripple carry and look-ahead carry parallel adders
- ☐ Use the magnitude comparator to determine the relationship between two binary numbers and use cascaded comparators to handle the comparison of larger numbers
- ☐ Implement a basic binary decoder
- ☐ Use BCD-to-7-segment decoders in display systems
- ☐ Apply a decimal-to-BCD priority encoder in a simple keyboard application
- ☐ Convert from BCD to binary, binary to Gray code, and Gray code to binary by using logic devices
- ☐ Apply multiplexers in data selection, multiplexed displays, logic function generation, and simple communications systems
- ☐ Use decoders as demultiplexers
- ☐ Explain the meaning of parity
- ☐ Use parity generators and checkers to detect bit errors in digital systems
- ☐ Implement a simple data communications system
- ☐ Identify glitches, common bugs in digital systems

■ CHAPTER OVERVIEW

In this chapter, several types of MSI combinational logic circuits are introduced including adders, comparators, decoders, encoders, code converters, multiplexers (data selectors), demultiplexers, and parity generators/checkers. Examples of system applications of many of these devices are included to demonstrate how the devices can be used in practical situations.

■ SPECIFIC DEVICES

74LS00	7442A	7447
74LS48	74LS83A	7485
74LS138	74LS139	74LS147
74148	74150	74151A
74154	74157	74180
74184	74185	74LS283

■ DIGITAL SYSTEM APPLICATION

This Digital System Application illustrates concepts taught in this chapter and deals with one portion of a traffic light control system. The system application in the next chapter presents another approach, while the system applications in Chapters 8, 9, and 11 focus on other parts of the system. Basically, this system controls the traffic light at the intersection of a busy street and a lightly traveled side street. The system includes a combinational logic portion to which the topics in this chapter and the previous chapter apply, a timing circuit portion to which Chapter 8 applies, and a sequential logic portion to which both Chapters 8 and 9 apply.

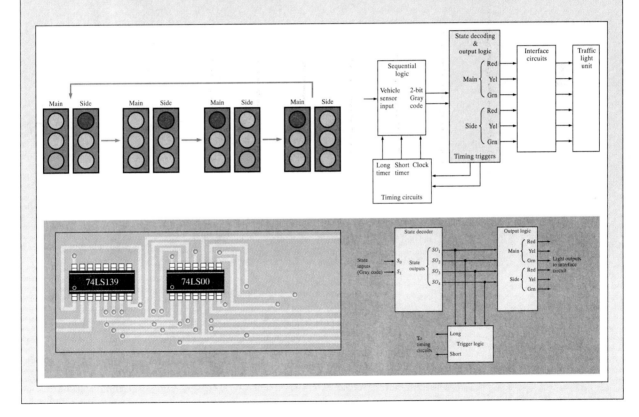

6–1 ■ BASIC ADDERS

Adders are important not only in computers, but in many types of digital systems in which numerical data are processed. An understanding of the basic adder operation is fundamental to the study of digital systems. In this section, the half-adder and the full-adder are introduced. After completing this section, you should be able to

☐ Describe the function of a half-adder   ☐ Draw a half-adder logic diagram   ☐ Describe the function of the full-adder   ☐ Draw a full-adder logic diagram using half-adders   ☐ Implement a full-adder using AND-OR logic

The Half-Adder

Recall the basic rules for binary addition as stated in Chapter 2.

$$0 + 0 = 0$$
$$0 + 1 = 1$$
$$1 + 0 = 1$$
$$1 + 1 = 10$$

These operations are performed by a logic circuit called a **half-adder.**

> **The half-adder accepts two binary digits on its inputs and produces two binary digits on its outputs, a sum bit and a carry bit.**

A half-adder is represented by the logic symbol in Figure 6–1.

FIGURE 6–1
Logic symbol for a half-adder.

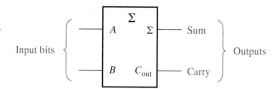

From the logical operation of the half-adder as stated in Table 6–1, expressions can be derived for the sum and the output carry as functions of the inputs. Notice that the output carry (C_{out}) is a 1 only when both A and B are 1s; therefore, C_{out} can be expressed as the AND of the input variables.

$$C_{out} = AB \tag{6–1}$$

Now observe that the sum output (Σ) is a 1 only if the input variables, A and B, are not equal. The sum can therefore be expressed as the exclusive-OR of the input variables.

$$\Sigma = A \oplus B \tag{6–2}$$

TABLE 6–1
Half-adder truth table

A	B	C_{out}	Σ
0	0	0	0
0	1	0	1
1	0	0	1
1	1	1	0

Σ = sum

C_{out} = output carry

A and B = input variables (operands)

FIGURE 6–2

Half-adder logic diagram.

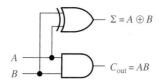

$\Sigma = A \oplus B$

$C_{out} = AB$

From Equations (6–1) and (6–2), the logic implementation required for the half-adder function can be developed. The output carry is produced with an AND gate with A and B on the inputs, and the sum output is generated with an exclusive-OR gate, as shown in Figure 6–2.

The Full-Adder

The second basic category of adder is the **full-adder.**

> **The full-adder accepts three inputs including an input carry and generates a sum output and an output carry.**

The basic difference between a full-adder and a half-adder is that the full-adder accepts an input carry. A logic symbol for a full-adder is shown in Figure 6–3, and the truth table in Table 6–2 shows the operation of a full-adder.

FIGURE 6–3

Logic symbol for a full-adder.

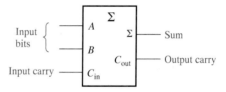

TABLE 6–2

Full-adder truth table

A	B	C_{in}	C_{out}	Σ
0	0	0	0	0
0	0	1	0	1
0	1	0	0	1
0	1	1	1	0
1	0	0	0	1
1	0	1	1	0
1	1	0	1	0
1	1	1	1	1

C_{in} = input carry, sometimes designated as CI

C_{out} = output carry, sometimes designated as CO

Σ = sum

A and B = input variables (operands)

The full-adder must add the two input bits and the input carry. From the half-adder we know that the sum of the input bits A and B is the exclusive-OR of those two variables, $A \oplus B$. For the input carry (C_{in}) to be added to the input bits, it must be exclusive-ORed with $A \oplus B$, yielding the equation for the sum output of the full-adder.

$$\Sigma = (A \oplus B) \oplus C_{in} \tag{6–3}$$

This means that to implement the full-adder sum function, two exclusive-OR gates can be used. The first must generate the term $A \oplus B$, and the second has as its inputs the output of the first XOR gate and the input carry, as illustrated in Figure 6–4(a).

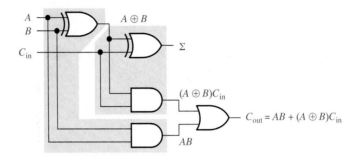

(a) Logic required to form the sum of three bits

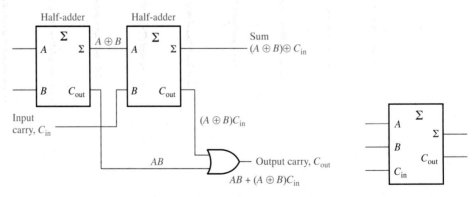

(b) Complete logic circuit for a full-adder (each half-adder is enclosed by a shaded area)

FIGURE 6–4
Full-adder logic.

The output carry is a 1 when both inputs to the first XOR gate are 1s or when both inputs to the second XOR gate are 1s. You can verify this fact by studying Table 6–2. The output carry of the full-adder is therefore produced by the inputs A ANDed with B and $A \oplus B$ ANDed with C_{in}. These two terms are ORed, as expressed in Equation (6-4). This function is implemented and combined with the sum logic to form a complete full-adder circuit, as shown in Figure 6–4(b).

$$C_{out} = AB + (A \oplus B)C_{in} \qquad \textbf{(6–4)}$$

Notice in Figure 6–4(b) there are two half-adders, connected as shown in the block diagram of Figure 6–5(a), with their output carries ORed. The logic symbol shown in Figure 6–5(b) will normally be used to represent the full-adder.

(a) Arrangement of two half-adders to form a full-adder   (b) Full-adder logic symbol

FIGURE 6–5
Full-adder implemented with half-adders.

EXAMPLE 6–1   Determine an alternative method for implementing the full-adder.

Solution   Referring to Table 6–2, you can write sum-of-products expressions for both Σ and C_{out} by observing the input conditions that make them 1s. The expressions are as follows:

$$\Sigma = \overline{A}\,\overline{B}C_{\text{in}} + \overline{A}B\overline{C}_{\text{in}} + A\overline{B}\,\overline{C}_{\text{in}} + ABC_{\text{in}}$$
$$C_{\text{out}} = \overline{A}BC_{\text{in}} + A\overline{B}C_{\text{in}} + AB\overline{C}_{\text{in}} + ABC_{\text{in}}$$

Mapping these two expressions on the Karnaugh maps in Figure 6–6, you find that the sum (Σ) expression cannot be simplified. The output carry expression (C_{out}) is reduced as indicated.

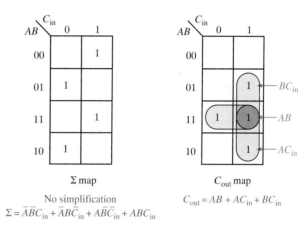

FIGURE 6–6

These two expressions are implemented with AND-OR logic as shown in Figure 6–7 to form a complete full-adder.

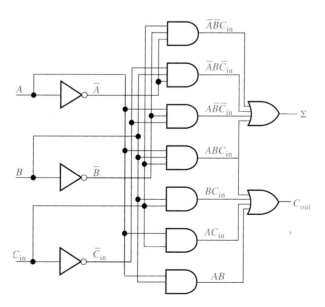

FIGURE 6–7

Related Exercise   Implement the full-adder in Figure 6–7 using only NAND gates.

6–2 ■ PARALLEL BINARY ADDERS

Adders that are available in integrated circuit form are parallel binary adders. In this section, you will learn the basics of this type of adder so that you will understand all the necessary input and output functions when working with these devices. After completing this section, you should be able to

☐ Use full-adders to implement a parallel binary adder   ☐ Explain the addition process in a parallel binary adder   ☐ Use the truth table for a 4-bit parallel adder
☐ Apply the 74LS83A and the 74LS283 for the addition of two 4-bit numbers
☐ Expand the 4-bit adder to accommodate 8-bit, 16-bit, or 32-bit addition

As you saw in Section 6–1, a single full-adder is capable of adding two 1-bit numbers and an input carry. To add binary numbers with more than one bit, additional full-adders must be used. When one binary number is added to another, each column generates a sum bit and a 1 or 0 carry bit to the next column to the left, as illustrated here with 2-bit numbers.

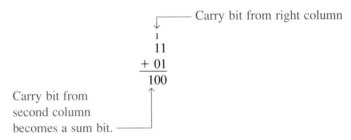

To implement the addition of binary numbers, a full-adder is required for each bit in the numbers. So for 2-bit numbers, two adders are needed; for 4-bit numbers, four adders are used; and so on. The carry output of each adder is connected to the carry input of the next higher-order adder, as shown in Figure 6–8 for a 2-bit adder. Notice that either a half-adder can be used for the least significant position or the carry input of a full-adder can be made 0 (grounded) because there is no carry input to the least significant bit position.

In Figure 6–8 the least significant bits (LSB) of the two numbers are represented by A_1 and B_1. The next higher-order bits are represented by A_2 and B_2. The three sum bits are Σ_1, Σ_2, and Σ_3. Notice that the output carry from the left-most full-adder becomes the most significant bit (MSB) in the sum, Σ_3.

FIGURE 6–8

Block diagram of a basic 2-bit parallel adder.

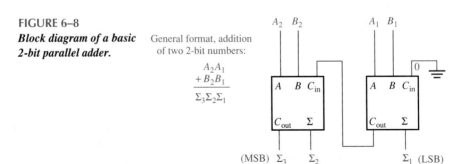

EXAMPLE 6–2   Verify that the 2-bit parallel adder in Figure 6–9 properly performs the following addition:

$$\begin{array}{r} 11 \\ + \ 10 \\ \hline 101 \end{array}$$

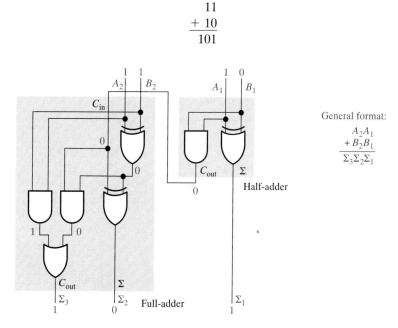

General format:
$$\begin{array}{r} A_2 A_1 \\ + B_2 B_1 \\ \hline \Sigma_3 \Sigma_2 \Sigma_1 \end{array}$$

FIGURE 6–9

Solution   The logic level at each point in the circuit for the given input numbers is determined from the truth table of the relevant gate. By following these levels through the circuit as indicated on the logic diagram, you find that the proper levels appear on the sum outputs.

Related Exercise   Replace the half-adder with a full-adder and repeat the addition process. Assume an input carry of 0.

Four-Bit Parallel Adders

A basic 4-bit parallel adder is implemented with four full-adders as shown in Figure 6–10, on the next page. Again, the LSBs (A_1 and B_1) in each number being added go into the right-most full-adder; the higher-order bits are applied as shown to the successively higher-order adders, with the MSBs (A_4 and B_4) in each number being applied to the left-most full-adder. The carry output of each adder is connected to the carry input of the next higher-order adder as indicated. These are called *internal carries.*

In keeping with most manufacturers' data sheets, the input labeled C_0 is the input carry to the least significant bit adder; C_4, in the case of four bits, is the output carry of the most significant bit adder; and Σ_1 (LSB) through Σ_4 (MSB) are the sum outputs. The logic symbol is shown in Figure 6–10(b).

Truth Table for a 4-Bit Parallel Adder

Table 6–3 is the truth table for a 4-bit adder. On some data sheets, truth tables may be called *function tables.* At first glance, this table may be confusing but, with a little practice, you will learn to interpret it easily. The internal carry, which is designated C_2 in the table, is the output carry from the addition, $A_2 A_1 + B_2 B_1$. This internal carry bit is then used to find the sum of $A_4 A_3 + B_4 B_3$, as well as the output carry, C_4.

FIGURE 6–10
A 4-bit parallel adder.

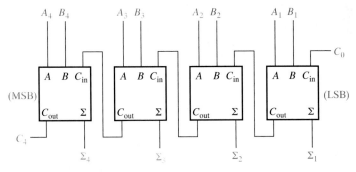

(a) Block diagram

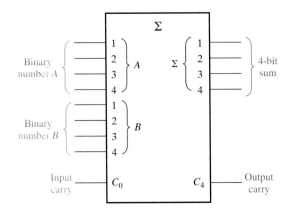

(b) Logic symbol

TABLE 6–3
Truth table for a 4-bit parallel adder

Inputs				Outputs When $C_0 = 0$		When $C_2 = 0$		When $C_0 = 1$		When $C_2 = 1$	
A_1 / A_3	B_1 / B_3	A_2 / A_4	B_2 / B_4	Σ_1 / Σ_3	Σ_2 / Σ_4	C_2 / C_4	Σ_1 / Σ_3	Σ_2 / Σ_4	C_2 / C_4
0	0	0	0	0	0	0	1	0	0
1	0	0	0	1	0	0	0	1	0
0	1	0	0	1	0	0	0	1	0
1	1	0	0	0	1	0	1	1	0
0	0	1	0	0	1	0	1	1	0
1	0	1	0	1	1	0	0	0	1
0	1	1	0	1	1	0	0	0	1
1	1	1	0	0	0	1	1	0	1
0	0	0	1	0	1	0	1	1	0
1	0	0	1	1	1	0	0	0	1
0	1	0	1	1	1	0	0	0	1
1	1	0	1	0	0	1	1	0	1
0	0	1	1	0	0	1	1	0	1
1	0	1	1	1	0	1	0	1	1
0	1	1	1	1	0	1	0	1	1
1	1	1	1	0	1	1	1	1	1

NOTE: Input conditions at A_1, B_1, A_2, B_2, and C_0 are used to determine outputs Σ_1 and Σ_2 and the value of the internal carry C_2. The values at C_2, A_3, B_3, A_4, and B_4 are then used to determine outputs Σ_3, Σ_4, and C_4.

EXAMPLE 6–3

Use the 4-bit parallel adder truth table (Table 6–3) to find the sum and output carry for the addition of the following two 4-bit numbers if the input carry is 0:

$$A_4A_3A_2A_1 = 1100 \qquad \text{and} \qquad B_4B_3B_2B_1 = 1100$$

Solution For $A_1 = 0$, $B_1 = 0$, $A_2 = 0$, $B_2 = 0$, and $C_0 = 0$, the first row in the output columns for $C_0 = 0$ in Table 6–3 shows that $\Sigma_1 = 0$, $\Sigma_2 = 0$, and $C_2 = 0$.

Next, for $A_3 = 1$, $B_3 = 1$, $A_4 = 1$, $B_4 = 1$, and $C_2 = 0$, the last row in the output columns for $C_2 = 0$ in the table shows that $\Sigma_3 = 0$, $\Sigma_4 = 1$, and $C_4 = 1$. The following addition agrees with the table:

$$
\begin{array}{r}
1100 \\
+\ 1100 \\
\hline
11000
\end{array}
$$

Related Exercise Use the truth table to find the result of adding the binary numbers 1011 and 1010.

The 74LS83A and 74LS283 MSI Adders

Examples of 4-bit parallel adders that are available as medium-scale integrated (MSI) circuits are the 74LS83A and the 74LS283 low-power Schottky TTL devices. These devices are also available in other logic families such as standard TTL (7483A and 74283) and CMOS (74HC283). The 74LS83A and the 74LS283 are functionally identical to each other but not pin compatible; that is, the pin numbers for the inputs and outputs are different due to different power and ground pin connections. For the 74LS83A, V_{CC} is pin 5 and ground is pin 12 on the 16-pin package. For the 74LS283, V_{CC} is pin 16 and ground is pin 8, which is a more standard configuration. Logic symbols for both of these devices are shown, with pin numbers in parentheses, in Figure 6–11.

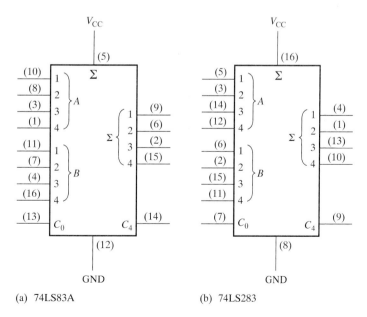

(a) 74LS83A   (b) 74LS283

FIGURE 6–11
MSI 4-bit parallel adders.

Adder Expansion

The 4-bit parallel adder can be expanded to handle the addition of two 8-bit numbers by using two 4-bit adders and connecting the carry input of the low-order adder (C_0) to ground because there is no carry into the least significant bit position and by connecting the carry output of the low-order adder to the carry input of the high-order adder as shown in Figure 6–12(a). This process is known as **cascading.** Notice that, in this case, the output carry is designated C_8 because it is generated from the eighth bit position. The low-order adder is the one that adds the lower or less significant four bits in the numbers and the high-order adder is the one that adds the higher or more significant four bits in the 8-bit numbers.

Similarly, four 4-bit adders can be cascaded to handle two 16-bit numbers as shown in Figure 6–12(b). Notice that the output carry is designated C_{16} because it is generated from the sixteenth bit position.

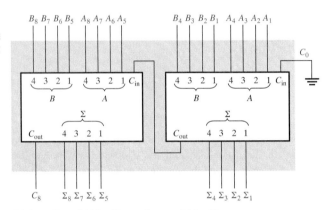

(a) Cascading of 4-bit adders to form an 8-bit adder

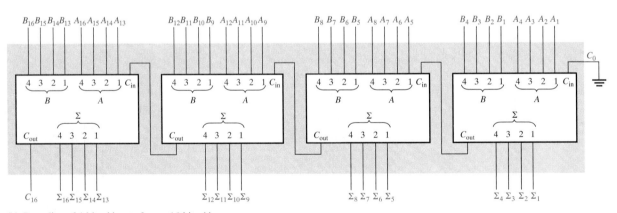

(b) Cascading of 4-bit adders to form a 16-bit adder

FIGURE 6–12
Examples of adder expansion.

EXAMPLE 6–4   Show how two 74LS83A adders can be connected to form an 8-bit parallel adder. Show output bits for the following 8-bit input numbers:

$$A_8A_7A_6A_5A_4A_3A_2A_1 = 10111001$$

and

$$B_8B_7B_6B_5B_4B_3B_2B_1 = 10011110$$

Solution Two 74LS83A 4-bit parallel adders are used to implement the 8-bit adder. The only connection between the two 74LS83As is the carry output (pin 14) of the low-order adder to the carry input (pin 13) of the high-order adder, as shown in Figure 6–13. Pin 13 of the low-order adder is grounded (no carry input).

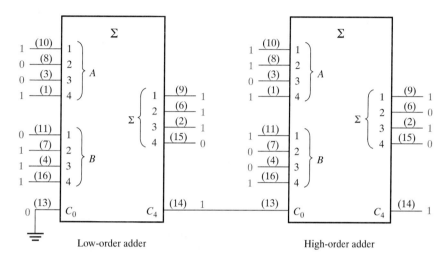

FIGURE 6–13
Two 74LS83A adders connected as an 8-bit parallel adder (pin numbers are in parentheses).

Related Exercise Use 74LS283 adders to implement a 12-bit parallel adder.

An Adder Application

An example of full-adder and parallel adder application is a simple voting system that can be used to simultaneously provide the number of "yes" votes and the number of "no" votes. For example, this type of system can be used where a group of people are assembled and there is a need for immediately determining opinions (for or against), making decisions, or voting on certain issues or other matters.

In its simplest form, the system includes a switch for "yes" or "no" selection at each position in the assembly and a digital display for the number of yes votes and one for the number of no votes. The basic system is shown in Figure 6–14 (on the next page) for a 6-position setup, but it can be expanded to any number of positions with additional 6-position modules and additional parallel adder and display circuits.

In Figure 6–14, each full-adder can produce the sum of up to three votes. The sum and output carry of each full-adder then goes to the two lower-order inputs of a parallel binary adder. The two higher-order inputs of the parallel adder are connected to ground (0) because there is never a case where the binary input exceeds 0011 (decimal 3). For this basic 6-position system, the outputs of the parallel adder go to a BCD-to-7-segment decoder which drives the 7-segment display. As mentioned, additional circuits must be included when the system is expanded.

The resistors from the inputs of each full-adder to ground assure that each input is LOW when the switch is in the neutral position (CMOS logic is used). When a switch is moved to the "yes" or to the "no" position, a HIGH level (V_{CC}) is applied to the associated full-adder input.

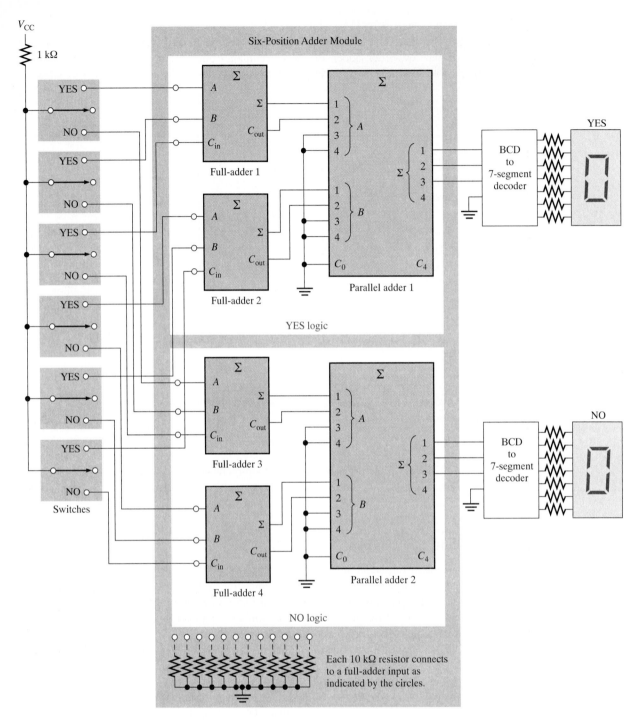

FIGURE 6–14
A voting system using full-adders and parallel binary adders.

1. Two 4-bit binary numbers (1101 and 1011) are applied to a 4-bit parallel adder. The input carry is 1. Determine the sum (Σ) and the output carry.
2. How many 74LS283 adders would be required to add two binary numbers each representing decimal numbers up through 1000_{10}?

6–3 ■ RIPPLE CARRY VERSUS LOOK-AHEAD CARRY ADDERS

Parallel adders can be placed into two categories based on the way in which internal carries from stage to stage are handled. Those categories are ripple carry and look-ahead carry. Externally, both types of adders are the same in terms of inputs and outputs. The difference is the speed at which they can add numbers. The look-ahead carry adder is much faster than the ripple carry adder. You will see why in this section. After completing this section, you should be able to

□ Discuss the difference between a ripple carry adder and a look-ahead carry adder
□ State the advantage of look-ahead carry addition □ Define *carry generation* and *carry propagation* and explain the difference □ Develop look-ahead carry logic
□ Explain why cascaded 74LS283s exhibit both ripple carry and look-ahead carry properties

The Ripple Carry Adder

A **ripple carry** adder is one in which the carry output of each full-adder is connected to the carry input of the next higher-order stage (a stage is one full-adder). The sum and the output carry of any stage cannot be produced until the input carry occurs; this causes a time delay in the addition process, as illustrated in Figure 6–15. The carry propagation delay for each full-adder is the time from the application of the input carry until the output carry occurs, assuming that the A and B inputs are already present.

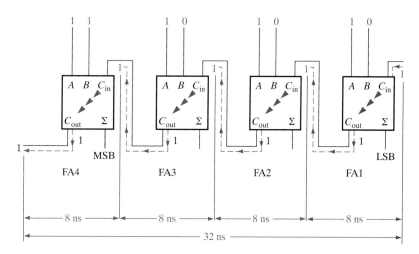

FIGURE 6–15

A 4-bit parallel ripple carry adder showing "worst-case" carry propagation delays.

Full-adder 1 (FA1) cannot produce a potential output carry until an input carry is applied. Full-adder 2 (FA2) cannot produce a potential output carry until full-adder 1 produces an output carry. Full-adder 3 (FA3) cannot produce a potential output carry until an output carry is produced by FA1 followed by an output carry from FA2, and so on. As you can see in Figure 6–15, the input carry to the least significant stage has to ripple through all the adders before a final sum is produced. The cumulative delay through all the adder stages is a "worst-case" addition time. The total delay can vary, depending on the carry bit produced by each full-adder. If two numbers are added such that no carries (0) occur between stages, the addition time is simply the propagation time through a single full-adder from the application of the data bits on the inputs to the occurrence of a sum output.

The Look-Ahead Carry Adder

The speed with which an addition can be performed is limited by the time required for the carries to propagate, or ripple, through all the stages of a parallel adder. One method of speeding up the addition process by eliminating this ripple carry delay is called **look-ahead carry** addition. The look-ahead carry adder anticipates the output carry of each stage, and based on the inputs, produces the output carry by either carry generation or carry propagation.

Carry generation occurs when an output carry is produced (generated) internally by the full-adder. A carry is generated only when both input bits are 1s. The generated carry, C_g, is expressed as the AND function of the two input bits, A and B.

$$C_g = AB \qquad\qquad (6\text{--}5)$$

Carry propagation occurs when the input carry is rippled to become the output carry. An input carry may be propagated by the full-adder when either or both of the input bits are 1s. The propagated carry, C_p, is expressed as the OR function of the input bits.

$$C_p = A + B \qquad\qquad (6\text{--}6)$$

The conditions for carry generation and carry propagation are illustrated in Figure 6–16. The three arrowheads symbolize ripple (propagation).

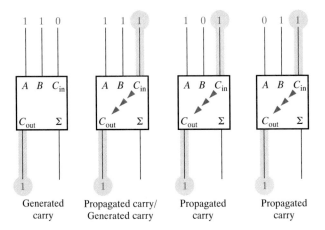

FIGURE 6–16
Illustration of conditions for carry generation and carry propagation.

The output carry of a full-adder can be expressed in terms of both the generated carry (C_g) and the propagated carry (C_p). The output carry (C_{out}) is a 1 if the generated carry is a 1 OR if the propagated carry is a 1 AND the input carry (C_{in}) is a 1. In other words, we get an output carry of 1 if it is generated by the full-adder ($A = 1$ AND $B = 1$) or if the adder propagates the input carry ($A = 1$ OR $B = 1$) AND $C_{in} = 1$. This relationship is expressed as

$$C_{out} = C_g + C_p C_{in} \qquad\qquad (6\text{--}7)$$

Now let's see how this concept can be applied to a parallel adder, whose individual stages are shown in Figure 6–17 for a 4-bit example. For each full-adder, the output carry is

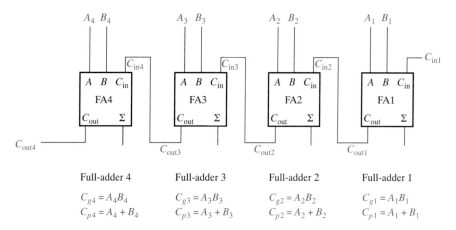

FIGURE 6–17
Carry generation and carry propagation in terms of the input bits to a 4-bit adder.

dependent on the generated carry (C_g), the propagated carry (C_p), and its input carry (C_{in}). The C_g and C_p functions for each stage are *immediately* available as soon as the input bits A and B and the input carry to the LSB adder are applied, because they are dependent only on these bits. The input carry to each stage is the output carry of the previous stage.

Based on this analysis, we can now develop expressions for the output carry, C_{out}, of each full-adder stage for the 4-bit example.

Full-adder 1:

$$C_{out1} = C_{g1} + C_{p1}C_{in1} \qquad (6\text{–}8)$$

Full-adder 2:

$$C_{in2} = C_{out1}$$
$$\begin{aligned} C_{out2} &= C_{g2} + C_{p2}C_{in2} \\ &= C_{g2} + C_{p2}C_{out1} \\ &= C_{g2} + C_{p2}(C_{g1} + C_{p1}C_{in1}) \end{aligned}$$
$$C_{out2} = C_{g2} + C_{p2}C_{g1} + C_{p2}C_{p1}C_{in1} \qquad (6\text{–}9)$$

Full-adder 3:

$$C_{in3} = C_{out2}$$
$$\begin{aligned} C_{out3} &= C_{g3} + C_{p3}C_{in3} \\ &= C_{g3} + C_{p3}C_{out2} \\ &= C_{g3} + C_{p3}(C_{g2} + C_{p2}C_{g1} + C_{p2}C_{p1}C_{in1}) \end{aligned}$$
$$C_{out3} = C_{g3} + C_{p3}C_{g2} + C_{p3}C_{p2}C_{g1} + C_{p3}C_{p2}C_{p1}C_{in1} \qquad (6\text{–}10)$$

Full-adder 4:

$$C_{in4} = C_{out3}$$
$$\begin{aligned} C_{out4} &= C_{g4} + C_{p4}C_{in4} \\ &= C_{g4} + C_{p4}C_{out3} \\ &= C_{g4} + C_{p4}(C_{g3} + C_{p3}C_{g2} + C_{p3}C_{p2}C_{g1} + C_{p3}C_{p2}C_{p1}C_{in1}) \end{aligned}$$
$$C_{out4} = C_{g4} + C_{p4}C_{g3} + C_{p4}C_{p3}C_{g2} + C_{p4}C_{p3}C_{p2}C_{g1} + C_{p4}C_{p3}C_{p2}C_{p1}C_{in1} \qquad (6\text{–}11)$$

Notice that in each of these expressions, the output carry for each full-adder stage is dependent only on the initial input carry (C_{in1}), the C_g and C_p functions of that stage, and the C_g and C_p functions of the preceding stages. Since each of the C_g and C_p functions can be expressed in terms of the A and B inputs to the full-adders, all the output carries are immediately available (except for gate delays), and you do not have to wait for a carry to ripple through all the stages before a final result is achieved. Thus, the look-ahead carry technique speeds up the addition process.

Equations (6–8) through (6–11) are implemented with logic gates and connected to the full-adders to create a 4-bit look-ahead carry adder, as shown in Figure 6–18.

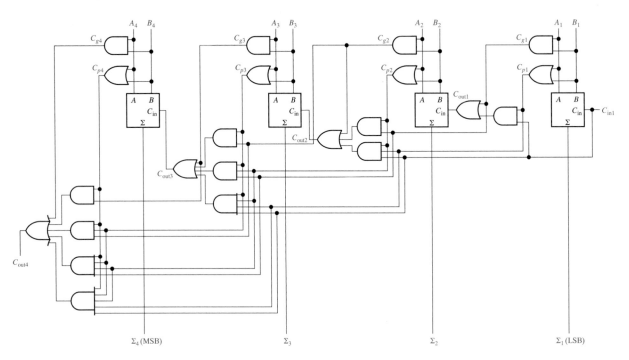

FIGURE 6–18
Logic diagram for a 4-stage look-ahead carry adder.

Combination Look-Ahead and Ripple Carry Adders

The 74LS83A and the 74LS283 4-bit adders that were introduced in Section 6–2 are look-ahead carry adders. When these adders are cascaded to expand their capability to handle binary numbers with more than four bits, the output carry of one adder is connected to the input carry of the next. This creates a ripple carry condition between the 4-bit adders so that when two or more 74LS83As or 74LS283s are cascaded, the resulting adder is actually a combination look-ahead and ripple carry adder. The look-ahead carry operation is internal to each MSI adder and the ripple carry feature comes into play when there is a carry out of one of the adders to the next one.

SECTION 6–3 REVIEW	1. The input bits to a full-adder are $A = 1$ and $B = 0$. Determine C_g and C_p.
	2. Determine the output carry of a full-adder when $C_{in} = 1$, $C_g = 0$, and $C_p = 1$.

6–4 ■ COMPARATORS

The basic function of a comparator is to compare the magnitudes of two binary quantities to determine the relationship of those quantities. In its simplest form, a comparator circuit determines whether two numbers are equal. After completing this section, you should be able to

☐ Use the exclusive-OR gate as a basic comparator   ☐ Analyze the internal logic of a magnitude comparator that has both equality and inequality outputs   ☐ Apply the 7485 magnitude comparator to the comparison of two 4-bit numbers   ☐ Cascade 7485s to expand a comparator to eight or more bits

As you learned in Chapter 3, the exclusive-OR gate can be used as a basic comparator because its output is a 1 if the two input bits are not equal and a 0 if the input bits are equal. Figure 6–19 shows the exclusive-OR gate as a 2-bit comparator.

0
0 ————⟫D—— 0   The input bits are equal.   1
0 ————⟫D—— 1   The input bits are not equal.

0
1 ————⟫D—— 1   The input bits are not equal.   1
1 ————⟫D—— 0   The input bits are equal.

FIGURE 6–19
Basic comparator operation.

In order to compare binary numbers containing two bits each, an additional exclusive-OR gate is necessary. The two least significant bits (LSBs) of the two numbers are compared by gate G_1, and the two most significant bits (MSBs) are compared by gate G_2, as shown in Figure 6–20. If the two numbers are equal, their corresponding bits are the same, and the output of each exclusive-OR gate is a 0. If the corresponding sets of bits are not equal, a 1 occurs on that exclusive-OR gate output. In order to produce a single output indicating an equality or inequality of two numbers, two inverters and an AND gate can be used, as shown in Figure 6–20. The output of each exclusive-OR gate is inverted and applied to the AND gate input. When the two input bits for each exclusive-OR are equal, the corresponding bits of the numbers are equal, producing a 1 on both inputs to the AND gate and thus a 1 on the output. When the two numbers are not equal, one or both sets of corresponding bits are unequal, and a 0 appears on at least one input to the AND gate to produce a 0 on its output. Thus, the output of the AND gate indicates equality (1) or inequality (0) of the two numbers. Example 6–5 illustrates this operation for two specific cases. The exclusive-OR gate and inverter are replaced by an exclusive-NOR symbol. Recall that in the system application for Chapter 3, you used this basic type of circuit to compare two 8-bit numbers.

FIGURE 6–20
Logic diagram for comparison of two 2-bit numbers.

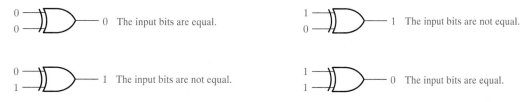

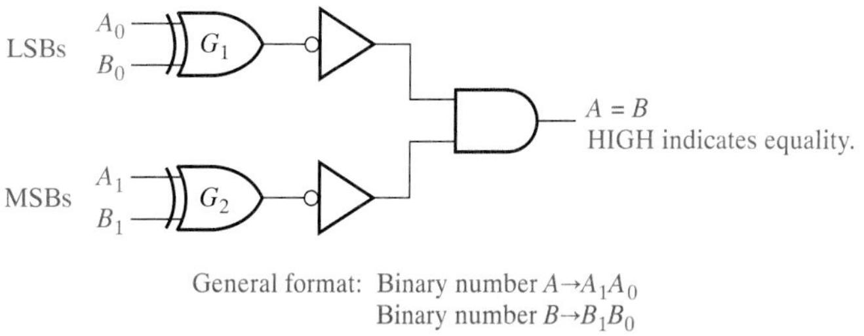

General format: Binary number $A \rightarrow A_1 A_0$
                 Binary number $B \rightarrow B_1 B_0$

EXAMPLE 6–5 Apply each of the following sets of binary numbers to the comparator inputs in Figure 6–21, and determine the output by following the logic levels through the circuit.
(a) 10 and 10   **(b)** 11 and 10

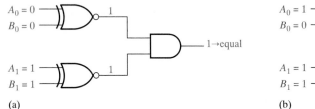

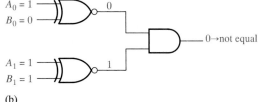

(a)

(b)

FIGURE 6–21

Solution
(a) The output is 1 for inputs 10 and 10, as shown in Figure 6–21(a).
(b) The output is 0 for inputs 11 and 10, as shown in Figure 6–21(b).

Related Exercise   Repeat the process for binary inputs of 01 and 10.

As you know from Chapter 3, the basic comparator circuit can be expanded to any number of bits, as illustrated in Figure 6–22 for two 4-bit numbers. The AND gate sets the condition that all corresponding bits of the two numbers must be equal if the two numbers themselves are equal.

Integrated Circuit Comparators

In addition to the equality output, many integrated circuit comparators provide additional outputs that indicate which of the two binary numbers being compared is the larger. That is, there is an output that indicates when number A is greater than number B ($A > B$) and an output that indicates when number A is less than number B ($A < B$), as shown in the logic symbol for a 4-bit comparator in Figure 6–23.

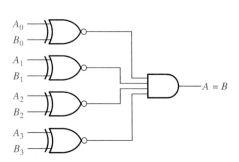

FIGURE 6–22
Logic diagram for the comparison of two 4-bit numbers, $A_3A_2A_1A_0$ and $B_3B_2B_1B_0$.

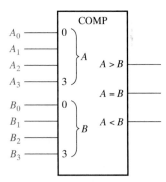

FIGURE 6–23
Logic symbol for a 4-bit comparator with inequality indication.

In order to understand the logic circuitry required for the $A > B$ and $A < B$ outputs, let's examine two general binary numbers and determine what characterizes an inequality of the numbers. For our purposes we use two 4-bit numbers with the general format $A_3A_2A_1A_0$ for one binary number, which we call number A, and $B_3B_2B_1B_0$ for the other binary number, which we call number B. This format is shown in Figure 6–24.

To determine an inequality of numbers A and B, we first examine the highest-order bit in each number. The following conditions are possible:

1. If $A_3 = 1$ and $B_3 = 0$, number A is greater than number B.
2. If $A_3 = 0$ and $B_3 = 1$, number A is less than number B.
3. If $A_3 = B_3$, then we must examine the next lower bit position for an inequality.

The three observations listed above are valid for each bit position in the numbers. The general procedure is to check for an inequality in a bit position, starting with the highest order bits (MSBs). When such an inequality is found, the relationship of the two numbers is established, and any other inequalities in lower-order bit positions must be ignored because it is possible for an opposite indication to occur; *the highest-order indication must take precedence.*

To illustrate, let's assume that number A is 0111 and number B is 1000. Comparison of bits A_3 and B_3 indicates that $A < B$ because $A_3 = 0$ and $B_3 = 1$. In this case, it doesn't

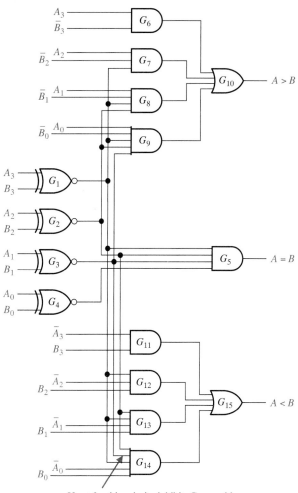

FIGURE 6–24
Logic diagram for a 4-bit comparator. Inverters at inputs and some interconnections are omitted for clarity.

If not for this priority inhibit, G_{14} would produce an invalid indication because all of its other inputs are HIGH (1).

matter how the remaining lower-order bits compare because the $A < B$ condition has been established by bits A_3 and B_3.

As a second example, number A is 1110 and number B is 1001. Since both A_3 and B_3 are 1, the next lower-order bits must be checked. $A_2 = 1$ and $B_2 = 0$ indicates that $A > B$. The remaining lower-order bits do not need to be checked.

As a final example, number A is 1011 and number B is 1010. In this case, all of the bits are equal except the least significant bits. $A_0 = 0$ and $B_0 = 1$ indicates that $A < B$.

Figure 6–24 shows a method of comparing two 4-bit numbers and generating an $A > B$, or an $A < B$, or an $A = B$ output. The $A > B$ condition is determined by gates G_6 through G_{10}. Gate G_6 checks for $A_3 = 1$ and $B_3 = 0$, and its function is expressed as $A_3\overline{B}_3$; gate G_7 checks for $A_2 = 1$ and $B_2 = 0$ ($A_2\overline{B}_2$); gate G_8 checks for $A_1 = 1$ and $B_1 = 0$ ($A_1\overline{B}_1$); gate G_9 checks for $A_0 = 1$ and $B_0 = 0$ ($A_0\overline{B}_0$). These conditions all indicate that number A is greater than number B. The outputs of all these gates are ORed by gate G_{10} to produce the $A > B$ output.

Notice that the output of gate G_1 is connected to inputs of gates G_7, G_8, and G_9. This provides a *priority inhibit* so that if the proper inequality ($A_3 < B_3$) occurs in bits A_3 and B_3, the lower-order bit checks will be inhibited. A priority inhibit is also provided by gate G_2 to gates G_8 and G_9, and by gate G_3 to gate G_9.

Gates G_{11} through G_{15} check for an $A < B$ condition. Each AND gate checks a given bit position for the occurrence of a 0 in number A and a 1 in number B. The AND gate outputs are ORed by gate G_{15} to provide the $A < B$ output. Priority inhibiting is provided as previously discussed. When all four bits in A equal the bits in B, the output of each exclusive-OR gate is 1. This enables G_5 to produce a 1 on the $A = B$ output.

Example 6–6 shows the comparison of specific binary numbers and indicates the logic levels throughout the comparator.

EXAMPLE 6–6

Analyze the comparator operation in Figure 6–24 for the numbers $A_3A_2A_1A_0 = 1010$ and $B_3B_2B_1B_0 = 1001$.

Solution Table 6–4 shows all logic levels at each gate output within the comparator for the specified inputs.

TABLE 6–4

Inputs		Gate Outputs													
$A_3A_2A_1A_0$	$B_3B_2B_1B_0$	G_1	G_2	G_3	G_4	G_5	G_6	G_7	G_8	G_9	G_{10}	G_{11}	G_{12}	G_{13}	G_{14}	G_{15}
1010	1001	1	1	0	0	0	0	0	1	0	1	0	0	0	0	0

Related Exercise Repeat the analysis of the comparator in Figure 6–24 for binary inputs 0010 and 1000.

The 7485 4-Bit Magnitude Comparator

The 7485 is an MSI comparator that is also available in the LS TTL family as well as others. The logic symbol is shown in Figure 6–25 with pin numbers in parentheses.

Notice that this device has all the inputs and outputs of the generalized comparator previously discussed and, in addition, has three cascading inputs: $A < B, A = B, A > B$. These inputs allow several comparators to be cascaded for comparison of any number of bits greater than four. To expand the comparator, the $A < B, A = B$, and $A > B$ outputs of the lower-order comparator are connected to the corresponding cascading inputs of the

FIGURE 6–25
Logic symbol for the 7485 4-bit magnitude comparator (pin numbers are in parentheses).

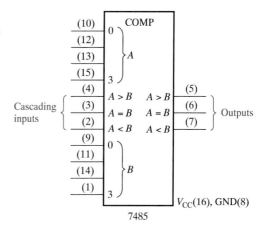

next higher-order comparator. The lowest-order comparator must have a HIGH on the $A = B$ input and LOWs on the $A < B$ and $A > B$ inputs.

EXAMPLE 6–7

Use 7485 comparators to compare the magnitudes of two 8-bit numbers. Show the comparators with proper interconnections.

Solution   Two 7485s are required to compare two 8-bit numbers. They are connected as shown in Figure 6–26, in a cascaded arrangement.

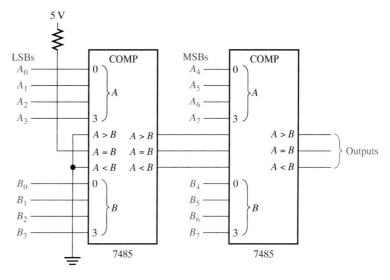

FIGURE 6–26
An 8-bit magnitude comparator using two 7485s.

Related Exercise   Expand the circuit in Figure 6–26 to a 16-bit comparator.

SECTION 6–4 REVIEW

1. The binary numbers $A = 1011$ and $B = 1010$ are applied to the inputs of a 7485. Determine the outputs.
2. The binary numbers $A = 11001011$ and $B = 11010100$ are applied to the 8-bit comparator in Figure 6–26. Determine the states of output pins 5, 6, and 7 on each 7485.

6–5 ■ DECODERS

The basic function of a decoder is to detect the presence of a specified combination of bits (code) on its inputs and to indicate the presence of that code by a specified output level. In its general form, a decoder has n input lines to handle n bits and from one to 2^n output lines to indicate the presence of one or more n-bit combinations. In this section, several decoders are introduced. The basic principles can be extended to other types of decoders. After completing this section, you should be able to

☐ Define *decoder*   ☐ Design a logic circuit to decode any combination of bits
☐ Describe the 74154 binary-to-decimal decoder   ☐ Describe the 7442A BCD-to-decimal decoder   ☐ Expand decoders to accommodate larger numbers of bits in a code   ☐ Describe the 7447 BCD-to-7-segment decoder   ☐ Discuss zero suppression in 7-segment displays   ☐ Apply decoders to specific applications

The Basic Binary Decoder

Suppose you need to determine when a binary 1001 occurs on the inputs of a digital circuit. An AND gate can be used as the basic decoding element because it produces a HIGH output only when all of its inputs are HIGH. Therefore, you must make sure that all of the inputs to the AND gate are HIGH when the binary number 1001 occurs; this can be done by inverting the two middle bits (the 0s), as shown in Figure 6–27.

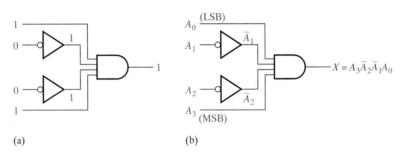

(a)   (b)

FIGURE 6–27
Decoding logic for the binary code 1001 with an active-HIGH output.

The logic equation for the **decoder** of Figure 6–27(a) is developed as illustrated in Figure 6–27(b). You should verify that the output is 0 except when $A_0 = 1, A_1 = 0, A_2 = 0$, and $A_3 = 1$ are applied to the inputs. A_0 is the LSB and A_3 is the MSB. *In the representation of a binary number or other weighted code in this book, the LSB is always the rightmost bit in a horizontal arrangement and the topmost bit in a vertical arrangement, unless specified otherwise.*

If a NAND gate is used in place of the AND gate, as shown in Figure 6–28, a LOW output will indicate the presence of the proper binary code, which is 1001 in this case.

FIGURE 6–28
Decoding logic for the binary code 1001 with an active-LOW output.

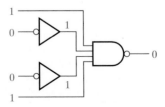

EXAMPLE 6–8

Determine the logic required to decode the binary number 1011 by producing a HIGH level on the output.

Solution The decoding function can be formed by complementing only the variables that appear as 0 in the binary number, as follows:

$$X = A_3\overline{A}_2A_1A_0 \qquad (1011)$$

This function can be implemented by connecting the true (uncomplemented) variables A_0, A_1, and A_3 directly to the inputs of an AND gate, and inverting the variable A_2 before applying it to the AND gate input. The decoding logic is shown in Figure 6–29.

FIGURE 6–29
Decoding logic for producing a HIGH output when 1011 is on the inputs.

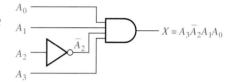

Related Exercise Develop the logic required to detect the binary code 10010 and produce an active-LOW output.

The 4-Bit Decoder

In order to decode all possible combinations of four bits, sixteen decoding gates are required ($2^4 = 16$). This type of decoder is commonly called a *4-line-to-16-line decoder* because there are four inputs and sixteen outputs or a *1-of-16 decoder* because for any given code on the inputs, one of the sixteen outputs is activated. A list of the sixteen binary codes and their corresponding decoding functions is given in Table 6–5.

TABLE 6–5
Decoding functions and truth table for a 4-line-to-16-line decoder with active-LOW outputs

Decimal Digit	Binary Inputs				Decoding Function	Outputs															
	A_3	A_2	A_1	A_0		0	1	2	3	4	5	6	7	8	9	10	11	12	13	14	15
0	0	0	0	0	$\overline{A}_3\overline{A}_2\overline{A}_1\overline{A}_0$	0	1	1	1	1	1	1	1	1	1	1	1	1	1	1	1
1	0	0	0	1	$\overline{A}_3\overline{A}_2\overline{A}_1A_0$	1	0	1	1	1	1	1	1	1	1	1	1	1	1	1	1
2	0	0	1	0	$\overline{A}_3\overline{A}_2A_1\overline{A}_0$	1	1	0	1	1	1	1	1	1	1	1	1	1	1	1	1
3	0	0	1	1	$\overline{A}_3\overline{A}_2A_1A_0$	1	1	1	0	1	1	1	1	1	1	1	1	1	1	1	1
4	0	1	0	0	$\overline{A}_3A_2\overline{A}_1\overline{A}_0$	1	1	1	1	0	1	1	1	1	1	1	1	1	1	1	1
5	0	1	0	1	$\overline{A}_3A_2\overline{A}_1A_0$	1	1	1	1	1	0	1	1	1	1	1	1	1	1	1	1
6	0	1	1	0	$\overline{A}_3A_2A_1\overline{A}_0$	1	1	1	1	1	1	0	1	1	1	1	1	1	1	1	1
7	0	1	1	1	$\overline{A}_3A_2A_1A_0$	1	1	1	1	1	1	1	0	1	1	1	1	1	1	1	1
8	1	0	0	0	$A_3\overline{A}_2\overline{A}_1\overline{A}_0$	1	1	1	1	1	1	1	1	0	1	1	1	1	1	1	1
9	1	0	0	1	$A_3\overline{A}_2\overline{A}_1A_0$	1	1	1	1	1	1	1	1	1	0	1	1	1	1	1	1
10	1	0	1	0	$A_3\overline{A}_2A_1\overline{A}_0$	1	1	1	1	1	1	1	1	1	1	0	1	1	1	1	1
11	1	0	1	1	$A_3\overline{A}_2A_1A_0$	1	1	1	1	1	1	1	1	1	1	1	0	1	1	1	1
12	1	1	0	0	$A_3A_2\overline{A}_1\overline{A}_0$	1	1	1	1	1	1	1	1	1	1	1	1	0	1	1	1
13	1	1	0	1	$A_3A_2\overline{A}_1A_0$	1	1	1	1	1	1	1	1	1	1	1	1	1	0	1	1
14	1	1	1	0	$A_3A_2A_1\overline{A}_0$	1	1	1	1	1	1	1	1	1	1	1	1	1	1	0	1
15	1	1	1	1	$A_3A_2A_1A_0$	1	1	1	1	1	1	1	1	1	1	1	1	1	1	1	0

If an active-LOW output is desired for each decoded number, the entire decoder can be implemented with NAND gates and inverters as follows. First, since each variable and its complement are required in the decoder, as seen from Table 6–5, each complement can be generated once and then used for all decoding gates as required, rather than be generated by a separate inverter for each place that complement is used. This arrangement is shown in Figure 6–30. In order to decode each of the sixteen binary codes, sixteen NAND gates are required (AND gates can be used to produce active-HIGH outputs). The general decoding gate arrangement is illustrated in Figure 6–30. Rather than reproducing the complex logic diagram for the decoder each time it is required in a schematic, a simpler representation is normally used.

A logic symbol for a 4-line-to-16-line decoder is shown in Figure 6–31. The BIN/DEC label indicates that a binary input makes the corresponding decimal output active. The input labels 1, 2, 4, and 8 represent the binary weights of the input bits ($2^3 2^2 2^1 2^0$).

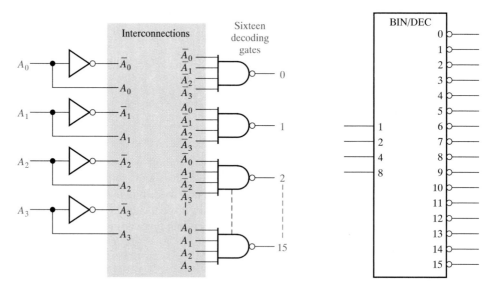

FIGURE 6–30
Abbreviated logic for a 4-line-to-16-line decoder.

FIGURE 6–31
Logic symbol for a 4-line-to-16-line decoder.

The 74154 4-Line-to-16-Line Decoder

The 74154 is a good example of an MSI decoder that is available in the LS TTL family as well as others. The CMOS equivalent is designated 74HC154. The logic diagram for the 74154 is shown in Figure 6–32(a), and the logic symbol is shown in Figure 6–32(b). The additional inverters on the inputs are required to prevent excessive loading of the driving source(s). Each input is connected to the input of only one inverter, rather than to the inputs of several NAND gates as in Figure 6–30. There is also an enable function provided on this particular device, which is implemented with a NOR gate used as a negative-AND. A LOW level on each input, $\overline{G}_1$ and $\overline{G}_2$, is required in order to make the enable gate output (EN) HIGH. The enable gate output is connected to an input of *each* NAND gate, so it must be HIGH for the NAND gates to be en-

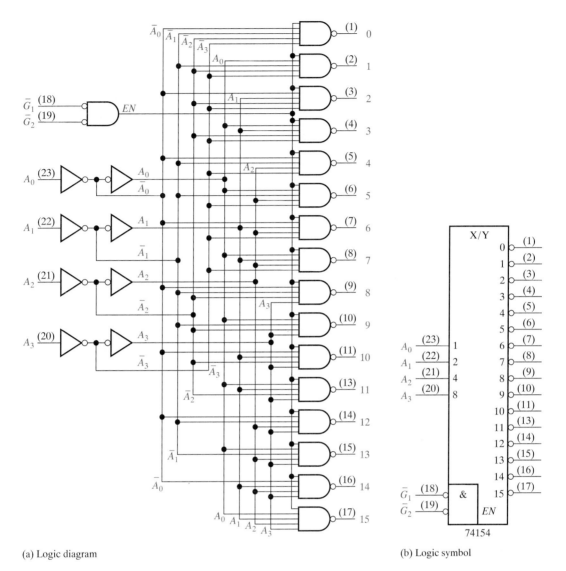

(a) Logic diagram

(b) Logic symbol

FIGURE 6–32

The 74154 4-line-to-16-line decoder.

abled. If the enable gate is not activated by a LOW on both inputs, then all sixteen decoder outputs will be HIGH regardless of the states of the four input variables, A_0, A_1, A_2, and A_3.

EXAMPLE 6–9

A certain application requires that a 5-bit number be decoded. Use 74154 decoders to implement the logic. The binary number is represented by the format $A_4A_3A_2A_1A_0$.

Solution Since the 74154 can handle only four bits, two decoders must be used to decode five bits. The fifth bit, A_4, is connected to the enable inputs, $\overline{G}_1$ and $\overline{G}_2$, of one decoder, and $\overline{A}_4$ is connected to the enable inputs of the other decoder, as shown in Figure 6–33. When the decimal number is 15 or less, $A_4 = 0$, and the low-order decoder is enabled and the high-order decoder is disabled. When the decimal number is greater than 15, $A_4 = 1$ so $\overline{A}_4 = 0$, and the high-order decoder is enabled and the low-order decoder is disabled.

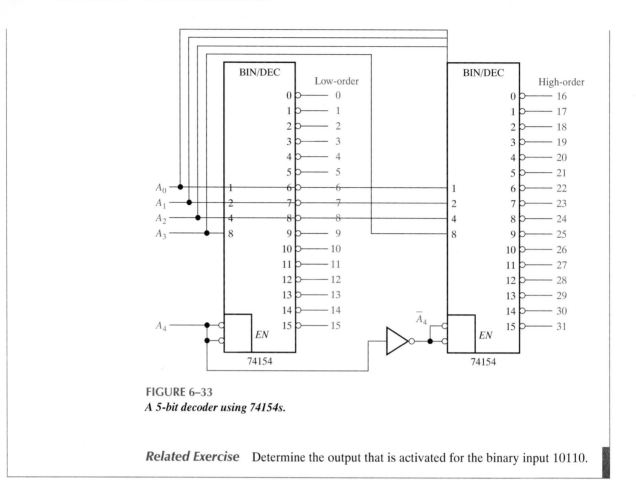

FIGURE 6–33
A 5-bit decoder using 74154s.

Related Exercise   Determine the output that is activated for the binary input 10110.

A Decoder Application

Decoders are used in many types of applications. One example is in computers for input/output selection as depicted in the general diagram of Figure 6–34.

Computers must communicate with a variety of external devices called *peripherals* by sending and/or receiving data through what is known as input/output (I/O) ports. These external devices include printers, modems, scanners, external disk drives, keyboard, video monitors, and other computers. As indicated in Figure 6–34, a decoder is used to select the I/O port as determined by the computer so that data can be sent or received from a specific external device.

Each I/O port has a number, called an address, which uniquely identifies it. When the computer wants to communicate with a particular device, it issues the appropriate address code for the I/O port to which that particular device is connected. This binary port address is decoded and the appropriate decoder output is activated to enable the I/O port.

As shown in Figure 6–34, binary data are transferred within the computer on a data bus, which is a set of parallel lines. For example, an 8-bit bus consists of eight parallel lines that can carry one byte of data at a time. The data bus goes to all of the I/O ports, but any data coming in or going out will only pass through the port that is enabled by the port address decoder.

The BCD-to-Decimal Decoder

The BCD-to-decimal decoder converts each BCD code (8421 code) into one of ten possible decimal digit indications. It is frequently referred to as a *4-line-to-10-line decoder* or a *1-of-10 decoder*.

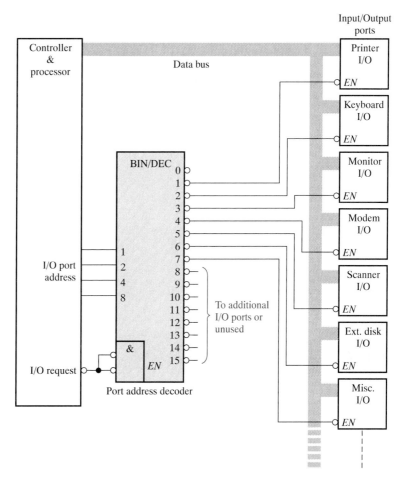

FIGURE 6–34
A generalized computer I/O port system with a port address decoder.

The method of implementation is the same as for the 4-line-to-16-line decoder previously discussed, except that only ten decoding gates are required because the BCD code represents only the ten decimal digits 0 through 9. A list of the ten BCD codes and their corresponding decoding functions is given in Table 6–6. Each of these decoding functions is implemented with NAND gates to provide active-LOW outputs. If an active-HIGH output is required, AND gates are used for decoding. The logic is identical to that of the first ten decoding gates in the 4-line-to-16-line decoder (see Table 6–5).

TABLE 6–6
BCD decoding functions

Decimal Digit	BCD Code				Decoding Function
	A_3	A_2	A_1	A_0	
0	0	0	0	0	$\overline{A}_3\overline{A}_2\overline{A}_1\overline{A}_0$
1	0	0	0	1	$\overline{A}_3\overline{A}_2\overline{A}_1 A_0$
2	0	0	1	0	$\overline{A}_3\overline{A}_2 A_1\overline{A}_0$
3	0	0	1	1	$\overline{A}_3\overline{A}_2 A_1 A_0$
4	0	1	0	0	$\overline{A}_3 A_2\overline{A}_1\overline{A}_0$
5	0	1	0	1	$\overline{A}_3 A_2\overline{A}_1 A_0$
6	0	1	1	0	$\overline{A}_3 A_2 A_1\overline{A}_0$
7	0	1	1	1	$\overline{A}_3 A_2 A_1 A_0$
8	1	0	0	0	$A_3\overline{A}_2\overline{A}_1\overline{A}_0$
9	1	0	0	1	$A_3\overline{A}_2\overline{A}_1 A_0$

EXAMPLE 6–10   The 7442A is an integrated circuit BCD-to-decimal decoder. The logic symbol is shown in Figure 6–35. If the input waveforms in Figure 6–36(a) are applied to the inputs of the 7442A, sketch the output waveforms.

FIGURE 6–35
The 7442A BCD-to-decimal decoder.

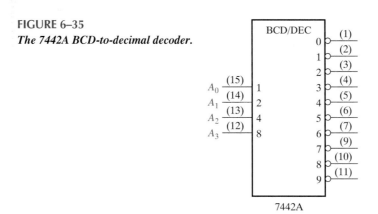

FIGURE 6–36

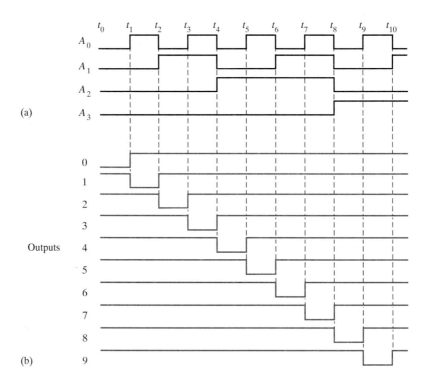

(a)

Outputs

(b)

Solution   The output waveforms are shown in Figure 6–36(b). As you can see, the inputs are sequenced through the BCD for digits 0 through 9. The output waveforms in the timing diagram indicate that sequence.

Related Exercise   Construct a timing diagram showing input and output waveforms for the case where the binary inputs sequence through the decimal numbers as follows: 0, 2, 4, 6, 8, 1, 3, 5, and 9.

The BCD-to-7-Segment Decoder

As you learned in the system application of Chapter 4, this type of decoder accepts the BCD code on its inputs and provides outputs to energize 7-segment display devices to produce a decimal readout. The logic diagram for a basic 7-segment decoder is shown in Figure 6–37.

FIGURE 6–37
Logic symbol for a BCD-to-7-segment decoder/driver with active-LOW outputs.

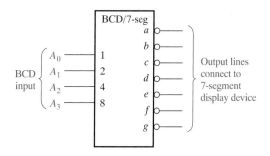

The 7447 BCD-to-7-Segment Decoder/Driver

The 7447 (also 74LS47) is an example of an MSI device that decodes a BCD input and drives a 7-segment display. In addition to its decoding and segment drive capability, the 7447 has several additional features as indicated by the *LT, RBI,* and *BI/RBO* functions in the logic symbol of Figure 6–38. As indicated by the bubbles on the logic symbol, all of the outputs (*a* through *g*) are active-LOW as are the *LT* (lamp test), *RBI* (ripple blanking input), and *BI/RBO* (blanking input/ripple blanking output) functions. The outputs can drive a common-anode 7-segment display directly. Recall that 7-segment displays were discussed in Chapter 4.

FIGURE 6–38
The 7447 BCD-to-7-segment decoder/driver.

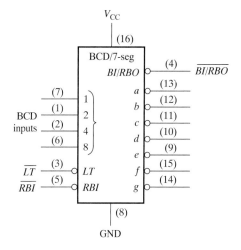

In addition to decoding a BCD input and producing the appropriate 7-segment outputs, the 7447 has lamp test and zero suppression capability.

Lamp Test When a LOW is applied to the *LT* input and the *BI/RBO* is HIGH, all of the 7 segments in the display are turned on. Lamp test is used to verify that no segments are burned out.

Zero Suppression **Zero suppression** is a feature used for multidigit displays to blank out unnecessary zeros. For example, in a 6-digit display the number 6.4 may be displayed as 006.400 if the zeros are not blanked out. Blanking the zeros at the front of a number is called *leading zero suppression* and blanking the zeros at the back of the number is called *trailing zero suppression.* Keep in mind that only nonessential zeros are blanked. With zero suppression, the number 030.080 will be displayed as 30.08 (the essential zeros remain).

Zero suppression in the 7447 is accomplished using the *RBI* and *BI/RBO* functions. *RBI* is the ripple blanking input and *RBO* is the ripple blanking output on the 7447; these are used for zero suppression. *BI* is the blanking input which shares the same pin with *RBO;* in other words, the *BI/RBO* pin can be used as an input or an output. When used as a *BI* (blanking input), all segment outputs are HIGH (nonactive) when *BI* is LOW, which overrides all other inputs. The *BI* function is not part of the zero suppression capability of the device.

All of the segment outputs of the decoder are nonactive (HIGH) if a zero code (0000) is on its BCD inputs and if its *RBI* is LOW. This causes the display to be blank and produces a LOW *RBO.*

The logic diagram in Figure 6–39(a) illustrates leading zero suppression for a whole number. The highest-order digit position (left-most) is always blanked if a zero code is on its BCD inputs because the *RBI* of the most-significant decoder is made LOW by connecting it to ground. The *RBO* of each decoder is connected to the *RBI* of the next lowest-order

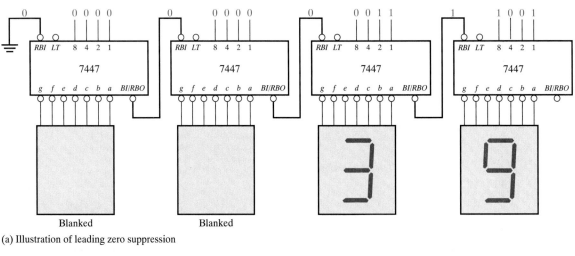

(a) Illustration of leading zero suppression

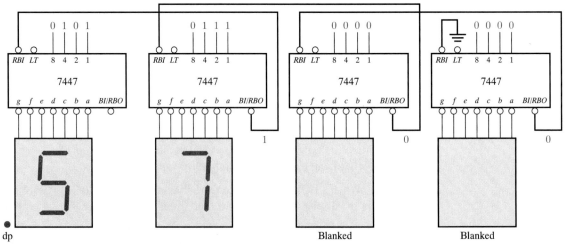

(b) Illustration of trailing zero suppression

FIGURE 6–39

Examples of zero suppression using the 7447 BCD to 7-segment decoder/driver.

decoder so that all zeros to the left of the first nonzero digit are blanked. For example, in part (a) of the figure the two highest-order digits are zeros and therefore are blanked. The remaining two digits, 3 and 9 are displayed.

The logic diagram in Figure 6–39(b) illustrates trailing zero suppression for a fractional number. The lowest-order digit (right-most) is always blanked if a zero code is on its BCD inputs because the *RBI* is connected to ground. The *RBO* of each decoder is connected to the *RBI* of the next highest-order decoder so that all zeros to the right of the first nonzero digit are blanked. In part (b) of the figure, the two lowest-order digits are zeros and therefore are blanked. The remaining two digits, 5 and 7 are displayed. To combine both leading and trailing zero suppression in one display and to have decimal point capability, additional logic is required.

SECTION 6–5 REVIEW

1. A 3-line-to-8-line decoder can be used for octal-to-decimal decoding. When a binary 101 is on the inputs, which output line is activated?
2. How many 74154 4-line-to-16-line decoders are necessary to decode a 6-bit binary number?
3. Would you select a decoder/driver with active-HIGH or active-LOW outputs to drive a common-cathode 7-segment LED display?

6–6 ■ ENCODERS

An encoder is a combinational logic circuit that essentially performs a "reverse" decoder function. An encoder accepts an active level on one of its inputs representing a digit, such as a decimal or octal digit, and converts it to a coded output, such as BCD or binary. Encoders can also be devised to encode various symbols and alphabetic characters. The process of converting from familiar symbols or numbers to a coded format is called encoding. After completing this section, you should be able to

☐ Determine the logic for a decimal encoder ☐ Explain the purpose of the priority feature in encoders ☐ Describe the 74LS147 decimal-to-BCD priority encoder ☐ Describe the 74148 octal-to-binary priority encoder ☐ Expand an encoder ☐ Apply the encoder to a specific application

The Decimal-to-BCD Encoder

This type of **encoder** has ten inputs—one for each decimal digit—and four outputs corresponding to the BCD code, as shown in Figure 6–40. This is a basic 10-line-to-4-line encoder.

FIGURE 6–40
Logic symbol for a decimal-to-BCD encoder.

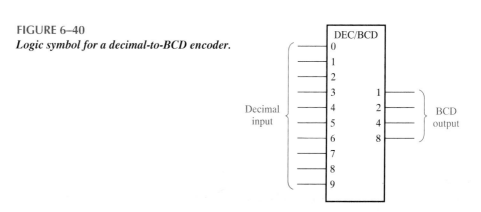

TABLE 6–7

Decimal Digit	BCD Code			
	A_3	A_2	A_1	A_0
0	0	0	0	0
1	0	0	0	1
2	0	0	1	0
3	0	0	1	1
4	0	1	0	0
5	0	1	0	1
6	0	1	1	0
7	0	1	1	1
8	1	0	0	0
9	1	0	0	1

The BCD (8421) code is listed in Table 6–7. From this table you can determine the relationship between each BCD bit and the decimal digits in order to analyze the logic. For instance, the most significant bit of the BCD code, A_3, is a 1 for decimal digit 8 or 9. The OR expression for bit A_3 in terms of the decimal digits can therefore be written

$$A_3 = 8 + 9$$

Bit A_2 is a 1 for decimal digit 4, 5, 6, or 7 and can be expressed as an OR function as follows:

$$A_2 = 4 + 5 + 6 + 7$$

Bit A_1 is a 1 for decimal digit 2, 3, 6, or 7 and can be expressed as

$$A_1 = 2 + 3 + 6 + 7$$

Finally, A_0 is a 1 for digit 1, 3, 5, 7, or 9. The expression for A_0 is

$$A_0 = 1 + 3 + 5 + 7 + 9$$

Now let's implement the logic circuitry required for encoding each decimal digit to a BCD code by using the logic expressions just developed. It is simply a matter of ORing the appropriate decimal digit input lines to form each BCD output. The basic encoder logic resulting from these expressions is shown in Figure 6–41.

FIGURE 6–41

Basic logic diagram of a decimal-to-BCD encoder. A 0-digit input is not needed because the BCD outputs are all LOW when there are no HIGH inputs.

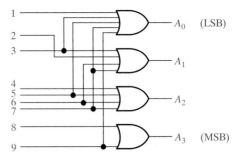

The basic operation of the circuit in Figure 6–41 is as follows: When a HIGH appears on *one* of the decimal digit input lines, the appropriate levels occur on the four BCD output lines. For instance, if input line 9 is HIGH (assuming all other input lines are LOW), this condition will produce a HIGH on outputs A_0 and A_3 and LOWs on outputs A_1 and A_2, which is the BCD code (1001) for decimal 9.

The Decimal-to-BCD Priority Encoder

The decimal-to-BCD **priority encoder** performs the same basic encoding function as previously discussed. It also offers additional flexibility in that it can be used in applications that require priority detection. The priority function means that the encoder will produce a BCD output corresponding to the highest-order decimal digit input that is active and will ignore any other active inputs. For instance, if the 6 and the 3 inputs are both active, the BCD output is 0110 (which represents decimal 6).

Priority Logic Now let's look at the requirements for the priority detection logic. This logic circuitry prevents a lower-order digit input from disrupting the encoding of a higher-order digit. We start by examining each BCD output (beginning with output A_0). Referring to Figure 6–41, notice that A_0 is HIGH when 1, 3, 5, 7, or 9 is HIGH. Digit input 1 should activate the A_0 output only if no higher-order digits other than those that also activate A_0 are HIGH. This requirement can be stated as follows:

1. A_0 is HIGH if 1 is HIGH *and* 2, 4, 6, and 8 are LOW.
2. A_0 is HIGH if 3 is HIGH *and* 4, 6, and 8 are LOW.
3. A_0 is HIGH if 5 is HIGH *and* 6 and 8 are LOW.
4. A_0 is HIGH if 7 is HIGH *and* 8 is LOW.
5. A_0 is HIGH if 9 is HIGH.

These five statements describe the priority of encoding for the BCD bit A_0. The A_0 output is HIGH if any of the conditions listed occur; that is, A_0 is true if statement 1, statement 2, statement 3, statement 4, or statement 5 is true. This can be expressed in the form of the following logic equation:

$$A_0 = (1 \cdot \overline{2} \cdot \overline{4} \cdot \overline{6} \cdot \overline{8}) + (3 \cdot \overline{4} \cdot \overline{6} \cdot \overline{8}) + (5 \cdot \overline{6} \cdot \overline{8}) + (7 \cdot \overline{8}) + 9$$

From this expression the logic circuitry required for the A_0 output with priority inhibits can be implemented, as shown in Figure 6–42.

The same reasoning process can be applied to output A_1, and the following logical statements can be made:

1. A_1 is HIGH if 2 is HIGH *and* 4, 5, 8, and 9 are LOW.
2. A_1 is HIGH if 3 is HIGH *and* 4, 5, 8, and 9 are LOW.
3. A_1 is HIGH if 6 is HIGH *and* 8 and 9 are LOW.
4. A_1 is HIGH if 7 is HIGH *and* 8 and 9 are LOW.

These statements are summarized in the following equation, and the logic implementation is shown in Figure 6–43.

$$A_1 = (2 \cdot \overline{4} \cdot \overline{5} \cdot \overline{8} \cdot \overline{9}) + (3 \cdot \overline{4} \cdot \overline{5} \cdot \overline{8} \cdot \overline{9}) + (6 \cdot \overline{8} \cdot \overline{9}) + (7 \cdot \overline{8} \cdot \overline{9})$$

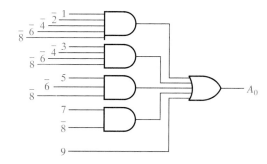

FIGURE 6–42
Logic for the A_0 output of a decimal-to-BCD priority encoder.

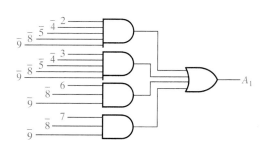

FIGURE 6–43
Logic for the A_1 output of a decimal-to-BCD priority encoder.

Next, output A_2 can be described as follows:

1. A_2 is HIGH if 4 is HIGH *and* 8 and 9 are LOW.
2. A_2 is HIGH if 5 is HIGH *and* 8 and 9 are LOW.
3. A_2 is HIGH if 6 is HIGH *and* 8 and 9 are LOW.
4. A_2 is HIGH if 7 is HIGH *and* 8 and 9 are LOW.

In equation form, output A_2 is

$$A_2 = (4 \cdot \overline{8} \cdot \overline{9}) + (5 \cdot \overline{8} \cdot \overline{9}) + (6 \cdot \overline{8} \cdot \overline{9}) + (7 \cdot \overline{8} \cdot \overline{9})$$

The logic circuitry for the A_2 output appears in Figure 6–44.
Finally, for the A_3 output,

$$A_3 \text{ is HIGH if 8 is HIGH } or \text{ if 9 is HIGH.}$$

This statement is expressed in equation form as follows:

$$A_3 = 8 + 9$$

The logic for the A_3 output is shown in Figure 6–45. No inhibits are required for this one.
We now have developed the basic logic for the decimal-to-BCD priority encoder. All the complements of the input digits are realized by inverting the inputs.

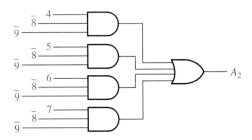

FIGURE 6–44
Logic for the A_2 output of a decimal-to-BCD priority encoder.

FIGURE 6–45
Logic for the A_3 output of a decimal-to-BCD priority encoder.

The 74LS147 Decimal-to-BCD Encoder

The 74LS147 is a priority encoder with active-LOW inputs for decimal digits 1 through 9 and active-LOW BCD outputs as indicated in the logic symbol in Figure 6–46. A BCD zero output is represented when none of the inputs is active. The device pin numbers are in parentheses.

FIGURE 6–46
Logic symbol for the 74LS147 decimal-to-BCD priority encoder (HPRI means highest value input has priority).

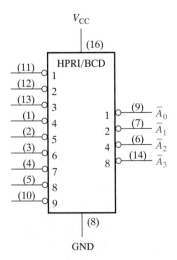

EXAMPLE 6–11

If LOW levels appear on pins 1, 4, and 13 of the 74LS147 illustrated in Figure 6–46, indicate the state of the four outputs. All other inputs are HIGH.

Solution Pin 4 is the highest-order decimal digit input having a LOW level and represents decimal 7. Therefore, the output levels indicate the BCD code for decimal 7 where $\overline{A}_0$ is the LSB and $\overline{A}_3$ is the MSB. Output $\overline{A}_0$ is LOW, $\overline{A}_1$ is LOW, $\overline{A}_2$ is LOW, and $\overline{A}_3$ is HIGH.

Related Exercise What are the outputs of the 74LS147 if all its inputs are LOW? If all its inputs are HIGH?

The 74148 Octal-to-Binary Encoder

The 74148 is a priority encoder that has eight active-LOW inputs and three active-LOW binary outputs, as shown in Figure 6–47. This device can be used for converting octal inputs (recall that the octal digits are 0 through 7) to a 3-bit binary code. To enable the device, the *EI* (enable input) must be LOW. It also has the *EO* (enable output) and *GS* output for expansion purposes. The *EO* is LOW when the *EI* is LOW and none of the inputs (0 through 7) is active. *GS* is LOW when *EI* is LOW and any of the inputs is active.

The 74148 can be expanded to a 16-line-to-4-line encoder by connecting the *EO* of the higher-order encoder to the *EI* of the lower-order encoder and negative-ORing the corresponding binary outputs as shown in Figure 6–48. The *EO* is used as the fourth and most-significant bit. This particular configuration produces active-HIGH outputs for the 4-bit binary number.

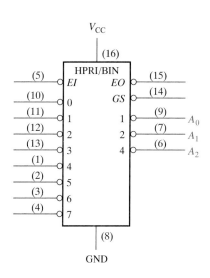

FIGURE 6–47
Logic symbol for the 74148 octal-to-binary encoder.

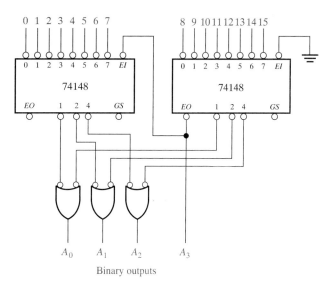

FIGURE 6–48
A 16-line-to-4 line encoder using 74148s and external logic.

An Encoder Application

A classic application example is a keyboard encoder. The ten decimal digits on the keyboard of a computer, for example, must be encoded for processing by the logic circuitry. When one of the keys is pressed, the decimal digit is encoded to the corresponding BCD code. Figure 6–49 shows a simple keyboard encoder arrangement using a 74LS147 priority encoder. The keys are represented by ten push-button switches, each with a **pull-up resistor** to $+V$. The pull-up resistor ensures that the line is HIGH when a key is not depressed. When a key is depressed, the line is connected to ground, and a LOW is applied to the corresponding encoder input. The zero key is not connected because the BCD output represents zero when none of the other keys is depressed.

The BCD output of the encoder goes into a storage device, and each successive BCD code is stored until the entire number has been entered. Methods of storing BCD numbers and binary data are covered in later chapters.

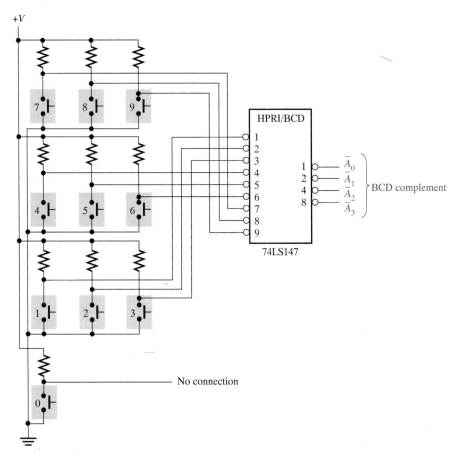

FIGURE 6–49

A simplified keyboard encoder.

SECTION 6–6 REVIEW

1. Suppose that HIGH levels are applied to the 2 input and the 9 input of the circuit in Figure 6–41 (page 288).
   (a) What are the states of the output lines?
   (b) Does this represent a valid BCD code?
   (c) What is the restriction on the encoder logic in Figure 6–41?
2. **(a)** What is the $\overline{A}_3\overline{A}_2\overline{A}_1\overline{A}_0$ output when LOWs are applied to pins 1 and 5 of the 74LS147 in Figure 6–46?
   (b) What does this output represent?

6–7 ■ CODE CONVERTERS

In this section, we will examine some methods of using combinational logic circuits to convert from one code to another. After completing this section, you should be able to

☐ Explain the process for converting BCD to binary   ☐ Apply the 74184 and the 74185 to BCD-to-binary and binary-to-BCD conversion, respectively   ☐ Expand the 74184 and the 74185 to handle larger binary and BCD numbers   ☐ Use exclusive-OR gates for conversions between binary and Gray codes

BCD-to-Binary Conversion

One method of BCD-to-binary code conversion uses adder circuits. The basic conversion process is as follows:

1. The value, or weight, of each bit in the BCD number is represented by a binary number.
2. All of the binary representations of the weights of bits that are 1s in the BCD number are added.
3. The result of this addition is the binary equivalent of the BCD number.

A more concise statement of this operation is

The binary numbers representing the weights of the BCD bits are summed to produce the total binary number.

Let's examine an 8-bit BCD code (one that represents a 2-digit decimal number) to understand the relationship between BCD and binary. For instance, we already know that the decimal number 87 can be expressed in BCD as

$$\underbrace{1000}_{8} \quad \underbrace{0111}_{7}$$

The left-most 4-bit group represents 80, and the right-most 4-bit group represents 7. That is, the left-most group has a weight of 10, and the right-most has a weight of 1. Within each group, the binary weight of each bit is as follows:

	Tens Digit				*Units Digit*		
Weight:	80	40	20	10	8	4	2	1
Bit designation:	B_3	B_2	B_1	B_0	A_3	A_2	A_1	A_0

The binary equivalent of each BCD bit is a binary number representing the weight of that bit within the total BCD number. This representation is given in Table 6–8.

TABLE 6–8
Binary representations of BCD bit weights

BCD Bit	BCD Weight	(MSB) 64	Binary Representation 32	16	8	4	2	(LSB) 1
A_0	1	0	0	0	0	0	0	1
A_1	2	0	0	0	0	0	1	0
A_2	4	0	0	0	0	1	0	0
A_3	8	0	0	0	1	0	0	0
B_0	10	0	0	0	1	0	1	0
B_1	20	0	0	1	0	1	0	0
B_2	40	0	1	0	1	0	0	0
B_3	80	1	0	1	0	0	0	0

If the binary representations for the weights of all the 1s in the BCD number are added, the result is the binary number that corresponds to the BCD number. Example 6–12 illustrates this.

EXAMPLE 6–12

Convert the BCD numbers 00100111 (decimal 27) and 10011000 (decimal 98) to binary.

Solution   Write the binary representations of the weights of all 1s appearing in the numbers, and then add them together.

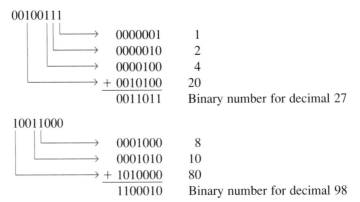

Related Exercise   Show the process of converting 01000001 in BCD to binary.

With this basic procedure in mind, let's determine how the process can be implemented with logic circuits. Once the binary representation for each 1 in the BCD number is determined, adder circuits can be used to add the 1s in each column of the binary representation. The 1s occur in a given column only when the corresponding BCD bit is a 1. The occurrence of a BCD 1 can therefore be used to generate the proper binary 1 in the appropriate column of the adder structure. To handle a two-decimal-digit (two-decade) BCD code, eight BCD input lines and seven binary outputs are required. (It takes seven binary bits to represent numbers through ninety-nine.)

Referring to Table 6–8, notice that the "1" (LSB) column of the binary representation has only a single 1 and no possibility of an input carry, so that a straight connection from the A_0 bit of the BCD input to the least significant binary output is sufficient. In the "2" column of the binary representation, the possible occurrence of the two 1s can be accommodated by adding the A_1 bit and the B_0 bit of the BCD number. In the "4" column of the binary representation, the possible occurrence of the two 1s is handled by adding the A_2 bit and the B_1 bit of the BCD number. In the "8" column of the binary representation, the possibility of the three 1s is handled by adding the A_3, B_0, and B_2 bits of the BCD number. In the "16" column, the B_1 and the B_3 bits are added. In the "32" column, only a single 1 is possible, so the B_2 bit is added to the carry from the "16" column. In the "64" column, only a single 1 can occur, so the B_3 bit is added only to the carry from the "32" column. A method of implementing these requirements with 4-bit adders is shown in Figure 6–50.

MSI BCD-to-Binary and Binary-to-BCD Converters

MSI **code converters** are generally implemented with preprogrammed read-only memories (ROMs). ROMs are covered in Chapter 12. Examples are the 74184, a ROM device programmed as a BCD-to-binary converter, and the 74185, a ROM device programmed as a binary-to-BCD converter. Logic symbols for these devices used as 6-bit converters are shown in Figure 6–51. Figure 6–52 and Figure 6–53 (page 296) show how these devices can be expanded for conversions involving larger numbers of bits.

FIGURE 6–50
Two-digit BCD-to-binary converter using 4-bit adders.

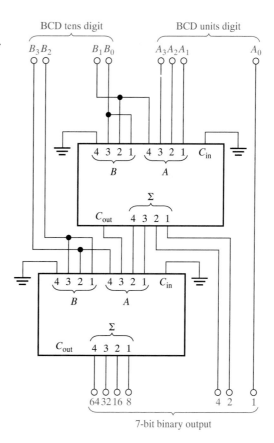

FIGURE 6–51

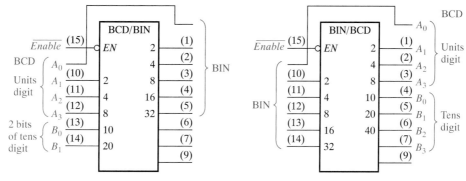

(a) The 74184 6-bit BCD-to-binary converter   (b) The 74185 6-bit binary-to-BCD converter

FIGURE 6–52
74184s expanded for conversion of two BCD digits to binary.

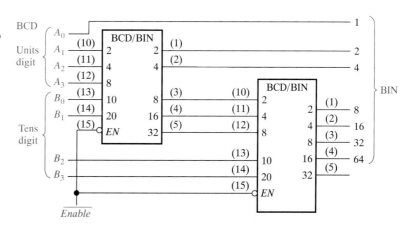

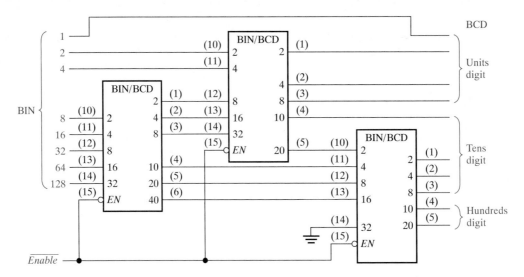

FIGURE 6–53
74185s expanded for 8-bit binary-to-BCD conversion.

Binary-to-Gray and Gray-to-Binary Conversion

The basic process for Gray-binary conversions was covered in Chapter 2. You will now see how exclusive-OR gates can be used for these conversions. Figure 6–54 shows a 4-bit binary-to-Gray code converter, and Figure 6–55 illustrates a 4-bit Gray-to-binary converter.

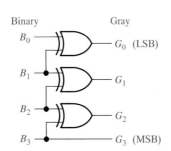

FIGURE 6–54
Four-bit binary-to-Gray conversion logic.

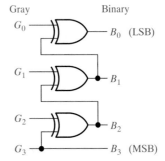

FIGURE 6–55
Four-bit Gray-to-binary conversion logic.

EXAMPLE 6–13

Convert the following binary numbers to Gray code by using exclusive-OR gates:
(a) 0101   (b) 00111   (c) 101011

Solution
(a) 0101_2 is 0111 Gray.
(b) 00111_2 is 00100 Gray.
(c) 101011_2 is 111110 Gray.
See Figure 6–56.

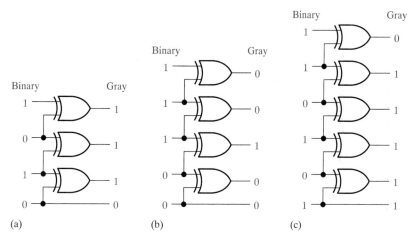

(a)   (b)   (c)

FIGURE 6–56

Related Exercise   How many exclusive-OR gates are required to convert 8-bit binary to Gray?

EXAMPLE 6–14   Convert the following Gray codes to binary by using exclusive-OR gates:
(a) 1011   (b) 11000   (c) 1001011

Solution
(a) 1011 Gray is 1101_2.
(b) 11000 Gray is 10000_2.
(c) 1001011 Gray is 1110010_2.
See Figure 6–57.

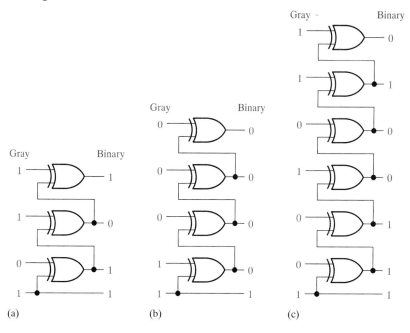

(a)   (b)   (c)

FIGURE 6–57

Related Exercise   Implement a 5-bit Gray-to-binary converter.

6–8 ■ MULTIPLEXERS (DATA SELECTORS)

A multiplexer (MUX) is a device that allows digital information from several sources to be routed onto a single line for transmission over that line to a common destination. The basic multiplexer has several data-input lines and a single output line. It also has data-select inputs, which permit digital data on any one of the inputs to be switched to the output line. Multiplexers are also known as data selectors. After completing this section, you should be able to

□ Explain the basic operation of a multiplexer   □ Describe the 74150, 74151A, and the 74157 MSI multiplexers   □ Expand a multiplexer to handle more data inputs
□ Use the multiplexer as a logic function generator

A logic symbol for a 4-input **multiplexer (MUX)** is shown in Figure 6–58. Notice that there are two data-select lines because with two select bits, any one of the four data-input lines can be selected.

FIGURE 6–58
Logic symbol for a 1-of-4 data selector/ multiplexer.

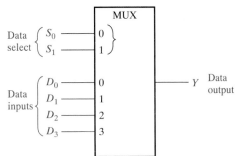

In Figure 6–58, a 2-bit binary code on the data-select (S) inputs will allow the data on the selected data input to pass through to the data output. If a binary 0 ($S_1 = 0$ and $S_0 = 0$) is applied to the data-select lines, the data on input D_0 appear on the data-output line. If a binary 1 ($S_1 = 0$ and $S_0 = 1$) is applied to the data-select lines, the data on input D_1 appear on the data output. If a binary 2 ($S_1 = 1$ and $S_0 = 0$) is applied, the data on D_2 appear on the output. If a binary 3 ($S_1 = 1$ and $S_0 = 1$) is applied, the data on D_3 are switched to the output line. A summary of this operation is given in Table 6–9.

TABLE 6–9
Data selection for a 1-of-4 multiplexer

Data-select Inputs		Input Selected
S_1	S_0	
0	0	D_0
0	1	D_1
1	0	D_2
1	1	D_3

Now let's look at the logic circuitry required to perform this multiplexing operation. The data output is equal to the state of the *selected* data input. We can therefore, derive a logic expression for the output in terms of the data input and the select inputs.

The data output is equal to D_0 if and only if $S_1 = 0$ and $S_0 = 0$: $Y = D_0\overline{S_1}\overline{S_0}$.

The data output is equal to D_1 if and only if $S_1 = 0$ and $S_0 = 1$: $Y = D_1\overline{S_1}S_0$.

The data output is equal to D_2 if and only if $S_1 = 1$ and $S_0 = 0$: $Y = D_2 S_1\overline{S_0}$.

The data output is equal to D_3 if and only if $S_1 = 1$ and $S_0 = 1$: $Y = D_3 S_1 S_0$.

When these terms are ORed, the total expression for the data output is

$$Y = D_0\overline{S_1}\overline{S_0} + D_1\overline{S_1}S_0 + D_2 S_1\overline{S_0} + D_3 S_1 S_0$$

The implementation of this equation requires four 3-input AND gates, a 4-input OR gate, and two inverters to generate the complements of S_1 and S_0, as shown in Figure 6–59. Because data can be selected from any one of the input lines, this circuit is also referred to as a **data selector.**

FIGURE 6–59
Logic diagram for a 4-input multiplexer.

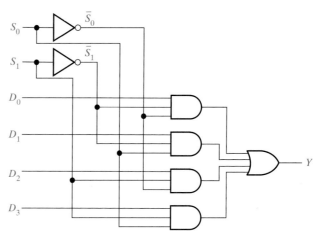

EXAMPLE 6–15

The data-input and data-select waveforms in Figure 6–60(a) are applied to the multiplexer in Figure 6–59. Determine the output waveform in relation to the inputs.

FIGURE 6–60

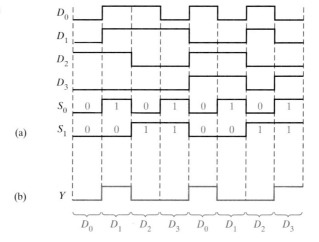

(a)

(b)

Solution   The binary state of the data-select inputs during each interval determines which data input is selected. Notice that the data-select inputs go through a repetitive binary sequence 00, 01, 10, 11, 00, 01, 10, 11, and so on. The resulting output waveform is shown in Figure 6–60(b).

Related Exercise   Construct a timing diagram showing all inputs and the output if the S_0 and S_1 waveforms in Figure 6–60 are interchanged.

The 74157 Quadruple 2-Input Data Selector/Multiplexer

The 74157, as well as its LS and CMOS versions, consists of four separate 2-input multiplexers. Each of the four multiplexers shares a common data-select line and a common *Enable*, as shown in Figure 6–61(a). Because there are only two inputs to be selected in each multiplexer, a single data-select input is sufficient.

   As you can see in the logic diagram, the data-select input is ANDed with the *B* input of each 2-input multiplexer, and the complement of data-select is ANDed with each *A* input.

   A LOW on the $\overline{Enable}$ input allows the selected input data to pass through to the output. A HIGH on the $\overline{Enable}$ input prevents data from going through to the output; that is, it disables the multiplexers.

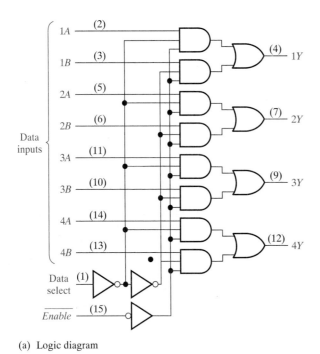

(a) Logic diagram

(b) Logic symbol

FIGURE 6–61
The 74157 quadruple 2-input data selector/multiplexer.

The ANSI/IEEE Logic Symbol

The ANSI/IEEE logic symbol for the 74157 is shown in Figure 6–61(b). Notice that the four multiplexers are indicated by the partitioned outline and that the inputs common to all four multiplexers are indicated as inputs to the notched block at the top, which is called the *common control block*. All labels within the upper MUX block apply to the other blocks below it.

Notice the 1 and $\overline{1}$ labels in the MUX blocks and the G1 label in the common control block. These labels are an example of the **dependency notation** system specified in the ANSI/IEEE Standard 91–1984. In this case G1 indicates an AND relationship between the data-select input and the data inputs with 1 or $\overline{1}$ labels. (The $\overline{1}$ means that the AND relationship applies to the complement of the G1 input.) In other words, when the data-select input is HIGH, the *B* inputs of the multiplexers are selected; and when the data-select input is LOW, the *A* inputs are selected. A "G" is always used to denote AND dependency. Other aspects of dependency notation are introduced as appropriate throughout the book.

The 74151A 8-Input Data Selector/Multiplexer

The 74151A has eight data inputs and, therefore, three data-select input lines. Three bits are required to select any one of the eight data inputs ($2^3 = 8$). A LOW on the $\overline{Enable}$ input allows the selected input data to pass through to the output. Notice that the data output and its complement are both available. The logic diagram is shown in Figure 6–62(a), and the ANSI/IEEE logic symbol is shown in part (b). In this case there is no

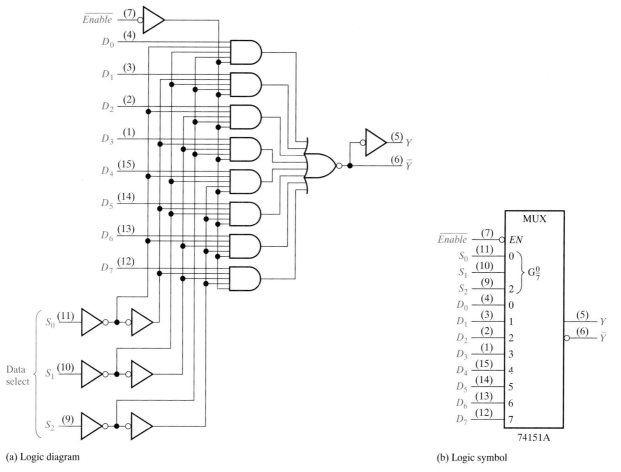

(a) Logic diagram

(b) Logic symbol

FIGURE 6–62
The 74151A 8-input data selector/multiplexer.

need for a common control block on the logic symbol because there is only one multi-plexer to be controlled, not four as in the 74157. The G_7^0 label within the logic symbol indicates the AND relationship between the data-select inputs and each of the data inputs 0 through 7.

The 74150 16-Input Data Selector/Multiplexer

The 74150 has sixteen data inputs and four data-select lines. In this case four bits are required to select any one of the sixteen data inputs ($2^4 = 16$). There is also an active-LOW $\overline{Enable}$ input. On this particular device, only the complement of the output is available. The logic symbol is shown in Figure 6–63.

FIGURE 6–63
The 74150 16-input data selector/multiplexer.

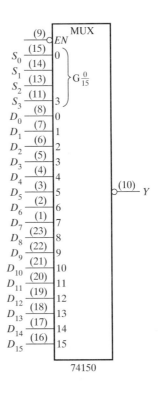

EXAMPLE 6–16   Use 74150s and any other logic necessary to multiplex 32 data lines onto a single data-output line.

Solution   An implementation of this system is shown in Figure 6–64. Five bits are required to select one of 32 data inputs ($2^5 = 32$). In this application the $\overline{Enable}$ input is used as the most significant data-select bit. When the MSB in the data-select code is LOW, the left 74150 is enabled, and one of the data inputs (D_0 through D_{15}) is selected by the other four data-select bits. When the data-select MSB is HIGH, the right 74150 is enabled, and one of the data inputs (D_{16} through D_{31}) is selected. The selected input data are then passed through to the negative-OR gate and onto the single output line.

Related Exercise   Determine the codes on the select inputs required to select each of the following data inputs: D_0, D_8, D_{21}, and D_{29}.

FIGURE 6–64

A 32-input multiplexer.

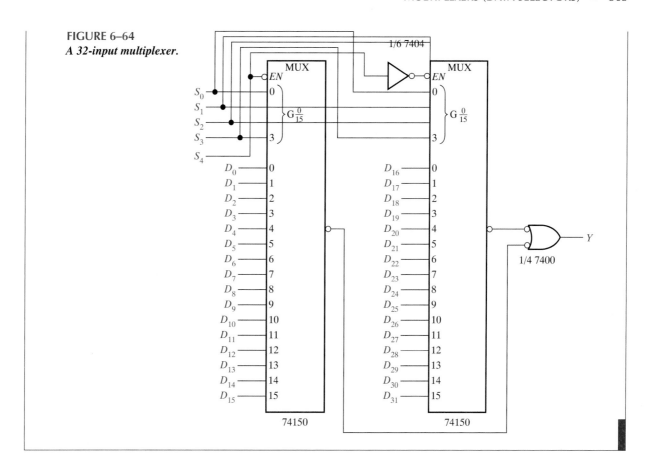

Data Selector/Multiplexer Applications

A 7-Segment Display Multiplexer   Figure 6–65 (page 304) shows a simplified method of multiplexing BCD numbers to a 7-segment display. In this example, 2-digit numbers are displayed on the 7-segment readout by the use of a single BCD-to-7-segment decoder. This basic method of display multiplexing can be extended to displays with any number of digits. The basic operation is as follows.

Two BCD digits ($A_3A_2A_1A_0$ and $B_3B_2B_1B_0$) are applied to the multiplexer inputs. A square wave is applied to the data-select line, and when it is LOW, the A bits ($A_3A_2A_1A_0$) are passed through to the inputs of the 74LS48 BCD-to-7-segment decoder. The LOW on the data-select also puts a LOW on the 1 input of the 74LS139 2-line-to-4-line decoder, thus activating its 0 output and enabling the A-digit display by effectively connecting its common terminal to ground. The A digit is now on and the B digit is off.

When the data-select line goes HIGH, the B bits ($B_3B_2B_1B_0$) are passed through to the inputs of the BCD-to-7-segment decoder. Also, the 74LS139 decoder's 1 output is activated, thus enabling the B-digit display. The B digit is now on and the A digit is off. The cycle repeats at the frequency of the data-select square wave. This frequency must be high enough (about 30 Hz) to prevent visual flicker as the digit displays are multiplexed.

A Logic Function Generator   A useful application of the data selector/multiplexer is in the generation of combinational logic functions in sum-of-products form. When used in this way, the device can replace discrete gates, can often greatly reduce the number of ICs, and can make design changes much easier.

To illustrate, a 74151A 8-input data selector/multiplexer can be used to implement any specified 3-variable logic function if the variables are connected to the data-select inputs and each data input is set to the logic level required in the truth table for that function. For example, if the function is a 1 when the variable combination is $\overline{A}_2A_1\overline{A}_0$, the 2 input (selected by 010) is connected to a HIGH. This HIGH is passed through to the output when

FIGURE 6–65
Simplified 7-segment display multiplexing logic.

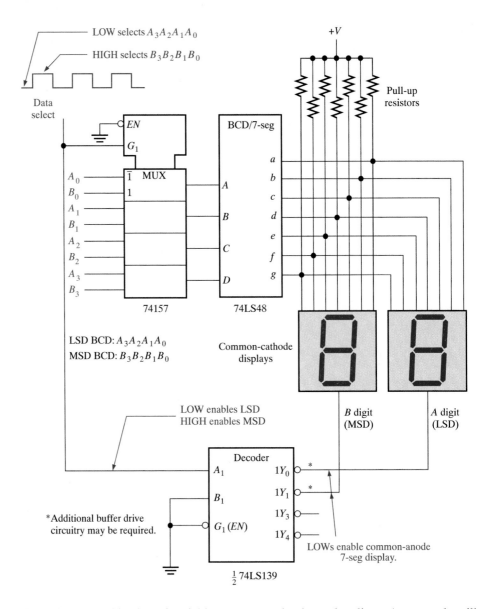

this particular combination of variables occurs on the data-select lines. An example will help clarify this application.

EXAMPLE 6–17

Implement the logic function specified in Table 6–10 by using a 74151A 8-input data selector/multiplexer. Compare this method with a discrete logic gate implementation.

TABLE 6–10

	Inputs		Output
A_2	A_1	A_0	Y
0	0	0	0
0	0	1	1
0	1	0	0
0	1	1	1
1	0	0	0
1	0	1	1
1	1	0	1
1	1	1	0

Solution Notice from the truth table that Y is a 1 for the following input variable combinations: 001, 011, 101, and 110. For all other combinations, Y is 0. For this function to be implemented with the data selector, the data input selected by each of the above-mentioned combinations must be connected to a HIGH (5 V) through a pull-up resistor. All the other data inputs must be connected to a LOW (ground), as shown in Figure 6–66.

The implementation of this function with logic gates would require four 3-input AND gates, one 4-input OR gate, and three inverters unless the expression can be simplified.

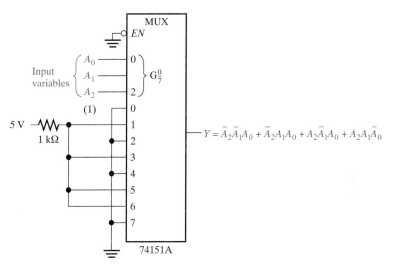

$$Y = \bar{A}_2\bar{A}_1A_0 + \bar{A}_2A_1A_0 + A_2\bar{A}_1A_0 + A_2A_1\bar{A}_0$$

FIGURE 6–66
Data selector/multiplexer connected as a 3-variable logic function generator.

Related Exercise Use the 74151A to implement the following expression:

$$Y = \bar{A}_2\bar{A}_1\bar{A}_0 + A_2\bar{A}_1\bar{A}_0 + \bar{A}_2A_1\bar{A}_0$$

Example 6–17 illustrated how the 8-input data selector can be used as a logic function generator for three variables. Actually, this device can be also used as a 4-variable logic function generator by the utilization of one of the bits (A_0) in conjunction with the data inputs.

A 4-variable truth table has sixteen combinations of input variables. When an 8-bit data selector is used, each input is selected twice: the first time when A_0 is 0 and the second time when A_0 is 1. With this in mind, the following rules can be applied (Y is the output, and A_0 is the least significant bit):

1. If $Y = 0$ both times a given data input is selected by a certain combination of the input variables, $A_3A_2A_1$, connect that data input to ground (0).
2. If $Y = 1$ both times a given data input is selected by a certain combination of the input variables, $A_3A_2A_1$, connect that data input to $+V$ (1).
3. If Y is different the two times a given data input is selected by a certain combination of the input variables, $A_3A_2A_1$, and if $Y = A_0$, connect that data input to A_0.
4. If Y is different the two times a given data input is selected by a certain combination of the input variables, $A_3A_2A_1$, and if $Y = \bar{A}_0$, connect that data input to $\bar{A}_0$.

The following example illustrates this method.

EXAMPLE 6–18   Implement the logic function in Table 6–11 by using a 74151A 8-input data selector/multiplexer. Compare this method with a discrete logic gate implementation.

TABLE 6–11

Decimal Digit	Inputs				Output
	A_3	A_2	A_1	A_0	Y
0	0	0	0	0	0
1	0	0	0	1	1
2	0	0	1	0	1
3	0	0	1	1	0
4	0	1	0	0	0
5	0	1	0	1	1
6	0	1	1	0	1
7	0	1	1	1	1
8	1	0	0	0	1
9	1	0	0	1	0
10	1	0	1	0	1
11	1	0	1	1	0
12	1	1	0	0	1
13	1	1	0	1	1
14	1	1	1	0	0
15	1	1	1	1	1

Solution   The data-select inputs are $A_3 A_2 A_1$. In the first row of the table, $A_3 A_2 A_1 = 000$ and $Y = A_0$. In the second row, where $A_3 A_2 A_1$ again is 000, $Y = A_0$. Thus, A_0 is connected to the 0 input. In the third row of the table, $A_3 A_2 A_1 = 001$ and $Y = \overline{A}_0$. Also, in the fourth row, when $A_3 A_2 A_1$ again is 001, $Y = \overline{A}_0$. Thus, A_0 is inverted and connected to the 1 input. This analysis is continued until each input is properly connected according to the specified rules. The implementation is shown in Figure 6–67.

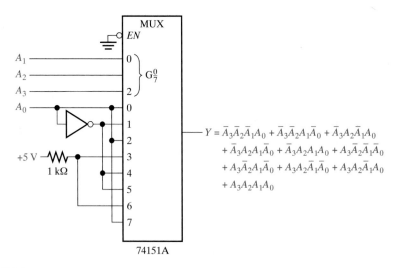

$$Y = \overline{A}_3 \overline{A}_2 \overline{A}_1 A_0 + \overline{A}_3 \overline{A}_2 A_1 \overline{A}_0 + \overline{A}_3 A_2 \overline{A}_1 A_0$$
$$+ \overline{A}_3 A_2 A_1 \overline{A}_0 + \overline{A}_3 A_2 A_1 A_0 + A_3 \overline{A}_2 \overline{A}_1 \overline{A}_0$$
$$+ A_3 \overline{A}_2 A_1 \overline{A}_0 + A_3 A_2 \overline{A}_1 \overline{A}_0 + A_3 A_2 \overline{A}_1 A_0$$
$$+ A_3 A_2 A_1 A_0$$

FIGURE 6–67
Data selector/multiplexer connected as a 4-variable logic function generator.

If implemented with logic gates, the function would require as many as ten 4-input AND gates, one 10-input OR gate, and four inverters, although possible simplification would reduce this requirement.

Related Exercise In Table 6–11, if $Y = 0$ when the inputs are all zeros and is alternately a 1 and a 0 for the remaining rows in the table, use a 74151A to implement the resulting logic function.

SECTION 6–8 REVIEW

1. In Figure 6–59 (page 299), $D_0 = 1$, $D_1 = 0$, $D_2 = 1$, $D_3 = 0$, $S_0 = 1$, and $S_1 = 0$. What is the output?
2. Identify each MSI device.
   (a) 74157   **(b)** 74151A   **(c)** 74150
3. A 74151A has alternating LOW and HIGH levels on its data inputs beginning with $D_0 = 0$. The data-select lines are sequenced through a binary count (000, 001, 010, and so on) at a frequency of 1 kHz. The enable input is LOW. Describe the data output waveform.
4. Briefly describe the purpose of each of the following devices in Figure 6–65:
   (a) 74157   **(b)** 74LS48   **(c)** 74LS139

6–9 ■ DEMULTIPLEXERS

A demultiplexer (DEMUX) basically reverses the multiplexing function. It takes data from one line and distributes them to a given number of output lines. For this reason, the demultiplexer is also known as a data distributor. As you will learn, decoders can also be used as demultiplexers. After completing this section, you should be able to

☐ Explain the basic operation of a demultiplexer   ☐ Describe how the 74154 4-line-to-16-line decoder can be used as a demultiplexer   ☐ Develop the timing diagram for a demultiplexer with specified data and data selection inputs

Figure 6–68 shows a 1-line-to-4-line **demultiplexer (DEMUX)** circuit. The data-input line goes to all of the AND gates. The two data-select lines enable only one gate at a time, and the data appearing on the data-input line will pass through the selected gate to the associated data-output line.

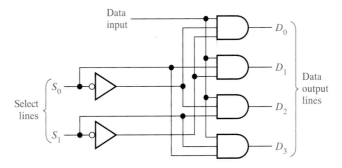

FIGURE 6–68
A 1-line-to-4-line demultiplexer.

EXAMPLE 6–19

The serial data-input waveform (Data in) and data-select inputs (S_0 and S_1) are shown in Figure 6–69. Determine the data-output waveforms on D_0 through D_3 for the demultiplexer in Figure 6–68.

FIGURE 6–69

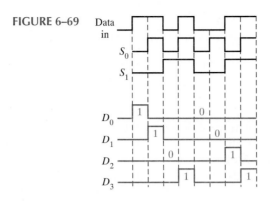

Solution   Notice that the select lines go through a binary sequence so that each successive input bit is routed to D_0, D_1, D_2, and D_3 in sequence, as shown by the output waveforms in Figure 6–69.

Related Exercise   Develop the timing diagram for the demultiplexer if the S_0 and S_1 waveforms are both inverted.

The 74154 as a Demultiplexer

We have already discussed the 74154 in its application as a 4-line-to-16-line decoder (Section 6–5). This device and other decoders are also used in demultiplexing applications. The logic symbol for this device when used as a demultiplexer is shown in Figure 6–70. In demultiplexer applications, the input lines are used as the data-select lines. One of the *Enable* inputs is used as the data-input line, with the other *Enable* input held LOW to enable the internal negative-AND gate at the bottom of the diagram.

FIGURE 6–70

The 74154 used as a demultiplexer.

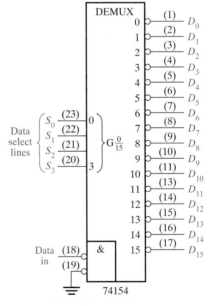

SECTION 6–9 REVIEW

1. Generally, how can an MSI decoder be used as a demultiplexer?
2. The 74154 demultiplexer in Figure 6–70 has a binary code of 1010 on the data-select lines, and the data-input line is LOW. What are the states of the output lines?

6–10 ■ PARITY GENERATORS/CHECKERS

Errors can occur as digital codes are being transferred from one point to another within a digital system or while codes are being transmitted from one system to another. The errors take the form of undesired changes in the bits that make up the coded information; that is, a 1 can change to a 0, or a 0 to a 1, because of component malfunctions or electrical noise. In most digital systems, the probability that even a single bit error will occur is very small, and the likelihood that more than one will occur is even smaller. Nevertheless, when an error occurs undetected, it can cause serious problems in a digital system. After completing this section, you should be able to

☐ Explain the concept of parity ☐ Implement a basic parity circuit with exclusive-OR gates ☐ Describe the operation of basic parity generating and checking logic ☐ Discuss the 74180 9-bit parity generator/checker ☐ Discuss how error detection can be implemented in a data transmission

The parity method of error detection in which a **parity bit** is attached to a group of information bits in order to make the total number of 1s either even or odd (depending on the system) was covered in Chapter 2. In addition to parity bits, several specific codes also provide inherent error detection; a few of the most important, as well as an error-correcting code, are discussed in Appendix B.

Parity Logic

In order to check for or to generate the proper parity in a given code, a very basic principle can be used:

The sum (disregarding carries) of an even number of 1s is always 0, and the sum of an odd number of 1s is always 1.

Therefore, to determine if a given code has **even parity** or **odd parity,** all the bits in that code are summed. As you know, the sum of two bits can be generated by an exclusive-OR gate, as shown in Figure 6–71(a); the sum of three bits can be formed by two exclusive-OR gates connected as shown in Figure 6–71(b); and so on.

FIGURE 6–71

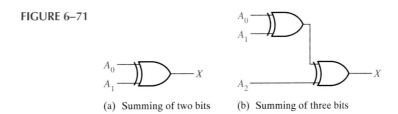

(a) Summing of two bits   (b) Summing of three bits

Parity Generator/Checker Logic

A typical 5-bit generator/checker circuit is shown in Figure 6–72. It can be used for either odd or even parity. When used as an odd parity checker, as shown, the operation is as follows: A 5-bit code (four data bits and one parity bit) is applied to the inputs. The four data bits are on the exclusive-OR inputs, and the parity bit is applied to the ODD input line. When the number of 1s in the 5-bit code is odd, the Σ ODD output is LOW, indicating proper parity. When there is an even number of 1s, the Σ ODD output is HIGH, indicating incorrect parity. Both of these conditions are illustrated in Figure 6–72.

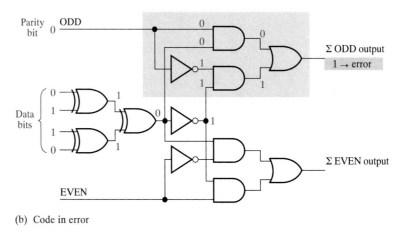

(a) Code correct

(b) Code in error

FIGURE 6–72
Examples of odd parity detection.

Similarly, even parity checks are illustrated for both nonerror and error conditions in Figure 6–73.

The same circuit can be used as a parity generator, as shown in Figure 6–74 (page 312). The operation for odd parity generation is illustrated in part (a). A 4-bit code is applied to the inputs, and the ODD line is held LOW (by grounding in this case). When the 4-bit code has an even number of 1s, as shown, the Σ ODD output is HIGH (1). This 1 output is the odd parity bit and is combined with the 4-bit code to form a 5-bit odd parity code as shown. Similarly, a 0 parity bit is produced when there is an odd number of 1s in the input code.

Figure 6–74(b) shows an example of the same circuit used in an even parity system. Notice that the EVEN line is grounded in this case.

The basic logic shown in Figure 6–74 can be expanded to accommodate any number of input bits by adding more exclusive-OR gates.

The 74180 9-Bit Parity Generator/Checker

The 74180 is represented in Figure 6–75 (page 312). This particular MSI device can be used to check for odd or even parity on a 9-bit code (eight data bits and one parity bit), or it can be used to generate a 9-bit odd or even parity code. The truth table operation varies slightly from the simpler basic circuits just discussed, but the principle is the same.

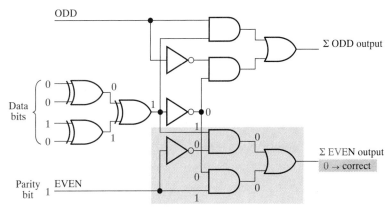

(a) Code correct

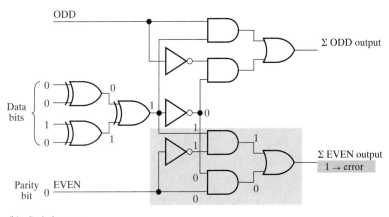

(b) Code in error

FIGURE 6–73
Examples of even parity detection.

A Data Transmission System with Error Detection

A simplified data transmission system is shown in Figure 6–76 (page 313) to illustrate an application of parity generators/checkers, as well as multiplexers and demultiplexers, and to illustrate the need for data storage in some applications.

In this application, digital data from seven sources are multiplexed onto a single line for transmission to a distant point. The seven data bits (D_0 through D_6) are applied to the multiplexer data inputs and, at the same time, to the even parity generator inputs. The Σ ODD output of the parity generator is used as the even parity bit. This bit is 0 if the number of 1s on the inputs A through H is even and is a 1 if the number of 1s on A through H is odd. This bit is D_7 of the transmitted code.

The data-select inputs are repeatedly cycled through a binary sequence, and each data bit, beginning with D_0, is serially passed through and onto the transmission line ($\overline{Y}$). In this example, the transmission line consists of four conductors: one carries the serial data and three carry the timing signals (data selects). There are more sophisticated ways of sending the timing information, but we are using this direct method to illustrate a basic application.

At the demultiplexer end of the system, the data-select signals and the serial data stream are applied to the demultiplexer. The data bits are distributed by the demultiplexer onto the output lines in the order in which they occurred on the multiplexer inputs. That is, D_0 comes out on the D_0 output, D_1 comes out on the D_1 output, and so on. The parity bit comes out on the D_7 output. These eight bits are temporarily stored and applied to the even

FIGURE 6–74
Examples of parity generation.

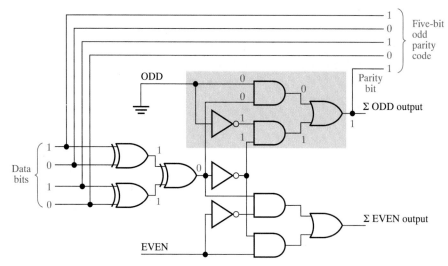

(a) Odd parity generation

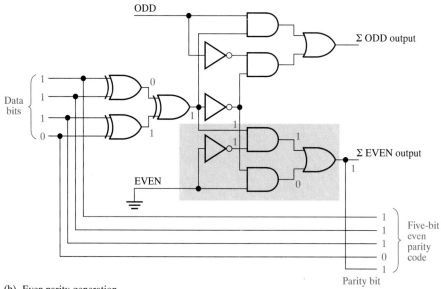

(b) Even parity generation

FIGURE 6–75
The 74180 9-bit parity generator/checker.

(a) Traditional logic symbol

	Inputs		Outputs	
Σ of 1s at A through H	EVEN	ODD	Σ EVEN	Σ ODD
EVEN	1	0	1	0
ODD	1	0	0	1
EVEN	0	1	0	1
ODD	0	1	1	0
X	1	1	0	0
X	0	0	1	1

(b) Truth table (X ≡ don't care)

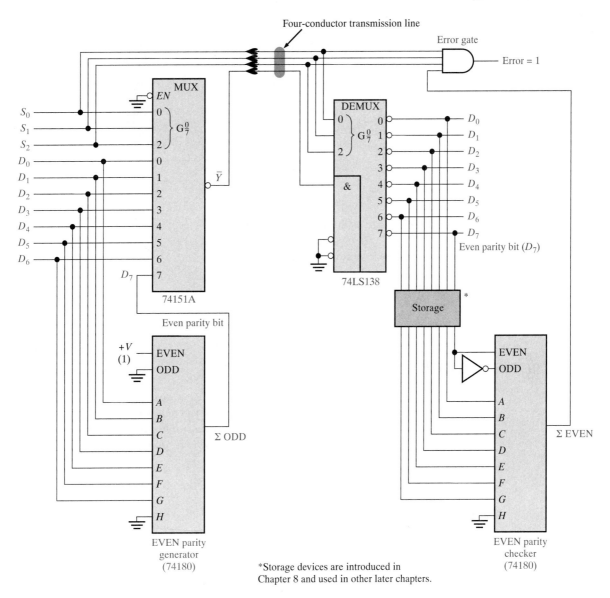

FIGURE 6–76
Simplified data transmission system with error detection.

parity checker. Not all of the bits are present on the parity checker inputs until the parity bit D_7 comes out and is stored. At this time, the error gate is enabled by the 111 data-select code. If the parity is correct, a 0 appears on the Σ EVEN output, keeping the ERROR output at 0. If the parity is incorrect, all 1s appear on the error gate inputs, and a 1 on the ERROR output results.

This particular application has demonstrated the need for data storage so that you will be better able to appreciate the usefulness of the storage devices introduced in Chapter 8 and used in other later chapters.

The timing diagram in Figure 6–77 illustrates a specific case in which two 8-bit words are transmitted, one with correct parity and one with an error.

FIGURE 6–77

Example of data transmission with and without error for the system in Figure 6–76.

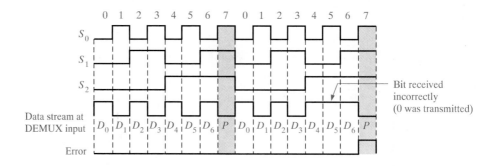

SECTION 6–10
REVIEW

1. Add an even parity bit to each of the following codes:
   (a) 110100   (b) 01100011
2. Add an odd parity bit to each of the following codes:
   (a) 1010101   (b) 1000001
3. Check each of the even parity codes for an error.
   (a) 100010101   (b) 1110111001

6–11 ■ TROUBLESHOOTING

In this section, the problem of decoder glitches is introduced and examined from a troubleshooting standpoint. A glitch is any undesired voltage or current spike (pulse) of very short duration. A glitch can be interpreted as a valid signal by a logic circuit and may cause improper operation. After completing this section, you should be able to

□ Explain what a glitch is   □ Determine the cause of glitches in a decoder application   □ Use the method of output strobing to eliminate glitches

A 74LS138 3-line-to-8-line decoder (binary-to-octal) is used in Figure 6–78 to illustrate how **glitches** occur and how to identify their cause. The $A_2A_1A_0$ inputs of the decoder are sequenced through a binary count, and the resulting waveforms of the inputs and outputs can be displayed on the screen of a logic analyzer as shown in Figure 6–78.

The output waveforms are correct except for the glitches that occur on some of the output signals. An oscilloscope or the oscilloscope function of a logic analyzer can be used to examine the critical timing details of the input waveforms in an effort to pinpoint the cause of the glitches. The oscilloscope is useful in this case because the waveforms can be "magnified" so that very small time differences between waveform transitions can be seen. A_2 transitions are delayed from A_1 transitions and A_1 transitions are delayed from A_0 transitions. This commonly occurs when waveforms are generated by a binary counter, as you will learn in Chapter 9.

The points of interest indicated by the highlighted areas on the input waveforms in Figure 6–78 are displayed on an oscilloscope as shown in Figure 6–79. At point 1 there is a transitional state of 000 due to delay differences in the waveforms. This causes the first glitch on the $\overline{0}$ output of the decoder. At point 2 there are two transitional states, 010 and 000. These cause the glitch on the $\overline{2}$ output of the decoder and the second glitch on the $\overline{0}$ output, respectively. At point 3 the transitional state is 100, which causes the first glitch on the $\overline{4}$ output of the decoder. At point 4 the two transitional states, 110 and 100, result in the glitch on the $\overline{6}$ output and the second glitch on the $\overline{4}$ output, respectively.

One way to eliminate the glitch problem is a method called **strobing**, in which the decoder is enabled by a strobe pulse only during the times when the waveforms are not in transition. This method is illustrated in Figure 6–80 (page 316).

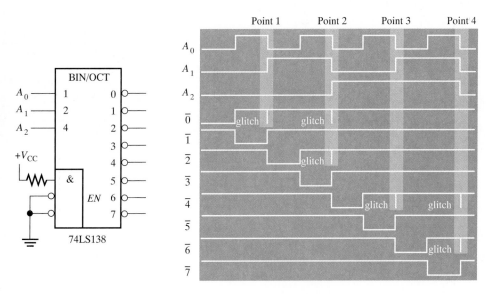

FIGURE 6–78
Decoder waveforms with output glitches.

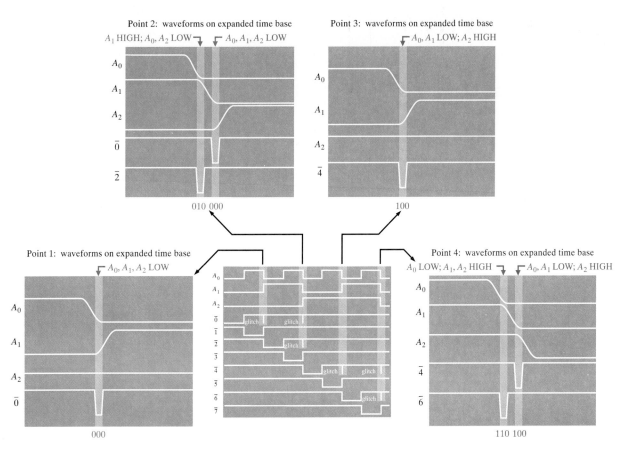

FIGURE 6–79
Decoder waveform displays showing how transitional input states produce glitches in the output waveforms.

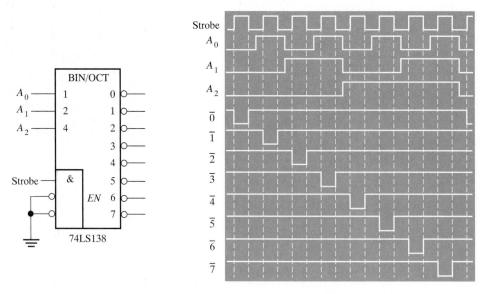

FIGURE 6–80

Application of a strobe waveform to eliminate glitches on decoder outputs.

SECTION 6–11 REVIEW	1. Define the term *glitch*. 2. Explain the basic cause of glitches in decoder logic. 3. Define the term *strobe*.

6–12 ■ DIGITAL SYSTEM APPLICATION

In this system application, you will begin working with a traffic light control system. In this section, the system requirements are established, a general block diagram is developed, and a state diagram is created to help define the sequence of operation. A portion of the system involving combinational logic is designed and methods of testing are considered. The sequential and timing portions of the system will be dealt with in Chapters 8 and 9. After completing this section, you should be able to

□ Determine system specifications □ Develop a system block diagram from basic specifications □ Explain the purpose of a state diagram □ Use a state diagram and system specifications to design and implement the combinational logic portions of the system □ Develop a test procedure for testing the breadboarded combinational logic □ Specify a simple test setup

General System Requirements

A digital controller is required to control a traffic light at the intersection of a busy main street and an occasionally used side street. The main street is to have a green light for a minimum of 25 s or as long as there is no vehicle on the side street. The side street is to have a green light until there is no vehicle on the side street or for a maximum of 25 s. There is to be a 4 s caution light (yellow) between changes from green to red on both the main street and on the side street. These requirements are illustrated in the pictorial diagram in Figure 6–81.

FIGURE 6–81
Requirements for the traffic light sequence.

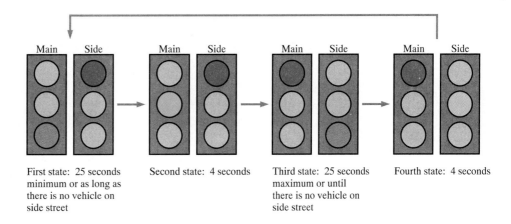

First state: 25 seconds minimum or as long as there is no vehicle on side street

Second state: 4 seconds

Third state: 25 seconds maximum or until there is no vehicle on side street

Fourth state: 4 seconds

Developing a Block Diagram of the System

From the requirements, you can develop a block diagram of the system. First, you know that the system must control six different pairs of lights. These are the red, yellow, and green lights for both directions on the main street and the red, yellow, and green lights for both directions on the side street. Also, you know that there is one external input (other than power) from a side street vehicle sensor. Figure 6–82 is a minimal block diagram showing these requirements.

FIGURE 6–82
A minimal system block diagram.

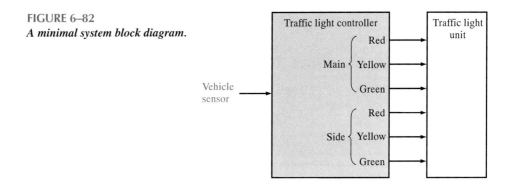

From the minimal system block diagram, you can begin to fill in the details. The system has four states, as indicated in Figure 6–81, so a logic circuit is needed to control the sequence of states (sequential logic). Also, circuits are needed to generate the proper time intervals of 25 s and 4 s that are required in the system and to generate a clock signal for cycling the system (timing circuits). The time intervals (long and short) and the vehicle sensor are inputs to the sequential logic because the sequencing of states is a function of these variables. Logic circuits are also needed to determine which of the four states the system is in at any given time, to generate the proper outputs to the lights (state decoding and output logic), and to initiate the long and short time intervals. Finally an interface circuit is needed to convert the logic levels of the decoding and output circuitry to the voltages and currents required to illuminate each of the lights. Figure 6–83 is a more detailed block diagram showing these essential elements.

The State Diagram

A state diagram graphically shows the sequence of states in a system and the conditions for each state and for transitions from one state to the next. Actually, Figure 6–81 is a form of state diagram because it shows the sequence of states and the conditions.

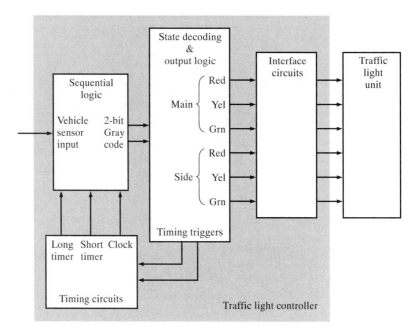

FIGURE 6–83
System block diagram showing essential elements.

Definition of Variables   Before a traditional state diagram can be developed, the variables that determine how the system sequences through its states must be defined. These variables and their symbols are listed as follows:

- Vehicle present on side street = V_s
- 25 s timer (long timer) is *on* = T_L
- 4 s timer (short timer) is *on* = T_S

The use of complemented variables indicates the opposite conditions. For example, $\overline{V_s}$ indicates that there is no vehicle on the side street, $\overline{T_L}$ indicates the long timer is *off*, and $\overline{T_S}$ indicates the short timer is *off*.

Description of the State Diagram   A state diagram is shown in Figure 6–84. Each of the four states is labeled according to the 2-bit Gray code sequence, as indicated by the circles. The looping arrow at each state indicates that the system remains in that state under the condition defined by the associated variable or expression. Each of the arrows going from one state to the next indicates a state transition under the condition defined by the associated variable or expression.

- ***First state***   The Gray code for this state is 00. The main street light is green and the side street light is red. The system remains in this state for at least 25 s when the long timer is *on* or as long as there is no vehicle on the side street ($T_L + \overline{V_s}$). The system goes to the next state when the 25 s timer is *off* and there is a vehicle on the side street ($\overline{T_L}V_s$).
- ***Second state***   The Gray code for this state is 01. The main street light is yellow (caution) and the side street light is red. The system remains in this state for 4 s when the short timer is *on* (T_S) and goes to the next state when the short timer goes *off* ($\overline{T_S}$).
- ***Third state***   The Gray code for this state is 11. The main street light is red and the side street light is green. The system remains in this state when the long timer is *on* and there is a vehicle on the side street ($T_L V_s$). The system goes to the next state when the 25 s have elapsed or when there is no vehicle on the side street, whichever comes first ($\overline{T_L} + \overline{V_s}$).

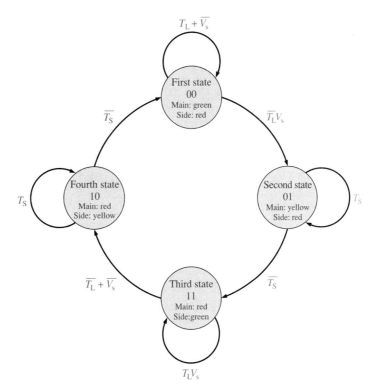

FIGURE 6–84
State diagram for the traffic light control system showing the Gray code sequence.

■ *Fourth state* The Gray code for this state is 10. The main street light is red and the side street light is yellow. The system remains in this state for 4 s when the short timer is *on* (T_S) and goes back to the first state when the short timer goes *off* ($\overline{T_S}$).

The State Decoding and Output Logic

The focus here is the state decoding and output logic portion of the block diagram of Figure 6–83. The timing and the sequential logic circuits will be the subjects of the system application sections in Chapters 8 and 9.

A block diagram for just the state decoding and output logic portion of the system should be developed as a first step in designing the logic. To do that, the three functions that this logic must perform are defined as follows and the resulting diagram with a block for each of the three functions is shown in Figure 6–85 (page 320):

■ Decode the 2-bit Gray code from the sequential logic to determine which of the four states the system is in.
■ Use the decoded state to activate (through the interface circuits) the appropriate two pairs of traffic lights out of the six pairs of traffic lights in the light unit.
■ Use the decoded states to produce signals for properly initiating (triggering) the long timer and the short timer.

Implementing the Logic The state decoder has two inputs (2-bit Gray code) and must have an output for each of the four states. A 2-line-to-4-line decoder is needed and the 74LS139 should do the job nicely. The 74LS139 contains two separate decoders but only one of them will be used.

The output logic must take the four active-LOW state outputs from the 74LS139 decoder and produce six outputs for activating the traffic lights. A truth table for the decoding and output logic is given in Table 6–12. The state outputs ($\overline{SO_1}$ through $\overline{SO_4}$) are active-LOW (0), and it is required that the light outputs from the output logic be active-LOW (0).

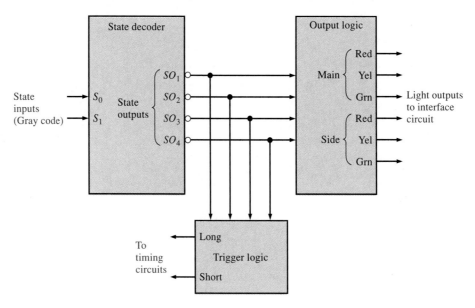

FIGURE 6–85
Block diagram of the state decoding and output logic.

TABLE 6–12
Truth table for the decoding and output logic

| State Inputs | State Outputs | | | | Light Outputs | | | | | | Trigger Outputs | |
$S_1 S_0$	$\overline{SO_1}$	$\overline{SO_2}$	$\overline{SO_3}$	$\overline{SO_4}$	$\overline{MR}$	$\overline{MY}$	$\overline{MG}$	$\overline{SR}$	$\overline{SY}$	$\overline{SG}$	*Long*	*Short*
00	0	1	1	1	1	1	0	0	1	1	1	0
01	1	0	1	1	1	0	1	0	1	1	0	1
11	1	1	0	1	0	1	1	1	1	0	1	0
10	1	1	1	0	0	1	1	1	0	1	0	1

State outputs are active-LOW and light outputs are active-LOW. *MR* stands for main street red, *SG* for side street green, etc.

The trigger logic produces two outputs. The long output produces a LOW-to-HIGH transition to trigger the 25 s timing circuit (long timer) when the system goes into the first or third states. The short output produces is a LOW-to-HIGH transition to trigger the 4 s timing circuit (short timer) when the system goes into the second or fourth states. This requirement can be seen by examining the system requirements and the state diagram in Figure 6–84.

■ THE DIGITAL WORKBENCH

Digital Workbench 1 involves the design activities for the state decoder and output logic portion of the system. Make every effort to design the logic diagram required for this part of the system and then compare your results with the portion of the printed circuit board shown in Digital Workbench 2. Only the portion of the circuit board containing the state decoder and output logic is shown. Keep in mind that you will be adding to your logic diagram as we progress through the rest of this system in Chapters 8 and 9.

In Digital Workbench 2, all of the connections to the V_{CC} and ground pads are on the back side of the board. Circuit interconnections on the back side of the board feed through to the pads on the component side. When you trace out the circuit, interconnected common pads are generally oriented vertically or horizontally and not at an angle with respect to each other.

■ DIGITAL WORKBENCH 1: Design

■ *Activity 1* Using the information from Table 6–12 and the pin diagram from a 74LS08 data sheet shown below, design the output logic required to generate the six light outputs to the interface circuit. Draw the logic diagram. *Hint: An AND gate can be treated as a negative-NOR gate where a LOW on either or both inputs produces a LOW output as shown below.*

■ *Activity 2* Using the state diagram and the 74LS08, design the trigger logic to meet the previously stated requirements. Add this to your logic diagram from Activity 1.

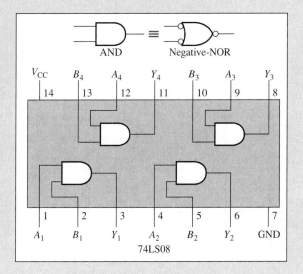

■ *Activity 3* Combine your circuit from Activities 1 and 2 with the 74LS139 decoder for which the pin diagram and truth table from the data sheet are shown below to create a complete logic diagram for the state decoder and output portion of the system. Make sure your diagram includes pin numbers.

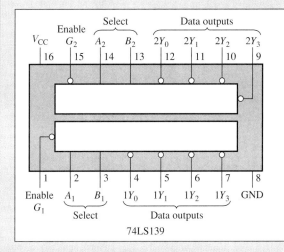

74LS139

Inputs			Outputs			
Enable	**Select**					
G	B	A	Y_0	Y_1	Y_2	Y_3
H	X	X	H	H	H	H
L	L	L	L	H	H	H
L	L	H	H	L	H	H
L	H	L	H	H	L	H
L	H	H	H	H	H	L

NOTE: H = High Level, L = Low Level, X = Don't Care

■ DIGITAL WORKBENCH 2: Verification and Testing

■ *Activity 1* Compare the logic diagram that you developed in Workbench 1 with the portion of the system PC board shown and make sure they agree.

■ *Activity 2* Identify the function of each of the points on the PC board labeled with a circled number.

■ *Activity 3* Using the switches connected as shown to simulate the Gray code input from the sequential portion of the system, develop a test procedure to check out this portion of the system board.

■ *Activity 4* When the switches are thrown into each of the positions shown in the table, determine from the indications at the test points represented by the circled numbers if the circuit is operating properly for each case and, if not, what the most likely problem is. Analyze each case separately.

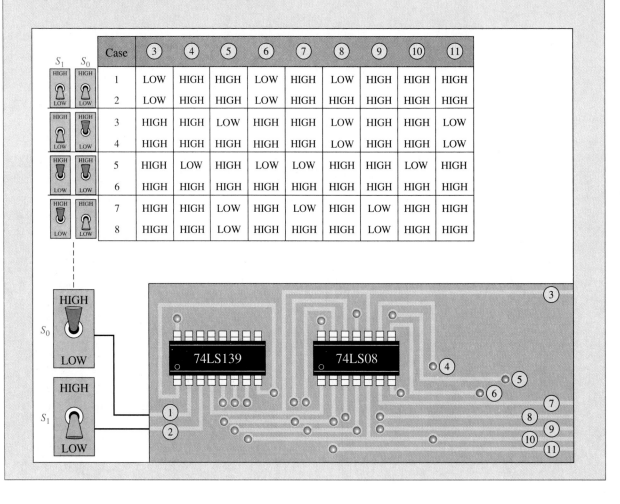

S_1	S_0	Case	③	④	⑤	⑥	⑦	⑧	⑨	⑩	⑪
HIGH / LOW	HIGH / LOW	1	LOW	HIGH	HIGH	LOW	HIGH	LOW	HIGH	HIGH	HIGH
		2	LOW	HIGH	HIGH	LOW	HIGH	HIGH	HIGH	HIGH	HIGH
HIGH / LOW	HIGH / LOW	3	HIGH	HIGH	LOW	HIGH	HIGH	LOW	HIGH	HIGH	LOW
		4	HIGH	HIGH	HIGH	HIGH	HIGH	LOW	HIGH	HIGH	LOW
HIGH / LOW	HIGH / LOW	5	HIGH	LOW	HIGH	LOW	LOW	HIGH	HIGH	LOW	HIGH
		6	HIGH	HIGH	HIGH	HIGH	HIGH	HIGH	HIGH	HIGH	HIGH
HIGH / LOW	HIGH / LOW	7	HIGH	HIGH	LOW	HIGH	LOW	HIGH	LOW	HIGH	HIGH
		8	HIGH	HIGH	LOW	HIGH	HIGH	HIGH	LOW	HIGH	HIGH

SECTION 6–12 REVIEW

1. List the sequence of states on the state decoder inputs.
2. Explain how the NAND gates used for the output logic and the trigger logic on the breadboard in the Digital Workbench act as negative-OR gates.
3. When do the timing trigger output transitions occur for the long timer? For the short timer?

■ SUMMARY

■ Half-adder and full-adder operations are summarized in Figure 6–86.

FIGURE 6–86

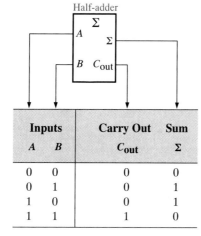

Half-adder			
Inputs		**Carry Out**	**Sum**
A	*B*	C_{out}	Σ
0	0	0	0
0	1	0	1
1	0	0	1
1	1	1	0

Full-adder				
Inputs		**Carry In**	**Carry Out**	**Sum**
A	*B*	C_{in}	C_{out}	Σ
0	0	0	0	0
0	0	1	0	1
0	1	0	0	1
0	1	1	1	0
1	0	0	0	1
1	0	1	1	0
1	1	0	1	0
1	1	1	1	1

■ Logic symbols with pin numbers for the ICs used in this chapter are shown in Figure 6–87, which continues on the next page. Pin labeling may differ from some manufacturers' data sheets.

FIGURE 6–87

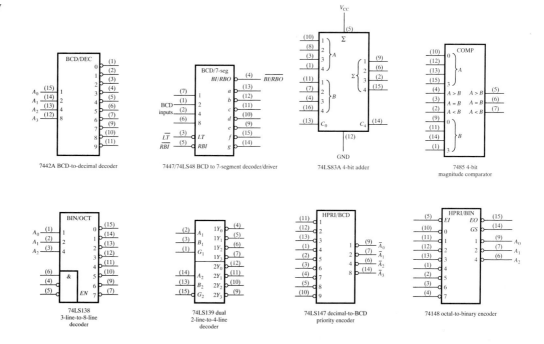

FIGURE 6–87
(Continued)

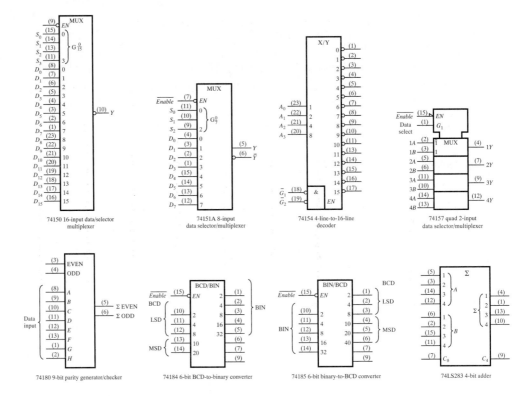

■ **SELF-TEST**

1. A half-adder is characterized by
   (a) two inputs and two outputs   (b) three inputs and two outputs
   (c) two inputs and three outputs   (d) two inputs and one output

2. A full-adder is characterized by
   (a) two inputs and two outputs   (b) three inputs and two outputs
   (c) two inputs and three outputs   (d) two inputs and one output

3. The inputs to a full-adder are $A = 1$, $B = 1$ and $C_{in} = 0$. The outputs are
   (a) $\Sigma = 1$, $C_{out} = 1$ (b) $\Sigma = 1$, $C_{out} = 0$
   (c) $\Sigma = 0$, $C_{out} = 1$ (d) $\Sigma = 0$, $C_{out} = 0$

4. A 4-bit parallel adder can add
   (a) two 4-bit binary numbers   (b) two 2-bit binary numbers
   (c) four bits at a time   (d) four bits in sequence

5. The 74LS83A is an example of a 4-bit parallel adder. To expand this device to an 8-bit adder, you must
   (a) use four adders with no interconnections
   (b) use two adders and connect the sum outputs of one to the bit inputs of the other
   (c) use eight adders with no interconnections
   (d) use two adders with the carry output of one connected to the carry input of the other

6. If a 7485 magnitude comparator has $A = 1011$ and $B = 1001$ on its inputs, the outputs are
   (a) $A > B = 0$, $A < B = 1$, $A = B = 0$ (b) $A > B = 1$, $A < B = 0$, $A = B = 0$
   (c) $A > B = 1$, $A < B = 1$, $A = B = 0$ (d) $A > B = 0$, $A < B = 0$, $A = B = 1$

7. If a 4-line-to-16-line decoder with active-LOW outputs exhibits a LOW on the decimal 12 output, what are the inputs?
   (a) $A_3A_2A_1A_0 = 1010$ (b) $A_3A_2A_1A_0 = 1110$
   (c) $A_3A_2A_1A_0 = 1100$ (d) $A_3A_2A_1A_0 = 0100$

8. A BCD-to-7-segment decoder has 0100 on its inputs. The active outputs are
   (a) a, c, f, g (b) b, c, f, g (c) b, c, e, f (d) b, d, e, g

9. If an octal-to-binary priority encoder has its 0, 2, 5, and 6 inputs at the active level, the active-HIGH binary output is
   (a) 110   (b) 010   (c) 101   (d) 000

10. In general, a multiplexer has
    (a) one data input, several data outputs, and selection inputs
    (b) one data input, one data output, and one selection input
    (c) several data inputs, several data outputs, and selection inputs
    (d) several data inputs, one data output, and selection inputs

11. Data selectors are basically the same as
    (a) decoders   (b) demultiplexers
    (c) multiplexers   (d) encoders

12. Which of the following codes exhibit even parity?
    (a) 10011000   (b) 01111000   (c) 11111111
    (d) 11010101   (e) all   (f) both answers (b) and (c)

■ PROBLEMS

SECTION 6–1   Basic Adders

1. For the full-adder of Figure 6–4 (p. 260), determine the logic state (1 or 0) at each gate output for the following inputs:
   (a) $A = 1, B = 1, C_{in} = 1$ (b) $A = 0, B = 1, C_{in} = 1$ (c) $A = 0, B = 1, C_{in} = 0$

2. What are the full-adder inputs that will produce each of the following outputs?
   (a) $\Sigma = 0, C_{out} = 0$ (b) $\Sigma = 1, C_{out} = 0$
   (c) $\Sigma = 1, C_{out} = 1$ (d) $\Sigma = 0, C_{out} = 1$

3. Simplify, if possible, the full-adder circuit of Example 6–1 (Figure 6–7, p. 261) by using the Karnaugh map method.

SECTION 6–2   Parallel Binary Adders

4. For the parallel adder in Figure 6–88, determine the complete sum by analysis of the logical operation of the circuit. Verify your result by longhand addition of the two input numbers.

5. Repeat Problem 4 for the circuit and input conditions in Figure 6–89.

FIGURE 6–88
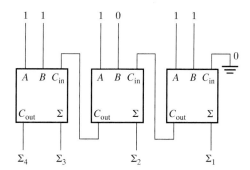

FIGURE 6–89
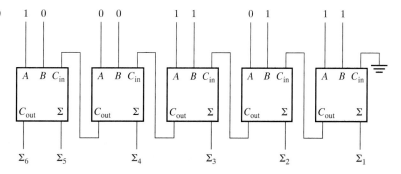

FIGURE 6–90

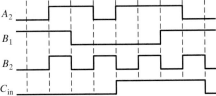

6. The input waveforms in Figure 6–90 are applied to a 2-bit adder. Determine the waveforms for the sum and the output carry in relation to the inputs by constructing a timing diagram.

7. The following sequences of bits (right-most bit first) appear on the inputs to a 74LS83A adder. Determine the resulting sequence of bits on each output.

$$
\begin{array}{ll}
A_1 & 10010110 \\
A_2 & 11101000 \\
A_3 & 00001010 \\
A_4 & 10111010 \\
B_1 & 11111000 \\
B_2 & 11001100 \\
B_3 & 10101010 \\
B_4 & 00100100 \\
\end{array}
$$

8. In the process of checking a 74LS83A 4-bit full-adder, the following voltage levels are observed on its pins: 1-LOW, 2-HIGH, 3-HIGH, 4-HIGH, 6-HIGH, 7-HIGH, 8-LOW, 9-LOW, 10-LOW, 11-LOW, 13-LOW, 14-HIGH, 15-LOW, and 16-HIGH. Determine if the IC is functioning properly.

SECTION 6–3 Ripple Carry Versus Look-Ahead Carry Adders

9. Each of the eight full-adders in an 8-bit parallel ripple carry adder exhibits the following propagation delays:

$$
\begin{array}{ll}
A \text{ to } \Sigma \text{ and } C_{out}: & 40 \text{ ns} \\
B \text{ to } \Sigma \text{ and } C_{out}: & 40 \text{ ns} \\
C_{in} \text{ to } \Sigma: & 35 \text{ ns} \\
C_{in} \text{ to } C_{out}: & 25 \text{ ns} \\
\end{array}
$$

Determine the maximum total time for the addition of two 8-bit numbers.

10. Show the additional logic circuitry necessary to make the 4-bit look-ahead carry adder in Figure 6–18 (p. 272) into a 5-bit adder.

SECTION 6–4 Comparators

11. The waveforms in Figure 6–91 are applied to the comparator as shown. Determine the output $(A = B)$ waveform.

FIGURE 6–91

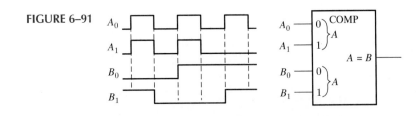

12. For the 4-bit comparator in Figure 6–92, plot each output waveform for the inputs shown. The outputs are active-HIGH.

13. For each set of binary numbers, determine the logic states at each gate output in the comparator circuit of Figure 6–24 (p. 275), and verify that the output indications are correct.

(a) $A_3A_2A_1A_0 = 1100$ (b) $A_3A_2A_1A_0 = 1000$ (c) $A_3A_2A_1A_0 = 0100$
    $B_3B_2B_1B_0 = 1001$ $B_3B_2B_1B_0 = 1011$ $B_3B_2B_1B_0 = 0100$

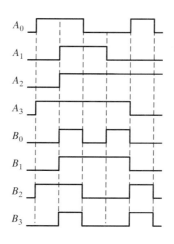
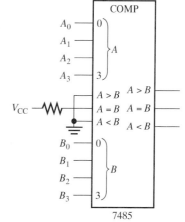

FIGURE 6–92

SECTION 6–5   Decoders

14. When a HIGH is on the output of each of the decoding gates in Figure 6–93, what is the binary code appearing on the inputs? The MSB is A_3.

15. Show the decoding logic for each of the following codes if an active-HIGH (1) output is required:

(a) 1101   (b) 1000   (c) 11011   (d) 11100
(e) 101010   (f) 111110   (g) 000101   (h) 1110110

16. Solve Problem 15, given that an active-LOW (0) output is required.

17. You wish to detect only the presence of the codes 1010, 1100, 0001, and 1011. An active-HIGH output is required to indicate their presence. Develop the minimum decoding logic with a single output that will indicate when any one of these codes is on the inputs. For any other code, the output must be LOW.

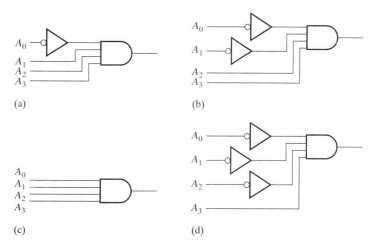

FIGURE 6–93

18. If the input waveforms are applied to the decoding logic as indicated in Figure 6–94, sketch the output waveform in proper relation to the inputs.

19. BCD numbers are applied sequentially to the BCD-to-decimal decoder in Figure 6–95. Draw a timing diagram, showing each output in the proper relationship with the others and with the inputs.

20. A 7-segment decoder/driver drives the display in Figure 6–96. If the waveforms are applied as indicated, determine the sequence of digits that appears on the display.

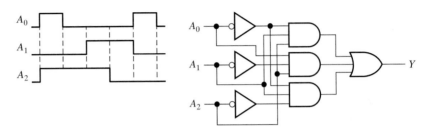

FIGURE 6–94

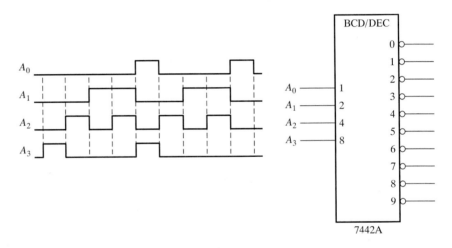

FIGURE 6–95

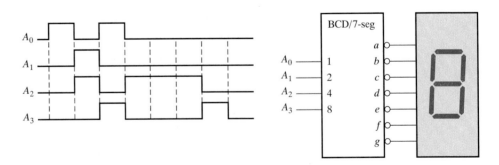

FIGURE 6–96

SECTION 6–6   Encoders

21. For the decimal-to-BCD encoder logic of Figure 6–41 on p. 288, assume that the 9 input and the 3 input are both HIGH. What is the output code? Is it a valid BCD (8421) code?

22. A 74147 encoder has LOW levels on pins 2, 5, and 12. What BCD code appears on the outputs if all the other inputs are HIGH?

SECTION 6–7 Code Converters

23. Convert the following decimal numbers first to BCD. Then, using the logical operation of the BCD-to-binary converter of Figure 6–50 (p. 295), convert the BCD to binary. Verify the result in each case.

    (a) 2   **(b)** 8   **(c)** 13   **(d)** 26   **(e)** 33

24. Show the logic required to convert a 10-bit binary number to Gray code, and use that logic to convert the following binary numbers to Gray code:

    (a) 1010101010   **(b)** 1111100000   **(c)** 0000001110   **(d)** 1111111111

25. Show the logic required to convert a 10-bit Gray code to binary, and use that logic to convert the following Gray code words to binary:

    (a) 1010000000   **(b)** 0011001100   **(c)** 1111000111   **(d)** 0000000001

SECTION 6–8 Multiplexers (Data Selectors)

26. For the multiplexer in Figure 6–97, determine the output for the following input states: $D_0 = 0, D_1 = 1, D_2 = 1, D_3 = 0, S_0 = 1, S_1 = 0$.

27. If the data-select inputs to the multiplexer in Figure 6–97 are sequenced as shown by the waveforms in Figure 6–98, determine the output waveform with the data inputs specified in Problem 26.

28. The waveforms in Figure 6–99 are observed on the inputs of a 74151A 8-input multiplexer. Sketch the Y output waveform.

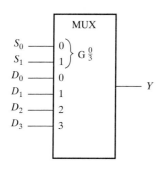

FIGURE 6–97

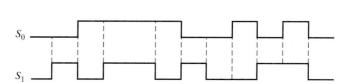

FIGURE 6–98

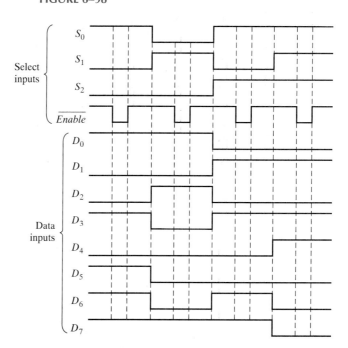

FIGURE 6–99

SECTION 6–9 Demultiplexers

29. Develop the total timing diagram (inputs and outputs) for a 74154 used in a demultiplexing application in which the inputs are as follows: The data-select inputs are repetitively sequenced through a straight binary count beginning with 0000, and the data input is a serial data stream carrying BCD data representing the decimal number 2468. The least significant digit (8) is first in the sequence, with its LSB first, and it should appear in the first 4-bit positions of the output.

SECTION 6–10 Parity Generators/Checkers

30. The waveforms in Figure 6–100 are applied to the 4-bit parity logic. Determine the output waveform in proper relation to the inputs. For how many bit times does even parity occur, and how is it indicated? The timing diagram includes eight bit times.

31. Determine the Σ EVEN and the Σ ODD outputs of a 74180 9-bit parity generator/checker for the inputs in Figure 6–101. Refer to the truth table in Figure 6–75 (p. 312).

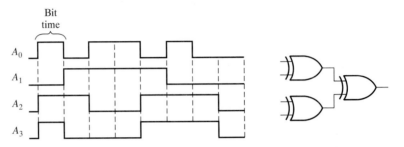

FIGURE 6–100

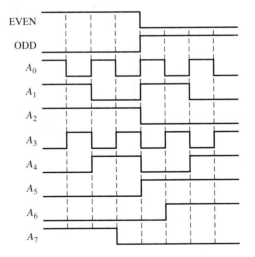

FIGURE 6–101

SECTION 6–11 Troubleshooting

32. The full-adder in Figure 6–102 is tested under all input conditions with the input waveforms shown. From your observation of the Σ and C_{out} waveforms, is it operating properly, and if not, what is the most likely fault?

33. List the possible faults for each decoder/display in Figure 6–103.

34. Develop a systematic test procedure to check out the complete operation of the keyboard encoder in Figure 6–49 (p. 292).

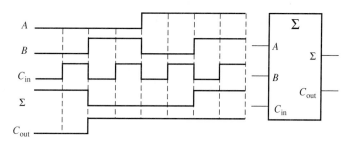

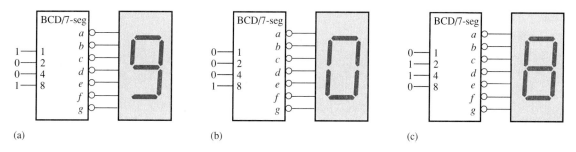

FIGURE 6–102

(a)   (b)   (c)

FIGURE 6–103

35. You are testing the BCD-to-binary converter in Figure 6–50 (p. 295). The procedure calls for applying BCD numbers in sequential order beginning with 0_{10} and checking for the correct binary output. What symptom or symptoms will appear on the binary outputs in the event of each of the following faults? For what BCD number is each fault *first* detected?

(a) The A_1 input is open (top adder).

(b) The C_{out} is open (top adder).

(c) The Σ_4 output is shorted to ground (top adder).

(d) The 32 output is shorted to ground (bottom adder).

36. For the display multiplexing system in Figure 6–65 (p. 304), determine the most likely cause or causes for each of the following symptoms:

(a) The *B*-digit (MSD) display does not turn on at all.

(b) Neither 7-segment display turns on.

(c) The *f*-segment of both displays appears to be on all the time.

(d) There is a visible flicker on the displays.

37. Develop a systematic procedure to fully test the 74151A data selector IC.

38. During the testing of the data transmission system in Figure 6–76 (p. 313), a code is applied to the D_0 through D_6 inputs that contains an odd number of 1s. A single bit error is deliberately introduced on the serial data transmission line between the MUX and the DEMUX, but the system does not indicate an error (error output = 0). After some investigation, you check the inputs to the even parity checker and find that D_0 through D_6 contain an even number of 1s, as you would expect. Also, you find that the D_7 parity bit is a 1. What are the possible reasons for the system not indicating the error?

39. In general, describe how you would fully test the data transmission system in Figure 6–76, and specify a method for the introduction of parity errors.

SECTION 6–12   Digital System Application

■ System Applications

40. The output logic block is implemented on the workbench using a 74LS08 with the AND gates operating as negative-NOR gates. Use a 74LS00 (quad NAND gates) and any other device that may be required to produce the active-HIGH outputs.

41. Implement the output logic if active-LOW outputs are required.

Special Design Problems

42. Modify the design of the 7-segment display multiplexing system in Figure 6–65 (p. 304) to accommodate two additional digits.

43. Implement the sum logic in Figure 6–7 (p. 261), by using a 74151A data selector. The three input variables are A, B, and C_{in}.

44. Implement the logic function specified in Table 6–13 by using a 74151A data selector.

TABLE 6–13

	Inputs			Output
A_3	A_2	A_1	A_0	Y
0	0	0	0	0
0	0	0	1	0
0	0	1	0	1
0	0	1	1	1
0	1	0	0	0
0	1	0	1	0
0	1	1	0	1
0	1	1	1	1
1	0	0	0	1
1	0	0	1	0
1	0	1	0	1
1	0	1	1	1
1	1	0	0	0
1	1	0	1	1
1	1	1	0	0
1	1	1	1	1

45. Using two of the 6-position adder modules from Figure 6–14 (p. 268), design a 12-position voting system.

46. The adder block in the tablet counting and control system in Figure 6–104 performs the addition of the 8-bit binary number from the counter and the 16-bit binary number from Register B. The result from the adder goes back into Register B. Use 74LS283s to implement this function and draw a complete logic diagram including pin numbers. Refer to Chapters 1 and 2 to review the system operation.

47. The comparator block in the tablet counting and control system in Figure 6–104 was implemented in Chapter 3 using discrete gates. Now you are prepared to use MSI devices to simplify the design. The comparator compares the 8-bit binary number (actually only seven bits are required) from the BCD-to-binary converter with the 8-bit binary number from the counter. Use 7485s to implement this function and draw a complete logic diagram including pin numbers.

48. Two BCD-to-7-segment decoders are used in the tablet counting and control system in Figure 6–104. One is required to drive the 2-digit *tablets/bottle* display and the other to drive the 5-digit *total tablets bottled* display. Use 7447s to implement each decoder and draw a complete logic diagram including pin numbers.

49. The encoder shown in the system block diagram of Figure 6–104 encodes each decimal key closure and converts it to BCD. Use a 74LS147 to implement this function and draw a complete logic diagram including pin numbers.

50. The system in Figure 6–104 requires two code converters. The BCD-to-binary converter changes the 2-digit BCD number in Register A to an 8-bit binary code (actually only 7 bits are required because the MSB is always 0). Use appropriate MSI code converters to implement the BCD-to-binary converter function and draw a complete logic diagram including pin numbers.

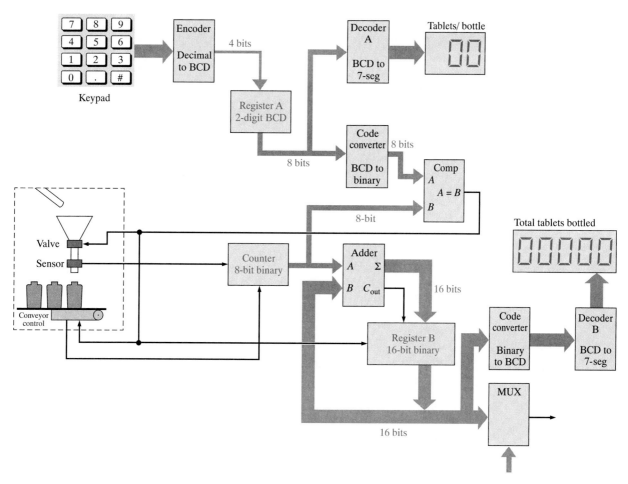

FIGURE 6–104

■ **ANSWERS TO SECTION REVIEWS**

SECTION 6–1

1. (a) $\Sigma = 1, C_{out} = 0$ (b) $\Sigma = 0, C_{out} = 0$ (c) $\Sigma = 1, C_{out} = 0$ (d) $\Sigma = 0, C_{out} = 1$
2. $\Sigma = 1, C_{out} = 1$

SECTION 6–2

1. $C_{out}\Sigma_4\Sigma_3\Sigma_2\Sigma_1 = 11001$ 2. Three 74LS283s are required to add two 10-bit numbers.

SECTION 6–3

1. $C_g = 0, C_p = 1$ 2. $C_{out} = 1$

SECTION 6–4

1. $A > B = 1, A < B = 0, A = B = 0$ when $A = 1011$ and $B = 1010$
2. Right comparator: pin 7: $A < B = 1$; pin 6: $A = B = 0$; pin 5: $A > B = 0$
   Left comparator: pin 7: $A < B = 0$; pin 6: $A = B = 0$; pin 5: $A > B = 1$

SECTION 6–5

1. Output 5 is active when 101 is on the inputs.
2. Four 74154s are used to decode a 6-bit binary number.
3. Active-LOW output drives a common-cathode LED display.

SECTION 6–6

1. (a) $A_0 = 1, A_1 = 1, A_2 = 0, A_3 = 1$

   (b) No, this is not a valid BCD code.

   (c) Only one input can be active for a valid output.

2. (a) $\overline{A}_3 = 0, \overline{A}_2 = 1, \overline{A}_1 = 1, \overline{A}_0 = 1$

   (b) The output is 0111 which is the complement of 1000 (8).

SECTION 6–7

1. 10000101 (BCD) = 1010101_2

2. An 8-bit binary to Gray converter consists of seven exclusive-OR gates in an arrangement like that in Figure 6–54.

SECTION 6–8

1. The output is 0.

2. (a) 74157: Quad 2-input data selector

   (b) 74151A: 8-input data selector

   (c) 74150: 16-input data selector

3. The data output alternates between LOW and HIGH as the data-select inputs sequence through the binary states.

4. (a) The 74157 multiplexes the two BCD codes to the 7-segment decoder.

   (b) The 74LS48 decodes the BCD to energize the display.

   (c) The 74LS139 enables the 7-segment displays alternately.

SECTION 6–9

1. A decoder can be used as a multiplexer by using the input lines for data selection and an Enable line for data input.

2. The outputs are all HIGH except D_{10}, which is LOW.

SECTION 6–10

1. (a) Even parity: 1110100   **(b)** Even parity: 001100011

2. (a) Odd parity: 11010101   **(b)** Odd parity: 11000001

3. (a) Code is correct, four 1s.

   (b) Code is in error, seven 1s

SECTION 6–11

1. A glitch is a very short-duration voltage spike (usually unwanted).

2. Glitches are caused by transition states.

3. Strobe is the enabling of a device for a specified period of time when the device is not in transition.

SECTION 6–12

1. Decoder input state sequence is 00, 01, 11, 10.

2. Negative-OR: $\overline{A} + \overline{B} = \overline{AB}$

3. Long timer triggers when going into the first or third states.
Short timer triggers when going into the second or fourth states.

7

INTRODUCTION TO PROGRAMMABLE LOGIC DEVICES

■ **CHAPTER OBJECTIVES**

☐ Discuss programmable arrays and describe the major types of PLDs
☐ Describe the basic PAL structure and explain how programming the AND array produces standard logic functions
☐ Describe the basic GAL structure and explain how programming the AND array produces standard logic functions
☐ Describe the basic GAL22V10 architecture
☐ Describe the basic GAL16V8 architecture
☐ Discuss the basic procedure of PLD programming
☐ Write simple ABEL programs to implement combinational logic in a PLD
☐ Implement simple combinational logic using a GAL

■ CHAPTER OVERVIEW

Programmable logic devices (PLDs) are introduced in this chapter. The emphasis is on using PLDs to implement combinational logic. Sequential logic applications of PLDs are discussed in Chapter 11. PLDs, particularly programmable array logic (PAL) and generic array logic (GAL) devices, can be used to replace SSI and MSI logic devices in many types of applications, resulting in fewer parts and lower cost.

Programming software, such as ABEL and CUPL, are used to implement logic designs in PLDs. In this book, ABEL is used to illustrate basic programming principles. Once you are familiar with one PLD programming language, the others are easily learned because of the similarities. The coverage of ABEL is limited and is not intended as a comprehensive treatment. Materials from the manufacturer and from other sources are recommended to supplement this coverage.

■ SPECIFIC DEVICES

PAL16L8   GAL22V10   GAL16V8
GAL20V8

■ DIGITAL SYSTEM APPLICATION

This Digital System Application illustrates concepts taught in this chapter. The combinational logic portion of the traffic light control system from Chapter 6 is redesigned using a programmable logic device. A program is developed using ABEL to implement the state decoding and output logic.

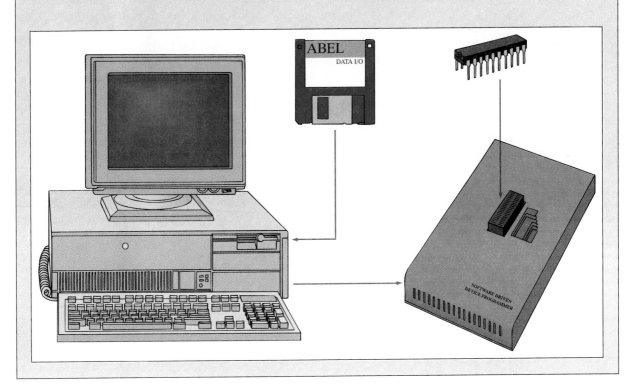

7–1 ■ PLD ARRAYS AND CLASSIFICATIONS

Programmable logic devices (PLDs) are used in many applications to replace SSI and MSI circuits; they save space and reduce the actual number and cost of devices in a given design. A PLD consists of a large array of AND gates and OR gates that can be programmed to achieve specified logic functions. Four types of devices that are classified as PLDs are the programmable read-only memory (PROM), the programmable logic array (PLA), the programmable array logic (PAL), and the generic array logic (GAL). After completing this section you should be able to

□ Explain a basic OR array   □ Explain a basic AND array   □ Describe the structure of a PROM   □ Describe the structure of a PLA   □ Describe the basic PAL   □ Describe the basic GAL   □ Discuss the differences between a PAL and a GAL

Programmable Arrays

All **PLDs** consist of programmable arrays. A programmable **array** is essentially a grid of conductors that form rows and columns with a fusible link at each cross point. Arrays can be either fixed or programmable. The earliest type of programmable array, dating back to the 1960s, was a diode matrix with a fuse at each cross point of the matrix.

The OR Array   The original diode array evolved into the integrated OR array, which consists of an array of OR gates connected to a programmable matrix with fusible links at each cross point of a row and column, as shown in Figure 7–1(a). The array is programmed by blowing **fuses** to eliminate selected variables from the output functions, as illustrated in part (b) for a specific case. For each input to an OR gate, only one fuse is left intact in order to connect the desired variable to the gate input. Once a fuse is blown, it cannot be reconnected.

The AND Array   This type of array consists of AND gates connected to a programmable matrix with fusible links at each cross point, as shown in Figure 7–2(a). Like the OR array, the AND array is programmed by blowing fuses to eliminate variables from the output function, as illustrated in part (b). For each input to an AND gate, only one fuse is left intact in order to connect the desired variable to the gate input. Also like the OR array, the AND array with fusible links is one-time programmable.

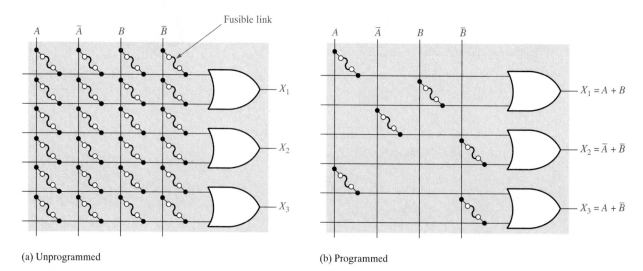

(a) Unprogrammed   (b) Programmed

FIGURE 7–1
An example of a basic OR array.

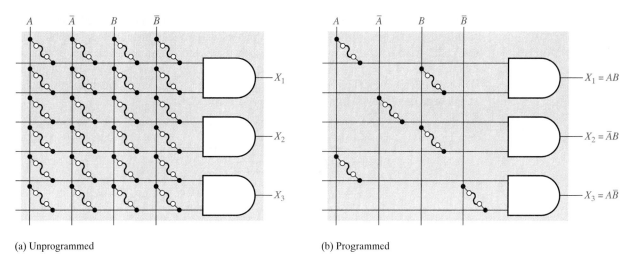

(a) Unprogrammed   (b) Programmed

FIGURE 7–2
An example of a basic AND array.

Classifications of PLDs

PLDs are classified according to their **architecture,** which is basically the functional arrangement of internal elements that give a device its unique characteristic.

Programmable Read-Only Memory   The **PROM** consists of a set of fixed (nonprogrammable) AND gates connected as a decoder and a programmable OR array, as shown in the generalized block diagram of Figure 7–3. The PROM is used primarily as an addressable memory and not as a logic device because of limitations imposed by the fixed AND gates. PROMs are covered in detail in Chapter 12.

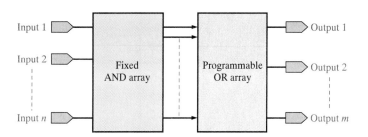

FIGURE 7–3
Block diagram of a PROM (programmable read-only memory).

Programmable Logic Array (PLA)   The **PLA** is a PLD that consists of a programmable AND array and a programmable OR array, as shown in Figure 7–4. The PLA was developed to overcome some of the limitations of the PROM. The PLA is also called an FPLA (field-programmable logic array) because the user in the field, not the manufacturer, programs it.

FIGURE 7–4
Block diagram of a PLA (programmable logic array).

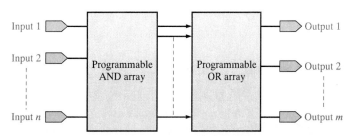

Programmable Array Logic (PAL) The **PAL** is a PLD that was developed to overcome certain disadvantages of the PLA, such as longer delays due to the additional fusible links that result from using two programmable arrays and more circuit complexity. The basic PAL consists of a programmable AND array and a fixed OR array with output logic, as shown in Figure 7–5. The PAL is the most common one-time programmable logic device and is implemented with bipolar technology (TTL or ECL).

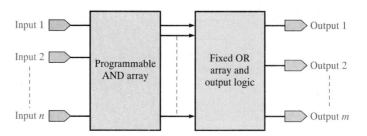

FIGURE 7–5
Block diagram of a PAL (programmable array logic).

Generic Array Logic (GAL) The most recent development in PLDs is the GAL. The **GAL,** like the PAL, has a programmable AND array and a fixed OR array with programmable output logic. The two main differences between GAL and PAL devices are (a) the GAL is reprogrammable and (b) the GAL has programmable output configurations.

The GAL can be reprogrammed again and again because it uses E^2CMOS (Electrically Erasable CMOS) technology instead of bipolar technology and fusible links. The block diagram of a GAL is shown in Figure 7–6.

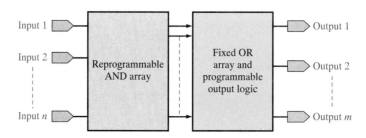

FIGURE 7–6
Block diagram of a GAL (generic array logic).

SECTION 7–1 REVIEW	1. List four PLDs. 2. What is the difference between a PLA and a PAL? 3. Describe the differences between a PAL and a GAL?

7–2 ■ PROGRAMMABLE ARRAY LOGIC (PAL)

The PAL and the GAL are the most common PLDs used for logic implementation. PAL is a designation originally used by Monolithic Memories, Inc. (now a part of Advanced Micro Devices) and later licensed to other manufacturers. As you learned in the last section, the PAL in its basic form is a PLD with a one-time programmable AND array and a fixed OR array. In this section, you will learn how PALs are used to produce specified combinational logic functions and examine a specific PAL. After completing this section, you should be able to

☐ Describe basic PAL operation ☐ Show how a sum-of-products expression is implemented in a PAL ☐ Discuss simplified PAL logic diagrams ☐ Explain the three basic types of PAL output combinational logic ☐ Interpret standard PAL numbers ☐ Discuss the PAL16L8

PAL Operation

The PAL consists of a programmable array of AND gates that connects to a fixed array of OR gates. This structure allows any sum-of-products (SOP) logic expression with a defined number of variables to be implemented. Recall from Chapter 3 that any logic function can be expressed in SOP form.

The basic structure of a PAL is illustrated in Figure 7–7 for two input variables and one output although most PALs have many inputs and many outputs. As you know, a programmable array is essentially a grid of conductors forming rows and columns with a fusible link at each cross point. Each fused cross point of a row and column is called a **cell** and is the programmable element of a PAL. Each row is connected to the input of an AND gate and each column is connected to an input variable or its complement. By using the presence or absence of fused connections created by programming, any combination of input variables or complements can be applied to an AND gate to form any desired product term.

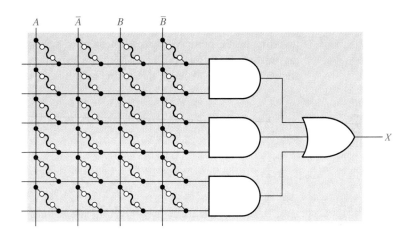

FIGURE 7–7
Basic structure of a PAL.

Implementing a Sum-of-Products Expression In its simplest form, each cell in a basic AND array consists of a fusible link connecting a row and a column as represented in Figure 7–7. When the connection between a row and column is required, the fuse is left intact. When no connection between a row and column is required, the fuse is blown open during the programming process.

As an example, a simple array is programmed as shown in Figure 7–8 so that the product term AB is produced by the top AND gate, $A\overline{B}$ by the middle AND gate, and $\overline{A}\,\overline{B}$ by the bottom AND gate. As you can see, the fusible links are left intact to connect the desired variables or their complements to the appropriate AND gate inputs. The fusible links are opened where a variable or its complement is not used in a given product term. The final output from the OR gate is the SOP expression,

$$X = AB + A\overline{B} + \overline{A}\,\overline{B}$$

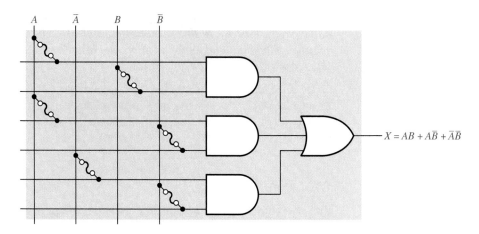

FIGURE 7–8
PAL implementation of a sum-of-products expression.

Simplified Symbols

What you have seen so far represents a small segment of a typical PAL. Actual PALs have many AND gates and many OR gates in addition to other circuitry and are capable of handling many input variables and their complements. Since PALs are very complex integrated circuit devices, manufacturers have adopted a simplified notation for the logic diagrams to keep them from being overwhelmingly complicated.

Input Buffers   The input variables to a PAL are buffered to prevent loading by the large number of AND gate inputs to which a variable or its complement may be connected. An inverting buffer produces the complement of an input variable. The symbol representing the **buffer** circuit that produces both the variable and its complement on its outputs is shown in Figure 7–9 where the bubble output is the complement.

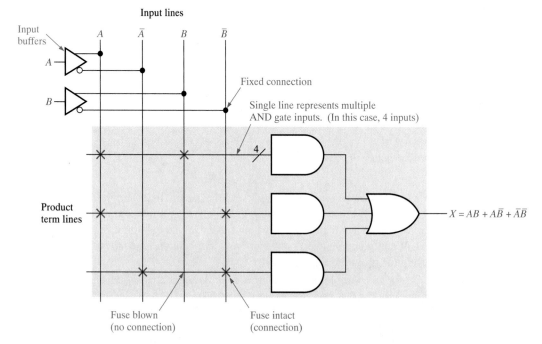

FIGURE 7–9
Simplified diagram of a programmed PAL.

AND Gates   A typical PAL AND array has an extremely large number of interconnecting lines, and each AND gate has multiple inputs. PAL logic diagrams show an AND gate that actually has several inputs by using an AND gate symbol with a single input line representing all of its input lines, as indicated Figure 7–9. Also, multiple input lines are sometimes indicated by a slash and the number of lines as shown in the top AND gate for the case of four lines.

PAL Connections   To keep a logic diagram as simple as possible, the fusible links in a programmed AND array are indicated by an X at the cross point if the fuse is left intact and by the absence of an X if the fuse is blown, as indicated in Figure 7–9. Fixed connections use the standard dot notation, as also indicated.

EXAMPLE 7–1   Indicate how a PAL is programmed for the following 3-variable logic function:

$$X = A\bar{B}C + \bar{A}B\bar{C} + \bar{A}\,\bar{B} + AC$$

Solution   The programmed array is shown in Figure 7–10. The intact fusible links are indicated by Xs. The absence of an X means that the fuse has been blown.

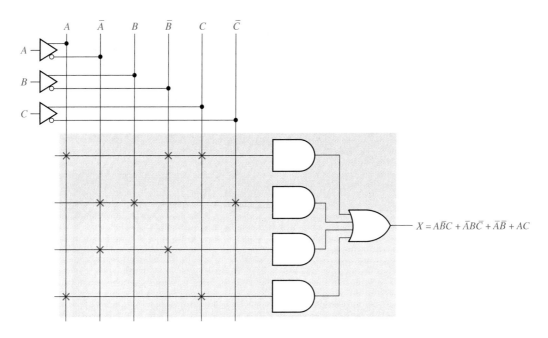

FIGURE 7–10

Related Exercise   Write the expression for the output if the fusible links connecting input A to the top row and to the bottom row in Figure 7–10 are also blown.

The PAL Block Diagram

A block diagram of a PAL is shown in Figure 7–11. The AND array outputs go to the OR array, and the output of each OR gate goes to its associated output logic. A typical PAL has eight or more inputs to its AND array and up to eight outputs from its output logic as indicated, where $n \geq 8$ and $m \leq 8$. Some PALS provide a combined input and output **(I/O)** pin that can be programmed as either an output or an input. The symbol ◄▷ means that a pin can be either an input or an output.

FIGURE 7–11
Block diagram of a PAL.

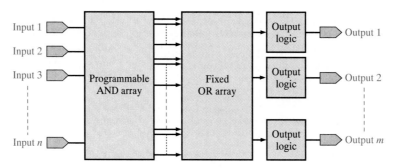

PAL Output Combinational Logic

There are several basic types of PAL output logic which allows you to configure the device for a specific application. In this chapter, only the aspects of the output logic related to combinational logic functions are discussed. Additional features are covered in Chapter 11 after you have studied flip-flops, counters, and registers. Figure 7–12 shows three basic types of combinational output logic with tristate outputs and the associated OR gate. The following are types of PAL output logic:

- *Combinational output*   This output is used for an SOP function and is usually available as either an active-LOW or an active-HIGH output.
- *Combinational input/output (I/O)*   This output is used when the output function must feed back to be an input to the array or be used to make the I/O pin an input only.
- *Programmable polarity output*   This output is used for selecting either the output function or its complement by programming the exclusive-OR gate for inversion or noninversion. The fusible link on the exclusive-OR input is blown open for inversion and left intact for noninversion.

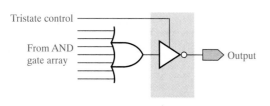

(a) Combinational output (active-LOW). An active-HIGH output would be shown without the bubble on the tristate gate symbol.

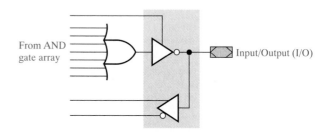

(b) Combinational input/output (active-LOW).

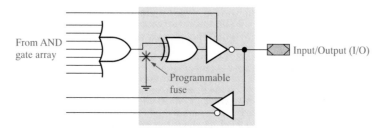

(c) Programmable polarity output.

FIGURE 7–12
Basic types of PAL combinational output logic.

Standard PAL Numbering

Standard PALs come in a variety of configurations, each of which is identified by a unique part number. This part number always begins with the prefix PAL. The first two digits following PAL indicate the number of inputs, which includes outputs that can be configured as inputs. The letter following the number of inputs designates the type of output: L—active-LOW, H—active-HIGH, or P—programmable polarity. The one or two digits that follow the output type is the number of outputs. The following number is an example.

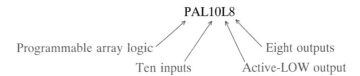

PAL10L8

Programmable array logic

Ten inputs

Active-LOW output

Eight outputs

In addition, a PAL part number may carry suffixes that specify speed, package type, and temperature range.

As an example of a PAL configuration, a block diagram of the PAL16L8 is shown in Figure 7–13. This particular device has 10 dedicated inputs (I) plus 6 input/outputs (I/O) that can be used as inputs. There are 8 outputs including 2 dedicated outputs (O) plus 6 input/outputs that can be used as outputs. Each of the outputs is active-LOW.

FIGURE 7–13
Block diagram of the PAL16L8.

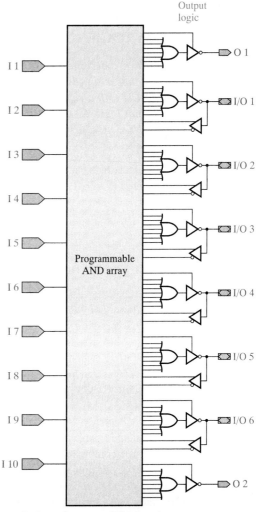

I = input, O = output, I/O = input/output

EXAMPLE 7–2

Determine the number of inputs, the number of outputs, and the type of output for each of the following part numbers:

(a) PAL12H6   (b) PAL16L2   (c) PAL20P8

Solution

(a) 12 inputs, 6 outputs, active-HIGH outputs
(b) 16 inputs, 2 outputs, active-LOW outputs
(c) 20 inputs, 8 outputs, programmable polarity outputs

Related Exercise   Describe a PAL with the part number PAL14H4.

SECTION 7–2 REVIEW

1. What is a PAL?
2. Explain the purpose of a fusible link in a PAL.
3. Can any combinational logic function be implemented with a PAL?
4. List three types of PAL output logic.
5. What does the designation PAL12H6 mean?

7–3 ■ GENERIC ARRAY LOGIC (GAL)

GAL is a designation originally used by Lattice Semiconductor and later licensed to other manufacturers. As you learned in Section 7–1, the GAL in its basic form is a PLD with a reprogrammable AND array, a fixed OR array, and programmable output logic. In this section, basic concepts are introduced and in Sections 7–4 and 7–5 specific GALs are examined. After completing this section, you should be able to

□ Describe basic GAL operation   □ Show how a sum-of-products expression is implemented in a GAL   □ Compare the E^2CMOS cell in a GAL with the fusible link cell in a PAL   □ Define OLMC and explain its purpose   □ Interpret standard GAL numbers

GAL Operation

The GAL basically consists of a reprogrammable array of AND gates that connects to a fixed array of OR gates. Just as in a PAL, this structure allows any sum-of-products (SOP) logic expression with a defined number of variables to be implemented.

The basic structure of a GAL is illustrated in Figure 7–14 for two input variables and one output although most GALs have many inputs and many outputs. The reprogrammable array is essentially a grid of conductors forming rows and columns with an electrically-erasable CMOS (E^2CMOS) cell at each cross point, rather than a fuse as in a PAL. These cells are shown as blocks in the figure.

Each row is connected to the input of an AND gate, and each column is connected to an input variable or its complement. By programming each E^2CMOS cell to be either *on* or *off,* any combination of input variables or complements can be applied to an AND gate to form any desired product term. A cell that is *on* effectively connects its corresponding row and column, and a cell that is *off* disconnects the row and column. The cells can be electrically erased and reprogrammed. A typical E^2CMOS cell can retain its programmed state for 20 years or more. See Chapter 15 for a discussion of how an E^2CMOS cell works.

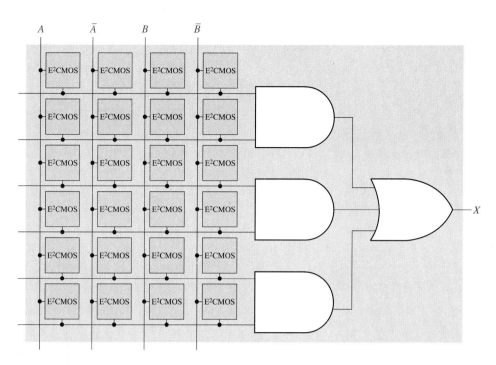

FIGURE 7–14
Basic E²CMOS array structure of a GAL.

Implementing a Sum-of-Products Expression As an example, a simple GAL array is programmed as shown in Figure 7–15 so that the product term $\overline{A}B$ is produced by the top AND gate, AB by the middle AND gate, and $\overline{A}\,\overline{B}$ by the bottom AND gate. As shown, the E²CMOS cells are *on* to connect the desired variables or their complements to the appropriate AND gate inputs. The E²CMOS cells are *off* where a variable or its complement is not used in a given product term. The final output from the OR gate is an SOP expression.

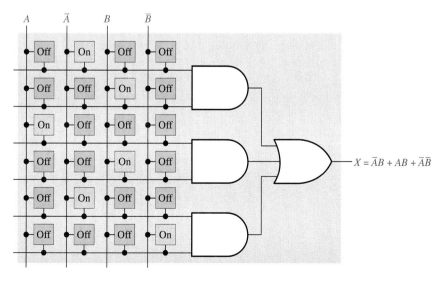

FIGURE 7–15
GAL implementation of a sum-of-products expression.

EXAMPLE 7–3   Indicate how a GAL is programmed for the following 3-variable logic function:

$$X = \overline{A}B\overline{C} + \overline{A}BC + BC + A\overline{B}$$

Solution   The programmed array using simplified notation is shown in Figure 7–16. Cells that are *on* are indicated by Xs. The absence of an X means that the cell is *off*.

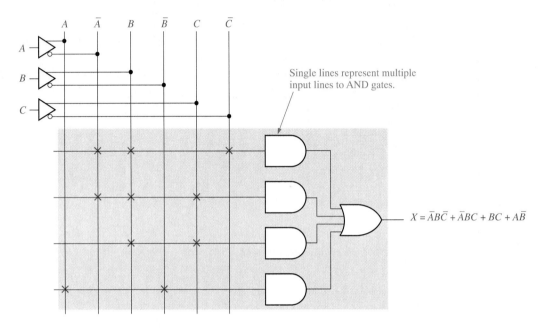

FIGURE 7–16

Related Exercise   Write the expression for the output if the cells connecting input $\overline{A}$ to the top row and to the second row in Figure 7–16 are also *off*.

The GAL Block Diagram

A block diagram of a GAL is shown in Figure 7–17. The AND array outputs go to the output logic macrocells **(OLMC),** which contain the OR gates and programmable logic. A typical GAL has eight or more inputs to its AND array and eight or more input/outputs (I/Os) from its OLMCs as indicated, where $n \geq 8$ and $m \geq 8$.

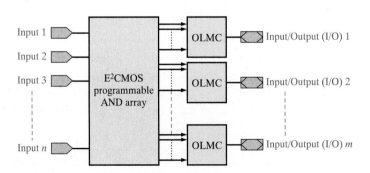

FIGURE 7–17
Block diagram of a GAL.

The OLMC is made up of logic circuits that can be programmed as either *combinational logic* or as *registered logic*. The OLMC provides much more flexibility than the fixed output logic in a PAL. In this chapter, the combinational output configurations of OLMCs are considered; the registered configurations are discussed in Chapter 11 after you have studied flip-flops, counters, and registers.

Standard GAL Numbering

GALs come in a variety of configurations, each of which is identified by a unique part number. This part number always begins with the prefix GAL. The first two digits following the prefix indicate the number of inputs, which includes outputs that can be configured as inputs. The letter V following the number of inputs designates a variable-output configuration. The one or two digits that follow the output type is the number of outputs. The following number is an example.

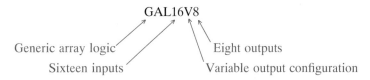

GAL16V8

Generic array logic

Sixteen inputs

Eight outputs

Variable output configuration

EXAMPLE 7–4

Determine the number of inputs and the number of outputs for each of the following part numbers:

(a) GAL20V8   (b) GAL22V10

Solution

(a) 20 inputs, 8 outputs
(b) 22 inputs, 10 outputs

Related Exercise   Describe a GAL with the part number GAL18V10.

SECTION 7–3 REVIEW

1. What is a GAL?
2. Explain the purpose of the E^2CMOS cell in a GAL.
3. Can any combinational logic function be implemented with a GAL?
4. What is the OLMC?

7–4 ■ THE GAL22V10

The various GALs all have the same type of programmable array. They differ in the size of the array, in the type of OLMCs, and in operating parameters such as speed and power dissipation. In this section, a popular generic array logic device, the GAL22V10, is discussed in order to illustrate principles that are common to all types of GALs. In the next section, we will look at the GAL16V8. After completing this section, you should be able to

□ Explain the operation of the OLMC and list the four configurations   □ Describe the combinational mode   □ Discuss the operation and purpose of the tristate buffer
□ Explain how the OLMC can operate as either an input or an output   □ Describe how the E^2CMOS cells are organized   □ Illustrate how an SOP function is implemented

Block Diagram of the GAL22V10

The GAL22V10 contains twelve dedicated inputs and ten input/outputs (I/Os) as shown in the block diagram of Figure 7–18. This device is available in either a 24-pin DIP (dual-in-line package) or a 28-pin PLCC (plastic chip carrier), as shown in Figure 7–19.

FIGURE 7–18
GAL22V10 block diagram.

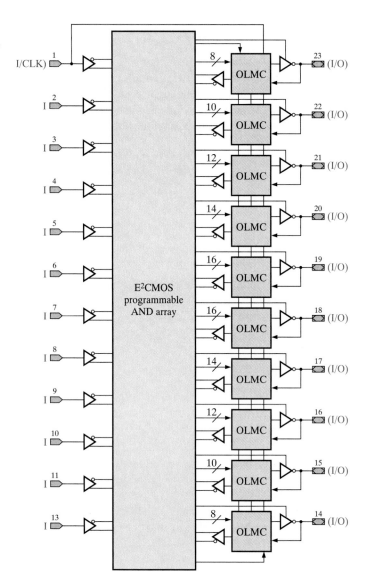

FIGURE 7–19
GAL22V10 package diagrams.

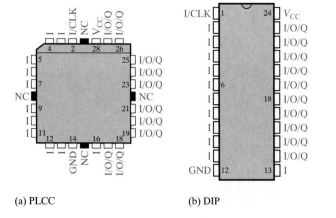

(a) PLCC   (b) DIP

The Output Logic Macrocells (OLMCs)

As stated in the discussion of GALs, an OLMC contains programmable logic circuits that can be configured either for a combinational output or input or for a registered output. In the registered mode (discussed in Chapter 11), the output comes from a flip-flop. In this chapter, the focus is on the combinational mode. OLMC combinational mode configurations are automatically set by programming.

As indicated by the notation in the block diagram of Figure 7–18, of the ten available GAL22V10 OLMCs, two have eight product terms (lines from the AND array to the OR gate), two have ten product terms, two have twelve product terms, two have fourteen product terms, and two have sixteen product terms. Each OLMC can be programmed for either an active-HIGH or an active-LOW output. Also, each OLMC can be programmed as an input.

Logic Diagram   A basic logic diagram for the GAL22V10 OLMC is shown in Figure 7–20. The inputs from the AND gates to the OR gate vary from ten to sixteen, as mentioned. The logic inside the blue box consists of a flip-flop (defined in Chapter 1 and not used in this chapter) and two multiplexers (covered in Chapter 6).

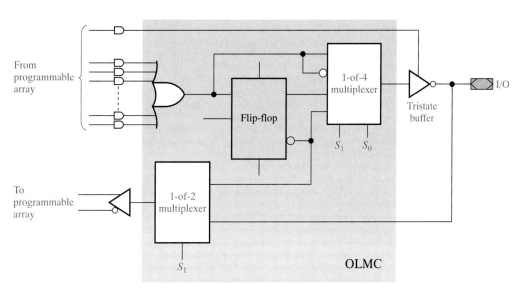

FIGURE 7–20
The GAL22V10 OLMC.

The 1-of-4 multiplexer connects one of its four input lines to the tristate output buffer based on the states of two select inputs, S_0 and S_1. The inputs to the 1-of-4 multiplexer are the OR gate output, the complement of the OR gate output, the flip-flop output, and the complement of the flip-flop output. The 1-of-2 multiplexer connects either the output of the tristate buffer or the flip-flop back through a buffer to the AND array based on the state of S_1. The select bits, S_0 and S_1, for each OLMC are programmed into a dedicated group of cells in the array.

The four OLMC configurations are

- Combinational mode with active-LOW output
- Combinational mode with active-HIGH output
- Registered mode with active-LOW output
- Registered mode with active-HIGH output

Combinational Mode   When $S_0 = 0$ and $S_1 = 1$, the multiplexer selects the OR gate output. The effective logic of the OLMC is shown in Figure 7–21(a) where the output polarity is active-LOW because of the inversion of the tristate buffer. When $S_0 = 1$ and $S_1 = 1$, the complement of the OR gate output is selected, and the effective logic of the OLMC is as shown in part (b). The output is active-HIGH because of the double inversion (complement of the OR and tristate inversion). The S_0 and S_1 bits for each of the ten OLMCs are set by programming a set of dedicated cells in the array separate from the logic array cells. The OLMC can be configured as an output or an input by controlling the tristate buffer.

FIGURE 7–21
OLMC combinational mode.

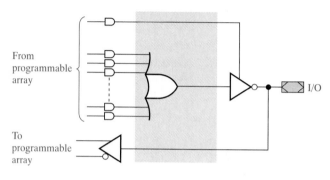

(a) Active-LOW output: $S_1 = 1$, $S_0 = 0$

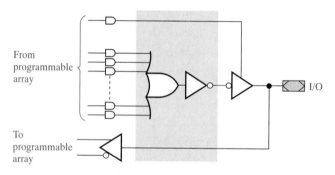

(b) Active-HIGH output: $S_1 = 1$, $S_0 = 1$

Tristate Buffer   The **tristate buffer,** shown in Figure 7–22(a), is a two-state logic inverter with a control line that permits the output to be disconnected from the input. The three output states are *LOW* (when the input is HIGH), *HIGH* (when the input is LOW), and *high-impedance* (when the output is disconnected from the input). When the control

FIGURE 7–22
Tristate buffer operation.

Tristate
control

Input ——▷o—— Output

(a)

HIGH ——┐
HIGH ——▷o—— LOW
Active state

HIGH ——┐
LOW ——▷o—— HIGH
Active state

LOW ——┐
——▷o——
High-impedance state

(b)

line is HIGH, the buffer is active; and when the control line is LOW, the buffer is in the high-impedance state. These three states are illustrated in Figure 7–22(b). Notice that in the high-impedance state the buffer effectively acts as an open switch.

Output or Input Selection   In the OLMC combinational mode logic of Figure 7–21, you can see that the control line for the tristate buffer comes from an AND gate in the array. When this control line is HIGH, the OLMC produces a combinational output. When this control line is LOW, the buffer is open (high-impedance) and the pin becomes an input. This is indicated in Figure 7–23. The state of the tristate control line is set by programming.

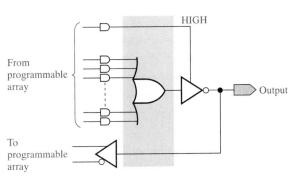

(a) Output (active-LOW shown for illustration)

(b) Input

FIGURE 7–23
OLMC combinational output and input configurations.

EXAMPLE 7–5

Determine the output expression of an OLMC in a GAL22V10 for the product terms from the AND array and the select bits as shown in Figure 7–24.

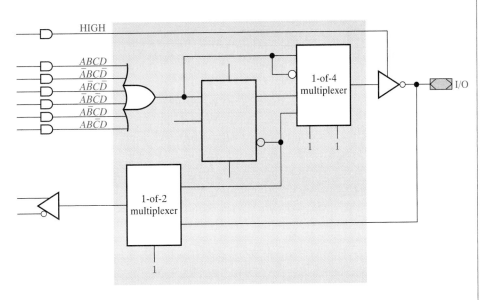

FIGURE 7–24

Solution   The OLMC is configured for an active-HIGH combinational output; therefore, the SOP expression is

$$X = ABCD + \overline{A}BC\overline{D} + A\overline{B}C\overline{D} + \overline{A}B\overline{C}D + A\overline{B}CD + AB\overline{C}D$$

Related Exercise   Write the SOP expression for the output if $S_0 = 0$ and $S_1 = 1$.

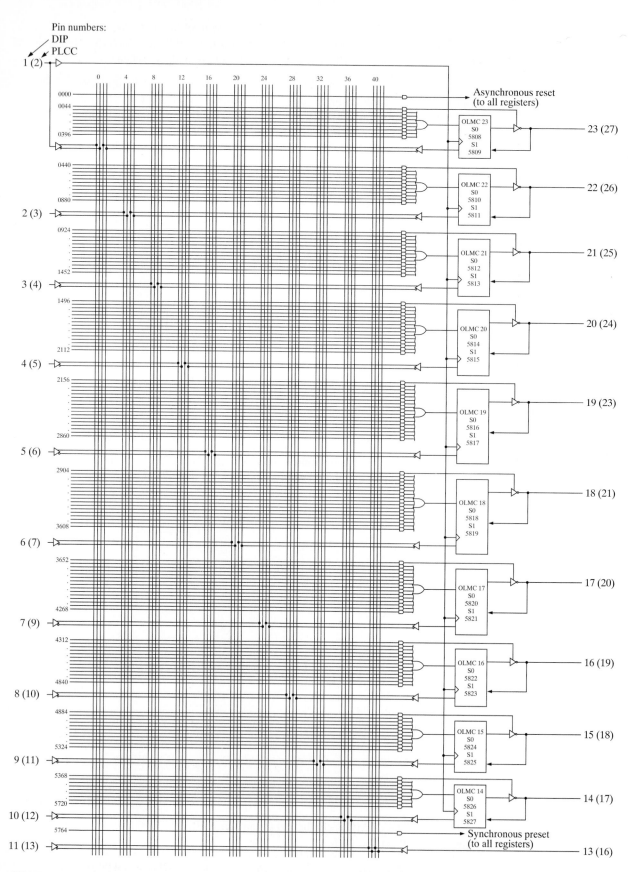

FIGURE 7–25
GAL22V10 array diagram.

The GAL22V10 Array

The diagram for the GAL22V10 in Figure 7–25 is similar to the block diagram, shown in Figure 7–18, except that details of the programmable AND array are shown.

The GAL22V10 programmable array is organized as 22 input lines and complements crossing 132 product term lines. There is a programmable E^2CMOS cell at each of the 5808 intersections. Each cell can be programmed in the *on* state to connect an input variable or its complement to a product term line.

A single line represents the product term line in the diagram of Figure 7–25. However, in the GAL22V10 there are actually 44 lines to each AND gate. So, each product term line consists of 44 AND gate inputs, one for each input line and its complement. The numbers shown for each product term line are the beginning cell numbers for the 44 cells associated with that line. The number in each OLMC for S_0 and S_1 are the cell numbers in the array into which the bits are programmed. These twenty special cells (5808 through 5827) are not shown in the array diagram.

Figure 7–26 shows details of the programmable array for the first OLMC section. The OLMC has a capacity for eight product terms that can be used in an SOP function. One product term is reserved for the tristate control, and one product term is used for a reset function in the registered mode for all of the OLMCs. The latter product term is called a *global* function because it is common to all OLMCs.

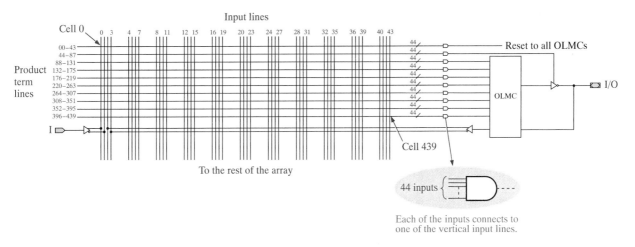

FIGURE 7–26
Organization of the programmable array showing one OLMC portion.

Implementing an SOP Function

As mentioned, a cell is programmed to the *on* state where a variable or its complement is part of a product term. Programming a cell to the *on* state simply connects an input line to a product term line. Each AND gate in the array produces one product term. When all of the input variables that are connected to an AND gate are HIGH, the output of the AND gate is HIGH.

The product terms from all of the AND gates for a given OLMC are ORed to form the sum-of-products function. The OLMC select bits are programmed to route the OR gate output, either complemented or uncomplemented, to the output pin.

With the GAL22V10, up to ten separate SOP functions can be programmed, the largest of which can contain sixteen product terms. If an SOP expression with more than sixteen terms is required, two OLMCs are necessary. The OR gate output of one OLMC is fed back to the array and connected into the other OLMC by programming.

EXAMPLE 7–6   Show how the following 6-variable SOP function is implemented with the GAL22V10.

$$X = ABCDEF + \overline{A}B\overline{C}\overline{D}E\overline{F} + \overline{A}B\overline{C}D\overline{E}F + \overline{A}BCD\overline{E}\overline{F} + A\overline{B}\overline{C}D\overline{E}F + \overline{A}BCDEF + \overline{A}B\overline{C}\overline{D}E\overline{F}$$

Solution   The programmed portion of the array with the associated OLMC is shown in Figure 7–27. The rest of the array remains unused in this case. The Xs represent E^{2}CMOS cells that are programmed to the *on* state.

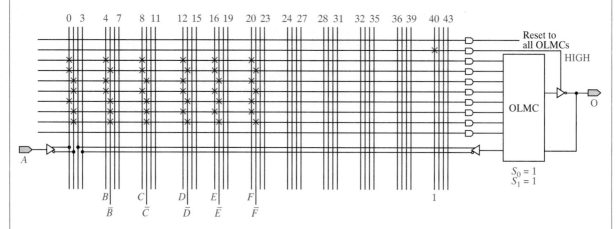

FIGURE 7–27

Related Exercise   How many more product terms can be added to the SOP expression programmed in Figure 7–27?

**SECTION 7–4
REVIEW**

1. What are the two modes for the GAL22V10?
2. How is the mode determined?
3. How is the polarity of the output set?
4. What is the maximum number of product terms in an SOP expression that can be implemented with a single GAL22V10 OLMC?

7–5 ■ THE GAL16V8

In this section, another type of generic array logic device, the GAL16V8, is discussed. The main feature of the GAL16V8 is that it can be programmed to emulate many types of PALs. After completing this section, you should be able to

□ List the modes for the OLMC   □ Discuss the simple mode   □ Discuss the complex mode   □ State how the GAL16V8 can replace various types of PALs

Block Diagram of the GAL16V8

The GAL16V8 contains eight dedicated inputs and eight input/outputs (I/Os) as shown in the block diagram of Figure 7–28(a) with the package pin numbers indicated. This device is available in either a 20-pin DIP (dual-in-line package) or a 20-pin PLCC (plastic chip carrier) as shown in part (b).

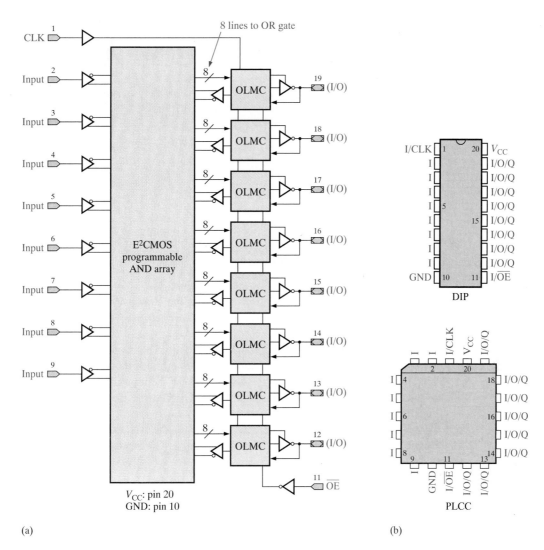

FIGURE 7–28
GAL16V8 block diagram and packaging.

The OLMC

The OLMC in the GAL16V8 is somewhat different from that in the GAL22V10. The GAL16V8 is designed to be programmed in one of three available modes to emulate most of the existing PALs; thus, it may replace the PAL for which it is programmed.

PAL Emulation All of the possible OLMC configurations of the GAL16V8 are classified into three basic modes: *simple, complex,* and *registered.* Each of these modes allows the GAL16V8 to be programmed to emulate a number of PALs as specified in Table 7–1, on the next page. The simple and complex modes are associated only with combinational outputs.

Simple Mode In this mode, the OLMCs are configured as dedicated active combinational outputs or as dedicated inputs (limited to six). The three possible configurations in the simple mode are

- Combinational output
- Combinational output with feedback to AND array
- Dedicated input

TABLE 7–1
PALs which can be emulated by each OLMC mode of the GAL16V8

GAL16V8 Simple Mode	GAL16V8 Complex Mode	GAL16V8 Registered Mode
PAL10L8	PAL16L8	PAL16R8
PAL12L6	PAL16H6	PAL16R6
PAL14L4	PAL16P8	PAL16R4
PAL16L2		PAL16RP8
PAL10H8		PAL16RP6
PAL12H6		PAL16RP4
PAL14H4		
PAL16H2		
PAL10P8		
PAL12P6		
PAL14P4		
PAL16P2		

Each of these configurations is illustrated in Figure 7–29. Notice that the tristate control line is HIGH (V_{CC}) making the buffer active for the combinational output configurations and LOW (ground) making the buffer open (high-impedance) for the dedicated input configuration.

The feedback allows the combinational output to be connected back to the AND array. The *XOR* input to the exclusive-OR gate determines the active state of the output. When $XOR = 0$, the output is active-LOW because the input from the OR gate is not inverted. When $XOR = 1$, the output is active-HIGH because the input from the OR gate is inverted.

The eight inputs to the OR gate come from the AND array and indicate that the OLMC can produce up to eight product terms in an SOP expression.

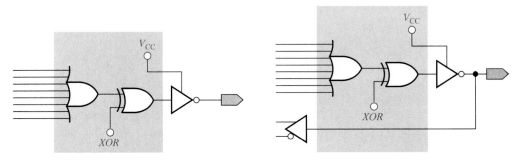

(a) Combinational output   (b) Combinational output with feedback to AND array

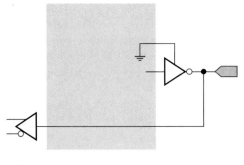

(c) Dedicated input

FIGURE 7–29
OLMC configurations in the simple mode.

Complex Mode   In this mode, the OLMCs can be configured two ways:

- Combinational output
- Combinational input/output (I/O)

These two configurations are illustrated in Figure 7–30. The I/Os are limited to six in this mode. Notice that there are only seven inputs to the OR gate from the AND array because the eighth input is used for the tristate control. This means that an OLMC can produce up to seven product terms in an SOP expression.

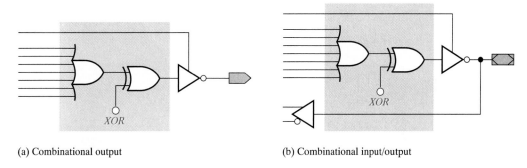

(a) Combinational output   (b) Combinational input/output

FIGURE 7–30
OLMC configurations in the complex mode.

SECTION 7–5 REVIEW	1. State the basic differences between a GAL16V8 and a GAL22V10. 2. How does the GAL16V8 emulate a PAL device? 3. List the three OLMC modes.

7–6 ■ PLD PROGRAMMING

As you have learned, PALs are programmed by leaving specified fusible links intact and blowing open all others. GALs are programmed in a similar way except the E^2CMOS cells are turned on or off. The logic function(s) to be implemented determines which cells are affected. In order to program a PAL or GAL, the following items are required: a computer, programming software, and a PLD programmer. After completing this section, you should be able to

☐ Name the components required to program a PLD   ☐ Describe the programming process in terms of a flow chart   ☐ Define in-system programming (ISP)

The three components required to program PLDs are

1. Programming software (logic compiler)
2. Personal computer that meets the software requirements
3. Software-driven programmer

The computer must meet the **software** requirements. The **programmer** is a hardware device that accepts programming data from the computer and implements a specified logic design in the PLD that is inserted in a socket in the programmer. These elements are illustrated in Figure 7–31.

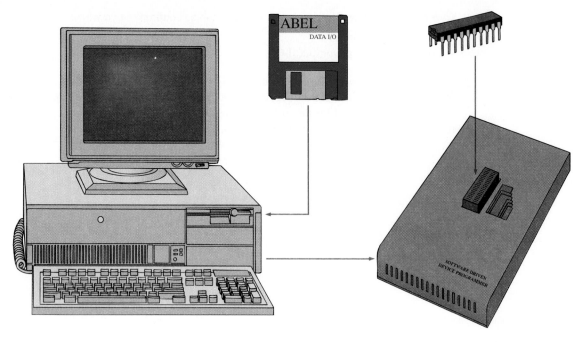

FIGURE 7–31
Components for PLD programming.

Computer

Any personal computer that meets the software and programmer specifications can be used. These specifications normally include the type of microprocessor on which the computer is based, the amount of memory, and the operating system (DOS, Windows, and Macintosh are examples).

Software

The software packages for PLD programming are called logic **compilers.** Various software packages are available including, but not limited to, the following: ABEL, CUPL, OrCAD-PLD, and LOG/iC. All of these software packages perform similar functions; they process and synthesize the logic design entered by a specified method, convert the entered data into an intermediate file, and then generate a final output file called a **JEDEC file** (also known as a *cell map* or *fuse map*) for the device programmer. Also, with the software, a logic design can be fully simulated and debugged before any hardware is built.

A key feature of PLD software is the method for entering a logic design. Three basic design entry methods are *Boolean equation, truth table,* and *state machine.* Additional entry methods are schematic, timing waveform, and hardware description. All available software packages include some or all of these design entry methods.

The Programmer

The PLD device is inserted into the programmer socket which is usually a **ZIF socket** (zero insertion force socket). Using the JEDEC file generated by the logic compiler, the programmer programs the PLD by applying required voltages to specified pins to alter the specified cells in the array as directed by the fuse map. All PLD software and programmers, regardless of the manufacturer, conform to a standard for generation of the JEDEC file established by the Joint Electronic Device Engineering Council **(JEDEC).**

The Programming Process

The general flow chart in Figure 7–32 shows the sequence for programming a PLD. First, the logic circuit is designed and expressed in terms of Boolean equations, truth tables, logic schematic, or other acceptable form, and the software is loaded into the computer.

Entering the Design The logic design is entered into the computer by creating an input or source file. Typically, some preliminary information is entered, such as user's name, company, data, and description of the design. Then, the type of PLD device, the input and output pin numbers, and variable labels are entered. Finally, the logic function(s) is entered in the form of a Boolean equation, truth table, schematic diagram, or the like. Any syntax

FIGURE 7–32
Flow chart of a PLD programming sequence.

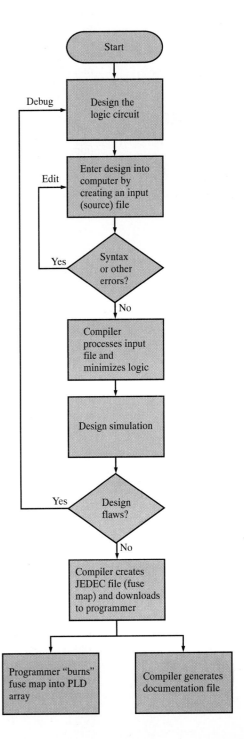

errors or other errors made in entering data into the input file are indicated and must be corrected. **Syntax** is the prescribed format and symbology used in describing a category of functions.

Running the Software   The software compiler processes and translates the input file and minimizes the logic. The logic design is then simulated using a set of hypothetical inputs known as *test vectors*. This process effectively "exercises" the design in software to determine if it works properly before the PLD is actually programmed. If any design flaws are discovered during software simulation, the design must be debugged and modified to correct the flaw.

Once the design is finalized, the compiler generates a **documentation file,** which includes the final logic equations, the JEDEC file, and if desired a pinout diagram of the PLD. The documentation file essentially provides a "hard copy" of the final design.

Programming the Device   When the design is finalized, the compiler creates a fuse map (JEDEC file) and downloads it to the programmer. The fuse map tells the programmer which fuse links to blow and which to leave intact or, in the case of a GAL, which E^2CMOS cells to turn *on* or *off*. The programmer then patterns the PLD device according to the fuse map.

An introduction to PLD programming using ABEL software is provided in Section 7–7.

In-System Programmability

Standard PLDs including EEPROMs, PALs, and GALs are programmed using the hardware programmer, as you have just learned. A variety of programmable logic including GALs is also available as in-system programmable (ISP) devices; for example, an in-system programmable version of the GAL22V10 is designated ispGAL22V10. The ISP device is the same as the standard device except that it has additional logic for in-system programming.

An ISP device can be programmed as a standard device prior to being placed on a PC board, or it can be programmed in several ways after it has been installed on a printed circuit board. ISP devices can be programmed with other circuits in a system as follows:

- Using a programmer before assembly on a printed circuit board
- Using automatic test equipment (ATE) after assembly on a board
- Using a personal computer after assembly on a board
- Using an in-system microprocessor (embedded) after assembly on a board

SECTION 7–6 REVIEW	**1.** What is required to program a PLD? **2.** What are the software packages for PLDs called? **3.** What is a JEDEC file? **4.** Define ISP. **5.** How does an ispGAL differ from a standard GAL?

7–7 ■ PLD SOFTWARE

As mentioned earlier, there are several software packages for implementing logic designs in PLDs. ABEL and CUPL are two of the most commonly used hardware description languages (HDLs). Because both languages are similar and produce the same results in terms of programming a PLD, usage is often a matter of preference and availability. This section introduces basic concepts of PLD software using the ABEL (trademark of Data I/O Corp.) language and is it not intended to provide a comprehen-

sive coverage. **Digital Design Using ABEL,** *by Pellerin and Holley (Prentice Hall, 1994), is recommended for a thorough coverage. After completing this section, you should be able to*

□ List the three formats for design entry   □ Define *test vector*   □ Identify the ABEL logic operator symbols   □ Express logic functions using the ABEL equation format   □ Express logic functions using the ABEL truth table format   □ Develop test vectors for a given logic design   □ Write an ABEL input file for a given logic design   □ Describe a documentation file   □ Discuss the JEDEC file

The ABEL Language

ABEL, which is the acronym for *Advanced Boolean Expression Language,* allows logic designs to be implemented in programmable logic devices. ABEL can be used to program any type of PLD and is therefore a device-independent language. As discussed in the previous section, ABEL is run on a computer connected to a language-independent device programmer into which the PLD is inserted.

Logic Design Entry   ABEL provides three different formats for describing and entering a logic design from the computer keyboard: *equations, truth tables,* and *state diagrams.* The equation and truth table formats are used for combinational logic design, which is the focus of this chapter. The state diagram format, as well as the other two formats, can be used for sequential logic design (state machines) and is not discussed until Chapter 11.

Design Simulation   Once a logic circuit design has been entered, its operation can be simulated using **test vectors** to make sure there are no design errors. Basically, a test vectors section lists a sample of input values and the corresponding output values. The software essentially "exercises" the logic design to make sure it works as expected by taking it through various combinations of inputs and checking the resulting outputs.

Logic Synthesis   The software process of converting a circuit description in the form of equations, truth tables, or state diagrams to a standard JEDEC file format required to actually implement the design in a PLD is called logic **synthesis.** Logic synthesis usually includes optimization and minimization of a design.

Boolean Operations

The NOT (inversion), AND, OR, and exclusive-OR operations have special symbols in ABEL as shown in Table 7–2 in order of precedence.

TABLE 7–2
ABEL symbols for the logic operations

Logic Operation	ABEL Symbol
NOT	!
AND	&
OR	#
XOR	$

Examples of the basic operations in standard Boolean notation and in ABEL notation are as follows:

Standard Boolean	ABEL
$\overline{A}$	!A
AB	A & B
$A + B$	A # B
$A \oplus B$	A $ B

Equations

One of the ABEL entry methods uses logic equations. The equal sign (=) in ABEL is the same as in standard equations, and any letters or combinations of letters and numbers can be used as variable identifiers. With respect to variable identifiers, ABEL is case-sensitive. For example, uppercase A and lowercase a are two different variables in ABEL.

All equations in ABEL must end with a semicolon (;). For example, the sum-of-products expression $X = AB + \overline{A}B + A\overline{B}$ is written in the ABEL language as

$$X = A \& B \# !A \& B \# A \& !B;$$

EXAMPLE 7–7

Write each of the following logic expressions in ABEL:
(a) $X = A\overline{B}C + \overline{A}\,\overline{B}\,\overline{C} + AB + \overline{B}C$ (b) $Y = (\overline{A} + B + \overline{C} + D)(A + B + C)$

Solution
(a) $X = A \& !B \& C \# !A \& !B \& !C \# A \& B \# !B \& C;$
(b) $Y = (!A \# B \# !C \# D) \& (A \# B \# C);$

Note that the parentheses in part (b) are used to change the order of precedence as defined in Table 7–2. That is, the variables within the parentheses are first ORed and then the two resulting terms are ANDed.

Related Exercise Express $W = X + Y$ in the ABEL format, where $X = \overline{A}B\overline{C}$ and $Y = A + \overline{B} + C$.

Sets In some cases, multiple input or output variables may be grouped as a set using brackets in order to simplify an equation. For example, a group of inputs labeled A_3, A_2, A_1, A_0 can be defined as a single variable A using ABEL set notation as follows:

$$A = [A3, A2, A1, A0];$$

EXAMPLE 7–8

Develop the logic equations to describe the 12-input, 4-output multiplexer shown in Figure 7–33(a) using ABEL. The select inputs, S_0 and S_1, select one of the three sets of four data inputs and route it to the four output lines. The function table is shown in part (b).

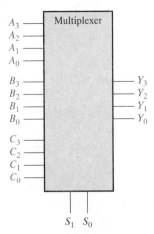

(a) Block diagram

Select Inputs		Data Outputs			
S_1	S_0	Y_3	Y_2	Y_1	Y_0
0	1	A_3	A_2	A_1	A_0
1	0	B_3	B_2	B_1	B_0
1	1	C_3	C_2	C_1	C_0

(b) Function table

FIGURE 7–33

Solution   The sum-of-products equations in standard Boolean form are

$$Y_3 = A_3\overline{S}_1 S_0 + B_3 S_1 \overline{S}_0 + C_3 S_1 S_0$$
$$Y_2 = A_2\overline{S}_1 S_0 + B_2 S_1 \overline{S}_0 + C_2 S_1 S_0$$
$$Y_1 = A_1\overline{S}_1 S_0 + B_1 S_1 \overline{S}_0 + C_1 S_1 S_0$$
$$Y_0 = A_0\overline{S}_1 S_0 + B_0 S_1 \overline{S}_0 + C_0 S_1 S_0$$

The equations in ABEL format are

```
Y3 = A3 & !S1 & S0 # B3 & S1 & !S0 # C3 & S1 & S0;

Y2 = A2 & !S1 & S0 # B2 & S1 & !S0 # C2 & S1 & S0;

Y1 = A1 & !S1 & S0 # B1 & S1 & !S0 # C1 & S1 & S0;

Y0 = A0 & !S1 & S0 # B0 & S1 & !S0 # C0 & S1 & S0;
```

The four equations can be combined into one ABEL equation by grouping the inputs and outputs into sets with the following declaration:

```
A = [A3, A2, A1, A0];

B = [B3, B2, B1, B0];

C = [C3, C2, C1, C0];

Y = [Y3, Y2, Y1, Y0];

S = [S1, S0];
```

The sets are then used as follows to produce a single multiplexer output equation:

```
Y = (S == 1) & A # (S == 2) & B # (S == 3) & C;
```

The symbol == is a relational operator that provides an equality comparison for the sets. For example, the first term in the right side of the equation is (S == 1). This means that the select inputs, S_1 and S_0, are equal to 1 or binary 01, which is expressed in standard Boolean notation as $\overline{S}_1 S_0$.

With the set declarations and the single ABEL equation, the compiler software will implement each of the four original logic equations.

Related Exercise   Rewrite the single multiplexer ABEL equation if the order of selection in the multiplexer is reversed. That is, the C inputs are selected when the select inputs are equal to binary 01, etc.

Truth Tables

Another ABEL entry method uses truth tables. A specified logic function can be entered in either equation form, as you have seen, or in truth table form. Both methods are equivalent in terms of the result. Truth tables are usually more convenient than equations for describing certain types of circuits such as decoders.

The ABEL truth table format includes a header and the truth table entries. The header format is illustrated by the following sample header:

```
TRUTH_TABLE ([A, B, C, D] -> [X1, X2])
```

where A, B, C, and D are the inputs and X1 and X2 are the outputs. The symbol -> is an operator that indicates the inputs produce combinational outputs. To illustrate how an ABEL truth table is constructed, a simple example of an exclusive-OR gate with inputs *A* and *B* and output *X* is written as follows:

```
TRUTH_TABLE ([A, B] -> [X])
            [1, 0] -> [1];
            [0, 1] -> [1];
            [0, 0] -> [0];
            [1, 1] -> [0];
```

EXAMPLE 7–9   Develop an ABEL truth table for a BCD to 7-segment decoder. Refer to the truth table in Table 4–10. The block diagram and 7-segment format are shown in Figure 7–34. The outputs are to be active-HIGH.

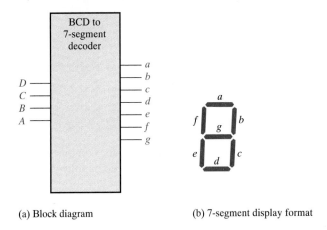

(a) Block diagram   (b) 7-segment display format

FIGURE 7–34

Solution   The most significant input is *D* and the least significant input is *A*. The segment outputs are *a* through *g*. An output of 1 means the segment is *on* and 0 means the segment is *off*. The ABEL truth table is

```
TRUTH_TABLE ([D, C, B, A] -> [a, b, c, d, e, f, g])
           [0, 0, 0, 0] -> [1, 1, 1, 1, 1, 1, 0];
           [0, 0, 0, 1] -> [0, 1, 1, 0, 0, 0, 0];
           [0, 0, 1, 0] -> [1, 1, 0, 1, 1, 0, 1];
           [0, 0, 1, 1] -> [1, 1, 1, 1, 0, 0, 1];
           [0, 1, 0, 0] -> [0, 1, 1, 0, 0, 1, 1];
           [0, 1, 0, 1] -> [1, 0, 1, 1, 0, 1, 1];
           [0, 1, 1, 0] -> [1, 0, 1, 1, 1, 1, 1];
           [0, 1, 1, 1] -> [1, 1, 1, 0, 0, 0, 0];
           [1, 0, 0, 0] -> [1, 1, 1, 1, 1, 1, 1];
           [1, 0, 0, 1] -> [1, 1, 1, 1, 0, 1, 1];
```

By using a set declaration for the inputs and decimal numbers for the BCD input values, the table can be simplified.

```
BCD = [D, C, B, A];

TRUTH_TABLE (BCD  -> [a, b, c, d, e, f, g])
           0    -> [1, 1, 1, 1, 1, 1, 0];
           1    -> [0, 1, 1, 0, 0, 0, 0];
           2    -> [1, 1, 0, 1, 1, 0, 1];
           3    -> [1, 1, 1, 1, 0, 0, 1];
           4    -> [0, 1, 1, 0, 0, 1, 1];
           5    -> [1, 0, 1, 1, 0, 1, 1];
           6    -> [1, 0, 1, 1, 1, 1, 1];
           7    -> [1, 1, 1, 0, 0, 0, 0];
           8    -> [1, 1, 1, 1, 1, 1, 1];
           9    -> [1, 1, 1, 1, 0, 1, 1];
```

Related Exercise   Rework the ABEL truth table for active-low outputs (0).

Test Vectors

As you know, test vectors are used in ABEL to describe input signals and expected outputs for logic simulation. In other words, the test vectors examine a logic design before it is committed to hardware by applying all possible input combinations and checking for the proper outputs.

Test vectors are essentially the same as truth tables and are entered in the same format. For example, the test vectors section for the BCD to 7-segment decoder is

```
TEST_VECTORS ([D, C, B, A] -> [a, b, c, d, e, f, g])
            [0, 0, 0, 0] -> [1, 1, 1, 1, 1, 1, 0];
            [0, 0, 0, 1] -> [0, 1, 1, 0, 0, 0, 0];
            [0, 0, 1, 0] -> [1, 1, 0, 1, 1, 0, 1];
            [0, 0, 1, 1] -> [1, 1, 1, 1, 0, 0, 1];
            [0, 1, 0, 0] -> [0, 1, 1, 0, 0, 1, 1];
            [0, 1, 0, 1] -> [1, 0, 1, 1, 0, 1, 1];
            [0, 1, 1, 0] -> [1, 0, 1, 1, 1, 1, 1];
            [0, 1, 1, 1] -> [1, 1, 1, 0, 0, 0, 0];
            [1, 0, 0, 0] -> [1, 1, 1, 1, 1, 1, 1];
            [1, 0, 0, 1] -> [1, 1, 1, 1, 0, 1, 1];
```

The test vectors can be simplified using sets, the same as truth tables.

The ABEL Input File

When you create an **input** (source) **file** in ABEL, you create a module which contains three basic sections following the module and title statements: *declarations, logic descriptions,* and *test vectors.*

Declarations   The declarations section generally includes *device declaration, pin declarations,* and *set declarations.* We used set declarations in Examples 7–8 and 7–9.

Device declarations are used to specify the PLD that is to be programmed, sometimes referred to as a *target device.* An example of a device declaration statement is

```
decoder device 'P22V10';
```

The first word is simply a description and can be anything you wish. The second word is an essential keyword and can appear in the statement in either uppercase or lowercase letters. The specific device must be in the form shown with the prefix P used for both PALs and GALs.

An example of a pin declaration statement is

```
A0, A1, A2, A3, PIN 1, 2, 3, 4;
```

The keyword PIN can also be written in lowercase letters. The numbers following the keyword are pin numbers on the PLD device that is to be programmed. A0 goes to pin 1, A1 goes to pin 2, A2 goes to pin 3, and A3 goes to pin 4.

Logic Descriptions   The two methods of entry that we have already discussed, equations and truth tables, are two types of logic descriptions. The third type is the state diagram which is covered in Chapter 11.

Test Vectors   As stated before, the test vector section tests a logic design before it is finalized and implemented in hardware. Test vectors have the same form as truth tables.

EXAMPLE 7–10

Create an ABEL input file for implementing the quad 1-of-4 multiplexer shown as a block diagram in Figure 7–35(a). The function table is shown in part (b), and the pin assignment diagram of the GAL20V8 target device is shown in part (c).

Solution   The ABEL input file for the multiplexer logic design is as follows:

```
Module    quad_1of4_mux

Title     'Quad 1 of 4 multiplexer in a GAL20V8'

    mux        device      'P20V8';

    A0, A1, A2, A3          pin 1, 2, 3, 4;
    B0, B1, B2, B3          pin 5, 6, 7, 8;
    C0, C1, C2, C3          pin 9, 10, 11, 13;
    D0, D1, D2, D3          pin 14, 15, 16, 17;
    Aout, Bout, Cout, Dout  pin 21, 20, 19, 18;
    S0, S1                  pin 22, 23;
```

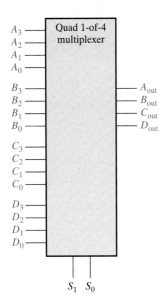

(a) Block diagram

Select Inputs		Data Outputs			
S_1	S_0	D_{out}	C_{out}	B_{out}	A_{out}
0	0	D_0	C_0	B_0	A_0
0	1	D_1	C_1	B_1	A_1
1	0	D_2	C_2	B_2	A_2
1	1	D_3	C_3	B_3	A_3

(b) Function table

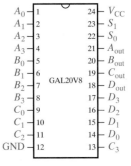

(c) Pin assignments

FIGURE 7–35

```
Equations

    Aout = !S1 & !S0 & A0 # !S1 & S0 & A1 # S1 & !S0 & A2 # S1 & S0 & A3;

    Bout = !S1 & !S0 & B0 # !S1 & S0 & B1 # S1 & !S0 & B2 # S1 & S0 & B3;

    Cout = !S1 & !S0 & C0 # !S1 & S0 & C1 # S1 & !S0 & C2 # S1 & S0 & C3;

    Dout = !S1 & !S0 & D0 # !S1 & S0 & D1 # S1 & !S0 & D2 # S1 & S0 & D3;

Test_vectors
([S1,S0,A0,A1,A2,A3,B0,B1,B2,B3,C0,C1,C2,C3,D0,D1,D2,D3] ->
[Aout,Bout,Cout,Dout])

"S  S  A  A  A  A  B  B  B  B  C  C  C  C  D  D  D  D      outputs
"1  0  0  1  2  3  0  1  2  3  0  1  2  3  0  1  2  3      A  B  C  D

[0, 0, 1, 0, 0, 0, 0, 1, 0, 0, 0, 0, 1, 0, 0, 0, 0, 1] -> [1, 0, 0, 0];
[0, 1, 1, 0, 0, 0, 0, 1, 0, 0, 0, 0, 1, 0, 0, 0, 0, 1] -> [0, 1, 0, 0];
[1, 0, 1, 0, 0, 0, 0, 1, 0, 0, 0, 0, 1, 0, 0, 0, 0, 1] -> [0, 0, 1, 0];
[1, 1, 1, 0, 0, 0, 0, 1, 0, 0, 0, 0, 1, 0, 0, 0, 0, 1] -> [0, 0, 0, 1];

[0, 0, 1, 1, 1, 0, 1, 1, 0, 1, 1, 0, 1, 1, 0, 1, 1, 1] -> [1, 1, 1, 0];
[0, 1, 1, 1, 1, 0, 1, 1, 0, 1, 1, 0, 1, 1, 0, 1, 1, 1] -> [1, 1, 0, 1];
[1, 0, 1, 1, 1, 0, 1, 1, 0, 1, 1, 0, 1, 1, 0, 1, 1, 1] -> [1, 0, 1, 1];
[1, 1, 1, 1, 1, 0, 1, 1, 0, 1, 1, 0, 1, 1, 0, 1, 1, 1] -> [0, 1, 1, 1];

END
```

Notice that comment lines can be inserted anyplace but must be preceded by the quotation symbol " as we did in the test vectors section to clarify the inputs and outputs in the entries. Comment lines in general are used to clarify or identify parts of the input file for later use by anyone who reads the file. Comment lines are ignored by the software and do not affect the final design.

Related Exercise Use set declarations to write the logic equations for the multiplexer.

The Documentation File

After an input file is processed, ABEL generates a documentation file. The documentation file provides a hard copy of the final reduced equations, a JEDEC file such as the one shown in Figure 7-36, and a device pin diagram.

```
QP24* QF2706*
L0000
0000000000000000000000000000000000000000000
0000000000000000000000000000000000000000000
0000000000000000000000000000000000000000000
0000000000000000000000000000000000000000000
0000000000000000000000000000000000000000000
0000000000000000000000000000000000000000000
0000000000000000000000000000000000000000000
0000000000000000000000000000000000000000000
```

A_{out} logic
```
1101111011101111111111111111111111111111
0111111011011111111111111111111111111111
1111010111101111111111111111111111111111
1111110101011111111111111111111111111111
```
```
0000000000000000000000000000000000000000000
0000000000000000000000000000000000000000000
0000000000000000000000000000000000000000000
0000000000000000000000000000000000000000000
```

B_{out} logic
```
1111110111001111111111111111111111111111
1111110110111101111111111111111111111111
1111101111101111111011111111111111111111
1111110111011111111111101111111111111111
```
```
0000000000000000000000000000000000000000000
0000000000000000000000000000000000000000000
0000000000000000000000000000000000000000000
0000000000000000000000000000000000000000000
```

C_{out} logic
```
1111110111011111111111111110111111111111
1111110110101111111111111111111101111111
1111101111101111111111111111111111110111
1111110111011111111111111111111111111101
```
```
0000000000000000000000000000000000000000000
0000000000000000000000000000000000000000000
0000000000000000000000000000000000000000000
0000000000000000000000000000000000000000000
```

D_{out} logic
```
1111110111011111111111111111111011111111
1111110110111111111111111111110111111111
1111101111101111111111111011111111111111
1111110111011111111111101111111111111111
```
```
0000000000000000000000000000000000000000000
0000000000000000000000000000000000000000000
0000000000000000000000000000000000000000000
00000000000000000000000000000000000000000000*
```

Used for configuring OLMCs and other special programming
```
L2560
01111000*
L2568
0000000000000000000000000000000000000000000000000000000000000000*
L2632
1*
L2633
0*
L2634
0*
L2635
0*
L2636
0*
L2637
1*
L2638
1*
L2639
1*
L2640
11111111111111111111111111111111111111111111111111111111111111111*
L2704
10*
```

Test vectors section (not programmed into array)
```
V0001 10000100001N00001LLLH00N*
V0002 10000100001N00001LLHL10N*
V0003 10000100001N00001LHLL01N*
V0004 10000100001N00001HLLL11N*
V0005 11101101101N10111LHHH00N*
V0006 11101101101N10111HLHH10N*
V0007 11101101101N10111HHLH01N*
V0008 11101101101N10111HHHL11N*
C5127*
```

FIGURE 7–36

JEDEC file for the quad 1-of-4 multiplexer from Example 7–10. The red labels are not part of the file.

The JEDEC File As mentioned in the last section, this part of the documentation file is also called a *cell map* or a *fuse map*. The JEDEC file is down-loaded to the programmer for programming the target device. JEDEC files are standarized for all programming languages.

The cell map shows which cells in the device array are disconnected (*off* or blown) and which are connected (*on* or intact). A 0 in a cell map indicates a cell that is connected and a 1 indicates a cell that is disconnected.

Figure 7–36 (p. 369) illustrates the cell map for the quad 1-of-4 multiplexer from Example 7–10. The four digits preceded by an "L" correspond to the product line numbers in

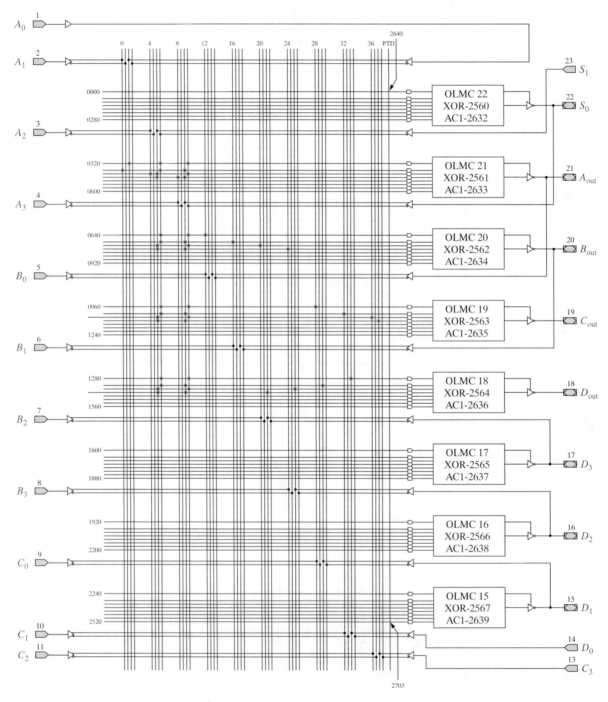

FIGURE 7–37
GAL20V8 for Example 7–11.

the array. The four sum-of-product equations for multiplexer outputs are implemented in the four sections of the array as indicated. An asterisk indicates the end of the logic implementation. All of the lines that start with an "L" are for special cell programming for user identification, configuring the OLMCs, and other functions. The lines in the cell map beginning with "V" are used for the test vectors, and the last line beginning with a "C" is a checksum.

EXAMPLE 7–11

Show the GAL20V8 logic array with the input and output assignments and the array programming for the quad 1-of-4 multiplexer based on the JEDEC file in Figure 7–36.

Solution The programmed logic diagram for the GAL20V8 is shown in Figure 7–37 (on facing page) where the Xs indicate the cells that are *on*.

Related Exercise How many I/O pins in the GAL20V8 in Figure 7–37 are used as inputs and how many are used as outputs?

SECTION 7–7 REVIEW

1. What is ABEL?
2. List the three formats for design entry in ABEL.
3. Write the ABEL symbols for NOT, AND, OR, and XOR.
4. Express the following Boolean equation using ABEL:

$$Y = \overline{DEF} + \overline{D}\,\overline{E} + F$$

5. What are test vectors?
6. List the three parts of an ABEL input file.
7. What types of declarations are usually found in an input file?
8. What is a documentation file?
9. Describe a JEDEC file and give two other names used for it.

7–8 ■ DIGITAL SYSTEM APPLICATION

In the last chapter the combinational logic portion of the traffic light control system was implemented using SSI and MSI devices. You will now apply your knowledge of PLDs and PLD programming to implement the state decoding and output logic with a GAL22V10. You should review Section 6–12 before continuing. After completing this section, you should be able to

☐ Program the GAL22V10   ☐ Explain how the GAL replaces the logic previously implemented in Chapter 6

State Decoding and Output Logic

The block diagram of the state decoding and output logic is shown in Figure 7–38, on the next page. The traffic light sequence and the corresponding state diagram are given in Figure 7–39 for reference.

The state decoding and output logic portion of the system has two inputs, S_1 and S_0, and eight outputs, *MR, MY, MG, SR, SY, SG, Long,* and *Short*. These input and output variables are defined as follows:

S_1 State input, Gray code MSB
S_0 State input, Gray code LSB
MR   Main street red light output
MY   Main street yellow light output

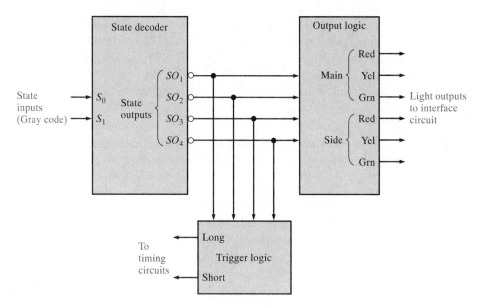

FIGURE 7–38
Block diagram of the state decoding and output logic.

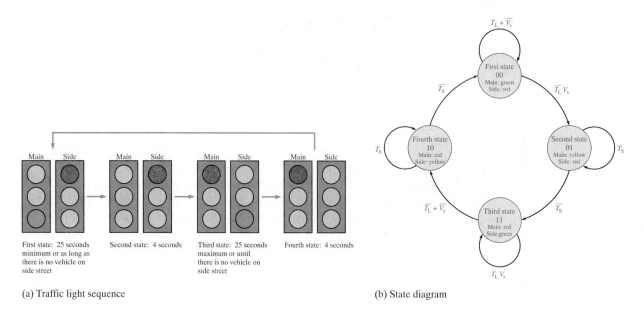

(a) Traffic light sequence

(b) State diagram

FIGURE 7–39
Traffic light sequence and state diagram.

MG	Main street green light output
SR	Side street red light output
SY	Side street yellow light output
SG	Side street green light output
Long	Trigger output to long timer
Short	Trigger output to short timer

■ DIGITAL WORKBENCH 1: GAL22V10 Programming

■ *Activity 1* Using the truth table and the pin assignment diagram for a GAL22V10 shown here, write an ABEL input file to program the device with the state decoding and output logic.

State Inputs		Light Outputs						Trigger Outputs	
S_1	S_0	$\overline{MR}$	$\overline{MY}$	$\overline{MG}$	$\overline{SR}$	$\overline{SY}$	$\overline{SG}$	*Long*	*Short*
0	0	1	1	0	0	1	1	1	0
0	1	1	0	1	0	1	1	0	1
1	1	0	1	1	1	1	0	1	0
1	0	0	1	1	1	0	1	0	1

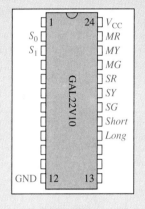

■ DIGITAL WORKBENCH 2: Verification and Testing

The single GAL22V10 package replaces the 74LS139 and the 74LS08 that were used to implement the same logic in the Chapter 6 system application. However, the GAL can be reprogrammed for additional logic functions required in this system, as you will see in Chapter 11.

■ *Activity 1* Identify the function of each of the points labeled with a circled number on the portion of the system PC board shown.

■ *Activity 2* Using the switches connected as shown to simulate the Gray code sequence on the inputs, develop a test procedure to check out this portion of the system board.

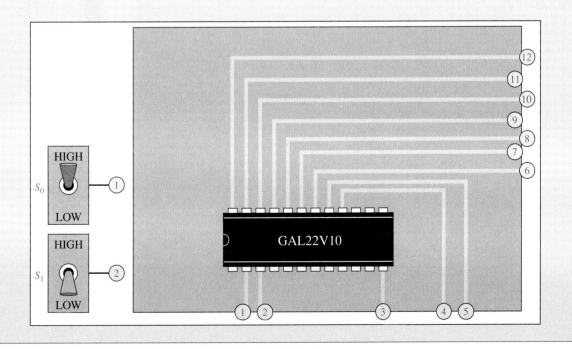

SECTION 7–8 REVIEW

1. After programming the GAL22V10 in this section, how many outputs are still unused and available?
2. How many inputs are unused and available?
3. Compare the GAL implementation to the implementation in Chapter 6 in terms of complexity?

■ SUMMARY

- A PLD is an integrated circuit programmable logic device into which a logic design can be programmed.
- All PLDs have programmable arrays.
- Four types of PLDs are Programmable Read-only Memory (PROM), Programable Logic Array (PLA), Programmable Array Logic (PAL), and Generic Array Logic (GAL).
- PROMs, PLAs, and PALs are one-time programmable devices.
- GALs are reprogrammable devices that can be programmed many times.
- The array of programmable cells in PROMs, PLAs, and PALs consists of fusible links.
- The array of reprogrammable cells in a GAL consists of E^2CMOS circuits.
- A PAL has fixed output logic.
- A GAL has programmable output logic circuits called OLMCs.
- PLD programming requires a computer, programming software, and a programmer.
- ABEL is one PLD programming language for describing a logic circuit design on a computer. CUPL is another.
- A logic design can be entered as equations, truth tables, or state diagrams into a computer which is running ABEL, CUPL, or other hardware description language (HDL) software.
- Test vectors are part of the software that provides for testing a logic design before it is committed to hardware.
- A JEDEC file is the final computer output that goes to the programmer to implement a design in a PLD. A JEDEC file is also called a *fuse map* or *cell map*.

■ SELF-TEST

1. The basic types of programmable arrays are
   (a) NOT and XOR   (b) NAND and NOR
   (c) AND and OR   (d) diode and transistor
2. An array can best be described as
   (a) a group of programmable logic gates
   (b) a matrix of rows and columns with a programmable cell at each intersection
   (c) rows of programmable cells
   (d) columns of fusible links
3. PLDs can be classified as
   (a) combinational and registered
   (b) programmable and nonprogrammable
   (c) one-time programmable and reprogrammable
   (d) PROMs, PLAs, PALs, and GALs
   (e) answers (a) and (c)
   (f) answers (c) and (d)
4. A PLD that has two programmable arrays is
   (a) PAL   (b) PROM   (c) PLA   (d) GAL
5. A PLD that has a programmable AND array and a fixed OR array is
   (a) PROM   (b) PLA   (c) PAL   (d) GAL
6. A PLD that has a reprogrammable AND array is
   (a) PROM   (b) PLA   (c) PAL   (d) GAL
7. The two types of lines in a PLD array are
   (a) vertical and horizontal   (b) input and product term
   (c) fusible and nonfusible   (d) buffered and tristate

8. A connection between a row and column in a PAL array is made by
   (a) blowing a fusible link
   (b) leaving a fusible link intact
   (c) connecting an input variable to the input line
   (d) connecting an input variable to the product term line

9. The device number PAL14H4 indicates
   (a) a PAL with fourteen active-HIGH outputs and four inputs
   (b) a PAL that implements fourteen AND gates and four OR gates
   (c) a PAL with fourteen inputs and four active-HIGH outputs
   (d) who the manufacturer is

10. A GAL is different from a PAL because
   (a) a GAL has more inputs and outputs
   (b) a GAL is implemented with a different technology
   (c) a GAL can replace several different PALs
   (d) a GAL can be reprogrammed and a PAL cannot
   (e) all of the above answers
   (f) all except answer (a)
   (g) all except answer (c)

11. The reprogrammable cells in a GAL array are
   (a) TTL   **(b)** E^2CMOS **(c)** ECL   **(d)** bipolar fuses

12. OLMC is an acronym for
   (a) Output Logic Main Cell
   (b) Optimum Logic Multiple Channel
   (c) Output Logic Macrocell
   (d) Odd-parity Logic Master Check

13. Two ways in which a GAL output can be configured are
   (a) combinational and I/O
   (b) simple and complex
   (c) fixed and variable
   (d) combinational and registered

14. The device number GAL22V10 means that
   (a) the device has 22 dedicated inputs and 10 dedicated outputs
   (b) the device has 22 inputs including dedicated inputs and I/Os and 10 outputs either dedicated or I/Os
   (c) the device has a variable number of inputs from a maximum of 22 to a minimum of 10

15. To program a PLD, you need only
   (a) a special fixture
   (b) a special fixture and a master PLD that has been preprogrammed at the factory
   (c) a computer and a programmer
   (d) a computer, a programmer, and HDL software
   (e) a computer, a programmer, and BASIC software

16. ISP stands for
   (a) In-System Programmable
   (b) Integrated System Program
   (c) Integrated Silicon Programmer

17. ABEL and CUPL are types of
   (a) programmers   **(b)** PLDs
   (c) HDL software   **(d)** disk operating systems

18. The symbols for NOT, AND, OR, and XOR in ABEL are
   (a) !, &, +, $ **(b)** !, @, #, $ **(c)** !, &, #, $ **(d)** *, &, +, !

19. The sum-of-products expression, $X = \overline{A}BC + A\overline{B}C + AB\overline{C}$ is written in ABEL as

    (a) X = !ABC # A!BC # AB!C;

    (b) X == !A & B & C # A & !B & C # A & B & !C;

    (c) X = !A & B & C # A & !B & C # A & B & !C;

    (d) X = (!A & B & C) + (A & !B & C) + (A & B & !C);

20. A logic circuit design to be implemented in a PLD is first described in

    (a) a declaration   **(b)** an input file   **(c)** a JEDEC file   **(d)** a documentation file

---

■ **PROBLEMS**

SECTION 7–1   PLD Arrays and Classifications

1. Name four types of PLDs and describe each.

2. In the simple programmed AND array with fusible links in Figure 7–40, determine the Boolean output expressions.

FIGURE 7–40

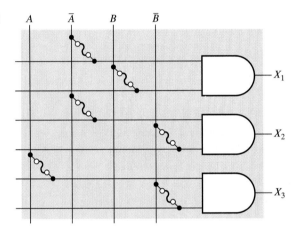

3. Determine by row and column number which fusible links must be blown in the programmable AND array of Figure 7–41 to implement each of the following product terms: $X_1 = \overline{A}BC$, $X_2 = AB\overline{C}$, $X_3 = \overline{A}B\overline{C}$.

FIGURE 7–41

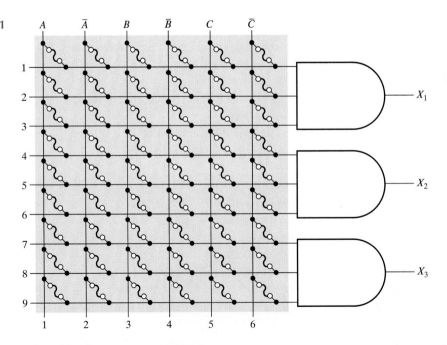

SECTION 7–2 Programmable Array Logic (PAL)

4. Determine the Boolean output expression for the simple PAL shown in Figure 7–42, where the Xs represent intact fusible links.

5. Show how the PAL in Figure 7–42 should be programmed in order to implement each of the following SOP expressions. Use an X to indicate an intact fuse. Simplify the expressions, if necessary, to fit the PAL shown.

   (a) $Y = A\overline{B}C + \overline{A}B\overline{C} + ABC$

   (b) $Y = A\overline{B}C + \overline{A}\,\overline{B}C + A\overline{B}\,\overline{C} + \overline{A}BC$

FIGURE 7–42

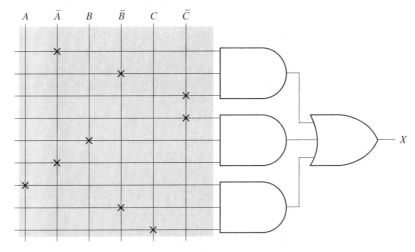

6. Interpret each of the PAL device numbers.

   (a) PAL16L2   (b) PAL12H6   (c) PAL10P8   (d) PAL16R6

7. Explain how a programmed polarity output in a PAL works.

SECTION 7–3 Generic Array Logic (GAL)

8. Determine the output expression for the GAL array shown in Figure 7–43. The Xs represent an *on* cell.

FIGURE 7–43

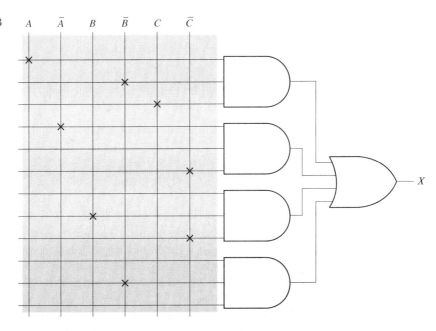

9. Show how the GAL in Figure 7–43 should be reprogrammed in order to implement each of the following expressions. Use Xs to indicate *on* cells. Modify the expressions, if necessary.

(a) $X = A\overline{B}C + \overline{A}B\overline{C} + A\overline{B} + BC$ (b) $X = (A + \overline{B} + \overline{C})(\overline{A} + B)$

SECTION 7–4   The GAL22V10

10. Describe a GAL22V10 in terms of

    (a) the number of dedicated inputs

    (b) the number of I/Os

    (c) the number of dedicated inputs plus the number of I/Os

11. If fifteen inputs are required for a particular GAL22V10 implementation, what is the maximum number of available outputs?

12. What is the largest single SOP expression that can be implemented with a GAL22V10 in terms of the number of variables and the number of product terms? Assume that only one output is used.

13. State the purpose of the select inputs to each OLMC in a GAL22V10.

14. Determine the SOP output of an OLMC in a GAL22V10 for the product terms and the select bits shown in Figure 7–44.

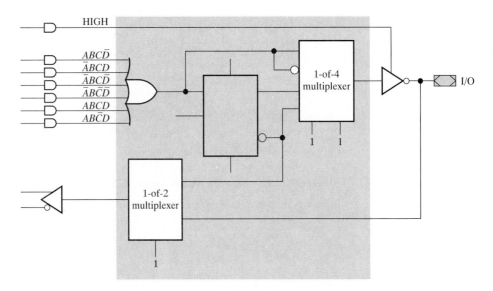

FIGURE 7–44

SECTION 7–5   The GAL16V8

15. Describe a GAL16V8 in terms of

    (a) the number of dedicated inputs

    (b) the number of I/Os

    (c) the number of dedicated inputs plus the number of I/Os

16. If there are ten inputs to a GAL16V8, what is the maximum number of available outputs?

17. What is the largest single SOP expression that can be implemented with a GAL16V8 in terms of the number of variables and the number of product terms? Assume that only one output is used.

18. PAL emulation with a GAL16V8 can be achieved by programming the device in any one of three modes. What are these modes?

SECTION 7–6 PLD Programming

19. List the items you must have before programming a PLD.

20. List the sequence of events that must occur to program a PLD.

21. How does an ispGAL22V10 differ from a GAL22V10?

SECTION 7–7 PLD Software

22. Express each of the following Boolean operations using ABEL symbols:
  (a) NOT **(b)** AND **(c)** OR **(d)** XOR

23. Convert each of the following Boolean expressions to ABEL notation:
  (a) $\overline{A}\,\overline{B}\,\overline{C}$ **(b)** $A + B + \overline{C}$ **(c)** $A(\overline{B + C})$

24. Determine the standard Boolean expression for each of the ABEL equations.
  (a) W = X & Y & Z # !X & !Y & !Z;
  (b) X = !(A & B) # !(!A & B & !C);
  (c) !Y = !(A # A & !B # !A & B);

25. Write each of the logic equations using ABEL:
  (a) $X = A\overline{B}C\overline{D}E + A\overline{B}C + \overline{A}\,\overline{B}$
  (b) $Y = AB\overline{C} + DEF + \overline{GH\overline{I}} + \overline{JK}L$
  (c) $Z = (\overline{A} + \overline{B} + C + D)(E + \overline{F}) + G$

26. Write a set declaration for the following ABEL equations and then express the set of equations with a single equation using the net notation.

```
X1 = A1 & P1 & Q1 # B1 & !P1 & Q1 # C1 & P1 & !Q1;
X2 = A2 & P1 & Q1 # B2 & !P1 & Q1 # C2 & P1 & !Q1;
X3 = A3 & P1 & Q1 # B3 & !P1 & Q1 # C3 & P1 & !Q1;
```

27. Write an ABEL truth table for an exclusive-NOR gate.

28. Write an ABEL truth table for a BCD-to-decimal decoder with active-HIGH outputs.

29. Write the test vectors for the BCD-to-decimal decoder in Problem 28.

SECTION 7–8 Digital System Application

30. For the traffic light control system, the long time interval is to be increased from 25 s to 45 s and the short time interval from 4 s to 5 s. How does this change affect the truth table for the state decoding and output logic?

31. Can a GAL16V8 be used to implement the state decoding and output logic? Give the reason for your answer.

Special Design Problems

32. Write a complete ABEL input file for a BCD-to-decimal decoder with active-HIGH outputs to be implemented with a GAL22V10.

33. Write an ABEL input file to implement a decimal-to-BCD priority encoder with active-HIGH outputs using a GAL16V8. Assume that only one input at a time will be active.

34. Use a GAL22V10 to replace the three 74185s used for the 8-bit binary-to-BCD converter in Chapter 6 (Figure 6–53, page 296). Develop the input file. Discuss any potential problems in this implementation.

■ ANSWERS TO SECTION REVIEWS

SECTION 7–1

1. Four types of PLDs are PROM, PLA, PAL, and GAL.

2. In a PLA, both arrays are programmable. In a PAL, only the AND array is programmable.

3. A PAL is one-time programmable; a GAL is reprogrammable and as programmable output sections.

SECTION 7–2

1. A PAL is a PLD with a programmable AND array and a fixed OR array.
2. The fusible link is the programmable element in an array.
3. Any combinational logic can be implemented with a PAL within the limitations of available inputs and outputs.
4. Three types of PAL output logic are combinational output, combinational I/O, and programmable polarity output.
5. A PAL12H6 is a PAL that has 12 inputs, including I/Os, and 6 active-HIGH outputs.

SECTION 7–3

1. A GAL is a PLD with a reprogrammable AND array, a fixed OR array, and programmable output logic.
2. An E^2CMOS cell can be programmed repeatedly to connect or disconnect an input line and a product term line.
3. Any combinational logic can be implemented with a GAL within the limitations of available inputs and outputs.
4. OLMC (output logic marocell) is the programmable output logic in a GAL, it can be programmed for several output configurations.

SECTION 7–4

1. The GAL22V10 has a combinational mode and a registered mode.
2. The mode is determined by the programmable select bits.
3. The polarity is set by the programmable select bits.
4. A maximum of sixteen product terms can be implemented in an SOP expression with a GAL22V10 OLMC.

SECTION 7–5

1. A GAL22V10 has more inputs and outputs than a GAL16V8, but the GAL16V8 can be readily programmed to emulate many PALs.
2. PAL emulation is done by programming of the OLMCs.
3. GAL16V8 modes are simple, complex, and registered.

SECTION 7–6

1. Programming of a PLD requires a computer, a programmer, and PLD software.
2. The software for PLDs are compilers.
3. A JEDEC (Joint Electronic Device Engineering Council) file is the final output from the PLD software that actually programs the device.
4. ISP means in-system programmable.
5. An ispGAL can be programmed after being connected on a printed circuit board; a standard GAL must be programmed in a programmer.

SECTION 7–7

1. ABEL (Advanced Boolean Expression Language) is a PLD hardware description language.
2. Design entry formats are equations, truth tables, and state diagrams.
3. NOT: !, AND: &, OR: #, XOR: $
4. Y = !D & E & !F # !D & !E # F
5. Test vectors list input and output values for testing a logic design.
6. Three parts of an ABEL input file are declarations, logic description, and test vectors.
7. An input file usually contains a device declaration, pin declarations, and set declarations.
8. A documentation file is the output file containing reduced equations, the JEDEC file, and a pin diagram.
9. A JEDEC file is the file that goes to the programmer to actually program the PLD. It is also called a fuse map or cell map.

SECTION 7–8

1. Two outputs are unused after programming the GAL22V10.
2. Ten inputs are unused and available.
3. The GAL replaces two IC packages.

8

FLIP-FLOPS AND RELATED DEVICES

■ CHAPTER OBJECTIVES

☐ Use logic gates to construct basic latches
☐ Explain the difference between an S-R latch and a D latch
☐ Recognize the difference between a latch and a flip-flop
☐ Explain how S-R, D, and J-K flip-flops differ
☐ Explain how edge-triggered and master-slave flip-flops differ
☐ Understand the significance of propagation delays, set-up time, hold time, maximum operating frequency, minimum clock pulse widths, and power dissipation in the application of flip-flops
☐ Apply flip-flops in basic applications
☐ Analyze circuits for race conditions and the occurrence of glitches
☐ Explain how retriggerable and nonretriggerable one-shots differ
☐ Connect a 555 timer to operate as either an astable multivibrator or a one-shot
☐ Approach the debugging of a new design
☐ Troubleshoot basic flip-flop and one-shot circuits
☐ Apply one-shots in a system application

■ CHAPTER OVERVIEW

In this chapter, we cover bistable, monostable, and astable logic devices called *multivibrators.* Two categories of bistable devices are the latch and the flip-flop. Bistable devices have two stable states, called SET and RESET; they can retain either of these states indefinitely, making them useful as storage devices. The basic difference between latches and flip-flops is the way in which they are changed from one state to the other. The flip-flop is a basic building block for counters, registers, and other sequential control logic. The monostable multivibrator, commonly known as the one-shot, has only one stable state. A one-shot produces a single controlled-width pulse when activated or triggered. The astable multivibrator has no stable state and is used primarily as an oscillator, which is a self-sustained waveform generator. Pulse oscillators are used as the sources for timing waveforms in digital systems.

■ SPECIFIC DEVICES

74LS74A	74LS279	74LS139
74122	555	74121
74LS75	74LS76A	74LS00

■ DIGITAL SYSTEM APPLICATION

This Digital System Application illustrates the concepts taught in this chapter. The system application in Section 8–9 continues with the traffic light control system from Chapter 6. The focus of this chapter is the timing circuit portion of the system which produces the clock, the long time interval for the red and green lights, and the short time interval for the caution light. The clock is used as the basic system timing signal for advancing the sequential logic through its states. The sequential logic will be developed in Chapter 9.

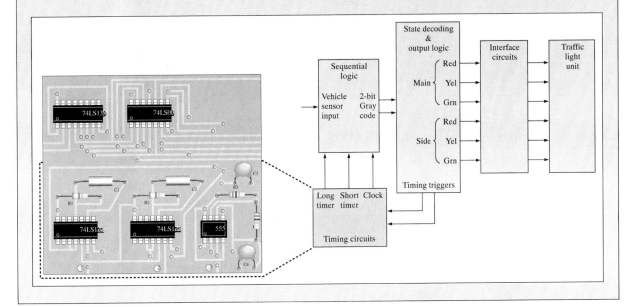

8–1 ▪ LATCHES

The latch is a type of bistable storage device that is normally placed in a category sepa-rate from that of flip-flops. Latches are basically similar to flip-flops because they are bistable devices that can reside in either of two states by virtue of a feedback arrange-ment, in which the outputs are connected back to the opposite inputs. The main differ-ence between latches and flip-flops is in the method used for changing their state. After completing this section, you should be able to

☐ Explain the operation of a basic S-R latch ☐ Explain the operation of a gated S-R latch ☐ Explain the operation of a gated D latch ☐ Implement an S-R or D latch with logic gates ☐ Describe the 74LS279 and 74LS75 quad latches

The S-R Latch

A **latch** is a type of **bistable** multivibrator. An active-HIGH input S-R (SET-RESET) latch is formed with two cross-coupled NOR gates as shown in Figure 8–1(a); an active-LOW input $\overline{S}$-$\overline{R}$ latch is formed with two cross-coupled NAND gates as shown in Figure 8–1(b). Notice that the output of each gate is connected to an input of the opposite gate. This pro-duces the regenerative **feedback** that is characteristic of all **multivibrators.**

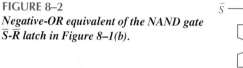

(a) Active-HIGH input S-R latch   (b) Active-LOW input $\overline{S}$-$\overline{R}$ latch

FIGURE 8–1
Two versions of SET-RESET (S-R) latches.

To understand the operation of the latch, we will use the NAND gate $\overline{S}$-$\overline{R}$ latch in Fig-ure 8–1(b). This latch is redrawn in Figure 8–2 with the negative-OR equivalents used for the NAND gates. This is done because LOWs on the $\overline{S}$ and $\overline{R}$ lines are the activating inputs.

The latch in Figure 8–2 has two inputs, $\overline{S}$ and $\overline{R}$, and two outputs, Q and $\overline{Q}$. Let's start by assuming that both inputs and the Q output are HIGH. Since the Q output is connected back to an input of gate G_2, and the $\overline{R}$ input is HIGH, the output of G_2 must be LOW. This LOW output is coupled back to an input of gate G_1, ensuring that its output is HIGH.

FIGURE 8–2
Negative-OR equivalent of the NAND gate $\overline{S}$-$\overline{R}$ latch in Figure 8–1(b).

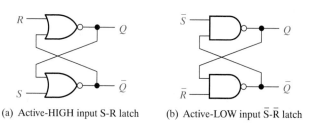

When the Q output is HIGH, the latch is in the **SET** state. It will remain in this state indefinitely until a LOW is temporarily applied to the $\overline{R}$ input. With a LOW on the $\overline{R}$ input and a HIGH on $\overline{S}$, the output of gate G_2 is forced HIGH. This HIGH on the $\overline{Q}$ output is cou-pled back to an input of G_1, and since the $\overline{S}$ input is HIGH, the output of G_1 goes LOW. This LOW on the Q output is then coupled back to an input of G_2, ensuring that the $\overline{Q}$ out-put remains HIGH even when the LOW on the $\overline{R}$ input is removed. When the Q output is

LOW, the latch is in the **RESET** state. Now the latch remains indefinitely in the RESET state until a LOW is applied to the $\bar{S}$ input.

The outputs of a latch are always complements of each other: When Q is HIGH, $\bar{Q}$ is LOW, and when Q is LOW, $\bar{Q}$ is HIGH.

An invalid condition in the operation of an active-LOW input $\overline{S}$-$\overline{R}$ latch occurs when LOWs are applied to both $\bar{S}$ and $\bar{R}$ at the same time. As long as the LOW levels are simultaneously held on the inputs, both the Q and $\bar{Q}$ outputs are forced HIGH, thus violating the basic complementary operation of the outputs. Also, if the LOWs are released simultaneously, both outputs will attempt to go LOW. Since there is always some small difference in the propagation delay time of the gates, one of the gates will dominate in its transition to the LOW output state. This, in turn, forces the output of the slower gate to remain HIGH. In this situation, you cannot reliably predict the next state of the latch.

Figure 8–3 illustrates the active-LOW input $\overline{S}$-$\overline{R}$ latch operation for each of the four possible combinations of levels on the inputs. (The first three combinations are valid, but the last is not.) Table 8–1 summarizes the logical operation in truth table form. Operation of the active-HIGH input NOR gate latch in Figure 8–1(a) is similar but requires the use of opposite logic levels.

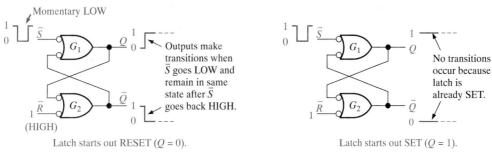

(a) Two possibilities for the SET operation

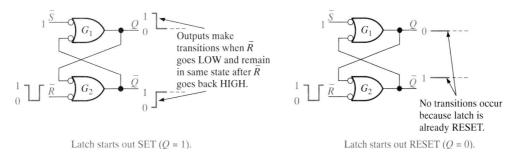

(b) Two possibilities for the RESET operation

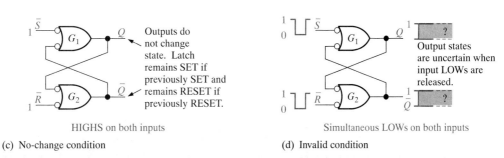

(c) No-change condition   (d) Invalid condition

FIGURE 8–3

The three modes of basic S-R latch operation (SET, RESET, No-Change) and the invalid condition.

TABLE 8–1

Truth table for an active-LOW input $\overline{S}$-$\overline{R}$ latch

Inputs		Outputs		
$\overline{S}$	$\overline{R}$	Q	$\overline{Q}$	Comments
1	1	NC	NC	No change. Latch remains in present state.
0	1	1	0	Latch SET.
1	0	0	1	Latch RESET.
0	0	1	1	Invalid condition

FIGURE 8–4

Logic symbols for the S-R latch.

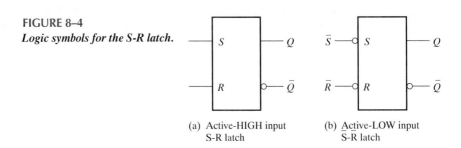

(a) Active-HIGH input
   S-R latch

(b) Active-LOW input
   S-R latch

   Logic symbols for both the active-HIGH input and the active-LOW input latches are shown in Figure 8–4.

   Example 8–1 illustrates how an active-LOW input $\overline{S}$-$\overline{R}$ latch responds to conditions on its inputs. LOW levels are pulsed on each input in a certain sequence and the resulting Q output waveform is observed. The $\overline{S} = 0, \overline{R} = 0$ condition is avoided because it results in an invalid mode of operation and is a major drawback of any SET-RESET type of latch.

EXAMPLE 8–1

If the $\overline{S}$ and $\overline{R}$ waveforms in Figure 8–5(a) are applied to the inputs of the latch in Figure 8–4(b), determine the waveform that will be observed on the Q output. Assume that Q is initially LOW.

FIGURE 8–5

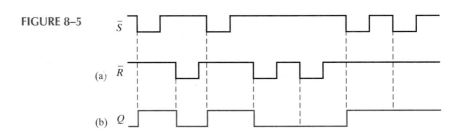

(a)

(b)

Solution See Figure 8–5(b).

Related Exercise Determine the Q output of an active-HIGH input S-R latch if the waveforms in Figure 8–5(a) are inverted and applied to the inputs.

The Latch as a Contact-Bounce Eliminator

A good example of an application of an $\overline{S}$-$\overline{R}$ latch is in the elimination of mechanical switch contact "bounce." When the pole of a switch strikes the contact upon switch closure, it physically vibrates or bounces several times before finally making a solid contact. Although these bounces are minute, they produce voltage spikes that are often not acceptable in a digital system. This situation is illustrated in Figure 8–6(a).

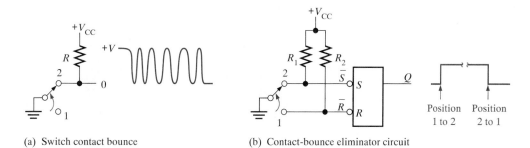

(a) Switch contact bounce   (b) Contact-bounce eliminator circuit

FIGURE 8–6
The $\overline{S}$-$\overline{R}$ latch used to eliminate switch contact bounce.

An $\overline{S}$-$\overline{R}$ latch can be used to eliminate the effects of switch bounce as shown in Figure 8–6(b). The switch is normally in position 1, keeping the $\overline{R}$ input LOW and the latch RESET. When the switch is thrown to position 2, $\overline{R}$ goes HIGH because of the pull-up resistor to V_{CC}, and $\overline{S}$ goes LOW on the first contact. Although $\overline{S}$ remains LOW for only a very short time before the switch bounces, this is sufficient to SET the latch. Any further voltage spikes on the $\overline{S}$ input due to switch bounce do not affect the latch, and it remains SET. Notice that the Q output of the latch provides a clean transition from LOW to HIGH, thus eliminating the voltage spikes caused by contact bounce. Similarly, a clean transition from HIGH to LOW is made when the switch is thrown back to position 1.

The 74LS279 Quad $\overline{S}$-$\overline{R}$ Latch

The 74LS279 is an example of a quad $\overline{S}$-$\overline{R}$ latch represented by the diagram of Figure 8–7. Notice that two of the latches each have two $\overline{S}$ inputs.

FIGURE 8–7

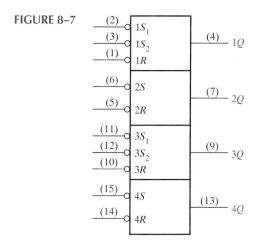

The Gated S-R Latch

A gated latch requires an enable input, *EN*. The logic diagram and logic symbol for a gated S-R latch are shown in Figure 8–8. The *S* and *R* inputs control the state to which the latch will go when a HIGH level is applied to the *EN* input. The latch will not change until *EN* is HIGH, but as long as it remains HIGH, the output is controlled by the state of the *S* and *R* inputs. In this circuit the invalid state occurs when both *S* and *R* are simultaneously HIGH.

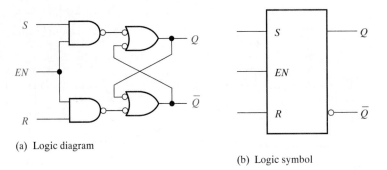

(a) Logic diagram

(b) Logic symbol

FIGURE 8–8
A gated S-R latch.

EXAMPLE 8–2

Determine the Q output waveform if the inputs shown in Figure 8–9(a) are applied to a gated S-R latch that is initially RESET.

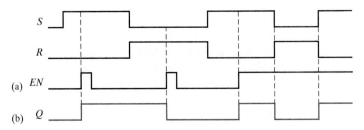

(a)

(b)

FIGURE 8–9

Solution   The Q waveform is shown in Figure 8–9(b). Anytime S is HIGH and R is LOW, a HIGH on the *EN* input SETS the latch. Anytime S is LOW and R is HIGH, a HIGH on the *EN* input RESETS the latch.

Related Exercise   Determine the Q output of a gated S-R latch if the S and R inputs in Figure 8–9(a) are inverted.

The Gated D Latch

Another type of gated latch is called the D latch. It differs from the S-R latch because it has only one input in addition to *EN*. This input is called the D (data) input. Figure 8–10 contains a logic diagram and logic symbol of a D latch. When the D input is HIGH and the *EN*

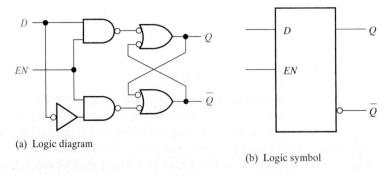

(a) Logic diagram

(b) Logic symbol

FIGURE 8–10
A gated D latch.

input is HIGH, the latch will SET. When the D input is LOW, and EN is HIGH, the latch will RESET. Stated another way, the output Q follows the input D when EN is HIGH.

EXAMPLE 8–3

Determine the Q output waveform if the inputs shown in Figure 8–11(a) are applied to a gated D latch, which is initially RESET.

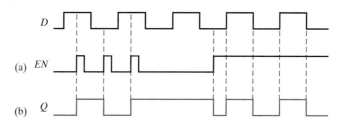

FIGURE 8–11

Solution   The Q waveform is shown in Figure 8–11(b). Whenever D is HIGH and EN is HIGH, Q goes HIGH. Whenever D is LOW and EN is HIGH, Q goes LOW. When EN is LOW, the state of the latch is not affected by the D input.

Related Exercise   Determine the Q output of the gated D latch if the D input in Figure 8–11(a) is inverted.

The 74LS75 Quad Latches

An example of an IC gated D latch is the 74LS75 represented by the logic symbol in Figure 8–12(a). This device has four latches. Notice that each active-HIGH EN input is shared by two latches and is designated as a control input (C). The truth table for each latch is shown in Figure 8–12(b). The X in the truth table represents a "don't care" condition. In this case, when the EN input is LOW, it does not matter what the D input is because the outputs are unaffected and remain in their prior states.

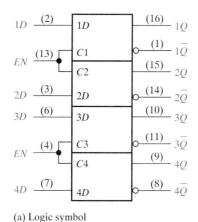

(a) Logic symbol

Inputs		Outputs		
D	EN	Q	$\bar{Q}$	Comments
0	1	0	1	RESET
1	1	1	0	SET
X	0	Q_0	$\bar{Q}_0$	No change

NOTE: Q_0 is the prior output level before the indicated input conditions were established.

(b) Truth table (each latch)

FIGURE 8–12
The 74LS75 quad gated D latches.

SECTION 8–1 REVIEW

1. List three types of latches.
2. Develop the truth table for the active-HIGH input S-R latch in Figure 8–1(a).
3. What is the Q output of a D latch when $EN = 1$ and $D = 1$?

8–2 ■ EDGE-TRIGGERED FLIP-FLOPS

Flip-flops are synchronous bistable devices. In this case the term synchronous means that the output changes state only at a specified point on a triggering input called the clock (CLK) which is designated as a control input, C; that is, changes in the output occur in synchronization with the clock. After completing this section, you should be able to

☐ Define *clock* ☐ Define *edge-triggered flip-flop* ☐ Explain the difference between a flip-flop and a latch ☐ Identify an edge-triggered flip-flop by its logic symbol ☐ Discuss the difference between a positive and a negative edge-triggered flip-flop ☐ Discuss and compare the operation of S-R, D, and J-K edge-triggered flip-flops and explain the differences in their truth tables ☐ Discuss the asynchronous inputs of a flip-flop ☐ Describe the 74LS74A and the 74LS76A flip-flops

A **flip-flop** is a synchronous bistable device. An **edge-triggered flip-flop** changes state either at the positive edge (rising edge) or at the negative edge (falling edge) of the clock pulse and is sensitive to its inputs only at this transition of the clock. Three types of edge-triggered flip-flops are covered in this section: S-R, D, and J-K. The logic symbols for all of these are shown in Figure 8–13. Notice that each type can be either positive edge-triggered (no bubble at *C* input) or negative edge-triggered (bubble at *C* input). The key to identifying an edge-triggered flip-flop by its logic symbol is the small triangle inside the block at the clock (*C*) input. This triangle is called the *dynamic input indicator*.

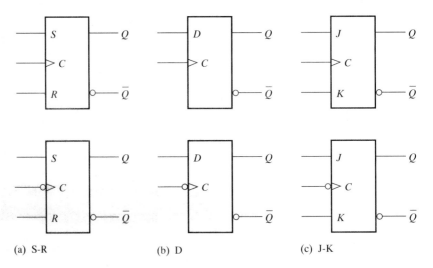

(a) S-R (b) D (c) J-K

FIGURE 8–13
Edge-triggered flip-flop logic symbols (top—positive edge-triggered; bottom—negative edge-triggered).

The Edge-Triggered S-R Flip-Flop

The *S* and *R* inputs of the **S-R flip-flop** are called **synchronous** inputs because data on these inputs are transferred to the flip-flop's output only on the triggering edge of the clock pulse. When *S* is HIGH and *R* is LOW, the *Q* output goes HIGH on the triggering edge of the clock pulse, and the flip-flop is SET. When *S* is LOW and *R* is HIGH, the *Q* output goes LOW on the triggering edge of the clock pulse, and the flip-flop is RESET. When both *S* and *R* are LOW, the output does not change from its prior state. An invalid condition exists when both *S* and *R* are HIGH.

This basic operation of a positive edge-triggered flip-flop is illustrated in Figure 8–14, and Table 8–2 is the truth table for this type of flip-flop. Remember, *the flip-flop cannot change state except on the triggering edge of a clock pulse.* The S and R inputs can be changed at any time when the clock input is LOW or HIGH (except for a very short interval around the triggering transition of the clock) without affecting the output.

The operation and truth table for a negative edge-triggered S-R flip-flop are the same as those for a positive edge-triggered device except that the falling edge of the clock pulse is the triggering edge.

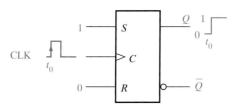

(a) $S = 1, R = 0$ flip-flop SETS on positive clock edge. (If already SET, it remains SET.)

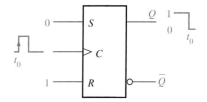

(b) $S = 0, R = 1$ flip-flop RESETS on positive clock edge. (If already RESET, it remains RESET.)

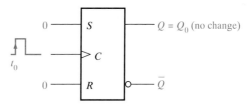

(c) $S = 0, R = 0$ flip-flop does not change. (If SET, it remains SET; if RESET, it remains RESET.)

FIGURE 8–14
Operation of a positive edge-triggered S-R flip-flop.

TABLE 8–2
Truth table for a positive edge-triggered S-R flip-flop

Inputs			Outputs		
S	R	CLK	Q	$\bar{Q}$	Comments
0	0	X	Q_0	$\bar{Q}_0$	No change
0	1	↑	0	1	RESET
1	0	↑	1	0	SET
1	1	↑	?	?	Invalid

↑ = clock transition LOW to HIGH
X = irrelevant ("don't care")
Q_0 = output level prior to clock transition

EXAMPLE 8–4

Determine the Q and $\bar{Q}$ output waveforms of the flip-flop in Figure 8–15 for the S, R, and CLK inputs in Figure 8–16(a). Assume that the positive edge-triggered flip-flop is initially RESET.

FIGURE 8–15

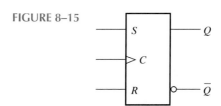

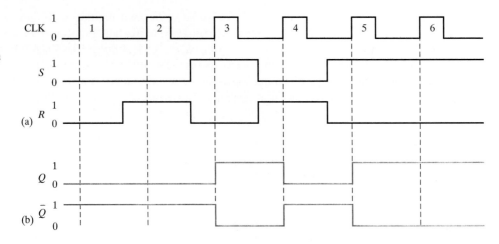

FIGURE 8–16

Solution

1. At clock pulse 1, S is LOW and R is LOW, so Q does not change.
2. At clock pulse 2, S is LOW and R is HIGH, so Q remains LOW (RESET).
3. At clock pulse 3, S is HIGH and R is LOW, so Q goes HIGH (SET).
4. At clock pulse 4, S is LOW and R is HIGH, so Q goes LOW (RESET).
5. At clock pulse 5, S is HIGH and R is LOW, so Q goes HIGH (SET).
6. At clock pulse 6, S is HIGH and R is LOW, so Q stays HIGH.

Once Q is determined, $\overline{Q}$ is easily found since it is simply the complement of Q. The resulting waveforms for Q and $\overline{Q}$ are shown in Figure 8–16(b) for the input waveforms in part (a).

Related Exercise Determine Q and $\overline{Q}$ for the S and R inputs in Figure 8–16(a) if the flip-flop is a negative edge-triggered device.

A Method of Edge-Triggering

A simplified implementation of an edge-triggered S-R flip-flop is illustrated in Figure 8–17(a) and is used to demonstrate the concept of edge-triggering. This coverage of the S-R flip-flop does not imply that it is the most important type. Actually, the D flip-flop and the J-K flip-flop are more widely used and more available in IC form than is the S-R type. However, understanding the S-R is important because both the D and the J-K flip-flops are derived from the S-R flip-flop. Notice that the S-R flip-flop differs from the gated S-R latch only in that it has a pulse transition detector. This circuit produces a very short-duration spike on the positive-going transition of the clock pulse.

One basic type of pulse transition detector is shown in Figure 8–17(b). As you can see, there is a small delay on one input to the NAND gate so that the inverted clock pulse arrives at the gate input a few nanoseconds after the true clock pulse. This produces an output spike with a duration of only a few nanoseconds. In a negative edge-triggered flip-flop the clock pulse is inverted first, thus producing a narrow spike on the negative-going edge.

Notice that the circuit in Figure 8–17 is partitioned into two sections, one labeled Steering gates, and the other Latch. The steering gates direct, or steer, the clock spike either to the input to gate G_3 or to the input to gate G_4, depending on the state of the S and R inputs. To understand the operation of this flip-flop, begin with the assumptions that it is in the RESET state ($Q = 0$) and that the S, R, and CLK inputs are all LOW. For this condition, the outputs of gate G_1 and gate G_2 are both HIGH. The LOW on the Q output is coupled back into one input of gate G_4, making the $\overline{Q}$ output HIGH. Because $\overline{Q}$ is HIGH, both inputs to gate G_3 are HIGH (remember, the output of gate G_1 is HIGH), holding the Q out-

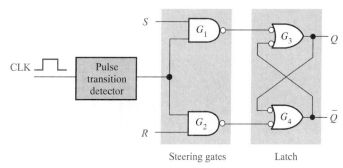

(a) A simplified logic diagram for a positive edge-triggered S-R flip-flop

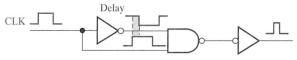

(b) A type of pulse transition detector

FIGURE 8–17
Edge triggering.

put LOW. If a pulse is applied to the CLK input, the outputs of gates G_1 and G_2 remain HIGH because they are disabled by the LOWs on the S input and the R input; therefore, there is no change in the state of the flip-flop—it remains RESET.

Let's now make S HIGH, leave R LOW and apply a clock pulse. Because the S input to gate G_1 is now HIGH, the output of gate G_1 goes LOW for a very short time (spike) when CLK goes HIGH, causing the Q output to go HIGH. Both inputs to gate G_4 are now HIGH (remember, gate G_2 output is HIGH because R is LOW), forcing the $\overline{Q}$ output LOW. This LOW on $\overline{Q}$ is coupled back into one input of gate G_3, ensuring that the Q output will remain HIGH. The flip-flop is now in the SET state. Figure 8–18 illustrates the logic level transitions that take place within the flip-flop for this condition.

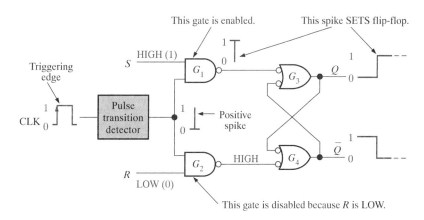

FIGURE 8–18
Flip-flop making a transition from the RESET state to the SET state on the positive-going edge of the clock pulse.

Next, let's make S LOW and R HIGH and apply a clock pulse. Because the R input is now HIGH, the positive-going edge of the clock produces a negative-going spike on the output of gate G_2, causing the $\overline{Q}$ output to go HIGH. Because of this HIGH on $\overline{Q}$, both inputs to gate G_3 are now HIGH (remember, the output of gate G_1 is HIGH because of the LOW on S), forcing the Q output to go LOW. This LOW on Q is coupled back into one input of gate G_4,

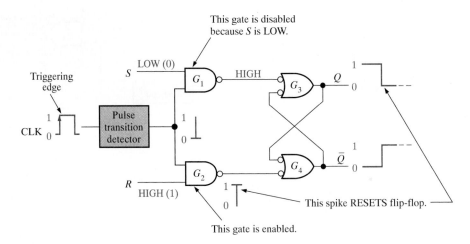

FIGURE 8–19
Flip-flop making a transition from the SET state to the RESET state on the positive-going edge of the clock pulse.

ensuring that $\overline{Q}$ will remain HIGH. The flip-flop is now in the RESET state. Figure 8–19 illustrates the logic level transitions that occur within the flip-flop for this condition. As with the gated latch, an invalid condition exists when both S and R are HIGH at the same time. This is the major drawback of the S-R flip-flop.

The Edge-Triggered D Flip-Flop

The **D flip-flop** is useful when a single data bit (1 or 0) is to be stored. The addition of an inverter to an S-R flip-flop creates a basic D flip-flop, as in Figure 8–20, where a positive edge-triggered type is shown.

Notice that the flip-flop in Figure 8–20 has only one input, the D input, in addition to the clock. If there is a HIGH on the D input when a clock pulse is applied, the flip-flop will SET, and the HIGH on the D input is stored by the flip-flop on the positive-going edge of the clock pulse. If there is a LOW on the D input when the clock pulse is applied, the flip-flop will RESET, and the LOW on the D input is stored by the flip-flop on the leading edge of the clock pulse. In the SET state the flip-flop is storing a 1, and in the RESET state it is storing a 0.

The operation of the positive edge-triggered D flip-flop is summarized in Table 8–3. The operation of a negative edge-triggered device is, of course, the same, except that triggering occurs on the falling edge of the clock pulse. *Remember, Q follows D at the clock edge.*

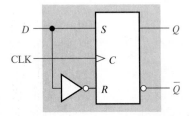

FIGURE 8–20
A positive edge-triggered D flip-flop formed with an S-R flip-flop and an inverter.

TABLE 8–3
Truth table for a positive edge-triggered D flip-flop

Inputs		Outputs		
D	CLK	Q	$\overline{Q}$	Comments
1	↑	1	0	SET (stores a 1)
0	↑	0	1	RESET (stores a 0)

↑ = clock transition LOW to HIGH

EXAMPLE 8–5   Given the waveforms in Figure 8–21(a) for the D input and the clock, determine the Q output waveform if the flip-flop starts out RESET.

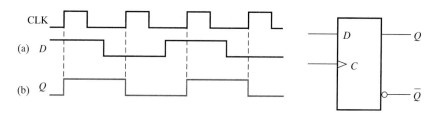

FIGURE 8–21

Solution   The Q output goes to the state of the D input at the time of the positive-going clock edge. The resultant output is shown in Figure 8–21(b).

Related Exercise   Determine the Q output for the D flip-flop if the D input in Figure 8–21(a) is inverted.

The Edge-Triggered J-K Flip-Flop

The **J-K flip-flop** is versatile and is perhaps the most widely used type of flip-flop. The J and K designations for the inputs have no known significance except that they are adjacent letters in the alphabet.

The functioning of the J-K flip-flop is identical to that of the S-R flip-flop in the SET, RESET, and no-change conditions of operation. *The difference is that the J-K flip-flop has no invalid state as does the S-R flip-flop.*

Figure 8–22 shows the basic internal logic for a positive edge-triggered J-K flip-flop. Notice that it differs from the S-R edge-triggered flip-flop in that the Q output is connected back to the input of gate G_2, and the $\overline{Q}$ output is connected back to the input of gate G_1. The two inputs are labeled J and K. A J-K flip-flop can also be of the negative edge-triggered type, in which case the clock input is inverted.

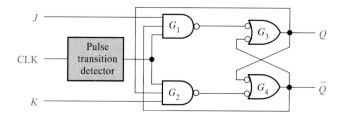

FIGURE 8–22
A simplified logic diagram for a positive edge-triggered J-K flip-flop.

Let's assume that the flip-flop in Figure 8–23 is RESET and that the J input is HIGH and the K input is LOW rather than as shown. When a clock pulse occurs, a leading-edge spike indicated by ① is passed through gate G_1 because $\overline{Q}$ is HIGH and J is HIGH. This will cause the latch portion of the flip-flop to change to the SET state.

The flip-flop is now SET. If we now make J LOW and K HIGH, the next clock spike indicated by ② will pass through gate G_2 because Q is HIGH and K is HIGH. This will cause the latch portion of the flip-flop to change to the RESET state.

Now if a LOW is applied to both the J and K inputs, the flip-flop will stay in its present state when a clock pulse occurs. So, a LOW on both J and K results in a *no-change* condition.

FIGURE 8–23
Transitions illustrating the toggle operation when J = 1 and K = 1.

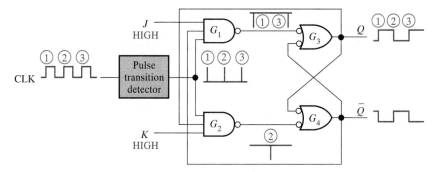

So far, the logical operation of the J-K flip-flop is the same as that of the S-R type in the SET, RESET, and no-change modes. The difference in operation occurs when both the J and K inputs are HIGH. To see this, assume that the flip-flop is RESET. The HIGH on the $\overline{Q}$ enables gate G_1, so the clock spike indicated by ③ passes through to SET the flip-flop. Now there is a HIGH on Q, which allows the next clock spike to pass through gate G_2 and RESET the flip-flop.

As you can see, on each successive clock spike, the flip-flop changes to the opposite state. This mode is called **toggle** operation. Figure 8–23 illustrates the transitions when the flip-flop is in the toggle mode.

Table 8–4 summarizes the operation of the edge-triggered J-K flip-flop in truth table form. *Notice that there is no invalid state as there is with an S-R flip-flop.* The truth table for a negative edge-triggered device is identical except that it is triggered on the falling edge of the clock pulse.

TABLE 8–4
Truth table for a positive edge-triggered J-K flip-flop

Inputs			Outputs		
J	K	**CLK**	Q	$\overline{Q}$	**Comments**
0	0	↑	Q_0	$\overline{Q}_0$	No change
0	1	↑	0	1	RESET
1	0	↑	1	0	SET
1	1	↑	$\overline{Q}_0$	Q_0	Toggle

↑ = clock transition LOW to HIGH
Q_0 = output level prior to clock transition

EXAMPLE 8–6

The waveforms in Figure 8–24(a) are applied to the J, K, and clock inputs as indicated. Determine the Q output, assuming that the flip-flop is initially RESET.

FIGURE 8–24

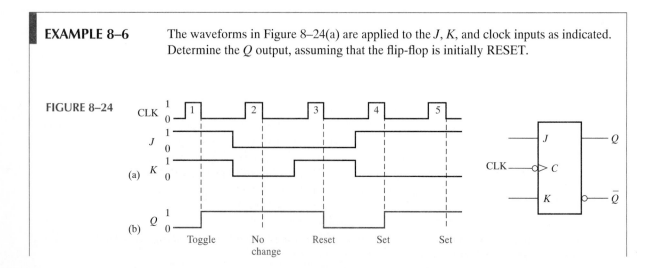

Solution

1. First, since this is a negative edge-triggered flip-flop, as indicated by the "bubble" at the clock input, the Q output will change only on the negative-going edge of the clock pulse.
2. At the first clock pulse, both J and K are HIGH; and because this is a toggle condition, Q goes HIGH.
3. At clock pulse 2, a no-change condition exists on the inputs, keeping Q at a HIGH level.
4. When clock pulse 3 occurs, J is LOW and K is HIGH, resulting in a RESET condition; Q goes LOW.
5. At clock pulse 4, J is HIGH and K is LOW, resulting in a SET condition; Q goes HIGH.
6. A SET condition still exists on J and K when clock pulse 5 occurs, so Q will remain HIGH.

The resulting Q waveform is indicated in Figure 8–24(b).

Related Exercise Determine the Q output of the J-K flip-flop if the J and K inputs in Figure 8–24(a) are inverted.

Asynchronous Inputs

For the flip-flops just discussed, the *S-R, D,* and *J-K* inputs are called *synchronous inputs* because data on these inputs are transferred to the flip-flop's output only on the triggering edge of the clock pulse; that is, the data are transferred synchronously with the clock.

Most integrated circuit flip-flops also have **asynchronous** inputs. These are inputs that affect the state of the flip-flop independent of the clock. They are normally labeled **preset** (*PRE*) and **clear** (*CLR*), or *direct set* (S_D) and *direct reset* (R_D) by some manufacturers. An active level on the preset input will SET the flip-flop, and an active level on the clear input will RESET it. A logic symbol for a J-K flip-flop with preset and clear inputs is shown in Figure 8–25. These inputs are active-LOW, as indicated by the bubbles. These preset and clear inputs must both be kept HIGH for synchronous operation.

Figure 8–26 shows the logic diagram for an edge-triggered J-K flip-flop with active-LOW preset ($\overline{PRE}$) and clear ($\overline{CLR}$) inputs. This figure illustrates basically how these inputs work. As you can see, they are connected so that they override the effect of the synchronous inputs, $J, K,$ and the clock.

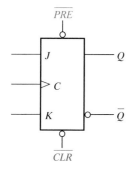

FIGURE 8–25

Logic symbol for a J-K flip-flop with active-LOW preset and clear inputs.

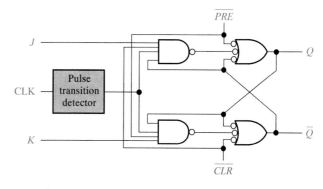

FIGURE 8–26

Logic diagram for a basic J-K flip-flop with active-LOW preset and clear.

EXAMPLE 8–7   For the positive edge-triggered J-K flip-flop with preset and clear inputs in Figure 8–27(a), determine the Q output for the inputs shown in the timing diagram if Q is initially LOW.

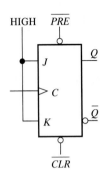

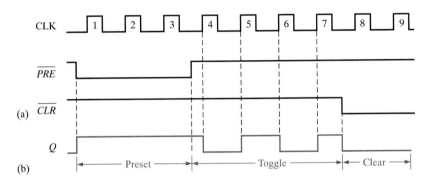

(a)

(b)

FIGURE 8–27

Solution

1. During clock pulses 1, 2, and 3, the preset ($\overline{PRE}$) is LOW, keeping the flip-flop SET regardless of the synchronous J and K inputs.
2. For clock pulses 4, 5, 6, and 7, toggle operation occurs because J is HIGH, K is HIGH, and both $\overline{PRE}$ and $\overline{CLR}$ are HIGH.
3. For clock pulses 8 and 9, the clear ($\overline{CLR}$) input is LOW, keeping the flip-flop RESET regardless of the synchronous inputs. The resulting Q output is shown in Figure 8–27(b).

Related Exercise   If you interchange the $\overline{PRE}$ and $\overline{CLR}$ waveforms in Figure 8–27(a), what will the Q output look like?

Specific Devices

We will now look at two specific IC edge-triggered flip-flops. They are representative of the various types of flip-flops available in IC form and, like most other devices, are available in TTL and in CMOS.

74LS74A Dual D Flip-Flops   This TTL device contains two identical flip-flops that are independent of each other except for sharing V_{CC} and ground. The flip-flops are positive edge-triggered and have active-LOW asynchronous preset and clear inputs. The logic sym-

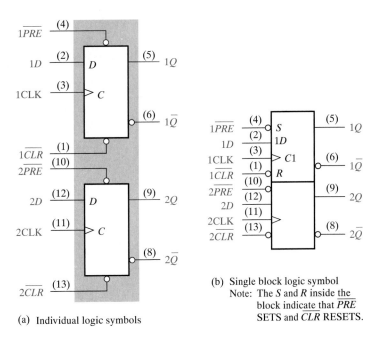

(a) Individual logic symbols

(b) Single block logic symbol
Note: The *S* and *R* inside the block indicate that $\overline{PRE}$ SETS and $\overline{CLR}$ RESETS.

FIGURE 8–28
Logic symbols for the 74LS74A dual positive edge-triggered D flip-flops.

bols for the individual flip-flops within the package are shown in Figure 8–28(a), and an ANSI/IEEE standard single block symbol representing the entire device is shown in part (b) of the figure. The pin numbers are shown in parentheses.

74LS76A Dual J-K Flip-Flops This TTL device also has two identical flip-flops that are negative edge-triggered and have active-LOW asynchronous preset and clear inputs. The logic symbols are shown in Figure 8–29.

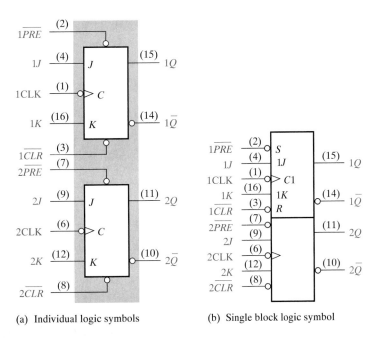

(a) Individual logic symbols

(b) Single block logic symbol

FIGURE 8–29
Logic symbols for the 74LS76A dual negative edge-triggered J-K flip-flops.

EXAMPLE 8–8

The $1J$, $1K$, 1CLK, $1\overline{PRE}$, and $1\overline{CLR}$ waveforms in Figure 8–30(a) are applied to one of the negative edge-triggered flip-flops in a 74LS76A package. Determine the $1Q$ output waveform.

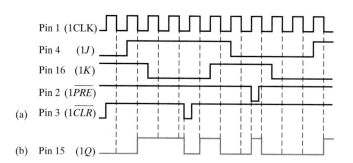

FIGURE 8–30

Solution The resulting $1Q$ waveform is shown in Figure 8–30(b). Notice that each time a LOW is applied to the $1\overline{PRE}$ or $1\overline{CLR}$, the flip-flop is SET or RESET regardless of the states of the other inputs.

Related Exercise Determine the $1Q$ output waveform if the waveforms for $1\overline{PRE}$ and $1\overline{CLR}$ are interchanged.

SECTION 8–2 REVIEW

1. Describe the main difference between a gated S-R latch and an edge-triggered S-R flip-flop.
2. How does a J-K flip-flop differ from an S-R flip-flop in its basic operation?
3. Assume that the flip-flop in Figure 8–21 is negative edge-triggered. Describe the output waveform for the same CLK and D waveforms.

8–3 ■ MASTER-SLAVE FLIP-FLOPS

Another class of flip-flop is the master-slave. Although this type of flip-flop has largely been replaced by the edge-triggered devices, a limited selection is still available from IC manufacturers and you may encounter this type of flip-flop in some existing equipment. Two basic types of master-slave flip-flops are the pulse-triggered and the data lock-out versions. In both types, data are entered into the flip-flop on the leading edge of the clock pulse, but the output does not reflect the input state until the trailing edge. The pulse-triggered version does not allow data to change while the clock pulse is active, whereas the data lock-out version does not have this restriction. After completing this section, you should be able to

☐ Identify a master-slave flip-flop by its logic symbol   ☐ Explain how the master-slave flip-flop differs from the edge-triggered devices   ☐ Discuss the basic operation of the pulse-triggered and the data lock-out types of master-slave flip-flop

The Pulse-Triggered Master-Slave J-K Flip-Flop

A basic **master-slave** J-K flip-flop is shown in Figure 8–31. The truth table operation is the same as that for the edge-triggered J-K flip-flop except for the way it is clocked. Internally, though, it is quite different.

This type of flip-flop is composed of two sections, the master section and the slave section. The master section is basically a gated latch, and the slave section is the same except that it is clocked on the inverted clock pulse and is controlled by the outputs of the master section rather than by the external $J\text{-}K$ inputs.

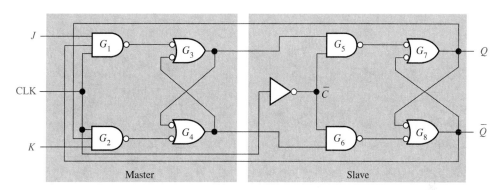

FIGURE 8–31
Logic diagram for a master-slave J-K flip-flop.

The master section will assume the state determined by the J and K inputs at the leading (positive-going) edge of the clock pulse. The state of the master section is then transferred to the slave section on the trailing (negative-going) edge of the clock pulse, because the outputs of the master are applied to the inputs of the slave and the clock pulse to the slave is inverted. The state of the slave then immediately appears on the Q and $\overline{Q}$ outputs. The Q output is connected back to an input of gate G_2 and the $\overline{Q}$ output is connected back to an input of gate G_1 to produce the characteristic toggle operation when $J = 1$ and $K = 1$. The logical operation is summarized in Table 8–5.

TABLE 8–5
Truth table for the master-slave J-K flip-flop

Inputs			Outputs		
J	K	CLK	Q	$\overline{Q}$	Comments
0	0	⊓	Q_0	$\overline{Q}_0$	No change
0	1	⊓	0	1	RESET
1	0	⊓	1	0	SET
1	1	⊓	$\overline{Q}_0$	Q_0	Toggle

⊓ = clock pulse
Q_0 = output level before clock pulse

The logic symbols for the J-K master-slave flip-flop are shown in Figure 8–32. The key to identifying a pulse-triggered (master-slave) flip-flop by its logic symbol is the ANSI/IEEE *postponed output* symbol (⌐) at the outputs. This symbol means that the output does not reflect the $J\text{-}K$ input data until the occurrence of the clock edge (either positive-going or negative-going) following the triggering edge. Notice that there is no dynamic input indicator (▷) at the clock (C) input as there is for an edge-triggered flip-flop.

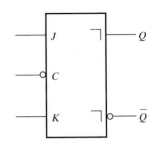

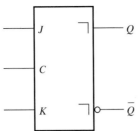

(a) Data are clocked in on positive-going edge of clock pulse and transferred to output on the following negative-going edge.

(b) Data are clocked in on negative-going edge of clock pulse and transferred to output on the following positive-going edge.

FIGURE 8–32
Pulse-triggered (master-slave) J-K flip-flop logic symbols.

EXAMPLE 8–9

Determine the Q output of the master-slave J-K flip-flop for the input waveforms shown in Figure 8–33(a). The flip-flop starts out RESET.

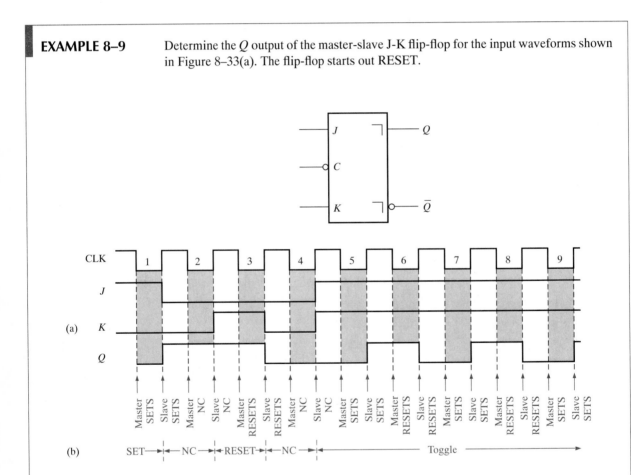

FIGURE 8–33

Solution The Q waveform is shown in Figure 8–33(b). The input states and events at the beginning and end of each clock pulse are labeled to demonstrate the operation. NC means no change.

Related Exercise What would the Q output look like if the J and K waveforms were inverted?

The Data Lock-Out Master-Slave J-K Flip-Flop

The data lock-out flip-flop is similar to the pulse-triggered master-slave flip-flop except that it has a *dynamic* clock input, which makes it sensitive to the data inputs only during a clock transition. In a data lock-out flip-flop, after the leading-edge clock transition, the data inputs are disabled (locked out) and do not have to be held constant while the clock pulse is HIGH. In essence, the master portion of this flip-flop is like an edge-triggered device, and the slave portion performs as in a pulse-triggered device to produce a postponed output on the trailing-edge transition of the clock.

A logic symbol for a data lock-out J-K flip-flop is shown in Figure 8–34. Notice that this symbol has both the dynamic input indicator for the clock and the postponed output indicators. This type of flip-flop is actually classified as a master-slave with a special lock-out feature.

FIGURE 8–34
Logic symbol for a data lock-out master-slave J-K flip-flop.

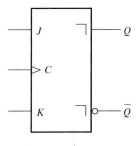

EXAMPLE 8–10

The waveforms in Figure 8–35(a) are applied to the data lock-out flip-flop as shown. Determine the Q output, starting in the RESET state.

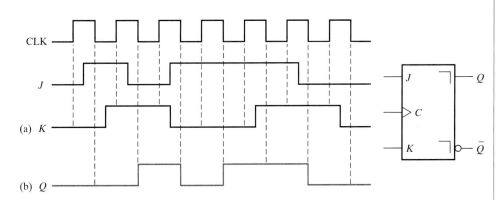

FIGURE 8–35

Solution The states of the J and K inputs are clocked into the master on the positive-going edge (triggering edge) of the clock pulse and into the slave (and thus onto the outputs) on the negative-going edge. A change in J or K after the triggering edge of the clock has no effect on the output, as shown in Figure 8–35, where J and K change several times while the clock is HIGH.

Related Exercise Interchange the J and K inputs and determine the resulting Q output.

1. Describe the basic difference between pulse-triggered and edge-triggered flip-flops.
2. Suppose that the D input of a flip-flop changes from LOW to HIGH in the middle of a positive-going clock pulse.
   (a) Describe what happens if the flip-flop is a positive edge-triggered type.
   (b) Describe what happens if the flip-flop is a pulse-triggered master-slave type.
3. Describe how a data lock-out master-slave flip-flop differs from a pulse-triggered master-slave flip-flop.

8–4 ■ FLIP-FLOP OPERATING CHARACTERISTICS

The performance, operating requirements, and limitations of flip-flops are specified by several operating characteristics or parameters found on the data sheet for the device. Generally, the specifications are applicable to all flip-flops in a given series (standard, LS, S, AS TTL, or CMOS). After completing this section, you should be able to

☐ Define *propagation delay time*   ☐ Explain the various propagation delay time specifications   ☐ Define *set-up time* and discuss how it limits flip-flop operation ☐ Define *hold time* and discuss how it limits flip-flop operation   ☐ Discuss the significance of maximum clock frequency   ☐ Discuss the various pulse width specifications   ☐ Define *power dissipation* and calculate its value for a specific device ☐ Compare various series of flip-flops in terms of their operating parameters

Propagation Delay Times

A **propagation delay time** is the interval of time required after an input signal has been applied for the resulting output change to occur. Several categories of propagation delay are important in the operation of a flip-flop:

1. Propagation delay t_{PLH} as measured from the triggering edge of the clock pulse to the LOW-to-HIGH transition of the output. This delay is illustrated in Figure 8–36(a).
2. Propagation delay t_{PHL} as measured from the triggering edge of the clock pulse to the HIGH-to-LOW transition of the output. This delay is illustrated in Figure 8–36(b).

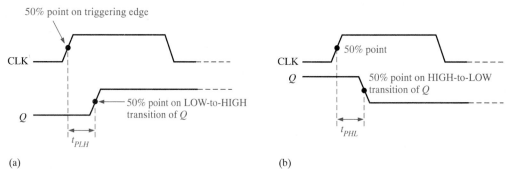

(a)   (b)

FIGURE 8–36
Propagation delays, clock to output.

3. Propagation delay t_{PLH} as measured from the preset input to the LOW-to-HIGH transition of the output. This delay is illustrated in Figure 8–37(a) for an active-LOW preset input.
4. Propagation delay t_{PHL} as measured from the clear input to the HIGH-to-LOW transition of the output. This delay is illustrated in Figure 8–37(b) for an active-LOW clear input.

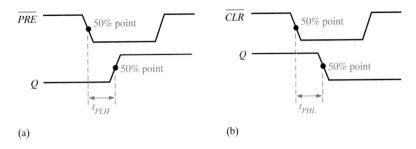

FIGURE 8–37

Propagation delays, preset input to output and clear input to output.

Set-up Time

The **set-up time** (t_s) is the minimum interval required for the logic levels to be maintained constantly on the inputs (J and K, or S and R, or D) prior to the triggering edge of the clock pulse in order for the levels to be reliably clocked into the flip-flop. This interval is illustrated in Figure 8–38 for a D flip-flop.

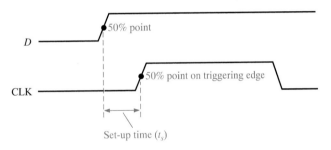

FIGURE 8–38
Set-up time (t_s).

Hold Time

The **hold time** (t_h) is the minimum interval required for the logic levels to remain on the inputs after the triggering edge of the clock pulse in order for the levels to be reliably clocked into the flip-flop. This is illustrated in Figure 8–39 for a D flip-flop.

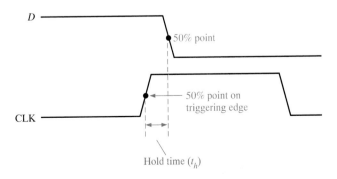

FIGURE 8–39
Hold time (t_h).

Maximum Clock Frequency

The maximum clock frequency (f_{max}) is the highest rate at which a flip-flop can be reliably triggered. At clock frequencies above the maximum, the flip-flop would be unable to respond quickly enough, and its operation would be impaired.

Pulse Widths

Minimum pulse widths (t_W) for reliable operation are usually specified by the manufacturer for the clock, preset, and clear inputs. Typically, the clock is specified by its minimum HIGH time and its minimum LOW time.

Power Dissipation

The **power dissipation** of any digital circuit is the total power consumption of the device. For example, if the flip-flop operates on a +5 V dc source and draws 50 mA of current, the power dissipation is

$$P = V_{CC} \times I_{CC} = 5 \text{ V} \times 50 \text{ mA} = 250 \text{ mW}$$

The power dissipation is very important in most applications in which the capacity of the dc supply is a concern. As an example, let's assume that we have a digital system that requires a total of ten flip-flops, and each flip-flop dissipates 250 mW of power. The total power requirement is

$$P_T = 10 \times 250 \text{ mW} = 2500 \text{ mW} = 2.5 \text{ W}$$

This tells us the output capacity required of the dc supply. If the flip-flops operate on +5 V dc, then the amount of current that the supply must provide is as follows:

$$I = \frac{2.5 \text{ W}}{5 \text{ V}} = 0.5 \text{ A}$$

We must use a +5 V dc supply that is capable of providing at least 0.5 A of current.

Comparison of Specific Flip-Flops

Table 8–6 provides a comparison, in terms of the operating parameters discussed in this section, of several TTL devices, as well as a CMOS device.

TABLE 8–6

Comparison of operating parameters for several types of flip-flops

Parameter (Times in ns)	TTL					CMOS
	7474	74LS74A	74S74	74LS76A	74LS112A	74HC112
t_{PHL} (CLK to Q)	40	30	9	20	20	21
t_{PLH} (CLK to Q)	25	25	9	20	20	21
t_{PHL} ($\overline{CLR}$ to Q)	40	30	13.5	20	20	26
t_{PLH} ($\overline{PRE}$ to Q)	25	25	6	20	20	28
t_s (set-up time)	20	20	3	20	20	20
t_h (hold time)	5	0	2	0	0	0
t_W (CLK HIGH)	30	18	6	20	20	16
t_W (CLK LOW)	37	—	7.3	—	25	—
t_W $\overline{CLR/PRE}$)	30	15	7	25	30	—
f_{max} (MHz)	15	25	75	45	30	30
Power (mW/F-F)	60	14	105	10	14	0.14

SECTION 8–4
REVIEW

1. Define the following:
   (a) set-up time   (b) hold time
2. Which specific flip-flop in Table 8–6 can be operated at the highest frequency?

8–5 ■ FLIP-FLOP APPLICATIONS

In this section, several general applications of flip-flops are discussed to give you a basic idea of how they can be used. In Chapters 9 and 10 there are detailed coverages of flip-flop applications in counters and registers. After completing this section, you should be able to

☐ Discuss the application of flip-flops to data storage   ☐ Describe how flip-flops are used for frequency division   ☐ Explain how flip-flops are used in basic counter applications

Parallel Data Storage

A common requirement in digital systems is to store several bits of data from parallel lines simultaneously in a group of flip-flops. This operation is illustrated in Figure 8–40(a) using four flip-flops. Each of the four parallel data lines is connected to the D input of a flip-flop. The clock inputs of the flip-flops are connected together, so that each flip-flop is triggered

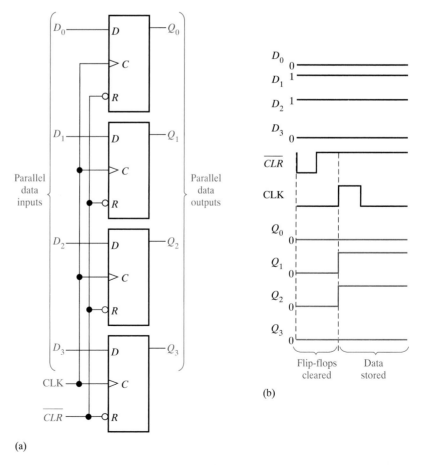

(a)

(b)

FIGURE 8–40
Example of flip-flops used in a basic register for parallel data storage.

by the same clock pulse. In this example, positive edge-triggered flip-flops are used, so the data on the *D* inputs are stored simultaneously by the flip-flops on the positive edge of the clock, as indicated in the timing diagram in Figure 8–40(b). Also, the asynchronous reset (*R*) inputs are connected to a common $\overline{CLR}$ line, which initially resets all the flip-flops.

This group of four flip-flops is an example of a basic register used for data storage. In digital systems, data are normally stored in groups of bits (usually eight or multiples thereof) that represent numbers, codes, or other information. Registers are covered in detail in Chapter 10.

Frequency Division

Another application of a flip-flop is dividing (reducing) the frequency of a periodic waveform. When a pulse waveform is applied to the clock input of a J-K flip-flop that is connected to toggle (*J* = *K* = 1), the *Q* output is a square wave with one-half the frequency of the clock input. Thus, a single flip-flop can be applied as a divide-by-2 device, as is illustrated in Figure 8–41. As you can see, the flip-flop changes state on each triggering clock edge (positive edge-triggered in this case). This results in an output that changes at half the frequency of the clock waveform.

Further division of a clock frequency can be achieved by using the output of one flip-flop as the clock input to a second flip-flop, as shown in Figure 8–42. The frequency of the Q_A output is divided by 2 by flip-flop B. The Q_B output is, therefore, one-fourth the frequency of the original clock input. Propagation delay times are not shown on the timing diagrams.

FIGURE 8–41
The J-K flip-flop as a divide-by-2 device.

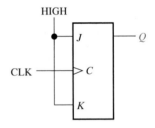

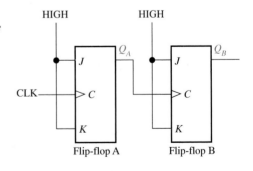

FIGURE 8–42
Example of two J-K flip-flops used to divide the clock frequency by 4.

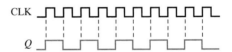

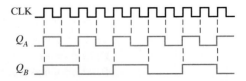

By connecting flip-flops in this way, a frequency division of 2^n is achieved, where n is the number of flip-flops. For example, three flip-flops divide the clock frequency by $2^3 = 8$; four flip-flops divide the clock frequency by $2^4 = 16$; and so on.

EXAMPLE 8–11

Develop the f_{out} waveform for the circuit in Figure 8–43 when an 8 kHz square wave input is applied to the clock input of flip-flop A.

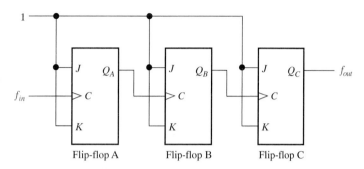

FIGURE 8–43

Solution The three flip-flops are connected to divide the input frequency by eight ($2^3 = 8$) and the f_{out} waveform is shown in Figure 8–44. Since these are positive edge-triggered flip-flops, the outputs change on the positive-going clock edge. There is one output pulse for every eight input pulses, so the output frequency is 1 kHz. Waveforms of Q_A and Q_B are also shown.

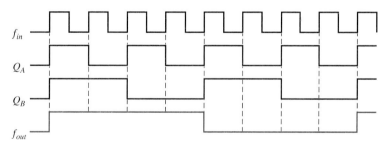

FIGURE 8–44

Related Exercise How many flip-flops are required to divide a frequency by thirty-two?

Counting

Another important application of flip-flops is in digital counters, which are covered in detail in the next chapter. The concept is illustrated in Figure 8–45. The flip-flops are negative edge-triggered J-Ks. Both flip-flops are initially RESET. Flip-flop A toggles on the negative-going transition of each clock pulse. The Q output of flip-flop A clocks flip-flop B, so each time Q_A makes a HIGH-to-LOW transition, flip-flop B toggles. The resulting Q_A and Q_B waveforms are shown in the figure.

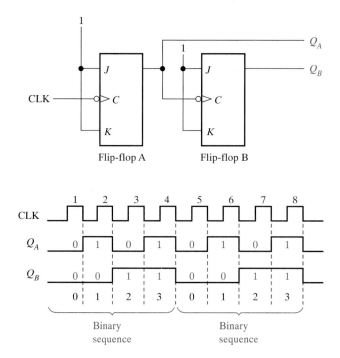

FIGURE 8–45

Flip-flops used to generate a binary count sequence. Two repetitions (00, 01, 10, 11) are shown.

Observe the sequence of Q_A and Q_B in Figure 8–45. Prior to clock pulse 1, $Q_A = 0$ and $Q_B = 0$; after clock pulse 1, $Q_A = 1$ and $Q_B = 0$; after clock pulse 2, $Q_A = 0$ and $Q_B = 1$; and after clock pulse 3, $Q_A = 1$ and $Q_B = 1$. If we take Q_A as the least significant bit, a 2-bit binary sequence is produced as the flip-flops are clocked. This binary sequence repeats every four clock pulses, as shown in the timing diagram of Figure 8–45. Thus, the flip-flops are counting in sequence from 0 to 3 (00, 01, 10, 11) and then recycling back to 0 to begin the sequence again.

EXAMPLE 8–12

Determine the output waveforms in relation to the clock for Q_A, Q_B, and Q_C in the circuit of Figure 8–46 and show the binary sequence represented by these waveforms.

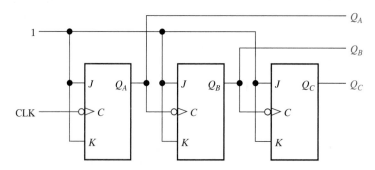

FIGURE 8–46

Solution The output timing diagram is shown in Figure 8–47. Notice that the outputs change on the negative-going edge of the clock pulses. The outputs go through the binary sequence 000, 001, 010, 011, 100, 101, 110, and 111 as indicated.

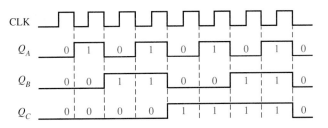

FIGURE 8–47

Related Exercise How many flip-flops are required to produce a binary sequence representing decimal numbers 0 through 15?

SECTION 8–5 REVIEW
1. A group of flip-flops used for data storage is called a _____.
2. How must a J-K flip-flop be connected to function as a divide-by-2 device?
3. How many flip-flops are required to produce a divide-by-64 device?

8–6 ■ ONE-SHOTS

The one-shot is a monostable multivibrator, a device with only one stable state. A one-shot is normally in its stable state and will change to its unstable state only when triggered. Once it is triggered, the one-shot remains in its unstable state for a predetermined length of time and then automatically returns to its stable state. The time that the device stays in its unstable state determines the pulse width of its output. After completing this section, you should be able to

□ Describe the basic operation of a one-shot □ Explain how a nonretriggerable one-shot works □ Explain how a retriggerable one-shot works □ Set up the 74121 and the 74122 one-shots to obtain a specified output pulse width □ Recognize a Schmitt trigger symbol and explain basically what it means

A **one-shot** is a type of **monostable** multivibrator. Figure 8–48 shows a basic one-shot that is composed of a logic gate and an inverter. When a pulse is applied to the **trigger** input, the output of gate G_1 goes LOW. This HIGH-to-LOW transition is coupled through the capacitor to the input of inverter G_2. The apparent LOW on G_2 makes its output go HIGH. This HIGH is connected back into G_1, keeping its output LOW. Up to this point the trigger pulse has caused the output of the one-shot, Q, to go HIGH.

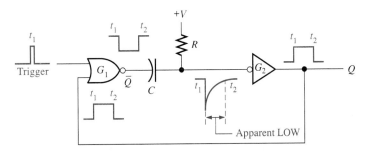

FIGURE 8–48
A simple one-shot circuit.

The capacitor immediately begins to charge through R toward the high voltage level. The rate at which it charges is determined by the RC time constant. When the capacitor charges to a certain level, which appears as a HIGH to G_2, the output goes back LOW.

To summarize, the output of inverter G_2 goes HIGH in response to the trigger input. It remains HIGH for a time set by the RC time constant. At the end of this time, it goes LOW. So *a single narrow trigger pulse produces a single output pulse whose time duration is controlled by the RC time constant.* This operation is illustrated in Figure 8–48.

A typical one-shot logic symbol is shown in Figure 8–49(a), and the same symbol with an external R and C is shown in Figure 8–49(b). The two basic types of IC one-shots are nonretriggerable and retriggerable.

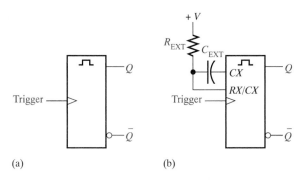

(a)   (b)

FIGURE 8–49
Basic one-shot logic symbols. CX and RX stand for external components.

Nonretriggerable One-Shots

A nonretriggerable one-shot will not respond to any additional trigger pulses from the time it is triggered into its unstable state (fired) until it returns to its stable state. In other words, it will ignore any trigger pulses occuring before it times out. The time that the one-shot remains in its unstable state is the pulse width of the output.

Figure 8–50 shows the nonretriggerable one-shot being triggered at intervals greater than its pulse width and at intervals less than the pulse width. Notice that in the second case, the additional pulses are ignored.

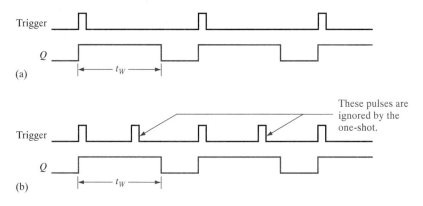

FIGURE 8–50
Nonretriggerable one-shot action.

The 74121 Nonretriggerable One-Shot The 74121 is an example of a nonretriggerable IC one-shot. It has provisions for external R and C, as shown in Figure 8–51. The inputs labeled A_1, A_2, and B are gated trigger inputs. The R_{INT} input connects to a 2 kΩ internal timing resistor.

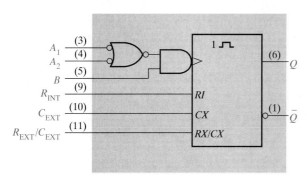

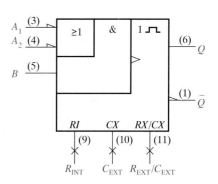

(a) Traditional logic symbol

(b) *ANSI/IEEE* std. 91–1984 logic symbol (✕ = nonlogic connection). "1 ⌐⌐" is the qualifying symbol for a nonretriggerable one-shot.

FIGURE 8–51
Logic symbols for the 74121 nonretriggerable one-shot.

Setting the Pulse Width   A typical pulse width of about 30 ns is produced when no external timing components are used and the internal timing resistor (R_{INT}) is connected to V_{CC}, as shown in Figure 8–52(a). The pulse width can be set anywhere between about 30 ns and 28 s by the use of external components. Figure 8–52(b) shows connection of the internal resistor (2 kΩ) and an external capacitor. Part (c) illustrates connection of an external resistor and an external capacitor. The output pulse width is set by the values of the resistor (R_{INT} = 2 kΩ, and R_{EXT} is selected) and the capacitor according to the following formula:

$$t_W = 0.7RC_{EXT} \qquad (8\text{–}1)$$

where R is either R_{INT} or R_{EXT}. When R is in kilohms (kΩ) and C_{EXT} is in picofarads (pF), the ouput pulse width t_W is in nanoseconds (ns).

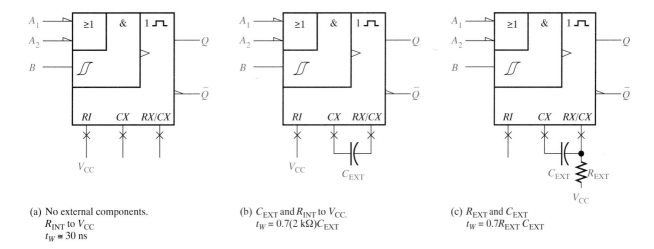

(a) No external components.
   R_{INT} to V_{CC}
   $t_W \cong 30$ ns

(b) C_{EXT} and R_{INT} to V_{CC}.
   $t_W = 0.7(2 \text{ k}\Omega)C_{EXT}$

(c) R_{EXT} and C_{EXT}
   $t_W = 0.7R_{EXT}C_{EXT}$

FIGURE 8–52
Three ways to set the pulse width of a 74121.

The Schmitt-Trigger Symbol   The symbol $\int\!\!\!\int$ indicates a Schmitt-trigger input. This type of input uses a special threshold circuit that produces **hysteresis,** a characteristic that prevents erratic switching between states when a slow-changing trigger voltage hovers around the critical input level. This allows reliable triggering to occur even when the input is changing as slowly as 1 volt/second.

EXAMPLE 8–13

A certain application requires a one-shot with a pulse width of approximately 100 ms. Using a 74121, show the connections and the component values.

Solution   Arbitrarily select $R_{EXT} = 39$ kΩ and calculate the necessary capacitance.

$$t_W = 0.7R_{EXT}C_{EXT}$$

$$C_{EXT} = \frac{t_W}{0.7R_{EXT}}$$

where C_{EXT} is in pF, R_{EXT} is in kΩ, and t_W is in ns. Since 100 ms $= 1 \times 10^8$ ns,

$$C_{EXT} = \frac{1 \times 10^8 \text{ ns}}{0.7(39 \text{ k}\Omega)} = 3.66 \times 10^6 \text{ pF} = 3.66 \text{ }\mu\text{F}$$

Use a standard 3.3 μF capacitor. The proper connections are shown in Figure 8–53.

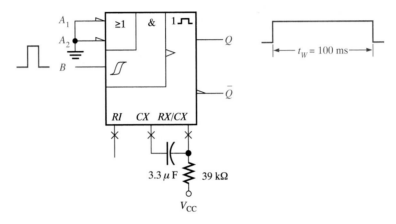

FIGURE 8–53

Related Exercise   Use an external capacitor in conjunction with R_{INT} to produce an output pulse width of 10 μs from the 74121.

Retriggerable One-Shots

A retriggerable one-shot can be triggered before it times out. The result of retriggering is an extension of the pulse width as illustrated in Figure 8–54.

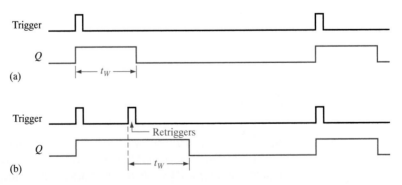

FIGURE 8–54
Retriggerable one-shot action.

The 74122 Retriggerable One-Shot The 74122 is an example of a retriggerable IC one-shot with a clear input. It also has provisions for external R and C, as shown in Figure 8–55. The inputs labeled A_1, A_2, B_1, and B_2 are the gated trigger inputs.

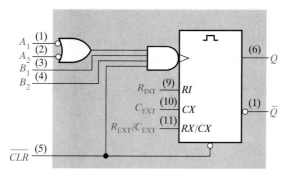

(a) Traditional logic symbol

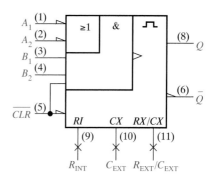

(b) *ANSI/IEEE* std. 91–1984 logic symbol ($\times$ = nonlogic connection). ⊓ is the qualifying symbol for a retriggerable one-shot.

FIGURE 8–55
Logic symbol for the 74122 retriggerable one-shot.

A minimum pulse width of approximately 45 ns is obtained with no external components. Wider pulse widths are achieved by using external components. A general formula for calculating the values of these components for a specified pulse width (t_W) is

$$t_W = 0.32RC_{EXT}\left(1 + \frac{0.7}{R}\right) \tag{8-2}$$

where 0.32 is a constant determined by the particular type of one-shot, R is in kΩ and is either the internal or the external resistor, C_{EXT} is in pF, and t_W is in ns. The internal resistance is 10 kΩ and can be used instead of an external resistor. (Notice the difference between this formula and that for the 74121, shown in Equation (8–1).)

EXAMPLE 8–14

Determine the values of R_{EXT} and C_{EXT} that will produce a pulse width of 1 μs when connected to a 74122.

Solution Assume a value of $C_{EXT} = 560$ pF and then solve for R_{EXT}. t_W must be expressed in ns and C_{EXT} in pF. R_{EXT} will be in kΩ.

$$t_W = 0.32R_{EXT}C_{EXT}\left(1 + \frac{0.7}{R_{EXT}}\right) = 0.32R_{EXT}C_{EXT} + 0.7\left(\frac{0.32R_{EXT}C_{EXT}}{R_{EXT}}\right)$$

$$= 0.32R_{EXT}C_{EXT} + (0.7)(0.32)C_{EXT}$$

$$R_{EXT} = \frac{t_W - (0.7)(0.32)C_{EXT}}{0.32C_{EXT}} = \frac{t_W}{0.32C_{EXT}} - 0.7$$

$$= \frac{1000 \text{ ns}}{(0.32)560 \text{ pF}} - 0.7 = 4.88 \text{ k}\Omega$$

Use a standard value of 4.7 kΩ.

Related Exercise Show the connections and component values for a 74122 one-shot with an output pulse width of 5 μs. Assume $C_{EXT} = 560$ pF.

A One-Shot Application

One practical one-shot application is a sequential timer that can be used to illuminate a series of lights. This type of circuit can be used, for example, in a lane change directional indicator for highway construction projects or in sequential turn signals on automobiles.

Figure 8–56 shows three 74122 one-shots connected as a sequential timer. This particular circuit produces a sequence of three 1 s pulses. The first one-shot is triggered by a switch closure or a low-frequency pulse input producing a 1 s output pulse. When the first one-shot (OS 1) times out and the 1 s pulse goes LOW, the second one-shot (OS 2) is triggered also producing a 1 s output pulse. When this second pulse goes LOW, the third one-shot (OS 3) is triggered and the third 1 s pulse is produced. The output timing is illustrated in the figure. Variations of this basic arrangement can be used to produce a variety of timed outputs.

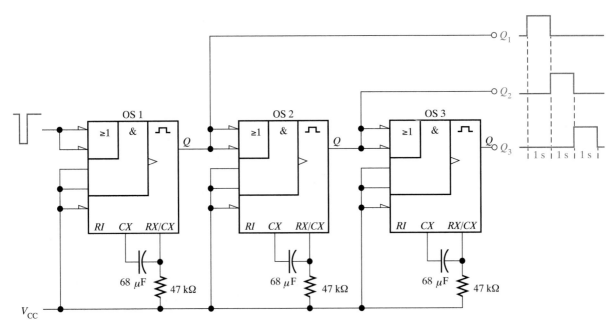

FIGURE 8–56
A sequential timing circuit using three 74122 one-shots.

SECTION 8–6 REVIEW	1. Describe the difference between a nonretriggerable and a retriggerable one-shot.
	2. How is the output pulse width set in most IC one-shots?

8–7 ■ THE 555 TIMER

The 555 timer is a versatile and widely used device because it can be configured in two different modes as either a monostable multivibrator (one-shot) or as an astable multivibrator (oscillator). An astable multivibrator has no stable states and therefore changes back and forth (oscillates) between two unstable states without any external triggering. After completing this section, you should be able to

☐ Describe the basic elements in a 555 timer   ☐ Set up a 555 timer as a one-shot
☐ Set up a 555 timer as an oscillator

Basic Operation

A functional diagram showing the internal components of a 555 **timer** is given in Figure 8–57. The comparators are devices whose outputs are HIGH when the voltage on the positive (+) input is greater than the voltage on the negative (−) input and LOW when the − input voltage is greater than the + input voltage. The voltage divider consisting of three 5 kΩ resistors provides a trigger level of $\frac{1}{3}V_{CC}$ and a threshold level of $\frac{2}{3}V_{CC}$. The control voltage input (pin 5) can be used to externally adjust the trigger and threshold levels to other values if necessary. When the normally HIGH trigger input momentarily goes below $\frac{1}{3}V_{CC}$, the output of comparator B switches from LOW to HIGH and SETS the S-R latch, causing the output (pin 3) to go HIGH and turning the discharge transistor Q_1 off. The output will stay HIGH until the normally LOW threshold input goes above $\frac{2}{3}V_{CC}$ and causes the output of comparator A to switch from LOW to HIGH. This RESETS the latch causing the output to go back LOW and turning the discharge transistor on. The external reset input can be used to RESET the latch independent of the threshold circuit. The trigger and threshold inputs (pins 2 and 6) are controlled by external components connected to produce either monostable or astable action.

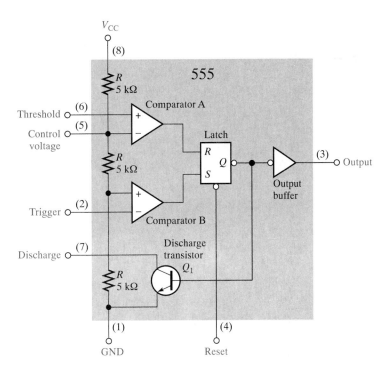

FIGURE 8–57
Internal functional diagram of a 555 timer (pin numbers are in parentheses).

Monostable (One-Shot) Operation

An external resistor and capacitor connected as shown in Figure 8–58 are used to set up the 555 timer as a nonretriggerable one-shot. The pulse width of the output is determined by the time constant of R_1 and C_1 according to the following formula:

$$t_W = 1.1R_1C_1 \qquad\qquad \text{(8–3)}$$

The control voltage input is not used and is connected to a decoupling capacitor C_2 to prevent noise from affecting the trigger and threshold levels.

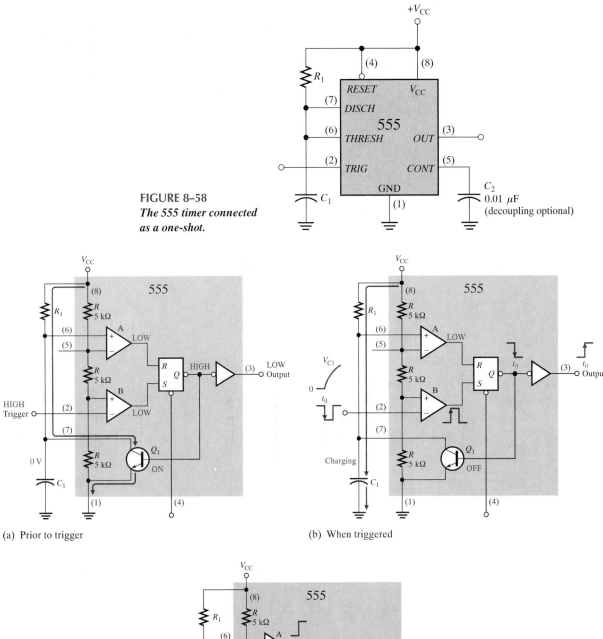

FIGURE 8–58
The 555 timer connected as a one-shot.

(a) Prior to trigger

(b) When triggered

(c) At end of charging interval

FIGURE 8–59
One-shot operation of the 555 timer.

Before a trigger pulse is applied, the output is LOW and the discharge transistor Q_1 is *on*, keeping C_1 discharged as shown in Figure 8–59(a). When a negative-going trigger pulse is applied, the output goes HIGH and the discharge transistor turns *off*, allowing capacitor C_1 to begin charging through R_1 as shown in part (b). When C_1 charges to ⅓V_{CC}, the output goes back LOW and Q_1 turns *on* immediately, discharging C_1 as shown in part (c). As you can see, the charging rate of C_1 determines how long the output is HIGH.

EXAMPLE 8–15

What is the output pulse width for a 555 monostable circuit with $R_1 = 2.2$ kΩ and $C_1 = 0.01 \, \mu F$?

Solution   From Equation (8–3) the pulse width is

$$t_W = 1.1 R_1 C_1 = 1.1(2.2 \text{ k}\Omega)(0.01 \, \mu F) = 24.2 \, \mu s$$

Related Exercise   For $C_1 = 0.01 \, \mu F$, determine the value of R_1 for a pulse width of 1 ms.

Astable Operation

A 555 timer connected to operate as an **astable** multivibrator, which is a free-running non-sinusoidal **oscillator,** is shown in Figure 8–60. Notice that the threshold input *(THRESH)* is now connected to the trigger input *(TRIG)*. The external components R_1, R_2, and C_1 form the timing network that sets the frequency of oscillation. The 0.01 μF capacitor C_2 connected to the control *(CONT)* input is strictly for decoupling and has no effect on the operation; in some cases it can be left off.

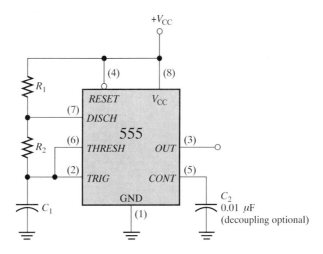

FIGURE 8–60
The 555 timer connected as an astable multivibrator.

Initially, when the power is turned on, the capacitor C_1 is uncharged and thus the trigger voltage (pin 2) is at 0 V. This causes the output of comparator B to be HIGH and the output of comparator A to be LOW, forcing the output of the latch, and thus the base of Q_1, LOW and keeping the transistor off. Now, C_1 begins charging through R_1 and R_2 as indicated in Figure 8–61. When the capacitor voltage reaches ⅓V_{CC}, comparator B switches to its LOW output state, and when the capacitor voltage reaches ⅔V_{CC}, comparator A switches to its HIGH output state. This RESETS the latch, causing the base of Q_1 to go HIGH, and turns on the transistor. This sequence creates a discharge path for the capacitor through R_2 and the transistor, as indicated. The capacitor now begins to discharge, causing

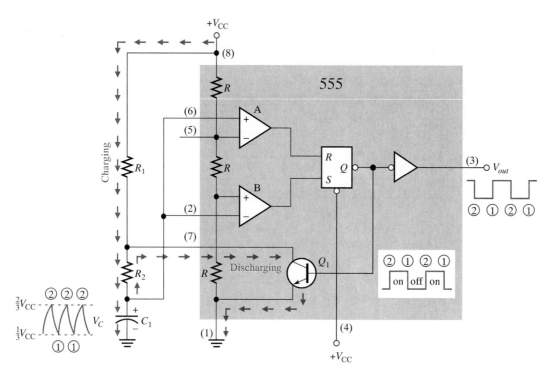

FIGURE 8–61

Operation of the 555 timer in the astable mode.

comparator A to go LOW. At the point where the capacitor discharges down to $\frac{1}{3}V_{CC}$, comparator B switches HIGH; this SETS the latch, which makes the base of Q_1 LOW and turns off the transistor. Another charging cycle begins, and the entire process repeats. The result is a rectangular wave output whose duty cycle depends on the values of R_1 and R_2. The frequency of oscillation is given by the following formula, or it can be found using the graph in Figure 8–62.

$$f = \frac{1.44}{(R_1 + 2R_2)C_1} \tag{8–4}$$

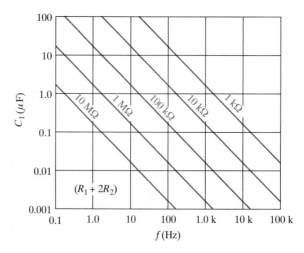

FIGURE 8–62

Frequency of oscillation as a function of C_1 and $R_1 + 2R_2$. The sloped lines are values of $R_1 + 2R_2$.

By selecting R_1 and R_2, the duty cycle of the output can be adjusted. Since C_1 charges through $R_1 + R_2$ and discharges only through R_2, duty cycles approaching a minimum of 50 percent can be achieved if $R_2 >> R_1$ so that the charging and discharging times are approximately equal.

An expression for the duty cycle is developed as follows. The time that the output is HIGH (t_H) is how long it takes C_1 to charge from $\frac{1}{3}V_{CC}$ to $\frac{2}{3}V_{CC}$. It is expressed as

$$t_H = 0.7(R_1 + R_2)C_1 \tag{8--5}$$

The time that the output is LOW (t_L) is how long it takes C_1 to discharge from $\frac{1}{3}V_{CC}$ to $\frac{2}{3}V_{CC}$. It is expressed as

$$t_L = 0.7R_2C_1 \tag{8--6}$$

The period, T, of the output waveform is the sum of t_H and t_L.

$$T = t_H + t_L = 0.7(R_1 + 2R_2)C_1$$

This is the reciprocal of f in Equation (8–4). Finally, the duty cycle is

$$\text{Duty cycle} = \frac{t_H}{T} = \frac{t_H}{t_H + t_L}$$

$$\text{Duty cycle} = \left(\frac{R_1 + R_2}{R_1 + 2R_2}\right)100\% \tag{8--7}$$

To achieve duty cycles of less than 50 percent, the circuit in Figure 8–60 can be modified so that C_1 charges through only R_1 and discharges through R_2. This is achieved with a diode D_1 placed as shown in Figure 8–63. The duty cycle can be made less than 50 percent by making R_1 less than R_2. Under this condition, the expression for the duty cycle is

$$\text{Duty cycle} = \left(\frac{R_1}{R_1 + R_2}\right)100\% \tag{8--8}$$

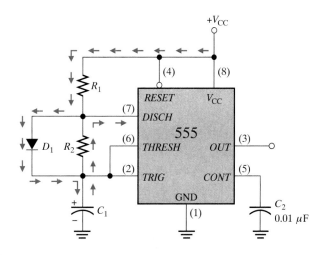

FIGURE 8–63
The addition of diode D_1 allows the duty cycle of the output to be adjusted to less than 50 percent by making $R_1 < R_2$.

EXAMPLE 8–16   A 555 timer configured to run in the astable mode (oscillator) is shown in Figure 8–64. Determine the frequency of the output and the duty cycle.

FIGURE 8–64

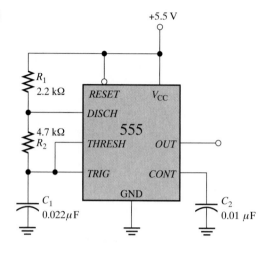

Solution   Use Equations (8–4) and (8–7).

$$f = \frac{1.44}{(R_1 + 2R_2)C_1} = \frac{1.44}{(2.2 \text{ k}\Omega + 9.4 \text{ k}\Omega)0.022 \ \mu\text{F}} = 5.64 \text{ kHz}$$

$$\text{Duty cycle} = \left(\frac{R_1 + R_2}{R_1 + 2R_2}\right)100\% = \left(\frac{2.2 \text{ k}\Omega + 4.7 \text{ k}\Omega}{2.2 \text{ k}\Omega + 9.4 \text{ k}\Omega}\right)100\% = 59.5\%$$

Related Exercise   Determine the duty cycle in Figure 8–64 if a diode is connected across R_2 as indicated in Figure 8–63.

SECTION 8–7 REVIEW

1. Explain the difference in operation between an astable multivibrator and a monostable multivibrator.
2. For a certain astable multivibrator, $t_H = 15$ ms and $T = 20$ ms. What is the duty cycle of the output?

8–8 ■ TROUBLESHOOTING

In industry, it is standard practice to test a new circuit design to be sure that it is operating as specified. New designs are usually "breadboarded" and tested before the design is finalized. The term breadboard refers to a method of temporarily hooking up a circuit so that its operation can be verified and any faults (bugs) worked out before a prototype unit is built. In this section, we will consider an example case. After completing this section, you should be able to

□ Describe how the timing of a circuit can produce erroneous glitches   □ Approach the debugging of a new design with greater insight and awareness of potential problems

---

The circuit shown in Figure 8–65(a) generates two clock waveforms (CLK A and CLK B) having an alternating occurrence of pulses. Each waveform is to be one-half the frequency of the original clock (CLK), as shown in the ideal timing diagram in part (b).

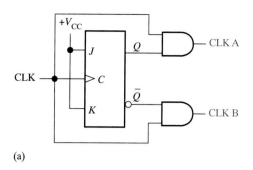

(a)

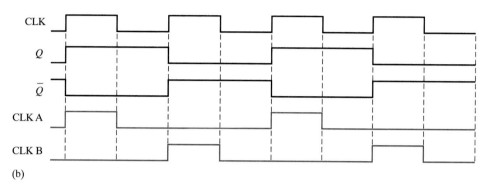

(b)

FIGURE 8–65
Two-phase clock generator with ideal waveforms.

When the circuit is tested, the CLK A and CLK B waveforms appear on the oscilloscope as shown in Figure 8–66(a). Since glitches are observed on both waveforms, something is wrong with the circuit either in its basic design or in the way it is connected. Further investigation reveals that the glitches are caused by a **race** condition between the CLK signal and the Q and $\overline{Q}$ signals at the inputs of the AND gates. As displayed in Figure 8–66(b), the propagation delays between CLK and Q and $\overline{Q}$ create a short-duration coincidence of HIGH levels at the leading edges of alternate clock pulses. Thus, there is a basic design flaw.

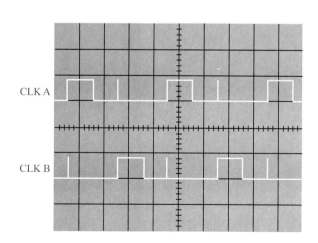

(a) Oscilloscope display of CLK A and CLK B waveforms

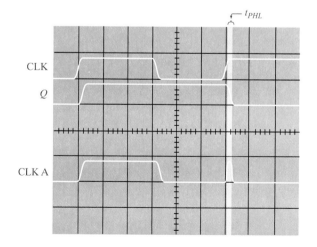

(b) Oscilloscope display showing propagation delay that creates glitch on CLK A waveform

FIGURE 8–66
Oscilloscope displays for the circuit in Figure 8–65.

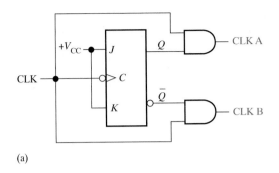

(a)

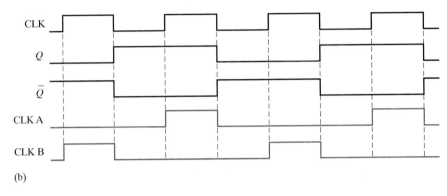

(b)

FIGURE 8–67

Two-phase clock generator using negative edge-triggered flip-flop to eliminate glitches.

The problem can be corrected by using a negative edge-triggered flip-flop in place of the positive edge-triggered device, as shown in Figure 8–67(a). Although the propagation delays between CLK and Q and $\overline{Q}$ still exist, they are initiated on the trailing edges of the clock (CLK), thus eliminating the glitches, as shown in the timing diagram of Figure 8–67(b).

SECTION 8–8
REVIEW

1. Can a negative edge-triggered D flip-flop be used in the circuit of Figure 8–67?
2. What device can be used to provide the clock for the circuit in Figure 8–67?

8–9 ■ DIGITAL SYSTEM APPLICATION

In this system application, you will continue working with the traffic light control system that was started in Chapter 6. You should review Section 6–12 before continuing. In this chapter, the focus is on the timing circuits that produce the 4 s interval for the caution light, the 25 s interval for the red and green lights, and the clock. After completing this section, you should be able to

☐ Develop a block diagram for the timing circuits portion of the system   ☐ Implement the short and long timers to meet specifications   ☐ Implement the clock oscillator to meet specifications   ☐ Develop a schematic from the printed circuit board   ☐ Develop a test procedure and troubleshoot the circuit board

General Requirements for the Timing Circuits

The overall system block diagram was discussed in the system application for Chapter 6 and is shown in Figure 8–68 again for reference. The state decoding and output logic block has been developed, and now we are going to concentrate on the timing circuits that must

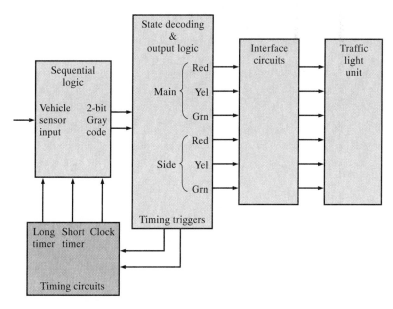

FIGURE 8–68
Traffic light control system block diagram.

produce a 4 s positive pulse or a 25 s positive pulse when triggered from the state decoding and output logic and a 10 kHz clock signal. All three of these outputs are used by the sequential logic that will be developed in Chapter 9.

A Block Diagram of the Timing Circuits

The timing circuits consist of the long timer, the short timer, and the clock oscillator as shown in Figure 8–69. The long timer and the short timer are implemented with one-shots and the clock oscillator is implemented with a 555 timer connected in the astable mode.

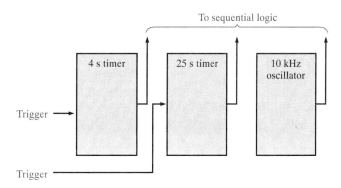

FIGURE 8–69
Block diagram of the timing circuits.

■ THE DIGITAL WORKBENCH

In the Chapter 6 workbench you focused on the portion of the traffic light controller board containing the state decoder and output logic. In this application, the portion of the board containing the timing circuits is now considered. As before, all of the connections to the V_{CC} and ground pads are on the back side of the board. Circuit interconnections on the back side of the board feed through to the pads on the component side. When you trace out the circuit, common pads are generally oriented vertically or horizontally and not at an angle with respect to each other.

■ DIGITAL WORKBENCH 1: Circuit Verification and Design

■ *Activity 1* Using the PC board, add the timing circuit portion of the system to the schematic that you began in Section 6–12.

■ *Activity 2* Verify that the timing circuit portion of the PC board is correct and properly connected to the state decoder and output logic.

■ *Activity 3* Identify the function of each of the points on the PC board to the right of each of the three circled numbers at the bottom of the board.

■ *Activity 4* Determine a value for each resistor (ignore the color bands shown) and capacitor on the board that will produce a 4 s positive pulse, a 25 s positive pulse, and a 10 kHz pulse waveform (duty cycle is not critical).

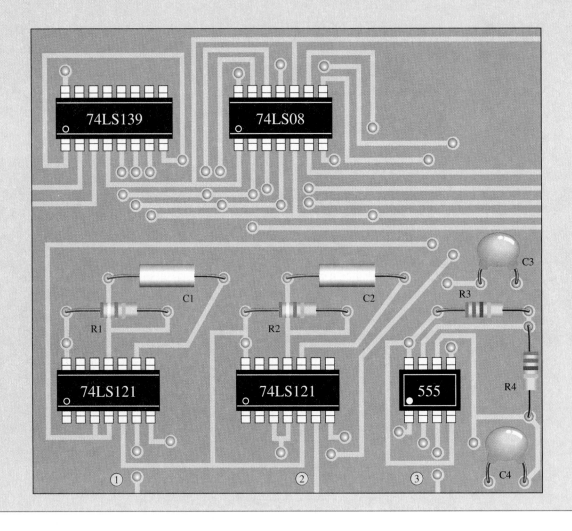

■ DIGITAL WORKBENCH 2: Testing and Troubleshooting

■ *Activity 1* Using the switches connected as shown to simulate the Gray code inputs from the sequential portion of the system, develop a test procedure to check out this portion of the system board.

■ *Activity 2* When the switches are thrown into each of the positions shown in the table, determine from the indications at the test points represented by the circled numbers if the circuit is operating properly for each case and, if not, what the most likely problem is.

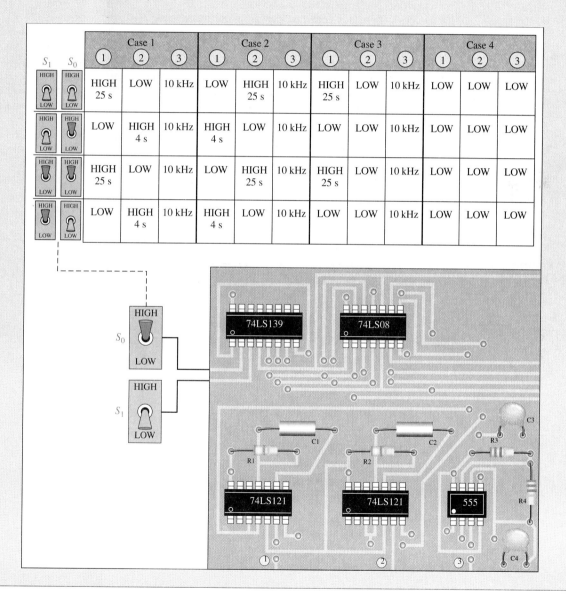

S_1	S_0	Case 1 (1)	Case 1 (2)	Case 1 (3)	Case 2 (1)	Case 2 (2)	Case 2 (3)	Case 3 (1)	Case 3 (2)	Case 3 (3)	Case 4 (1)	Case 4 (2)	Case 4 (3)
HIGH/LOW (HIGH)	HIGH/LOW (HIGH)	HIGH 25 s	LOW	10 kHz	LOW	HIGH 25 s	10 kHz	HIGH 25 s	LOW	10 kHz	LOW	LOW	LOW
HIGH/LOW (LOW)	HIGH/LOW (HIGH)	LOW	HIGH 4 s	10 kHz	HIGH 4 s	LOW	10 kHz	LOW	LOW	10 kHz	LOW	LOW	LOW
HIGH/LOW (HIGH)	HIGH/LOW (HIGH)	HIGH 25 s	LOW	10 kHz	LOW	HIGH 25 s	10 kHz	HIGH 25 s	LOW	10 kHz	LOW	LOW	LOW
HIGH/LOW (HIGH)	HIGH/LOW (LOW)	LOW	HIGH 4 s	10 kHz	HIGH 4 s	LOW	10 kHz	LOW	LOW	10 kHz	LOW	LOW	LOW

SECTION 8–9 REVIEW

1. Explain the purpose of the two switches in Workbench 2.
2. Why are the 4 s one-shot and the 25 s one-shot triggered separately?
3. Upon what other parts of the system does the operation of the clock oscillator depend?

■ SUMMARY

- The symbols for all devices in this summary conform with ANSI/IEEE Standard 91–1984. (Preset and clear inputs are not shown on devices.)
- Latches are bistable elements whose state normally depends on asynchronous inputs.
- Edge-triggered flip-flops are bistable elements with synchronous inputs whose state depends on the inputs only at the triggering transition of a clock pulse. Changes in the outputs occur at the triggering transition of the clock.
- Pulse-triggered or master-slave flip-flops are bistable elements with synchronous inputs whose state depends on the inputs at the leading edge of the clock pulse, but whose output is postponed and does not reflect the internal state until the trailing edge of the clock pulse. The synchronous inputs should not be allowed to change while the clock is HIGH.
- Data lock-out flip-flops are similar to master-slave types in that they are sensitive to the inputs only at the leading clock edge but the output is postponed until the trailing edge of the clock pulse. However, inputs can be changed while the clock is HIGH without affecting the operation.
- Monostable multivibrators (one-shots) have one stable state. When the one-shot is triggered, the output goes to its unstable state for a time determined by an RC circuit.
- Astable multivibrators have no stable states and are used as oscillators to generate timing waveforms in digital systems.

■ SELF-TEST

1. If an S-R latch has a 1 on the S input and a 0 on the R input and then the S input goes to 0, the latch will be
   (a) set   **(b)** reset   **(c)** invalid   **(d)** clear

2. The invalid state of an S-R latch occurs when
   (a) $S = 1, R = 0$ **(b)** $S = 0, R = 1$
   (c) $S = 1, R = 1$ **(d)** $S = 0, R = 0$

3. For a gated D latch, the Q output always equals the D input
   (a) before the enable pulse   **(b)** during the enable pulse
   (c) immediately after the enable pulse   **(d)** answers (b) and (c)

4. Like the latch, the flip-flop belongs to a category of logic circuits known as
   (a) monostable multivibrators   **(b)** bistable multivibrators
   (c) astable multivibrators   **(d)** one-shots

5. The purpose of the clock input to a flip-flop is to
   (a) clear the device
   (b) set the device
   (c) always cause the output to change states
   (d) cause the output to assume a state dependent on the controlling (S-R, J-K, or D) inputs.

6. For an edge-triggered D flip-flop,
   (a) a change in the state of the flip-flop can occur only at a clock pulse edge
   (b) the state that the flip-flop goes to depends on the D input
   (c) the output follows the input at each clock pulse
   (d) all of these answers

7. A feature that distinguishes the J-K flip-flop from the S-R flip-flop is the
   (a) toggle condition   **(b)** preset input
   (c) type of clock   **(d)** clear input

8. A flip-flop is in the toggle condition when
   (a) $J = 1, K = 0$ **(b)** $J = 1, K = 1$
   (c) $J = 0, K = 0$ **(d)** $J = 0, K = 1$

9. A J-K flip-flop with $J = 1$ and $K = 1$ has a 10 kHz clock input. The Q output is
   - (a) constantly HIGH
   - (b) constantly LOW
   - (c) a 10 kHz square wave
   - (d) a 5 kHz square wave

10. A one-shot is a type of
   - (a) monostable multivibrator
   - (b) astable multivibrator
   - (c) timer
   - (d) answers (a) and (c)
   - (e) answers (b) and (c)

11. The output pulse width of a nonretriggerable one-shot depends on
   - (a) the trigger intervals
   - (b) the supply voltage
   - (c) a resistor and capacitor
   - (d) the threshold voltage

12. An astable multivibrator
   - (a) requires a periodic trigger input
   - (b) has no stable state
   - (c) is an oscillator
   - (d) produces a periodic pulse output
   - (e) answers (a), (b), (c), and (d)
   - (f) answers (b), (c), and (d) only

■ PROBLEMS

SECTION 8–1   Latches

1. If the waveforms in Figure 8–70 are applied to an active-LOW input S-R latch, sketch the resulting Q output waveform in relation to the inputs. Assume that Q starts LOW.

2. Solve Problem 1 for the input waveforms in Figure 8–71 applied to an active-HIGH S-R latch.

FIGURE 8–70

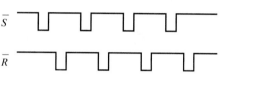
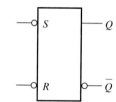

FIGURE 8–71

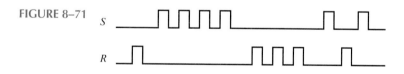

3. Solve Problem 1 for the input waveforms in Figure 8–72.

4. For a gated S-R latch, determine the Q and $\overline{Q}$ outputs for the inputs in Figure 8–73. Show them in proper relation to the Enable. Assume that Q starts LOW.

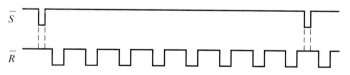

FIGURE 8–72

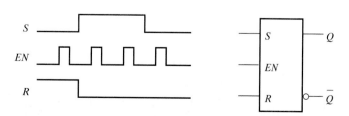

FIGURE 8–73

5. Solve Problem 4 for the inputs in Figure 8–74.

6. Solve Problem 4 for the inputs in Figure 8–75.

7. For a gated D latch, the waveforms shown in Figure 8–76 are observed on its inputs. Sketch the timing diagram showing the output waveform you would expect to see at Q if the latch is initially RESET.

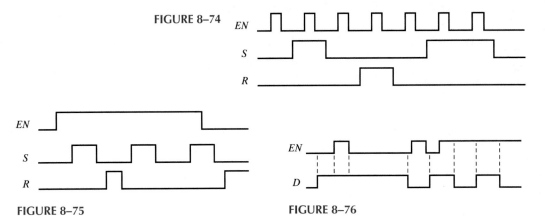

FIGURE 8–74

FIGURE 8–75

FIGURE 8–76

SECTION 8–2   Edge-Triggered Flip-Flops

8. Two edge-triggered S-R flip-flops are shown in Figure 8–77. If the inputs are as shown, sketch the Q output of each flip-flop relative to the clock, and explain the difference between the two. The flip-flops are initially RESET.

9. The Q output of an edge-triggered S-R flip-flop is shown in relation to the clock signal in Figure 8–78. Determine the input waveforms on the S and R inputs that are required to produce this output if the flip-flop is a positive edge-triggered type.

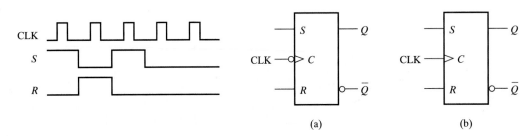

(a)   (b)

FIGURE 8–77

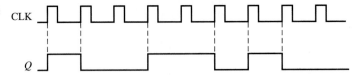

FIGURE 8–78

10. Sketch the Q output relative to the clock for a D flip-flop with the inputs as shown in Figure 8–79. Assume positive edge-triggering and Q initially LOW.

11. Solve Problem 10 for the inputs in Figure 8–80.

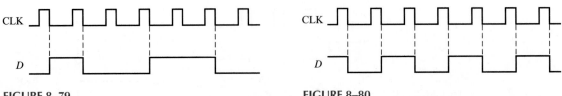

FIGURE 8–79

FIGURE 8–80

12. For a positive edge-triggered J-K flip-flop with inputs as shown in Figure 8–81, determine the Q output relative to the clock. Assume that Q starts LOW.

13. Solve Problem 12 for the inputs in Figure 8–82.

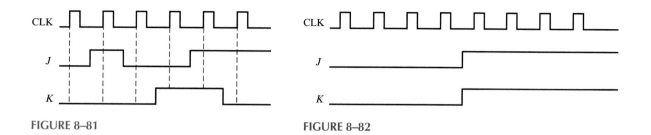

FIGURE 8–81

FIGURE 8–82

14. Determine the Q waveform relative to the clock if the signals shown in Figure 8–83 are applied to the inputs of the J-K flip-flop. Assume that Q is initially LOW.

15. For a negative edge-triggered J-K flip-flop with the inputs in Figure 8–84, sketch the Q output waveform relative to the clock. Assume that Q is initially LOW.

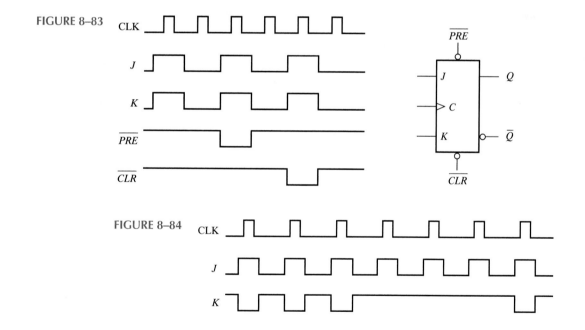

FIGURE 8–83

FIGURE 8–84

16. The following serial data are applied to the flip-flop through the AND gates as indicated in Figure 8–85. Determine the resulting serial data that appear on the Q output. There is one clock pulse for each bit time. Assume that Q is initially 0 and that $\overline{PRE}$ and $\overline{CLR}$ are HIGH. Right-most bits are applied first.

J_1: 1010011
J_2: 0111010
J_3: 1111000
K_1: 0001110
K_2: 1101100
K_3: 1010101

FIGURE 8–85

17. For the circuit in Figure 8–85, complete the timing diagram in Figure 8–86 by sketching the Q output (which is initially LOW). Assume $\overline{PRE}$ and $\overline{CLR}$ remain HIGH.

FIGURE 8–86

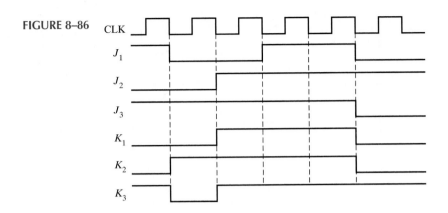

18. Solve Problem 17 with the same J and K inputs but with the $\overline{PRE}$ and $\overline{CLR}$ inputs as shown in Figure 8–87 in relation to the clock.

FIGURE 8–87

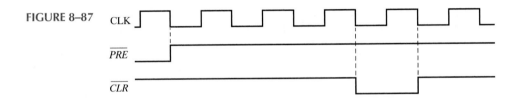

SECTION 8–3 Master-Slave Flip-Flops

19. Determine the Q output in Figure 8–88. Assume that Q is initially LOW.

20. Sketch the Q output of flip-flop B in Figure 8–89 in proper relation to the clock. The flip-flops are initially RESET.

FIGURE 8–88

FIGURE 8–89

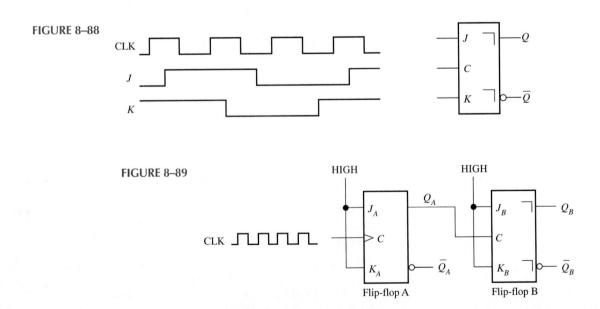

21. Sketch the Q output of flip-flop B in Figure 8–90 in proper relation to the clock. The flip-flops are initially RESET.

22. For a master-slave data lock-out J-K flip-flop with the inputs in Figure 8–84, draw the timing diagram showing Q starting in the LOW state. Assume triggering on the positive-going clock edge.

23. The D input and a single clock pulse are shown in Figure 8–91. Compare the resulting Q outputs for positive edge-triggered, negative edge-triggered, pulse-triggered master-slave and data lock-out master-slave flip-flops. The flip-flops are initially RESET.

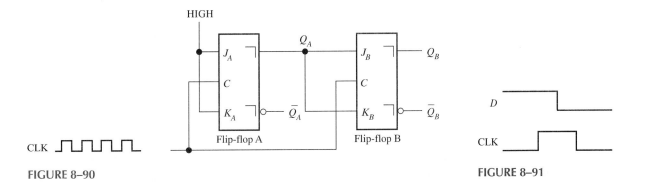

FIGURE 8–90   FIGURE 8–91

24. Solve Problem 19 for a master-slave data lock-out J-K flip-flop with triggering on the positive-going clock edge.

25. Solve Problem 24 for a device that triggers on the negative-going clock edge.

SECTION 8–4   Flip-Flop Operating Characteristics

26. Typically, a manufacturer's data sheet specifies four different propagation delay times associated with a flip-flop. Name and describe each one.

27. The data sheet of a certain flip-flop specifies that the minimum HIGH time for the clock pulse is 30 ns and the minimum LOW time is 37 ns. What is the maximum operating frequency?

28. The flip-flop in Figure 8–92 is initially RESET. Show the relation between the Q output and the clock pulse if propagation delay t_{PLH} (clock to Q) is 8 ns.

29. The direct current required by a particular flip-flop that operates on a $+5$ V dc source is found to be 10 mA. A certain digital device uses 15 of these flip-flops. Determine the current capacity required for the $+5$ V dc supply and the total power dissipation of the system.

30. For the circuit in Figure 8–93, determine the maximum frequency of the clock signal for reliable operation if the set-up time for each flip-flop is 20 ns and the propagation delays (t_{PLH} and t_{PHL}) from clock to output are 50 ns for each flip-flop.

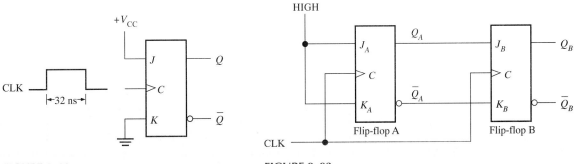

FIGURE 8–92   FIGURE 8–93

SECTION 8–5   Flip-Flop Applications

31. An S-R flip-flop is connected as shown in Figure 8–94. Determine the Q output in relation to the clock. What specific function does this device perform?

32. For the circuit in Figure 8–93, develop a timing diagram for eight clock pulses, showing the Q_A and Q_B outputs in relation to the clock.

FIGURE 8–94

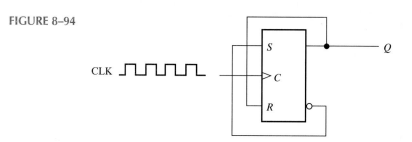

SECTION 8–6   One-Shots

33. Determine the pulse width of a 74121 one-shot if the external resistor is 3.3 kΩ and the external capacitor is 2000 pF.

34. An output pulse of 5 μs duration is to be generated by a 74122 one-shot. Using a capacitor of 10,000 pF, determine the value of external resistance required.

SECTION 8–7   The 555 Timer

35. Create a one-shot, using a 555 timer that will produce a 0.25 s output pulse.

36. A 555 timer is configured to run as an astable multivibrator as shown in Figure 8–95. Determine its frequency.

FIGURE 8–95

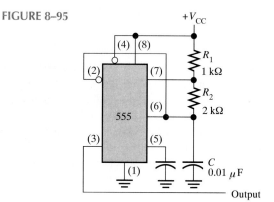

37. Determine the values of the external resistors for a 555 timer used as an astable multivibrator with an output frequency of 20 kHz, if the external capacitor C is 0.002 μF and the duty cycle is to be approximately 75%.

SECTION 8–8   Troubleshooting

38. The flip-flop in Figure 8–96 is tested under all input conditions as shown. Is it operating properly? If not, what is the most likely fault?

39. A 7400 quad NAND gate IC is used to construct a gated S-R latch on a protoboard in the lab as shown in Figure 8–97. The schematic in part (a) is used to connect the circuit in part (b). When you try to operate the latch, you find that the Q output stays HIGH no matter what the inputs are. Determine the problem.

40. Determine if the flip-flop in Figure 8–98 is operating properly, and if not, identify the most probable fault.

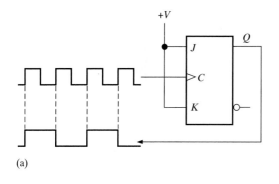

(a)

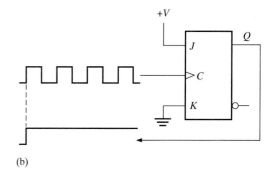

(b)

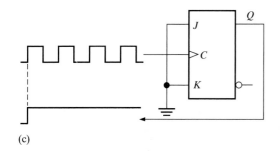

(c)

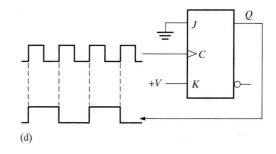

(d)

FIGURE 8–96

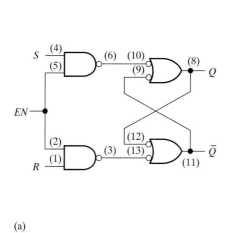

(a)

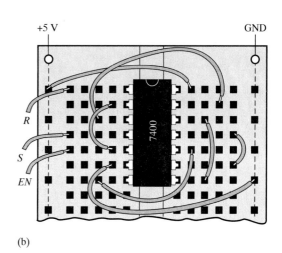

(b)

FIGURE 8–97

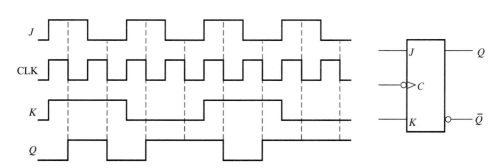

FIGURE 8–98

41. The parallel data storage circuit in Figure 8–40 (page 407) does not operate properly. To check it out, you first make sure that V_{CC} and ground are connected, and then you apply LOW levels to all the D inputs and pulse the clock line. You check the Q outputs and find them all to be LOW; so far, so good. Next you apply HIGHs to all the D inputs and again pulse the clock line. When you check the Q outputs, they are still all LOW. What is the problem, and what procedure will you use to isolate the fault to a single device?

42. The flip-flop circuit in Figure 8–99(a) is used to generate a binary count sequence. The gates form a decoder that is supposed to produce a HIGH when a binary zero or a binary three state occurs (00 or 11). When you check the Q_A and Q_B outputs, you get the display shown in part (b), which reveals glitches on the decoder output (X) in addition to the correct pulses. What is causing these glitches, and how can you eliminate them?

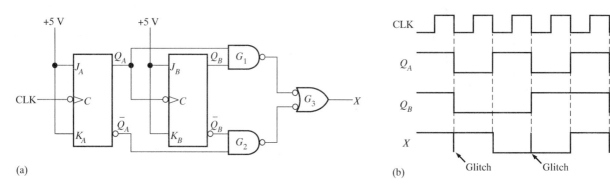

(a)   (b)

FIGURE 8–99

43. Determine the Q_A, Q_B, and X outputs over six clock pulses in Figure 8–99(a) for each of the following faults in the TTL circuits. Start with both Q_A and Q_B LOW.

(a) J_A input open   (b) K_B input open

(c) Q_B output open   (d) clock input to flip-flop B shorted

(e) gate G_2 output open

44. Two 74121 one-shots are connected on a circuit board as shown in Figure 8–100. After observing the oscilloscope display, do you conclude that the circuit is operating properly? If not, what is the most likely problem?

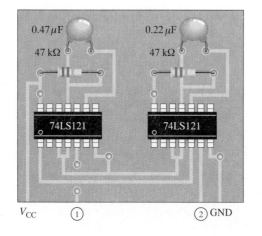 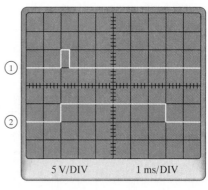

FIGURE 8–100

SECTION 8–9   Digital System Application

45. Use 555 timers to implement the 4 s and the 25 s one-shots for the timing circuits portion of the system. The trigger input to the 555 cannot stay LOW after its negative-going transition, so you will have to develop a circuit to produce very short negative-going pulses to trigger the long and short timers when the system goes into each state.

46. Modify the 10 kHz oscillator circuit in the timing portion of the traffic light control system to produce an output with a 25% duty cycle.

Special Design Problems

47. Devise a basic counting circuit that produces a binary sequence from zero through seven by using negative edge-triggered J-K flip-flops.

48. In the shipping department of a softball factory, the balls roll down a conveyor and through a chute single file into boxes for shipment. Each ball passing through the chute activates a switch circuit that produces an electrical pulse. The capacity of each box is 32 balls. Design a logic circuit to indicate when a box is full so that an empty box can be moved into position.

49. List the design changes that would be necessary in the traffic light control system to add a 15 s left turn arrow for the main street. The turn arrow will occur after the red light and prior to the green light. Modify the state diagram from Chapter 6 (Figure 6–84, p. 319) to show these changes.

■ ANSWERS TO SECTION REVIEWS

SECTION 8–1

1. Three types of latches are S-R, gated S-R, and gated D.

2. $SR = 00$, NC; $SR = 01$, $Q = 0$; $SR = 10$, $Q = 1$; $SR = 11$, invalid

3. $Q = 1$

SECTION 8–2

1. The output of a gated S-R latch can change any time the gate enable *(EN)* input is active. The output of an edge-triggered S-R flip-flop can change only on the triggering edge of a clock pulse.

2. The J-K flip-flop does not have an invalid state as does the S-R flip-flop.

3. Output Q goes HIGH on the trailing edge of the first clock pulse, LOW on the trailing edge of the second pulse, HIGH on the trailing edge of the third pulse, and LOW on the trailing edge of the fourth pulse.

SECTION 8–3

1. In the pulse-triggered flip-flop (master-slave), a data bit goes into the master section on the leading edge of a clock pulse and is transferred to the output (slave) on the trailing edge. In the edge-triggered flip-flop, a data bit goes into the flip-flop and appears on the output on the same clock edge.

2. **(a)** Nothing happens for the positive edge-triggered flip-flop.

   (b) The output of a pulse-triggered MS flip-flop will change.

3. A data lock-out is sensitive to the input data only during a clock transition and not while the clock is in its active state.

SECTION 8–4

1. **(a)** Set-up time is the time required for input data to be present before the triggering edge of the clock pulse.

   (b) Hold time is the time required for data to remain on input after the triggering edge of the clock pulse.

2. The 74S74 can be operated at the highest frequency, according to Table 8–6.

SECTION 8–5

1. A group of data storage flip-flops is a register.
2. For divide-by-2 operation, the flip-flop must toggle ($J = 1, K = 1$).
3. Six flip-flops are used in a divide-by-64 device.

SECTION 8–6

1. A nonretriggerable one-shot times out before it can respond to another trigger input. A retriggerable one-shot responds to each trigger input.
2. Pulse width is set with external R and C components.

SECTION 8–7

1. An astable has no stable state. A monostable has one stable state.
2. dc = (15 ms/20 ms)100% = 75%

SECTION 8–8

1. Yes, a negative edge-triggered D flip-flop can be used.
2. An astable multivibrator using a 555 timer can be used to provide the clock.

SECTION 8–9

1. The switches are used to simulate the 2-bit Gray code input.
2. The 4 s interval and the 25 s interval occur at different states.
3. none

9

COUNTERS

■ **CHAPTER OBJECTIVES**

☐ Describe the difference between an asynchronous and a synchronous counter
☐ Analyze counter timing diagrams
☐ Analyze counter circuits
☐ Explain how propagation delays affect the operation of a counter
☐ Determine the modulus of a counter
☐ Modify the modulus of a counter
☐ Recognize the difference between a 4-bit binary counter and a decade counter
☐ Use an up/down counter to generate forward and reverse binary sequences
☐ Determine the sequence of a counter
☐ Use IC counters in various applications
☐ Design a counter that will have any specified sequence of states
☐ Use cascaded counters to achieve a higher modulus
☐ Use logic gates to decode any given state of a counter
☐ Eliminate glitches in counter decoding
☐ Explain how a digital clock operates
☐ Troubleshoot counters for various types of faults
☐ Interpret counter logic symbols that use dependency notation
☐ Apply a counter in a system application

■ CHAPTER OVERVIEW

As you learned in Chapter 7, flip-flops can be connected together to perform counting operations. Such a group of flip-flops is a counter. The number of flip-flops used and the way in which they are connected determine the number of states (called the modulus) and also the specific sequence of states that the counter goes through during each complete cycle.

Counters are classified into two broad categories according to the way they are clocked: asynchronous and synchronous. In asynchronous counters, commonly called *ripple counters,* the first flip-flop is clocked by the external clock pulse and then each successive flip-flop is clocked by the output of the preceding flip-flop. In synchronous counters, the clock input is connected to all of the flip-flops so that they are clocked simultaneously. Within each of these two categories, counters are classified primarily by the type of sequence, the number of states, or the number of flip-flops in the counter.

■ SPECIFIC DEVICES

74LS93A	74LS153	74LS160A
74LS249	74LS161A	74LS163A
74LS190	74LS00	74LS74A
74LS08		

■ DIGITAL SYSTEM APPLICATION

This Digital System Application illustrates the concepts taught in this chapter. The system application in Section 9–10 continues the traffic light control system from the last three chapters. The focus of this chapter is the sequential logic portion of the system that produces the traffic light sequence based on inputs from the timing circuits and the vehicle sensor.

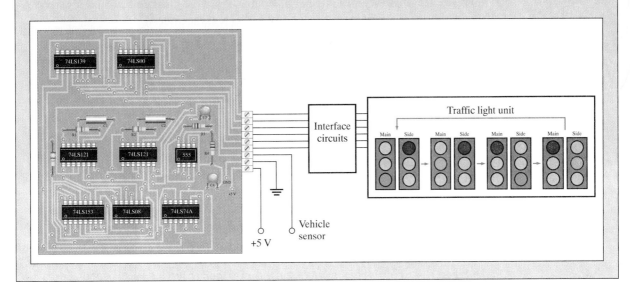

9–1 ■ ASYNCHRONOUS COUNTER OPERATION

The term asynchronous refers to events that do not have a fixed time relationship with each other and, generally, do not occur at the same time. An asynchronous counter is one in which the flip-flops (FF) within the counter do not change states at exactly the same time because they do not have a common clock pulse. After completing this section, you should be able to

□ Describe the operation of a 2-bit asynchronous binary counter □ Describe the operation of a 3-bit asynchronous binary counter □ Define *ripple* in relation to counters □ Describe the operation of an asynchronous decade counter □ Develop counter timing diagrams □ Discuss the 74LS93A 4-bit binary counter

A 2-Bit Asynchronous Binary Counter

Figure 9–1 shows a 2-bit counter connected for **asynchronous** operation. Notice that the clock line (CLK) is connected to the clock input (*C*) of only the first flip-flop, FF0. The second flip-flop, FF1, is triggered by the $\overline{Q}_0$ output of FF0. FF0 changes state at the positive-going edge of each clock pulse, but FF1 changes only when triggered by a positive-going transition of the $\overline{Q}_0$ output of FF0. Because of the inherent propagation delay time through a flip-flop, a transition of the input clock pulse and a transition of the $\overline{Q}_0$ output of FF0 can never occur at exactly the same time. Therefore, the two flip-flops are never simultaneously triggered, so the counter operation is asynchronous.

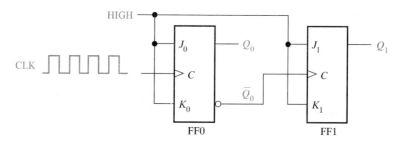

FIGURE 9–1

A 2-bit asynchronous binary counter.

The Timing Diagram   Let's examine the basic operation of the **asynchronous counter** of Figure 9–1 by applying four clock pulses to FF0 and observing the *Q* output of each flip-flop. Figure 9–2 illustrates the changes in the state of the flip-flop outputs in response to the clock pulses. Both flip-flops are connected for toggle operation ($J = 1$, $K = 1$) and are assumed to be initially RESET (*Q* LOW).

The positive-going edge of CLK1 (clock pulse 1) causes the Q_0 output of FF0 to go HIGH, as shown in Figure 9–2. At the same time the $\overline{Q}_0$ output goes LOW, but it has no effect on FF1, because a positive-going transition must occur to trigger the flip-flop. After

FIGURE 9–2

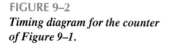

Timing diagram for the counter of Figure 9–1.

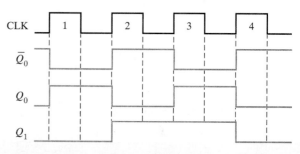

the leading edge of CLK1, $Q_0 = 1$ and $Q_1 = 0$. The positive-going edge of CLK2 causes Q_0 to go LOW. Output $\overline{Q_0}$ goes HIGH and triggers FF1, causing Q_1 to go HIGH. After the leading edge of CLK2, $Q_0 = 0$ and $Q_1 = 1$. The positive-going edge of CLK3 causes Q_0 to go HIGH again. Output $\overline{Q_0}$ goes LOW and has no effect on FF1. Thus, after the leading edge of CLK3, $Q_0 = 1$ and $Q_1 = 1$. The positive-going edge of CLK4 causes Q_0 to go LOW, while $\overline{Q_0}$ goes HIGH and triggers FF1, causing Q_1 to go LOW. After the leading edge of CLK4, $Q_0 = 0$ and $Q_1 = 0$. The counter has now recycled to its original state (both flip-flops are RESET).

In the timing diagram, the waveforms of the Q_0 and Q_1 outputs are shown relative to the clock pulses as illustrated in Figure 9–2. For simplicity, the transitions of Q_0, Q_1, and the clock pulses are shown as simultaneous even though this is an asynchronous counter. There is, of course, some small delay between the CLK, Q_0, and Q_1 transitions.

Note in Figure 9–2 that the 2-bit counter exhibits four different states, as you would expect with two flip-flops ($2^2 = 4$). Also, notice that if Q_0 represents the least significant bit (LSB) and Q_1 represents the most significant bit (MSB), the sequence of counter states is actually a sequence of binary numbers as listed in Table 9–1. *In this book, Q_0 is always the LSB unless otherwise specified.*

TABLE 9–1

Clock Pulse	Q_1	Q_0
Initially	0	0
1	0	1
2	1	0
3	1	1
4 (recycles)	0	0

Since it goes through a binary **sequence,** the counter in Figure 9–1 is a binary counter. It actually counts the number of clock pulses up to three, and on the fourth pulse it recycles to its original state ($Q_0 = 0$, $Q_1 = 0$). The term **recycle** is commonly applied to counter operation; it refers to the transition of the counter from its final state back to its original state.

A 3-Bit Asynchronous Binary Counter

A 3-bit asynchronous binary counter is shown in Figure 9–3(a), on the next page. The basic operation is the same as that of the 2-bit counter just discussed, except that the 3-bit counter has eight states, due to its three flip-flops. A timing diagram is shown in Figure 9–3(b) for eight clock pulses. Notice that the counter in Figure 9–3 progresses through a binary count of zero through seven and then recycles to the zero state. This counter sequence is listed in Table 9–2.

TABLE 9–2
State sequence for a 3-stage binary counter

Clock Pulse	Q_2	Q_1	Q_0
Initially	0	0	0
1	0	0	1
2	0	1	0
3	0	1	1
4	1	0	0
5	1	0	1
6	1	1	0
7	1	1	1
8 (recycles)	0	0	0

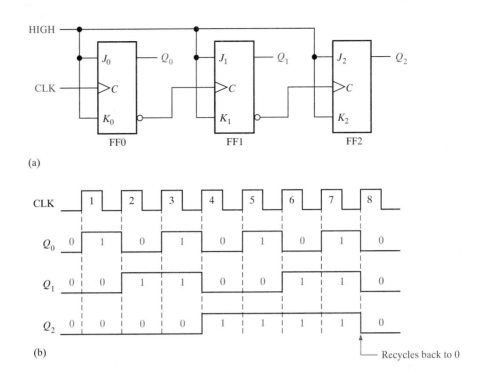

(a)

(b)

Recycles back to 0

FIGURE 9–3

Three-bit asynchronous binary counter and its timing diagram for one cycle.

Propagation Delay   Asynchronous counters are commonly referred to as **ripple coun-ters** for the following reason: The effect of the input clock pulse is first "felt" by FF0. This effect cannot get to FF1 immediately because of the propagation delay through FF0. Then there is the propagation delay through FF1 before FF2 can be triggered. Thus, the effect of an input clock pulse "ripples" through the counter, taking some time, due to propagation delays, to reach the last flip-flop.

To illustrate, notice that all three flip-flops in the counter of Figure 9–3 change state as a result of CLK4. This ripple clocking effect is shown in Figure 9–4 for the first four clock pulses, with the propagation delays indicated. The HIGH-to-LOW transition of Q_0 occurs one delay time (t_{PHL}) after the positive-going transition of the clock pulse. The HIGH-to-LOW transition of Q_1 occurs one delay time (t_{PHL}) after the positive-going transition of $\overline{Q}_0$. The LOW-to-HIGH transition of Q_2 occurs one delay time (t_{PHL}) after the positive-going transition of $\overline{Q}_1$. As you can see, FF2 is not triggered until two delay times after the positive-going edge of the clock pulse, CLK4. Thus, it takes three propagation de-

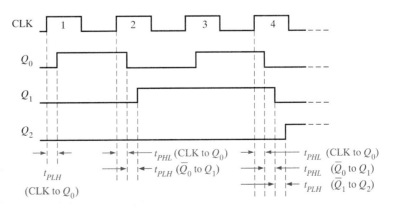

FIGURE 9–4

Propagation delays in an asynchronous (ripple-clocked) binary counter.

lay times for the effect of the clock pulse, CLK4, to ripple through the counter and change Q_2 from LOW to HIGH.

This cumulative delay of an asynchronous counter is a major disadvantage in many applications because it limits the rate at which the counter can be clocked and creates decoding problems. The maximum cumulative delay in a counter must be less than the period of the clock waveform.

EXAMPLE 9–1

A 4-bit asynchronous binary counter is shown in Figure 9–5(a). Each flip-flop is negative edge-triggered and has a propagation delay of 10 nanoseconds (ns). Draw a timing diagram showing the Q output of each flip-flop, and determine the total propagation delay time from the triggering edge of a clock pulse until a corresponding change can occur in the state of Q_3. Also determine the maximum clock frequency at which the counter can be operated.

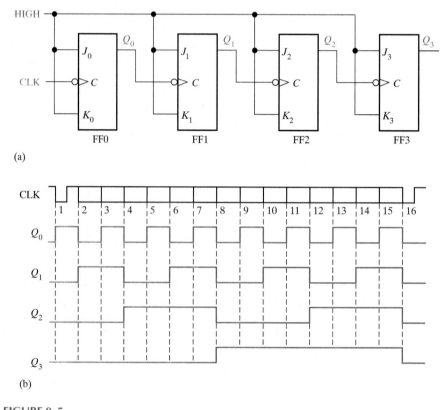

(a)

(b)

FIGURE 9–5
Four-bit asynchronous binary counter and its timing diagram.

Solution The timing diagram with delays omitted is as shown in Figure 9–5(b). For the total delay time, the effect of CLK8 or CLK16 must propagate through four flip-flops before Q_3 changes, so

$$t_{p(tot)} = 4 \times 10 \text{ ns} = 40 \text{ ns}$$

The maximum clock frequency is

$$f_{max} = \frac{1}{t_{p(tot)}} = \frac{1}{40 \text{ ns}} = 25 \text{ MHz}$$

Related Exercise Draw the timing diagram if all of the flip-flops in Figure 9–5(a) are positive edge-triggered.

Asynchronous Decade Counters

Regular binary counters, such as those previously introduced, have a maximum **modulus,** which means they progress through *all* of their possible states. The maximum possible number of states (maximum modulus) of a counter is 2^n, where n is the number of flip-flops in the counter.

Counters can also be designed to have a number of states in their sequence that is less than the maximum of 2^n. The resulting sequence is called a **truncated sequence.**

One common modulus for counters with **truncated** sequences is ten. Counters with ten states in their sequence (modulus-10) are called **decade counters.** A **decade** counter with a count sequence of zero (0000) through nine (1001) is a BCD decade counter because its ten-state sequence is the BCD code. This type of counter is useful in display applications in which BCD is required for conversion to a decimal readout.

To obtain a truncated sequence, it is necessary to force the counter to recycle before going through all of its normal states. For example, the BCD decade counter must recycle back to the 0000 state after the 1001 state. A decade counter requires four flip-flops (three flip-flops are insufficient because $2^3 = 8$).

We will use a 4-bit asynchronous counter such as the one in Figure 9–5(a) and modify its sequence to illustrate the principle of truncated counters. One way to make the counter recycle after the count of nine (1001) is to decode count ten (1010) with a NAND gate and connect the output of the NAND gate to the clear ($\overline{CLR}$) inputs of the flip-flops, as shown in Figure 9–6(a).

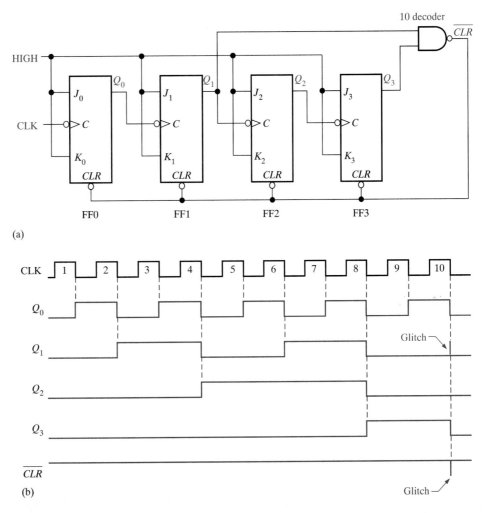

(a)

(b)

FIGURE 9–6
An asynchronously clocked decade counter with asynchronous recycling.

Notice in Figure 9–6(a) that only Q_1 and Q_3 are connected to the NAND gate inputs. This arrangement is an example of *partial decoding,* in which the two unique states ($Q_1 = 1$ and $Q_3 = 1$) are sufficient to decode the count of ten, because none of the other states (zero through nine) have both Q_1 and Q_3 HIGH at the same time. When the counter goes into count ten (1010), the decoding gate output goes LOW and asynchronously resets all the flip-flops.

The resulting timing diagram is shown in Figure 9–6(b). Notice that there is a glitch on the Q_1 waveform. The reason for this glitch is that Q_1 must first go HIGH before the count of ten can be decoded. Not until several nanoseconds after the counter goes to the count of ten does the output of the decoding gate go LOW (both inputs are HIGH). Thus, the counter is in the 1010 state for a short time before it is reset to 0000, thus producing the glitch on Q_1 and the resulting glitch on the $\overline{CLR}$ line which resets the counter.

Other truncated sequences can be implemented in a similar way, as Example 9–2 shows.

EXAMPLE 9–2

Show how an asynchronous counter can be implemented having a modulus of twelve with a straight binary sequence from 0000 through 1011.

Solution   Since three flip-flops can produce a maximum of eight states, four flip-flops are required to produce any modulus greater than eight but less than or equal to sixteen.

When the counter gets to its last state, 1011, it must recycle back to 0000 rather than going to its normal next state of 1100, as illustrated in the following sequence chart:

Q_3	Q_2	Q_1	Q_0	
0	0	0	0	←
.	.	.	.	
.	.	.	.	Recycles
.	.	.	.	
1	0	1	1	
1	1	0	0	← Normal next state

Observe that Q_0 and Q_1 both go to 0 anyway, but Q_2 and Q_3 must be forced to 0 on the twelfth clock pulse. Figure 9–7(a) shows the modulus-12 counter. The NAND gate partially decodes count twelve (1100) and resets flip-flop 2 and flip-flop 3. Thus, on the twelfth clock pulse, the counter is forced to recycle from count eleven to count zero, as shown in the timing diagram of Figure 9–7(b) ,on the next page. (It is in count twelve for only a few nanoseconds before it is reset by the glitch on $\overline{CLR}$.)

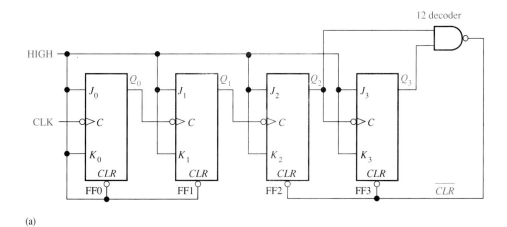

(a)

FIGURE 9–7

Asynchronously clocked modulus-12 counter with asynchronous recycling.

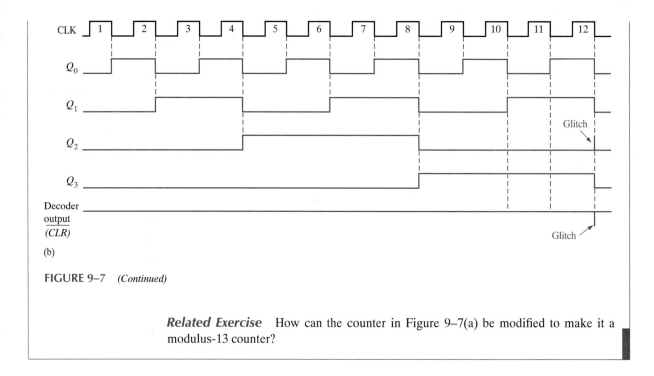

(b)

FIGURE 9–7 *(Continued)*

Related Exercise How can the counter in Figure 9–7(a) be modified to make it a modulus-13 counter?

The 74LS93A 4-Bit Binary Counter

The 74LS93A is an example of a specific integrated circuit asynchronous counter. As the logic diagram in Figure 9–8 shows, this device actually consists of a single flip-flop and a 3-bit asynchronous counter. This arrangement is for flexibility. It can be used as a divide-by-2 device if only the single flip-flop is used, or it can be used as a modulus-8 counter if only the 3-bit counter portion is used. This device also provides gated reset inputs, $RO(1)$ and $RO(2)$. When both of these inputs are HIGH, the counter is reset to the 0000 state by $\overline{CLR}$.

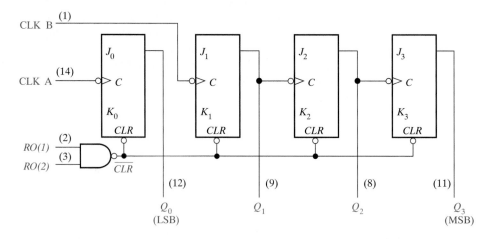

FIGURE 9–8
The 74LS93A 4-bit binary counter logic diagram. (Pin numbers are in parentheses, and all J and K inputs are internally connected HIGH.)

Additionally, the 74LS93A can be used as a 4-bit modulus-16 counter (counts 0 through 15) by connecting the Q_0 output to the CLK B input as shown in Figure 9–9(a). It can also be configured as a decade counter (counts 0 through 9) with asynchronous recycling by using the gated reset inputs for partial decoding of count ten, as shown in Figure 9–9(b).

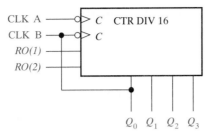

(a) 74LS93A connected as a modulus-16 counter

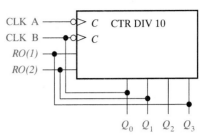

(b) 74LS93A connected as a decade counter

FIGURE 9–9

Two configurations of the 74LS93A asynchronous counter. (The qualifying label, CTR DIV n, indicates a counter with n states.)

EXAMPLE 9–3

Show how the 74LS93A can be used as a modulus-12 counter.

Solution Use the gated reset inputs, *RO(1)* and *RO(2)*, to partially decode count 12 (remember, there is an internal NAND gate associated with these inputs). The count-12 decoding is accomplished by connecting Q_3 to *RO(1)* and Q_2 to *RO(2)*, as shown in Figure 9–10. Output Q_0 is connected to CLK B to create a 4-bit counter.

FIGURE 9–10
74LS93A connected as a modulus-12 counter.

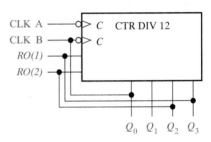

Immediately after the counter goes to count 12 (1100), it is reset to 0000. The recycling, however, results in a glitch on Q_2 because the counter must go into the 1100 state for several nanoseconds before recycling.

Related Exercise Show how the 74LS93A can be connected as a modulus-13 counter.

SECTION 9–1 REVIEW

1. What does the term *asynchronous* mean in relation to counters?
2. How many states does a modulus-14 counter have? What is the minimum number of flip-flops required?

9–2 ■ SYNCHRONOUS COUNTER OPERATION

The term synchronous refers to events that have a fixed time relationship with each other. With respect to counter operation, synchronous means that the flip-flops within the counter are all clocked at the same time by a common clock pulse. After completing this section, you should be able to

☐ Describe the operation of a 2-bit synchronous binary counter   ☐ Describe the operation of a 3-bit synchronous binary counter   ☐ Describe the operation of a 4-bit synchronous binary counter □ Describe the operation of a synchronous decade counter □ Develop counter timing diagrams □ Discuss the 74LS163A 4-bit binary counter and the 74LS160A BCD decade counter

A 2-Bit Synchronous Binary Counter

Figure 9–11 shows a 2-bit **synchronous** counter. Notice that an arrangement different from that for the asynchronous counter must be used for the J_1 and K_1 inputs of FF1 in order to achieve a binary sequence.

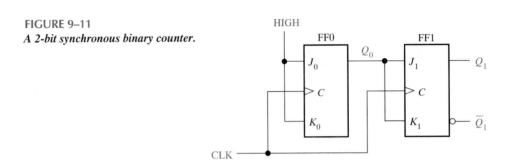

The operation of this **synchronous counter** is as follows: First, assume that the counter is initially in the binary 0 state; that is, both flip-flops are RESET. When the positive edge of the first clock pulse is applied, FF0 will toggle and Q_0 will therefore go HIGH. What happens to FF1 at the positive-going edge of CLK1? To find out, let's look at the input conditions of FF1. Inputs J_1 and K_1 are both LOW because Q_0, to which they are connected, has not yet gone HIGH. Remember, there is a propagation delay from the triggering edge of the clock pulse until the Q output actually makes a transition. So, $J = 0$ and $K = 0$ when the leading edge of the first clock pulse is applied. This is a no-change condition, and therefore FF1 does not change state. A timing detail of this portion of the counter operation is shown in Figure 9–12(a).

After CLK1, $Q_0 = 1$ and $Q_1 = 0$ (which is the binary 1 state). When the leading edge of CLK2 occurs, FF0 will toggle and Q_0 will go LOW. Since FF1 has a HIGH ($Q_0 = 1$) on

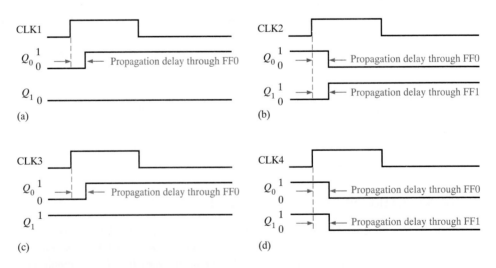

FIGURE 9–12
Timing details for the 2-bit synchronous counter operation (the propagation delays of both flip-flops are assumed to be equal).

its J_1 and K_1 inputs at the triggering edge of this clock pulse, the flip-flop toggles and Q_1 goes HIGH. Thus, after CLK2, $Q_0 = 0$ and $Q_1 = 1$ (which is a binary 2 state). The timing detail for this condition is shown in Figure 9–12(b).

When the leading edge of CLK3 occurs, FF0 again toggles to the SET state ($Q_0 = 1$), and FF1 remains SET ($Q_1 = 1$) because its J_1 and K_1 inputs are both LOW ($Q_0 = 0$). After this triggering edge, $Q_0 = 1$ and $Q_1 = 1$ (which is a binary 3 state). The timing detail is shown in Figure 9–12(c).

Finally, at the leading edge of CLK4, Q_0 and Q_1 go LOW because they both have a toggle condition on their J and K inputs. The timing detail is shown in Figure 9–12(d). The counter has now recycled to its original state, binary 0.

The complete timing diagram for the counter in Figure 9–11 is shown in Figure 9–13. Notice that all the waveform transitions appear coincident; that is, the propagation delays are not indicated. Although the delays are an important factor in the synchronous counter operation, in an overall timing diagram they are normally omitted for simplicity. Major waveform relationships resulting from the logical operation of a circuit can be conveyed completely without showing small delay and timing differences.

FIGURE 9–13
Timing diagram for the counter of Figure 9–11.

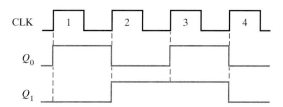

A 3-Bit Synchronous Binary Counter

A 3-bit synchronous binary counter is shown in Figure 9–14, and its timing diagram is shown in Figure 9–15 (next page). An understanding of this counter can be achieved by a careful examination of its sequence of states as shown in Table 9–3.

First, let's look at Q_0. Notice that Q_0 changes on each clock pulse as the counter progresses from its original state to its final state and then back to its original state. To produce this operation, FF0 must be held in the toggle mode by constant HIGHs on its J_0 and K_0 inputs. Notice that Q_1 goes to the opposite state following each time Q_0 is a 1. This change occurs at CLK2, CLK4, CLK6, and CLK8. The CLK8 pulse causes the counter to recycle. To produce this operation, Q_0 is connected to the J_1 and K_1 inputs of FF1. When Q_0 is a 1 and a clock pulse occurs, FF1 is in the toggle mode and therefore changes state. The other times, when Q_0 is a 0, FF1 is in the no-change mode and remains in its present state.

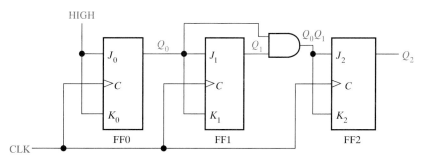

FIGURE 9–14
A 3-bit synchronous binary counter.

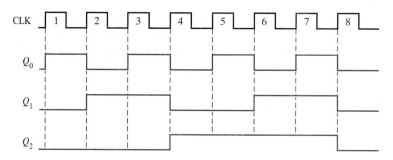

FIGURE 9–15
Timing diagram for the counter of Figure 9–14.

TABLE 9–3
State sequence for a 3-bit binary counter

Clock Pulse	Q_2	Q_1	Q_0
Initially	0	0	0
1	0	0	1
2	0	1	0
3	0	1	1
4	1	0	0
5	1	0	1
6	1	1	0
7	1	1	1
8 (recycles)	0	0	0

Next, let's see how FF2 is made to change at the proper times according to the binary sequence. Notice that both times Q_2 changes state, it is preceded by the unique condition in which both Q_0 and Q_1 are HIGH. This condition is detected by the AND gate and applied to the J_2 and K_2 inputs of FF2. Whenever both Q_0 and Q_1 are HIGH, the output of the AND gate makes the J_2 and K_2 inputs of FF2 HIGH, and FF2 toggles on the following clock pulse. At all other times, the J_2 and K_2 inputs of FF2 are held LOW by the AND gate output, and FF2 does not change state.

A 4-Bit Synchronous Binary Counter

Figure 9–16(a) shows a 4-bit binary counter, and Figure 9–16(b) shows its timing diagram. This particular counter is implemented with negative edge-triggered flip-flops. The reasoning behind the J and K input control for the first three flip-flops is the same as previously discussed for the 3-bit counter. The fourth stage, FF3, changes only twice in the sequence. Notice that both of these transitions occur following the times that Q_0, Q_1, and Q_2 are all HIGH. This condition is decoded by AND gate G_2 so that when a clock pulse occurs, FF3 will change state. For all other times the J_3 and K_3 inputs of FF3 are LOW, and it is in a no-change condition.

Synchronous Decade Counters

As you know, a BCD decade counter exhibits a truncated binary sequence and goes from 0000 through the 1001 state. Rather than going to the 1010 state, it recycles to the 0000 state. A synchronous BCD decade counter is shown in Figure 9–17.

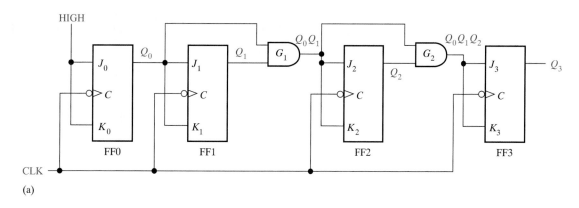

(a)

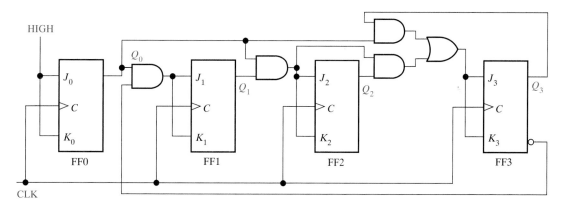

(b)

FIGURE 9–16

A 4-bit synchronous binary counter and timing diagram. Points where the AND gate outputs are HIGH are indicated by the shaded areas.

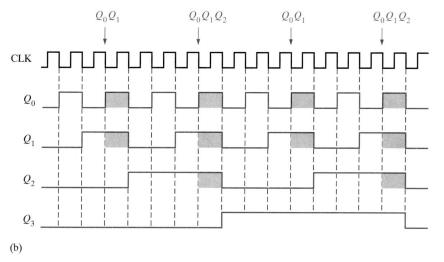

FIGURE 9–17

A synchronous BCD decade counter.

TABLE 9–4

States of a BCD decade counter

Clock Pulse	Q_3	Q_2	Q_1	Q_0
Initially	0	0	0	0
1	0	0	0	1
2	0	0	1	0
3	0	0	1	1
4	0	1	0	0
5	0	1	0	1
6	0	1	1	0
7	0	1	1	1
8	1	0	0	0
9	1	0	0	1
10 (recycles)	0	0	0	0

The counter operation can be understood by examining the sequence of states in Table 9–4 and by following the implementation in Figure 9–17. First, notice that FF0 (Q_0) toggles on each clock pulse, so the logic equation for its J_0 and K_0 inputs is

$$J_0 = K_0 = 1$$

This equation is implemented by connecting J_0 and K_0 to a constant HIGH level.

Next, notice in Table 9–4 that FF1 (Q_0) changes on the next clock pulse each time $Q_0 = 1$ and $Q_3 = 0$, so the logic equation for the J_1 and K_1 inputs is

$$J_1 = K_1 = Q_0\overline{Q}_3$$

This equation is implemented by ANDing Q_0 and $\overline{Q}_3$ and connecting the gate output to the J_1 and K_1 inputs of FF1.

Flip-flop 2 (Q_2) changes on the next clock pulse each time both $Q_0 = 1$ and $Q_1 = 1$. This requires an input logic equation as follows:

$$J_2 = K_2 = Q_0Q_1$$

This equation is implemented by ANDing Q_0 and Q_1 and connecting the gate output to the J_2 and K_2 inputs of FF2.

Finally, FF3 (Q_3) changes to the opposite state on the next clock pulse each time $Q_0 = 1$, $Q_1 = 1$, and $Q_2 = 1$ (state 7), or when $Q_0 = 1$ and $Q_3 = 1$ (state 9). The equation for this is as follows:

$$J_3 = K_3 = Q_0Q_1Q_2 + Q_0Q_3$$

This function is implemented with the AND/OR logic connected to the J_3 and K_3 inputs of FF3 as shown in the logic diagram in Figure 9–17. Notice that the only difference between this decade counter and the modulus-16 binary counter in Figure 9–16 is the Q_0Q_3 AND gate and the OR gate; this arrangement detects the occurrence of the 1001 state and causes the counter to recycle properly on the next clock pulse. The timing diagram for the decade counter is shown in Figure 9–18.

The 74LS163A Synchronous 4-Bit Binary Counter

The 74LS163A is an example of an integrated circuit synchronous 4-bit binary counter. A logic symbol is shown in Figure 9–19 with pin numbers in parentheses. This counter has several features in addition to the basic functions previously discussed for the general synchronous binary counter.

First, the counter can be synchronously preset to any 4-bit binary number by applying the proper levels to the parallel data inputs. When a LOW is applied to the $\overline{LOAD}$ in-

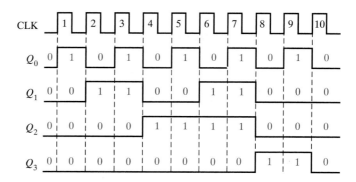

FIGURE 9–18

Timing diagram for the BCD decade counter (Q_0 is the LSB).

FIGURE 9–19

The 74LS163A 4-bit synchronous binary counter. (The qualifying label CTR DIV 16 indicates a counter with sixteen states.)

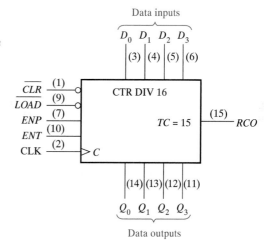

put, the counter will assume the state of the data inputs on the next clock pulse. Thus, the counter sequence can be started with any 4-bit binary number.

Also, there is an active-LOW clear input ($\overline{CLR}$), which synchronously RESETS all four flip-flops in the counter. There are two enable inputs, *ENP* and *ENT*. These inputs must both be HIGH for the counter to sequence through its binary states. When at least one input is LOW, the counter is disabled. The ripple clock output (*RCO*) goes HIGH when the counter reaches a **terminal count** of fifteen (*TC* = 15). This output, in conjunction with the enable inputs, allows these counters to be cascaded for higher count sequences, as will be discussed later.

Figure 9–20, on the next page, shows a timing diagram of this counter being preset to twelve (1100) and then counting up to its terminal count, fifteen (1111). Input D_0 is the least significant input bit, and Q_0 is the least significant output bit.

Let's examine this timing diagram in detail. This will aid you in interpreting timing diagrams found later in this chapter or in manufacturers' data sheets. To begin, the LOW level pulse on the $\overline{CLR}$ input causes all the outputs (Q_0, Q_1, Q_2, and Q_3) to go LOW.

Next, the LOW level pulse on the $\overline{LOAD}$ input synchronously enters the data on the data inputs (D_0, D_1, D_2, and D_3) into the counter. These data appear on the Q outputs at the time of the first positive-going clock edge after $\overline{LOAD}$ goes LOW. This is the preset operation. In this particular example, Q_0 is LOW, Q_1 is LOW, Q_2 is HIGH, and Q_3 is HIGH. This, of course, is a binary 12 (Q_0 is the LSB).

The counter now advances through states 13, 14, and 15 on the next three positive-going clock edges. It then recycles to 0, 1, 2 on the following clock pulses. Notice that both

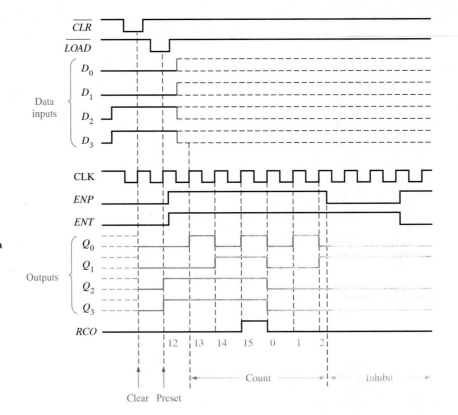

FIGURE 9–20
Timing example for a 74LS163A.

ENP and *ENT* inputs are HIGH during the state sequence. When *ENP* goes LOW, the counter is inhibited and remains in the binary 2 state.

The 74LS160A Synchronous BCD Decade Counter

This device has the same inputs and outputs as the 74LS163A binary counter previously discussed. It can be preset to any BCD count by the use of the data inputs and a LOW on the $\overline{LOAD}$ input. A LOW on the $\overline{CLR}$ will RESET the counter. The enable inputs *ENP* and *ENT* must both be HIGH for the counter to advance through its sequence of states in response to a positive transition on the CLK input. As in the 74LS163A, the enable inputs in conjunction with the ripple clock output *RCO* (terminal count of 1001) provide for cascading several decade counters. Figure 9–21 shows the logic symbol for the 74LS160A

FIGURE 9–21

The 74LS160A synchronous BCD decade counter. (The qualifying label CTR DIV 10 indicates a counter with ten states.)

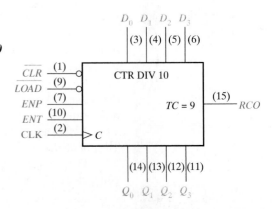

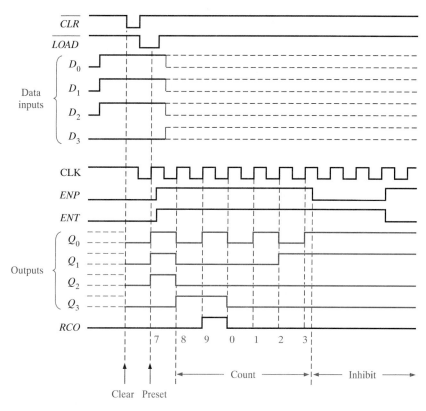

FIGURE 9–22
Timing example for a 74LS160A.

counter, and Figure 9–22 is a timing diagram showing the counter being preset to count 7 (0111). Cascaded counters will be discussed in Section 9–5.

SECTION 9–2 REVIEW

1. How does a synchronous counter differ from an asynchronous counter?
2. Explain the function of the preset feature of counters such as the 74LS160A and the 74LS163A.
3. Describe the purpose of the *ENP* and *ENT* inputs and the *RCO* output for the two specific counters introduced in this section.

9–3 ■ UP/DOWN SYNCHRONOUS COUNTERS

An up/down counter is one that is capable of progressing in either direction through a certain sequence. An up/down counter, sometimes called a bidirectional counter, can have any specified sequence of states. A 3-bit binary counter that advances upward through its sequence (0, 1, 2, 3, 4, 5, 6, 7) and then can be reversed so that it goes through the sequence in the opposite direction (7, 6, 5, 4, 3, 2, 1, 0) is an illustration of up/down sequential operation. After completing this section, you should be able to

☐ Explain the basic operation of an up/down counter   ☐ Discuss the 74LS190 up/down decade counter

In general, most **up/down counters** can be reversed at any point in their sequence. For instance, the 3-bit binary counter can be made to go through the following sequence:

$$\overbrace{0, 1, 2, 3, 4, 5,}^{\text{UP}} \underbrace{4, 3, 2,}_{\text{DOWN}} \overbrace{3, 4, 5, 6, 7,}^{\text{UP}} \underbrace{6, 5,}_{\text{DOWN}} \text{etc.}$$

Table 9–5 shows the complete up/down sequence for a 3-bit binary counter. The arrows indicate the state-to-state movement of the counter for both its UP and its DOWN modes of operation. An examination of Q_0 for both the up and down sequences shows that FF0 toggles on each clock pulse. So the J_0 and K_0 inputs of FF0 are

$$J_0 = K_0 = 1$$

For the up sequence, Q_1 changes state on the next clock pulse when $Q_0 = 1$. For the down sequence, Q_1 changes on the next clock pulse when $Q_0 = 0$. Thus, the J_1 and K_1 inputs of FF1 must equal 1 under the conditions expressed by the following equation:

$$J_1 = K_1 = (Q_0 \cdot \text{UP}) + (\overline{Q}_0 \cdot \text{DOWN})$$

For the up sequence, Q_2 changes state on the next clock pulse when $Q_0 = Q_1 = 1$. For the down sequence, Q_2 changes on the next clock pulse when $Q_0 = Q_1 = 0$. Thus, the J_2 and K_2 inputs of FF2 must equal 1 under the conditions expressed by the following equation:

$$J_2 = K_2 = (Q_0 \cdot Q_1 \cdot \text{UP}) + (\overline{Q}_0 \cdot \overline{Q}_1 \cdot \text{DOWN})$$

TABLE 9–5
Up/Down sequence for a 3-bit binary counter

Clock Pulse	UP	Q_2	Q_1	Q_0	DOWN
0		0	0	0	
1		0	0	1	
2		0	1	0	
3		0	1	1	
4		1	0	0	
5		1	0	1	
6		1	1	0	
7		1	1	1	

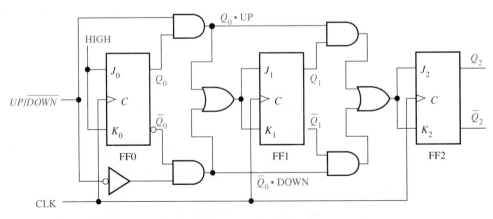

FIGURE 9–23
A basic 3-bit up/down synchronous counter.

Each of the conditions for the J and K inputs of each flip-flop produces a toggle at the appropriate point in the counter sequence.

Figure 9–23 shows a basic implementation of a 3-bit up/down binary counter using the logic equations just developed for the J and K inputs of each flip-flop. Notice that the $UP/\overline{DOWN}$ control input is HIGH for UP and LOW for DOWN.

EXAMPLE 9–4

Sketch the timing diagram and determine the sequence of a synchronous 4-bit binary up/down counter if the clock and $UP/\overline{DOWN}$ control inputs have waveforms as shown in Figure 9–24(a). The counter starts in the all 0s state and is positive edge-triggered.

FIGURE 9–24

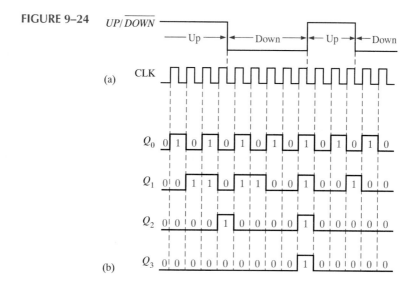

(a)

(b)

Solution The timing diagram showing the Q outputs is shown in Figure 9–24(b). From these waveforms, the counter sequence is as shown in Table 9–6.

TABLE 9–6

Q_3	Q_2	Q_1	Q_0	
0	0	0	0	
0	0	0	1	
0	0	1	0	UP
0	0	1	1	
0	1	0	0	
0	0	1	1	
0	0	1	0	
0	0	0	1	DOWN
0	0	0	0	
1	1	1	1	
0	0	0	0	
0	0	0	1	UP
0	0	1	0	
0	0	0	1	
0	0	0	0	DOWN

Related Exercise Draw the timing diagram if the $UP/\overline{DOWN}$ control waveform in Figure 9–24(a) is inverted.

The 74LS190 Up/Down Decade Counter

Figure 9–25 shows a logic diagram for the 74LS190, a good example of an integrated circuit up/down counter. The direction of the count is determined by the level of the up/down input ($D/\overline{U}$). When this input is HIGH, the counter counts down; when it is LOW, the counter counts up. Also, this device can be preset to any desired BCD digit as determined by the states of the data inputs when the *LOAD* input is LOW.

FIGURE 9–25
The 74LS190 up/down synchronous decade counter.

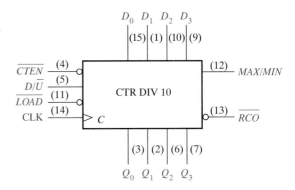

The *MAX/MIN* output produces a HIGH pulse when the terminal count nine (1001) is reached in the UP mode or when the terminal count zero (0000) is reached in the DOWN mode. This *MAX/MIN* output, along with the ripple clock output ($\overline{RCO}$) and the count enable input ($\overline{CTEN}$), is used when cascading counters. (Cascaded counters are discussed in Section 9–5.)

Figure 9–26 is an example timing diagram showing the 74LS190 counter preset to seven (0111) and then going through a count-up sequence followed by a count-down sequence. The *MAX/MIN* output is HIGH when the counter is in either the all-0s state (*MIN*) or the 1001 state (*MAX*).

FIGURE 9–26
Timing example for a 74LS190.

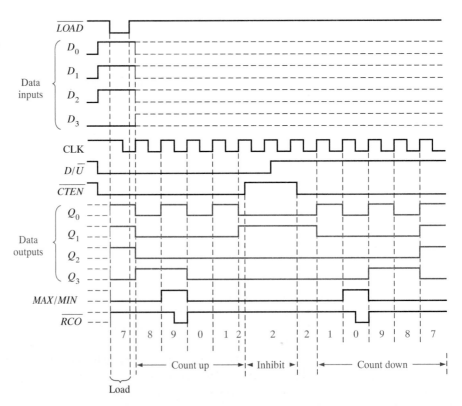

9–4 ■ DESIGN OF SYNCHRONOUS COUNTERS

This section is optional and may be omitted without affecting the flow of material in the remainder of the book. Inclusion of this material is recommended for those who want an introduction to counter design or to state machine design in general. After completing this section, you should be able to

☐ Describe a general sequential circuit in terms of its basic parts and its input and outputs ☐ Develop a state diagram for a given sequence ☐ Develop a next-state table for a specified counter sequence ☐ Create a flip-flop transition table ☐ Use the Karnaugh map method to derive the logic requirements for a synchronous counter ☐ Implement a counter to produce a specified sequence of states

General Model of a Sequential Circuit

Before proceeding with a specific design technique, let's begin with a general definition of a **sequential circuit (state machine):** A general sequential circuit consists of a combinational logic section and a memory section (flip-flops), as shown in Figure 9–27. In a clocked sequential circuit, there is a clock input to the memory section as indicated.

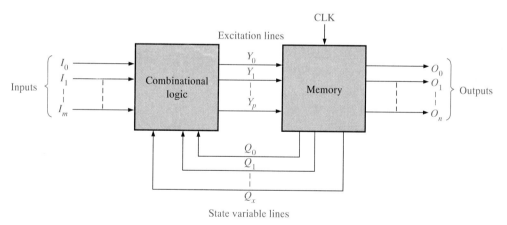

FIGURE 9–27
General clocked sequential circuit.

The information stored in the memory section, as well as the inputs to the combinational logic ($I_0, I_1, \ldots, I_m$), is required for proper operation of the circuit. At any given time, the memory is in a state called the *present state* and will advance to a *next state* on a clock pulse as determined by conditions on the excitation lines ($Y_0, Y_1, \ldots, Y_p$). The present state of the memory is represented by the state variables ($Q_0, Q_1, \ldots, Q_x$). These state variables, along with the inputs ($I_0, I_1, \ldots, I_m$), determine the system outputs ($O_0, O_1, \ldots, O_n$).

Not all sequential circuits have input and output variables as in the general model just discussed. However, all have excitation variables and state variables. Counters are a special case of clocked sequential circuits. In this section, a general design procedure for sequential circuits is applied to synchronous counters in a series of steps.

Step 1: State Diagram

A counter is first described by a **state diagram,** which shows the progression of states through which the counter advances when it is clocked. As an example, Figure 9–28 is a state diagram for a basic 3-bit Gray code counter. This particular circuit has no inputs other than the clock and no outputs other than the outputs taken off each flip-flop in the counter. You may wish to review the coverage of the Gray code in Chapter 2 at this time.

FIGURE 9–28

State diagram for a 3-bit Gray code counter.

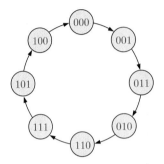

Step 2: Next-State Table

Once the sequential circuit is defined by a state diagram, the second step is to derive a next-state table, which lists each state of the counter (present state) along with the corresponding next state. *The next state is the state that the counter goes to from its present state upon application of a clock pulse.* The next-state table is derived from the state diagram and is shown in Table 9–7 for the 3-bit Gray code counter. Q_0 is the least significant bit.

TABLE 9–7

Next-state table for 3-bit Gray code counter

Present State			Next State		
Q_2	Q_1	Q_0	Q_2	Q_1	Q_0
0	0	0	0	0	1
0	0	1	0	1	1
0	1	1	0	1	0
0	1	0	1	1	0
1	1	0	1	1	1
1	1	1	1	0	1
1	0	1	1	0	0
1	0	0	0	0	0

Step 3: Flip-Flop Transition Table

Table 9–8 is a transition table for the J-K flip-flop. All possible output transitions are listed by showing the Q output of the flip-flop going from present states to next states. Q_N is the present state of the flip-flop (before a clock pulse) and Q_{N+1} is the next state (after a clock pulse). For each output transition, the J and K inputs that will cause the transition to occur are listed. The Xs indicate a "don't care" (the input can be either a 1 or a 0).

In designing the counter, the transition table is applied to each of the flip-flops in the counter, based on the next-state table (Table 9–7). For example, for the present state 000, Q_0 goes from a present state of 0 to a next state of 1. To make this happen, J_0 must be a 1 and we

TABLE 9–8
Transition table for a J-K flip-flop

Output Transitions		Flip-Flop Inputs	
Q_N	Q_{N+1}	J	K
0 $\longrightarrow$ 0		0	X
0 $\longrightarrow$ 1		1	X
1 $\longrightarrow$ 0		X	1
1 $\longrightarrow$ 1		X	0

Q_N: present state
Q_{N+1}: next state
X: "don't care"

don't care what K_0 is ($J_0 = 1$, $K_0 = $ X), as you can see in the transition table
(Table 9–8). Next, Q_1 is 0 in the present state and remains a 0 in the next state. For this transi-
tion, $J_1 = 0$ and $K_1 = $ X. Finally, Q_2 is 0 in the present state and remains a 0 in the next state.
Therefore, $J_2 = 0$ and $K_2 = $ X. This analysis in repeated for each present state in Table 9–7.

Step 4: Karnaugh Maps

Karnaugh maps are used to determine the logic required for the J and K inputs of each flip-
flop in the counter. There is a Karnaugh map for the J input and a Karnaugh map for the K
input of each flip-flop. In this design procedure, each cell in a Karnaugh map represents
one of the present states in the counter sequence listed in Table 9–7.

From the J and K states in the transition table (Table 9–8) a 1, 0, or X is entered into
each present state cell on the maps depending on the transition of the Q output for a particular flip-flop. To illustrate this procedure, two sample entries are shown for the J_0 and the
K_0 inputs to the least significant flip-flop (Q_0) in Figure 9–29.

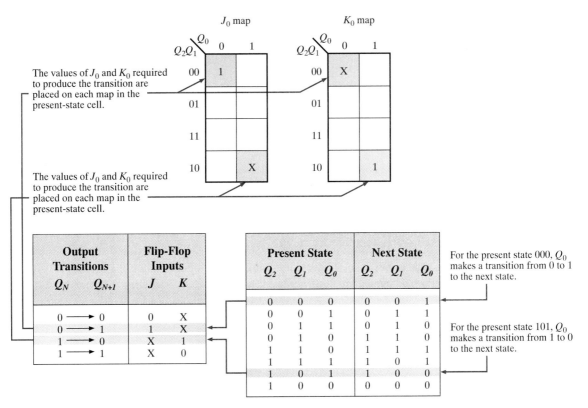

FIGURE 9–29
Examples of the mapping procedure for the counter sequence in Table 9–7.

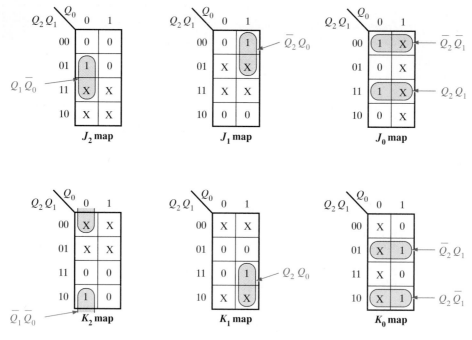

FIGURE 9–30
Karnaugh maps for present-state J and K inputs.

The completed Karnaugh maps for all three flip-flops in the counter are shown in Figure 9–30. The cells are grouped as indicated and the corresponding Boolean expressions for each group are derived.

Step 5: Logic Expressions for Flip-Flop Inputs

From the Karnaugh maps of Figure 9–30 we obtain the following expressions for the J and K inputs of each flip-flop:

$$J_0 = Q_2Q_1 + \overline{Q_2}\,\overline{Q_1} = \overline{Q_2 \oplus Q_1}$$
$$K_0 = Q_2\overline{Q_1} + \overline{Q_2}Q_1 = Q_2 \oplus Q_1$$
$$J_1 = \overline{Q_2}Q_0$$
$$K_1 = Q_2Q_0$$
$$J_2 = Q_1\overline{Q_0}$$
$$K_2 = \overline{Q_1}\,\overline{Q_0}$$

Step 6: Counter Implementation

The final step is to implement the combinational logic from the expressions for the J and K inputs and connect the flip-flops to form the complete 3-bit Gray code counter as shown in Figure 9–31.

A summary of steps used in the design of this counter follows. In general, these steps can be applied to any sequential circuit.

1. Specify the counter sequence and draw a state diagram.
2. Derive a next-state table from the state diagram.
3. Develop a transition table showing the flip-flop inputs required for each transition. The transition table is always the same for a given type of flip-flop.
4. Transfer the J and K states from the transition table to Karnaugh maps. There is a Karnaugh map for each input of each flip-flop.

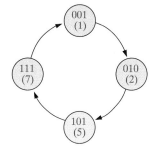

FIGURE 9–31
Three-bit Gray code counter.

5. Factor the maps to generate a logic expression for each flip-flop input.
6. Implement the expressions with combinational logic, and combine with the flip-flops to create the counter.

This procedure is now applied to the design of other synchronous counters in Examples 9–5 and 9–6.

EXAMPLE 9–5

Design a counter with the irregular binary count sequence shown in the state diagram of Figure 9–32. Use J-K flip-flops.

FIGURE 9–32

```
        001
        (1)
   111       010
   (7)       (2)
        101
        (5)
```

Solution
Step 1. The state diagram is as shown. Although there are only four states, a 3-bit counter is required to implement this sequence because the maximum binary count is seven. Since the required sequence does not include all the possible binary states, the invalid states (0, 3, 4, and 6) can be treated as "don't cares" in the design. However, if the counter should erroneously get into an invalid state, you must make sure that it goes back to a valid state.
Step 2. The next-state table is developed from the state diagram and is given in Table 9–9.

TABLE 9–9
Next-state table

Present State			Next State		
Q_2	Q_1	Q_0	Q_2	Q_1	Q_0
0	0	1	0	1	0
0	1	0	1	0	1
1	0	1	1	1	1
1	1	1	0	0	1

TABLE 9–10
Transition table for a J-K flip-flop

Output Transitions			Flip-Flop Inputs	
Q_N		Q_{N+1}	J	K
0	$\longrightarrow$	0	0	X
0	$\longrightarrow$	1	1	X
1	$\longrightarrow$	0	X	1
1	$\longrightarrow$	1	X	0

Step 3. The transition table for the J-K flip-flop is repeated in Table 9–10.

Step 4. The J and K inputs are plotted on the present-state Karnaugh maps in Figure 9–33. Also "don't cares" can be placed in the cells corresponding to the in-valid states of 000, 011, 100, and 110, as indicated by the red Xs.

Step 5. Group the 1s, taking advantage of as many of the "don't care" states as possible for maximum simplification, as shown in Figure 9–33. Notice that when *all* cells in a map are grouped, the expression is simply equal to 1. The expression for each J and K input taken from the maps is as follows:

$$J_0 = 1, K_0 = \overline{Q}_2$$
$$J_1 = K_1 = 1$$
$$J_2 = K_2 = Q_1$$

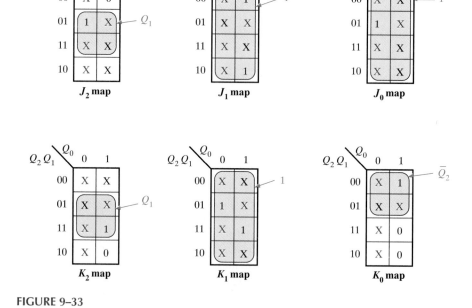

FIGURE 9–33

Step 6. The implementation of the counter is shown in Figure 9–34.

An analysis shows that if the counter, by accident, gets into one of the invalid states (0, 3, 4, 6), it will always return to a valid state according to the following se-quences: $0 \rightarrow 3 \rightarrow 4 \rightarrow 7$, and $6 \rightarrow 1$.

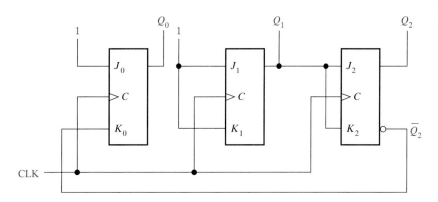

FIGURE 9–34

Related Exercise Verify the analysis that proves the counter will always return (eventually) to a valid state from an invalid state.

EXAMPLE 9–6

Develop a synchronous 3-bit up/down counter with a Gray code sequence. The counter should count up when an UP/$\overline{\text{DOWN}}$ control input is 1 and count down when the control input is 0.

Solution

Step 1. The state diagram is shown in Figure 9–35. The 1 or 0 beside each arrow indicates the state of the UP/$\overline{\text{DOWN}}$ control input, Y.

FIGURE 9–35
State diagram for a 3-bit up/down Gray code counter.

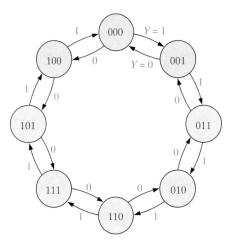

Step 2. The next-state table is derived from the state diagram and is shown in Table 9–11 (next page). Notice that for each present state there are two possible next states, depending on the UP/$\overline{\text{DOWN}}$ control variable, Y.

Step 3. The transition table for the J-K flip-flops is repeated in Table 9–12.

Step 4. The Karnaugh maps for the J and K inputs of the flip-flops are shown in Figure 9–36. The UP/$\overline{\text{DOWN}}$ control input, Y, is considered one of the state variables along with Q_0, Q_1, and Q_2. Using the next-state table, the information in the "Flip-Flop Inputs" column of Table 9–12 is transferred onto the maps as indicated for each present state of the counter.

TABLE 9–11

Next-state table for 3-bit up/down Gray code counter

Present State			Next State					
			Y = 0 (DOWN)			Y = 1 (UP)		
Q_2	Q_1	Q_0	Q_2	Q_1	Q_0	Q_2	Q_1	Q_0
0	0	0	1	0	0	0	0	1
0	0	1	0	0	0	0	1	1
0	1	1	0	0	1	0	1	0
0	1	0	0	1	1	1	1	0
1	1	0	0	1	0	1	1	1
1	1	1	1	1	0	1	0	1
1	0	1	1	1	1	1	0	0
1	0	0	1	0	1	0	0	0

$Y = \overline{\text{UP/DOWN}}$ control input.

TABLE 9–12

Transition table for a J-K flip-flop

Output Transitions		Flip-Flop Inputs	
Q_N	Q_{N+1}	J	K
0 → 0		0	X
0 → 1		1	X
1 → 0		X	1
1 → 1		X	0

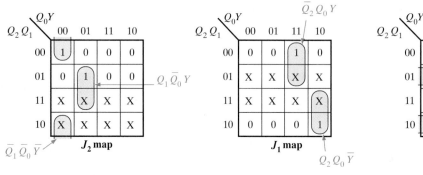

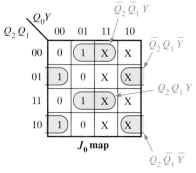

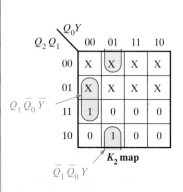

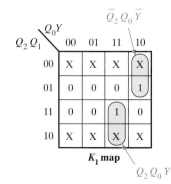

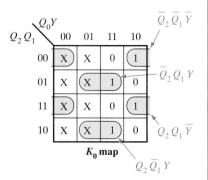

FIGURE 9–36

J and K maps for Table 9–11. The UP/$\overline{\text{DOWN}}$ control input, Y, is treated as a fourth variable.

Step 5. The 1s are combined in the largest possible groupings, with "don't cares" (Xs) used where possible. The groups are factored, and the expressions for the J and K inputs are as follows:

$$J_0 = Q_2Q_1Y + Q_2\overline{Q}_1\overline{Y} + \overline{Q}_2Q_1\overline{Y} + \overline{Q}_2\overline{Q}_1Y \qquad K_0 = \overline{Q}_2\overline{Q}_1\overline{Y} + \overline{Q}_2Q_1Y + Q_2Q_1\overline{Y} + Q_2\overline{Q}_1Y$$

$$J_1 = \overline{Q}_2Q_0Y + Q_2Q_0\overline{Y} \qquad\qquad\qquad K_1 = \overline{Q}_2Q_0\overline{Y} + Q_2Q_0Y$$

$$J_2 = Q_1\overline{Q}_0Y + \overline{Q}_1\overline{Q}_0\overline{Y} \qquad\qquad\qquad K_2 = Q_1\overline{Q}_0\overline{Y} + \overline{Q}_1\overline{Q}_0Y$$

Step 6. The J and K equations are implemented with combinational logic, and the complete counter is shown in Figure 9–37.

FIGURE 9–37
Three-bit up/down Gray code counter.

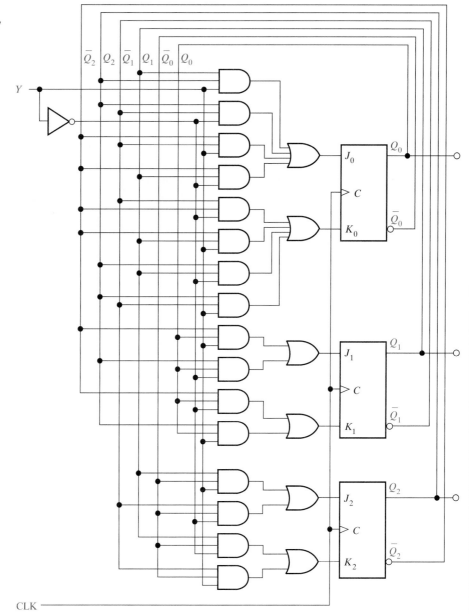

Related Exercise Verify that the logic in Figure 9–37 agrees with the expressions in Step 5.

In this section, you have seen how sequential circuit design techniques can be applied specifically to counter design. In general, sequential circuits can be classified into two types: (1) those in which the output or outputs depend only on the present internal state (called *Moore circuits*) and (2) those in which the output or outputs depend on both the present state and the input or inputs (called *Mealy circuits*).

SECTION 9–4 REVIEW

1. A flip-flop is presently in the RESET state and must go to the SET state on the next clock pulse. What must J and K be?
2. A flip-flop is presently in the SET state and must remain SET on the next clock pulse. What must J and K be?
3. A binary counter is in the 1010 state.
   (a) What is its next state?
   (b) What condition must exist on each flip-flop input to ensure that it goes to the proper next state on the clock pulse?

9–5 ■ CASCADED COUNTERS

Counters can be connected in cascade to achieve higher-modulus operation. In essence, cascading means that the last-stage output of one counter drives the input of the next counter. After completing this section, you should be able to

□ Determine the overall modulus of cascaded counters   □ Analyze the timing diagram of a cascaded counter configuration   □ Use cascaded counters as a frequency divider   □ Use cascaded counters to achieve specified truncated sequences

An example of two counters connected in **cascade** is shown for a 2-bit and a 3-bit ripple counter in Figure 9–38. The timing diagram is in Figure 9–39.

Notice that in the timing diagram in Figure 9–39, the final output of the modulus-8 counter, Q_4, occurs once for every 32 input clock pulses. The overall modulus of the cascaded counters is 32; that is, they act as a divide-by-32 counter. In general, the overall modulus of cascaded counters is equal to the product of the individual moduli. For instance, for the counter in Figure 9–38, the overall modulus is $4 \times 8 = 32$.

When operating synchronous counters in a cascaded configuration, it is necessary to use the count enable and the terminal count functions to achieve higher-modulus operation. On some devices the count enable is labeled simply *CTEN* or some other desig-

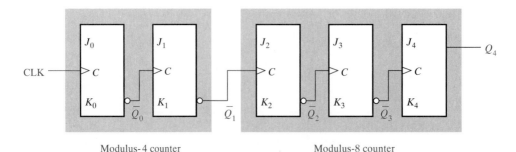

Modulus-4 counter   Modulus-8 counter

FIGURE 9–38
Two cascaded counters (all J and K inputs are HIGH).

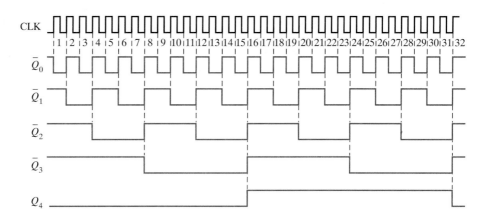

FIGURE 9–39
Timing diagram for the cascaded counter configuration of Figure 9–38.

nation, and terminal count (*TC*) is analogous to ripple clock output (*RCO*) on some IC counters.

Figure 9–40 shows two decade counters connected in cascade. The terminal count (*TC*) output of counter 1 is connected to the count enable (*CTEN*) input of counter 2. Counter 2 is inhibited by the LOW on its *CTEN* input until counter 1 reaches its last, or terminal, state and its terminal count output goes HIGH. This HIGH now enables counter 2, so that when the first clock pulse after counter 1 reaches its terminal count (CLK10), counter 2 goes from its initial state to its second state. Upon completion of the entire second cycle of counter 1 (when counter 1 reaches terminal count the second time), counter 2 is again enabled and advances to its next state. This sequence continues. Since these are decade counters, counter 1 must go through ten complete cycles before counter 2 completes its first cycle. In other words, for every ten cycles of counter 1, counter 2 goes through one cycle. Thus, counter 2 will complete one cycle after one hundred clock pulses. The overall modulus of these two cascaded counters is $10 \times 10 = 100$.

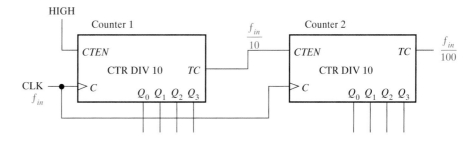

FIGURE 9–40
A modulus-100 counter using two cascaded decade counters.

When viewed as a frequency divider, the circuit of Figure 9–40 divides the input clock frequency by 100. Cascaded counters are often used to divide a high-frequency clock signal to obtain highly accurate pulse frequencies. Cascaded counter configurations used for such purposes are sometimes called *countdown chains.*

For example, suppose that you have a basic clock frequency of 1 MHz and you wish to obtain 100 kHz, 10 kHz, and 1 kHz; a series of cascaded decade counters can be used.

If the 1 MHz signal is divided by 10, the output is 100 kHz. Then if the 100 kHz signal is divided by 10, the output is 10 kHz. Another division by 10 produces the 1 kHz frequency. The general implementation of this countdown chain is shown in Figure 9–41.

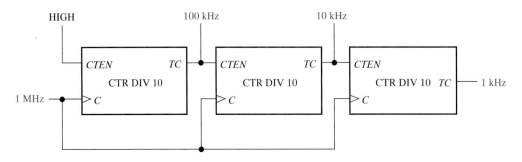

FIGURE 9–41

Three cascaded decade counters forming a divide-by-1000 frequency divider with intermediate divide-by-10 and divide-by-100 outputs.

EXAMPLE 9–7

Determine the overall modulus of the two cascaded counter configurations in Figure 9–42.

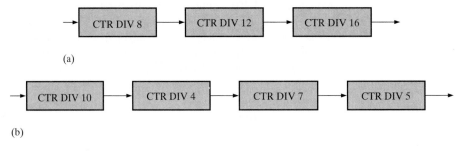

(a)

(b)

FIGURE 9–42

Solution In Figure 9–42(a), the overall modulus for the 3-counter configuration is

$$8 \times 12 \times 16 = 1536$$

In Figure 9–42(b), the overall modulus for the 4 counter configuration is

$$10 \times 4 \times 7 \times 5 = 1400$$

Related Exercise How many cascaded decade counters are required to divide a clock frequency by 100,000?

EXAMPLE 9–8

Use 74LS160A counters to obtain a 10 kHz waveform from a 1 MHz clock. Show the logic diagram.

Solution To obtain 10 kHz from a 1 MHz clock requires a division factor of 100. Two 74LS160A counters must be cascaded as shown in Figure 9–43. The left counter produces an *RCO* pulse for every 10 clock pulses. The right counter produces an *RCO* pulse for every 100 clock pulses.

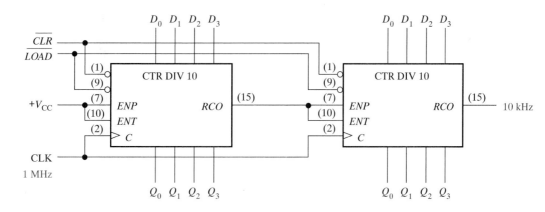

FIGURE 9–43

A divide-by-100 counter using two 74LS160A decade counters.

Related Exercise Determine the frequency of the waveform at the Q_0 output of the second counter in Figure 9–43.

Cascaded IC Counters with Truncated Sequences

The preceding discussion has shown how to achieve an overall modulus (divide-by factor) that is the product of the individual moduli of all the cascaded counters. This can be considered *full-modulus cascading.*

Often an application requires an overall modulus that is less than that achieved by full-modulus cascading. That is, a truncated sequence must be implemented with cascaded counters. To illustrate this method, the cascaded counter configuration in Figure 9–44 is used. This particular circuit uses four 74LS161A 4-bit binary counters. If these four counters (sixteen bits total) were cascaded in a full-modulus arrangement, the modulus would be

$$2^{16} = 65,536$$

Let's assume that a certain application requires a divide-by-40,000 counter (modulus 40,000). The difference between 65,536 and 40,000 is 25,536, which is the number of states that must be *deleted* from the full-modulus sequence. The technique used in the circuit of Figure 9–44 is to preset the cascaded counter to 25,536 (63C0 in hexadecimal) each

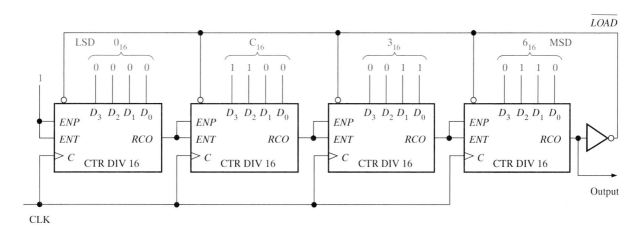

FIGURE 9–44

A divide-by-40,000 counter using 74LS161A 4-bit binary counters. Note that each of the parallel data inputs is shown in binary order (the right-most bit D_0 is the LSB in each counter).

time it recycles, so that it will count from 25,536 up to 65,536 on each full cycle. Therefore, each full cycle of the counter consists of 40,000 states.

Notice in Figure 9–44 that the RCO output of the right-most counter is inverted and applied to the $\overline{LOAD}$ input of each 4-bit counter. Each time the count reaches its terminal value of 65,536, RCO goes HIGH and causes the number on the parallel data inputs ($63C0_{16}$) to be preset into the counter. Thus, there is one RCO pulse from the right-most 4-bit counter for every 40,000 clock pulses.

With this technique any modulus can be achieved by the presetting of the counter to the appropriate initial state on each cycle.

**SECTION 9–5
REVIEW**

1. How many decade counters are necessary to implement a divide-by-1000 (modulus-1000) counter? A divide-by-10,000?
2. Show with general block diagrams how to achieve each of the following, using a flip-flop, a decade counter, and a 4-bit binary counter, or any combination of these:
   (a) Divide-by-20 counter   (b) Divide-by-32 counter
   (c) Divide-by-160 counter   (d) Divide-by-320 counter

9–6 ▪ COUNTER DECODING

In many applications, it is necessary that some or all of the counter states be decoded. The decoding of a counter involves using decoders or logic gates to determine when the counter is in a certain binary state in its sequence. For instance, the terminal count function previously discussed is a single decoded state (the last state) in the counter sequence. After completing this section, you should be able to

☐ Implement the decoding logic for any given state in a counter sequence   ☐ Explain why glitches occur in counter decoding logic   ☐ Use the method of strobing to eliminate decoding glitches

Suppose that you wish to decode binary state 6 (110) of a 3-bit binary counter. This can be done as shown in Figure 9–45. When $Q_2 = 1, Q_1 = 1$, and $Q_0 = 0$, a HIGH appears on the output of the decoding gate, indicating that the counter is at state 6. This is called *active-HIGH decoding*. Replacing the AND gate with a NAND gate provides active-LOW decoding.

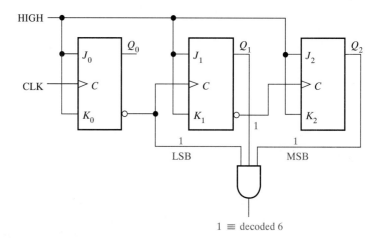

FIGURE 9–45
Decoding of state 6 (110).

EXAMPLE 9–9
Implement the decoding of binary state 2 and binary state 7 of a 3-bit synchronous counter. Show the entire counter timing diagram and the output waveforms of the decoding gates. Binary $2 = \overline{Q}_2 Q_1 \overline{Q}_0$, and binary $7 = Q_2 Q_1 Q_0$.

Solution See Figure 9–46. The 3-bit counter was originally discussed in Section 9–2 (Figure 9–14).

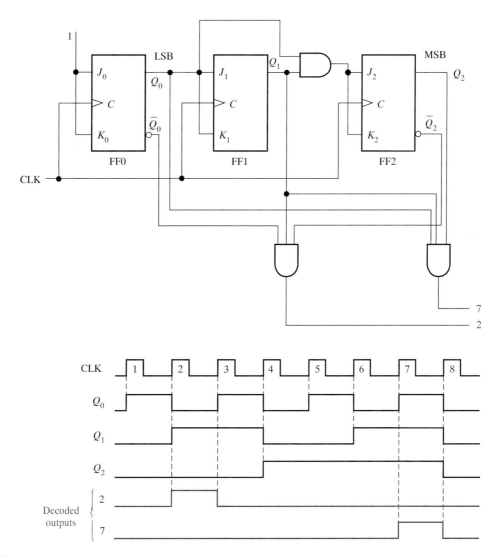

FIGURE 9–46
A 3-bit counter with active-HIGH decoding of count 2 and count 7.

Related Exercise Show the logic for decoding state 5 in the 3-bit counter.

Decoding Glitches

The problem of glitches produced by the decoding process was introduced in Chapter 6. As you have learned, the propagation delays due to the ripple effect in asynchronous counters create transitional states in which the counter outputs are changing at slightly different times. These transitional states produce undesired voltage spikes of short duration (glitches) on the outputs of a decoder connected to the counter. The glitch problem can also

FIGURE 9–47
A basic decade (BCD) counter and decoder.

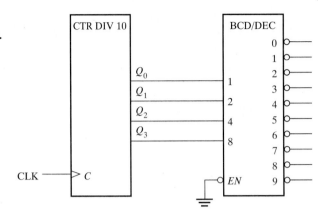

occur to some degree with synchronous counters because the propagation delays from clock to Q outputs of each flip-flop in a counter can vary slightly.

Figure 9–47 shows a basic asynchronous BCD decade counter connected to a BCD-to-decimal decoder. To see what happens in this case, we will look at a timing diagram in which the propagation delays are taken into account, as in Figure 9–48. Notice that

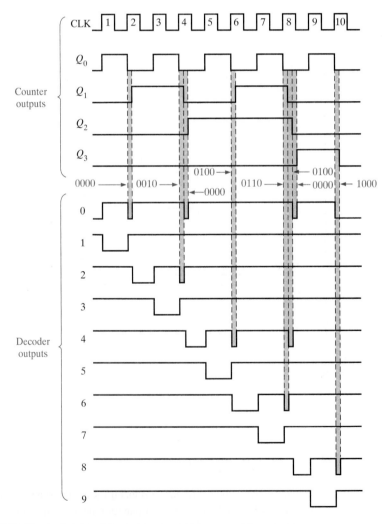

FIGURE 9–48
Outputs with glitches from the decoder in Figure 9–47. Glitch widths are exaggerated for illustration and are usually only several nanoseconds wide.

these delays cause false states of short duration. The value of the false binary state at each critical transition is indicated on the diagram. The resulting glitches can be seen on the decoder outputs.

One way to eliminate the glitches is to enable the decoded outputs at a time after the glitches have had time to disappear. This method is known as *strobing* and can be accomplished in this case by using the LOW level of the clock to enable the decoder, as shown in Figure 9–49. The resulting improved timing diagram is shown in Figure 9–50.

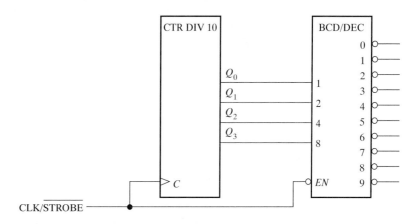

FIGURE 9–49
The basic decade counter and decoder with strobing to eliminate glitches.

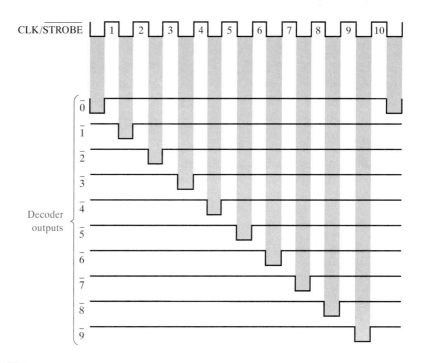

FIGURE 9–50
Strobed decoder outputs for the circuit of Figure 9–49.

SECTION 9–6 REVIEW

1. What transitional states are possible when an asynchronous 4-bit binary counter changes from
   (a) count 2 to count 3   (b) count 3 to count 4
   (c) count 10_{10} to count 11_{10} (d) count 15 to count 0

9–7 ■ COUNTER APPLICATIONS

The digital counter is a useful and versatile device that is found in many applications. In this section, some representative counter applications are presented. After completing this section, you should be able to

□ Describe how counters are used in a basic digital clock system   □ Explain how a divide-by-60 counter is implemented and how it is used in a digital clock   □ Explain how the hours counter is implemented   □ Discuss the application of a counter in an automobile parking control system   □ Describe how a counter is used in the process of parallel-to-serial data conversion

The Digital Clock

A common example of a counter application is in timekeeping systems. Figure 9–51 is a simplified logic diagram of a digital clock that displays seconds, minutes, and hours. First, a 60 Hz sinusoidal ac voltage is converted to a 60 Hz pulse waveform and divided down to a 1 Hz pulse waveform by a divide-by-60 counter formed by a divide-by-10 counter followed by a divide-by-6 counter. Both the *seconds* and *minutes* counts are also produced by divide-by-60 counters, the details of which are shown in Figure 9–52. These counters count from 0 to 59 and then recycle to 0; 74LS160A synchronous decade counters are used in this particular implementation. Notice that the divide-by-6 portion is formed with a

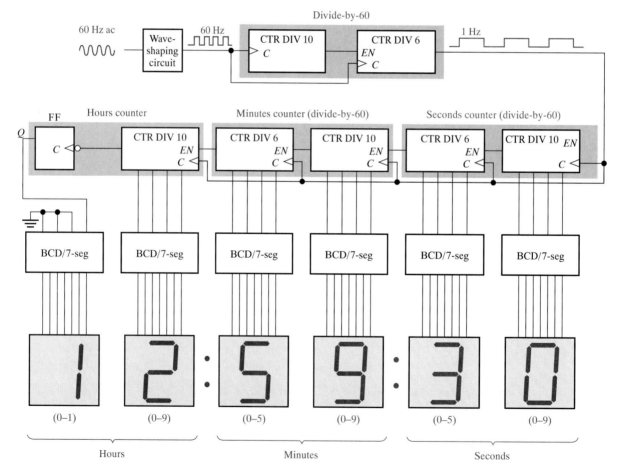

FIGURE 9–51
Simplified logic diagram for a 12-hour digital clock. Logic details using specific devices are shown in the following diagrams.

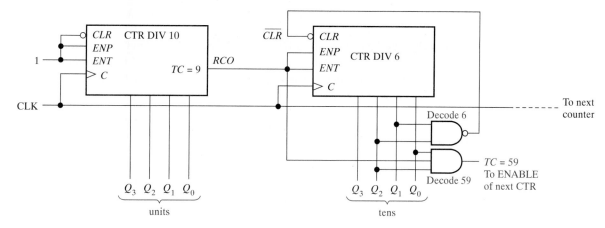

FIGURE 9–52

Logic diagram of typical divide-by-60 counter using 74LS160A synchronous decade counters.
Note that the outputs are in binary order (the right-most bit is the LSB).

decade counter with a truncated sequence achieved by using the decoded count 6 to asynchronously clear the counter. The terminal count, 59, is also decoded to enable the next
counter in the chain.

The *hours* counter is implemented with a decade counter and a flip-flop as shown in
Figure 9–53. Consider that initially both the decade counter and the flip-flop are RESET,
and the decode-12 gate and decode-9 gate outputs are HIGH. The decade counter advances
through all of its states from zero to nine, and on the clock pulse that recycles it from nine
back to zero, the flip-flop goes to the SET state ($J = 1$, $K = 0$). This illuminates a 1 on the
tens-of-hours display. The total count is now ten (the decade counter is in the zero state and
the flip-flop is SET).

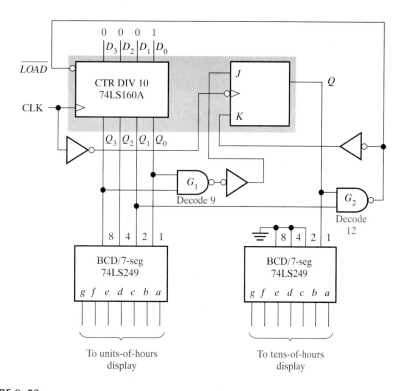

FIGURE 9–53

Logic diagram for hours counter and decoders. Note that on the counter inputs and outputs, the
right-most bit is the LSB.

Next, the total count advances to eleven and then to twelve. In state 12 the Q_2 output of the decade counter is HIGH, the flip-flop is still SET, and thus the decode-12 gate output is LOW. This activates the $\overline{LOAD}$ input of the decade counter. On the next clock pulse, the decade counter is preset to state 1 by the data inputs, and the flip-flop is RESET ($J = 0, K = 1$). As you can see, this logic always causes the counter to recycle from twelve back to one rather than back to zero.

Automobile Parking Control

Now a simple application example illustrates the use of an up/down counter to solve an everyday problem. The problem is to devise a means of monitoring available spaces in a one-hundred-space parking garage and provide for an indication of a full condition by illuminating a display sign and lowering a gate bar at the entrance.

A system that solves this problem consists of (1) optoelectronic sensors at the entrance and exit of the garage, (2) an up/down counter and associated circuitry, and (3) an interface circuit that uses the counter output to turn the FULL sign on or off as required and lower or raise the gate bar at the entrance. A general block diagram of this system is shown in Figure 9–54.

A logic diagram of the up/down counter is shown in Figure 9–55. It consists of two cascaded 74LS190 up/down decade counters. The operation is described in the following paragraphs.

The counter is initially preset to 0 using the parallel data inputs which are not shown. Each automobile entering the garage breaks a light beam, activating a sensor that produces an electrical pulse. This positive pulse sets the S-R latch on its leading edge. The LOW on the $\overline{Q}$ output of the latch puts the counter in the UP mode. Also, the sensor pulse goes through the NOR gate and clocks the counter on the LOW-to-HIGH transition of its trailing edge. Each time an automobile enters the garage, the counter is advanced by

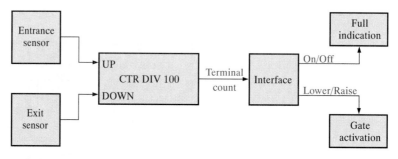

FIGURE 9–54
Functional block diagram for parking garage control.

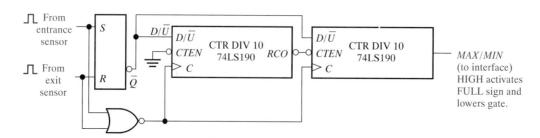

FIGURE 9–55
Logic diagram for modulus-100 up/down counter for automobile parking control.

one (**incremented**). When the one-hundredth automobile enters, the counter goes to its last state (100_{10}). The *MAX/MIN* output goes HIGH and activates the interface circuit (no detail), which lights the FULL sign and lowers the gate bar to prevent further entry.

When an automobile exits, an optoelectronic sensor produces a positive pulse, which resets the S-R latch and puts the counter in the DOWN mode. The trailing edge of the clock decreases the count by one (**decrements**). If the garage is full and an automobile leaves, the *MAX/MIN* output of the counter goes LOW, turning off the FULL sign and raising the gate.

Parallel-to-Serial Data Conversion (Multiplexing)

A simplified example of data transmission using multiplexing and demultiplexing techniques was introduced in Chapter 6. Essentially, the parallel data bits on the multiplexer inputs are converted to serial data bits on the single transmission line. A group of bits appearing simultaneously on parallel lines is called *parallel data.* A group of bits appearing on a single line in a time sequence is called *serial data.*

Parallel-to-serial conversion is normally accomplished by the use of a counter to provide a binary sequence for the data-select inputs of a data selector/multiplexer, as illustrated in Figure 9–56. The Q outputs of the modulus-8 counter are connected to the data-select inputs of an 8-bit multiplexer.

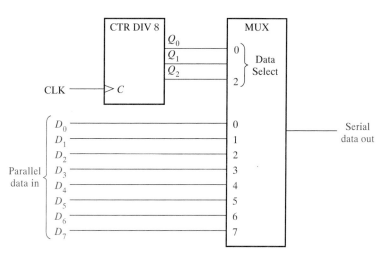

FIGURE 9–56
Parallel-to-serial data conversion logic.

Figure 9–57 is a timing diagram illustrating the operation of this circuit. The first byte (eight-bit group) of parallel data is applied to the multiplexer inputs. As the counter goes through a binary sequence from zero to seven, each bit, beginning with D_0, is sequentially selected and passed through the multiplexer to the output line. After eight clock pulses the data byte has been converted to a serial format and sent out on the transmission line. When the counter recycles back to 0, the next byte is applied to the data inputs and is sequentially converted to serial form as the counter cycles through its eight states. This process continues repeatedly as each parallel byte is converted to a serial byte. An end-of-chapter problem (Problem 52) requires that this circuit be implemented with specific devices.

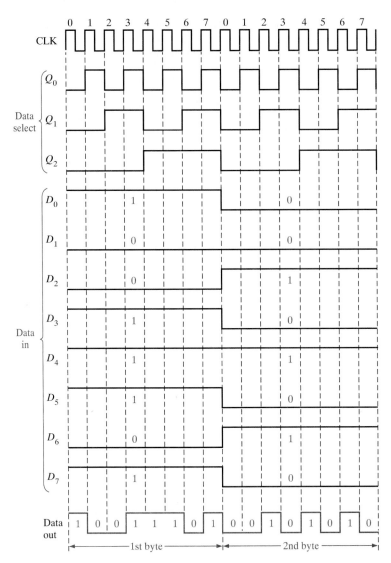

FIGURE 9–57
Example of parallel-to-serial conversion timing for the circuit in Figure 9–56.

SECTION 9–7 REVIEW

1. Explain the purpose of each NAND gate in Figure 9–53, on p. 479.
2. Identify the two recycle conditions for the hours counter in Figure 9–51, on p. 478, and explain the reason for each.

9–8 ■ TROUBLESHOOTING

The troubleshooting of counters can be simple or quite involved, depending on the type of counter and the type of fault. This section will give you some good practical insight into how to approach the troubleshooting of sequential circuits. After completing this section, you should be able to

□ Detect a faulty IC counter □ Isolate faults in maximum-modulus cascaded counters
□ Isolate faults in cascaded counters with truncated sequences □ Determine faults in counters implemented with individual flip-flops

IC Counters

For an IC counter with a straightforward sequence that is not controlled by external logic, about the only thing to check (other than V_{CC} and ground) is the possibility of open or shorted inputs or outputs. An IC counter almost never alters its sequence of states because of an internal fault, so you need only check for pulse activity on the Q outputs to detect the existence of an open or a short. The absence of pulse activity on one of the Q outputs indicates an internal short or open. Absence of pulse activity on all the Q outputs indicates that the clock input is faulty or the clear input is stuck in its active state.

To check the clear input, apply a constant active level while the counter is clocked. You will observe a LOW on each of the Q outputs if it is functioning properly.

The parallel load feature on a counter can be checked by activating the $\overline{LOAD}$ input and exercising each state as follows: Apply LOWs to the parallel data inputs, pulse the clock input, and check for LOWs on all the Q outputs. Next, apply HIGHs to all the parallel data inputs, pulse the clock input, and check for HIGHs on all the Q outputs.

Cascaded IC Counters with Maximum Modulus

A failure in one of the counters in a chain of cascaded counters can affect all the counters that follow it. For example, if a count enable input opens, it effectively acts as a HIGH (TTL), and the counter is always enabled. This type of failure in one of the counters will cause that counter to run at the full clock rate and will also cause all the succeeding counters to run at higher than normal rates. This is illustrated in Figure 9–58 for a divide-by-1000 cascaded counter arrangement where an open enable (*CTEN*) input acts as a TTL HIGH and continuously enables the second counter. Other faults that can affect "downstream" counter stages are open or shorted clock inputs or terminal count outputs. In some of these situations, pulse activity can be observed, but it may be at the wrong frequency. Exact frequency measurements must be made.

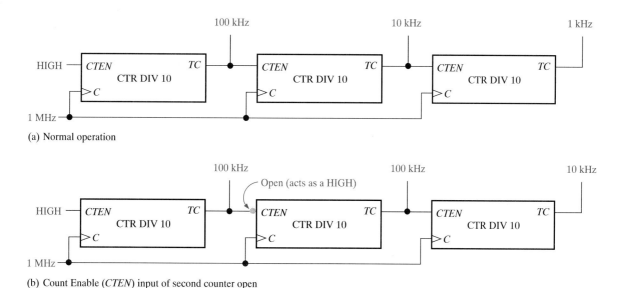

(a) Normal operation

(b) Count Enable (*CTEN*) input of second counter open

FIGURE 9–58

Example of a failure that affects following counters in a cascaded arrangement.

Cascaded Counters with Truncated Sequences

The count sequence of a cascaded counter with a truncated sequence, such as that in Figure 9–59, can be affected by other types of faults in addition to those mentioned for maximum-modulus cascaded counters. For example, a failure in one of the parallel data inputs,

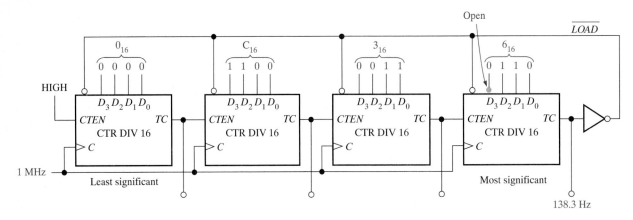

FIGURE 9–59

Example of a failure in a cascaded counter with a truncated sequence.

the $\overline{LOAD}$ input, or the inverter can alter the preset count and thus change the modulus of the counter.

For example, suppose the D_3 input of the most significant counter in Figure 9–59 is open and acts as a HIGH. Instead of 6_{16} (0110) being preset into the counter, E_{16} (1110) is preset in. So, instead of beginning with $63C0_{16}$ ($25,536_{10}$) each time the counter recycles, the sequence will begin with $E3C0_{16}$ ($58,304_{10}$). This changes the modulus of the counter from 40,000 to $65,536 - 58,304 = 7232$.

To check this counter, apply a known clock frequency, say 1 MHz, and measure the output frequency at the final terminal count output. If the counter is operating properly, the output frequency is

$$f_{out} = \frac{f_{in}}{modulus} = \frac{1 \text{ MHz}}{40,000} = 25 \text{ Hz}$$

In this case, the specific failure described in the preceding paragraph will cause the output frequency to be

$$f_{out} = \frac{f_{in}}{modulus} = \frac{1 \text{ MHz}}{7232} = 138.3 \text{ Hz}$$

EXAMPLE 9–10

Frequency measurements are made on the truncated counter in Figure 9–60 as indicated. Determine if the counter is working properly, and if not, isolate the fault.

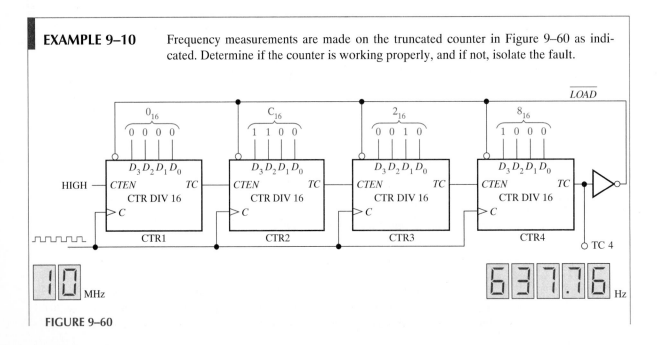

FIGURE 9–60

Solution Check to see if the frequency measured at TC 4 is correct. If it is, the counter is working properly.

$$\text{truncated modulus} = \text{full modulus} - \text{preset count}$$
$$= 16^4 - 82C0_{16}$$
$$= 65,536 - 33,472 = 32,064$$

The correct frequency at TC 4 is

$$f_4 = \frac{10 \text{ MHz}}{32,064} = 311.88 \text{ Hz}$$

Uh oh! There is a problem. The measured frequency of 637.76 Hz does not agree with the correct calculated frequency of 311.88 Hz.

To find the faulty counter, determine the actual truncated modulus as follows:

$$\text{modulus} = \frac{f_{out}}{f_{in}} = \frac{10 \text{ MHz}}{637.76 \text{ Hz}} = 15,680$$

Because the truncated modulus should be 32,064, most likely the counter is being preset to the wrong count when it recycles. The actual preset count is determined as follows:

$$\text{truncated modulus} = \text{full modulus} - \text{preset count}$$
$$\text{preset count} = \text{full modulus} - \text{truncated modulus}$$
$$= 65,536 - 15,680$$
$$= 49,856$$
$$= C2C0_{16}$$

This shows that the counter is being preset to $C2C0_{16}$ instead of $82C0_{16}$ each time it recycles.

Counters 1, 2, and 3 are being preset properly but counter 4 is not. Since $C_{16} = 1100_2$, the D_2 input to counter 4 is HIGH when it should be LOW. This is most likely caused by an open input. Check for an external open caused by a bad solder connection or a broken conductor; if none can be found, replace the IC and the counter should work properly.

Related Exercise Determine what the output frequency at TC 4 would be if the D_3 input of counter 3 were open.

Counters Implemented with Individual Flip-Flops

Counters implemented with individual flip-flop and gate ICs are sometimes more difficult to troubleshoot because there are many more inputs and outputs with external connections than there are in an IC counter. The sequence of a counter can be altered by a single open or short on an input or output, as Example 9–11 illustrates.

EXAMPLE 9–11

Suppose that you observe the output waveforms that are indicated for the counter in Figure 9–61 (next page). Determine if there is a problem with the counter.

Solution The Q_2 waveform is incorrect. The correct waveform is shown as a colored dashed line. You can see that the Q_2 waveform looks exactly like the Q_1 waveform. So whatever is causing FF1 to toggle appears to also be controlling FF2.

Checking the J and K inputs to FF2, you find a waveform that looks like Q_0. This result indicates that Q_0 is somehow getting through the AND gate. The only way this can happen is if the Q_1 input to the AND gate is always HIGH. But you have seen that

FIGURE 9-61

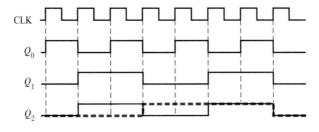

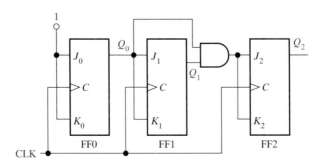

Q_1 has a correct waveform. This observation leads to the conclusion that the lower input to the AND gate must be internally open and acting as a HIGH. Replace the AND gate and retest the circuit.

Related Exercise Describe the Q_2 output of the counter in Figure 9-61 if the Q_1 output of FF1 is open.

SECTION 9-8 REVIEW

1. What failures can cause the counter in Figure 9-58 (p. 483) to have no pulse activity on any of the *TC* outputs?
2. What happens if the inverter in Figure 9-60 develops an open output?

9-9 ■ LOGIC SYMBOLS WITH DEPENDENCY NOTATION

Up to this point, the logic symbols with dependency notation specified in ANSI/IEEE Standard 91-1984 have been introduced on a limited basis. In many cases, the new symbols do not deviate greatly from the traditional symbols. A significant departure from what we are accustomed to does occur, however, for some devices, including counters and other more complex devices. After completing this section, you should be able to

□ Interpret logic symbols that include dependency notation □ Identify the common block and the individual elements of a counter symbol □ Interpret the qualifying symbol □ Discuss control dependency □ Discuss mode dependency □ Discuss AND dependency

Dependency notation is fundamental to the ANSI/IEEE standard. Dependency notation is used in conjunction with the logic symbols to specify the relationships of inputs and outputs so that the logical operation of a given device can be determined entirely from its logic symbol without a prior knowledge of the details of its internal structure and without a detailed logic diagram for reference.

Although we will continue to use primarily the more traditional and familiar symbols throughout this book, a brief coverage of logic symbols with dependency notation is provided. This section can be treated as optional, and its omission will not affect the remainder of the book.

The 74LS163A 4-bit binary counter is used for illustration. For comparison, Figure 9–62 shows a traditional block symbol and the ANSI/IEEE symbol with dependency notation. Basic descriptions of the symbol and the dependency notation follow.

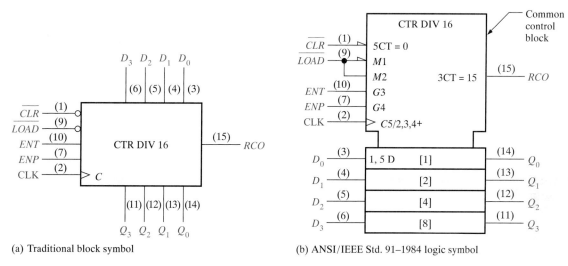

(a) Traditional block symbol   (b) ANSI/IEEE Std. 91–1984 logic symbol

FIGURE 9–62
The 74LS163A 4-bit synchronous counter.

Common Control Block   The upper block with notched corners in Figure 9–62(b) has inputs and an output that are considered common to all elements in the device and not unique to any one of the elements.

Individual Elements   The lower block in Figure 9–62(b), which is partitioned into four abutted sections, represents the four storage elements (D flip-flops) in the counter, with inputs D_0, D_1, D_2, and D_3 and outputs Q_0, Q_1, Q_2, and Q_3.

Qualifying Symbol   The label "CTR DIV 16" in Figure 9–62(b) identifies the device as a counter (CTR) with sixteen states (DIV 16).

Control Dependency (C)   As shown in Figure 9–62(b), the letter C denotes control dependency. Control inputs usually enable or disable the data inputs *(D, J, K, S,* and *R)* of a storage element. The C input is usually the clock input. In this case the digit 5 following C (C5/2,3,4+) indicates that the inputs labeled with a 5 prefix are dependent on the clock (synchronous with the clock). For example, $5CT = 0$ on the $\overline{CLR}$ input indicates that the clear function is dependent on the clock; that is, it is a synchronous clear. When the $\overline{CLR}$ input is LOW (0), the counter is reset to zero $(CT = 0)$ on the triggering edge of the clock pulse. Also, the 5 D label at the input of storage element [1] indicates that the data storage is dependent on (synchronous with) the clock. All labels in the [1] storage element apply to the [2], [4], and [8] elements below it, since they are not labeled differently.

Mode Dependency (M)   As shown in Figure 9–62(b), the letter M denotes mode dependency. This label is used to indicate how the functions of various inputs or outputs depend on the mode in which the device is operating. In this case the device has two modes of operation. When the $\overline{LOAD}$ input is LOW (0), as indicated by the triangle input, the counter is in a preset mode ($M1$) in which the input data (D_0, D_1, D_2, and D_3) are synchronously loaded into the four flip-flops. The digit 1 following M in $M1$ and the 1 in 1, 5 D show a dependency relationship and indicate that input data are stored only when the device is in the preset mode ($M1$), in which $\overline{LOAD} = 0$. When the $\overline{LOAD}$ input is HIGH (1), the counter advances through its normal binary sequence, as indicated by $M2$ and the 2 in $C5/2,3,4+$.

AND Dependency (G)   As shown in Figure 9–62(b), the letter G denotes AND dependency, indicating that an input designated with G followed by a digit is ANDed with any other input or output having the same digit as a prefix in its label. In this particular example, the $G3$ at the $\overline{ENT}$ input and the $3CT = 15$ at the RCO output are related, as indicated by the 3, and that relationship is an AND dependency, indicated by the G. This tells us that ENT must be HIGH (no triangle on the input) AND the count must be fifteen ($CT = 15$) for the RCO output to be HIGH.

Also, the digits 2, 3, and 4 in the label $C5/2,3,4+$ indicate that the counter advances through its states when $\overline{LOAD} = 1$, as indicated by the mode dependency label $M2$, and when $ENT = 1$ AND $ENP = 1$, as indicated by the AND dependency labels $G3$ and $G4$. The $+$ indicates that the counter advances by one count when these conditions exist.

This coverage of a specific logic symbol with dependency notation is intended to aid in the interpretation of other such symbols that you may encounter in the future.

SECTION 9–9 REVIEW

1. In dependency notation, what do the letters C, M, and G stand for?
2. By what letter is data storage denoted?

9–10 ■ DIGITAL SYSTEM APPLICATION

In this system application, you will continue working with the traffic light control system. You should review Section 6–12 and Section 8–9 before continuing. In this section, the focus is on the sequential logic portion of the system which is the last block of the system to be developed. After completing this section, you should be able to

☐ Develop a block diagram for the sequential logic portion of the system   ☐ Implement the counter and input logic and complete the system logic diagram   ☐ Compare the logic diagram to the printed circuit board   ☐ Develop a test procedure and troubleshoot the circuit board

General Requirements for the Sequential Logic

The overall system block diagram with the sequential logic block highlighted in blue and the light sequence indicated is shown in Figure 9–63. The state decoding and output logic block and the timing circuits block have been developed in Chapters 6 and 8. In this section, we are going to concentrate on the sequential logic that controls the sequencing of the traffic lights based on inputs from the timing circuits and from the vehicle sensor. The sequential logic produces a 2-bit Gray code sequence for the four states of the system that are indicated in Figure 9–63.

A Block Diagram of the Sequential Logic

The sequential logic consists of a 2-bit Gray code counter and associated input logic as shown in Figure 9–64. The counter produces a sequence of four states. Transitions from one state to the next are determined by the 4 s interval (short timer) and the 25 s interval

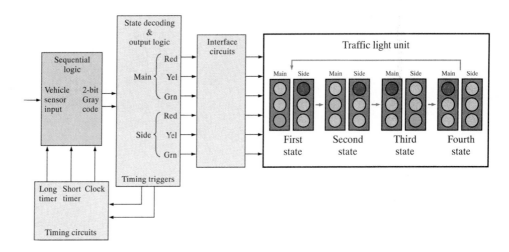

FIGURE 9–63
Traffic light control system block diagram and light sequence.

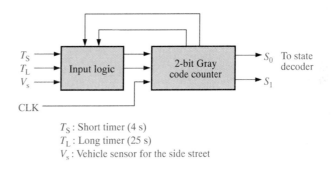

T_S : Short timer (4 s)
T_L : Long timer (25 s)
V_s : Vehicle sensor for the side street

FIGURE 9–64
Block diagram of the sequential logic.

(long timer) produced by the timing circuits and the vehicle sensor input. The clock for the counter is the 10 kHz signal produced by the 555 timer in the timing circuits portion of the system.

The State Diagram and the Sequential Operation

The state diagram for the system was introduced in Chapter 6 and is shown again in Figure 9–65, on the next page. Based on this state diagram, the sequential Gray code operation is described as follows.

First state: The Gray code for this state is 00. The main street light is green and the side street light is red. The system remains in this state for at least 25 s when the long timer is *on* or as long as there is no vehicle on the side street $(T_L + \overline{V}_s)$. The system goes to the next state when the 25 s timer is *off* and there is a vehicle on the side street $(\overline{T}_L V_s)$.

Second state: The Gray code for this state is 01. The main street light is yellow (caution) and the side street light is red. The system remains in this state for 4 s when the short timer is *on* (T_S) and goes to the next state when the short timer goes *off* $(\overline{T}_S)$.

Third state: The Gray code for this state is 11. The main street light is red and the side street light is green. The system remains in this state when the long timer is *on* and there is a vehicle on the side street $(T_L V_s)$. The system goes to the next state when the 25 s have elapsed or when there is no vehicle on the side street, whichever comes first $(\overline{T}_L + \overline{V}_s)$.

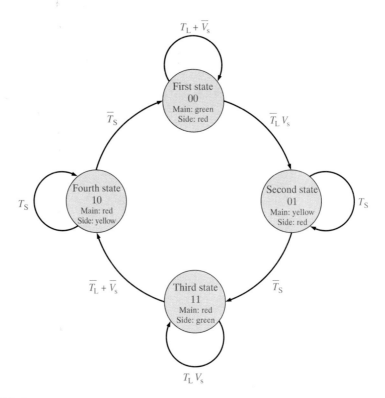

FIGURE 9–65
State diagram showing the 2-bit Gray code sequence.

Fourth state: The Gray code for this state is 10. The main street light is red and the side street light is yellow. The system remains in this state for 4 s when the short timer is *on* (T_S) and goes back to the first state when the short timer goes *off* ($\overline{T}_S$).

Implementing the Logic

The diagram in Figure 9–66 further defines the sequential logic. The counter consists of two synchronously clocked D flip-flops for which the D inputs come from the input logic. The clock is the 10 kHz output from the 555 timer. The input logic consists of five input variables: Q_1, Q_0, T_L, T_S, and V_s.

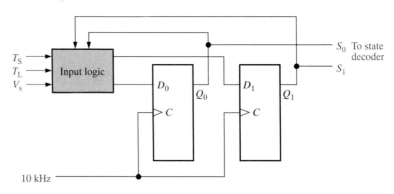

FIGURE 9–66
Sequential logic.

TABLE 9–13
Next-state table for the sequential logic transition

Present State Q_1	Present State Q_0	Next State Q_1	Next State Q_0	Input Conditions
0	0	0	0	$T_L + \overline{V}_s$
0	0	0	1	$\overline{T}_L V_s$
0	1	0	1	T_S
0	1	1	1	$\overline{T}_S$
1	1	1	1	$T_L V_s$
1	1	1	0	$\overline{T}_L + \overline{V}_s$
1	0	1	0	T_S
1	0	0	0	$\overline{T}_S$

TABLE 9–14
D flip-flop transition table

Output Transitions Q_N		Output Transitions Q_{N+1}	Flip-Flop Input D
0	⟶	0	0
0	⟶	1	1
1	⟶	0	0
1	⟶	1	1

From the state diagram, a next-state table can be developed as shown in Table 9–13. The input conditions for T_L, T_S, and V_s for each present-state/next-state combination are also listed in the table. The D flip-flop transition table is shown in Table 9–14.

■ THE DIGITAL WORKBENCH

In the Chapter 6 workbench, you worked with the portion of the traffic light controller board containing the state decoder and output logic. In Chapter 8, the timing circuits were developed. In this section, the entire board including the sequential logic portion is completed. As before, all of the connections to the V_{CC} and ground pads are on the back side of the board. Circuit interconnections on the back side of the board feed through to the pads on the component side. When you trace out the circuit, common pads are generally oriented vertically or horizontally and not at an angle with respect to each other.

■ DIGITAL WORKBENCH 1: Design

From Table 9–13 (next-state table) and Table 9–14 (D flip-flop transition table), you can determine the logic conditions required for each flip-flop to have a next state of 1. For example, Q_0 goes from 0 to 1 when the present state is 00 and the input condition is $\overline{T}_L V_s$, as indicated on the second row of the table shown here. D_0 must be a 1 to make Q_0 go to a 1 or to remain a 1. This condition for D_0 can be written as a logic expression as follows:

$$D_0 = \overline{Q}_1 \overline{Q}_0 \overline{T}_L V_s$$

Present State		Next State		Input Conditions	Flip-Flop Inputs	
Q_1	Q_0	Q_1	Q_0		D_1	D_0
0	0	0	0	$T_L + \overline{V}_s$		
0	0	0	1	$\overline{T}_L V_s$	0	1
0	1	0	1	T_s		
0	1	1	1	$\overline{T}_s$		
1	1	1	1	$T_L V_s$		
1	1	1	0	$\overline{T}_L + \overline{V}_s$		
1	0	1	0	T_s		
1	0	0	0	$\overline{T}_s$		

■ **Activity 1** Complete the *flop-flop inputs* part of the table. Determine the rest of the logic conditions that make $D_0 = 1$ and develop a logic expression for D_0. Keep in mind that you are looking for conditions for each present state that make the flip-flop go to a 1 or stay a 1 in the next state. The resulting expression should be a sum-of-products.

■ **Activity 2** Determine all of the logic conditions that make $D_1 = 1$ and develop a logic expression for D_1. Keep in mind that you are looking for conditions for each present state that make the flip-flop go to a 1 or stay a 1 in the next state. The resulting expression should be a sum-of-products.

■ **Activity 3** Using the SOP logic expressions developed for D_0 and D_1, implement the sequential logic based on the block diagram in Figure 9–64. Refer to the following data sheets:

■ 74LS74A dual D flip-flops
■ 74LS08 quad AND gates
■ 74LS153 data selector (multiplexer)

Use these devices to implement the logic. The 74LS153 is a dual 4-input data selector to be used as a logic function generator. This topic was discussed in Section 6–8.

■ DIGITAL WORKBENCH 2: Verification

■ *Activity 1* The complete system board shown here includes the sequential logic, state decoding and output logic, and the timing circuits. As the system block diagram shows, the outputs from this board go to the interface circuits which are a separate part of the system. Develop the complete system logic diagram from this board.

■ *Activity 2* Verify that your sequential logic design from Workench 1 agrees with the circuitry on this board.

■ *Activity 3* Identify and label the outputs and input on the circuit board connector.

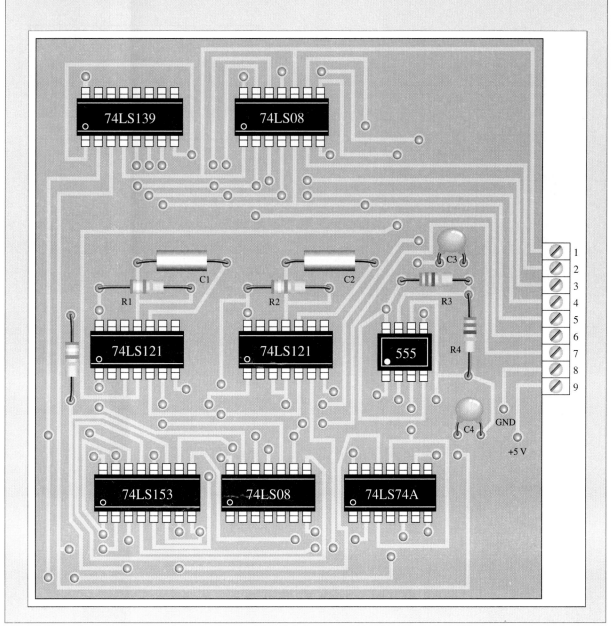

■ DIGITAL WORKBENCH 3: Troubleshooting

■ *Activity 1* Develop a test procedure for testing the system board in the laboratory by simulating the vehicle sensor input, the interface circuit, and the traffic lights with available components.

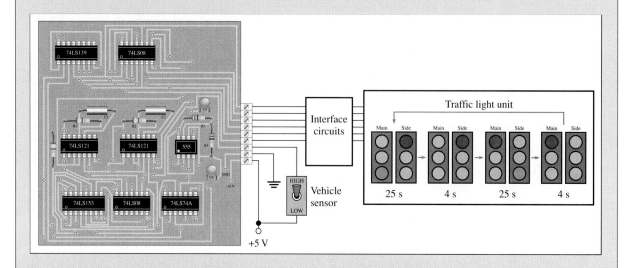

■ *Activity 2* The total system is shown in the diagram above with the correct sequence of lights indicated. For each of the following cases, determine if there is a problem, and if so, identify the most likely fault or faults on the system circuit board. Assume that the interface circuits and the lights are working properly so that any problem is confined to the system board.

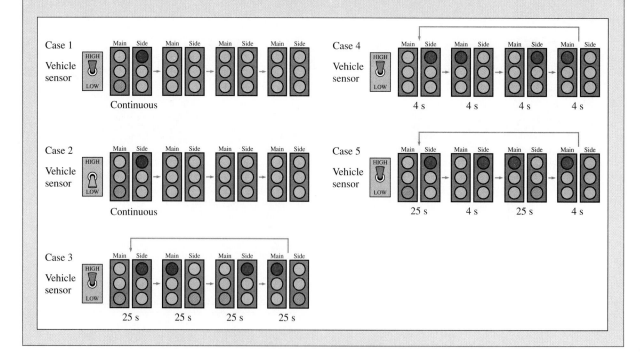

SECTION 9–10 REVIEW

1. Explain the purpose of the sequential logic portion of the system.
2. What controls the transitions of the sequential logic from one state to the next?
3. Explain why the frequency of the clock could be greater or less than 10 kHz and not visually affect the light sequence.

SUMMARY

■ Asynchronous and synchronous counters differ only in the way in which they are clocked, as shown in Figure 9–67. Synchronous counters can run at faster clock rates than asynchronous counters.

■ Connection diagrams for the IC counters introduced in this chapter are shown in Figure 9–68.

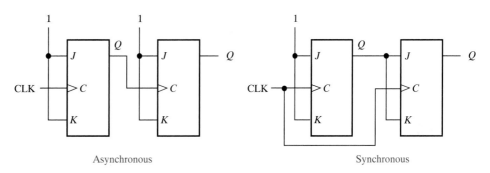

FIGURE 9–67
Comparison of asynchronous and synchronous counters.

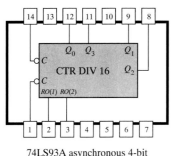

74LS93A asynchronous 4-bit binary counter

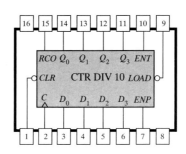

74LS160A synchronous decade counter with asynchronous clear

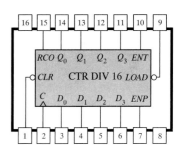

74LS161A synchronous 4-bit binary counter with asynchronous clear

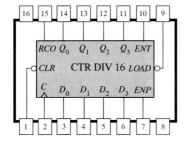

74LS163A synchronous 4-bit binary counter with synchronous clear

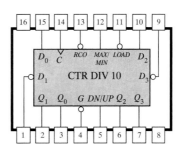

74LS190 synchronous up/down decade counter

FIGURE 9–68
Note that the labels (names of inputs and outputs) are consistent with text but may differ from the particular manufacturer's data book you are using.

■ The maximum modulus of a counter is the maximum number of possible states and is a function of the number of stages (flip-flops). Thus,

$$\text{Maximum modulus} = 2^n$$

where n is the number of stages in the counter. The modulus of a counter is the *actual* number of states in its sequence and can be equal to or less than the maximum modulus.

■ The overall modulus of cascaded counters is equal to the product of the moduli of the individual counters.

■ **SELF-TEST**

1. Asynchronous counters are known as
   (a) ripple counters   (b) multiple clock counters
   (c) decade counters   (d) modulus counters

2. An asynchronous counter differs from a synchronous counter in
   (a) the number of states in its sequence   (b) the method of clocking
   (c) the type of flip-flops used   (d) the value of the modulus

3. The modulus of a counter is
   (a) the number of flip-flops
   (b) the actual number of states in its sequence
   (c) the number of times it recycles in a second
   (d) the maximum possible number of states

4. A 3-bit binary counter has a maximum modulus of
   (a) 3   (b) 6   (c) 8   (d) 16

5. A 4-bit binary counter has a maximum modulus of
   (a) 16   (b) 32   (c) 8   (d) 4

6. A modulus-12 counter must have
   (a) 12 flip-flops   (b) 3 flip-flops
   (c) 4 flip-flops   (d) synchronous clocking

7. Which one of the following is an example of a counter with a truncated modulus?
   (a) Modulus 8   (b) Modulus 14
   (c) Modulus 16   (d) Modulus 32

8. A 4-bit ripple counter consists of flip-flops which each have a propagation delay from clock to Q output of 12 ns. For the counter to recycle from 1111 to 0000, it takes a total of
   (a) 12 ns   (b) 24 ns   (c) 48 ns   (d) 36 ns

9. A BCD counter is an example of
   (a) a full-modulus counter   (b) a decade counter
   (c) a truncated-modulus counter   (d) answers (b) and (c)

10. Which of the following is an invalid state in an 8421 BCD counter?
    (a) 1100   (b) 0010   (c) 0101   (d) 1000

11. Three cascaded modulus-10 counters have an overall modulus of
    (a) 30   (b) 100   (c) 1000   (d) 10,000

12. A 10 MHz clock frequency is applied to a cascaded counter consisting of a modulus-5 counter, and modulus-8 counter, and two modulus-10 counters. The lowest output frequency possible is
    (a) 10 kHz   (b) 2.5 kHz   (c) 5 kHz   (d) 25 kHz

13. A 4-bit binary up/down counter is in the binary state of zero. The next state in the DOWN mode is
    (a) 0001   (b) 1111   (c) 1000   (d) 1110

14. The terminal count of a modulus-13 binary counter is
    (a) 0000   (b) 1111   (c) 1101   (d) 1100

■ PROBLEMS

SECTION 9–1 Asynchronous Counter Operation

1. For the ripple counter shown in Figure 9–69, draw the complete timing diagram for eight clock pulses, showing the clock, Q_0 and Q_1 waveforms.

2. For the ripple counter in Figure 9–70, draw the complete timing diagram for sixteen clock pulses. Show the clock, Q_0, Q_1, and Q_2 waveforms.

3. In the counter of Problem 2, assume that each flip-flop has a propagation delay from the triggering edge of the clock to a change in the Q output of 8 ns. Determine the worst-case (longest) delay time from a clock pulse to the arrival of the counter in a given state. Specify the state or states for which this worst-case delay occurs.

4. Show how to connect a 74LS93A 4-bit asynchronous counter for each of the following moduli:
   (a) 9   (b) 11   (c) 13   (d) 14   (e) 15

FIGURE 9–69

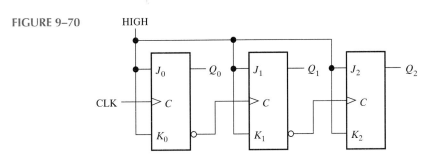

FIGURE 9–70

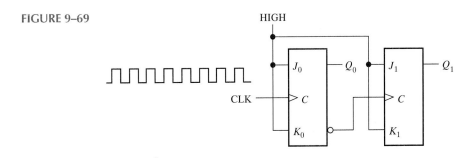

SECTION 9–2 Synchronous Counter Operation

5. If the counter of Problem 3 were synchronous rather than asynchronous, what would be the longest delay time?

6. Draw the complete timing diagram for the 5-stage synchronous binary counter in Figure 9–71. Verify that the waveforms of the Q outputs represent the proper binary number after each clock pulse.

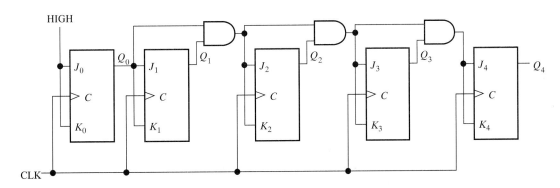

FIGURE 9–71

7. By analyzing the J and K inputs to each flip-flop prior to each clock pulse, prove that the decade counter in Figure 9–72 progresses through a BCD sequence. Explain how these conditions in each case cause the counter to go to the next proper state.

8. The waveforms in Figure 9–73 are applied to the count enable, clear, and clock inputs as indicated. Sketch the counter output waveforms in proper relation to these inputs. The clear input is asynchronous.

9. A BCD decade counter is shown in Figure 9–74. The waveforms are applied to the clock and clear inputs as indicated. Determine the waveforms for each of the counter outputs (Q_0, Q_1, Q_2, and Q_3). The clear is synchronous, and the counter is initially in the binary 1000 state.

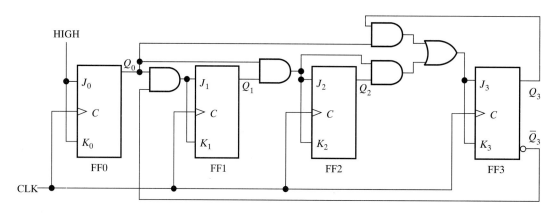

FIGURE 9–72

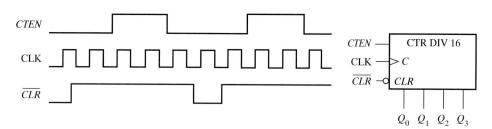

FIGURE 9–73

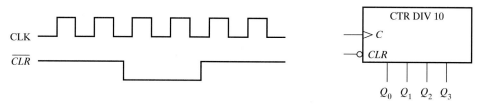

FIGURE 9–74

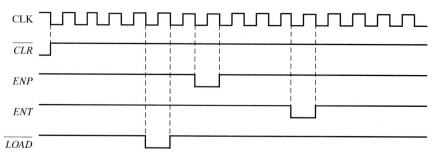

FIGURE 9–75

10. The waveforms in Figure 9–75 are applied to a 74LS163A counter. Determine the Q outputs and the RCO. The inputs are $D_0 = 1, D_1 = 1, D_2 = 0,$ and $D_3 = 1.$

11. The waveforms in Figure 9–75 are applied to a 74LS160A counter. Determine the Q outputs and the RCO. The inputs are $D_0 = 1, D_1 = 0, D_2 = 0,$ and $D_3 = 1.$

SECTION 9–3   Up/Down Synchronous Counters

12. Draw a complete timing diagram for a 3-bit up/down counter that goes through the following sequence. Indicate when the counter is in the UP mode and when it is in the DOWN mode. Assume positive edge-triggering.

$$0, 1, 2, 3, 2, 1, 2, 3, 4, 5, 6, 5, 4, 3, 2, 1, 0$$

13. Sketch the Q output waveforms for a 74LS190 up/down counter with the input waveforms shown in Figure 9–76. A binary 0 is on the data inputs. Start with a count of 0000.

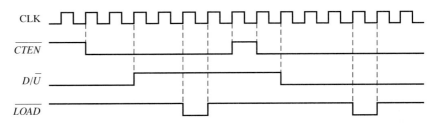

FIGURE 9–76

SECTION 9–4   Design of Synchronous Counters

14. Determine the sequence of the counter in Figure 9–77.

15. Determine the sequence of the counter in Figure 9–78. Begin with the counter cleared.

16. Design a counter to produce the following sequence. Use J-K flip-flops.

$$00, 10, 01, 11, 00, \ldots$$

FIGURE 9–77

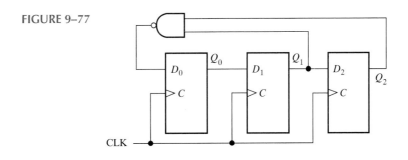

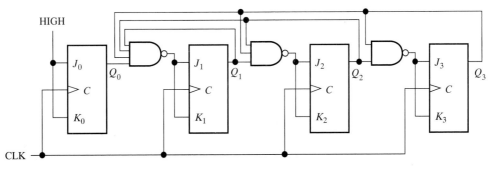

FIGURE 9–78

17. Design a counter to produce the following binary sequence. Use J-K flip-flops.

$$1, 4, 3, 5, 7, 6, 2, 1, \ldots$$

18. Design a counter to produce the following binary sequence. Use J-K flip-flops.

$$0, 9, 1, 8, 2, 7, 3, 6, 4, 5, 0, \ldots$$

19. Design a binary counter with the sequence shown in the state diagram of Figure 9–79.

FIGURE 9–79

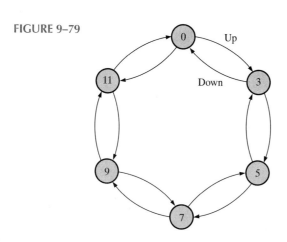

SECTION 9–5   Cascaded Counters

20. For each of the cascaded counter configurations in Figure 9–80, determine the frequency of the waveform at each point indicated by a circled number, and determine the overall modulus.

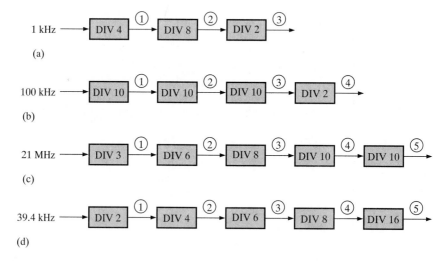

FIGURE 9–80

21. Expand the counter in Figure 9–41 (p. 472) to create a divide-by-10,000 counter and a divide-by-100,000 counter.

22. With general block diagrams, show how to obtain the following frequencies from a 10 MHz clock by using single flip-flops, modulus-5 counters, and decade counters:

   (a) 5 MHz   **(b)** 2.5 MHz   **(c)** 2 MHz   **(d)** 1 MHz   **(e)** 500 kHz

   (f) 250 kHz   **(g)** 62.5 kHz   **(h)** 40 kHz   **(i)** 10 kHz   **(j)** 1 kHz

SECTION 9–6 Counter Decoding

23. Given a BCD decade counter with only the Q outputs available, show what decoding logic is required to decode each of the following states and how it should be connected to the counter. A HIGH output indication is required for each decoded state. The MSB is to the left.

(a) 0001   (b) 0011   (c) 0101   (d) 0111   (e) 1000

24. For the 4-bit binary counter connected to the decoder in Figure 9–81, determine each of the decoder output waveforms in relation to the clock pulses.

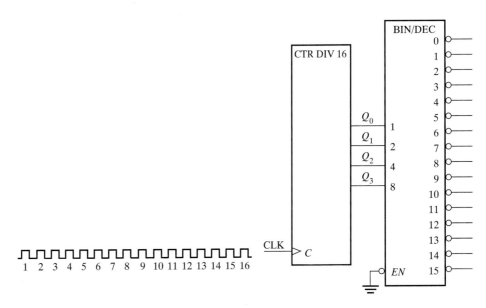

FIGURE 9–81

25. If the counter in Figure 9–81 is asynchronous, determine where the decoding glitches occur on the decoder output waveforms.

26. Modify the circuit in Figure 9–81 to eliminate decoding glitches.

27. Analyze the counter in Figure 9–45 (p. 474) for the occurrence of glitches on the decode gate output. If glitches occur, suggest a way to eliminate them.

28. Analyze the counter in Figure 9–46 (p. 475) for the occurrence of glitches on the outputs of the decoding gates. If glitches occur, make a design change that will eliminate them.

SECTION 9–7 Counter Applications

29. Assume that the digital clock of Figure 9–51 (p. 478) is initially reset to 12 o'clock. Determine the binary state of each counter after sixty-two 60 Hz pulses have occurred.

30. What is the output frequency of each counter in the digital clock circuit of Figure 9–51?

31. For the automobile parking control system in Figure 9–54 (p. 480) a pattern of entrance and exit sensor pulses during a given 24-hour period are shown in Figure 9–82. If there were 53 cars already in the garage at the beginning of the period, what is the state of the counter at the end of the 24 hours?

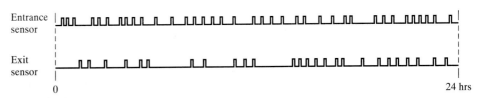

FIGURE 9–82

32. The binary number for decimal 57 appears on the parallel data inputs of the parallel-to-serial converter in Figure 9–56 (D_0 is the LSB). The counter initially contains all zeros and a 10 kHz clock is applied. Develop the timing diagram showing the clock, the counter outputs, and the serial data output.

SECTION 9–8 Troubleshooting

33. For the counter in Figure 9–1 (p. 442) draw the timing diagram for the Q_0 and Q_1 waveforms for each of the following faults (assume Q_0 and Q_1 are initially LOW):

(a) clock input to FF0 shorted to ground

(b) Q_0 output open

(c) clock input to FF1 open

(d) J input to FF0 open

(e) K input to FF1 shorted to ground

34. Solve Problem 33 for the counter in Figure 9–11 (p. 450).

35. Isolate the fault in the counter in Figure 9–3 (p. 444) by analyzing the waveforms in Figure 9–83.

36. From the waveform diagram in Figure 9–84, determine the most likely fault in the counter of Figure 9–14 (p. 451).

37. Solve Problem 36 if the Q_2 output has the waveform observed in Figure 9–85. Outputs Q_0 and Q_1 are the same as in Figure 9–84.

FIGURE 9–83

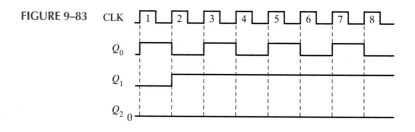

FIGURE 9–84

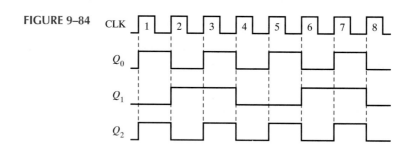

FIGURE 9–85

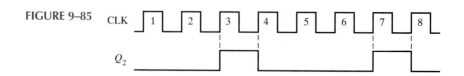

38. You apply a 5 MHz clock to the cascaded counter in Figure 9–44 (p. 473) and measure a frequency of 76.2939 Hz at the last RCO output. Is this correct, and if not, what is the most likely problem?

39. Develop a table for use in testing the counter in Figure 9–44 that will show the frequency at the final RCO output for all possible open failures of the parallel data inputs (D_0, D_1, D_2, and D_3) taken one at a time. Use 10 MHz as the test frequency for the clock.

40. The tens-of-hours 7-segment display in the digital clock system of Figure 9–51 (p. 478) continuously displays a 1. All the other digits work properly. What could be the problem?

41. What would be the visual indication of an open Q_1 output in the tens portion of the minutes counter in Figure 9–51 (p. 478)? Also see Figure 9–52 (p. 479).

42. One day (perhaps a Monday) complaints begin flooding in from patrons of a parking garage that has the control system depicted in Figures 9–54 and 9–55 (p. 480). The patrons say that they enter the garage because the gate is up and the FULL sign is off but that, once in, they can find no empty space. As the technician in charge of this facility, what do you think the problem is, and how will you troubleshoot and repair the system as quickly as possible?

SECTION 9–10 Digital System Application

43. Implement the input logic in the sequential circuit portion of the traffic light control system using only NAND gates.

44. Replace the D flip-flops in the 2-bit Gray code state counter in Figure 9–66 (p. 490) with specific J-K flip-flops.

45. Specify how you would change the time interval for the green light from 25 s to 60 s.

Special Design Problems

46. Design a modulus-1000 counter by using 74LS160A decade counters.

47. Modify the design of the counter in Figure 9–44 (p. 473) to achieve a modulus of 30,000.

48. Repeat Problem 47 for a modulus of 50,000.

49. Modify the digital clock in Figures 9–51, 9–52, and 9–53 (pp. 478–479) so that it can be preset to any desired time.

50. Design an alarm circuit for the digital clock that can detect a predetermined time (hours and minutes only) and produce a signal to activate an audio alarm.

51. Modify the design of the circuit in Figure 9–55 (p. 480) for a 1000-space parking garage and a 3000-space parking garage.

52. Implement the parallel-to-serial data conversion logic in Figure 9–56 (p. 481) with specific devices.

53. In Problem 15 you found that the counter locks up and alternates between two states. It turns out that this operation is the result of a design flaw. Redesign the counter so that when it goes into the second of the lock-up states, it will recycle to the all-0s state on the next clock pulse.

54. Modify the block diagram of the traffic light control system in Figure 9–63 (p. 489) to reflect the addition of a 15 s left turn signal on the main street immediately preceding the green light.

■ ANSWERS TO SECTION REVIEWS

SECTION 9–1

1. Asynchronous means that each flip-flop after the first one is enabled by the output of the preceding flip-flop.

2. A modulus-14 counter has fourteen states requiring four flip-flops.

SECTION 9–2

1. All flip-flops in a synchronous counter are clocked simultaneously.

2. The counter can be preset (initialized) to any given state.

3. Counter is enabled when *ENP* and *ENT* are both HIGH; *RCO* goes HIGH when final state in sequence is reached.

SECTION 9–3

1. The counter goes to 1001.   2. UP: 1111: DOWN: 0000; the next state is 1111

SECTION 9–4

1. $J = 1, K = $ X ("don't care")   **2.** $J = $ X ("don't care"), $K = 0$

3. (a) The next state is 1011.

(b) Q_3 (MSB): no-change or SET; Q_2: no-change or RESET; Q_1: no-change or SET; Q_0 (LSB): SET or toggle

SECTION 9–5

1. Three decade counters produce $\div 1000$; 4 decade counters produce $\div 10,000$.

2. (a) $\div 20$: flip-flop and DIV 10   **(b)** $\div 32$: flip-flop and DIV 16

(c) $\div 160$: DIV 16 and DIV 10   **(d)** $\div 320$: DIV 16 and DIV 10 and flip-flop

SECTION 9–6

1. (a) No transitional states because there is a single bit change

(b) 0000, 0001, 0010, 0101, 0110, 0111

(c) No transitional states because there is a single bit change

(d) 0001, 0010, 0011, 0100, 0101, 0110 , 0111, 1000, 1001, 1010, 1011, 1100, 1101, 1110

SECTION 9–7

1. Gate G_1 resets flip-flop on first clock pulse after count 12. Gate G_2 decodes count 12 to preset counter to 0001.

2. The hours decade counter advances through each state from zero to nine, and as it recycles from nine back to zero, the flip-flop is toggled to the SET state. This produces a ten (10) on the display. When the hours decade counter is in state 12, the decode NAND gate causes the counter to recycle to state 1 on the next clock pulse. The flip-flop RESETS. This results in a one (01) on the display.

SECTION 9–8

1. No pulses on *TC* outputs: *CTEN* of first counter shorted to ground or to a LOW; clock input of first counter open; clock line shorted to ground or to a LOW; *TC* output of first counter shorted to ground or to a LOW

2. With inverter output open, the counter does not recycle at the preset count but acts as a full-modulus counter.

SECTION 9–9

1. *C*: control, usually clock; *M*: mode; *G*: AND   **2.** *D* indicates data storage.

SECTION 9–10

1. The sequential logic produces the proper sequence of states to control the lights.

2. The long timer, the short timer, and/or the vehicle sensor controls sequential logic transitions.

3. The shortest light interval of 4 s is much greater than the period of a 10 kHz clock.

10

SHIFT REGISTERS

■ CHAPTER OBJECTIVES

☐ Identify the basic forms of data movement in shift registers

☐ Explain how serial in/serial out, serial in/parallel out, parallel in/serial out, and parallel in/parallel out shift registers operate

☐ Describe how a bidirectional shift register operates

☐ Determine the sequence of a Johnson counter

☐ Set up a ring counter to produce a specified sequence

☐ Construct a ring counter from a shift register

☐ Use a shift register as a time-delay device

☐ Use a shift register to implement a serial-to-parallel data converter

☐ Implement a basic shift-register-controlled keyboard encoder

☐ Troubleshoot digital systems by "exercising" the system with a known test pattern

☐ Interpret ANSI/IEEE Standard 91–1984 shift register symbols with dependency notation

☐ Use shift registers in a system application

■ CHAPTER OVERVIEW

Shift registers are a type of sequential logic circuit closely related to digital counters. Registers are used primarily for the storage of digital data and typically do not possess a characteristic internal sequence of states as do counters. There are exceptions, however, and these are covered in Section 10–7.

In this chapter, the basic types of shift registers are studied and several applications are presented. Also, an important troubleshooting method is introduced.

■ SPECIFIC DEVICES

74LS00	74LS02	74LS08
74LS85	74LS121	74LS133
74LS147	74LS164	74LS165
74LS174	74LS194A	74LS195A

■ DIGITAL SYSTEM APPLICATION

This Digital System Application illustrates the concepts taught in this chapter. The system application in Section 10–11 introduces a security entry system for controlling the alarms in a building. This system uses two types of shift registers as well as other types of devices covered in previous chapters. The system also includes a memory that will be the focus in Chapter 12.

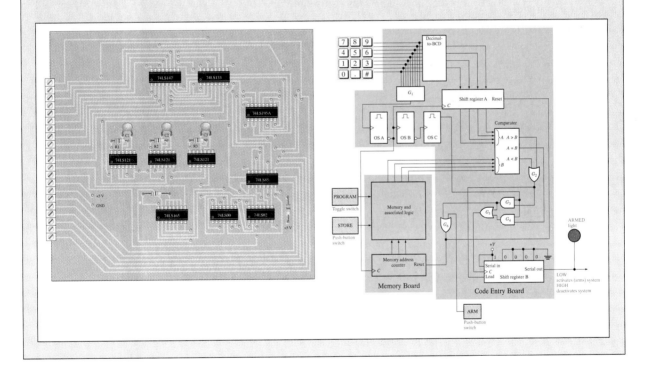

10–1 ■ BASIC SHIFT REGISTER FUNCTIONS

Shift registers consist of an arrangement of flip-flops and are important in applications involving the storage and transfer of data in a digital system. The basic difference between a register and a counter is that a register has no specified sequence of states, except in certain very specialized applications. A register, in general, is used solely for storing and shifting data (1s and 0s) entered into it from an external source and typically possesses no characteristic internal sequence of states. After completing this section, you should be able to

☐ Explain how a flip-flop stores a data bit   ☐ Define the storage capacity of a shift register   ☐ Define the shifting capability of a register

A **register** is a digital circuit with two basic functions: data storage and data movement. The storage capability of a register makes it an important type of memory device. Figure 10–1 illustrates the concept of storing a 1 or a 0 in a D flip-flop. A 1 is applied to the data input as shown, and a clock pulse is applied that stores the 1 by setting the flip-flop. When the 1 on the input is removed, the flip-flop remains in the SET state, thereby storing the 1. The same procedure applies to the storage of a 0, as also illustrated in Figure 10–1.

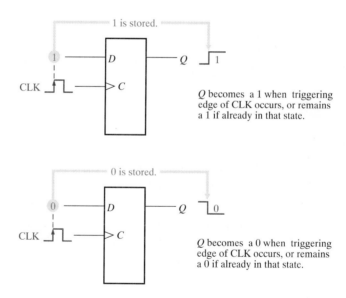

FIGURE 10–1
The flip-flop as a storage element.

The storage capacity of a register is the number of bits (1s and 0s) of digital data it can retain. Each **stage** (flip-flop) in a shift register represents one bit of storage capacity; therefore, the number of stages in a register determines its total storage capacity. Registers are implemented with flip-flops or other storage devices.

The **shifting** capability of a register permits the movement of data from stage to stage within the register or into or out of the register upon application of clock pulses. Figure 10–2 illustrates the types of data movement in shift registers. The block represents any arbitrary 4-bit register, and the arrows indicate the direction of data movement.

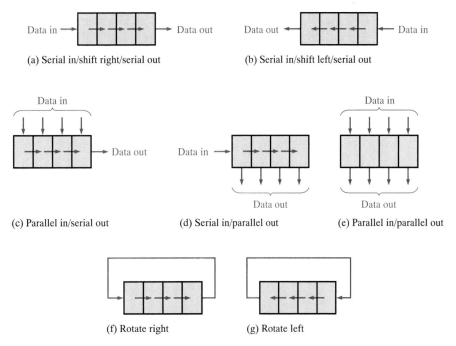

FIGURE 10–2
Basic data movement in shift registers (four bits are used for illustration).

SECTION 10–1 REVIEW	**1.** Generally, what is the difference between a counter and a shift register? **2.** What two principal functions are performed by a shift register?

10–2 ■ SERIAL IN/SERIAL OUT SHIFT REGISTERS

The serial in/serial out shift register accepts data serially—that is, one bit at a time on a single line. It produces the stored information on its output also in serial form. After completing this section, you should be able to

□ Explain how data bits are serially entered into a shift register □ Describe how data bits are shifted through the register   □ Explain how data bits are serially taken out of a shift register   □ Develop and analyze timing diagrams for serial in/serial out registers

Let's first look at the serial entry of data into a typical shift register. Figure 10–3 shows a 4-bit device implemented with D flip-flops. With four stages, this register can store up to four bits of data; its storage capacity is four bits.

FIGURE 10–3
Serial in/serial out shift register.

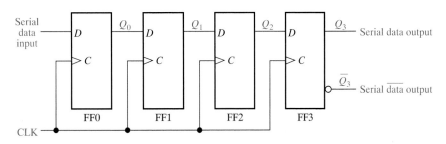

Figure 10–4 illustrates entry of the four bits 1010 into the register, beginning with the right-most bit. The register is initially clear. The 0 is put onto the data input line, making $D = 0$ for FF0. When the first clock pulse is applied, FF0 is RESET, thus storing the 0.

Next the second bit, which is a 1, is applied to the data input, making $D = 1$ for FF0 and $D = 0$ for FF1 because the D input of FF1 is connected to the Q_0 output. When the second clock pulse occurs, the 1 on the data input is shifted into FF0 because FF0 SETS, and the 0 that was in FF0 is shifted into FF1.

The third bit, a 0, is now put onto the data-input line, and a clock pulse is applied. The 0 is entered into FF0, the 1 stored in FF0 is shifted into FF1, and the 0 stored in FF1 is shifted into FF2.

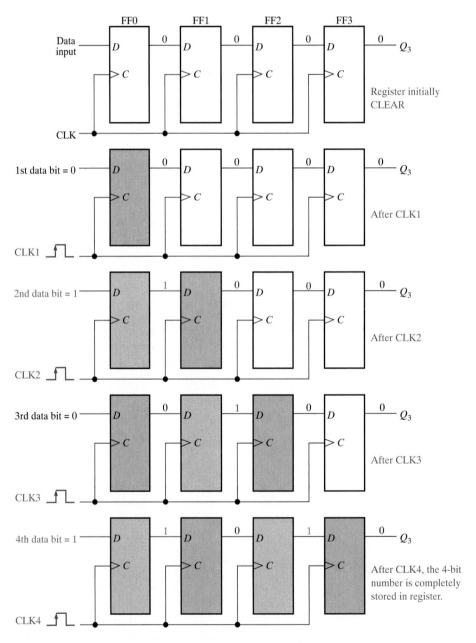

FIGURE 10–4
Four bits (1010) being entered serially into the register.

The last bit, a 1, is now applied to the data input, and a clock pulse is applied. This time the 1 is entered into FF0, the 0 stored in FF0 is shifted into FF1, the 1 stored in FF1 is shifted into FF2, and the 0 stored in FF2 is shifted into FF3. This completes the serial entry of the four bits into the shift register, where they can be stored for any length of time as long as the flip-flops have dc power.

If you want to get the data out of the register, the bits must be shifted out serially and taken off the Q_3 output, as Figure 10–5 illustrates. After CLK4 in the data-entry operation just described, the right-most bit, 0, appears on the Q_3 output. When clock pulse CLK5 is applied, the second bit appears on the Q_3 output. Clock pulse CLK6 shifts the

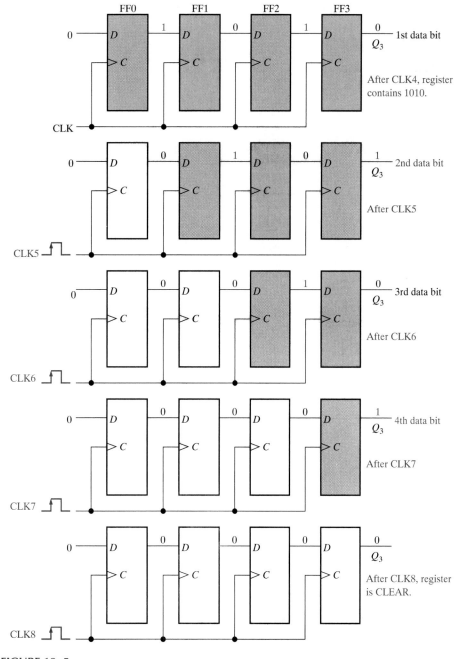

FIGURE 10–5
Four bits (1010) being serially shifted out of the register and replaced by all zeros.

third bit to the output, and CLK7 shifts the fourth bit to the output. Notice that while the original four bits are being shifted out, more bits can be shifted in. All zeros are shown being shifted in.

EXAMPLE 10–1

Show the states of the 5-bit register in Figure 10–6(a) for the specified data input and clock waveforms. Assume that the register is initially cleared (all 0s).

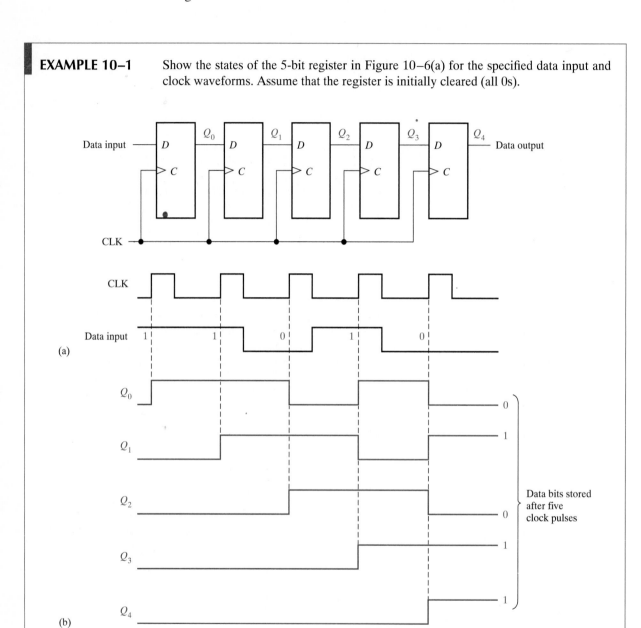

(a)

(b)

FIGURE 10–6

Solution   The first data bit (1) is entered into the register on the first clock pulse and then shifted from left to right as the remaining bits are entered and shifted. The register contains 11010 after five clock pulses. See Figure 10–6(b).

Related Exercise   Show the states of the register if the data input is inverted. The register is initially cleared.

A traditional logic block symbol for an 8-bit serial in/serial out shift register is shown in Figure 10–7. The "SRG 8" designation means a shift register (SRG) with an 8-bit capacity.

FIGURE 10–7
Logic symbol for an 8-bit serial in/serial out shift register.

Data in
SRG 8
Q_7
CLK
C
$\overline{Q}_7$

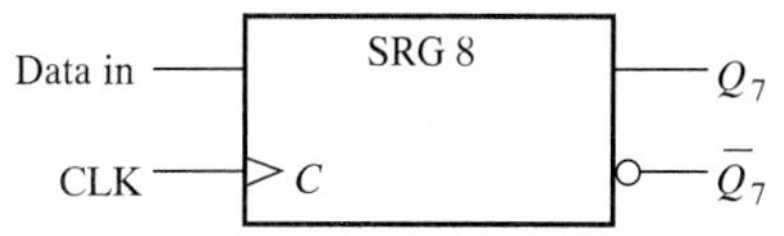

SECTION 10–2 REVIEW

1. Draw the logic diagram for the shift register in Figure 10–3, using J-K flip-flops to replace the D flip-flops.
2. How many clock pulses are required to enter a byte of data serially into an 8-bit shift register?

10–3 ■ SERIAL IN/PARALLEL OUT SHIFT REGISTERS

Data bits are entered serially (right-most bit first) into this type of register in the same manner as discussed in Section 10–2. The difference is the way in which the data bits are taken out of the register; in the parallel output register, the output of each stage is available. Once the data are stored, each bit appears on its respective output line, and all bits are available simultaneously, rather than on a bit-by-bit basis as with the serial output. After completing this section, you should be able to

☐ Explain how data bits are taken out of a shift register in parallel ☐ Compare serial output to parallel output ☐ Discuss the 74LS164 8-bit shift register ☐ Develop and analyze timing diagrams for serial in/parallel out registers

Figure 10–8 shows a 4-bit serial in/parallel out shift register and its logic block symbol.

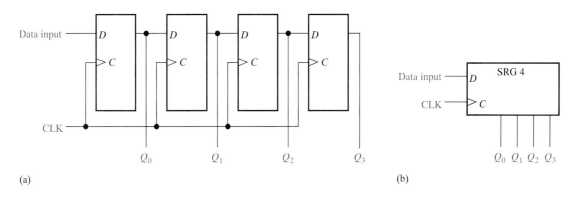

Data input
D
C
D
C
D
C
D
C
CLK
Q_0
Q_1
Q_2
Q_3
(a)
Data input
D
SRG 4
CLK
C
Q_0 Q_1 Q_2 Q_3
(b)

FIGURE 10–8
A serial in/parallel out shift register.

EXAMPLE 10–2

Show the states of the 4-bit register for the data input and clock waveforms in Figure 10–9(a). The register initially contains all 1s.

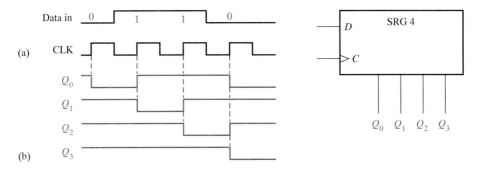

FIGURE 10–9

Solution The register contains 0110 after four clock pulses. See Figure 10–9(b).

Related Exercise If the data input remains 0 after the fourth clock pulse, what is the state of the register after three additional clock pulses?

The 74LS164 8-Bit Serial In/Parallel Out Shift Register

The 74LS164 is an example of an IC shift register having serial in/parallel out operation. The logic diagram is shown in Figure 10–10(a), and a typical logic block symbol is shown in part (b). Notice that this device has two gated serial inputs, A and B, and a clear ($\overline{CLR}$) input that is active-LOW. The parallel outputs are Q_0 through Q_7.

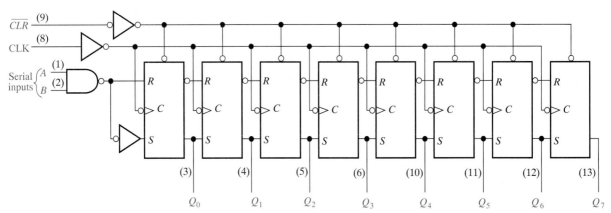

(a) Logic diagram

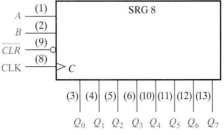

(b) Logic symbol

FIGURE 10–10
The 74LS164 8-bit serial in/parallel out shift register.

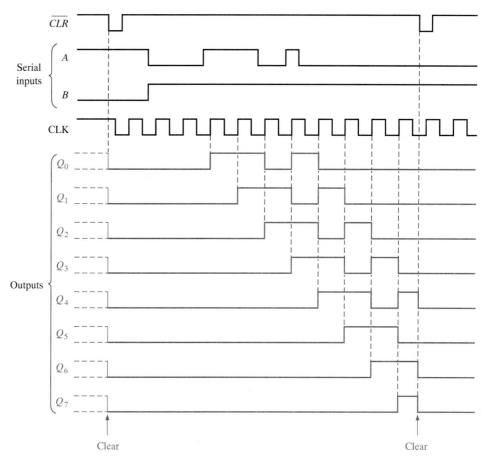

FIGURE 10–11
Sample timing diagram for a 74LS164 shift register.

A sample timing diagram for the 74LS164 is shown in Figure 10–11. Notice that the serial input data on input *A* are shifted into and through the register after input *B* goes HIGH.

SECTION 10–3 REVIEW

1. The bit sequence 1101 is serially entered (right-most bit first) into a 4-bit parallel out shift register that is initially clear. What are the *Q* outputs after two clock pulses?
2. How can a serial in/parallel out register be used as a serial in/serial out register?

10–4 ■ PARALLEL IN/SERIAL OUT SHIFT REGISTERS

For a register with parallel data inputs, the bits are entered simultaneously into their respective stages on parallel lines rather than on a bit-by-bit basis on one line as with serial data inputs. The serial output is the same as described in Section 10–2, once the data are completely stored in the register. After completing this section, you should be able to

□ Explain how data bits are entered into a shift register in parallel   □ Compare serial input to parallel input   □ Discuss the 74LS165 8-bit parallel-load shift register   □ Develop and analyze timing diagrams for parallel in/serial out registers

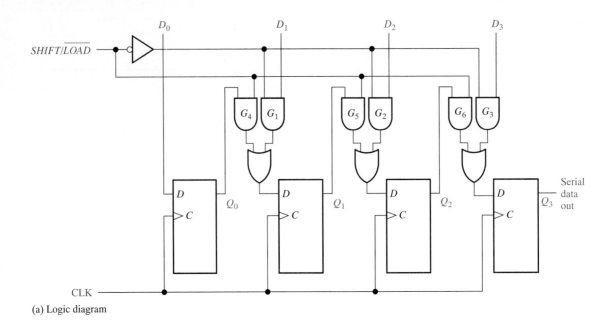

(a) Logic diagram

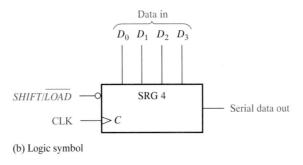

(b) Logic symbol

FIGURE 10–12

A 4-bit parallel in/serial out shift register.

Figure 10–12 illustrates a 4-bit parallel in/serial out shift register and a typical logic symbol. Notice that there are four data-input lines, D_0, D_1, D_2, and D_3, and a $SHIFT/\overline{LOAD}$ input, which allows four bits of data to be **loaded** in parallel into the register. When $SHIFT/\overline{LOAD}$ is LOW, gates G_1 through G_3 are enabled, allowing each data bit to be applied to the D input of its respective flip-flop. When a clock pulse is applied, the flip-flops with $D = 1$ will SET and those with $D = 0$ will RESET, thereby storing all four bits simultaneously.

When $SHIFT/\overline{LOAD}$ is HIGH, gates G_1 through G_3 are disabled and gates G_4 through G_6 are enabled, allowing the data bits to shift right from one stage to the next. The OR gates allow either the normal shifting operation or the parallel data-entry operation, depending on which AND gates are enabled by the level on the $SHIFT/\overline{LOAD}$ input.

EXAMPLE 10–3

Show the data-output waveform for a 4-bit register with the parallel input data and the clock and $SHIFT/\overline{LOAD}$ waveforms given in Figure 10–13(a). Refer to Figure 10–12(a) for the logic diagram.

Solution On clock pulse 1, the parallel data ($D_0D_1D_2D_3 = 1010$) are loaded into the register, making Q_3 a 0. On clock pulse 2 the 1 from Q_2 is shifted onto Q_3; on clock pulse 3 the 0 is shifted onto Q_3; on clock pulse 4 the last data bit (1) is shifted onto Q_3; and on clock pulse 5, all data bits have been shifted out, and only 1s remain in the register (assuming the D_0 input remains a 1). See Figure 10–13(b).

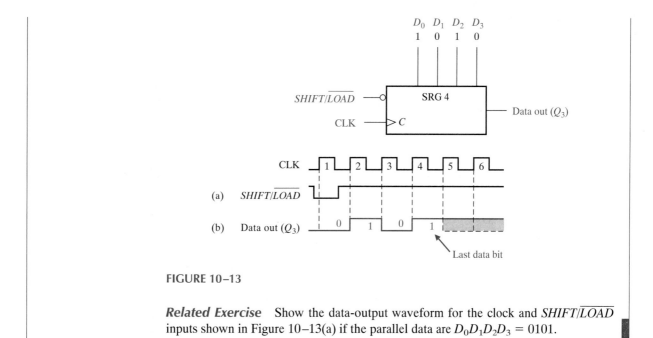

FIGURE 10–13

Related Exercise Show the data-output waveform for the clock and *SHIFT/$\overline{LOAD}$* inputs shown in Figure 10–13(a) if the parallel data are $D_0D_1D_2D_3 = 0101$.

The 74LS165 8-Bit Parallel Load Shift Register

The 74LS165 is an example of an IC shift register that has a parallel in/serial out operation (it can also be operated as serial in/serial out). Figure 10–14(a) shows the internal logic diagram for this device, and part (b) on the next page shows a typical logic block symbol. A

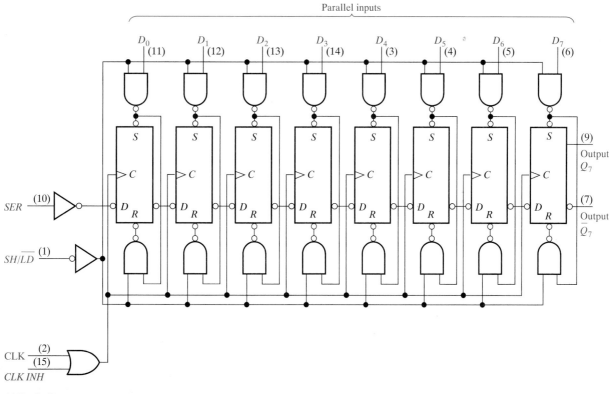

(a) Logic diagram

FIGURE 10–14
The 74LS165 8-bit parallel load shift register.

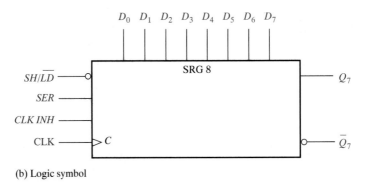

(b) Logic symbol

FIGURE 10–14 *(Continued.)*

LOW on the *SHIFT/LOAD* input (*SH/LD*) enables all the NAND gates for parallel loading. When an input data bit is a 1, the flip-flop is asynchronously SET by a LOW out of the upper gate. When an input data bit is a 0, the flip-flop is asynchronously RESET by a LOW out of the lower gate. Additionally, data can be entered serially on the *SER* input. Also, the clock can be inhibited anytime with a HIGH on the *CLK INH* input. The serial data outputs of the register are Q_7 and its complement $\overline{Q}_7$. This implementation is different from the synchronous method of parallel loading previously discussed, demonstrating that there are usually several ways to accomplish the same function.

Figure 10–15 is a timing diagram showing an example of the operation of a 74LS165 shift register.

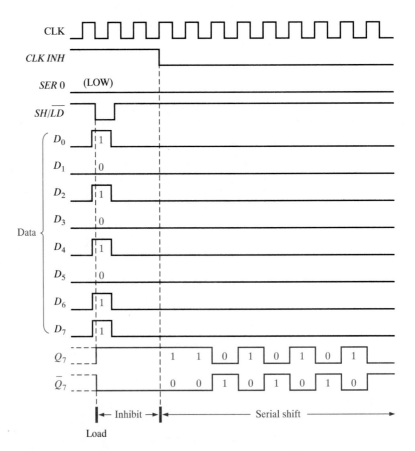

FIGURE 10–15
Sample timing diagram for a 74LS165 shift register.

1. Explain the function of the *SHIFT/$\overline{LOAD}$* input.
2. Is the parallel load operation in a 74LS165 shift register synchronous or asynchronous? What does this mean?

10–5 ■ PARALLEL IN/PARALLEL OUT SHIFT REGISTERS

Parallel entry of data was described in Section 10–4, and parallel output of data has also been discussed previously. The parallel in/parallel out register employs both methods. Immediately following the simultaneous entry of all data bits, the bits appear on the parallel outputs. After completing this section, you should be able to

□ Discuss the 74LS195A 4-bit parallel-access shift register □ Develop and analyze timing diagrams for parallel in/parallel out registers

Figure 10–16 shows a parallel in/parallel out register.

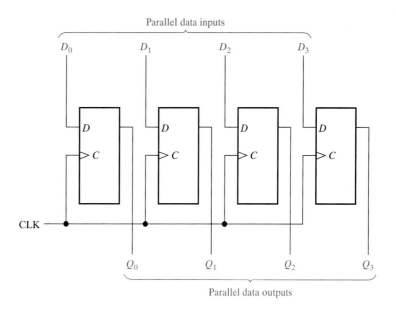

FIGURE 10–16
A parallel in/parallel out register.

The 74LS195A 4-Bit Parallel Access Shift Register

The 74LS195A can be used for parallel in/parallel out operation. Since it also has a serial input, it can be used for serial in/serial out and serial in/parallel out operations. It can be used for parallel in/serial out operation by using Q_3 as the output. A typical logic block symbol is shown in Figure 10–17 on the next page.

When the *SHIFT/$\overline{LOAD}$* input (*SH/$\overline{LD}$*) is LOW, the data on the parallel inputs are entered synchronously on the positive transition of the clock. When *SH/$\overline{LD}$* is HIGH, stored data will shift right (Q_0 to Q_3) synchronously with the clock. Inputs J and $\overline{K}$ are the serial data inputs to the first stage of the register (Q_0); Q_3 can be used for serial output data. The active-LOW clear input is asynchronous.

The timing diagram in Figure 10–18 illustrates the operation of this register.

FIGURE 10–17
The 74LS195A 4-bit parallel access shift register.

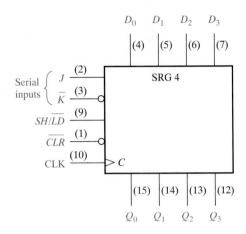

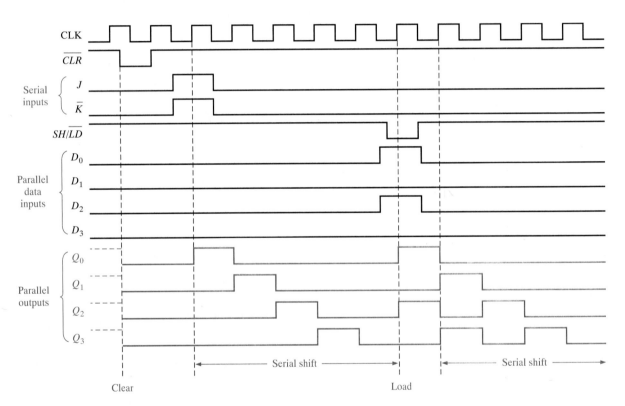

FIGURE 10–18
Sample timing diagram for a 74LS195A shift register.

SECTION 10–5 REVIEW

1. In Figure 10–16, $D_0 = 1, D_1 = 0, D_2 = 0$, and $D_3 = 1$. After three clock pulses, what are the data outputs?
2. For a 74LS195A, $SH/\overline{LD} = 1, J = 1$, and $\overline{K} = 1$. What is Q_0 after one clock pulse?

10–6 ■ BIDIRECTIONAL SHIFT REGISTERS

A bidirectional shift register is one in which the data can be shifted either left or right. It can be implemented by using gating logic that enables the transfer of a data bit from one stage to the next stage to the right or to the left, depending on the level of a control line. After completing this section, you should be able to

□ Explain the operation of a bidirectional shift register   □ Discuss the 74LS194A 4-bit bidirectional universal shift register   □ Develop and analyze timing diagrams for bidirectional shift registers

A 4-bit **bidirectional** shift register is shown in Figure 10–19. A HIGH on the $RIGHT/\overline{LEFT}$ control input allows data bits inside the register to be shifted to the right, and a LOW enables data bits inside the register to be shifted to the left. An examination of the gating logic will make the operation apparent. When the $RIGHT/\overline{LEFT}$ control input is HIGH, gates G_1 through G_4 are enabled, and the state of the Q output of each flip-flop is passed through to the D input of the *following* flip-flop. When a clock pulse occurs, the data bits are shifted one place to the *right*. When the $RIGHT/\overline{LEFT}$ control input is LOW, gates G_5 through G_8 are enabled, and the Q output of each flip-flop is passed through to the D input of the *preceding* flip-flop. When a clock pulse occurs, the data bits are then shifted one place to the *left*.

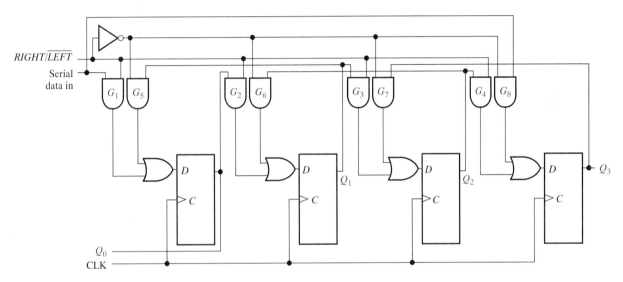

FIGURE 10–19
Four-bit bidirectional shift register.

EXAMPLE 10–4   Determine the state of the shift register of Figure 10–19 after each clock pulse for the given $RIGHT/\overline{LEFT}$ control input waveform in Figure 10–20(a). Assume that $Q_0 = 1$, $Q_1 = 1$, $Q_2 = 0$, and $Q_3 = 1$ and that the serial data-input line is LOW.

FIGURE 10–20

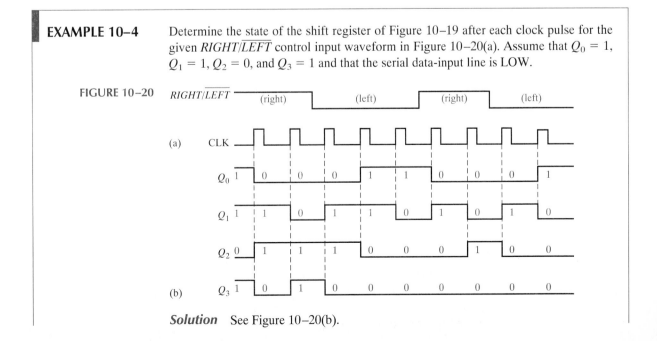

Solution   See Figure 10–20(b).

> **Related Exercise** Invert the *RIGHT/LEFT* waveform, and determine the state of the shift register in Figure 10–19 after each clock pulse.

The 74LS194A 4-Bit Bidirectional Universal Shift Register

The 74LS194A is an example of a universal bidirectional shift register in integrated circuit form. A **universal shift register** has both serial and parallel input and output capability. A logic block symbol is shown in Figure 10–21, and a sample timing diagram is shown in Figure 10–22.

FIGURE 10–21
The 74LS194A 4-bit bidirectional universal shift register.

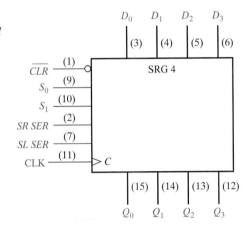

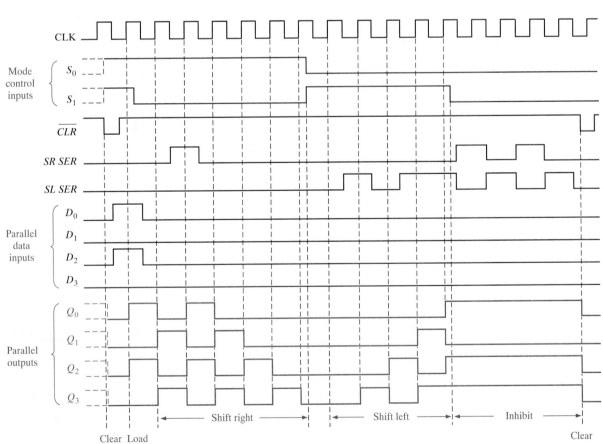

FIGURE 10–22
Sample timing diagram for a 74LS194A shift register.

Parallel loading, which is synchronous with a positive transition of the clock, is accomplished by applying the four bits of data to the parallel inputs and a HIGH to the S_0 and S_1 inputs. Shift right is accomplished synchronously with the positive edge of the clock when S_0 is HIGH and S_1 is LOW. Serial data in this mode are entered at the shift-right serial input (*SR SER*). When S_0 is LOW and S_1 is HIGH, data bits shift left synchronously with the clock, and new data are entered at the shift-left serial input *(SL SER)*. The sample timing diagram is shown in Figure 10–22. Input *SR SER* goes into the Q_0 stage, and *SL SER* goes into the Q_3 stage.

SECTION 10–6 REVIEW	**1.** Assume that the bidirectional shift register in Figure 10–19 has the following contents: $Q_0 = 1$, $Q_1 = 1$, $Q_2 = 0$, and $Q_3 = 0$. There is a 1 on the serial data-input line. If *RIGHT/$\overline{LEFT}$* is HIGH for three clock pulses and LOW for two more clock pulses, what are the contents after the fifth clock pulse?

10–7 ■ SHIFT REGISTER COUNTERS

A shift register counter is basically a shift register with the serial output connected back to the serial input to produce special sequences. These devices are often classified as counters because they exhibit a specified sequence of states. Two of the most common types of shift register counters, the Johnson counter and the ring counter, are introduced in this section. After completing this section, you should be able to

□ Discuss how a shift register counter differs from a basic shift register □ Explain the operation of a Johnson counter □ Specify a Johnson sequence for any number of bits □ Explain the operation of a ring counter and determine the sequence of any specific ring counter

The Johnson Counter

In a **Johnson counter** the complement of the output of the last flip-flop is connected back to the *D* input of the first flip-flop (it can be implemented with other types of flip-flops as well). This feedback arrangement produces a characteristic sequence of states, as shown in Table 10–1 for a 4-bit device and in Table 10–2 for a 5-bit device. Notice that the 4-bit se-

TABLE 10–1
Four-bit Johnson sequence

Clock Pulse	Q_0	Q_1	Q_2	Q_3
0	0	0	0	0
1	1	0	0	0
2	1	1	0	0
3	1	1	1	0
4	1	1	1	1
5	0	1	1	1
6	0	0	1	1
7	0	0	0	1

TABLE 10–2
Five-bit Johnson sequence

Clock Pulse	Q_0	Q_1	Q_2	Q_3	Q_4
0	0	0	0	0	0
1	1	0	0	0	0
2	1	1	0	0	0
3	1	1	1	0	0
4	1	1	1	1	0
5	1	1	1	1	1
6	0	1	1	1	1
7	0	0	1	1	1
8	0	0	0	1	1
9	0	0	0	0	1

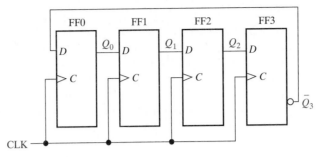

(a) Four-bit Johnson counter

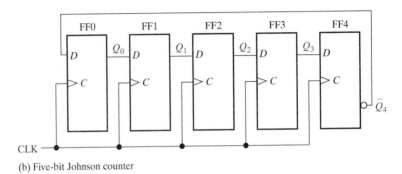

(b) Five-bit Johnson counter

FIGURE 10–23
Four-bit and 5-bit Johnson counters.

quence has a total of eight states, or bit patterns, and that the 5-bit sequence has a total of ten states. In general, a Johnson counter will produce a modulus of $2n$, where n is the number of stages in the counter.

The implementations of the 4-stage and 5-stage Johnson counters are shown in Figure 10–23. The implementation of a Johnson counter is very straightforward and is the same regardless of the number of stages. The Q output of each stage is connected to the D input of the next stage (assuming that D flip-flops are used). The single exception is that the $\overline{Q}$ output of the last stage is connected back to the D input of the first stage. As the sequences in Tables 10–1 and 10–2 show, the counter will "fill up" with 1s from left to right, and then it will "fill up" with 0s again.

Diagrams of the timing operations of the 4-bit and 5-bit counters are shown in Figures 10–24 and 10–25, respectively.

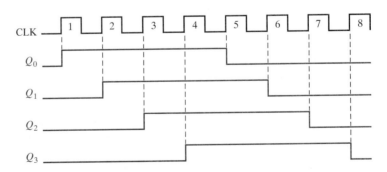

FIGURE 10–24
Timing sequence for a 4-bit Johnson counter.

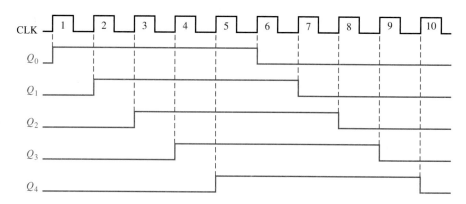

FIGURE 10–25
Timing sequence for a 5-bit Johnson counter.

The Ring Counter

The **ring counter** utilizes one flip-flop for each state in its sequence. It has the advantage that decoding gates are not required for decimal conversion, because there is an output for each decimal number.

A logic diagram for a 10-bit ring counter is shown in Figure 10–26. The sequence for this ring counter is given in Table 10–3. Initially, a 1 is preset into the first flip-flop, and the rest of the flip-flops are cleared. Notice that the interstage connections are the same as

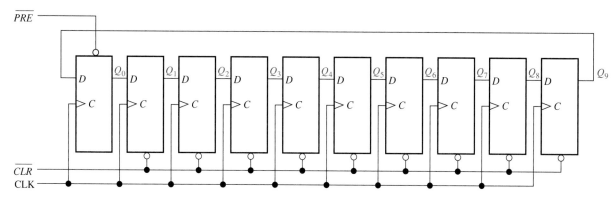

FIGURE 10–26
A 10-bit ring counter.

TABLE 10–3
Ten-bit ring counter sequence

Clock Pulse	Q_0	Q_1	Q_2	Q_3	Q_4	Q_5	Q_6	Q_7	Q_8	Q_9
0	1	0	0	0	0	0	0	0	0	0
1	0	1	0	0	0	0	0	0	0	0
2	0	0	1	0	0	0	0	0	0	0
3	0	0	0	1	0	0	0	0	0	0
4	0	0	0	0	1	0	0	0	0	0
5	0	0	0	0	0	1	0	0	0	0
6	0	0	0	0	0	0	1	0	0	0
7	0	0	0	0	0	0	0	1	0	0
8	0	0	0	0	0	0	0	0	1	0
9	0	0	0	0	0	0	0	0	0	1

those for a Johnson counter, except that Q rather than $\overline{Q}$ is fed back from the last stage. The ten outputs of the counter indicate directly the decimal count of the clock pulse. For instance, a 1 on Q_0 is a zero, a 1 on Q_1 is a one, a 1 on Q_2 is a two, a 1 on Q_3 is a three, and so on. You should verify for yourself that the 1 is always retained in the counter and simply shifted "around the ring," advancing one stage for each clock pulse.

Modified sequences can be achieved by having more than a single 1 in the counter, as illustrated in Example 10–5.

EXAMPLE 10–5

If a 10-bit ring counter similar to Figure 10–26 has the initial state 1010000000, determine the waveform for each of the Q outputs.

Solution See Figure 10–27.

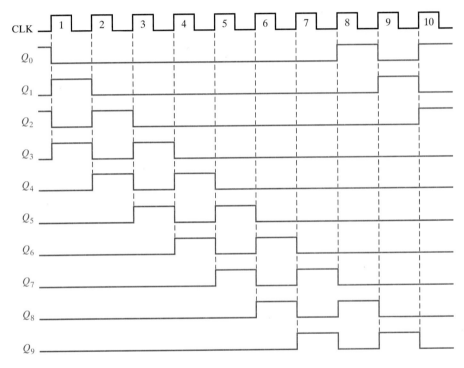

FIGURE 10–27

Related Exercise If a 10-bit ring counter has an initial state 0101001111, determine the waveform for each Q output.

SECTION 10–7 REVIEW

1. How many states are there in an 8-bit Johnson counter sequence?
2. Write the sequence of states for a 3-bit Johnson counter starting with 000.

10–8 ■ SHIFT REGISTER APPLICATIONS

Shift registers are found in many types of applications, a few of which are presented in this section. After completing this section, you should be able to

☐ Use a shift register to generate a time delay   ☐ Implement a specified ring counter sequence using a 74LS195A shift register   ☐ Discuss how shift registers are used for serial-to-parallel conversion of data   ☐ Define *UART*   ☐ Explain the operation of a keyboard encoder and how registers are used in this application

Time Delay

The serial in/serial out shift register can be used to provide a time delay from input to output that is a function of both the number of stages (n) in the register and the clock frequency.

When a data pulse is applied to the serial input in Figure 10–28 (A and B connected together), it enters the first stage on the triggering edge of the clock pulse. It is then shifted from stage to stage on each successive clock pulse until it appears on the serial output n clock periods later. This time-delay operation is illustrated in Figure 10–28, in which an 8-bit serial in/serial out shift register is used with a clock frequency of 1 MHz to achieve a time delay (t_d) of 8 μs (8×1 μs). This time can be adjusted up or down by changing the clock frequency. The time delay can also be increased by cascading shift registers and decreased by taking the outputs from successively lower stages in the register if the outputs are available, as illustrated in Example 10–6.

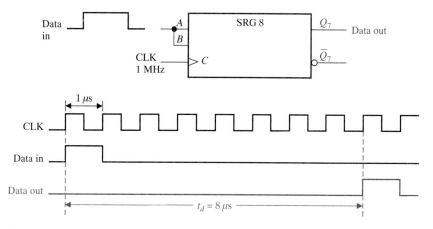

FIGURE 10–28
The shift register as a time-delay device.

EXAMPLE 10–6

Determine the amount of time delay between the serial input and each output in Figure 10–29. Show a timing diagram to illustrate.

FIGURE 10–29

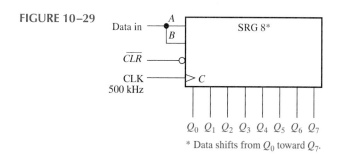

* Data shifts from Q_0 toward Q_7.

Solution The clock period is $2\,\mu s$. Thus, the time delay can be increased or decreased in $2\,\mu s$ increments from a minimum of $2\,\mu s$ to a maximum of $16\,\mu s$, as illustrated in Figure 10–30.

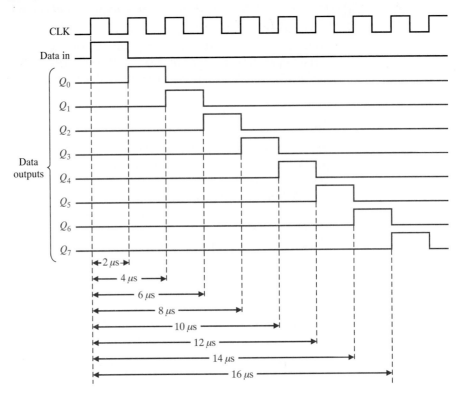

FIGURE 10–30
Timing diagram showing time delays for the register in Figure 10–29.

Related Exercise Determine the clock frequency required to obtain a time delay of $24\,\mu s$ to the Q_7 output in Figure 10–29.

A Ring Counter Using a 74LS195A Shift Register

If the output is connected back to the serial input, a shift register can be used as a ring counter. Figure 10–31 illustrates this application with a 74LS195A 4-bit shift register.

Initially, a bit pattern of 1000 (or any other pattern) can be synchronously preset into the counter by applying the bit pattern to the parallel data inputs, taking the $SH/\overline{LD}$ input LOW, and applying a clock pulse. After this initialization, the 1 continues to circulate through the ring counter, as the timing diagram in Figure 10–32 shows.

FIGURE 10–31
74LS195A connected as a ring counter.

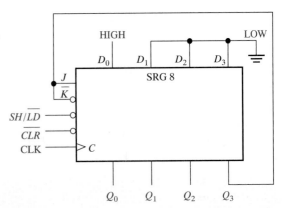

FIGURE 10–32
Timing diagram showing two complete cycles of the ring counter in Figure 10–31 when it is initially preset to 1000.

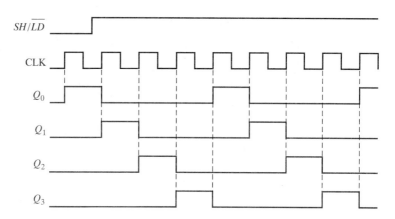

Serial-to-Parallel Data Converter

Serial data transmission from one digital system to another is commonly used to reduce the number of wires in the transmission line. For example, eight bits can be sent serially over one wire, but it takes eight wires to send the same data in parallel.

A computer or microprocessor-based system commonly requires incoming data to be in parallel format, thus the requirement for serial-to-parallel conversion. A simplified serial-to-parallel data converter, in which two types of shift registers are used, is shown in Figure 10–33.

To illustrate the operation of this serial-to-parallel converter, the serial data format shown in Figure 10–34 is used. It consists of eleven bits. The first bit (start bit) is always 0 and always begins with a HIGH-to-LOW transition. The next eight bits (D_7 through D_0) are the data bits (one of the bits can be parity), and the last two bits (stop bits) are always 1s. When no data are being sent, there is a continuous 1 on the serial data line.

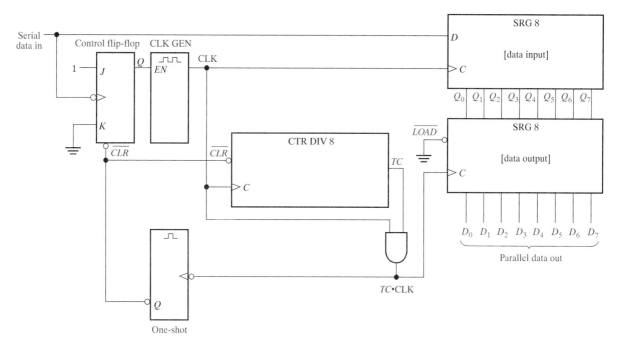

FIGURE 10–33
Simplified logic diagram of a serial-to-parallel converter.

FIGURE 10–34
Serial data format.

The HIGH-to-LOW transition of the start bit SETS the control flip-flop, which enables the clock generator. After a fixed delay time, the clock generator begins producing a pulse waveform, which is applied to the data-input register and to the divide-by-8 counter. The clock has a frequency precisely equal to that of the incoming serial data, and the first clock pulse after the start bit occurs simultaneously with the first data bit.

The timing diagram in Figure 10–35 illustrates the following basic operation: The eight data bits (D_7 through D_0) are serially shifted into the data-input register. After the eighth clock pulse, a HIGH-to-LOW transition of the terminal count (TC) output of the counter ANDed with the clock ($TC \cdot CLK$) loads the eight bits that are in the data-input register into the data-output register. This same transition also triggers the one-shot, which produces a short-duration pulse to clear the counter and RESET the control flip-flop and thus disable the clock generator. The system is now ready for the next group of

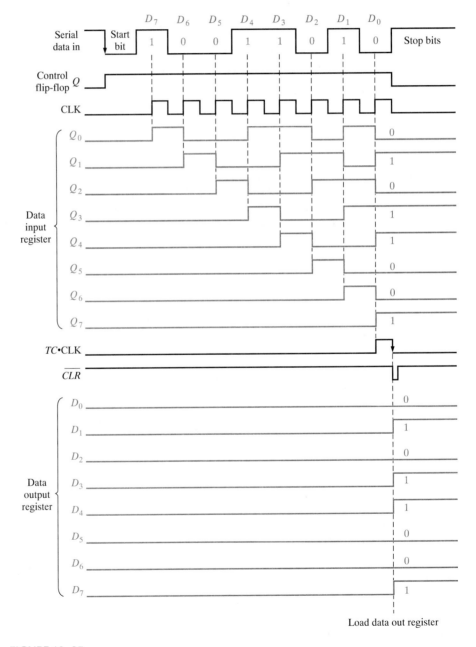

FIGURE 10–35

Timing diagram illustrating the operation of the serial-to-parallel data converter in Figure 10–33.

eleven bits, and it waits for the next HIGH-to-LOW transition at the beginning of the start bit.

By reversing the process just stated, parallel-to-serial data conversion can be accomplished. However, since the serial data format must be produced, additional requirements must be taken into consideration. Problem 43 at the end of the chapter relates to the design of an 8-bit parallel-to-serial data converter.

Universal Asynchronous Receiver Transmitter (UART)

As mentioned, computers and microprocessor-based systems often send and receive data in a parallel format. Frequently, these systems must communicate with external devices that send and/or receive serial data. An interfacing device used to accomplish these conversions is the UART. Figure 10–36 illustrates the UART in a general microprocessor-based system application.

A UART includes a serial-to-parallel data converter such as we have discussed and a parallel-to-serial converter, as shown in Figure 10–37. The data bus is basically a set of parallel conductors along which data move between the UART and the microprocessor system. Buffers interface the data registers with the data bus.

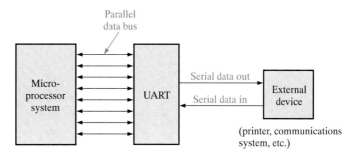

FIGURE 10–36
UART interface.

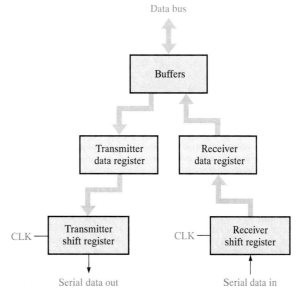

FIGURE 10–37
Basic UART block diagram.

The UART receives data in serial format, converts the data to parallel format, and places them on the data bus. The UART also accepts parallel data from the data bus, converts the data to serial format, and transmits them to an external device.

Keyboard Encoder

The keyboard encoder is a good example of the application of a shift register used as a ring counter in conjunction with other devices. Recall that a simplified computer keyboard encoder without data storage was presented in Chapter 6.

Figure 10–38 shows a simplified keyboard encoder for encoding a key closure in a 64-key matrix organized in eight rows and eight columns. Two 74LS195A 4-bit shift reg-

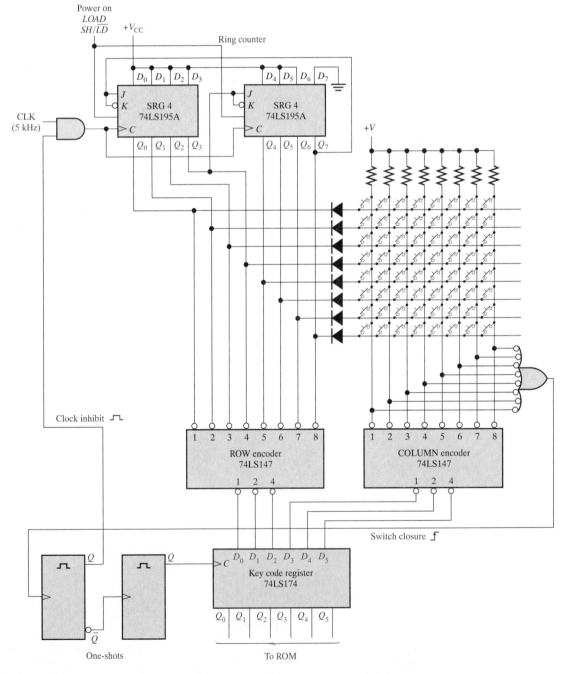

FIGURE 10–38
Simplified keyboard encoding circuit.

isters are connected as an 8-bit ring counter with a fixed bit pattern of seven 1s and one 0 preset into it when the power is turned on. Two 74LS147 priority encoders (introduced in Chapter 6) are used as eight-line-to-three-line encoders (9 input HIGH, 8 output unused) to encode the ROW and COLUMN lines of the keyboard matrix. The 74LS174 (hex flip-flops) is used as a parallel in/parallel out register in which the ROW/COLUMN code from the priority encoders is stored.

The basic operation of the keyboard encoder in Figure 10–38 is as follows: The ring counter "scans" the rows for a key closure as the clock signal shifts the 0 around the counter at a 5 kHz rate. The 0 (LOW) is sequentially applied to each ROW line, while all other ROW lines are HIGH. All the ROW lines are connected to the ROW encoder inputs, so the 3-bit output of the ROW encoder at any time is the binary representation of the ROW line that is LOW. When there is a key closure, one COLUMN line is connected to one ROW line. When the ROW line is taken LOW by the ring counter, that particular COLUMN line is also pulled LOW. The COLUMN encoder produces a binary output corresponding to the COLUMN in which the key is closed. The 3-bit ROW code plus the 3-bit COLUMN code uniquely identifies the key that is closed. This 6-bit code is applied to the inputs of the key code register. When a key is closed, the two one-shots produce a delayed clock pulse to parallel-load the 6-bit code into the key code register. This delay allows the contact bounce to die out. Also, the first one-shot output inhibits the ring counter to prevent it from scanning while the data are being loaded into the key code register.

The 6-bit code in the key code register is now applied to a ROM (read-only memory) to be converted to ASCII or some other appropriate alphanumeric code that identifies the keyboard character. ROMs are studied in Chapter 12.

SECTION 10–8 REVIEW	1. In the keyboard encoder, how many times per second does the ring counter scan the keyboard? 2. What is the 6-bit ROW/COLUMN code (key code) for the top row and the left-most column in the keyboard encoder? 3. What is the purpose of the diodes in the keyboard encoder? What is the purpose of the resistors?

10–9 ■ TROUBLESHOOTING

One basic method of troubleshooting sequential logic and other more complex digital systems uses a procedure of "exercising" the circuit under test with a known input waveform (stimulus) and then observing the output for the proper bit pattern. After completing this section, you should be able to

☐ Explain the procedure of "exercising" as a troubleshooting technique ☐ Discuss exercising of a serial-to-parallel converter ☐ Discuss the basics of signature analysis

The serial-to-parallel data converter in Figure 10–33 (p. 529) is used to illustrate the "exercising" procedure. The main objective in exercising the circuit is to force all elements (flip-flops and gates) into all of their states to be certain that nothing is stuck in a given state as a result of a fault. The input test pattern, in this case, must be designed to force each flip-flop in the registers into both states, to clock the counter through all of its eight states, and to take the control flip-flop, the clock generator, the one-shot, and the AND gate through their paces.

The input test pattern that accomplishes this objective for the serial-to-parallel data converter is based on the serial data format in Figure 10–34 (p. 530). It consists of the pattern 10101010 in one serial group of data bits followed by 01010101 in the next group, as shown in Figure 10–39. These patterns are generated on a repetitive basis by a special test-pattern generator. The basic test setup is shown in Figure 10–40.

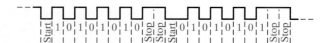

FIGURE 10–39
Sample test pattern.

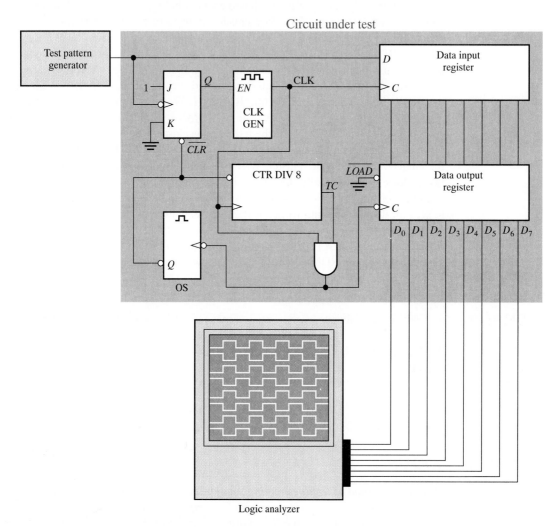

FIGURE 10–40
Basic test setup for the serial-to-parallel data converter of Figure 10–33.

After both patterns have been run through the circuit under test, all the flip-flops in the data-input register and in the data-output register have resided in both SET and RESET states, the counter has gone through its sequence (once for each bit pattern), and all the other devices have been exercised.

To check for proper operation, each of the parallel data outputs is observed for an alternating pattern of 1s and 0s as the input test patterns are repetitively shifted into the data-input register and then loaded into the data-output register. The proper timing diagram is shown in Figure 10–41. Each output can be observed individually, or the outputs can be observed in pairs with a dual-trace oscilloscope, or all eight outputs can be observed simultaneously with a logic analyzer configured for timing analysis.

If one or more outputs of the data-output register are incorrect, then we must back up to the outputs of the data-input register. If these outputs are correct, then probably the data-

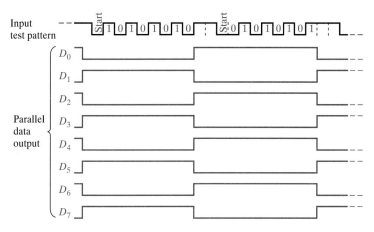

FIGURE 10–41

Proper outputs for the circuit under test in Figure 10–40. The input test patterns are shown.

output register is defective. If the data-input register outputs are also incorrect, the fault could be with the register itself or with any of the other logic, and additional investigation is necessary to isolate the problem.

SECTION 10–9 REVIEW	**1.** What is the purpose of providing a test input to a sequential logic circuit? **2.** Generally, when an output waveform is found to be incorrect, what is the next step to be taken?

10–10 ■ LOGIC SYMBOLS WITH DEPENDENCY NOTATION

This section can be omitted without affecting other coverage. Two examples of ANSI/IEEE Standard 91–1984 symbols with dependency notation for shift registers are presented. After completing this section, you should be able to

☐ Understand and interpret the logic symbols with dependency notation for the 74LS164 and the 74LS194A shift registers

The logic symbol for a 74LS164 8-bit parallel output serial shift register is shown in Figure 10–42 (next page). The common control inputs are shown on the notched block. The clear ($\overline{CLR}$) input is indicated by an R (for RESET) inside the block. Since there is no dependency prefix to link R with the clock ($C1$), the clear function is asynchronous. The right arrow symbol after $C1$ indicates data flow from Q_0 to Q_7. The A and B inputs are ANDed, as indicated by the embedded AND symbol, to provide the synchronous data input, $1D$, to the first stage (Q_0). Note the dependency of D on C, as indicated by the 1 suffix on C and the 1 prefix on D.

Figure 10–43 is the logic symbol for the 74LS194A 4-bit bidirectional universal shift register. Starting at the top left side of the control block, note that the $\overline{CLR}$ input is active-LOW and is asynchronous (no prefix link with C). Inputs S_0 and S_1 are mode inputs that determine the *shift-right, shift-left,* and *parallel load* modes of operation, as indicated by the $\frac{0}{3}$ dependency designation following the M. The $\frac{0}{3}$ represents the bianary states of 0, 1, 2, and 3 on the S_0 and S_1 inputs. When one of these digits is used as a prefix for another input, a dependency is established. The $1{\rightarrow}/2{\leftarrow}$ symbol on the clock input indicates the following: $1{\rightarrow}$ indicates that a right shift (Q_0 toward Q_3) occurs when the mode inputs (S_0, S_1) are in the binary 1 state ($S_0 = 1$, $S_1 = 0$), $2{\leftarrow}$ indicates that a left shift (Q_3 toward Q_0)

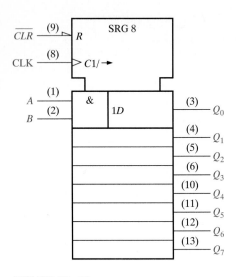

FIGURE 10–42
Logic symbol for the 74LS164.

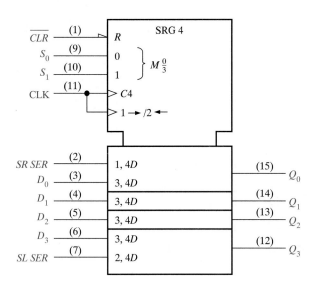

FIGURE 10–43
Logic symbol for the 74LS194A.

occurs when the mode inputs are in the binary 2 state ($S_0 = 0$, $S_1 = 1$). The shift-right serial input (*SR SER*) is both mode-dependent and clock-dependent, as indicated by 1, 4D. The parallel inputs (D_0, D_1, D_2, and D_3) are all mode-dependent (prefix 3 indicates parallel load mode) and clock-dependent, as indicated by 3, 4D. The shift-left serial input (*SL SER*) is both mode-dependent and clock-dependent, as indicated by 2, 4D.

The four modes for the 74LS194A are summarized as follows:

Do nothing:	$S_0 = 0, S_1 = 0$	(mode 0)
Shift right:	$S_0 = 1, S_1 = 0$	(mode 1, as in 1, 4D)
Shift left:	$S_0 = 0, S_1 = 1$	(mode 2, as in 2, 4D)
Parallel load:	$S_0 = 1, S_1 = 1$	(mode 3, as in 3, 4D)

SECTION 10–10 REVIEW

1. In Figure 10–43, are there any inputs that are dependent on the mode inputs being in the 0 state?
2. Is the parallel load synchronous with the clock?

10–11 ■ DIGITAL SYSTEM APPLICATION

In this system application, you will be working with a security entry system. This system allows you to deactivate (disarm) the alarms so that you can enter the building. Disarming the system is accomplished by entering a specified sequence of four digits on a keypad. The system is rearmed by pressing a switch when you leave the building. The system uses two shift registers as well as other devices that were covered in previous chapters. A memory is also included, but you will not get into specific details until you study memories in Chapter 12. After completing this section, you should be able to

□ Describe the overall function of the system   □ Explain the purposes of the shift registers as well as other types of devices   □ Determine the specific devices to be used in the system   □ Develop a detailed logic diagram from a general logic diagram   □ Develop a test procedure and troubleshoot the system

Basic Operation of the Security Entry System

A preliminary logic diagram of the system is shown in Figure 10–44. The system consists of two circuit boards, the panel switches, and the keypad. The focus in this section is on the code entry board and you should be familiar with all of the types of devices used. The memory board and complete system operation is the focus of the system application in Chapter 12.

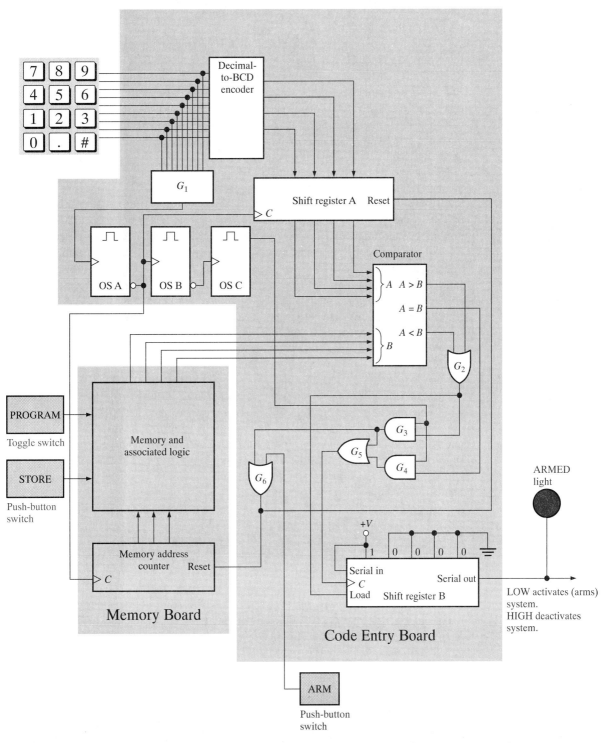

FIGURE 10–44
Basic logic diagram of the security entry system.

The security entry system controls the arming and disarming of the sensors and alarms in a given building. Arming or activation of the sensors and alarms is accomplished by pressing the single push-button switch labeled ARM and, while it is pressed, entering any digit on the keypad. Disarming or deactivation is accomplished by entering a predetermined 4-digit number on the keypad.

When the system is armed, shift register A and the memory counter are cleared to their all-zero state, while shift register B is preset to 10000. The resulting LOW on the serial output of shift register B enables the alarm system sensors and illuminates the ARMED light.

When the first digit of the entry number is entered on the keypad, the corresponding line from the keypad goes to its active level. The decimal-to-BCD encoder produces the 4-bit BCD code corresponding to the decimal digit. At the same time, one-shot A is triggered and produces a clock pulse that causes the 4-bit BCD digit from the encoder to be loaded into register A. These four bits then appear on the A inputs of the comparator. On the same clock pulse, the address counter on the memory board is advanced, causing the memory to place the first BCD code on the B inputs of the comparator. The memory stores the BCD codes for the correct 4-digit entry number.

If the correct digit has been entered on the keypad, the four bits on the A inputs of the comparator are the same as the four bits on the B inputs and the comparator produces a HIGH on its $A = B$ output. The clock pulse from one-shot A also triggers one-shot B to produce a delay interval before one-shot C is triggered. When one-shot C is triggered, it produces a pulse to the clock input of shift register B through AND gate G_4 and OR gate G_5. This delayed clock pulse shifts the 1 in shift register B to the right from the first stage to the second stage, so shift register B now contains 11000. Since there is still a 0 in the last stage of the register, the alarm system remains armed.

After the second correct digit is entered on the keypad, shift register B contains 11100. After the third correct digit is entered, the shift register contains 11110. After the fourth, and final, correct digit is entered, the shift register contains 11111. When the 1 is shifted into the last stage, the serial output of the register goes HIGH and disarms the system. The ARMED light goes off at this time and you can enter the building without setting off the alarm.

Now let's briefly see what happens if you enter an incorrect digit. If, at any point while entering the 4-digit number an incorrect digit is entered, the BCD code will not agree with the code stored in the memory. As a result, the comparator will produce a HIGH on one of its inequality outputs $A < B$ or $A > B$, depending on the digit entered. This will result in the parallel loading of shift register B with 10000 and the clearing of shift register A and the counter to all zeros. At this point, you must start over and reenter the 4-digit number.

■ THE DIGITAL WORKBENCH

The basic logic diagram in Figure 10–44 indicates the general logic for the system and does not represent the detailed logic required to implement the system. In the following three workbenches, you will analyze the basic operation of the code entry board logic to make sure that you understand the functional requirements and the basic system operation. You will then complete the design of the logic by selecting appropriate devices and then developing a detailed logic diagram incorporating the selected devices. Finally, you will compare your logic diagram to an implemented circuit board and then develop a test procedure.

On the circuit board, interconnections on the back side of the board feed through to the pads on the component side. When you trace out the circuit in Workbench 3, common pads that are connected on the back side are generally oriented vertically or horizontally with respect to each other and not at an angle.

■ DIGITAL WORKBENCH 1: Analysis

■ **Activity 1   The Decimal-to-BCD Encoder**
  (a) State the purpose of the encoder.
  (b) Describe what happens on the encoder inputs when you enter a digit on the keypad.
  (c) Describe what happens on the encoder outputs when you enter a digit on the keypad.

■ **Activity 2   The Timing Circuits**
  (a) State the purpose of the block labeled G_1.
  (b) Explain when and how one-shot A is triggered.
  (c) State the purpose of each of the one-shots.

■ **Activity 3   Shift Register A**
  (a) State the purpose of shift register A.
  (b) Determine the capacity of shift register A (number of bits).
  (c) Describe when and how the parallel data are entered into shift register A.

■ **Activity 4   Comparator**
  (a) State the purpose of the comparator.

  (b) Define each of the two sets (A and B) of four input bits.
  (c) Describe what happens on the outputs when the two inputs are the same.
  (d) Describe what happens on the outputs when the two inputs are different.

■ **Activity 5   Shift Register B and Gating Logic**
  (a) State the purpose of shift register B.
  (b) Discuss the bit pattern that is initially loaded into the register and state its purpose.
  (c) Describe what happens when the serial output is LOW. When the serial output is HIGH.
  (d) Describe what happens to shift register B each time a correct digit is entered on the keypad.
  (e) Explain how the logic gates provide for proper clocking and loading of shift register B.
  (f) Explain how the logic gates provide for resetting shift register A and the memory address counter.

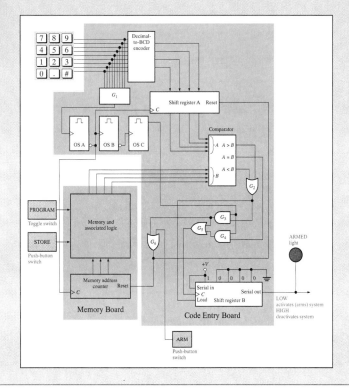

■ DIGITAL WORKBENCH 2: Design

This workbench focuses on the design of the code entry circuit board. For information on devices, reference to earlier chapters and/or device data sheets is necessary.

■ *Activity 1   Decimal-to-BCD Encoder*
The particular keypad used in this system produces active-LOW outputs. When a key is pressed, the corresponding line is LOW and the other lines are HIGH. Specify an IC encoder to properly interface with the keypad.

■ *Activity 2   The Timing Circuits*
The basic logic diagram shows a block (G_1) for the gate used to trigger one-shot A when a key is pressed. One-shot A is to produce a 1 ms pulse. One-shot B is to produce a 1 ms pulse triggered at the trailing edge of the one-shot A pulse. One-shot C is to produce a 1 ms pulse triggered at the trailing edge of the one-shot B pulse.
(a) Specify the type of gate to be used to implement gate G_1.

(b) Specify the type of IC one-shots and determine the external components required for each one-shot to produce a 1 ms pulse.

■ *Activity 3   Shift Register A*
Specify an IC shift register that can accept 4 bits in parallel and provide a parallel output. The shift register must be triggered at the trailing edge of the clock pulse from one-shot A.

■ *Activity 4   Comparator*
Specify an IC comparator that will compare two 4-bit numbers and produce the three required outputs.

■ *Activity 5   Shift Register B*
This shift register must store an initial bit pattern of 10000 which is parallel loaded. It must have a serial input and shift-right capability so that the initial 1 is shifted right on each clock pulse and the register fills with 1s from the serial input. The serial output is used to control the alarm system and

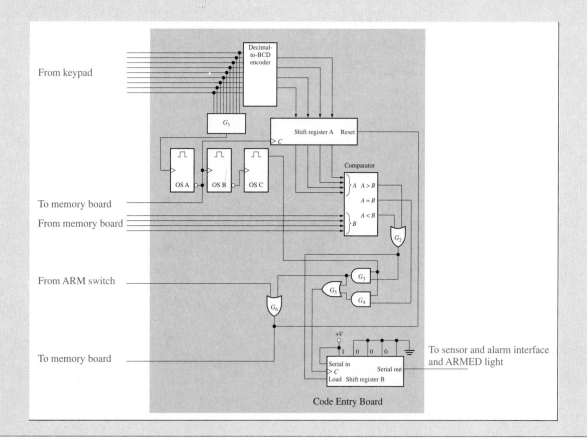

Code Entry Board

■ DIGITAL WORKBENCH 2: Design *(continued)*

ARMED light. An off-board drive circuit is required for this purpose.

Specify the IC device or devices for implementing this shift register.

■ *Activity 6   Gating Logic*

AND gates G_3 and G_4 and OR gates G_2, G_5, and G_6 are used for properly loading and clocking shift register B and for clearing shift register A and the memory counter. Based on your selection of ICs for shift registers A and B, specify the types of logic gates required to provide the proper active levels. Although the basic logic diagram indicates the basic gate function, you may have to specify NAND gates or negative-AND gates for the AND functions and NOR gates or negative-OR gates for the OR functions.

■ *Activity 7*   Develop a detailed logic diagram using the devices specified in preceding activities. Show all necessary details including device pin numbers.

■ DIGITAL WORKBENCH 3: Verification and Testing

■ *Activity 1* Draw the logic diagram for the entry code logic board below and compare to your logic diagram from Activity 7 of Workbench 2. Discuss any differences between your design and the design of this logic board and verify that both work.

■ *Activity 2* Identify the outputs and inputs on the circuit board connector. If there are more inputs or outputs than shown in the basic diagram in Workbench 2, they are for additional requirements for interfacing the memory board and will be discussed in Chapter 12.

■ *Activity 3* Develop a test procedure for checking the board shown.

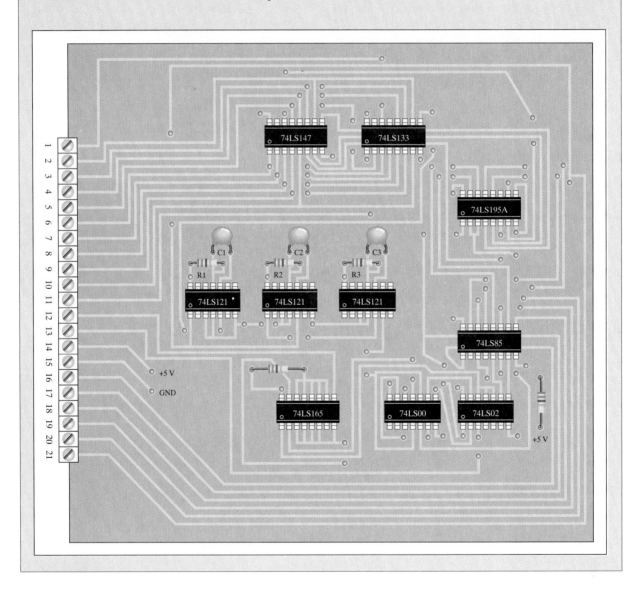

SECTION 10–11 REVIEW	1. What steps are required to arm the system?
	2. How is the system disarmed?
	3. What happens if an incorrect digit is entered after two correct digits?

■ SUMMARY

- The basic types of registers, classified by input and output, are
  1. serial in/serial out   3. parallel in/serial out
  2. serial in/parallel out   4. parallel in/parallel out
- The basic types of data movement in a register are
  1. shift right   4. parallel shift out
  2. shift left   5. rotate right
  3. parallel shift in (load)   6. rotate left
- Shift register counters are shift registers with feedback that exhibit special sequences. Examples are the Johnson counter and the ring counter.
- The Johnson counter has $2n$ states in its sequence, where n is the number of stages.
- The ring counter has n states in its sequence.

■ SELF-TEST

1. A stage in a shift register consists of
   (a) a latch   (b) a flip-flop   (c) a byte of storage   (d) four bits of storage

2. To serially shift a byte of data into a shift register, there must be
   (a) one clock pulse   (b) one load pulse
   (c) eight clock pulses   (d) one clock pulse for each 1 in the data

3. To parallel load a byte of data into a shift register, there must be
   (a) one clock pulse   (b) one clock pulse for each 1 in the data
   (c) eight clock pulses   (d) one clock pulse for each 0 in the data

4. The group of bits 10110101 is serially shifted (right-most bit first) into an 8-bit parallel output shift register with an initial state of 11100100. After two clock pulses, the register contains
   (a) 01011110   (b) 10110101   (c) 01111001   (d) 00101101

5. With a 100 kHz clock frequency, eight bits can be serially entered into a shift register in
   (a) $80\,\mu s$ (b) $8\,\mu s$ (c) 80 ms   (d) $10\,\mu s$

6. With a 1 MHz clock frequency, eight bits can be parallel entered into a shift register in
   (a) $8\,\mu s$ (b) in the propagation delay time of eight flip-flops
   (c) $1\,\mu s$ (d) in the propagation delay time of one flip-flop

7. A divide-by-10 Johnson counter requires
   (a) ten flip-flops   (b) four flip-flops   (c) five flip-flops   (d) twelve flip-flops

8. A divide-by-10 ring counter requires a minimum of
   (a) ten flip-flops   (b) five flip-flops   (c) four flip-flops   (d) twelve flip-flops

9. When an 8-bit serial in/serial out shift register is used for a $24\,\mu s$ time delay, the clock frequency must be
   (a) 41.67 kHz   (b) 333 kHz   (c) 125 kHz   (d) 8 MHz

10. The purpose of the ring counter in the keyboard encoding circuit of Figure 10–38 is
    (a) to sequentially apply a HIGH to each row for detection of key closure
    (b) to provide trigger pulses for the key code register
    (c) to sequentially apply a LOW to each row for detection of key closure
    (d) to sequentially reverse bias the diodes in each row

■ PROBLEMS

SECTION 10–1   Basic Shift Register Functions

1. Why are shift registers considered basic memory devices?

2. What is the storage capacity of a register that can retain two bytes of data?

SECTION 10–2   Serial In/Serial Out Shift Registers

3. For the data input and clock in Figure 10–45, determine the states of each flip-flop in the shift register of Figure 10–3 (p. 509) and draw the Q waveforms. Assume that the register contains all 1s initially.

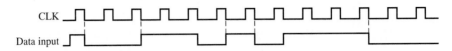

FIGURE 10–45

4. Solve Problem 3 for the waveforms in Figure 10–46.

FIGURE 10–46

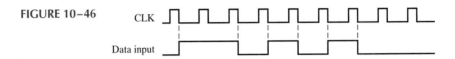

5. What is the state of the register in Figure 10–47 after each clock pulse if it starts in the 101001111000 state?

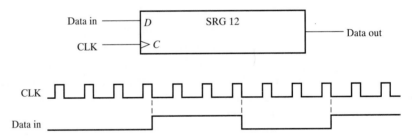

FIGURE 10–47

6. For the serial in/serial out shift register, determine the data-output waveform for the data-input and clock waveforms in Figure 10–48. Assume that the register is initially cleared.

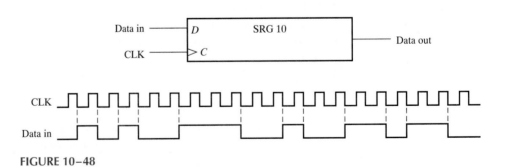

FIGURE 10–48

7. Solve Problem 6 for the waveforms in Figure 10–49.

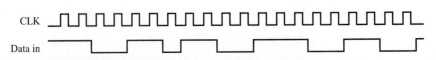

FIGURE 10–49

8. The data-output waveform in Figure 10–50 is related to the clock as indicated. What binary number is stored in an 8-bit serial in/serial out register if the first data bit out (left most) is the LSB?

FIGURE 10–50

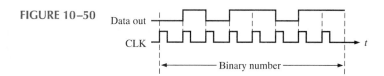

SECTION 10–3 Serial In/Parallel Out Shift Registers

9. Draw a complete timing diagram showing the parallel outputs for the shift register in Figure 10–8 (p. 513). Use the waveforms in Figure 10–48 with the register initially clear.

10. Solve Problem 9 for the input waveforms in Figure 10–49.

11. Sketch the Q_0 through Q_7 outputs for a 74LS164 shift register with the input waveforms shown in Figure 10–51.

FIGURE 10–51

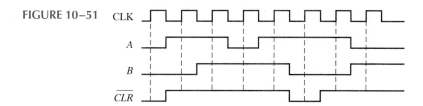

SECTION 10–4 Parallel In/Serial Out Shift Registers

12. The shift register in Figure 10–52(a) has $SHIFT/\overline{LOAD}$ and CLK inputs as shown in part (b). The serial data input (SER) is a 0. The parallel data inputs are $D_0 = 1$, $D_1 = 0$, $D_2 = 1$, and $D_3 = 0$. Sketch the data-output waveform in relation to the inputs.

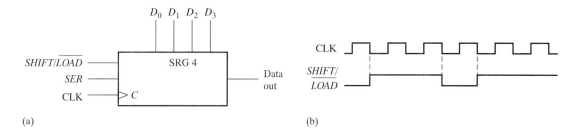

(a) (b)

FIGURE 10–52

13. The waveforms in Figure 10–53 are applied to a 74LS165 shift register. The parallel inputs are all 0. Determine the Q_7 waveform.

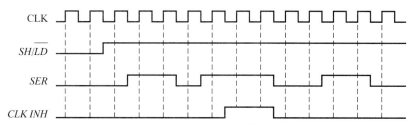

FIGURE 10–53

14. Solve Problem 13 if the parallel inputs are all 1.

15. Solve Problem 13 if the SER input is inverted.

SECTION 10–5   Parallel In/Parallel Out Shift Registers

16. Determine all the Q output waveforms for a 74LS195A 4-bit shift register when the inputs are as shown in Figure 10–54.

FIGURE 10–54

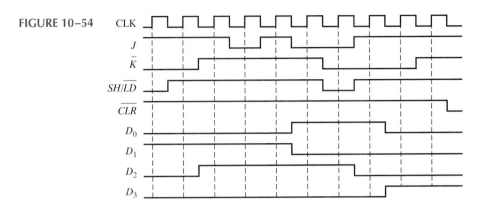

17. Solve Problem 16 if the $SH/\overline{LD}$ input is inverted and the register is initially clear.

18. Use two 74LS195A shift registers to form an 8-bit shift register. Show the required connections.

SECTION 10–6   Bidirectional Shift Registers

19. For the 8-bit bidirectional register in Figure 10–55, determine the state of the register after each clock pulse for the $RIGHT/\overline{LEFT}$ control waveform given. A HIGH on this input enables a shift to the right, and a LOW enables a shift to the left. Assume that the register is initially storing the decimal number seventy-six in binary, with the right-most position being the LSB. There is a LOW on the data-input line.

FIGURE 10–55

20. Solve Problem 19 for the waveforms in Figure 10–56.

21. Use two 74LS194A 4-bit bidirectional shift registers to create an 8-bit bidirectional shift register. Show the connections.

22. Determine the Q outputs of a 74LS194A with the inputs shown in Figure 10–57. Inputs D_0, D_1, D_2, and D_3 are all HIGH.

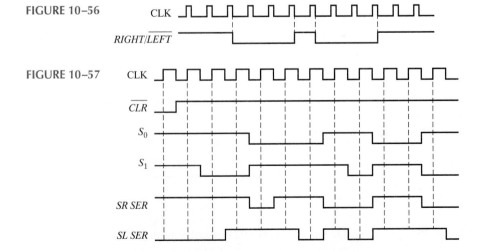

FIGURE 10–56

FIGURE 10–57

SECTION 10–7 Shift Register Counters

23. How many flip-flops are required to implement each of the following in a Johnson counter configuration:

(a) divide-by-6   (b) divide-by-10   (c) divide-by-14   (d) divide-by-16

(e) divide-by-20   (f) divide-by-24   (g) divide-by-36

24. Draw the logic diagram for a divide-by-18 Johnson counter. Sketch the timing diagram and write the sequence in tabular form.

25. For the ring counter in Figure 10–58, draw the waveforms for each flip-flop output with respect to the clock. Assume that FF0 is initially SET and that the rest are RESET. Show at least ten clock pulses.

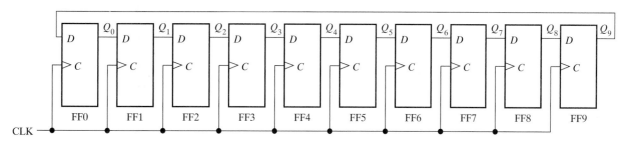

FIGURE 10–58

26. The waveform pattern in Figure 10–59 is required. Devise a ring counter, and indicate how it can be preset to produce this waveform on its Q_9 output. At CLK16 the pattern begins to repeat.

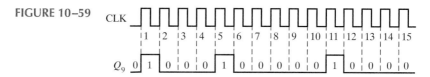

FIGURE 10–59

SECTION 10–8 Shift Register Applications

27. Use 74LS195A 4-bit shift registers to implement a 16-bit ring counter. Show the connections.

28. What is the purpose of the power-on $\overline{LOAD}$ input in Figure 10–38 (p. 532)?

29. What happens when two keys are pressed simultaneously in Figure 10–38?

SECTION 10–9 Troubleshooting

30. Based on the waveforms in Figure 10–60(a), determine the most likely problem with the register in part (b) of the figure.

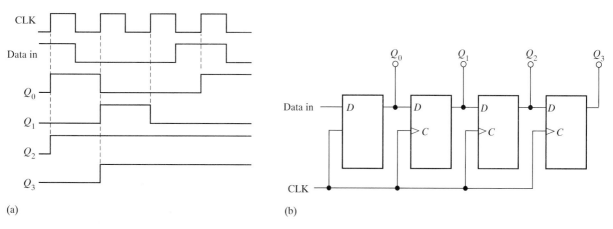

FIGURE 10–60

31. Refer to the parallel in/serial out shift register in Figure 10–12 (p. 516). The register is in the state where $Q_0 Q_1 Q_2 Q_3 = 1001$, and $D_0 D_1 D_2 D_3 = 1010$ is loaded in. When the $SHIFT/\overline{LOAD}$ input is taken HIGH, the data shown in Figure 10–61 are shifted out. Is this operation correct? If not, what is the most likely problem?

FIGURE 10–61

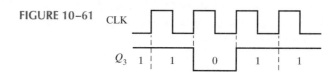

32. You have found that the bidirectional register in Figure 10–19 (p. 521) will shift data right but not left. What is the most likely fault?

33. For the keyboard encoder in Figure 10–38 (p. 532), list the possible faults for each of the following symptoms:

    (a) The state of the key code register does not change for any key closure.

    (b) The state of the key code register does not change when any key in the third row is closed. A proper code occurs for all other key closures.

    (c) The state of the key code register does not change when any key in the first column is closed. A proper code occurs for all other key closures.

    (d) When any key in the second column is closed, the left three bits of the key code ($Q_0 Q_1 Q_2$) are correct, but the right three bits are all 1s.

34. Develop a test procedure for exercising the keyboard encoder in Figure 10–38 (p. 532). Specify the procedure on a step-by-step basis, indicating the output code from the key code register that should be observed at each step in the test.

35. What symptoms are observed for the following failures in the serial-to-parallel converter in Figure 10–33:

    (a) AND gate output stuck in HIGH state

    (b) clock generator output stuck in LOW state

    (c) third stage of data-input register stuck in SET state

    (d) terminal count output of counter stuck in HIGH state

SECTION 10–11 Digital System Application

36. Sketch a waveform timing diagram of the three one-shot outputs as specified in Workbench 2.

37. Assume the entry code is 1939. Determine the states of shift register A and shift register B after the second correct digit has been entered.

38. Assume the entry code is 7646 and the digits 7645 are entered. Determine the states of shift register A and shift register B after each of the digits is entered.

39. A problem has developed with the security entry system such that it cannot be deactivated (the serial output of shift register B stays LOW). Upon checking several points, you find that one-shot A is never triggered. What are the possible faults?

40. As in Problem 39, the security entry system cannot be deactivated. This time, however, you find that the one-shots are all triggering properly. What devices could possibly be faulty in this case?

Special Design Problems

41. Specify the devices that can be used to implement the serial-to-parallel data converter in Figure 10–33. Draw the complete logic diagram, showing any modifications necessary to accommodate the specific devices used.

42. Modify the serial-to-parallel converter in Figure 10–33 (p. 529) to provide 16-bit conversion.

43. Design an 8-bit parallel-to-serial data converter that produces the data format in Figure 10–34 (p. 530). Show a logic diagram and specify the devices.

44. Design a power-on $\overline{LOAD}$ circuit for the keyboard encoder in Figure 10–38 (p. 532). This circuit must generate a short-duration LOW pulse when the power switch is turned on.

45. Implement the test pattern generator used in Figure 10–40 (p. 534) to troubleshoot the serial-to-parallel converter.

46. Review the tablet counting and control system that was introduced in Chapter 1 and was the topic of the system applications in Chapters 2, 3, and 4. Utilizing the knowledge gained in this chapter, implement registers A and B in that system using specific IC devices.

■ ANSWERS TO SECTION REVIEWS

SECTION 10–1

1. A counter has a specified sequence of states, but a shift register does not.

2. Storage and data movement are two functions of a shift register.

SECTION 10–2

1. FF0: data input to J_0, $\overline{\text{data input}}$ to K_0; FF1: Q_0 to J_1, $\overline{Q}_0$ to K_1; FF2: Q_1 to J_2, $\overline{Q}_1$ to K_2; FF3: Q_2 to J_3, $\overline{Q}_2$ to K_3

2. Eight clock pulses

SECTION 10–3

1. 0100 after 2 clock pulses

2. Take the serial output from the right-most flip-flop for serial out operation.

SECTION 10–4

1. When *SHIFT/$\overline{LOAD}$* is HIGH, the data are shifted right one bit per clock pulse. When *SHIFT/$\overline{LOAD}$* is LOW, the data on the parallel inputs are loaded into the register.

2. The parallel load operation is asynchronous, so it is not dependent on the clock.

SECTION 10–5

1. The data outputs are 1001.   **2.** $Q_0 = 1$ after one clock pulse

SECTION 10–6

1. 1111 after the fifth clock pulse

SECTION 10–7

1. Sixteen states are in an 8-bit Johnson counter sequence.

2. For a 3-bit Johnson counter: 000, 100, 110, 111, 011, 001, 000

SECTION 10–8

1. 625 scans/second   **2.** $Q_5Q_4Q_3Q_2Q_1Q_0 = 011011$

3. The diodes provide unidirectional paths for pulling the ROWs LOW and preventing HIGHs on the ROW lines from being connected to the switch matrix. The resistors pull the COLUMN lines HIGH.

SECTION 10–9

1. A test input is used to sequence the circuit through all of its states.

2. Check the input to that portion of the circuit. If the signal on that input is correct, the fault is isolated to the circuitry between the good input and the bad output.

SECTION 10–10

1. No, inputs are dependent on the mode inputs being in the 0 state.

2. Yes, the parallel lead is synchronous with the clock as indicated by the 4D label.

SECTION 10–11

1. Press ARM switch; then press any key to arm system.

2. Enter correct 4-digit number to disarm system.

3. The system is reinitialized.

11

SEQUENTIAL LOGIC APPLICATIONS OF PLDs

■ **CHAPTER OBJECTIVES**

☐ Describe the OLMCs in the GAL22V10 and the GAL16V8
☐ Explain the difference between the registered mode and the combinational mode
☐ Develop ABEL input files for the design of shift registers
☐ Use ABEL to implement counters in a PLD
☐ Use ABEL state diagrams to implement logic systems in PLDs
☐ Implement sequential logic in the traffic light control system using a GAL

CHAPTER OVERVIEW

The basics of programmable logic devices were covered in Chapter 7 with emphasis on using PLDs to implement combinational logic. The coverage of PLDs is continued in this chapter with an introduction to the registered mode and sequential logic applications. If necessary, review Chapter 7 during or prior to your study of this chapter.

As in Chapter 7, the ABEL programming language is used to illustrate the implementation of sequential logic, but this approach can readily be transferred to CUPL or other similar PLD languages.

SPECIFIC DEVICES

GAL16V8   GAL22V10

DIGITAL SYSTEM APPLICATION

This Digital System Application illustrates concepts taught in this chapter. The sequential logic portion of the traffic light control system from Chapter 9 is redesigned using a programmable logic device, and the complete system except for the timing circuits is implemented with a single GAL22V10.

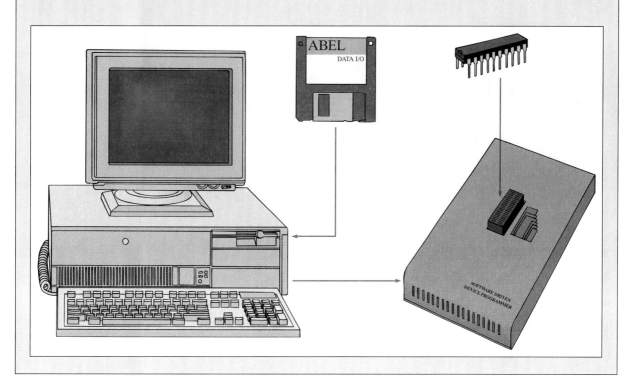

11–1 ■ THE COMPLETE OLMC

In Chapter 7 the output logic macrocells (OLMC) for the GAL22V10 and the GAL16V8 were examined in terms of their combinational modes, and the flip-flop contained in the OLMC was only briefly mentioned. In this section, we will discuss the complete OLMC which includes a flip-flop, again using the GAL22V10 and the GAL16V8 as representative PLD devices. After completing this section, you should be able to

☐ Discuss the basic operation of an OLMC   ☐ Discuss local and global cells

The block diagrams of the GAL22V10 and the GAL16V8 that were introduced in Chapter 7 are repeated in Figure 11–1. The GAL22V10 has ten output logic macrocells (OLMCs) and the GAL16V8 has eight.

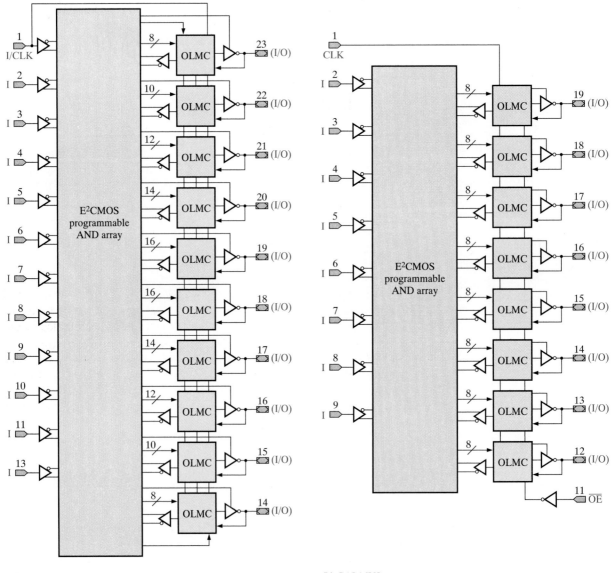

(a) GAL22V10   (b) GAL16V8

FIGURE 11–1
GAL block diagrams.

The Output Logic Macrocell (OLMC) in the GAL22V10

As stated in Chapter 7, an OLMC contains programmable logic circuits that can be configured for either a combinational output or input or for a **registered** output. In the registered mode, the output comes from a flip-flop. As you know, the OLMCs are internally configured automatically by programming a set of cells that are separate from the logic array cells.

Logic Diagram   Figure 11–2 shows a basic logic diagram for the GAL22V10 OLMC. The inputs from the AND gates in the array to the OR gate vary from ten to sixteen. The logic consists of a flip-flop and two multiplexers.

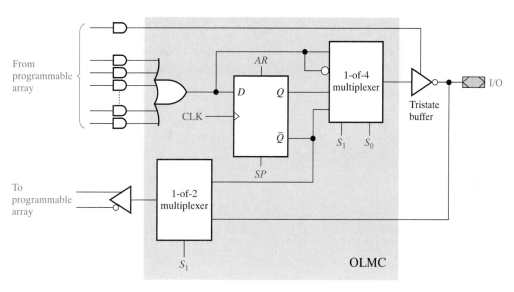

FIGURE 11–2
The GAL22V10 OLMC.

The 1-of-4 multiplexer connects one of its four input lines to the output tristate buffer based on the states of two select inputs, S_0 and S_1. The inputs to the 1-of-4 multiplexer are the OR gate output, the complement of the OR gate output, the flip-flop output $Q,$ and the complement of the flip-flop output $\overline{Q}$. This allows the output of the OLMC to be either active-HIGH or active-LOW in each mode. The 1-of-2 multiplexer connects either the output of the tristate buffer or the $\overline{Q}$ output of the flip-flop back through a buffer to the AND array, based on the state of S_1. The select bits, S_0 and S_1, for each OLMC are programmed into a dedicated group of cells in the array by the compiler software, so the user does not have to directly manipulate these bits.

The flip-flop is a positive edge-triggered D type. The asynchronous reset (AR) input resets the flip-flop to the logic 0 state ($Q = 0$) independent of the clock. The synchronous preset (SP) input sets the flip-flop to the logic 1 state ($Q = 1$) on the rising edge of the clock pulse. The GAL22V10 has product terms for the AR and the SP inputs.

The four OLMC configurations are

- Combinational mode with active-LOW output
- Combinational mode with active-HIGH output
- Registered mode with active-LOW output
- Registered mode with active-HIGH output

The combinational modes were covered in Chapter 7. The registered modes are covered in Section 11–2.

The OLMC in the GAL16V8

As you learned in Chapter 7, the OLMC in the GAL16V8 is somewhat different from that in the GAL22V10. A GAL16V8 OLMC can be programmed in one of three available modes to emulate most of the existing PALs; thus, a GAL16V8 can replace the PAL that it is programmed to emulate. There are two combinational modes, simple and complex, and one registered mode.

Logic Diagram Figure 11–3 shows a basic logic diagram for the GAL16V8 OLMC. Each GAL16V8 OLMC contains two **global cells**, $\overline{SYN}$ and $AC0$, which affect all of the OLMCs. Also, there are two **local cells** for each OLMC, $AC1(n)$ and XOR, which affect only the associated OLMC. Both global and local cells select various paths in the OLMC and determine the combinational or registered mode.

The $\overline{SYN}$ bit and the $AC0$ bit control the mode configuration. $\overline{SYN} = 0$ and $AC0 = 1$ set the registered mode, and $\overline{SYN} = 1$ and $AC0 = 0$ set the combinational modes. The $AC1$ bit controls the input/output configuration. The XOR bit selects either active-LOW ($XOR = 0$) or active-HIGH ($XOR = 1$) output polarity.

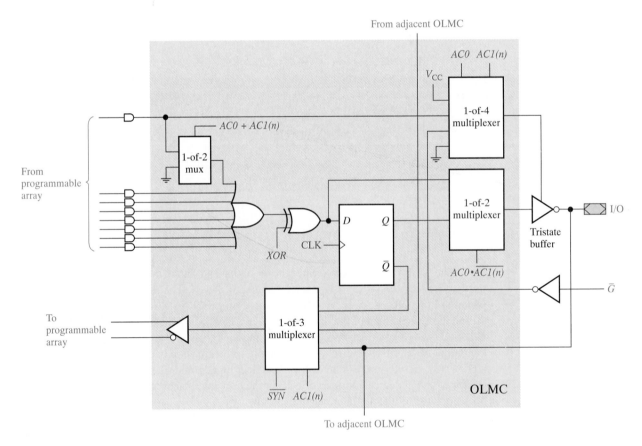

FIGURE 11–3
The GAL16V8 OLMC.

11–2 ■ OLMC MODE SELECTION

In this section, we will examine the mode configurations in the GAL22V10 and discuss the difference between the registered mode and the combinational mode. As in Chapter 7, ABEL is used to illustrate the software aspects of PLD applications. After completing this section, you should be able to

□ Discuss the role of the control bits S_0 and S_1 in mode selection   □ Show how active-LOW and active-HIGH outputs are achieved

The Combinational Mode

The modes in the GAL22V10 are determined by the S_0 and S_1 bits, which are controlled by programming. For the combinational mode, $S_1S_0 = 10$ or $S_1S_0 = 11$. Figure 11–4 (next page) shows the logic paths through the OLMC for active-LOW and active-HIGH combinational outputs along with the resulting effective OLMC logic in each case. The flip-flop is not used in the combinational mode.

In Figure 11–4(a), the inversion of the tristate buffer produces the active-LOW output. In part (b), the inversions of the 1-of-4 multiplexer and the tristate buffer cancel, resulting in the active-HIGH output.

The Registered Mode

For the registered mode, $S_1S_0 = 00$ or $S_1S_0 = 01$. Figure 11–5 (p. 557) shows the logic paths through the OLMC for active-LOW and active-HIGH registered outputs along with the resulting effective OLMC logic in each case. In part (a), the inversion of the tristate buffer produces the active-LOW output. In part (b), the inversions of the flip-flop ($\overline{Q}$ output) and the tristate buffer cancel, resulting in the active-HIGH output. Notice that the feedback to the AND array through a buffer is from the $\overline{Q}$ output of the flip-flop and not from the output pin as in the combinational mode, so a registered output cannot be used as an input.

Software Mode Specification

The modes of an OLMC are established with software. This is done using statements in the declarations part of the input file and in the way logic descriptions are written. Several basic elements of ABEL declarations and expressions are important in programming a given output as either combinational or registered.

1. The ISTYPE statement is used to declare an output as either combinational or registered with the use of the attributes 'com' or 'reg'. For example, to specify a given output as registered, an ISTYPE statement can be included in a pin declaration as follows:

   ```
   Q0 PIN 23 ISTYPE 'reg';
   ```

   Also, the output can be specified as either not inverted or inverted (active-HIGH or active-LOW) using the attributes 'buffer' or 'invert' in the ISTYPE statement. For example,

   ```
   Q0 PIN 23 ISTYPE 'reg,buffer';
   ```

   indicates that the registered output Q_0 is not inverted.
2. The assignment operators := and :> are used in logic descriptions to indicate a registered output. The registered assignment operator := is analogous to the operator = used in combinational logic expressions. The registered assignment operator :> is analogous

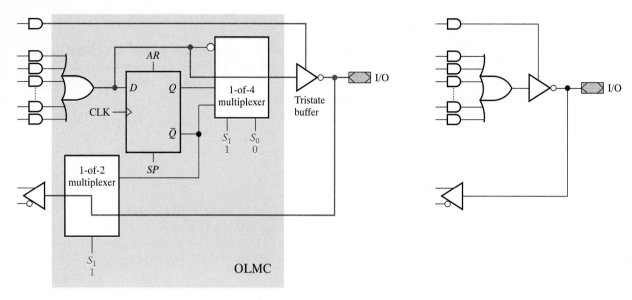

(a) OLMC in the active-LOW combinational mode and the effective logic diagram

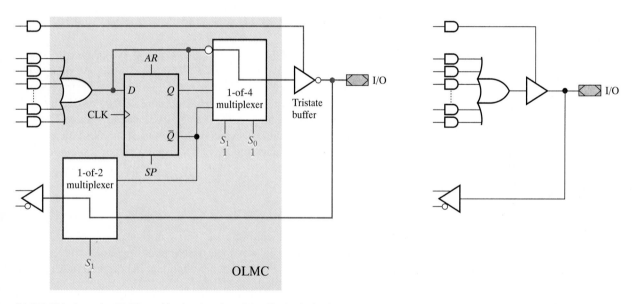

(b) OLMC in the active-HIGH combinational mode and the effective logic diagram

FIGURE 11–4

Combinational mode for active-LOW and active-HIGH outputs. The red lines show the logic paths in each case.

to the operator –> used in combinational logic truth tables. For example, the registered output expression

```
Q0 := D0;
```

indicates that the output Q_0 will assume the value of the D_0 input on the clock pulse (in the case of a clocked flip-flop) and will hold that value until the next clock pulse. As a comparison, the combinational output expression

```
X = A;
```

indicates that the output X is equal to the input A.

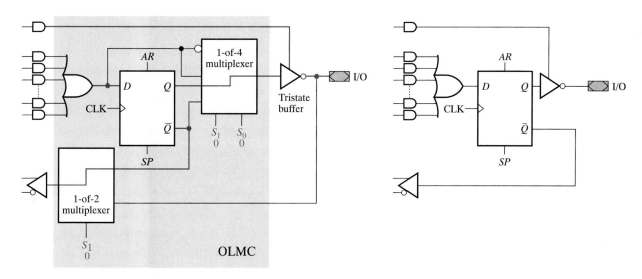

(a) OLMC in the active-LOW registered mode and the effective logic diagram

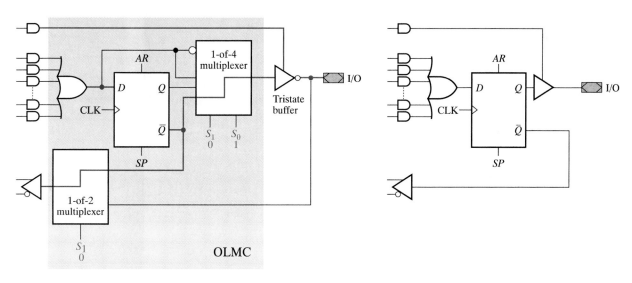

(b) OLMC in the active-HIGH registered mode and the effective logic diagram

FIGURE 11–5
Registered mode for active-LOW and active-HIGH outputs. The red lines show the logic paths in each case.

3. The dot extension .CLK is used to indicate that the register device is a clocked flip-flop. An equation using this dot extension must accompany the output equation. For example, the output equation

```
QO := DO;
```

must be followed by a clock equation such as

```
QO.CLK = Clock;
```

ABEL is an extensive language and the elements just introduced represent only a few of the many that are available. However, these few are essential in programming a PLD for operation as a sequential circuit.

EXAMPLE 11–1   Write the ABEL pin declarations and the equations to specify the GAL22V10 in Figure 11–6 as a parallel input/parallel output clocked register with D_0 through D_7 as inputs and Q_0 through Q_7 as registered outputs that are not inverted.

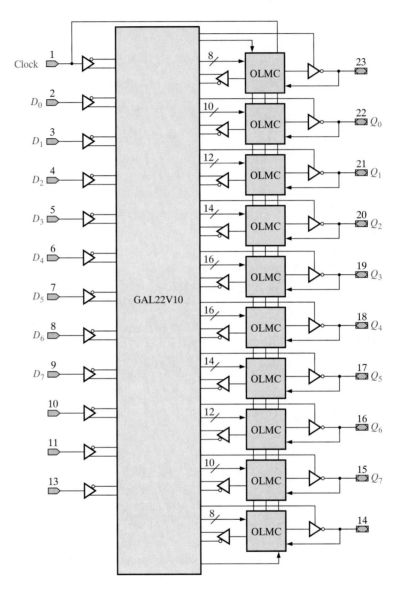

FIGURE 11–6

Solution   The pin declaration statements can be written as

```
Clock,D0,D1,D2,D3,D4,D5,D6,D7  PIN 1,2,3,4,5,6,7,8,9;
Q0,Q1,Q2,Q3,Q4,Q5,Q6,Q7  PIN 22,21,20,19,18,17,16,15 ISTYPE 'reg,buffer';
```

The equations can be written using set notation as follows:

```
[Q0, Q1, Q2, Q3, Q4, Q5, Q6, Q7] := [D0, D1, D2, D3, D4, D5, D6, D7];
[Q0, Q1, Q2, Q3, Q4, Q5, Q6, Q7].CLK = Clock;
```

This is equivalent to, but much simpler than, writing individual equations such as

```
Q0 := D0;
Q0.CLK = Clock;
Q1 := D1;
Q1.CLK = Clock;
Q2 := D2;
Q2.CLK = Clock;
Q3 := D3;
Q3.CLK = Clock;
Q4 := D4;
Q4.CLK = Clock;
Q5 := D5;
Q5.CLK = Clock;
Q6 := D6;
Q6.CLK = Clock;
Q7 := D7;
Q7.CLK = Clock;
```

Related Exercise Change the pin declaration statements to make the outputs inverted.

SECTION 11–2 REVIEW

1. What is the basic difference between the registered mode and the combinational mode?
2. How can ISTYPE statements be used?
3. What is the difference between the operator $:=$ and the operator $=$ in an equation?
4. What is the purpose of the .CLK extension?

11–3 ■ IMPLEMENTING SHIFT REGISTERS WITH PLDs

In the last section, you learned about the registered mode and how the OLMC flip-flop is used. In this section, you will see how shift registers can be implemented with a PLD. Again, the GAL22V10 is used for illustration. After completing this section, you should be able to

□ Use the .CLK extension   □ Use the .AR extension   □ Use the .c. special constant in test vectors   □ Implement an 8-bit serial in/parallel out shift register   □ Implement a 4-bit parallel in/serial out shift register

An 8-Bit Serial In/Parallel Out Shift Register

Figure 11–7 shows an 8-bit shift register in which the data are entered serially. When the Enable input to the AND gate is HIGH, the data are clocked into the left-most flip-flop and then shifted from left to right. The Q output of each flip-flop is available. The Clear input is active-LOW and resets all of the flip-flops ($Q = 0$) independent of the clock.

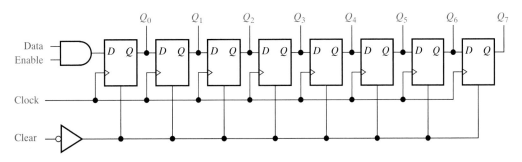

FIGURE 11–7
Eight-bit serial in/parallel out shift register.

Implementing the Shift Register with a PLD

The shift register inputs and outputs are assigned to pins on the GAL22V10 as shown in Figure 11–8. Since the D input of the first flip-flop comes from the AND gate, the output of the first flip-flop, Q_0, is expressed as the following ABEL registered equation:

```
Q0 := Data & Enable;
```

Since the Q output of each flip-flop is connected to the D input of the next flip-flop, the remaining flip-flop outputs are expressed as follows:

```
Q1 := Q0;
Q2 := Q1;
Q3 := Q2;
Q4 := Q3;
Q5 := Q4;
Q6 := Q5;
Q7 := Q6;
```

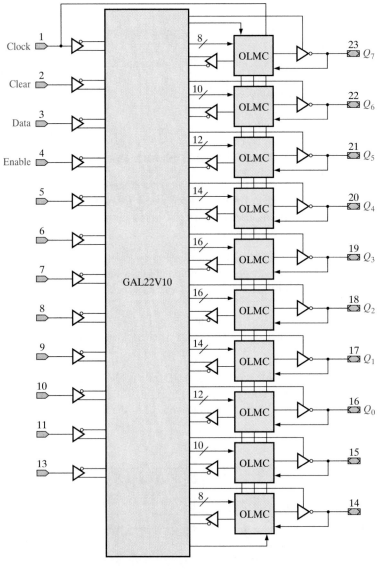

FIGURE 11–8
GAL22V10 input/output assignment for the 8-bit shift register.

Developing the ABEL Input File

Recall that an ABEL input file module typically contains three parts: declarations, logic descriptions, and test vectors. In the input file for the 8-bit shift register, the declarations section makes the pin assignments for the inputs and outputs.

The equations section expresses the output logic equations, utilizing the set notation that was introduced in Chapter 7. For the Clock equation, the dot extension .CLK specifies the clock input for each of the flip-flops associated with the specified outputs. For the Clear equation, the dot extension .AR specifies an asynchronous reset for each of the flip-flops associated with the specified outputs.

Entries in the test vectors section consist of numeric values and the special constant .c. that is used to indicate the clock pulse. Recall that test vectors are used to test the PLD to make sure that it works as expected when programmed. The ABEL input file is as follows:

```
Module   Eight_bit_shift_register

Title    '8-bit shift register in a GAL22V10'

"Device Declaration

    Register        device       'P22V10'

"Pin Declaration

    Clock, Clear               pin 1,2;
    Data, Enable               pin 3,4;
    Q0,Q1,Q2,Q3,Q4,Q5,Q6,Q7    pin 16,17,18,19,20,21,22,23 ISTYPE 'reg,buffer';

Equations

    Q0 := Data & Enable;

    [Q1,Q2,Q3,Q4,Q5,Q6,Q7] := [Q0,Q1,Q2,Q3,Q4,Q5,Q6];

    [Q0,Q1,Q2,Q3,Q4,Q5,Q6,Q7].CLK = Clock;

    [Q0,Q1,Q2,Q3,Q4,Q5,Q6,Q7].AR = Clear;

Test_Vectors

    ([Clock, Clear, Data, Enable] -> [Q0,Q1,Q2,Q3,Q4,Q5,Q6,Q7])
    [  0  ,  0  ,  1  ,  0  ] -> [ 0, 0, 0, 0, 0, 0, 0, 0];
    [ .c. ,  1  ,  1  ,  0  ] -> [ 0, 0, 0, 0, 0, 0, 0, 0];
    [ .c. ,  1  ,  0  ,  1  ] -> [ 0, 0, 0, 0, 0, 0, 0, 0];
    [ .c. ,  1  ,  1  ,  1  ] -> [ 1, 0, 0, 0, 0, 0, 0, 0];
    [ .c. ,  1  ,  0  ,  1  ] -> [ 0, 1, 0, 0, 0, 0, 0, 0];
    [ .c. ,  1  ,  1  ,  1  ] -> [ 1, 0, 1, 0, 0, 0, 0, 0];
    [ .c. ,  1  ,  0  ,  1  ] -> [ 0, 1, 0, 1, 0, 0, 0, 0];
    [ .c. ,  1  ,  1  ,  1  ] -> [ 1, 0, 1, 0, 1, 0, 0, 0];
    [ .c. ,  1  ,  0  ,  1  ] -> [ 0, 1, 0, 1, 0, 1, 0, 0];
    [ .c. ,  1  ,  1  ,  1  ] -> [ 1, 0, 1, 0, 1, 0, 1, 0];
    [ .c. ,  1  ,  0  ,  1  ] -> [ 0, 1, 0, 1, 0, 1, 0, 1];
    [ .c. ,  1  ,  1  ,  1  ] -> [ 1, 0, 1, 0, 1, 0, 1, 0];
    [ .c. ,  0  ,  1  ,  1  ] -> [ 0, 0, 0, 0, 0, 0, 0, 0];

END
```

EXAMPLE 11–2

Implement the 4-bit parallel in/serial out shift register in Figure 11–9 by developing an ABEL input file for a GAL22V10. The SHLD input is the Shift/Load. When this input is LOW, the parallel data are loaded into the register, and when it is HIGH, data are shifted from left to right. D_0 through D_3 are the parallel data inputs and Q_3 is the serial data output.

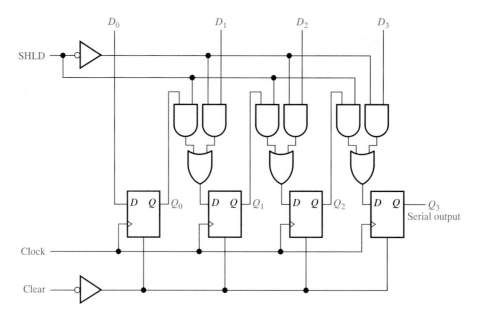

FIGURE 11–9
Four-bit parallel in/serial out shift register.

Solution The ABEL input file is as follows:

```
Module   Four_bit_shift_register

Title    '4-bit shift register in a GAL22V10'

"Device Declaration

    Register      device      'P22V10'

"Pin Declaration

    Clock, Clear    pin 1,2;
    SHLD            pin 3;
    D0, D1, D2, D3  pin 4,5,6,7 ISTYPE 'reg,buffer';
    Q0, Q1, Q2, Q3  pin 14,15,16,17 ISTYPE 'reg,buffer';

Equations

    Q0 := D0;

    Q1 := Q0 & SHLD # D1 & !SHLD;

    Q2 := Q1 & SHLD # D2 & !SHLD;

    Q3 := Q2 & SHLD # D3 & !SHLD;

    [Q0, Q1, Q2, Q3].CLK = Clock;

    [Q0, Q1, Q2, Q3].AR = !Clear;

Test_Vectors

    ([Clock, Clear, SHLD, D0, D1, D2, D3] -> [Q3])
    [  0  ,   0  ,  1 ,  0,  0,  0,  0] -> [ 0];
    [ .c. ,   1  ,  0 ,  0,  1,  0,  1] -> [ 1];
    [ .c. ,   1  ,  0 ,  1,  0,  1,  0] -> [ 0];
    [ .c. ,   1  ,  1 ,  1,  0,  1,  0] -> [ 1];
    [ .c. ,   1  ,  1 ,  1,  0,  1,  0] -> [ 0];
    [ .c. ,   1  ,  1 ,  1,  0,  1,  0] -> [ 1];
    [ .c. ,   0  ,  0 ,  1,  0,  1,  0] -> [ 0];

END
```

Related Exercise Modify the pin declaration and the equation sections of the ABEL input file to change the shift register in Figure 11–9 from four bits to five bits.

1. What is the meaning of the ABEL equation Q0 := A & !B # !A & B;?
2. What is the purpose of the dot extension .AR?
3. When is the special constant .c. used?

11–4 ■ IMPLEMENTING COUNTERS WITH PLDs

As you have learned, counters consist of flip-flops just as shift registers do; but counters have a prescribed sequence of states, whereas registers do not. Counters are in a category of digital circuits known as state machines as mentioned briefly in Chapter 9. In this section, the general types of state machines are discussed and three approaches to implementing a counter in a PLD are covered. After completing this section, you should be able to

☐ Define the two categories of state machines   ☐ Use the .x. special constant in truth tables and test vectors   ☐ Implement a counter using equation entry   ☐ Implement a counter using truth table entry   ☐ Implement a counter using state diagram entry

Types of State Machines

Sequential logic circuits can be classified into a category of circuits known as **state machines** of which there are two basic types. In the Moore state machine shown in Figure 11–10(a), the outputs depend only on the internal state and any inputs that are synchronized with the circuit. Counters are examples of the Moore type of state machine, where the flip-flops are the memory portion.

In the Mealy state machine, the outputs are determined by both the internal state and by inputs that are not synchronized with the circuit, as indicated in part (b).

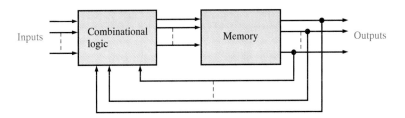

(a) Moore state machine

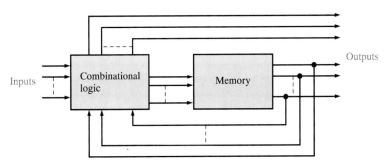

(b) Mealy state machine

FIGURE 11–10

Generalized block diagrams for the two categories of state machines.

A 3-Bit Up/Down Gray Code Counter Design

To illustrate the process of implementing a counter design in a PLD, we will use a 3-bit up/down Gray code counter.

The first step is to design the counter with D flip-flops. D flip-flops are used because most PLDs including the GAL22V10 and the GAL16V8 have this type of device in their OLMCs. A similar counter was designed in Chapter 9 with J-K flip-flops.

The State Diagram   Figure 11–11 shows the state diagram for a 3-bit up/down Gray code counter. The $Y = 1$ or $Y = 0$ beside each arrow indicates the state of the up/down control input.

FIGURE 11–11

State diagram for a 3-bit up/down Gray code counter.

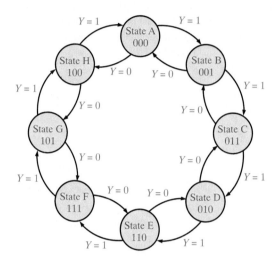

The Next-State Table   The next-state table is derived from the state diagram and is shown in Table 11–1. For each present state there are two possible next states, depending on the UP/DOWN control variable, Y.

TABLE 11–1

Next-state table for the 3-bit up/down Gray code counter

Present State			Next State					
			$Y = 0$ (DOWN)			$Y = 1$ (UP)		
Q_2	Q_1	Q_0	Q_2	Q_1	Q_0	Q_2	Q_1	Q_0
0	0	0	1	0	0	0	0	1
0	0	1	0	0	0	0	1	1
0	1	1	0	0	1	0	1	0
0	1	0	0	1	1	1	1	0
1	1	0	0	1	0	1	1	1
1	1	1	1	1	0	1	0	1
1	0	1	1	1	1	1	0	0
1	0	0	1	0	1	0	0	0

The Flip-Flop Transition Table   The transition table for the D flip-flop is shown in Table 11–2. This table indicates the required D input for each of the four possible state transitions. Q_N is the present state and Q_{N+1} is the next state

The Karnaugh Maps D inputs of the three flip-flops in the Gray code counter are plotted on the Karnaugh maps as shown in Figure 11–12. The up/down control input Y is

TABLE 11–2
Transition table for a D flip-flop

Output Transitions		Flip-Flop Input
Q_N	Q_{N+1}	D
0 $\longrightarrow$ 0		0
0 $\longrightarrow$ 1		1
1 $\longrightarrow$ 0		0
1 $\longrightarrow$ 1		1

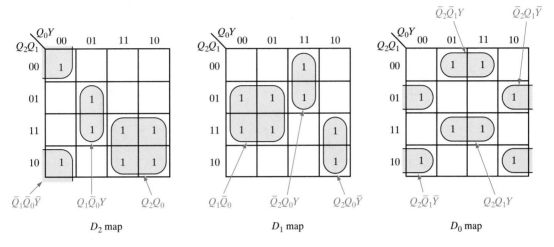

FIGURE 11–12
D input maps for the 3-bit up/down Gray code counter.

treated as one of the state variables along with Q_0, Q_1, and Q_2. Using the next-state table (Table 11–1), the flip-flop input from Table 11–2 is transferred onto the maps for each present state of the counter.

The Counter Circuit Figure 11–13 is the logic diagram for the 3-bit up/down Gray code counter. The combinational logic is used to implement the three equations for the D inputs to the flip-flops.

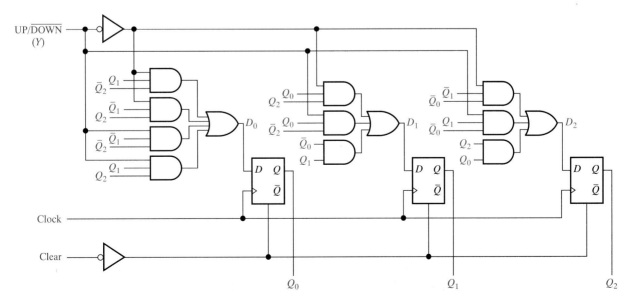

FIGURE 11–13
Logic diagram of the 3-bit up/down Gray code counter.

The Boolean Expressions for the D Inputs The 1s on the maps are combined in the largest possible groupings as illustrated in Figure 11–12 and the groups are factored, resulting in the following expressions:

$$D_0 = Q_2 Q_1 Y + \overline{Q}_2 \overline{Q}_1 Y + Q_2 \overline{Q}_1 \overline{Y} + \overline{Q}_2 Q_1 \overline{Y}$$
$$D_1 = Q_2 Q_0 \overline{Y} + \overline{Q}_2 Q_0 Y + Q_1 \overline{Q}_0$$
$$D_2 = \overline{Q}_1 \overline{Q}_0 \overline{Y} + Q_1 \overline{Q}_0 Y + Q_2 Q_0$$

Implementing the Counter with ABEL Equations

You are already familiar with using ABEL equations to describe a sequential circuit with registered outputs. The process for counters is basically the same as for shift registers in terms of the structure of the ABEL input file, as you will see in Example 11–3.

In addition to the .c. special constant for a clock pulse, another special constant is .x., which indicates a "don't care" for an input or output and can be used in the test vectors.

EXAMPLE 11–3

Develop an ABEL input file using the logic equation format to implement the 3-bit up/down Gray code counter in Figure 11–13. A GAL22V10 is the target device.

Solution Y represents the UP/$\overline{\text{DOWN}}$ control input. Since the Q output goes to the value of the D input on the next clock pulse, the ABEL equations are expressed in terms of the Q outputs instead of the D inputs. The input file is as follows:

```
Module    Three_bit_U/D_Gray_Code_Counter

Title     '3-bit Gray Code Counter in a GAL22V10'

"Device Declaration

    Counter      device      'P22V10'

"Pin Declaration

    Clock, Clear    pin 1,2;
    Y               pin 3;
    Q0,Q1,Q2        pin 21,22,23 ISTYPE 'reg,buffer';

Equations

    Q0 := Q2 & Q1 & Y # !Q2 & !Q1 & Y # Q2 & !Q1 & !Y # !Q2 & Q1 & !Y;

    Q1 := Q2 & Q0 & !Y # !Q2 & Q0 & Y # Q1 & !Q0;

    Q2 := !Q1 & !Q0 & !Y # Q1 & !Q0 & Y # Q2 & Q0;

    [Q0,Q1,Q2].CLK = Clock;

    [Q0,Q1,Q2].AR = !Clear;

Test_Vectors ([Clock, Clear,  Y ] -> [Q2,Q1,Q0])
             [ .c. ,   0  , .x.] -> [ 0, 0, 0];
             [ .c. ,   1  ,  0 ] -> [ 1, 0, 0];
             [ .c. ,   1  ,  0 ] -> [ 1, 0, 1];
             [ .c. ,   1  ,  0 ] -> [ 1, 1, 1];
             [ .c. ,   1  ,  0 ] -> [ 1, 1, 0];
             [ .c. ,   1  ,  0 ] -> [ 0, 1, 0];
             [ .c. ,   1  ,  0 ] -> [ 0, 1, 1];
             [ .c. ,   1  ,  0 ] -> [ 0, 0, 1];
             [ .c. ,   1  ,  0 ] -> [ 0, 0, 0];
             [ .c. ,   1  ,  1 ] -> [ 0, 0, 1];
             [ .c. ,   1  ,  1 ] -> [ 0, 1, 1];
             [ .c. ,   1  ,  1 ] -> [ 0, 1, 0];
             [ .c. ,   1  ,  1 ] -> [ 1, 1, 0];
             [ .c. ,   1  ,  1 ] -> [ 1, 1, 1];
             [ .c. ,   1  ,  1 ] -> [ 1, 0, 1];
             [ .c. ,   1  ,  1 ] -> [ 1, 0, 0];
             [ .c. ,   1  ,  1 ] -> [ 0, 0, 0];

END
```

Related Exercise Explain what the test vectors do line by line in this input file.

Implementing the Counter with an ABEL Truth Table

Instead of using equations, the counter can be described using the truth table entry method, which was covered in Chapter 7. The entry method used is basically a matter of personal preference.

EXAMPLE 11–4

Develop an ABEL input file using the truth table format to implement the 3-bit up/down Gray code counter in Figure 11–13. A GAL22V10 is the target device.

Solution In a truth table, the inputs are on the left of the assignment operator and the outputs are on the right. The assignment operator :> is used in truth tables for circuits with registered outputs and indicates that, for the given input values, the outputs will assume the given state after the next clock pulse. The ABEL input file is as follows. Notice that we still have the equations defining Clock and Clear. Also, the test vectors are not shown because they are the same as in Example 11–3.

```
Module   Three_bit_U/D_Gray_Code_Counter

Title   '3-bit Gray Code Counter in a GAL22V10'

"Device Declaration

   Counter     device     'P22V10'

"Pin Declaration

   Clock, Clear   pin 1,2;
   Y              pin 3;
   Q0,Q1,Q2       pin 21,22,23 ISTYPE 'reg,buffer';

Equations

   [Q0,Q1,Q2].CLK = Clock;

   [Q0,Q1,Q2].AR = !Clear;

Truth_Table ([Clear, Y , Q2 , Q1 , Q0 ] :> [Q2, Q1, Q0])
            [ 0 , .x. , .x. , .x. , .x.] :> [ 0,  0,  0];
            [ 1 , 0 , 0 , 0 , 0 ] :> [ 1,  0,  0];
            [ 1 , 0 , 1 , 0 , 0 ] :> [ 1,  0,  1];
            [ 1 , 0 , 1 , 0 , 1 ] :> [ 1,  1,  1];
            [ 1 , 0 , 1 , 1 , 1 ] :> [ 1,  1,  0];
            [ 1 , 0 , 1 , 1 , 0 ] :> [ 0,  1,  0];
            [ 1 , 0 , 0 , 1 , 0 ] :> [ 0,  1,  1];
            [ 1 , 0 , 0 , 1 , 1 ] :> [ 0,  0,  1];
            [ 1 , 0 , 0 , 0 , 1 ] :> [ 0,  0,  0];
            [ 1 , 1 , 0 , 0 , 0 ] :> [ 0,  0,  1];
            [ 1 , 1 , 0 , 0 , 1 ] :> [ 0,  1,  1];
            [ 1 , 1 , 0 , 1 , 1 ] :> [ 0,  1,  0];
            [ 1 , 1 , 0 , 1 , 0 ] :> [ 1,  1,  0];
            [ 1 , 1 , 1 , 1 , 0 ] :> [ 1,  1,  1];
            [ 1 , 1 , 1 , 1 , 1 ] :> [ 1,  0,  1];
            [ 1 , 1 , 1 , 0 , 1 ] :> [ 1,  0,  0];
            [ 1 , 1 , 1 , 0 , 0 ] :> [ 0,  0,  0];

END
```

Related Exercise Explain the use of the "don't care" special constant in the first line of the truth table.

Implementing the Counter with an ABEL State Diagram

A third alternative for ABEL design entry is the state diagram method, which is often better for more complex state machines. However, the 3-bit counter will serve to introduce this method.

The state diagram for the 3-bit up/down Gray code counter was shown in Figure 11–11. There are eight states labeled A through H, each representing a 3-bit Gray code. In each state, the counter can go to either to the next higher state or to the next lower state, depending on the value of the UP/$\overline{\text{DOWN}}$ control variable Y.

IF-THEN-ELSE Statements   ABEL provides a conditional statement to describe to which state the counter will go next based on certain conditions. In this case there are two conditions, $Y = 0$ and $Y = 1$. So, for example, when the counter is in State A (000), it will go to State B (001) if $Y = 1$ else it will go to State H ($Y = 0$). The ABEL IF-THEN-ELSE statement for State A is

```
IF Y THEN B ELSE H;
```

The statements can also be written using lowercase letters.

Defining the States   The use of a state diagram must include a definition of all the states in the counter. First, the variable *QSTATE* is defined as the set of Q outputs, which are the states of the flip-flops:

```
QSTATE = [Q2,Q1,Q0];
```

This is followed by the name and values for each of the eight states.

```
A = [0 , 0 , 0];
B = [0 , 0 , 1];
C = [0 , 1 , 1];
D = [0 , 1 , 0];
E = [1 , 1 , 0];
F = [1 , 1 , 1];
G = [1 , 0 , 1];
H = [1 , 0 , 0];
```

The next example shows how these basic ABEL state diagram elements can be used to describe the counter.

EXAMPLE 11–5

Develop an ABEL input file using the state diagram format to implement the 3-bit up/down Gray code counter in Figure 11–13. A GAL22V10 is the target device.

Solution   The ABEL input file is as follows. Notice that we still have the equations defining Clock and Clear. Again, the test vectors are not shown because they are the same as in Example 11–3.

```
Module    Three_bit_U/D_Gray_Code_Counter

Title    '3-bit Gray Code Counter in a GAL22V10'

"Device Declaration

    Counter      device      'P22V10'

"Pin Declaration

    Clock, Clear   pin 1,2;
    Y              pin 3;
    Q0,Q1,Q2       pin 21,22,23 ISTYPE 'reg,buffer';
```

```
"State Definitions

    QSTATE = [Q2,Q1,Q0];
    A       = [ 0, 0, 0];
    B       = [ 0, 0, 1];
    C       = [ 0, 1, 1];
    D       = [ 0, 1, 0];
    E       = [ 1, 1, 0];
    F       = [ 1, 1, 1];
    G       = [ 1, 0, 1];
    H       = [ 1, 0, 0];


Equations

    QSTATE.CLK = Clock;

    QSTATE.AR  = !Clear;


State_diagram QSTATE

    State A:    if Y then B else H;

    State B:    if Y then C else A;

    State C:    if Y then D else B;

    State D:    if Y then E else C;

    State E:    if Y then F else D;

    State F:    if Y then G else E;

    Stage G:    if Y then H else F;

    Stage H:    if Y then A else G;

END
```

Related Exercise Write the IF-THEN-ELSE statement for State A using $\overline{Y}$ instead of Y.

GOTO Statements In addition to the IF-THEN-ELSE statement for conditional state transitions, ABEL provides the GOTO statement for unconditional state transitions. If a counter goes to only one next state from a given present state, a GOTO statement can be used. For example, a 3-bit counter that sequences through a straight binary count can be described by the following ABEL state definitions and state diagram:

```
QSTATE = [Q2,Q1,Q0];
A       = [ 0, 0, 0];
B       = [ 0, 0, 1];
C       = [ 0, 1, 0];
D       = [ 0, 1, 1];
E       = [ 1, 0, 0];
F       = [ 1, 0, 1];
G       = [ 1, 1, 0];
H       = [ 1, 1, 1];


State_diagram QSTATE

    State A:    GOTO B;
    State B:    GOTO C;
    State C:    GOTO D;
    State D:    GOTO E;
    State E:    GOTO F;
    State F:    GOTO G;
    State G:    GOTO H;
    State H:    GOTO I;
    State I:    GOTO A;
```

11–5 ■ PLD SYSTEM IMPLEMENTATION

*In this section, a specific example is used to illustrate the use of the state diagrams
and other entry methods to implement a complete system with PLDs using ABEL. The
example system is an elevator control for a two-floor building and uses two GAL16V8
devices. This system is implemented from a description of the system requirements,
a block diagram, and a state diagram. After completing this section, you should be
able to*

☐ Describe an elevator control system   ☐ Implement a design from a state diagram
☐ Use the dot extension .FB

The System

A block diagram of a control system for a two-floor elevator is shown in Figure 11–14. The
control system monitors the status of the elevator push-button switches, responds to service calls, and displays the floor number and direction of the elevator. The system uses two
GAL16V8s, one for the control logic and one for the display logic, as indicated. The
amount of logic required could be implemented with a single GAL. However, one
GAL16V8 does not have enough outputs to accommodate the system requirements.

FIGURE 11–14
Block diagram of the elevator control system.

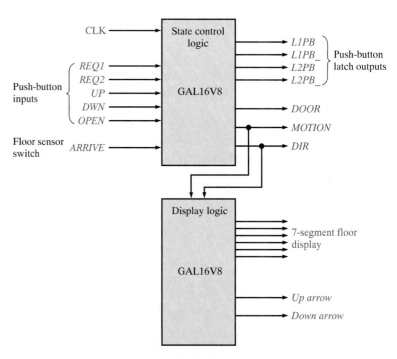

There is a "request" push-button on each floor beside the elevator door which you
press for service. When the elevator arrives at your floor, the doors open for a set period of
time to let passengers on or off and then close automatically. Inside the elevator you press
either the Up or the Down push-button switches. If you are on the first floor, you press the
Up push-button; and if you are on the second floor, you press the Down push-button.

Pressing the Open push-button inside the elevator will cause the door to reopen or stay open longer than the preset time as long as the elevator is not moving. If none of the push-button switches is activated, the elevator rests at the last floor serviced. A very slow clock signal controls both the movement of the elevator and the time that the door remains open when the elevator stops at a floor.

The state control logic consists of clock-independent latches that store a momentary closure of the push-button switches and a state machine that produces three elevator control outputs synchronized with the clock: *DOOR*, *MOTION*, and *DIR* (direction). Table 11–3 shows the states of these synchronized outputs.

TABLE 11–3

Output Variable	State	
DOOR	0 = Open	1 = Closed
MOTION	0 = Wait	1 = Move
DIR	0 = Up	1 = Down

The display unit is basically a decoder consisting only of combinational logic that produces six outputs to a 7-segment display (segment *f* is not used) to indicate either floor 1 or floor 2 and two outputs to an up/down arrow indicator light to indicate the direction that the elevator is moving.

Inputs to the State Control Logic

■ CLK   The clock signal establishes the length of time that the door remains open. The period of the clock is 10 s.
■ *REQ1*   Request for the elevator to come to the first floor. The request push-button is located outside of the elevator beside the door.
■ *REQ2*   Request for the elevator to come to the second floor. The request push-button is located outside of the elevator beside the door.
■ *UP*   Instruction for the elevator to go to the second floor. The Up push-button is located inside the elevator.
■ *DWN*   Instruction for the elevator to go to the first floor. The Down push-button is located inside the elevator.
■ *OPEN*   Instruction for the door to open when the elevator is not moving. The Open push-button is located inside the elevator.
■ *ARRIVE*   Indication that the elevator has arrived at a new floor. A sensor switch, activated by the elevator, on each floor provides this input.

Outputs from the State Control Logic

■ *L1PB* and *L1PB_*   First floor latch outputs for push-button switches. When one of the push-button switches for the first floor is pressed, the request is immediately stored in an S-R latch, independent of the clock. *L1PB* and *L1PB_* are complements.
■ *L2PB* and *L2PB_*   Second floor latch outputs for push-button switches. When one of the push-button switches for the second floor is pressed, the request is immediately stored in an S-R latch, independent of the clock. *L2PB* and *L2PB_* are complements.
■ *DOOR*   Indicates the status of the elevator door. When *DOOR* = 0, the door is open. When *DOOR* = 1, the door is closed.
■ *MOTION*   Indicates the movement of the elevator. When *MOTION* = 0, the elevator waits (does not move). When *MOTION* = 1, the elevator moves either up or down.
■ *DIR*   Indicates the direction of movement. When *DIR* = 0, the elevator moves up on the next call for second floor service. When *DIR* = 1, the elevator moves down on the next call for first floor service.

The State Diagram for the Elevator State Control Logic

The six states of the system are as follows:

- REST1 The elevator is waiting on the first floor with the door open.
- CLOSE1   The elevator is waiting on the first floor with the door closed.
- UP   The elevator is moving from first floor to second floor.
- REST2   The elevator is waiting on the second floor with the door open.
- CLOSE2   The elevator is waiting on the second floor with the door closed.
- DOWN   The elevator is moving from second floor to first floor.

The state diagram in Figure 11–15 shows the conditions for the system to change from one state to the next.

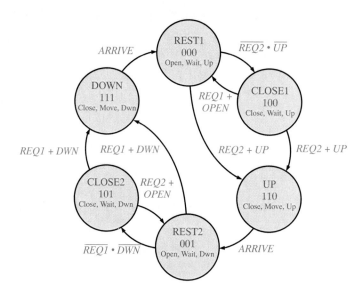

FIGURE 11–15
State diagram for the elevator control system.

Description of Operation

Refer to the state diagram in Figure 11–15. When the system is in the REST1 state, the elevator is waiting on the first floor with the door open. If, after a period of time established by the clock, there is no request from the second floor and the Up button is not pressed $(\overline{REQ2} \cdot \overline{UP})$, the system goes to the CLOSE1 state in which the elevator remains waiting on the first floor with the door closed. When the system is in the REST1 state and there is a request from the second floor or the Up button is pressed $(REQ2 + UP)$, the system bypasses the CLOSE1 state and goes to the UP state on the next clock pulse. In the UP state, the door closes and the elevator moves upward.

When the system is in the CLOSE1 state, it remains there until there is a request from the second floor or the Up button is pressed $(REQ2 + UP)$, which causes the system to go to the UP state. Alternatively, it remains in the CLOSE1 state until there is a request from the first floor or the Open button is pressed $(REQ1 + OPEN)$, which causes the system to return to the REST1 state.

The system remains in the UP state until the *ARRIVE* signal is received. The *ARRIVE* signal is produced when the elevator reaches the second floor and activates a switch, putting the system into the REST2 state. In the REST2 state, the elevator is waiting on the second floor with the door open. If, after a period of time established by the clock, there is no request from the first floor and the Down button is not pressed $(\overline{REQ1} \cdot \overline{DWN})$, the system goes to the CLOSE2 state in which the elevator remains waiting on the second floor with the door closed. When the system is in the REST2 state and there is a request from the first floor or the Down button is pressed (*REQ1* + *DWN*), the system bypasses the CLOSE2 state and goes to the DOWN state on the next clock pulse. In the DOWN state, the door closes and the elevator moves downward.

When the system is in the CLOSE2 state, it remains there until there is a request from the first floor or the Down button is pressed (*REQ1* + *DWN*), which causes the system to go to the DOWN state. Alternatively, it remains in the CLOSE2 state until there is a request from the second floor or the Open button is pressed (*REQ2* + *OPEN*), which causes the system to return to the REST2 state.

The system remains in the DOWN state until the *ARRIVE* signal is received when the elevator reaches the first floor, which activates a switch and puts the system back into the REST1 state.

Implementing the State Control Logic with a PLD

Refer to the block diagram for the GAL16V8 in Figure 11–1(b). Figure 11–16 shows the pin assignments for the GAL16V8 used in the state control logic portion of the system. Each of the elements required to describe the system with ABEL will be developed and then put together to form a complete input file. Notice that the state control logic consists of two parts, the latches for the push-button switch inputs and the state sequence logic.

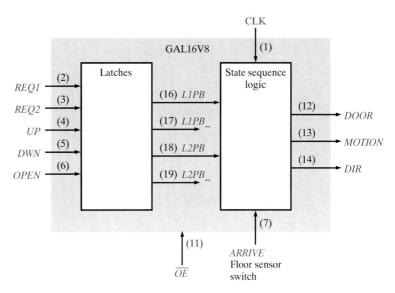

FIGURE 11–16
State control logic block diagram showing GAL16V8 pin assignments for inputs and outputs.

State Control Logic Inputs and Outputs

The statements defining the inputs can be written as follows. $\overline{\text{OE}}$ is the Output Enable that is required for the GAL16V8.

```
CLK, !OE              pin 1,11;
REQ1, REQ2            pin 2,3;
UP, DWN, OPEN, ARRIVE  pin 4,5,6,7;
```

Because of the very long clock period ($T = 10$ s), the latches for the first and second floor push-buttons (PB) are implemented with the logic array gates to achieve an immediate response instead of using the clock-dependent D flip-flops in the OLMCs. The statements defining the outputs can be written as follows:

```
L1PB, L1PB_           pin 16,17 ISTYPE 'com,buffer;
L2PB, L2PB_           pin 18,19 ISTYPE 'com,buffer;
DOOR, MOTION, DIR     pin 12,13,14 ISTYPE 'reg,buffer';
```

The variables, *L1PB, L1PB_, L2PB,* and *L2PB_,* produced by the latches are listed as outputs only because they are used as inputs by the sequential logic portion of the state control logic and must have pin designations in order to be fed back to the array logic. These outputs are not used for any external purpose.

State Definitions

The six states in the state control system must be defined in terms of their binary values. As you have seen, the state of the system is determined by the control output variables *DOOR, MOTION,* and *DIR.* First, these state control variables must be defined with a statement such as

```
CONSTATE = [DOOR, MOTION, DIR];
```

Next, the binary values must be stated and there are two ways to do this. The first way uses set notation. For example, the REST1 state is

```
REST1 = [ 0 , 0 , 0];
```

The second way uses ABEL binary notation where the binary number is prefixed with a $^\wedge$B as follows:

```
REST1  = ^B000;
CLOSE1 = ^B100;
UP     = ^B110;
REST2  = ^B001;
CLOSE2 = ^B101;
DOWN   = ^B111;
```

The Push-Button Latch Logic

The two latches for the push-button switch inputs are implemented with combinational logic. This is necessary because a push-button switch may be pressed and released at any time and, therefore, the clocked flip-flops in the OLMC cannot be used. For example, a button may be pressed and released between clock pulses (they are 10 seconds apart) and completely missed if a flip-flop is used to store the switch closure. An S-R latch, however, will immediately store a switch closure and hold it until it is reset.

The latch for the first floor is SET ($L1PB = 1$) if the first floor request button is pressed ($REQ1$) or if the elevator is on the first floor and the Open button is pressed ($\overline{DIR}$)($OPEN$) or if the elevator is on the second floor and the Down button is pressed (DIR)(DWN).

The elevator is on the first floor when the system is in the REST1 state or the CLOSE1 state. The elevator is on the second floor when the system is in the REST2 state or the CLOSE2 state. From the state diagram you can see that the *DIR* variable defines the floor because $DIR = 0$ when the system is in REST1 or CLOSE1 and $DIR = 1$ when the system is in REST2 or CLOSE2. The SET condition for the first-floor latch is expressed by the following Boolean equation:

$$L1PB = REQ1 + (\overline{DIR})(OPEN) + (DIR)(DWN) + \overline{L1PB\_}$$

The latch is RESET if the system is in the REST1 state (000). The RESET condition is expressed by the following Boolean equation:

$$L1PB\_ = (\overline{DOOR})(\overline{MOTION})(\overline{DIR}) + \overline{L1PB}$$

Equations for the second floor latch are similar.

$$L2PB = REQ2 + (DIR)(OPEN) + (\overline{DIR})(UP) + \overline{L2PB\_}$$
$$L2PB\_ = (\overline{DOOR})(\overline{MOTION})(DIR) + \overline{L2PB}$$

The combinational logic for the first-floor and second-floor latches is shown in Figure 11–17.

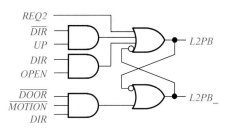

(a) Latch for push-button inputs, first floor   (b) Latch for push-button inputs, second floor

FIGURE 11–17
Latch circuits in the state control logic.

The Boolean equations for the latches are translated into ABEL as follows:

```
L1PB  = REQ1 # !DIR.FB & OPEN # DIR.FB & DWN # !L1PB_;
L1PB_ = (!DOOR.FB & !MOTION.FB & !DIR.FB) # !L1PB;
L2PB  = REQ2 # DIR.FB & OPEN # !DIR.FB & UP # !L2PB_;
L2PB_ = (!DOOR.FB & !MOTION.FB & DIR.FB) # !L2PB;
```

The use of the dot extension .FB with the state variables indicates that the variable is to be fed back to the array logic directly from the associated flip-flop. OLMC flip-flops produce the state variables *DOOR, MOTION,* and *DIR* in the sequential portion of the system.

The State Sequence Logic

The sequential portion of the state control logic produces the state variable outputs *DOOR, MOTION,* and *DIR.* This logic can be defined using state diagram entry without having to develop equations as we did for the combinational latch logic. The IF-THEN-ELSE statements for each of the six states are as follows. Input/output variables are enclosed in parentheses.

```
State REST1:   if (L2PB) then UP else CLOSE1;
State CLOSE1:  if (L2PB) then UP else if L1PB then REST1 else CLOSE1;
State UP:      if (ARRIVE) then REST2 else UP;
State REST2:   if (L1PB) then DOWN else CLOSE2;
State CLOSE2:  if (L1PB) then DOWN else if L2PB then REST2 else CLOSE2;
State DOWN:    if (ARRIVE) then REST1 else DOWN;
```

As you have seen, a sequential system can be implemented in a PLD without having to first design the logic because the software automatically implements the design based on a description of the state diagram.

EXAMPLE 11–6

Write a complete ABEL input file for the system control logic using the elements already developed.

Solution

```
Module   Elevator_State_Control

Title    'Control logic for two-floor elevator system'

"Device Declaration

    Elevator_cont      device      'P16V8'

"Pin Declaration
    CLK, !OE              pin 1,11;
    REQ1, REQ2           pin 2,3;
    UP, DWN, OPEN, ARRIVE pin 4,5,6,7;

    L1PB, L1PB_          pin 16,17 ISTYPE 'com,buffer';
    L2PB, L2PB_          pin 18,19 ISTYPE 'com,buffer';
    DOOR, MOTION, DIR    pin 12,13,14 ISTYPE 'reg,buffer';
```

```
"State Definitions

    CONSTATE = [DOOR, MOTION, DIR];
    REST1  = ^B000;
    CLOSE1 = ^B100;
    UP     = ^B110;
    REST2  = ^B001;
    CLOSE2 = ^B101;
    DOWN   = ^B111;

Equations

    CONSTATE.CLK = CLK;
    L1PB = REQ1 # !DIR.FB & OPEN # DIR.FB & DWN # !L1PB_;
    L1PB_ = (!DOOR.FB & !MOTION.FB & !DIR.FB) # !L1PB;
    L2PB = REQ2 # DIR.FB & OPEN # !DIR.FB & UP # !L2PB_;
    L2PB_ = (!DOOR.FB & !MOTION.FB & DIR.FB) # !L2PB;

State_diagram CONSTATE

    State REST1:   if (L2PB) then UP else CLOSE1;
    State CLOSE1:  if (L2PB) then UP else if L1PB then REST1 else CLOSE1;
    State UP:      if (ARRIVE) then REST2 else UP;
    State REST2:   if (L1PB) then DOWN else CLOSE2;
    State CLOSE2:  if (L1PB) then DOWN else if L2PB then REST2 else CLOSE2;
    State DOWN:    if (ARRIVE) then REST1 else DOWN;

Test_Vectors ([CLK, REQ1, REQ2, UP, DWN, OPEN, ARRIVE] -> [DOOR,MOTION,DIR])
             [.c.,  0 ,  0 , 0,  0,   0 ,   1  ] -> [ .x.,  .x.   ,.x.];
             [.c.,  1 ,  0 , 0,  0,   0 ,   1  ] -> [ .x.,  .x.   ,.x.];
             [.c.,  1 ,  0 , 0,  0,   0 ,   1  ] -> [  0 ,  0   , 0 ];
             [.c.,  0 ,  0 , 0,  0,   0 ,   0  ] -> [  1 ,  0   , 0 ];
             [.c.,  1 ,  0 , 0,  0,   0 ,   0  ] -> [  0 ,  0   , 0 ];
             [.c.,  0 ,  0 , 0,  0,   0 ,   1  ] -> [  1 ,  0   , 0 ];
             [.c.,  0 ,  0 , 0,  0,   1 ,   0  ] -> [  0 ,  0   , 0 ];
             [.c.,  0 ,  1 , 0,  0,   0 ,   0  ] -> [  1 ,  1   , 0 ];
             [.c.,  0 ,  0 , 0,  0,   0 ,   1  ] -> [  0 ,  0   , 1 ];
             [.c.,  0 ,  0 , 0,  0,   0 ,   0  ] -> [  1 ,  0   , 1 ];
             [.c.,  0 ,  1 , 0,  0,   0 ,   0  ] -> [  0 ,  0   , 1 ];
             [.c.,  0 ,  0 , 0,  0,   0 ,   0  ] -> [  1 ,  0   , 1 ];
             [.c.,  0 ,  0 , 0,  0,   1 ,   0  ] -> [  0 ,  0   , 1 ];
             [.c.,  1 ,  0 , 0,  0,   0 ,   0  ] -> [  1 ,  1   , 1 ];
             [.c.,  0 ,  0 , 0,  0,   0 ,   1  ] -> [  0 ,  0   , 0 ];
             [.c.,  0 ,  0 , 1,  0,   0 ,   0  ] -> [  1 ,  1   , 0 ];
             [.c.,  0 ,  0 , 0,  0,   0 ,   1  ] -> [  0 ,  0   , 1 ];
             [.c.,  0 ,  0 , 0,  1,   0 ,   0  ] -> [  1 ,  1   , 1 ];
             [.c.,  0 ,  0 , 0,  0,   0 ,   1  ] -> [  0 ,  0   , 0 ];
             [.c.,  0 ,  0 , 0,  0,   0 ,   0  ] -> [  1 ,  0   , 0 ];
             [.c.,  0 ,  1 , 0,  0,   0 ,   0  ] -> [  1 ,  1   , 0 ];
             [.c.,  0 ,  0 , 0,  0,   0 ,   1  ] -> [  0 ,  0   , 1 ];
             [.c.,  0 ,  0 , 0,  0,   0 ,   0  ] -> [  1 ,  0   , 1 ];
             [.c.,  1 ,  0 , 0,  0,   0 ,   0  ] -> [  1 ,  1   , 1 ];
             [.c.,  0 ,  0 , 0,  0,   0 ,   1  ] -> [  0 ,  0   , 0 ];
             [.c.,  0 ,  0 , 0,  0,   0 ,   0  ] -> [  1 ,  0   , 0 ];
             [.c.,  0 ,  0 , 1,  0,   0 ,   0  ] -> [  1 ,  1   , 0 ];
             [.c.,  0 ,  0 , 0,  0,   0 ,   1  ] -> [  0 ,  0   , 1 ];
             [.c.,  0 ,  0 , 0,  0,   0 ,   0  ] -> [  1 ,  0   , 1 ];
             [.c.,  0 ,  0 , 0,  1,   0 ,   0  ] -> [  1 ,  1   , 1 ];
             [.c.,  0 ,  0 , 0,  0,   0 ,   1  ] -> [  0 ,  0   , 0 ];

END
```

Related Exercise   The test vectors should take the system through all possible paths in the state diagram. Verify that this happens.

Implementing the Display Logic with a PLD

The pin assignments for the GAL16V8 used in the display logic portion of the system are shown in Figure 11–18. The inputs to the display logic come from the state control logic. Only six active-LOW outputs are required for the 7-segment readout because only the digit 1 and the digit 2 are displayed; therefore, segment *f* is not needed. Two other active-

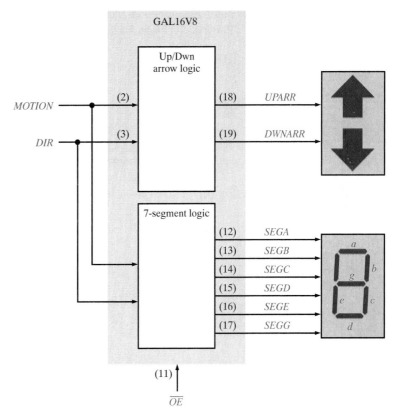

FIGURE 11–18
Display logic block diagram showing GAL16V8 pin assignments for inputs and outputs.

LOW outputs activate the up and down arrows. All of the logic in this portion of the system is combinational.

Input and Output Pin Declarations

The ABEL statement designating the input pins can be written as follows. $\overline{OE}$ is the Output Enable that is required for the GAL16V8.

```
!OE, MOTION, DIR    pin 11,2,3;
```

The ABEL statements designating the output pins can be written as follows:

```
SEGA, SEGB, SEGC    pin 12,13,14    ISTYPE 'com,invert';
SEGD, SEGE, SEGG    pin 15,16,17    ISTYPE 'com,invert';
UPARR, DWNARR       pin 18,19       ISTYPE 'com,invert';
```

The Up Arrow and Down Arrow Logic

The Up arrow must be illuminated when the elevator is moving ($MOTION = 1$) in the upward direction ($DIR = 0$). The Boolean equation is

$$UPARR = MOTION \cdot \overline{DIR}$$

The Down arrow must be illuminated when the elevator is moving ($MOTION = 1$) in the downward direction ($DIR = 1$). The Boolean equation is

$$DWNARR = MOTION \cdot DIR$$

As you can see, these functions can be implemented with 2-input AND gates.

The Boolean equations for the latches are translated into ABEL as follows:

```
UPARR = MOTION & !DIR;
DWNARR = MOTION & DIR;
```

The 7-Segment Logic

The display must show a "1" when the system is in the REST1 state and in the CLOSE1 state. The display must show a "2" in the REST2 state and in the CLOSE2 state.

Segments b and c are used for a "1" and segments a, b, d, e, and g are used for a "2", as shown in Figure 11–18. Segment b is common to both digits. Therefore, the ABEL equations for the segments are as follows:

```
SEGA = !MOTION & DIR;
SEGB = !MOTION;
SEGC = !MOTION & !DIR;
SEGD = !MOTION & DIR;
SEGE = !MOTION & DIR;
SEGG = !MOTION & DIR;
```

Notice that the equations for segments a, d, e, and g are the same, so one output could be used for all of these. However, this output will have to be buffered externally in order to drive the four segments.

EXAMPLE 11–7

Write a complete ABEL input file for the system display logic using the elements already developed.

Solution

```
Module    Elevator_Display_Logic

Title    'Display logic for two-floor elevator system'

"Device Declaration

    Elevator_disp    device    'P16V8'

"Pin Declaration

    !OE, MOTION, DIR    pin 11,2,3;

    SEGA, SEGB, SEGC    pin 12,13,14 ISTYPE 'com,invert';
    SEGD, SEGE, SEGG    pin 15,16,17 ISTYPE 'com,invert';
    UPARR, DWNARR       pin 18,19    ISTYPE 'com,invert';

Equations

    UPARR = MOTION & !DIR;
    DWNARR = MOTION & DIR;
    SEGA = !MOTION & DIR;
    SEGB = !MOTION;
    SEGC = !MOTION & !DIR;
    SEGD = !MOTION & DIR;
    SEGE = !MOTION & DIR;
    SEGG = !MOTION & DIR;

Test_Vectors
([!OE, MOTION, DIR] -> [SEGA,SEGB,SEGC,SEGD,SEGE,SEGG,UPARR,DWNARR])
 [ 0 ,   0  ,  0 ] -> [ 1 , 0 , 0 , 1 , 1 , 1 , 1 ,  1  ];
 [ 0 ,   0  ,  1 ] -> [ 0 , 0 , 1 , 0 , 0 , 0 , 1 ,  1  ];
 [ 0 ,   1  ,  0 ] -> [ 1 , 1 , 1 , 1 , 1 , 1 , 0 ,  1  ];
 [ 0 ,   1  ,  1 ] -> [ 1 , 1 , 1 , 1 , 1 , 1 , 1 ,  0  ];
 [ 1 ,  .x. , .x.] -> [ 1 , 1 , 1 , 1 , 1 , 1 , 1 ,  1  ];

END
```

Related Exercise   Redo the input file to combine the four equivalent segment outputs into one output called SEGADEG.

SECTION 11–5 REVIEW

1. Explain why the latches are required in the system control logic.
2. In the test vectors in Example 11–6, explain how the system is initialized to the REST1 state. Why are the output variables "don't cares" in the first two rows?
3. Can you use the state diagram entry method instead of equations for the display logic? Explain.
4. Can you use the truth table entry method for all of the logic design discussed in this section?

11–6 ■ DIGITAL SYSTEM APPLICATION

In Chapter 9 the sequential logic portion of the traffic light control system was implemented using SSI and MSI devices. You will now apply your knowledge of PLDs and PLD programming to implement the sequential logic with a GAL22V10. You should review Section 9–10 before proceeding. After completing this section, you should be able to

☐ Program the GAL22V10 from a state diagram   ☐ Explain how the GAL replaces the logic previously implemented in Chapter 9.   ☐ Develop an ABEL input file to implement the traffic light control system

Block Diagram of the Sequential Logic

The block diagram of the traffic light control system is shown in Figure 11–19, and the block diagram of the sequential logic is shown in Figure 11–20.

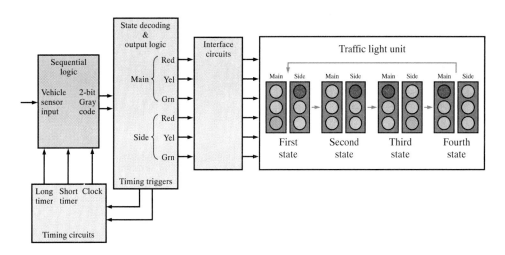

FIGURE 11–19
Traffic light control system block diagram.

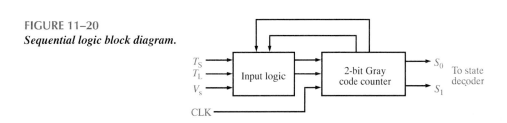

FIGURE 11–20
Sequential logic block diagram.

The sequential logic portion of the system uses a 2-bit Gray code counter to produce the sequence of states described by the state diagram in Figure 11–21. A review of the input variables is as follows:

V_s Vehicle on side street
$\overline{V}_s$ No vehicle on side street
T_L Long timer on (25 s)
$\overline{T}_L$ Long timer off (25 s)
T_S Short timer on (4 s)
$\overline{T}_S$ Short timer off (25 s)

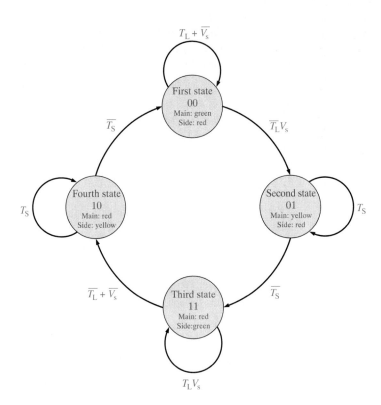

FIGURE 11–21
State diagram for the traffic light control system.

■ DIGITAL WORKBENCH 1: GAL22V10 Programming

The GAL22V10 that was used for the state decoding and output logic in the Chapter 7 system application can be programmed for the addition of the sequential logic because there are enough inputs and outputs still available.

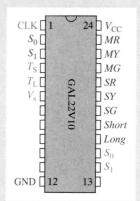

- *Activity 1*   From the state diagram in Figure 11–21 and the pin assignment diagram for the GAL22V10 shown here, write an ABEL input file to program the sequential logic into the device. The existing inputs and outputs for the previously programmed state decoding and output logic are shown in red. The inputs and outputs for the sequential logic to be added are shown in blue.

- *Activity 2*   Write a single input file for the complete system by properly combining the input file for the state decoding and output logic from Chapter 7 and the sequential logic input file that you just completed. The test vectors for this complete input file are developed in Digital Workbench 2.

■ DIGITAL WORKBENCH 2: Verification and Testing

The single GAL22V10 package replaces the 74LS74A, 74LS08, 74LS153 that were originally used to implement the sequential logic in the Chapter 9 system application as well as the 74LS139 and 74LS08 used for the state decoding and output logic in the Chapter 6 system application. The timing circuits are the same.

■ *Activity 1*   Develop test vectors to check the programmed GAL and add it to the input file from Workbench 1.

■ *Activity 2*   Identify the function of each of the input and output terminals on the system PC board shown below.

■ *Activity 3*   Draw a logic diagram of the system board.

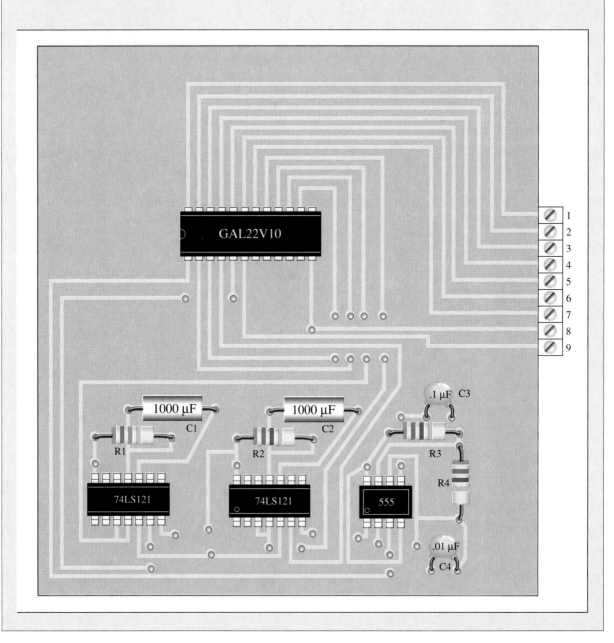

**SECTION 11–6
REVIEW**

1. After programming the GAL22V10 in this section, how many outputs are still unused and available?
2. How many inputs are unused and available?
3. Compare the GAL implementation to the implementation using SSI and MSI devices for all of the logic.

■ **SUMMARY**

- The GAL16V8 and the GAL22V10 are examples of reprogrammable logic devices.
- The OLMC in a GAL is the Output Logic Macrocell that can be configured by programming for either a combinational or a registered output.
- In the OLMC registered mode, the output comes from a D flip-flop.
- The ISTYPE statement in ABEL can be used to declare an output either combinational or registered as ISTYPE 'com'; or ISTYPE 'reg';.
- The ABEL assignment operators := and :> are used to indicate a registered output.
- The dot extension .CLK is used to indicate that the register device is a clocked flip-flop.
- The dot extension .AR is used to asynchronously reset a flip-flop.
- Two basic types of state machines are the Moore and the Mealy.
- In a Moore type of state machine, the outputs are determined only on the internal state and any synchronized inputs.
- In a Mealy type of state machine, the outputs are determined by both the internal state and by inputs that are not synchronized with the system.
- A counter is a simple example of a Moore type of state machine.
- State machines can be implemented in a PLD using the state diagram entry method in ABEL. The IF-THEN-ELSE statement is an important element of the state diagram syntax.

■ **SELF-TEST**

1. The GAL22V10 has
   (a) 10 inputs and 22 outputs   (b) 22 inputs and 10 outputs
   (c) 11 inputs and 10 input/outputs   (d) 11 inputs and 10 outputs

2. The GAL16V8 has
   (a) 8 inputs and 16 outputs   (b) 16 inputs and 8 outputs
   (c) 8 inputs and 8 input/outputs   (d) 8 inputs and 8 outputs

3. An OLMC is
   (a) Output Latch Main Cell   (b) Odd Logic Multiplexer Cell
   (c) Optimum Logic Minimization Method   (d) Output Logic Macrocell

4. A typical OLMC consists of
   (a) gates, multiplexers, and a flip-flop   (b) gates and a shift register
   (c) a Gray code counter   (d) a fixed logic array

5. Which of the following ABEL equations refers to a registered output?
   (a) X = A & B;   (b) !X = !(A & B);
   (c) X := A & B;   (d) X –> A & B;

6. A counter can be implemented in a PLD using ABEL by
   (a) equation entry   (b) truth table entry
   (c) state diagram entry   (d) all of the above
   (e) answers (b) and (c)

7. Combinational logic can be implemented in a PLD using ABEL by
   (a) equation entry   (b) truth table entry
   (c) state diagram entry   (d) all of the above
   (e) answers (a) and (b)

8. A valid IF-THEN-ELSE statement is
   (a) if A then C else D   (b) if A then C else D;
   (c) IF A THEN C ELSE D   (d) IF-A-THEN-C-ELSE-D:
   (e) answers (b) and (c)

9. The ABEL expression A = [0, 1, 0]

    (a) means that state A is defined by the binary value 010

    (b) means that variable A has a sequence of values 0,1,0

    (c) must be preceded by a state definition statement

    (d) answers (a) and (c)

10. .CLK and .AR are examples of

    (a) special constants   **(b)** ABEL variables

    (c) dot extensions   **(d)** none of the above

■ PROBLEMS

SECTION 11–1   The Complete OLMC

1. List the four GAL22V10 OLMC configurations.

2. Determine the GAL16V8 mode configuration for each combination of "cell" values.

    (a) $SYN = 0$, $AC0 = 0$, $XOR = 0$

    (b) $SYN = 1$, $AC0 = 1$, $XOR = 1$

    (c) $SYN = 0$, $AC0 = 1$, $XOR = 1$

    (d) $SYN = 1$, $AC0 = 0$, $XOR = 0$

3. Beside the flip-flop and logic gates, what type of device is used in an OLMC?

SECTION 11–2   OLMC Mode Selection

4. Determine the GAL22V10 mode for each of the bit combinations.

    (a) $S_1S_0 = 10$ **(b)** $S_1S_0 = 00$

    (c) $S_1S_0 = 11$ **(d)** $S_1S_0 = 01$

5. Explain each of the ABEL output pin declarations.

    (a) X pin 15 ISTYPE 'com';

    (b) Q2 pin 20 ISTYPE 'reg';

    (c) A PIN 18 ISTYPE 'com, invert';

    (d) Q1 pin 21 ISTYPE 'reg, buffer';

6. Identify each of the ABEL expressions as registered or combinational.

    (a) Y = A # B;   **(b)** X := A & !B;

    (b) Q1 := D1 # D0;   **(d)** Q2 = !C;

7. Write an ABEL clock equation to accompany the following expression: Q1 := (A & B) # C;

SECTION 11–3   Implementing Shift Registers with PLDs

8. Write an ABEL input file, excluding the test vectors, to implement the serial in/serial out shift register from Figure 11–22 in a GAL22V10.

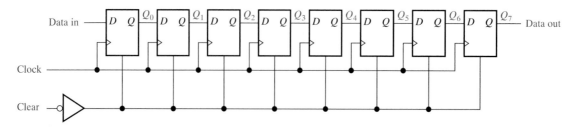

FIGURE 11–22

9. Develop the test vectors for the input file from Problem 8.

10. Repeat Problem 8 using a GAL16V8 as the target device.

11. Repeat Problem 9 using a GAL16V8 as the target device.

12. Write an ABEL input file to expand the 4-bit parallel in/serial out register in Figure 11–23 to an 8-bit register. The target device is to be a GAL22V10. Include test vectors.

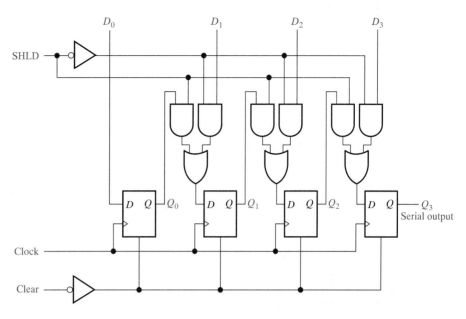

FIGURE 11–23

13. Determine the feasibility of implementing the 8-bit register in Problem 12 with A GAL16V8.

SECTION 11–4   Implementing Counters with PLDs

14. Explain the basic difference between a Moore and a Mealy state machine.

15. What is the largest counter in terms of the number of stages or bits that can be implemented with a single GAL16V8?

16. What is the largest counter in terms of the number of stages or bits that can be implemented with a single GAL22V10?

17. Convert the 3-bit up-down Gray code counter in Figure 11–24 to a 3-bit up only Gray code counter and implement it in a GAL22V10 using logic equation entry.

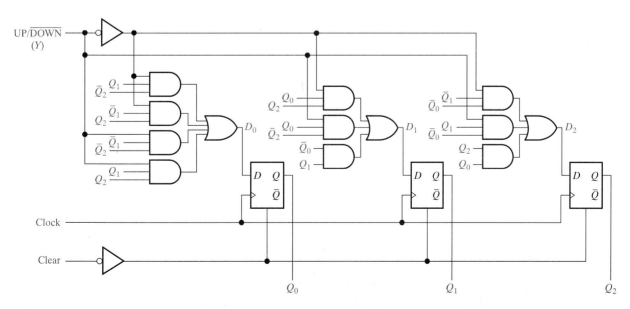

FIGURE 11–24

18. Repeat Problem 17 using truth table entry.

19. Repeat Problem 17 using state diagram entry.

20. Repeat Problem 17 with a GAL16V8 as the target device.

21. Repeat Problem 18 with a GAL16V8 as the target device.

22. Repeat problem 19 with a GAL16V8 as the target device.

SECTION 11–5 PLD System Implementation

23. Referring to the state diagram in Figure 11–25, if the Open push-button is pressed, what happens when the system is in the

(a) CLOSE1 state   (b) REST1 state

(c) Up state   '   (d) REST2 state

(e) CLOSE2 state   (f) DOWN state

FIGURE 11–25

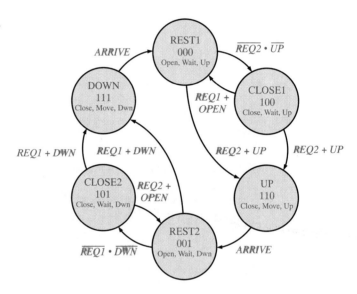

24. Repeat Problem 23 for each of the following cases:

(a) The Down push-button is pressed.

(b) The Up push-button is pressed.

(c) The REQ1 push-button is pressed.

(d) The REQ2 push-button is pressed.

25. Write input file to implement the state control logic portion of the elevator control system that was developed in this section with a GAL22V10 instead of a GAL16V8.

26. Repeat Problem 25 for the display logic in the elevator control system.

SECTION 11–6 Digital System Application

27. Can the traffic light control system be implemented with a single GAL16V8? If not, explain.

28. Assume the short timer input (T_S) to the sequential logic portion of the traffic light control system fails such that it is always HIGH ($T_S = 1$). How is the system operation affected?

29. Assume the long timer input (T_L) to the sequential logic portion of the traffic light control system fails such that it is always LOW ($T_L = 0$). How is the system operation affected?

30. If the vehicle sensor always produces a HIGH, how is the system operation affected?

Special Design Problems

31. Design an 8-bit serial in/serial out shift register and write an ABEL input file to implement it using GAL16V8s.

32. Modify the design in Problem 31 for a 16-bit serial in/serial out shift register.

33. Implement a 4-bit counter that will sequentially produce the excess-3 code using a GAL22V10 as the target device.

34. Modify the design in problems 33 so that the counter can be preloaded in parallel to any specified 4-bit excess-3 code number.

35. Develop a state diagram for a three-floor elevator control system.

36. Using the state diagram from Problem 35, write an ABEL input file to implement the state control logic for the three-floor elevator system using a GAL22V10.

■ ANSWERS TO SECTION REVIEWS

SECTION 11–1

1. The GAL22V10 modes are combinational and registered.
2. The mode is set by the select bits S_0 and S_1.
3. The GAL16V8 has two combinational modes (simple and complex) and a registered mode.

SECTION 11–2

1. In the registered mode, the output is from a flip-flop. In the combinational mode the output is not from a flip-flop.
2. An ISTYPE statement is used to declare an output either combinational or registered.
3. The := operator indicates a registered function while the = operator is for a combinational function.
4. The dot extension .CLK indicates that the register device is a clocked flip-flop.

SECTION 11–3

1. Q0 := A & !B # !A & B; means that output Q_0 is registered and equal to the exclusive-OR of A and B.
2. The dot extension .AR indicates asynchronous reset of a flip-flop.
3. The special constant .c. indicates a clock pulse.

SECTION 11–4

1. The categories of state machines are Moore and Mealy.
2. A counter is a Moore state machine.
3. Entry methods are equations, truth table, and state diagram.
4. The special constant .x. represents "don't care".

SECTION 11–5

1. The latches are required to store a temporary switch closure.
2. A maximum of three clock pulses are required to get the system to the REST1 state regardless of its initial state. Since three clock pulses are used, we don't care what the outputs are for the first two.
3. State diagram entry cannot be used for the display logic because the logic is all combinational.
4. Truth table entry can be used for all logic.

SECTION 11–6

1. No outputs are unused and available.
2. Six inputs are unused and available.
3. One GAL22V10 replaces five ICs in the original design.

12

MEMORIES

■ CHAPTER OBJECTIVES

☐ Define the basic memory characteristics
☐ Explain what a ROM is and how it works
☐ Explain how a ROM is programmed
☐ Describe the various types of PROMs
☐ Explain what a RAM is and how it works
☐ Explain how ROMs and RAMs are used in micro-processor-based systems
☐ Explain the difference between static RAMs (SRAMs) and dynamic RAMs (DRAMs)
☐ Discuss the characteristics of a flash memory
☐ Describe the expansion of ROMs and RAMs to increase word length and word capacity
☐ Describe the basic organization of magnetic disks and magnetic tapes
☐ Describe the basic operation of magneto-optic disks and discuss the CD-ROM and the WORM disk memory
☐ Describe basic methods for memory testing
☐ Develop flowcharts for memory testing
☐ Apply a memory device in a system application

■ CHAPTER OVERVIEW

Chapter 10 covered shift registers, which are a type of storage device; in fact, a shift register is essentially a small-scale memory. The memory devices covered in this chapter are generally used for longer-term storage of larger amounts of data than registers can provide.

Computers and other types of systems require the permanent or semipermanent storage of large amounts of binary data. Microprocessor-based systems rely on memories for their operation because of the necessity for storing programs and for retaining data during processing.

In this chapter semiconductor, magnetic, and optical memories are covered.

■ SPECIFIC DEVICES

2516	MCM6246	MCM6264C
MCM516100	MCM516400	

■ DIGITAL SYSTEM APPLICATION

This Digital System Application illustrates concepts taught in this chapter. The system application in Section 12–10 completes the security entry system from Chapter 10. The focus in this chapter is the memory logic portion of the system, which stores the entry code. Once the memory logic is developed, it is interfaced with the security code logic from Chapter 10 to form the complete system.

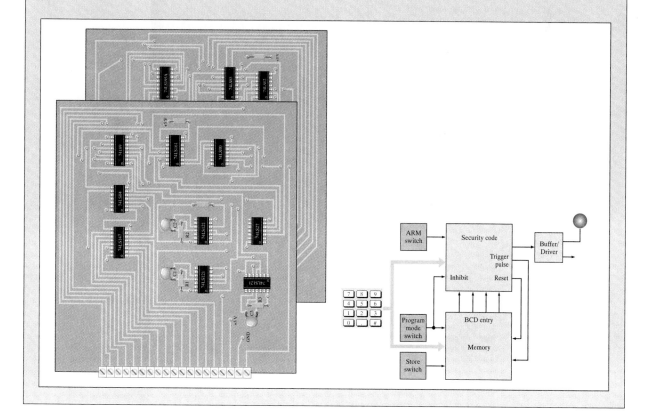

12–1 ■ BASICS OF SEMICONDUCTOR MEMORIES

The memory is a device for storing binary data on a long-term or on a short-term basis. Basic types of memories are semiconductor, magnetic and optical. Semiconductor memories consist of arrays of storage elements that are either latches, capacitors, or other charge-storage elements. After completing this chapter, you should be able to

☐ Explain how a memory stores binary data   ☐ Discuss the basic organization of a memory   ☐ Describe the write operation   ☐ Describe the read operation   ☐ Describe the addressing operation   ☐ Explain what RAMs and ROMs are

Units of Binary Data

As a rule, memories store data in units that have from one to eight bits. The smallest unit of binary data, as you know, is the bit. In many applications data are handled in an 8-bit unit called a **byte** or in multiples of 8-bit units. The byte can be split into two 4-bit units which are called **nibbles.** A complete unit of information is called a **word** and generally consists of one or more bytes, although a group of less than eight bits can also constitute a word.

The Basic Semiconductor Memory Array

Each storage element in a memory can retain either a 1 or a 0 and is called a **cell.** Memories are made up of arrays of cells as illustrated in Figure 12–1 using 64 cells as an example. Each block in the **memory array** represents one storage cell, and its location can be identified by specifying a row and a column.

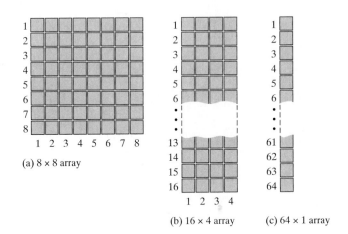

FIGURE 12–1
A 64-cell storage array organized in three different ways.

The 64-cell array can be organized in several ways based on units of data. Figure 12–1(a) shows an 8 × 8 array, which can be viewed as either a 64-bit memory or an 8-byte memory. Part (b) shows a 16 × 4 array, which is a 16-nibble memory, and part (c) shows a 64 × 1 array which is a 64-bit memory. A memory is identified by the number of words it can store times the word size. For example, a 16k × 4 memory can store 16,384 words of four bits each. The inconsistency here is common in memory terminology. The actual number of words is always a power of 2, which, in this case, is $2^{14}=16,384$. However, it is common practice to state the number to the nearest thousand, in this case, 16k.

Memory Address and Capacity

The location of a unit of data in a memory array is called its **address.** For example, in Figure 12–2(a), the address of a bit in the array is specified by the row and column as shown. In Figure 12–2(b), the address of a byte is specified only by the row. So, as you can see, the address depends on how the memory is organized into units of data.

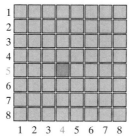
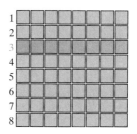

(a) The address of the bit highlighted in blue is row 5, column 4.

(b) The address of the byte highlighted in blue is row 3.

FIGURE 12–2
Examples of memory address.

The **capacity** of a memory is the total number of data units that can be stored. For example, in the bit-organized memory array in Figure 12–2(a), the capacity is 64 bits. In the byte-organized memory array in Figure 12–2(b), the capacity is 8 bytes, which is also 64 bits.

Basic Memory Operations

Since a memory stores binary data, data must be put into the memory and data must be taken out of the memory when needed. The **write** operation puts data into a specified address in the memory, and the **read** operation takes data out of a specified address in the memory. The addressing operation, which is part of both the write and the read operations, selects the specified memory address.

Data go into the memory during a write operation and come out of the memory during a read operation on a set of lines called the *data bus.* As indicated in Figure 12–3, the data bus is bidirectional, which means that data can go in either direction (into the memory or out of the memory). In the case of a byte-organized memory, the data bus has eight lines so that all eight bits in a selected byte of data are transferred in parallel. Also, for a write or a read operation, an address must be selected by placing a binary code representing the desired address on a set of lines called the *address bus.* The address code is decoded and the appropriate address is selected. The number of lines in the address bus depends on the ca-

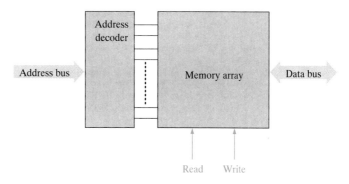

FIGURE 12–3
Block diagram of a memory showing address bus, address decoder, data bus, and read/write inputs.

pacity of the memory. For example, a 4-bit address code can select 16 locations (2^4) in the memory, an 8-bit address code can select 256 locations (2^8) in the memory, and so on.

The Write Operation The basic write operation is illustrated in Figure 12–4. To store a byte of data in the memory, a code held in the address register is placed on the address bus. Once the address code is on the bus, the address decoder decodes the address and selects the specified location in the memory. The memory then gets a write command, and the data byte held in the data register is placed on the data bus and stored in the selected memory address, thus completing the write operation. When a new data byte is written into a memory address, the current data byte stored at that address is destroyed.

FIGURE 12–4
Illustration of the write operation.

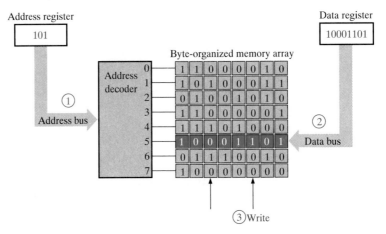

① Address code placed on address bus and address 5 selected

② Data byte placed on data bus

③ Write command causes data byte to be stored in address 5, replacing previous data

The Read Operation The basic read operation is illustrated in Figure 12–5. Again, a code held in the address register is placed on the address bus. Once the address code is on the bus, the address decoder decodes the address and selects the specified location in the memory. The memory then gets a read command, and a "copy" of the data byte that is stored in the selected memory address is placed on the data bus and loaded into the data register, thus completing the read operation. When a data byte is read from a memory address, it also remains stored at that address and is not destroyed.

FIGURE 12–5
Illustration of the read operation.

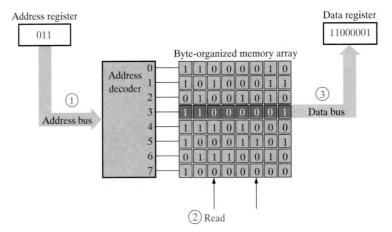

① Address code placed on address bus and address 3 selected

② Read command is applied

③ The contents of address 3 is placed on the data bus and shifted into data register. The contents of address 3 is not destroyed by the read operation.

RAMs and ROMs

The two major categories of semiconductor memories are the RAM and the ROM. **RAM** (random-access memory) is a type of memory in which all addresses are accessible in an equal amount of time and can be selected in any order for a read or write operation. All RAMs have both read and write capability. Because RAMs lose stored data when the power is turned off, they are **volatile** memories.

ROM (read-only memory) is a type of memory in which data are stored permanently or semipermanently. Data can be read from a ROM, but there is no write operation as in the RAM. The ROM, like the RAM, is a random-access memory but the term RAM traditionally means a random-access *read/write* memory. Several types of RAMs and ROMs will be covered in this chapter. Because ROMs retain stored data even if power is turned off, they are **nonvolatile** memories.

Memory Technologies

Semiconductor memories are manufactured using two basic technologies: bipolar and MOS. The MOS (metal-oxide semiconductor) technology, which uses MOSFETs, is the most dominant memory technology. Bipolar memories make up less and less of the market as time goes by, although BiMOS, a combination of bipolar and MOS, is used in many applications.

Three major MOS technologies are PMOS, NMOS, and CMOS. PMOS is implemented with *p*-channel FETs, NMOS uses *n*-channel FETs, and CMOS combines both. Most early semiconductor memories were made with PMOS technology. As higher speeds and greater densities were needed, NMOS became the dominant MOS technology. More recently, CMOS has become widely used in memory devices. BiCMOS memories incorporate bipolar transistor circuits (TTL or ECL) for input and output functions to make the devices compatible with bipolar logic.

SECTION 12–1 REVIEW	1. What is the smallest unit of data that can be stored in a memory? 2. What is the bit capacity of a memory that can store 256 bytes of data? 3. What is a write operation? 4. What is a read operation? 5. How is a given unit of data located in a memory? 6. Describe the difference between a RAM and a ROM. 7. Name two types of semiconductor technologies used in memory devices.

12–2 ■ READ-ONLY MEMORIES (ROMs)

As mentioned before, a ROM contains permanently or semipermanently stored data, which can be read from the memory but either cannot be changed at all or cannot be changed without specialized equipment. A ROM stores data that are used repeatedly in system applications, such as tables, conversions, or programmed instructions for system initialization and operation. ROMs retain stored data when the power is off. After completing this section, you should be able to

☐ List the types of ROMs   ☐ Describe a basic mask ROM storage cell   ☐ Explain how data are read from a ROM   ☐ Discuss internal organization of a typical ROM   ☐ Explain the purpose of a tristate output   ☐ Discuss some ROM applications

The ROM Family

Semiconductor ROMs are manufactured with bipolar technology (such as TTL) or with MOS (metal oxide semiconductor) technology. Figure 12–6 shows how ROMs are categorized. The mask ROM is the type in which data are permanently stored in the memory during the manufacturing process. The **PROM,** or programmable ROM, is the type in which the data are electrically stored by the user with the aid of specialized equipment. Notice that both the mask ROM and the PROM can be of either technology. The **EPROM,** or erasable PROM, is strictly an MOS device.

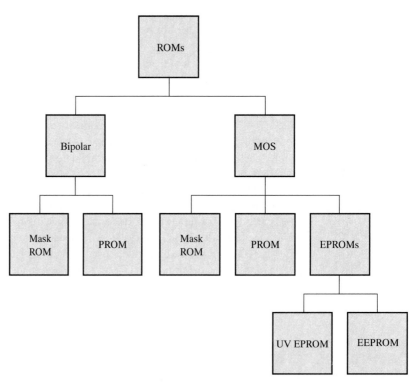

FIGURE 12–6
The semiconductor ROM family.

The **UV EPROM** is electrically programmable by the user, but the stored data must be erased by exposure to ultraviolet light over a period of several minutes. The electrically erasable PROM (**EEPROM** or E^2PROM) can be erased in a few milliseconds.

The Mask ROM

The mask ROM is usually referred to simply as a ROM. It is permanently programmed during the manufacturing process to provide widely used standard functions, such as popular conversions, or to provide user-specified functions. Once the memory is programmed, it cannot be changed. Most IC ROMs utilize the presence or absence of a transistor connection at a row/column junction to represent a 1 or a 0. A ROM can be either bipolar or MOS.

Figure 12–7(a) shows bipolar ROM cells. The presence of a connection from a row line to the base of a transistor represents a 1 at that location because when the

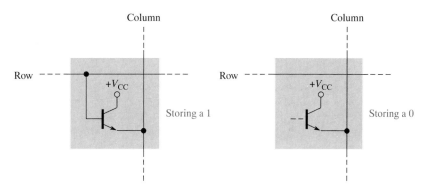

(a) Bipolar cells

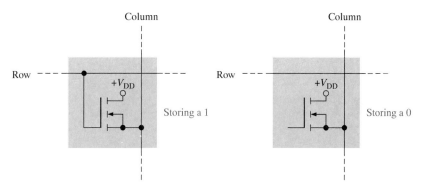

(b) MOS cells

FIGURE 12–7
ROM cells.

row line is taken HIGH, all transistors with a base connection to that row line turn on and connect the HIGH (1) to the associated column lines. At row/column junctions where there are no base connections, the column lines remain LOW (0) when the row is addressed.

Figure 12–7(b) illustrates MOS ROM cells. They are basically the same as the bipolar cells, except they are made with MOSFETs (metal oxide semiconductor field-effect transistors). The presence or absence of a gate connection at a junction permanently stores a 1 or a 0, as shown.

A Simple ROM

To illustrate the ROM concept, Figure 12–8 (next page) shows a small, simplified ROM array. The blue squares represent stored 1s, and the gray squares represent stored 0s. The basic read operation is as follows: When a binary address code is applied to the address input lines, the corresponding row line goes HIGH. This HIGH is connected to the column lines through the transistors at each junction (cell) where a 1 is stored. At each cell where a 0 is stored, the column line stays LOW because of the terminating resistor. The column lines form the data output. The eight data bits stored in the selected row appear on the output lines.

As you can see, the example ROM in Figure 12–8 is organized into 16 addresses, each of which stores 8 data bits. Thus, it is a 16 × 8 (16-by-8) ROM, and its total capacity is 128 bits or 16 bytes.

FIGURE 12–8
A 16 × 8-bit ROM array.

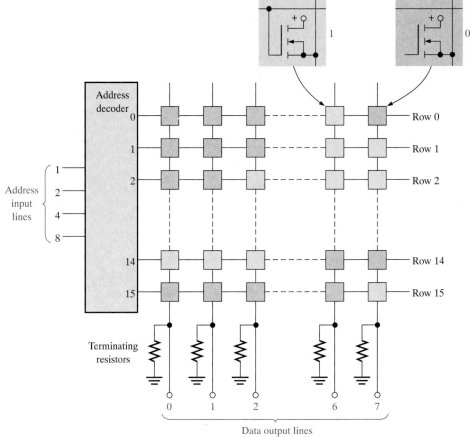

EXAMPLE 12–1   Show a basic ROM, similar to the one in Figure 12–8, programmed for 4-bit binary-to-Gray conversion.

Solution   Review Chapter 2 for the Gray code. Table 12–1 is developed for use in programming the ROM.

TABLE 12–1

Binary				Gray			
B_3	B_2	B_1	B_0	G_3	G_2	G_1	G_0
0	0	0	0	0	0	0	0
0	0	0	1	0	0	0	1
0	0	1	0	0	0	1	1
0	0	1	1	0	0	1	0
0	1	0	0	0	1	1	0
0	1	0	1	0	1	1	1
0	1	1	0	0	1	0	1
0	1	1	1	0	1	0	0
1	0	0	0	1	1	0	0
1	0	0	1	1	1	0	1
1	0	1	0	1	1	1	1
1	0	1	1	1	1	1	0
1	1	0	0	1	0	1	0
1	1	0	1	1	0	1	1
1	1	1	0	1	0	0	1
1	1	1	1	1	0	0	0

The resulting ROM array is shown in Figure 12–9. You can see that a binary code on the address input lines produces the corresponding Gray code on the output lines (columns). For example, when the binary number 0110 is applied to the address input lines, address 6, which stores the Gray code 0101, is selected.

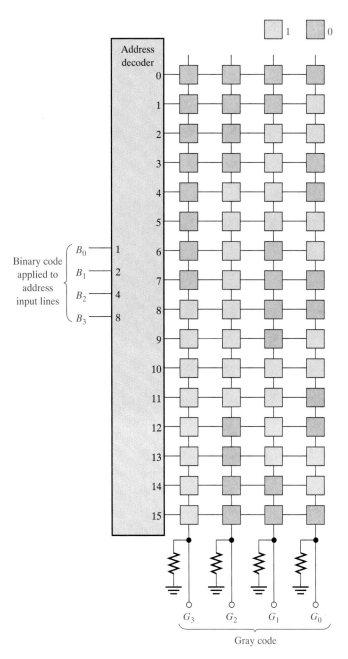

FIGURE 12–9
Representation of a ROM programmed as a binary-to-Gray code converter.

Related Exercise   Using Figure 12–9, determine the Gray code output when a binary code of 1011 is applied to the address input lines.

Internal ROM Organization

Most IC ROMs have a somewhat more complex internal organization than that in the basic simplified example just presented. To illustrate how an IC ROM is structured, a 1024-bit device with a 256 × 4 organization is used. The logic symbol is shown in Figure 12–10. When any one of 256 binary codes (eight bits) is applied to the address lines, four data bits appear on the outputs if the chip enable inputs are LOW. (There are eight address lines because $2^8 = 256$.)

Although the 256 × 4 organization of this device implies that there are 256 rows and 4 columns in the memory array, this is not the case. The memory cell array is actually a 32 × 32 matrix (32 rows and 32 columns), as shown in the block diagram in Figure 12–11.

FIGURE 12–10
A 256 × 4 ROM logic symbol. The A $\frac{0}{255}$ designator means that the 8-bit address code selects addresses 0 through 255.

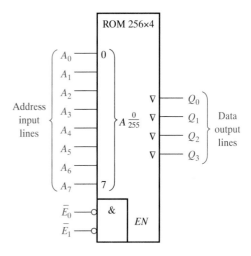

FIGURE 12–11
A typical ROM. This particular example is a 1024-bit ROM with a 256 × 4 organization based on a 32 × 32 array.

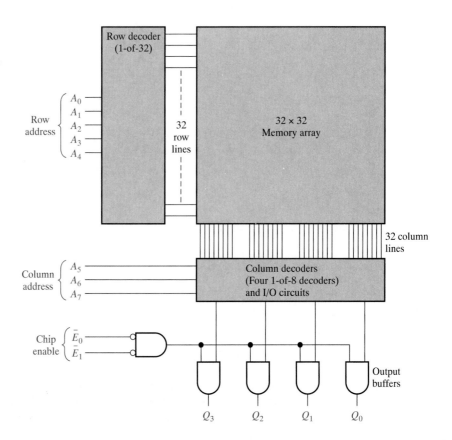

The ROM in Figure 12–11 works as follows: Five of the eight address lines (A_0 through A_4) are decoded by the row decoder (often called the Y decoder) to select one of the 32 rows. Three of the eight address lines (A_5 through A_7) are decoded by the column decoder (often called the X decoder) to select four of the 32 columns. Actually, the column decoder consists of four 1-of-8 decoders (data selectors), as shown in Figure 12–11.

The result of this structure is that when an 8-bit address code (A_0 through A_7) is applied, a 4-bit data word appears on the data outputs when the chip enable lines ($\overline{E}_0$ and $\overline{E}_1$) are LOW to enable the output buffers. This type of internal organization (architecture) is typical of IC ROMs of various capacities. In fact, the 74187 bipolar ROM has this configuration.

Tristate Outputs and Buses

The tristate outputs are indicated on logic symbols by a small inverted triangle (∇), as shown in Figure 12–10, and are used for compatibility with bus structures such as those found in microprocessor-based systems.

Physically, a **bus** is a set of conductive paths that serve to interconnect two or more functional components of a system or several diverse systems. Electrically, a bus is a collection of specified voltage levels and/or current levels and signals that allow the various devices connected to the bus to communicate and work properly together.

For example, a microprocessor is connected to memories and input/output devices by certain bus structures as illustrated in Figure 12–12. An address bus allows the microprocessor to address the memories, and the data bus provides for transfer of data between the microprocessor, the memories, and the input/output devices such as monitors, printers, keyboards, and modems. The control bus allows the microprocessor to control data transfers and timing for the various components.

FIGURE 12–12
Basic microprocessor-based system.

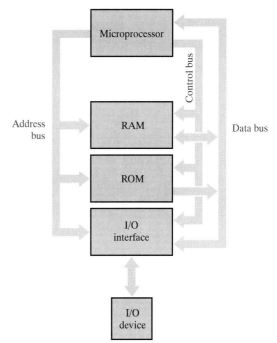

Tristate Interface to the Bus In a typical application, several devices—for example, a microprocessor, a RAM, and a ROM—are connected to one bus. For this reason, tristate logic circuits are used to interface digital devices such as memories to a bus. Figure 12–13(a) shows the logic symbol for a noninverting tristate buffer with an active-HIGH enable. Part (b) of the figure shows one with an active-LOW enable.

FIGURE 12–13
Tristate buffer symbols.

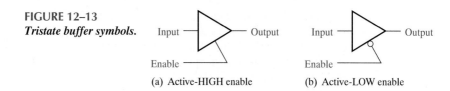

(a) Active-HIGH enable   (b) Active-LOW enable

The basic operation of a tristate buffer can be understood in terms of switching action as illustrated in Figure 12–14. When the enable input is active, the gate operates as a normal noninverting circuit. That is, the output is HIGH when the input is HIGH and LOW when the input is LOW, as shown in parts (a) and (b). The HIGH and LOW levels represent two of the states. The buffer operates in its third state when the enable input is not active. In this state the circuit acts as an open switch, and the output is completely disconnected from the input, as shown in part (c). This is sometimes called the *high-impedance* or *high-Z* state. A more detailed coverage of tristate circuitry is found in Chapter 15.

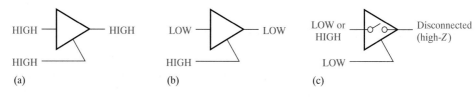

FIGURE 12–14
Tristate buffer operation.

Many microprocessors, memories, and other integrated circuit functions have tristate buffers that serve to interface with the buses. Such buffers are necessary when two or more devices are connected to a common bus. To prevent the devices from interfering with each other, the tristate buffers are used to disconnect all devices except the ones that are communicating at any given time.

ROM Access Time

A typical timing diagram that illustrates ROM access time is shown in Figure 12–15. The **access time,** t_a, of a ROM is the time from the application of a valid address code on the input lines until the appearance of valid output data. Access time can also be measured from the activation of the chip enable ($\overline{E}$) input to the occurrence of valid output data when a valid address is already on the input lines.

FIGURE 12–15
ROM access time (t_a) from address change to data output with chip enable already active.

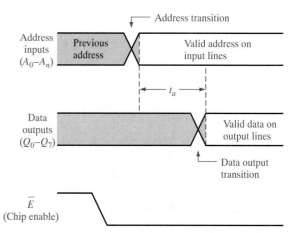

ROM in Computer Applications

ROM is used in the IBM personal computer, for example, to store what is referred to as the BIOS (Basic Input/Output Services). These are programs that are used to perform fundamental supervisory and support functions for the computer. For example, BIOS programs stored in the ROM control certain video monitor functions, provide for disk formatting, scan the keyboard for inputs, and control certain printer functions.

Another use of ROMs in computer systems is to store the language interpreter programs such as BASIC.

SECTION 12–2 REVIEW

1. What is the bit storage capacity of a ROM with a 512×8 organization?
2. Explain the purpose of tristate outputs.
3. How many address bits are required for a 2048-bit memory organized as a 256×8 memory?

12–3 ■ PROGRAMMABLE ROMs (PROMs AND EPROMs)

PROMs are basically the same as mask ROMs, once they have been programmed. The difference is that PROMs come from the manufacturer unprogrammed and are custom programmed in the field to meet the user's needs. After completing this section, you should be able to

☐ Distinguish between a mask ROM and a PROM   ☐ Describe a basic PROM storage cell   ☐ Discuss EPROMs including UV EPROMs and EEPROMs   ☐ Analyze a PROM programming cycle

PROMs

PROMs are available in both bipolar and MOS technologies and generally have 4-bit or 8-bit output word formats and bit capacities ranging in excess of 250,000. PROMs use some type of fusing process to store bits, whereby a memory link is fused open or left intact to represent a 0 or a 1. The fusing process is irreversible; once a PROM is programmed, it cannot be changed.

Figure 12–16 (next page) illustrates a MOS PROM array with fusible links. The fusible links are manufactured into the PROM between the source of each cell's transistor and its column line. In the programming process, a sufficient current is injected through the fusible link to burn it open to create a stored 0. The link is left intact for a stored 1.

Three basic fuse technologies used in PROMs are metal links, silicon links, and *pn* junctions. A brief description of each of these follows.

1. Metal links are made of a material such as nichrome. Each bit in the memory array is represented by a separate link. During programming, the link is either "blown" open or left intact. This is done basically by first addressing a given cell and then forcing a sufficient amount of current through the link to cause it to open.
2. Silicon links are formed by narrow, notched strips of polycrystalline silicon. Programming of these fuses requires melting of the links by passing a sufficient amount of current through them. This amount of current causes a high temperature at the fuse location that oxidizes the silicon and forms an insulation around the now-open link.
3. Shorted junction, or avalanche-induced migration, technology consists basically of two *pn* junctions arranged back-to-back. During programming, one of the diode junctions is avalanched, and the resulting voltage and heat cause aluminum ions to migrate and short the junction. The remaining junction is then used as a forward-biased diode to represent a data bit.

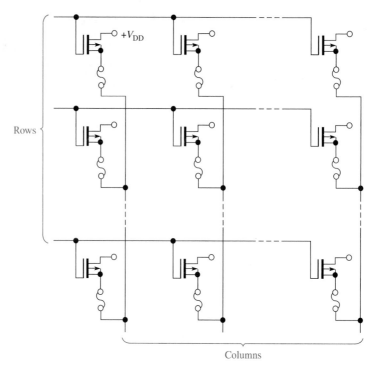

FIGURE 12–16
MOS PROM array with fusible links. (All drains are commonly connected to V_{DD}.)

PROM Programming

A PROM is normally programmed by inserting it into a special instrument called a *PROM programmer.* Basically, the programming is accomplished as shown by the simplified setup in Figure 12–17. An address is selected by the switch settings on the address lines, and then a pulse is applied to those output lines corresponding to bit locations where 0s are to be stored (the PROM starts out with all 1s). These pulses blow the fusible links, thus creating the desired bit pattern. The next address is then selected and the process is repeated. This sequence is done automatically by a software-driven PROM programmer.

FIGURE 12–17
Simplified PROM programming setup.

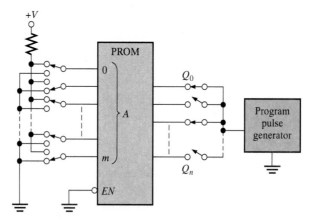

EPROMs

An EPROM is an erasable PROM. Unlike an ordinary PROM, an EPROM can be reprogrammed if an existing program in the memory array is erased first.

An EPROM uses an NMOSFET array with an isolated-gate structure. The isolated transistor gate has no electrical connections and can store an electrical charge for indefinite periods of time. The data bits in this type of array are represented by the presence or absence of a stored gate charge. Erasure of a data bit is a process that removes the gate charge.

Two basic types of erasable PROMs are the ultraviolet light erasable PROM (UV EPROM) and the electrically erasable PROM (EEPROM).

UV EPROMs   You can recognize the UV EPROM device by the transparent quartz lid on the package, as shown in Figure 12–18. The isolated gate in the FET of an ultraviolet EPROM is "floating" within an oxide insulating material. The programming process causes electrons to be removed from the floating gate. Erasure is done by exposure of the memory array chip to high-intensity ultraviolet radiation through the quartz window on top of the package. The positive charge stored on the gate is neutralized after several minutes to an hour of exposure time.

FIGURE 12–18
Ultraviolet erasable PROM package.

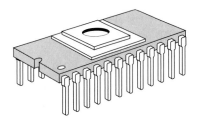

EEPROMs   An electrically erasable PROM can be both erased and programmed with electrical pulses. Since it can be both electrically written into and electrically erased, the EEPROM can be rapidly programmed and erased in circuit for reprogramming.

Two types of EEPROMs are the floating-gate MOS and the metal nitride-oxide silicon (MNOS). The application of a voltage on the control gate in the floating-gate structure permits the storage and removal of charge from the floating gate.

A Specific PROM

The 2516 is an example of a MOS EPROM device. Its operation is representative of that of other typical EPROMs. As the logic symbol in Figure 12–19 shows, this device has 2048 addresses ($2^{11} = 2048$), each with eight bits. Notice that the eight outputs are tristate (∇).

FIGURE 12–19
The logic symbol for a 2516 PROM.

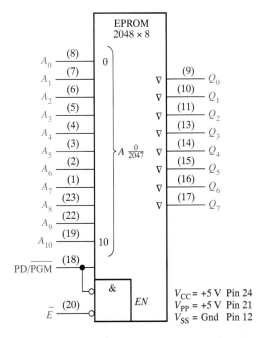

To read from the memory, the chip enable input ($\overline{E}$) must be LOW and the power-down/program (PD/$\overline{PGM}$) input LOW. When the PD/$\overline{PGM}$ input is HIGH, the device is in a low-power standby mode that reduces the current drain on the dc power supply. To erase the stored data, the device is exposed to high-intensity ultraviolet light through the transparent lid. A typical 12 mW/cm² filterless UV lamp will erase the data in about 20 to 25 minutes. As in most EPROMs after erasure, all bits are 1s. Normal ambient light contains the correct wavelength of UV light for erasure. Therefore, the transparent lid must be kept covered.

To program the device, +25 V dc is applied to V_{PP} (which is normally +5 V), and $\overline{E}$ is HIGH. The eight data bits to be programmed into a given address are applied to the outputs (Q_0 through Q_7), and the address is selected on inputs A_0 through A_{10}. Next, a 10 ms to 55 ms HIGH level pulse is applied to the PD/$\overline{PGM}$ input. The addresses can be programmed in any order.

A timing diagram for the programming mode is shown in Figure 12–20. These signals are normally produced by an EPROM programmer.

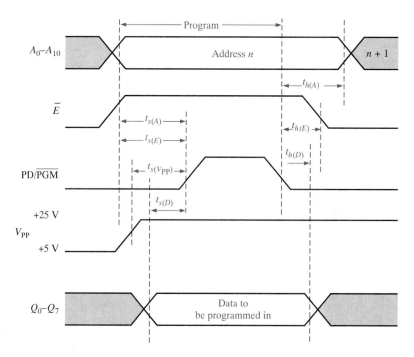

FIGURE 12–20

Timing diagram for 2516 PROM programming cycle, with critical setup times (t_s) and hold times (t_h) indicated.

SECTION 12–3 REVIEW

1. How do PROMs differ from ROMs?
2. After erasure, all bits are (1s, 0s) in a typical EPROM.
3. What is the normal mode of operation for a PROM?

12–4 ■ READ/WRITE RANDOM-ACCESS MEMORIES (RAMs)

Data can be readily written into and read from a RAM at any selected address in any sequence. When data are written into a given address in the RAM, the data previously stored at that address are destroyed and are replaced by the new data. When data are read from a given address in the RAM, the data at that address are not destroyed. This nondestructive read operation can be thought of as copying the contents of an address while leaving the contents intact. A RAM is used for short-term data storage and cannot retain stored data when power is off. After completing this section, you should be able to

☐ List the types of RAM   ☐ Describe how RAMs differ from ROMs   ☐ Explain what a SRAM is   ☐ Describe a basic SRAM storage cell   ☐ Discuss basic SRAM organization and read/write cycles   ☐ Explain what a DRAM is   ☐ Describe a basic DRAM storage cell   ☐ Discuss basic DRAM organization and the refresh operation

The RAM Family

Semiconductor RAMs are also manufactured with either bipolar or MOS technologies. Some memory devices manufactured using a combination of bipolar (TTL or ECL) and MOS are called BiMOS. Bipolar RAMs are all static RAMs (**SRAM**). **Static memory** uses storage elements such as latches, so data can be stored for an indefinite period of time as long as the power is on. Some MOS RAMs are of the static type and some are dynamic. A dynamic RAM (**DRAM**) memory is one in which data are stored on capacitors, which require periodic recharging (refreshing) to retain the data. Figure 12–21 shows the categories of RAMs.

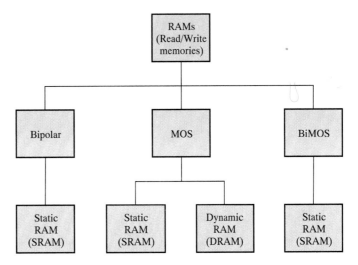

FIGURE 12–21
The semiconductor RAM family.

The Static RAM (SRAM) Cell

As mentioned before, the storage cells in a SRAM are either bipolar or MOS latches. Once a data bit is stored in a cell, it remains indefinitely (unless power is lost or a new data bit is written in). Because data are lost if the power to the memory is lost, a SRAM is a type of volatile memory.

Figure 12–22 shows a functional logic diagram of a SRAM cell. The basic operation is as follows: The cell (or a group of cells) is selected by HIGHs on the row and column lines. When the $\overline{WRITE}$ line is LOW (write), the input data bit is written into the cell by setting the cell for a 1 or resetting the cell for a 0. When the $\overline{WRITE}$ line is HIGH (read), the cell is unaffected, but the stored data bit (Q) is gated to the data-output line.

FIGURE 12–22
A generalized logic diagram for a SRAM cell (latch).

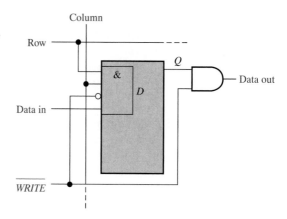

Basic Organization of a Static RAM (SRAM)

A SRAM is addressed in basically the same way as a ROM. The main difference between the organization of SRAMs and ROMs is that the SRAM has data inputs and a read/write control. To illustrate the general organization of a SRAM, a 32k × 8 bit memory is used. A logic symbol for this memory is shown in Figure 12–23.

FIGURE 12–23
Logic diagram for a 32k × 8 SRAM.

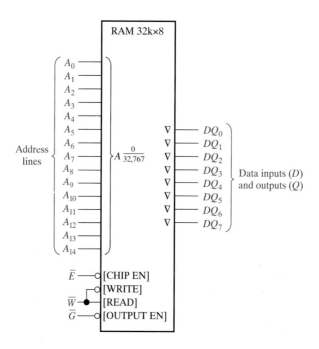

In the READ mode, the eight data bits that are stored in the selected address appear on the data output lines. In the WRITE mode, the eight data bits that are applied to the data input lines are stored at the selected address. The data input and data output lines (DQ_0 through DQ_7) are the same lines. During READ, they act as output lines (Q_0 through Q_7) and during WRITE they act as input lines (D_0 through D_7).

Memory Organization SRAMs are organized in single bits, nibbles (4 bits), bytes (8 bits), or multiple bytes (16, 24, or 32 bits) with capacities typically ranging up to 4 Mbits. Other organizations are available in some devices.

Figure 12–24 shows the organization of a typical 32k × 8 SRAM. The memory cell array is arranged in 256 rows and 128 × 8 columns (128 columns, each with 8 bits). There are actually $2^{15} = 32,768$ addresses and each address contains 8 bits. The capacity of this example memory is 32,768 bytes (typically expressed as 32 kbytes) or 262,144 bits.

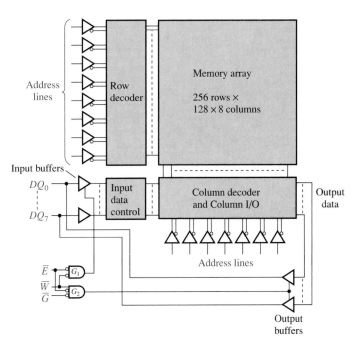

FIGURE 12–24
Basic organization of a 32k × 8 SRAM.

The SRAM in Figure 12–24 works as follows. First, the chip enable, $\overline{E}$, must be LOW for the memory to operate. Eight of the fifteen address lines are decoded by the row decoder to select one of the 256 rows. Seven of the fifteen address lines are decoded by the column decoder to select one of the 128 8-bit columns.

READ In the READ mode, the write enable input, $\overline{W}$, is inactive (HIGH) and the output enable, $\overline{G}$, is active (LOW). The input buffers are disabled by gate G_1, and the column output tristate buffers are enabled by gate G_2. Therefore, the eight data bits from the selected address are routed through the column I/O to the data lines (DQ_0 through DQ_7).

WRITE In the WRITE mode, $\overline{W}$ is active (LOW) and $\overline{G}$ is inactive (HIGH). The input buffers are enabled by gate G_1, and the output buffers are disabled by gate G_2. Therefore, the eight input data bits on the data lines are routed through the input data control and the column I/O to the selected address and stored.

Read and Write Cycles Figure 12–25 (next page) shows typical timing diagrams for a memory read cycle and a write cycle. For the read cycle shown in part (a), a valid address code is applied to the address lines for a specified time interval called the *read cycle time*, t_{RC}. Next, the chip enable ($\overline{E}$) and the output enable ($\overline{G}$) inputs go LOW. One time interval after the $\overline{G}$ input goes LOW, valid data from the selected address appear on the data lines. This time interval is called the *output enable access time*, t_{GQ}.

Two other access times for the read cycle are the *address access time*, t_{AQ}, measured from the beginning of a valid address to the appearance of valid data on the data lines and the *chip enable access time*, t_{EQ}, measured from the HIGH-to-LOW transition of $\overline{E}$ to the appearance of valid data on the data lines.

During each read cycle, one unit of data (single bit, nibble, byte, or multiple bytes depending on the memory) is read from the memory.

For the write cycle shown in Figure 12–25(b), a valid address code is applied to the address lines for a specified time interval called the *write cycle time, t_{WC}*. Next, the chip enable ($\overline{E}$) and the write enable ($\overline{W}$) inputs go LOW. The required time interval from the beginning of a valid address until the $\overline{W}$ input goes LOW is called the *address setup*

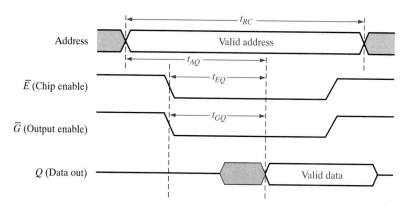

(a) Read cycle

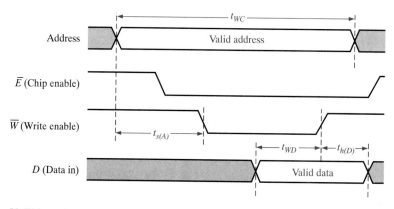

(b) Write cycle

FIGURE 12–25
Basic read and write cycle timing for the SRAM in Figure 12–24.

time, $t_{s(A)}$. The time that the $\overline{W}$ input must be LOW is the write pulse width. The time that the $\overline{W}$ input must remain LOW after valid data are applied to the data inputs is designated t_{WD}; the time that the valid input data must remain on the data lines after the $\overline{W}$ input goes HIGH is the *data hold time*, $t_{h(D)}$.

During each write cycle, one unit of data (single bit, nibble, byte, or multiple bytes depending on the memory) is written into the memory.

Specific Static RAMs

The MCM6264C This memory, shown in Figure 12–26, is an 8k × 8 bit SRAM with an organization and operation similar to the memory in Figure 12–24 that we previously discussed. The differences are that the 8k × 8 memory array consists of 256 rows and 32 8-bit columns. The 8,192 addresses are selected by thirteen address lines. Also, this memory has two chip enable inputs, $\overline{E}_1$ and E_2.

The operation of this SRAM is summarized in Table 12–2. The memory is enabled when the $\overline{E}_1$ input is LOW and the E_2 input is HIGH. During the READ mode, the $\overline{W}$ input is HIGH and the outputs are enabled by a LOW on the $\overline{G}$ input. During the WRITE mode, the $\overline{W}$ input is LOW and $\overline{G}$ can be either HIGH or LOW. Notice that the output buffers are always in the high-Z (high impedance) state except during a read cycle.

The pin assignments and package configurations for the MCM6264C are shown in Figure 12–27.

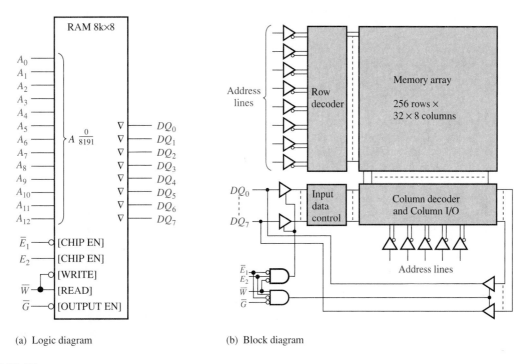

(a) Logic diagram   (b) Block diagram

FIGURE 12–26
Logic symbol and block diagram for the MCM6264C SRAM.

TABLE 12–2
Truth table for the MCM6264C SRAM (X = don't care, H = HIGH, L = LOW)

$\overline{E}_1$	E_2	$\overline{G}$	$\overline{W}$	Mode	Outputs	Cycle
H	X	X	X	Unselected	High-Z	None
X	L	X	X	Unselected	High-Z	None
L	H	H	H	Output disabled	High-Z	None
L	H	L	H	READ	D_0–D_7	Read
L	H	X	L	WRITE	High-Z	Write

FIGURE 12–27
The MCM6264C pin assignments and packages.

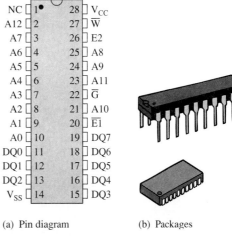

(a) Pin diagram   (b) Packages

The MCM6246   This memory, shown in Figure 12–28, is a 512k × 8 bit SRAM also with an organization and operation similar to the memory in Figure 12–24. The differences are that the 512k × 8 memory array consists of 1024 rows and 4096 columns and that the 524,288 byte addresses are selected by nineteen address lines.

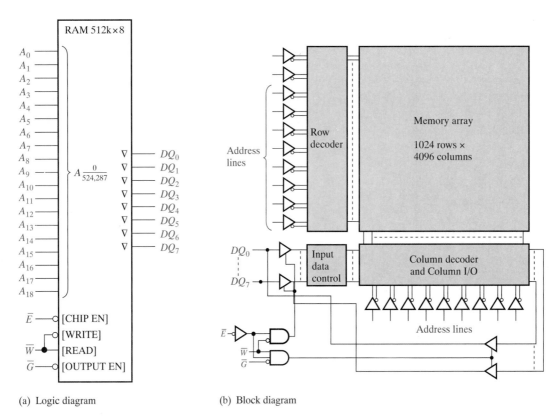

(a) Logic diagram

(b) Block diagram

FIGURE 12–28
Logic symbol and block diagram for the MCM6246 SRAM.

The operation of this SRAM is summarized in Table 12–3. The memory is enabled when the $\overline{E}$ input is LOW. During the READ mode, the $\overline{W}$ input is HIGH and the outputs are enabled by a LOW on the $\overline{G}$ input. During the WRITE mode, the $\overline{W}$ input is LOW and $\overline{G}$ can be either HIGH or LOW.

The pin assignments and package configuration for the MCM6246 are shown in Figure 12–29.

The read and write cycle timing diagrams for the MCM6264C and the MCM6246 are shown in Figure 12–30 with critical times indicated. The red labels apply to the 6264C and the blue labels to the 6246.

TABLE 12–3
Truth table for the MCM6246 SRAM.

$\overline{E}$	$\overline{G}$	$\overline{W}$	Mode	Outputs	Cycle
H	X	X	Unselected	High-Z	None
L	H	H	Output disabled	High-Z	None
L	L	H	READ	D_0–D_7	Read
L	X	L	WRITE	High-Z	Write

FIGURE 12–29
The MCM6246 pin assignments and package.

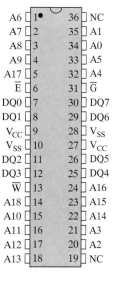

(a) Pin diagram   (b) Package

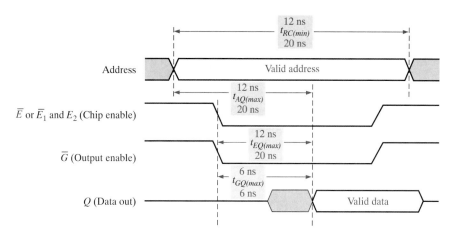

(a) Read cycle

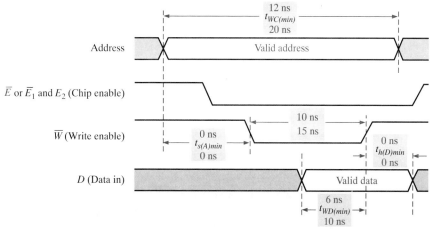

(b) Write cycle

FIGURE 12–30
Read and write cycles for the MCM6264C (red labels) and the MCM6246 (blue labels) SRAMs.

The Dynamic RAM (DRAM) Cell

Dynamic memory cells store a data bit in a small capacitor rather than in a latch. The advantage of this type of cell is that it is very simple, thus allowing very large memory arrays to be constructed on a chip at a lower cost per bit than in static memories. The disadvantage is that the storage capacitor cannot hold its charge over an extended period of time and will lose the stored data bit unless its charge is refreshed periodically. To **refresh** requires additional memory circuitry and complicates the operation of the DRAM. Figure 12–31 shows a typical DRAM cell consisting of a single MOS transistor and a capacitor.

FIGURE 12–31
A MOS dynamic RAM cell.

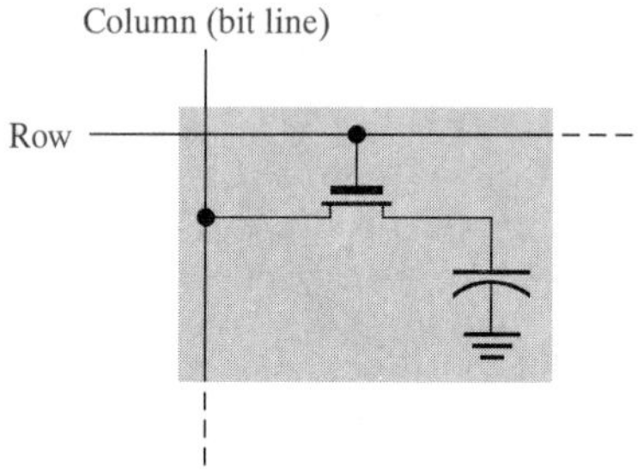

In this type of cell, the transistor acts as a switch. The basic simplified operation is illustrated in Figure 12–32 and is as follows. A LOW on the $\overline{W}$ line (WRITE mode) enables the tristate input buffer and disables the output buffer. For a 1 to be written into the cell, the D_{IN} line must be HIGH, and the transistor must be turned on by a HIGH on the row line. The transistor acts as a closed switch connecting the capacitor to the bit line. This connection allows the capacitor to charge to a positive voltage, as shown in Figure 12–32(a). When a 0 is to be stored, a LOW is applied to the D_{IN} line. If the capacitor is storing a 0, it remains uncharged, or if it is storing a 1, it discharges as indicated in Figure 12–32(b). When the row line is taken back LOW, the transistor turns off and disconnects the capacitor from the bit line, thus "trapping" the charge (1 or 0) on the capacitor.

To read from the cell, the $\overline{W}$ line is HIGH, enabling the output buffer and disabling the input buffer. When the row line is taken HIGH, the transistor turns on and connects the capacitor to the bit line and thus to the output buffer (sense amplifier), so the data bit appears on the data-output line (D_{OUT}). This process is illustrated in Figure 12–32(c).

For refreshing the memory cell, the $\overline{W}$ line is HIGH, the row line is HIGH, and the refresh line is HIGH. The transistor turns on, connecting the capacitor to the bit line. The output buffer is enabled, and the stored data bit is applied to the input of the refresh buffer, which is enabled by the HIGH on the refresh input. This produces a voltage on the bit line corresponding to the stored bit, thus replenishing the capacitor as illustrated in Figure 12–32(d).

Basic Organization of a Dynamic RAM (DRAM)

The main difference between DRAMs and SRAMs is the type of memory cell. As you have seen, the DRAM memory cell consists of one transistor and a capacitor and is much simpler than the SRAM cell. This allows much greater densities in DRAMs and results in greater bit capacities for a given chip area.

Because charge stored in a capacitor will leak off, the DRAM cell requires a frequent refresh operation to preserve the stored data bit. This requirement results in more complex circuitry than in a SRAM.

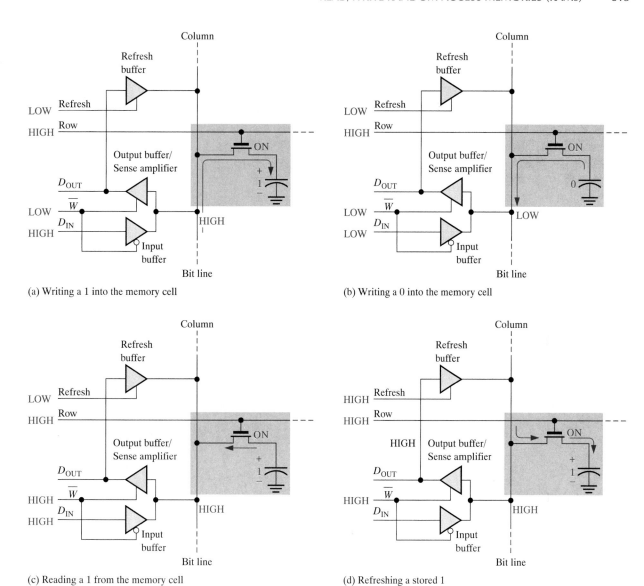

FIGURE 12–32
Basic operation of a DRAM cell.

Several features common to most DRAMs are now discussed using a generic 1M × 1 bit DRAM as an example.

Address Multiplexing DRAMs use a technique called *address multiplexing* to reduce the number of address lines. Figure 12–33 (next page) shows the block diagram of a 1,048,576-bit (1 Mbit) DRAM with a 1M × 1 organization. We will focus on the blue blocks to illustrate address multiplexing. The gray blocks represent the refresh logic.

The ten address lines are time multiplexed at the beginning of a memory cycle by the row address strobe ($\overline{RAS}$) and the column address strobe ($\overline{CAS}$) into two separate 10-bit address fields. First, the 10-bit row address is latched into the row address latch. Next, the 10-bit column address is latched into the column address latch. The row address and the column address are decoded to select one of the 1,048,576 addresses ($2^{20} = 1,048,576$) in the memory array. The basic timing for the address multiplexing operation is shown in Figure 12–34.

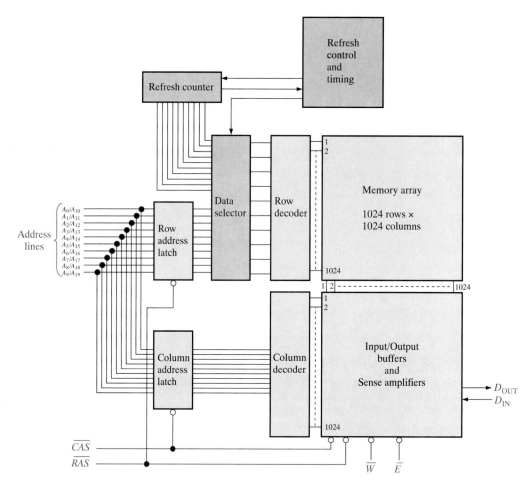

FIGURE 12–33
Simplified block diagram of a 1M × 1 DRAM.

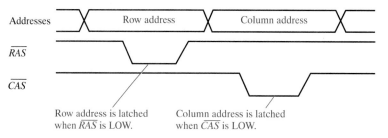

FIGURE 12–34
Basic timing for address multiplexing.

Read and Write Cycles At the beginning of each read or write memory cycle, $\overline{RAS}$ and $\overline{CAS}$ go active (LOW) to multiplex the row and column addresses into the latches and decoders. For a read cycle, the $\overline{W}$ (Write) input is HIGH. For a write cycle the $\overline{W}$ input is LOW. This is illustrated in Figure 12–35.

Page Mode Cycles In the normal read or write cycle described previously, the row address for a particular memory location is first loaded by an active-LOW $\overline{RAS}$ and then the column address for that location is loaded by an active-LOW $\overline{CAS}$. The next location is selected by another $\overline{RAS}$ followed by a $\overline{CAS}$, and so on.

Page mode allows fast successive read or write operations at each column address in a selected row. A row address is first loaded by $\overline{RAS}$ going LOW and remaining LOW

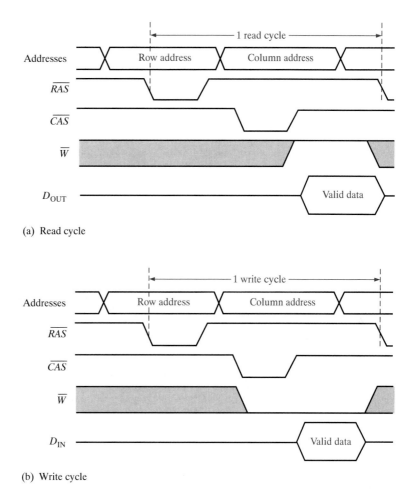

(a) Read cycle

(b) Write cycle

FIGURE 12–35
Normal read and write cycle timing.

while $\overline{CAS}$ is toggled between HIGH and LOW. A single row address is selected and remains selected while $\overline{RAS}$ is active. Each successive $\overline{CAS}$ selects another column in the selected row. So, after a page mode cycle, all of the addresses in the selected row have been read from or written into, depending on $\overline{W}$. For example, a page mode cycle for the DRAM in Figure 12–33 requires $\overline{CAS}$ to go active 1024 times for each row selected by $\overline{RAS}$. Basic page mode operation for read is illustrated by the timing diagram in Figure 12–36, p. 616.

Refresh Cycles As you know, DRAMs are based on capacitor charge storage for each bit in the memory array. This charge degrades (leaks off) with time and temperature, so each bit must be periodically refreshed (recharged) to maintain the correct bit state. Typically, a DRAM must be refreshed every 8 ms to 16 ms, although for some devices the refresh period can exceed 100 ms.

A read operation automatically refreshes all the addresses in the selected row. However, in typical applications, you cannot always predict how often there will be a read cycle and so you cannot depend on a read cycle to occur frequently enough to prevent data loss. Therefore, special refresh cycles must be implemented in DRAM systems.

Burst refresh and *distributed refresh* are the two basic refresh modes for refresh operations. In burst refresh, all rows in the memory array are refreshed consecutively each refresh period. For a memory with a refresh period of 8 ms, a burst refresh of all rows occurs once every 8 ms. The normal read and write operations are suspended during a burst refresh cycle.

In distributed refresh, each row is refreshed at intervals interspersed between normal read or write cycles. For example, the memory in Figure 12–33 has 1024 rows. For an 8 ms

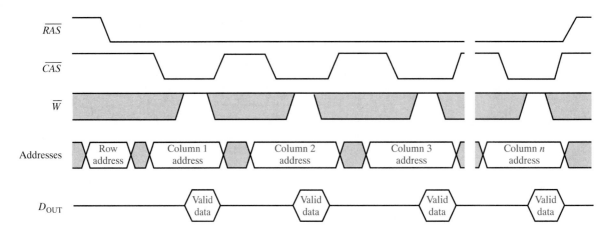

FIGURE 12–36
Basic page mode timing for read.

refresh period, each row must be refreshed every 8 ms/1024 = 7.8 μs when distributed re-fresh is used.

The two types of refresh operations are $\overline{RAS}$-only refresh and $\overline{CAS}$ before $\overline{RAS}$ refresh. $\overline{RAS}$-only refresh consists of a $\overline{RAS}$ transition to the LOW (active) state which latches the address of the row to be refreshed while $\overline{CAS}$ remains HIGH (inactive) throughout the cycle. An external counter is used to provide the row addresses for this type of operation.

The $\overline{CAS}$ before $\overline{RAS}$ refresh is initiated by $\overline{CAS}$ going LOW before $\overline{RAS}$ goes LOW. This sequence activates an internal refresh counter that generates the row address to be refreshed. This address is switched by the data selector into the row decoder.

Specific Dynamic RAMs

The MCM516100 This memory, shown in Figure 12–37(a), is a 16M × 1 bit DRAM. The 16,777,216-bit (16 Mbit) memory array consists of 4096 rows and 4096 columns. The addresses are selected by twelve multiplexed address lines. There is one data input line *(D)* and one data output line *(Q)*.

FIGURE 12–37
Logic symbols for the MCM516100 and the MCM516400 DRAMs.

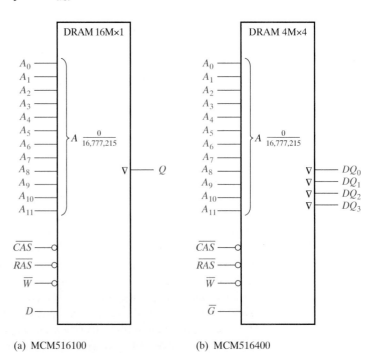

(a) MCM516100   (b) MCM516400

The MCM516400   This device, shown in Figure 12–37(b), also has a total capacity of 16,777,216 bits but is organized as a 4M $\times$ 4 memory, so that it has four data input/output lines.

SECTION 12–4
REVIEW
1. Explain how SRAMs and DRAMs differ.
2. Describe the refresh operation in a DRAM.

12–5 ■ FLASH MEMORIES

The ideal memory has high storage capacity, nonvolatility, in-system read and write capability, comparatively fast operation, and cost effectiveness. The traditional memory technologies such as ROM, PROM, EPROM, EEPROM, SRAM, and DRAM individually exhibit one or more of these characteristics, but no single technology has all of them except the flash memory. After completing this section, you should be able to

□ Discuss the basic characteristics of a flash memory   □ Compare flash memories with other types of memories

Flash memories are high-density read/write memories (high-density translates into large bit storage capacity) that are nonvolatile, which means that data can be stored indefinitely without power.

High-density means that a large number of cells can be packed into a given surface area on the chip; that is, the higher the density, the more bits that can be stored on a given size chip. This high density is achieved in flash memories with a storage cell that consists of a single floating-gate MOS transistor. A data bit is stored as charge or the absence of charge on the floating gate depending if a 0 or a 1 is stored.

Comparison of Flash Memories with Other Memories

Let's compare flash memories with other types of memories with which you are already familiar.

Flash vs. ROM, EPROM, and EEPROM   Read-only memories are high-density, non-volatile devices. However, once programmed the contents of a ROM can never be altered. Also, the initial programming is a time-consuming and costly process.

Although the EPROM is a high-density, nonvolatile memory, it can be erased only by removing it from the system and using ultraviolet light. It can be reprogrammed only with specialized equipment.

The EEPROM has a more complex cell structure than either the ROM or EPROM and so the density is not as high, although it can be reprogrammed without being removed from the system. Because of its lower density, the cost/bit is higher than ROMs or EPROMs.

A flash memory can be reprogrammed easily in the system because it is essentially a READ/WRITE device. The density of a flash memory compares with the ROM and EPROM because both have single transistor cells. A flash memory (like a ROM, EPROM, or EEPROM) is nonvolatile, which allows data to be stored indefinitely with power off.

Flash vs. SRAM   As you have learned, static random-access memories are volatile READ/WRITE devices. A SRAM requires constant power to retain the stored data. In many applications, a battery backup is used to prevent data loss if the main power source is turned off. However, since battery failure is always a possibility, indefinite retention of the stored data in a SRAM cannot be guaranteed. Because the memory cell in a SRAM is basically a latch consisting of several transistors, the density is relatively low.

A flash memory is also a READ/WRITE memory, but unlike the SRAM it is non-volatile. Also, a flash memory has a much higher density than a SRAM.

Flash vs. DRAM Dynamic random-access memories are volatile high-density READ/WRITE devices. DRAMs require not only constant power to retain data, but also the stored data must be refreshed frequently. In many applications, backup storage such as hard disk must be used with a DRAM.

Flash memories exhibit higher densities than DRAMs because a flash memory cell consists of one transistor and does not need refreshing, whereas a DRAM cell is one transistor plus a capacitor that has to be refreshed. Typically, a flash memory consumes much less power than an equivalent DRAM and can be used as a hard disk replacement in many applications.

Table 12–4 provides a summary of the comparison of the memory technologies.

TABLE 12–4

Comparison of types of memories

Memory Type	Nonvolatile	High-density	One-transistor Cell	In-system Writability
Flash	Yes	Yes	Yes	Yes
SRAM	No	No	No	Yes
DRAM	No	Yes	Yes	Yes
ROM	Yes	Yes	Yes	No
EPROM	Yes	Yes	Yes	No
EEPROM	Yes	No	No	Yes

SECTION 12–5 REVIEW

1. What types of memories are nonvolatile?
2. What is a major advantage of a flash memory over a SRAM or DRAM?

12–6 ■ MEMORY EXPANSION

Available memory can be expanded to increase the word length (number of bits in each address) or the word capacity (number of addresses) or both. Memory expansion is accomplished by adding an appropriate number of memory chips to the address, data, and control buses as explained in this section. After completing this section, you should be able to

☐ Define *word-length expansion*   ☐ Show how to expand the word length of a memory
☐ Define *word-capacity expansion*   ☐ Show how to expand the word capacity of a memory

Word-Length Expansion

To increase the **word length** of a memory, the number of bits in the data bus must be increased. For example, an 8-bit word length can be achieved by using two memories, each with 4-bit words as illustrated in Figure 12–38(a). As you can see in part (b), the 8-bit address bus is commonly connected to both memories so that the combination memory still has the same number of addresses ($2^8 = 256$) as each individual memory. The 4-bit data buses from the two memories are combined to form an 8-bit data bus. Now when an address is selected, eight bits are produced on the data bus—four from each memory.

The following example shows the details of 256 × 4 to 256 × 8 expansion.

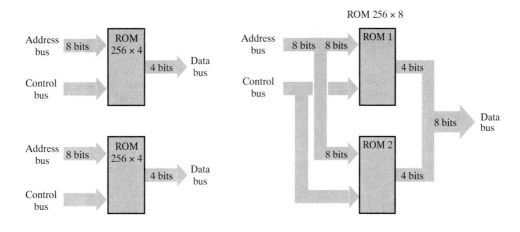

(a) Two separate 256 × 4 ROMs

(b) A 256 × 8 ROM from two 256 × 4 ROMs

FIGURE 12–38
Expansion of two 256 × 4 ROMs into a 256 × 8 ROM to illustrate word-length expansion.

EXAMPLE 12–2

Expand the 256 × 4 ROM in Figure 12–39 to form a 256 × 8 ROM.

FIGURE 12–39
A 256 × 4 ROM.

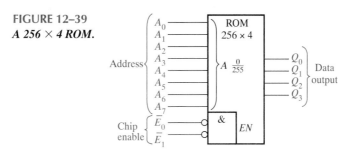

Solution Two 256 × 4 ROMs are connected as shown in Figure 12–40. Notice that a specific address is accessed in ROM 1 and ROM 2 at the same time. The four bits from a selected address in ROM 1 and the four bits from the corresponding address in ROM 2 go out in parallel to form an 8-bit word on the data bus. Also notice that a LOW on the chip enable line, $\overline{E}$, which forms a simple control bus, enables *both* memories.

FIGURE 12–40

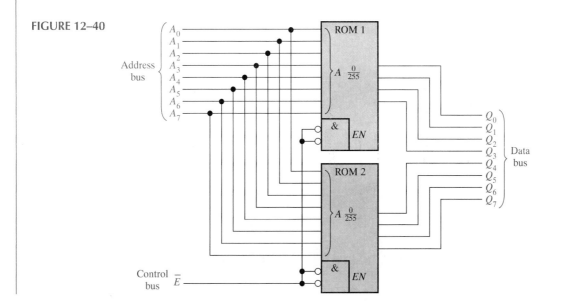

> **Related Exercise**   Describe how you would expand a 256 × 1 ROM to a 256 × 8 ROM.

EXAMPLE 12–3   Use the memories in Example 12–2 to form a 256 × 16 ROM.

Solution   In this case you need a memory that stores 256 16-bit words. Four 256 × 4 ROMs are required to do the job, as shown in Figure 12–41.

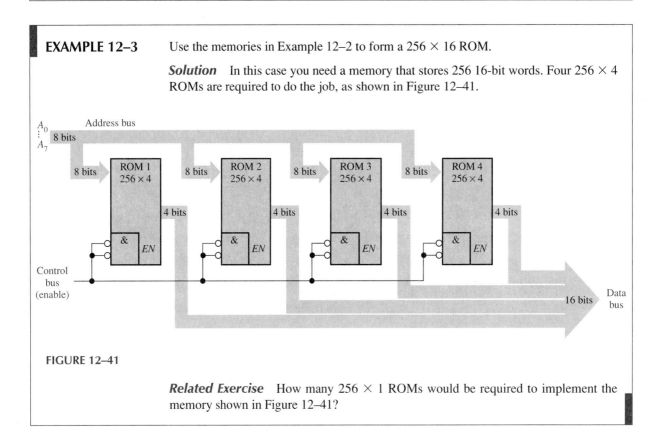

FIGURE 12–41

> **Related Exercise**   How many 256 × 1 ROMs would be required to implement the memory shown in Figure 12–41?

A ROM has only data outputs, but a RAM has both data inputs and data outputs. For word-length expansion in a RAM, the data inputs *and* data outputs form the data bus. Because data input lines and the corresponding data output lines must be connected together, tristate buffers are required. Most available IC RAMs provide internal tristate circuitry. Figure 12–42 illustrates RAM expansion to increase word length.

FIGURE 12–42
Illustration of word-length expansion with two 256 × 4 RAMs forming a 256 × 8 RAM.

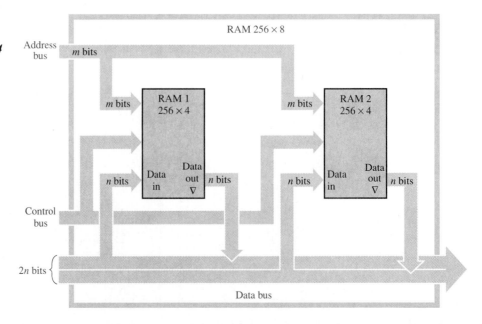

EXAMPLE 12–4   Use 16 × 4 SRAMs to create a 16 × 8 SRAM.

Solution   Two 16 × 4 SRAMs are connected as shown in Figure 12–43. When the $\overline{W}$ input is LOW, the tristate data outputs are disabled for the write operation.

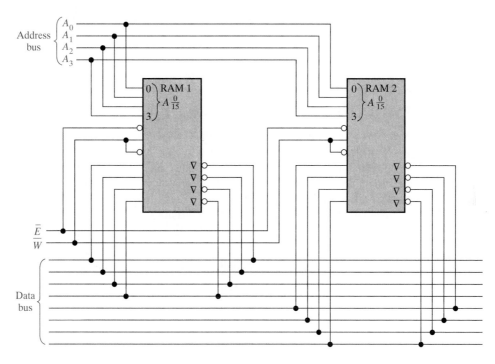

FIGURE 12–43

Related Exercise   Use MCM6264C RAMs to create an 8k × 16 RAM. See Figure 12–26.

FIGURE 12–44
Illustration of word-capacity expansion, discussed on the next page.

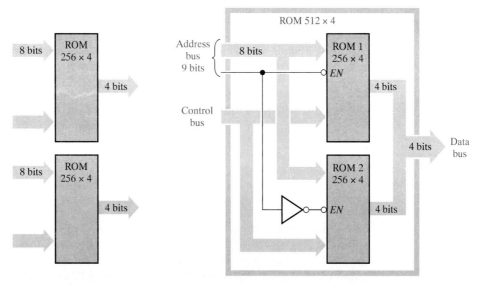

(a) Individual memories each store 256 4-bit words

(b) Memories expanded to form a 512 × 4 ROM

Word-Capacity Expansion

When memories are expanded to increase the **word capacity,** the number of addresses is increased. To achieve this increase, the number of address bits must be increased, as illustrated in Figure 12–44, p. 621, (where two 256 × 4 ROMs are expanded to form a 512 × 4 memory.

Each individual memory has eight address bits to select its 256 addresses, as shown in part (a). The expanded memory has 512 addresses and therefore requires nine address bits, as shown in part (b). The ninth address bit is used to enable the appropriate memory chip. The data bus for the expanded memory remains four bits wide. Details of this expansion are illustrated in Example 12–5.

EXAMPLE 12–5   Use 256 × 4 ROMs like the one in Figure 12–39 to implement a 512 × 4 memory.

Solution   The expanded addressing is achieved by connecting the chip enable $(\overline{E}_0)$ input to the ninth address bit (A_8), as shown in Figure 12–45. Input $\overline{E}_1$ is used as an enable input common to both memories. When the ninth address bit (A_8) is LOW, ROM 1 is selected (ROM 2 is disabled), and the eight lower-order address bits $(A_0–A_7)$ access each of the addresses in ROM 1. When the ninth address bit (A_8) is HIGH, ROM 2 is enabled by a LOW on the inverter output (ROM 1 is disabled), and the eight lower-order address bits $(A_0–A_7)$ access each of the ROM 2 addresses.

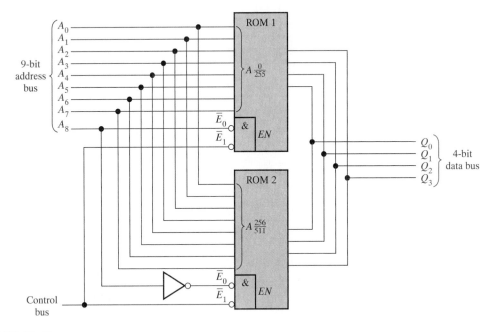

FIGURE 12–45

Related Exercise   What are the ranges of addresses in ROM 1 and in ROM 2 in Figure 12–45?

Single-In-Line Memory Modules (SIMMs)

When memory devices are combined to increase the word length, they are commonly packaged in a SIMM. A **SIMM** is essentially a small, rectangular-shaped PC board with a single row of input/output pins, as shown in Figure 12–46. SIMMs are typically available in sizes from 1M × 8 to 8M × 40. The SIMM shown in Figure 12–46 uses eight 4M × 1 bit memory chips to form a 4M × 8 bit memory module and is an example of word-length expansion.

FIGURE 12-46
A 4M × 8 single-in-line memory module (SIMM).

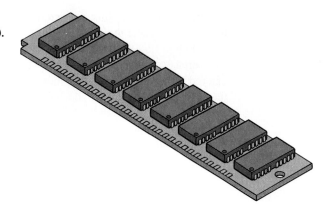

**SECTION 12–6
REVIEW**

1. How many 16k × 1 RAMs are required to achieve a memory with a word capacity of 16k and a word length of eight bits?
2. To expand the 16k × 8 memory in question 1 to a 32k × 8 organization, how many more 16k × 1 RAMs are required?
3. What does SIMM stand for?

12–7 ■ SPECIAL TYPES OF MEMORIES

In this section, the first in–first out (FIFO) memory, the last in–first out (LIFO) memory, the memory stack, and the charge-coupled device memory are covered. After completing this section, you should be able to

□ Describe a FIFO memory □ Describe a LIFO memory □ Discuss memory stacks
□ Explain how to use a portion of RAM as a memory stack □ Describe a basic CCD memory

First In–First Out (FIFO) Memories

This type of memory is formed by an arrangement of shift registers. The term **FIFO** refers to the basic operation of this type of memory, in which the first data bit written into the memory is the first to be read out.

One important difference between a conventional shift register and a FIFO register is illustrated in Figure 12–47. In a conventional register, a data bit moves through the regis-

Conventional shift register						FIFO shift register					
Input	X	X	X	X	Output	Input	—	—	—	—	Output
0	0	X	X	X	→	0	—	—	—	0	→
1	1	0	X	X	→	1	—	—	1	0	→
1	1	1	0	X	→	1	—	1	1	0	→
0	0	1	1	0	→	0	0	1	1	0	→

X = unknown data bits.
In a conventional shift register, data stay to the left until "forced" through by additional data.

— = empty positions.
In a FIFO shift register, data "fall" through (go right).

FIGURE 12–47
Comparison of conventional and FIFO register operation.

ter only as new data bits are entered; in a FIFO register, a data bit immediately goes through the register to the right-most bit location that is empty.

Figure 12–48 is a block diagram of a FIFO serial memory. This particular memory has four serial 64-bit data registers and a 64-bit control register (marker register). When data are entered by a shift-in pulse, they move automatically under control of the marker register to the empty location closest to the output. Data cannot advance into occupied positions. However, when a data bit is shifted out by a shift-out pulse, the data bits remaining in the registers automatically move to the next position toward the output. In an asynchronous FIFO, data are shifted out independent of data entry, with the use of two separate clocks.

FIGURE 12–48
Block diagram of a typical FIFO serial memory.

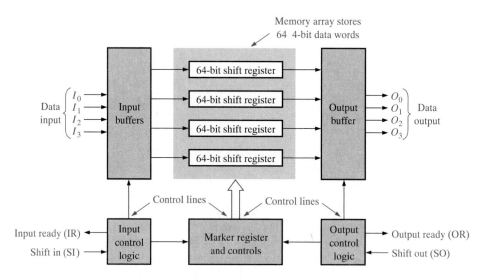

FIFO Applications

One important application area for the FIFO register is the case in which two systems of differing data rates must communicate. Data can be entered into a FIFO register at one rate and taken out at another rate. Figure 12–49 illustrates how a FIFO register might be used in these situations.

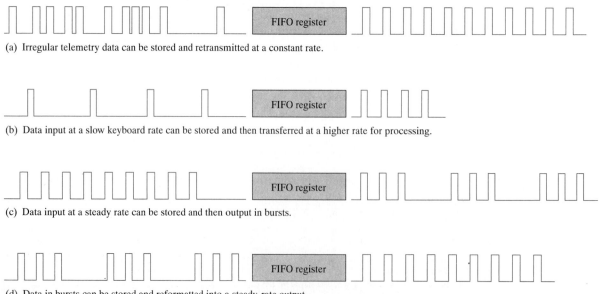

(a) Irregular telemetry data can be stored and retransmitted at a constant rate.

(b) Data input at a slow keyboard rate can be stored and then transferred at a higher rate for processing.

(c) Data input at a steady rate can be stored and then output in bursts.

(d) Data in bursts can be stored and reformatted into a steady-rate output.

FIGURE 12–49
The FIFO register in data-rate buffering applications.

Last In–First Out (LIFO) Memories

The last in–first out (**LIFO**) memory is found in applications involving microprocessors and other computing systems. It allows data to be stored and then recalled in reverse order; that is, the last data byte to be stored is the first data byte to be retrieved.

Stacks   A LIFO memory is commonly referred to as a push-down stack. In some systems, it is implemented with a group of registers as shown in Figure 12–50. A stack can consist of any number of registers, but the register at the top is called the *top-of-stack.*

FIGURE 12–50
Register stack.

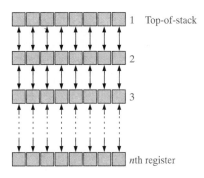

To illustrate the principle, a byte of data is loaded in parallel onto the top of the stack. Each successive byte pushes the previous one down into the next register. This process is illustrated in Figure 12–51. Notice that the new data byte is always loaded into the top register and the previously stored bytes are pushed deeper into the stack. The name *push-down stack* comes from this characteristic.

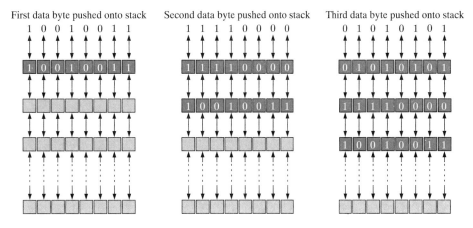

FIGURE 12–51
Pushing data onto the stack.

Data bytes are retrieved in the reverse order. The last byte entered is always at the top of the stack, so when it is pulled from the stack, the other bytes pop up into the next higher locations. This process is illustrated in Figure 12–52 (next page).

RAM Stack   Another approach to LIFO memory used in some microprocessor-based systems is the allocation of a section of RAM as the stack rather than the use of a dedicated set of registers.

Consider a random-access memory that is byte oriented—that is, one in which each address contains eight bits—as illustrated in Figure 12–53. The binary address 0000000000001111, for example, can be written as 000F in hexadecimal. A 16-bit address can have a *minimum* hexadecimal value of 0000_{16} and a *maximum* value of $FFFF_{16}$. With

Initially storing 3 data bytes. The last byte in is at top-of-stack.

After third byte is pulled from stack, the second byte that was stored pops up to the top-of-stack.

After second byte is pulled from stack, the first byte that was stored pops up to the top-of-stack.

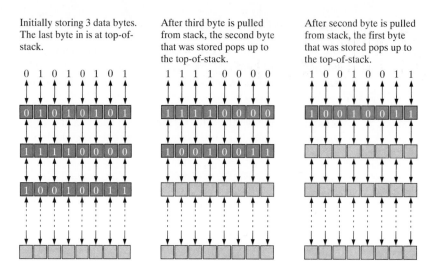

FIGURE 12–52
Pulling data from the stack.

FIGURE 12–53
Representation of a 64 kbyte memory with the 16-bit addresses expressed in hexadecimal.

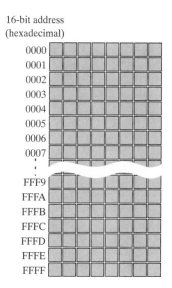

this notation, a 64 kbyte memory array can be represented as shown in Figure 12–53. The lowest memory address is 0000_{16} and the highest address is $FFFF_{16}$.

Now, consider a section of RAM set aside for use as a stack. A special separate register called a *stack pointer* contains the address of the top of the stack, as illustrated in Figure 12–54. A 4-digit hexadecimal representation is used for the binary addresses. In the figure, the addresses are chosen arbitrarily for purposes of illustration.

Now let's see how data are pushed onto the stack. A data byte is stored by a normal memory write operation at address $00FF_{16}$, which is the top of the stack as shown in Figure 12–54(a). The stack pointer is then decremented (decreased by 1) to $00FE_{16}$. This moves the top of the stack to the next lower memory address, as shown in Figure 12–54(b). Notice that the top of the stack is not stationary as in the fixed register stack but moves downward (to lower addresses) in the RAM as data bytes are stored. Figure 12–54(c) and (d) show two more bytes being pushed onto the stack. After the third byte is stored, the top of the stack is at $00FC_{16}$.

Figure 12–55 illustrates pulling data out of the RAM stack. The last data byte stored at the top-of-stack (address $00FC_{16}$) is read as indicated in part (a). The stack pointer is

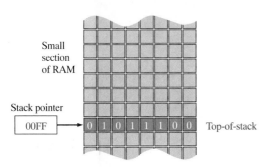

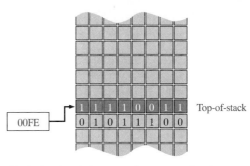

(a) Stack pointer addresses top-of-stack and data byte is pushed onto stack from the data bus.

(b) Stack pointer is decremented to next top-of-stack and second data byte is pushed onto stack.

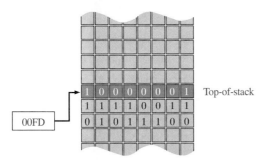

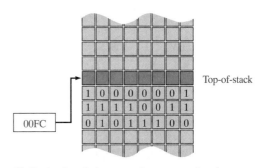

(c) Stack pointer is decremented to next top-of-stack and third data byte is pushed onto stack.

(d) Stack pointer is decremented to next top-of-stack and waits for fourth data byte.

FIGURE 12–54
Illustration of the pushing of data onto a RAM stack.

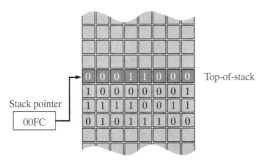

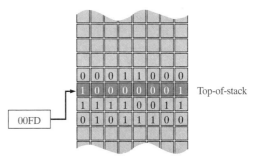

(a) Last (fourth) byte in is pulled (read) from top-of-stack.

(b) Stack pointer is incremented and third data byte in is pulled from top-of-stack .

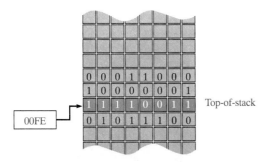

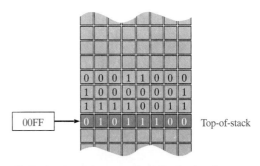

(c) Stack pointer is incremented and second data byte in is pulled from top-of-stack.

(d) Stack pointer is incremented and first data byte in is pulled from top-of-stack.

FIGURE 12–55
Illustration of the pulling of data out of the RAM stack.

then incremented (increased by 1) to address $00FD_{16}$ and a read operation is performed as shown in part (b). After this byte is read, the stack pointer is again incremented to $00FE_{16}$, and another read operation is performed, and so on, as shown in parts (c) and (d). Keep in mind that RAMs are nondestructive when read, so the data byte still remains in the memory after a read operation. A data byte is destroyed only when a new byte is written over it.

A RAM stack can be of any depth, depending on the number of continuous memory addresses not needed for any other use.

CCD Memories

The **CCD** (charge-coupled device) memory stores data as charges on capacitors. Unlike the dynamic RAM, however, the storage cell does not include a transistor. High density is the main advantage of CCDs.

The CCD memory consists of long rows of semiconductor capacitors, called *channels*. Data are entered into a channel serially by depositing a small charge for a 0 and a large charge for a 1 on the capacitors. These charge packets are then shifted along the channel by clock signals as more data are entered.

As with the DRAM, the charges must be refreshed periodically. This process is done by shifting the charge packets serially through a refresh circuit. Figure 12–56 shows the basic concept of a CCD channel. Because data are shifted serially through the channels, the CCD memory has a relatively long access time.

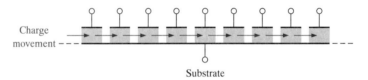

FIGURE 12–56
A CCD (charge-coupled device) channel.

SECTION 12–7 REVIEW	1. What is a FIFO memory?
	2. What is a LIFO memory?
	3. What does the term *CCD* stand for.

12–8 ■ MAGNETIC AND OPTICAL MEMORIES

In this section, we will cover the fundamentals of magnetic disks, tapes, magneto-optic disks, and other laser disks. These storage media are very important, particularly in computer applications, where they are used for nonvolatile mass storage. After completing this section, you should be able to

☐ List the types of magnetic memory ☐ Describe the floppy disk ☐ Describe the hard disk ☐ Describe how magnetic tape stores data ☐ Explain the basic principle of magneto-optic disks ☐ Define CD-ROM ☐ Define WORM memory

Types of Magnetic Memories

Figure 12–57 shows the categories of magnetic memory. Magnetic storage devices are used in computer systems primarily for mass data storage. Magnetic memories have a much slower access time (time to read data) than semiconductor memories; thus their ap-

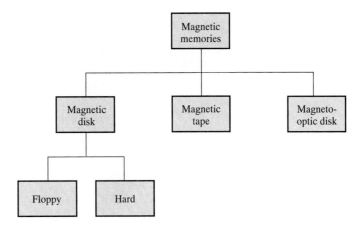

FIGURE 12–57
The magnetic memory family.

plications are limited to permanent or semipermanent storage of large amounts of data. Magnetic memories fall into three basic categories: magnetic disk, magneto-optic disk, and tape.

Two basic types of magnetic disks are used in computer systems: floppy, or flexible, disks and hard, or fixed, disks. Basically, the **floppy disk** provides less storage capacity than the hard disk, but the floppy is transportable, and the 3.5-in. variety can be carried in a shirt pocket. The **hard disk** provides a very large storage capacity that can be accessed relatively fast.

Two basic types of tape storage that are sometimes used in computer systems are the cassette tape and the reel-to-reel tape. Although they are not used as widely as they once were, tapes provide an inexpensive method for mass data storage. However, access time is extremely long compared to disks.

The **magneto-optic** disk is a technology that uses an electromagnet and laser beams to write and read data. This type of optical disk can store much more data than a magnetic disk and is less expensive in terms of cost per bit. Also, the optical disk is more durable than the magnetic disk. Other optical disks, although not magnetic, are related to the magneto-optic disk because of the use of laser beams. These are the **CD-ROM** and the **WORM** (write once–read many) disks.

Floppy Disks

There are two types of floppy disks: the older 5.25 in. floppy (minifloppy) and the more current 3.5 in. floppy (microfloppy). They are called *floppies* because the disks themselves are made of a flexible polyester material with a thin magnetic layer on both sides.

A protective flexible jacket encloses the 5.25 in. disk. There are cutout areas in the jacket for the drive spindle, the read/write head, and the index position sensor. The index hole establishes a reference point for all the tracks on the disk. As the disk rotates within the stationary jacket, the read/write head makes contact with the magnetic surface through the access window. Covering the write-protect notch with tape prevents data from being written or unintentionally erased so that the disk can be used for read-only applications if desired. Figure 12–58(a) illustrates the basic construction of the 5.25 in. disk.

The 3.5 in. flexible disk is enclosed in a rigid plastic jacket. A spring-loaded door covers the access window, and it remains closed until the disk is inserted into a disk drive to protect the disk from dust and fingerprints. A metal hub has one hole to center the disk and another for spinning it. Indexing is accomplished by a magnetic signal recorded on the disk rather than a hole as in the 5.25 in. disk. The write-protect tab can be moved to the open position to prevent unintentional writing or erasure. Figure 12–58(b) shows the basic construction of the 3.25 in. disk.

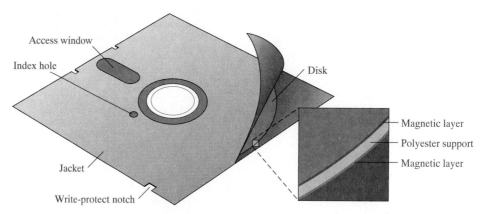

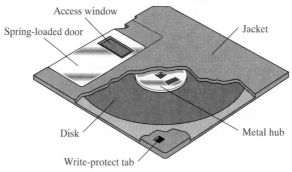

(a) 5.25 in. floppy disk

(b) 3.5 in. floppy disk

FIGURE 12–58
Magnetic floppy disks.

General Floppy-Disk Track and Sector Format

Floppy disks are organized into concentric tracks, which are divided into sectors. Figure 12–59 illustrates a general floppy-disk format for a nine-sector disk. Most disks have 80 tracks per side, although this figure varies. The blow-up of one track in one sector shows the data block followed by an error-detection code and an intersector gap. The data block is preceded by a predata gap, a data field header that identifies the sector, and a synchronizing marker. In terms of storage capacity, the more widely used 3.5 in. disk is available in a double density version (720 kbytes) and a high-density version (1.44 Mbytes). The double-density (DD) disk has 80 tracks per side with 9 sectors, and the high-density (HD) disk has 80 tracks per side with 18 sectors.

The Hard Disk

Hard disks are contained in a sealed drive package such as the one shown in Figure 12–60. The read/write head, unlike that of a floppy, floats above the disk surface on a layer of air. Because of the extremely small spacing between the disk and the head (about 10^{-6} in.), a very clean dust-free environment is essential to reliable operation. This requirement is the reason for using a sealed enclosure.

The storage capacities of typical hard disks vary from 20 megabytes to several gigabytes. Also, in part because of a higher rotational speed, access to the data is much faster than for a floppy disk drive.

Hard disks are found in both 3.5 in. and 5.25 in sizes. These disks are formatted into tracks and sections just as floppy disks are. The typical hard disk has over 300 tracks, and the number of sectors varies depending on the system.

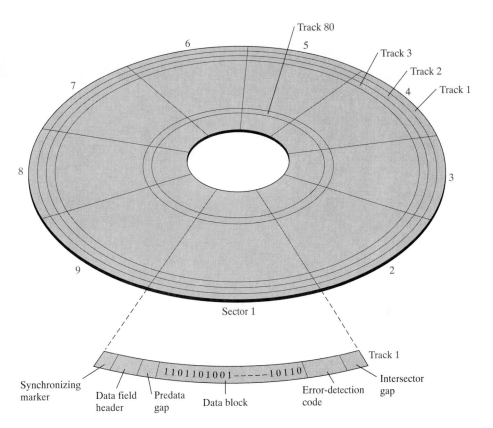

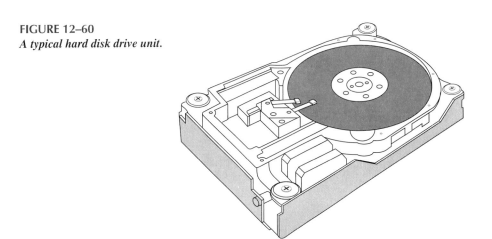

FIGURE 12–59
General format of a floppy disk.

FIGURE 12–60
A typical hard disk drive unit.

Magnetic Tape

Compared with disks, magnetic tape is a very slow storage medium in terms of access time because data are accessed serially rather than by random selection. That is, to get to a given block of data, all data preceding it must be read through.

On the nine-track tape format depicted in Figure 12–61, related bytes of data are grouped together in what is called a *record* and stored in one or more data blocks along the tape. Since the read/write head cannot jump from place to place on the tape as in a disk, many feet of tape may have to be scanned to locate a particular piece of data. Therefore, tapes require an average access time of up to 30 s to find a record.

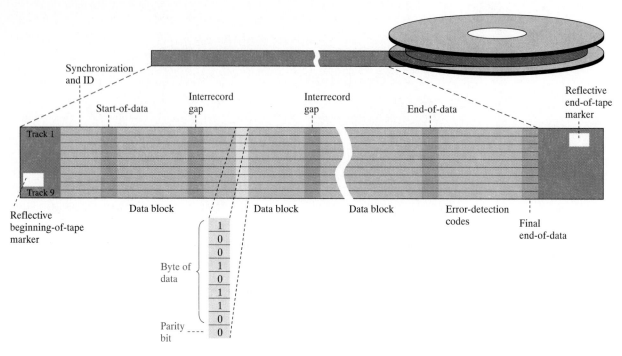

FIGURE 12–61
Basic magnetic tape format.

For the nine-track format shown in Figure 12–61 eight data bits, each in a separate track, are recorded simultaneously by eight write heads across the tape. A ninth write head records a parity bit for each data byte. The data are read by nine read heads, one for each track across the tape.

The tape illustrated in Figure 12–61 shows a typical record that is identified by a synchronization and identification (ID) code. A given record, of which there are many on a tape, typically contains numerous data blocks.

Read/Write Head Principles

A simplified diagram of the magnetic surface read/write operation is shown in Figure 12–62. A data bit (1 or 0) is written on the magnetic surface by the magnetization of a small segment of the surface as it moves by the write head. The direction of the magnetic flux lines is controlled by the direction of the current pulse in the winding, as shown in Figure 12–62(a). At the air gap in the write head, the magnetic flux takes a path through the surface of the storage device. This magnetizes a small spot on the surface in the direction

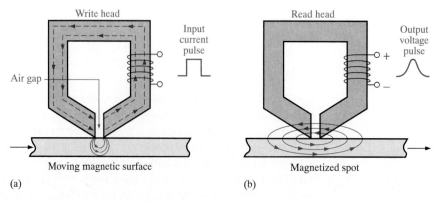

FIGURE 12–62
Read/write function on a magnetic surface.

of the field. A magnetized spot of one polarity represents a binary 1, and one of the opposite polarity represents a binary 0. Once a spot on the surface is magnetized, it remains until written over with an opposite magnetic field.

When the magnetic surface passes a read head, the magnetized spots produce magnetic fields in the read head, which induce voltage pulses in the winding. The polarity of these pulses depends on the direction of the magnetized spot and indicates whether the stored bit is a 1 or a 0. This process is illustrated in Figure 12–62(b). Often the read and write heads are combined into a single unit.

The Magneto-Optic Disk

The magneto-optic disk uses an electromagnet and laser beams to read and write (record) data on a magnetic surface. Magneto-optic disks are formatted in tracks and sectors similar to magnetic floppy disks and hard disks. However, because of the ability of a laser beam to be precisely directed to an extremely small spot, magneto-optic disks are capable of storing much more data than standard magnetic hard disks.

Figure 12–63(a) illustrates a small cross-sectional area of a disk before recording, with an electromagnet positioned above it. Tiny magnetic particles, represented by the arrows, are all magnetized in the same direction.

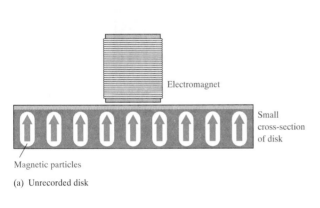

(a) Unrecorded disk

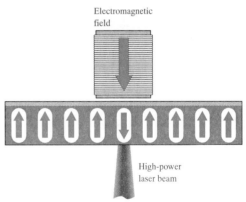

(b) Writing on the disk. High-power laser beam heats the spot and the magnetic particle aligns with the external electromagnetic field.

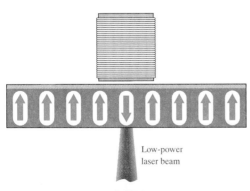

(c) Reading from the disk. Low-power laser beam reflects off of reversed polarity particle and its polarization shifts. If the particle is not reversed, the polarization of the reflected beam is unchanged.

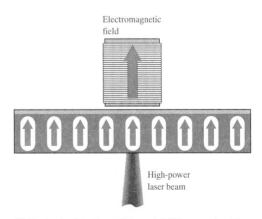

(d) Erasing the disk. External magnetic field is reversed and the high-power laser beam causes the magnetic particle to align with the external beam restoring its original polarity.

FIGURE 12–63

Basic concept of the magneto-optic disk.

Writing (recording) on the disk is accomplished by applying an external magnetic field opposite to the direction of the magnetic particles as indicated in Figure 12–63(b) and then directing a high-power laser beam to heat the disk at a precise point where a binary 1 is to be stored. The disk material, a magneto-optic alloy, is highly resistant to magnetization at room temperature; but at the spot where the laser beam heats the material, the inherent direction of magnetism is reversed by the external magnetic field produced by the electromagnet. At points where binary 0s are to be stored, the laser beam is not applied and the inherent upward direction of the magnetic particle remains.

As illustrated in Figure 12–63(c), reading data from the disk is accomplished by turning off the external magnetic field and directing a low-power laser beam at a spot where a bit is to be read. Basically, if a binary 1 is stored at the spot (reversed magnetization), the laser beam is reflected and its polarization is shifted; but if a binary 0 is stored, the polarization of the reflected laser beam is unchanged. A detector senses the difference in the polarity of the reflected laser beam to determine if the bit being read is a 1 or a 0.

Figure 12–63(d) shows that the disk is erased by restoring the original magnetic direction of each particle by reversing the external magnetic field and applying the high-power laser beam.

Two other types of optical disks are related to the magneto-optical disk because they also use laser beams to access data. However, these disks are not magnetic.

CD-ROM The compact-disk read-only memory comes from the factory with prerecorded data which cannot be erased. The data are stored on plastic disk in the form of microscopic pits in the surface and are read by a laser beam. When the laser beam is focused on a flat spot, most of it is reflected back but when it is focused on a pit, much of the light is scattered and very little reflected. A detector senses the difference in the reflected laser beam and decodes the data accordingly. In comparison with a floppy disk that may have up to 100 tracks per inch and a hard magnetic disk that may have several hundred tracks per inch, the CD-ROM has about 16,000 tracks per inch.

WORM The write once–read many disk comes from the factory unrecorded so that the user can record data one time, after which it is permanent and cannot be erased. To write data, a high-power laser beam is used to burn microscopic pits in the surface material. After that, a low-power beam is used to read the data based on beam reflection just as in the CD-ROM.

SECTION 12–8 REVIEW	**1.** List the major types of magnetic memories. **2.** What are the two physical sizes of floppy disks currently in use? **3.** Generally, how is a magnetic disk organized? **4.** What is the major disadvantage of magnetic tape compared to disk? **5.** How are data written on and read from a magneto-optic disk?

12–9 ■ TESTING AND TROUBLESHOOTING

Because memories can contain large numbers of storage cells, testing each cell can be a very lengthy and frustrating process. Fortunately, memory testing is usually an automated process performed with a programmable test instrument or with the aid of software for in-system testing. Most microprocessor-based systems provide automatic memory testing as part of their system software. After completing this section, you should be able to

☐ Discuss the checksum method of testing ROMs ☐ Discuss the checkerboard pattern method of testing RAMs

ROM Testing

Since ROMs contain known data, they can be checked for the correctness of the stored data by reading each data word from the memory and comparing it with a data word that is known to be correct. One way of doing this is illustrated in Figure 12–64. This process requires a reference ROM that contains the same data as the ROM to be tested. A special test instrument is programmed to read each address in both ROMs simultaneously and to compare the contents. A flowchart in Figure 12–65 illustrates the basic sequence.

Checksum Method   Although the previous method checks each ROM address for correct data, it has the disadvantage of requiring a reference ROM for each different ROM to be tested. Also, a failure in the reference ROM can produce a fault indication.

FIGURE 12–64
Block diagram for a complete contents check of a ROM.

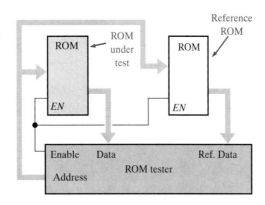

FIGURE 12–65
Flowchart for a complete contents check of a ROM.

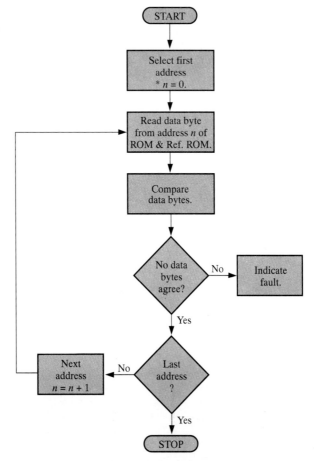

* *n* is the address number.

In the checksum method a number, the sum of the contents of all the ROM addresses, is stored in a designated ROM address when the ROM is programmed. To test the ROM, the contents of all the addresses except the checksum are added, and the result is compared with the checksum stored in the ROM. If there is a difference, there is definitely a fault. If the checksums compare, the ROM is most likely good. However, there is a remote possibility that a combination of bad memory cells could cause the checksums to compare.

This process is illustrated in Figure 12–66 with a simple example. The checksum in this case is produced by taking the sum of each column of data bits and discarding the carries. This is actually an XOR sum of each column. The flowchart in Figure 12–67 illustrates the basic checksum test.

The checksum test can be implemented with a special test instrument, or it can be incorporated as a test routine in the built-in (system) software of microprocessor-based systems. In that case, the ROM test routine is automatically run on system start-up.

FIGURE 12–66

Simplified illustration of a programmed ROM with the checksum stored at a designated address.

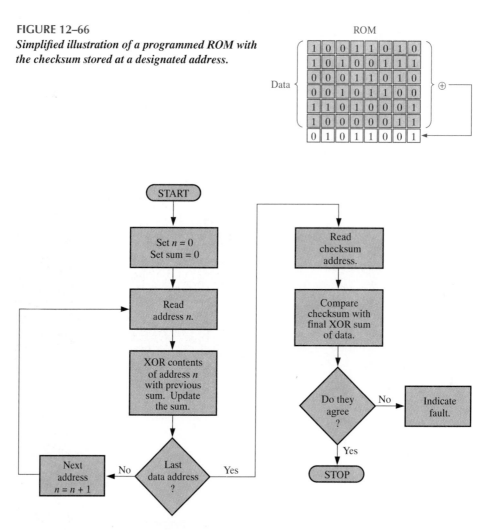

FIGURE 12–67

Flowchart for a basic checksum test.

RAM Testing

To test a RAM for its ability to store both 0s and 1s in each cell, first 0s are written into all the cells in each address and then read out and checked. Next, 1s are written into all the cells in each address and then read out and checked. This basic test will detect a cell that is stuck in either a 1 state or a 0 state.

Some memory faults cannot be detected with the all-0s–all-1s test. For example, if two adjacent memory cells are shorted, they will always be in the same state, both 0s or both 1s. Also, the all-0s–all-1s test is ineffective if there are internal noise problems such that the contents of one or more addresses are altered by a change in the contents of another address.

The Checkerboard Pattern Test One way to more fully test a RAM is by using a checkerboard pattern of 1s and 0s, as illustrated in Figure 12–68. Notice that all adjacent cells have opposite bits. This pattern checks for a short between two adjacent cells, because if there is a short, both cells will be in the same state.

After the RAM is checked with the pattern in Figure 12–68(a), the pattern is reversed, as shown in part (b). This reversal checks the ability of all cells to store both 1s and 0s.

A further test is to alternate the pattern one address at a time and check all the other addresses for the proper pattern. This test will catch a problem in which the contents of an address are dynamically altered when the contents of another address change.

A basic procedure for the checkerboard test is illustrated by the flowchart in Figure 12–69. The procedure can be implemented with the system software in microprocessor-based systems so that either the tests are automatic when the system is powered up or they can be initiated from the keyboard.

1	0	1	0	1	0	1	0
0	1	0	1	0	1	0	1
1	0	1	0	1	0	1	0
0	1	0	1	0	1	0	1
1	0	1	0	1	0	1	0
0	1	0	1	0	1	0	1
1	0	1	0	1	0	1	0
0	1	0	1	0	1	0	1
1	0	1	0	1	0	1	0
0	1	0	1	0	1	0	1
1	0	1	0	1	0	1	0
0	1	0	1	0	1	0	1

(a)

0	1	0	1	0	1	0	1
1	0	1	0	1	0	1	0
0	1	0	1	0	1	0	1
1	0	1	0	1	0	1	0
0	1	0	1	0	1	0	1
1	0	1	0	1	0	1	0
0	1	0	1	0	1	0	1
1	0	1	0	1	0	1	0
0	1	0	1	0	1	0	1
1	0	1	0	1	0	1	0
0	1	0	1	0	1	0	1
1	0	1	0	1	0	1	0

(b)

FIGURE 12–68
The RAM checkerboard test pattern.

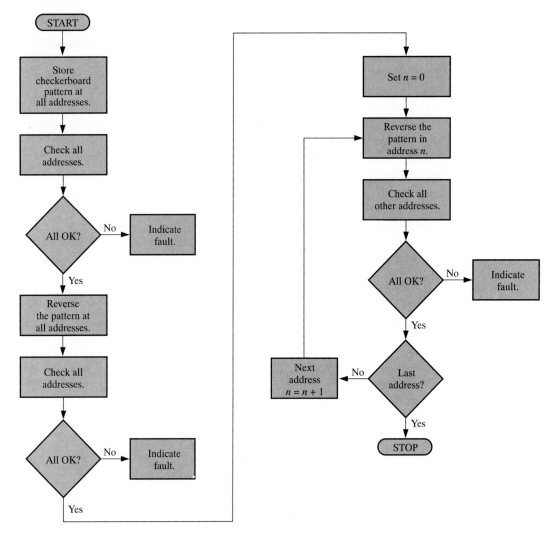

FIGURE 12-69
Flowchart for basic RAM checkerboard test.

SECTION 12-9 REVIEW	**1.** Describe the checksum method of ROM testing. **2.** Why can the checksum method not be applied to RAM testing? **3.** List the three basic faults that the checkerboard pattern test can detect in a RAM.

12-10 ■ DIGITAL SYSTEM APPLICATION

In this system application, you will complete the security entry system. In Chapter 10, the security code logic was developed. In this section, the memory logic will be developed and the two parts of the system interfaced. After completing this section, you should be able to

☐ Implement the memory portion of the system and develop a logic diagram
☐ Interface the memory logic with the security code logic and with the external switches ☐ Explain the overall operation of the system ☐ Develop a test procedure for the memory board ☐ Develop a test procedure for the complete system

General System Operation

A basic block diagram of the complete security entry system is shown in Figure 12–70. The security code logic was developed in Chapter 10 and you may want to review that material before proceeding.

The memory logic uses a RAM to store the 4-digit entry code in BCD format. In the switch-selected programming mode, four digits are entered into the memory from the keypad. Once stored in the memory, the four BCD digits become the permanent entry code. If it is necessary to change the entry code, the memory is reprogrammed with a new one. During the programming mode, the security code logic is inhibited.

The memory is programmed by first putting the system into the program mode with the program mode switch and using the store switch and keypad to enter the desired 4-digit code in the memory.

Once the memory is programmed with the desired code, the program mode switch is turned off. The system is then armed by pressing the arm switch while selecting any digit on the keypad.

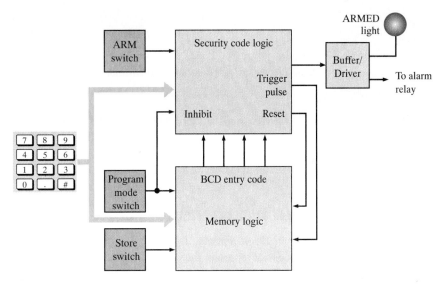

FIGURE 12–70
Basic block diagram of the security entry system.

Basic Operation of the Memory Logic

A simplified diagram of the memory logic is shown in Figure 12–71 (next page). To initially store a code in the RAM, the program mode switch is closed, triggering one-shot B, and it remains closed during programming. The pulse output of the one-shot goes through OR gate G_2 to reset (clear) the counter. In the RESET state, the counter contains all zeros (000) which selects address 0 in the RAM. Address 0 is not used for storing a code.

To enter the first security code digit, the desired key on the keypad is pressed and held for a few seconds while the store switch is closed momentarily, triggering one-shot A. The pulse output of the one-shot goes through OR gate G_1, to clock the counter to the 001 state, thus selecting address 1 in the RAM. At the end of the pulse from one-shot A, one-shot C is triggered and generates a write pulse to the memory, which stores the selected digit code in address 1.

To enter the second digit, the desired key is pressed and held, again triggering one-shot A and advancing the counter to the 010 state, which selects memory address 2. One-shot C is then triggered and generates a write pulse which stores the digit code in memory address 2.

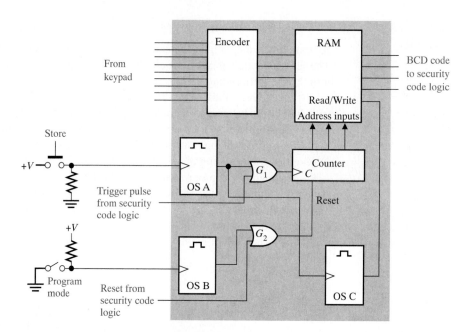

FIGURE 12–71
Simplified basic logic diagram of the memory logic.

Similar steps are repeated to store the third digit in address 3 and the fourth digit in address 4 in the RAM. Once the four digits are stored, the program mode switch is opened. The RAM stores the code as long as there is power. The system is assumed to have a battery backup in the event of power failure.

■ THE DIGITAL WORKBENCH

The basic logic diagram in Figure 12–71 indicates the general logic required for the system but does not represent the detailed logic necessary to implement the system. In the following workbench assignments, you will analyze the basic operation of the memory logic to make sure that you understand the functional requirements, complete the design of the logic by selecting appropriate devices, and then develop a detailed logic diagram incorporating the selected devices. You will compare your logic diagram to an implemented circuit board and then develop a test procedure for the memory logic board. Finally, you will interface the security code logic board, memory logic board, and the front panel components (switches, keypad, and ARMED light with its buffer/driver board) to form the complete system.

On the circuit board, interconnections on the back side of the board feed through to the pads on the component side. When you trace out the circuit in Workbench 2, common pads that are connected on the back side are generally oriented vertically or horizontally with respect to each other and not at an angle.

■ DIGITAL WORKBENCH 1: Design

The focus of this workbench is the design of the memory logic. For information on devices, reference to earlier chapters and/or device data sheets is necessary.

■ *Activity 1 Decimal-to-BCD Encoder*
The particular keypad used in this system produces active-LOW outputs. When a key is pressed, the corresponding line is LOW and the other lines are HIGH. Specify an IC encoder to properly interface with the keypad.

■ *Activity 2 The RAM*
The requirement for the memory to store 4-bit BCD codes establishes the word length. The requirement for the memory to store four BCD codes in addresses 1 through 4 and the requirement for address 0 to be reserved for the reset or armed conditions establishes the minimum word capacity. Select a RAM device of minimum capacity that meets the storage requirements in terms of its organization.

■ *Activity 3 Counter*
The counter must be capable of producing a 3-bit binary sequence for addressing the memory, and it must have a reset or clear input. Select an appropriate IC counter.

■ *Activity 4 One-shots*
One-shot A is to produce an output pulse with a width of approximately 10 ms. One-shots B and C are to each produce an output pulse with a width of approximately 1 ms. Select an appropriate type of IC one-shot and determine the values of the resistor and capacitor for each one.

■ *Activity 5 Gating Logic*
Based on your selection of the IC for the counter and the inputs from the security code logic, specify the types of logic gates required to provide the proper active levels. Although the basic logic diagram indicates OR gates, you may have to specify NOR gates or negative-OR gates.

■ *Activity 6* Develop a detailed logic diagram using the devices specified in the preceding activities. Show all necessary details including device pin numbers.

■ *Activity 7* Write a detailed technical report describing the operation of the memory logic and specifying a parts list.

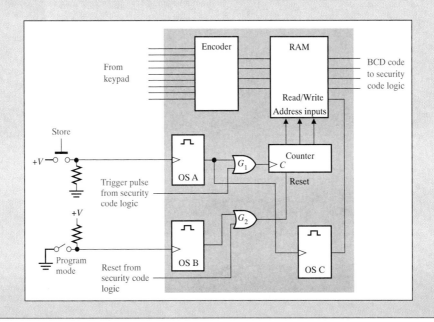

■ DIGITAL WORKBENCH 2: Verification and Testing

■ *Activity 1* Draw the logic diagram for the memory logic board below and compare to your logic diagram from Activity 6 of Workbench 1. Discuss any differences between your design and the design of this logic board and verify that both work.

■ *Activity 2* Identify the outputs and inputs on the circuit board connector.

■ *Activity 3* Develop a test procedure for checking the board shown.

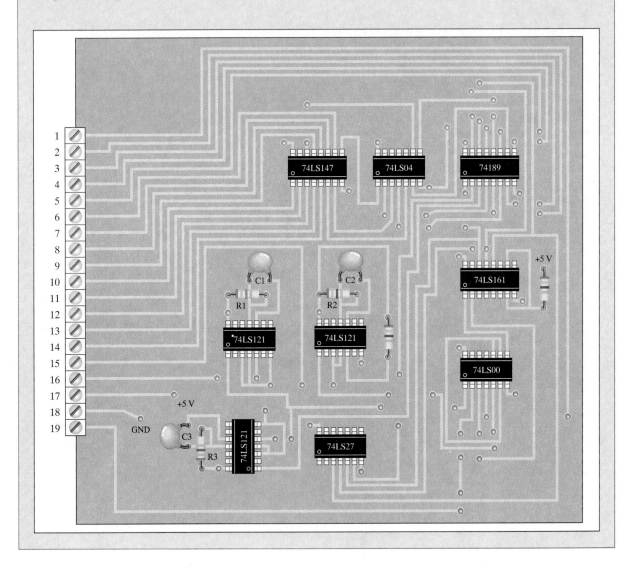

■ DIGITAL WORKBENCH 3: The Complete System

The components and logic boards for the security entry system are shown. When interconnected, these elements form the complete system for activating and deactivating the alarm system.

■ *Activity 1* Create a point-to-point list for properly interconnecting all of the components and boards.

■ *Activity 2* In Chapter 10 and in Workbench 2 of this chapter, you developed test procedures for checking each board individually. Now, develop a test procedure for the complete system that includes all of the parts shown below.

■ *Activity 3* Write a detailed technical report describing the overall system operation. Provide detailed step-by-step instructions for operating the system.

■ *Activity 4* List possible reasons for the following system malfunctions:
(a) ARMED light will not come on when the ARM switch is pressed.
(b) ARMED light will not turn off when proper code is entered.
(c) ARMED light turns off after the first code digit is entered.

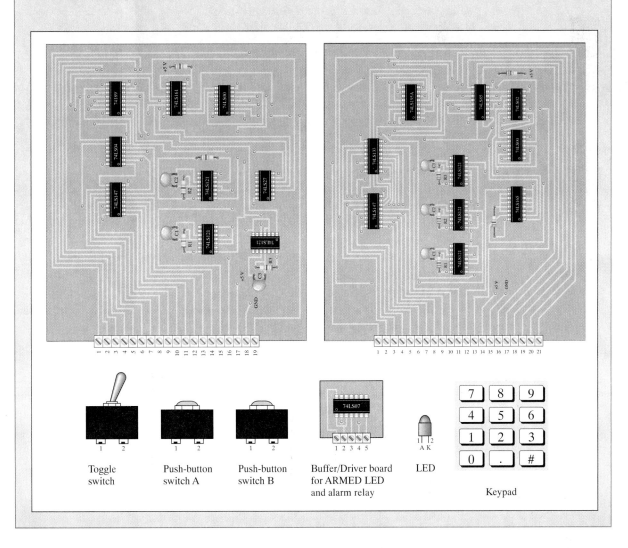

| Toggle switch | Push-button switch A | Push-button switch B | Buffer/Driver board for ARMED LED and alarm relay | LED | Keypad |

SECTION 12–10 REVIEW

1. Why does the system have to be reprogrammed with the entry code if there is a power failure and there is no battery backup?
2. How many addresses in the RAM on the memory logic board are unused?
3. How does the program mode switch inhibit the security code logic during the programming mode?

■ SUMMARY

- Semiconductor memories: ROM, SRAM, DRAM, Flash, CCD
- Magnetic memories: tape, floppy disk, hard disk, magneto-optic disk
- Random-access memories: ROM, SRAM, DRAM, flash, floppy disk, hard disk, magneto-optic disk
- Serial-access memories: CCD, tape
- Read/write memories: SRAM, DRAM, flash, CCD, tape, floppy disk, hard disk, magneto-optic disk
- Read-only memories: semiconductor ROM, CD-ROM
- Special function memories: FIFO, LIFO, WORM

■ SELF-TEST

1. The bit capacity of a memory that has 1024 addresses and can store 8 bits at each address is
   (a) 1024   (b) 8192   (c) 8   (d) 4096
2. A 32-bit data word consists of
   (a) 2 bytes   (b) 4 nibbles   (c) 4 bytes   (d) 3 bytes and 1 nibble
3. Data are stored in a random-access memory (RAM) during the
   (a) read operation   (b) enable operation
   (c) write operation   (d) addressing operation
4. Data that are stored at a given address in a random-access memory (RAM) is lost when
   (a) power goes off   (b) the data are read from the address
   (c) new data are written at the address   (d) answers (a) and (c)
5. A ROM is a
   (a) nonvolatile memory   (b) volatile memory
   (c) read/write memory   (d) byte-organized memory
6. A memory with 256 addresses has
   (a) 256 address lines   (b) 6 address lines   (c) 1 address line   (d) 8 address lines
7. A byte-organized memory has
   (a) 1 data output line   (b) 4 data output lines
   (c) 8 data output lines   (d) 16 data output lines
8. The storage cell in a SRAM is
   (a) a flip-flop   (b) a capacitor   (c) a fuse   (d) a magnetic domain
9. A DRAM must be
   (a) replaced periodically   (b) refreshed periodically
   (c) always enabled   (d) programmed before each use
10. A flash memory is
    (a) volatile   (b) a read-only memory
    (c) a read/write memory   (d) nonvolatile
    (e) answers (a) and (c)   (f) answers (c) and (d)

■ PROBLEMS

SECTION 12–1   Basics of Semiconductor Memories

1. Identify the ROM and the RAM in Figure 12–72.

2. Explain why RAMs and ROMs are both random-access memories.

3. Explain the purposes of the address bus, the data bus, and the control bus in a typical microprocessor-based system.

4. What memory address (0 through 256) is represented by each of the following hexadecimal numbers:

    (a) $0A_{16}$ **(b)** $3F_{16}$ **(c)** CD_{16}

FIGURE 12–72

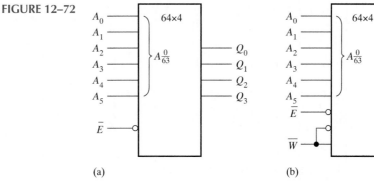

(a)   (b)

SECTION 12–2   Read-Only Memories (ROMs)

5. For the ROM array in Figure 12–73, determine the outputs for all possible input combinations, and summarize them in tabular form (light cell is a 1, dark cell is a 0).

6. Determine the truth table for the ROM in Figure 12–74.

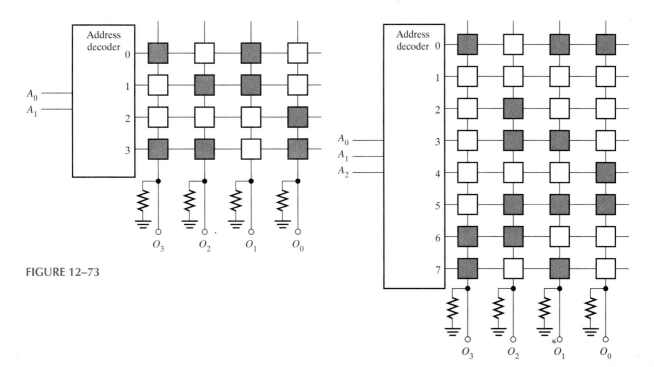

FIGURE 12–73

FIGURE 12–74

7. Using a procedure similar to that in Example 12–1, design a ROM for conversion of single-digit BCD to excess-3 code.

8. What is the total bit capacity of a ROM that has 14 address lines and 8 data outputs?

SECTION 12–3 Programmable ROMs (PROMs and EPROMs)

9. Assuming that the PROM matrix in Figure 12–75 is programmed by blowing a fuse link to create a 0, indicate the links to be blown to program an X^3 look-up table, where X is a number from 0 through 7.

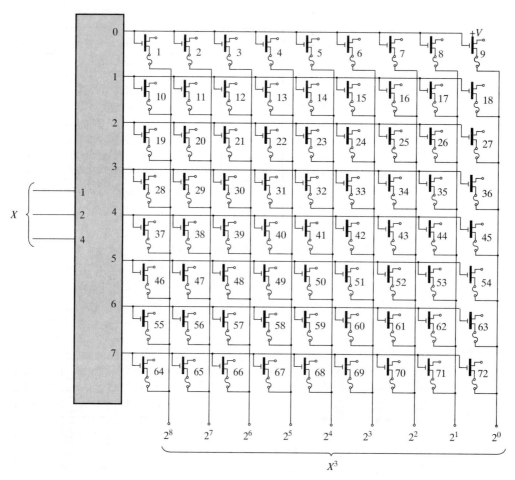

FIGURE 12–75

10. Determine the addresses that are programmed and the contents of each address after the programming sequence in Figure 12–76 has been applied to a 2516 PROM.

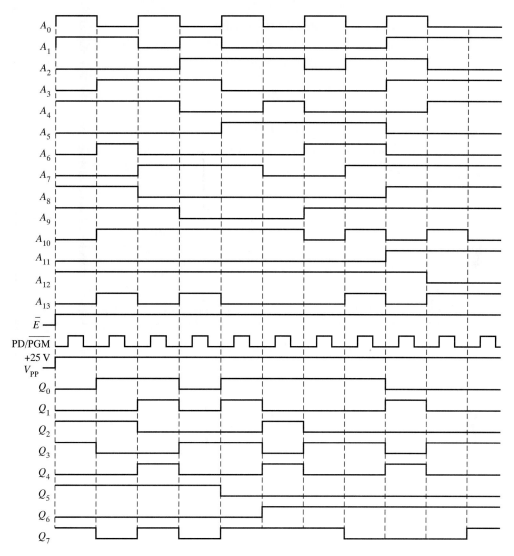

FIGURE 12–76

SECTION 12–4   Read/Write Random-Access Memories (RAMs)

11. A static memory cell such as the one in Figure 12–22 (p. 606) is storing a 0 (RESET). What is its state after each of the following conditions?

   (a) Row = 1, Column = 1, Data in = 1, $\overline{WRITE} = 1$

   (b) Row = 0, Column = 1, Data in = 1, $\overline{WRITE} = 1$

   (c) Row = 1, Column = 1, Data in = 1, $\overline{WRITE} = 0$

12. Draw a basic logic diagram for a 512 × 8-bit static RAM, showing all the inputs and outputs.

13. Assuming that a 64k × 8 SRAM has a structure similar to that of the SRAM in Figure 12–24 (p. 607), determine the number of rows and 8-bit columns in its memory cell array.

14. Redraw the block diagram in Figure 12–24 for a 64k × 8 memory.

15. Explain the difference between a SRAM and a DRAM.

16. What is the bit capacity of a DRAM that has twelve address lines?

SECTION 12–6   Memory Expansion

17. Use 16k × 4 DRAMs to build a 64k × 8 RAM. Show the logic diagram.

18. Using a block diagram, show how 64k × 1 dynamic RAMs can be expanded to build a 256k × 4 RAM.

19. What is the word length and the word capacity of the memory of Problem 17? Problem 18?

SECTION 12–7   Special Types of Memories

20. Complete the timing diagram in Figure 12–77 by showing the output waveforms that are initially all LOW for a FIFO serial memory like that shown in Figure 12–48 on p. 624.

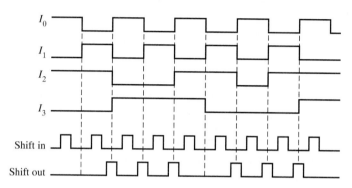

FIGURE 12–77

21. Consider a 4096 × 8 RAM in which the last 64 addresses are used as a LIFO stack. If the first address in the RAM is 000_{16}, designate the 64 addresses used for the stack.

22. In the memory of Problem 21, sixteen bytes are pushed into the stack. At what address is the first byte in located? At what address is the last byte in located?

SECTION 12–8   Magnetic and Optical Memories

23. Show the format of a track sector in a floppy disk.

24. Why does magnetic tape require a much longer access time than does a disk?

25. How many tracks are in a basic magnetic tape?

26. Explain the differences in a magneto-optic disk, a CD-ROM, and a WORM.

SECTION 12–9   Testing and Troubleshooting

27. Determine if the contents of the ROM in Figure 12–78 are correct.

FIGURE 12–78

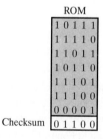

28. A 128 × 8 ROM is implemented as shown in Figure 12–79. The decoder decodes the two most significant address bits to enable the ROMs one at a time, depending on the address selected.

    (a) Express the lowest address and the highest address of each ROM as hexadecimal numbers.

    (b) Assume that a single checksum is used for the entire memory and it is stored at the highest address. Develop a flowchart for testing the complete memory system.

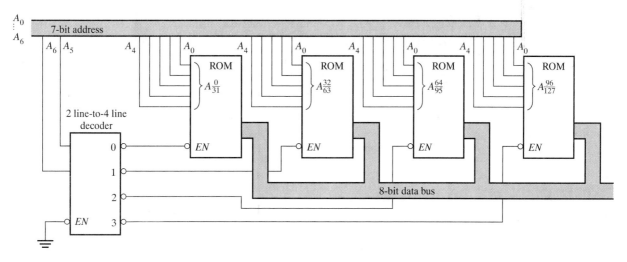

FIGURE 12–79

(c) Assume that each ROM has a checksum stored at its highest address. Modify the flowchart developed in part (b) to accommodate this change.

(d) What is the disadvantage of using a single checksum for the entire memory rather than a checksum for each individual ROM?

29. Suppose that a checksum test is run on the memory in Figure 12–79 and each individual ROM has a checksum at its highest address. What IC or ICs will you replace for each of the following error messages that appear on the system's video monitor?

(a) ADDRESSES 40 – 5F FAULTY

(b) ADDRESSES 20 – 3F FAULTY

(c) ADDRESSES 00 – 7F FAULTY

 SECTION 12–10 Digital System Application

30. Develop a timing diagram for the basic memory logic in Figure 12–71 (p. 640) to illustrate the entry of the digits 4321 into the RAM. Include all the inputs and outputs of each device.

31. In the programming mode, what is the state of the counter in Figure 12–71 (p. 640) after two code digits have been entered?

32. What is the purpose of the 74LS27 on the memory logic board in Workbench 2?

33. Discuss the advantages and the disadvantages of using a PROM instead of a RAM in the memory logic.

 Special Design Problems

34. Develop a basic block diagram for the ROM tester for the setup in Figure 12–64 (p. 635) to test the ROM in Figure 12–80.

FIGURE 12–80

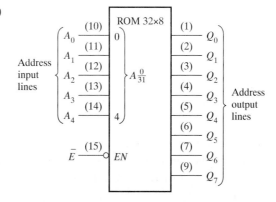

35. Repeat Problem 34 for the ROM in Figure 12–81.

FIGURE 12–81

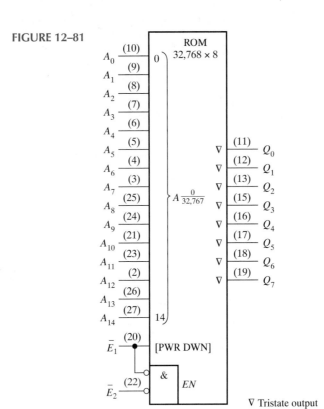

∇ Tristate output

36. Modify, if necessary, the design of the memory logic of the security entry system to accommodate a 5-digit entry code.

37. Make the appropriate modifications to the entry code logic of the security entry system for a 5-digit entry code. Refer to Chapter 10.

■ **ANSWERS TO SECTION REVIEWS**

SECTION 12–1

1. Bit is the smallest unit of data.   **2.** 2048 bits is 256 bytes.

3. A write operation stores data in memory.

4. A read operation takes a copy of data from memory.

5. A unit of data is located by its address.

6. A RAM is volatile and has read/write capability. A ROM is nonvolatile and has only read capability.

7. Memory technologies are bipolar and MOS.

SECTION 12–2

1. 512×8 equals 4096 bits.

2. Tristate logic allows outputs to be disconnected (high-Z) from bus lines.

3. Eight bits of address are required for 256 byte locations ($2^8 = 256$).

SECTION 12–3

1. PROMs are field-programmable; ROMs are not.   **2.** 1s are left after EPROM erasure.

3. Read is the normal mode of operation for a PROM.

SECTION 12–4

1. SRAMs have latch storage cells that can retain data indefinitely. DRAMs have capacitive storage cells that must be periodically refreshed.
2. The refresh operation prevents data from being lost because of capacitive discharge. A stored bit is restored periodically by recharging the capacitor.

SECTION 12–5

1. Flash, ROM, EPROM, and EEPROM are nonvolatile.
2. Flash is nonvolatile; SRAM and DRAM are volatile.

SECTION 12–6

1. Eight RAMs   2. Eight RAMs   3. SIMM: single-in-line memory module.

SECTION 12–7

1. In a FIFO memory the *first* bit (or word) *in* is the *first* bit (or word) *out.*
2. In a LIFO memory the *last* bit (or word) *in* is the *first* bit (or word) *out.* A stack is a LIFO.
3. CCD is a charge-coupled device.

SECTION 12–8

1. Magnetic memories are floppy disk, hard disk, tape, and magneto-optic disk.
2. Floppy disk sizes are 3.5 in. and 5.25 in.
3. A magnetic disk is organized in tracks and sectors.
4. Magnetic tape has a long access time.
5. A magneto-optic disk uses a laser beam and an electromagnet.

SECTION 12–9

1. The contents of the ROM are added and compared with a prestored checksum.
2. Checksum cannot be used because the contents of a RAM are not fixed.
3. (1) a short between adjacent cells; (2) an inability of some cells to store both 1s and 0s; (3) dynamic altering of the contents of one address when the contents of another address change.

SECTION 12–10

1. The system must be reprogrammed because the RAM is volatile and loses stored data when power is off.
2. Five of the sixteen addresses are used. Eleven addresses are unused.
3. The program mode switch applies a LOW to the negative-OR gate (NAND) that triggers one-shot A in the security code logic.

13

INTERFACING

■ CHAPTER OBJECTIVES

□ Describe applications of digital-to-analog and analog-to-digital conversion

□ Explain D/A conversion by the binary-weighted input method and by the $R/2R$ ladder method

□ Explain A/D conversion by any of the following processes: flash (simultaneous), digital ramp, tracking, single slope, dual slope, and successive approximation

□ Recognize such conversion errors as nonmonotonicity, differential nonlinearity, offset, improper gain, and missing or incorrect codes in DACs and ADCs

□ Troubleshoot DAC and ADC systems

□ Explain how digital devices are interfaced by the use of bus systems

□ Define the basic characteristics and applications of the Multibus and PC bus standards

□ Define the basic characteristics and applications of the GPIB, VXI, RS-232C, and SCSI bus standards

■ CHAPTER OVERVIEW

Interfacing is the process of making two or more devices or systems operationally compatible with each other so that they function together as required.

Many quantities are analog in nature; that is, they are continuous quantities. Physical quantities such as temperature, pressure, time, and velocity are examples of analog quantities. In this chapter, methods of converting from digital codes to analog quantities are introduced. This is called digital-to-analog conversion (D/A conversion). Also, methods for converting analog quantities to digital codes are covered. This is called analog-to-digital conversion (A/D conversion).

In addition to D/A and A/D conversion, the concept of buses is introduced and several types of bus structures are discussed.

■ SPECIFIC DEVICES

ADC0804   AD673   AD9300

■ DIGITAL SYSTEM APPLICATION

This Digital System Application illustrates concepts taught in this chapter and focuses on both the analog and the digital portions of an antenna positioning system. A/D conversion is required to convert analog voltages representing the antenna position into digital form and logic circuits are required to process and display the angular positions.

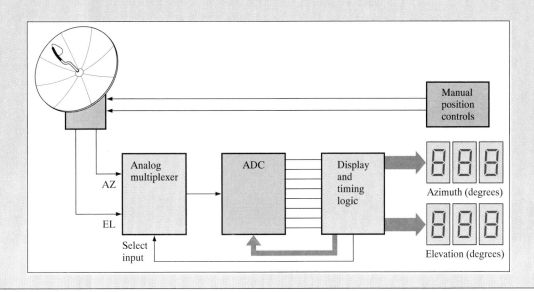

13–1 ■ INTERFACING THE DIGITAL AND ANALOG WORLDS

In Chapter 1 we briefly described analog and digital quantities and pointed out the difference between them. Analog quantities are sometimes called real-world quantities because most physical quantities are analog in nature. Many applications of computers and other digital systems require the input of real-world quantities, such as temperature, speed, position, pressure, and force. Real-world quantities can even include graphic images. Also, digital systems often must produce outputs to control real-world quantities. After completing this section, you should be able to

☐ Define an analog quantity   ☐ State the purpose of analog-to-digital conversion
☐ State the purpose of digital-to-analog conversion   ☐ Give examples of both analog-to-digital and digital-to-analog conversion

Digital and Analog Signals

An analog quantity is one that has continuous values over a given range, as contrasted with a discrete set of values for the digital case. To illustrate the difference between an analog and a digital representation of a quantity, let's take the case of a voltage that varies over a range from 0 V to +15 V. The analog representation of this quantity takes in *all* values between 0 and +15 V of which there is an infinite number.

In the case of a digital representation of the voltage using a 4-bit binary code, only sixteen values can be defined. More values between 0 and +15 can be represented by using more bits in the digital code. So an analog quantity can be represented to some degree of accuracy with a digital code that specifies discrete values within the range. This concept is illustrated in Figure 13–1, where the analog function shown is a smoothly changing curve that takes on values between 0 V and +15 V. If a 4-bit code is used to represent this curve, each binary number represents a discrete point on the curve.

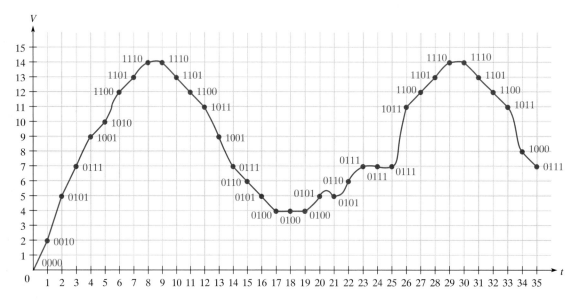

FIGURE 13–1
Discrete (digital) points on an analog curve.

In Figure 13–1 the voltage on the analog curve is measured, or sampled, at each of thirty-five equal intervals. The voltage at each of these intervals is represented by a 4-bit code as indicated. At this point, we have a series of binary numbers representing various voltage values along the analog curve. This is the basic idea of analog-to-digital (A/D) conversion.

An approximation of the analog function in Figure 13–1 can be reconstructed from the sequence of digital numbers that has been generated. Obviously, there will be some error in the reconstruction because only certain values are represented (thirty-six in this example) and not the continuous set of values. If the digital values at all of the thirty-six points are graphed as shown in Figure 13–2, we have a reconstructed function. As you can see, the graph only approximates the original curve because values between the points are not known. Accuracy can be increased by sampling the curve more often and by increasing the number of bits used to represent each sampled value.

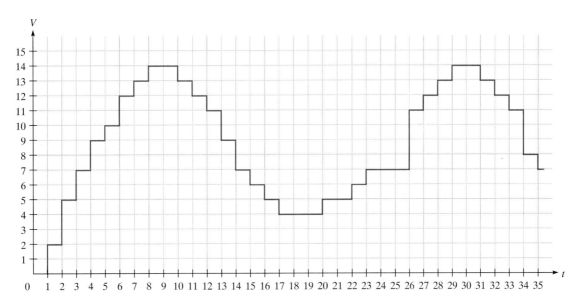

FIGURE 13–2
A rough digital reproduction of the analog curve in Figure 13–1.

Real-World Interfacing

To **interface** between the digital and analog worlds, two basic processes are required. These are **analog-to-digital (A/D) conversion** and **digital-to-analog (D/A) conversion.** The following three system examples illustrate applications of these conversion processes.

An Electronic Thermostat   A simplified block diagram of an electronic thermostat is shown in Figure 13–3 (next page). The room temperature sensor produces an analog voltage that is proportional to the temperature. This voltage is increased by the linear amplifier and applied to the **analog-to-digital converter (ADC),** where it is converted to a digital code and periodically sampled by the control logic. For example, suppose the room temperature is 67°F. A specific voltage value corresponding to this temperature appears on the ADC input and is converted to an 8-bit binary number, 01000011.

The control logic compares this binary number with a binary number representing the desired temperature (say 01001000 for 72°F). This desired value has been previously entered from the keypad and stored in a register. The comparison shows that the actual room temperature is less than the desired temperature. As a result, the control logic instructs the unit control circuit to turn the furnace on. As the furnace runs, the control logic continues to monitor the actual temperature via the ADC. When the actual temperature equals or exceeds the desired temperature, the control logic turns the furnace off. Both the actual temperature and the desired temperature are displayed.

CD Player   The compact-disk (CD) player is an example of a system that uses a **digital-to-analog converter (DAC).** A basic block diagram is shown in Figure 13–4. An audio signal is digitally recorded on the CD in the form of pits that are sensed by the laser pick-

FIGURE 13–3
Basic block diagram of an electronic thermostat that uses an ADC.

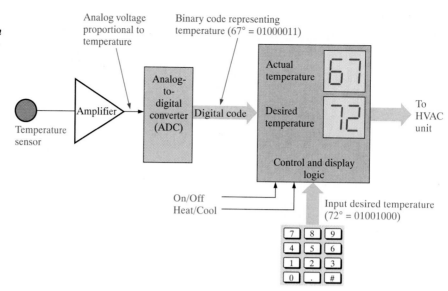

FIGURE 13–4
Basic block diagram of a CD player.

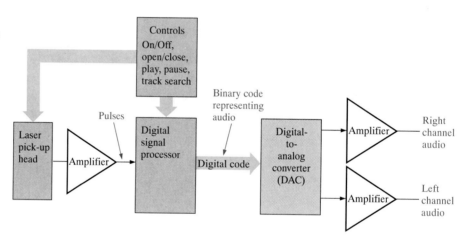

FIGURE 13–5
Basic block diagram of a DAT system.

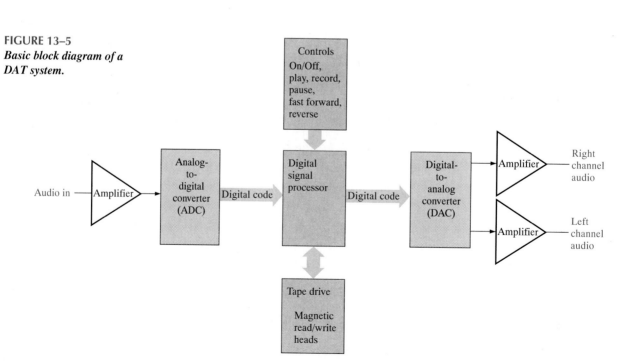

up and amplified. The amplified digital signal is converted to a sequence of binary codes that represent the originally recorded audio signal. The binary representation of the audio is converted to analog form by the DAC, amplified, and sent to the speakers.

A Digital Audiotape (DAT) Player/Recorder Another system example that includes both A/D and D/A conversion is the **DAT** player/recorder. A basic block diagram is shown in Figure 13–5.

An audio signal, of course, is an analog quantity. In the record mode, sound is picked up, amplified, and converted to digital form by the ADC. The digital codes representing the audio signal are processed and recorded on the tape.

In the play mode, the digitized audio signal is read from the tape, processed, and converted back to analog form by the DAC. It is then amplified and sent to the speaker system.

SECTION 13–1 **REVIEW**	**1.** In nature, quantities normally appear in what form? **2.** Explain the basic purpose of A/D conversion. **3.** Explain the basic purpose of D/A conversion.

13–2 ■ DIGITAL-TO-ANALOG (D/A) CONVERSION

D/A conversion is an important part of many systems. In this section, we will examine two basic types of digital-to-analog converters (DACs) and learn about their performance characteristics. After completing this section, you should be able to

□ Explain basically what an operational amplifier is □ Show how the op-amp can be used as an inverting amplifier or a comparator □ Explain the operation of a binary-weighted-input DAC □ Explain the operation of an $R/2R$ ladder DAC □ Discuss resolution, accuracy, linearity, monotonicity, and settling time in a DAC

The Operational Amplifier

Before getting into digital-to-analog converters (DACs), let's look briefly at an element that is common to most types of DACs and analog-to-digital converters (ADCs). This element is the operational amplifier, or op-amp for short.

An **op-amp** is a linear amplifier that has two inputs (inverting and noninverting) and one output. It has a very high voltage gain and a very high input impedance, as well as a very low output impedance. The op-amp symbol is shown in Figure 13–6(a). When used as an inverting amplifier, the op-amp is configured as shown in part (b). The feedback resistor, R_f, and the input resistor, R_{in}, control the voltage gain according to the formula in

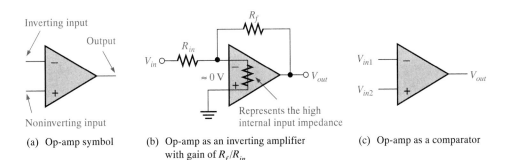

(a) Op-amp symbol

(b) Op-amp as an inverting amplifier with gain of R_f/R_{in}

(c) Op-amp as a comparator

FIGURE 13–6
The operational amplifier (op-amp).

Equation (13–1), where V_{out}/V_{in} is the closed-loop voltage gain (closed loop refers to the feedback from output to input provided by R_f). The negative sign indicates inversion.

$$\frac{V_{out}}{V_{in}} = -\frac{R_f}{R_{in}} \tag{13–1}$$

In the inverting amplifier configuration, the inverting input of the op-amp is approximately at ground potential (0 V) because feedback and the extremely high open-loop gain make the differential voltage between the two inputs extremely small. Since the noninverting input is grounded, the inverting input is at approximately 0 V, which is called *virtual ground*.

When the op-amp is used as a comparator, as shown in Figure 13–6(c), two voltages are applied to the inputs. When these input voltages differ by a very small amount, the op-amp is driven into one of its two saturated output states, either HIGH or LOW, depending on which input voltage is greater.

Binary-Weighted-Input Digital-to-Analog Converter

One method of D/A conversion uses a resistor network with resistance values that represent the binary weights of the input bits of the digital code. Figure 13–7 shows a 4-bit DAC of this type. Each of the input resistors will either have current or have no current, depending on the input voltage level. If the input voltage is zero (binary 0), the current is also zero. If the input voltage is HIGH (binary 1), the amount of current depends on the input resistor value and is different for each input resistor, as indicated in the figure.

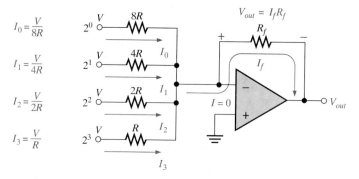

FIGURE 13–7
A 4-bit DAC with binary-weighted inputs.

Since there is practically no current into the op-amp inverting input, all of the input currents sum together and go through R_f. Since the inverting input is at 0 V (virtual ground), the drop across R_f is equal to the output voltage, so $V_{out} = I_f R_f$.

The values of the input resistors are chosen to be inversely proportional to the binary weights of the corresponding input bits. The lowest-value resistor (R) corresponds to the highest binary-weighted input (2^3). The other resistors are multiples of R(2R, 4R, and 8R) and correspond to the binary weights 2^2, 2^1, and 2^0, respectively. The input currents are also proportional to the binary weights. Thus, the output voltage is proportional to the sum of the binary weights because the sum of the input currents is through R_f.

One of the disadvantages of this type of DAC is the number of different resistor values. For example, an 8-bit converter requires eight resistors, ranging from some value of R to 128R in binary-weighted steps. This range of resistors requires tolerances of one part in 255 (less than 0.5%) to accurately convert the input, making this type of DAC very difficult to mass-produce.

EXAMPLE 13–1   Determine the output of the DAC in Figure 13–8(a) if the waveforms representing a sequence of 4-bit numbers in Figure 13–8(b) are applied to the inputs. Input D_0 is the least significant bit (LSB).

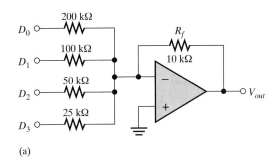

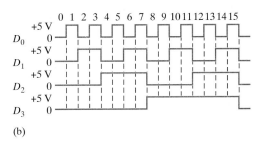

(a)   (b)

FIGURE 13–8

Solution   First, determine the current for each of the weighted inputs. Since the inverting input $(-)$ of the op-amp is at 0 V (virtual ground) and a binary 1 corresponds to $+5$ V, the current through any of the input resistors is 5 V divided by the resistance value:

$$I_0 = \frac{5\ V}{200\ k\Omega} = 0.025\ mA$$

$$I_1 = \frac{5\ V}{100\ k\Omega} = 0.05\ mA$$

$$I_2 = \frac{5\ V}{50\ k\Omega} = 0.1\ mA$$

$$I_3 = \frac{5\ V}{25\ k\Omega} = 0.2\ mA$$

Almost no current goes into the inverting op-amp input because of its extremely high impedance. Therefore, assume that all of the current goes through the feedback resistor R_f. Since one end of R_f is at 0 V (virtual ground), the drop across R_f equals the output voltage, which is negative with respect to virtual ground.

$$V_{out(D0)} = (10\ k\Omega)(-0.025\ mA) = -0.25\ V$$
$$V_{out(D1)} = (10\ k\Omega)(-0.05\ mA) = -0.5\ V$$
$$V_{out(D2)} = (10\ k\Omega)(-0.1\ mA) = -1\ V$$
$$V_{out(D3)} = (10\ k\Omega)(-0.2\ mA) = -2\ V$$

From Figure 13–8(b), the first binary input code is 0000, which produces an output voltage of 0 V. The next input code is 0001, which produces an output voltage of -0.25 V. For this, the output voltage is -0.25 V. The next code is 0010 which produces an output voltage of -0.5 V. The next code is 0011 which produces an output voltage of -0.25 V $+ -0.5$ V $= -0.75$ V. Each successive binary code increases the output voltage by -0.25 V, so for this particular straight binary sequence on the inputs, the output is a stairstep waveform going from 0 V to -3.75 V in -0.25 V steps. This is shown in Figure 13–9.

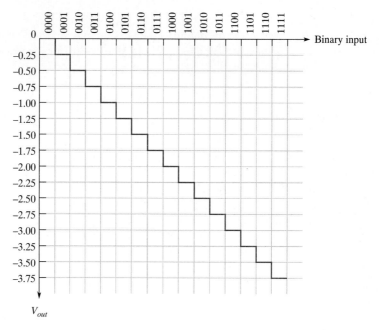

FIGURE 13–9
Output of the DAC in Figure 13–8.

Related Exercise Reverse the input waveforms to the DAC in Figure 13–8 (D_3 to D_0, D_2 to D_1, D_1 to D_2, D_0 to D_3) and determine the output.

The *R/2R* Ladder Digital-to-Analog Converter

Another method of D/A conversion is the *R/2R* ladder, as shown in Figure 13–10 for four bits. It overcomes one of the problems in the binary-weighted-input DAC in that it requires only two resistor values.

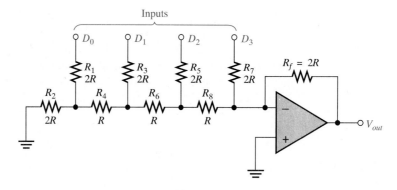

FIGURE 13–10
An R/2R ladder DAC.

Start by assuming that the D_3 input is HIGH (+5 V) and the others are LOW (ground, 0 V). This condition represents the binary number 1000. A circuit analysis will show that this reduces to the equivalent form shown in Figure 13–11(a). Essentially no current goes through the 2R equivalent resistance because the inverting input is at virtual ground. Thus, all of the current ($I = 5$ V/2R) through R_7 also goes through R_f, and the output voltage is −5 V. The operational amplifier keeps the inverting (−) input near zero volts (≈0 V) because of negative feedback. Therefore, all current goes through R_f rather than into the inverting input.

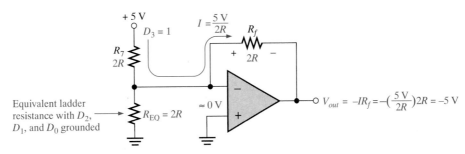

(a) Equivalent circuit for $D_3 = 1, D_2 = 0, D_1 = 0, D_0 = 0$

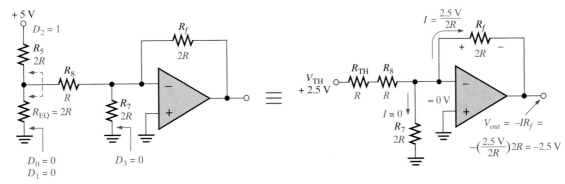

(b) Equivalent circuit for $D_3 = 0, D_2 = 1, D_1 = 0, D_0 = 0$

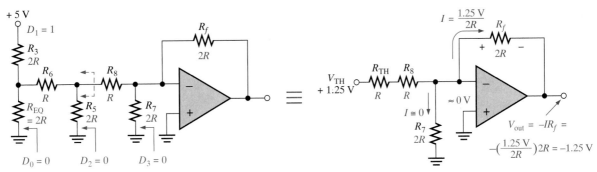

(c) Equivalent circuit for $D_3 = 0, D_2 = 0, D_1 = 1, D_0 = 0$

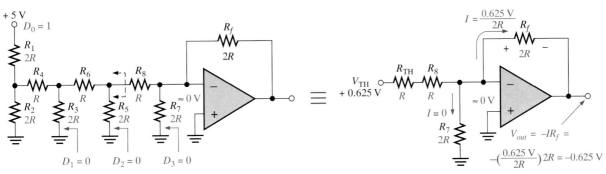

(d) Equivalent circuit for $D_3 = 0, D_2 = 0, D_1 = 0, D_0 = 1$

FIGURE 13–11

Analysis of the R/2R ladder DAC.

Figure 13–11(b) shows the equivalent circuit when the D_2 input is at $+5$ V and the others are at ground. This condition represents 0100. If we thevenize looking from R_8, we get 2.5 V in series with R, as shown. This results in a current through R_f of $I = 2.5$ V/$2R$, which gives an output voltage of -2.5 V. Keep in mind that there is no current into the op-amp inverting input and that there is no current through the equivalent resistance to ground because it has 0 V across it, due to the virtual ground.

Figure 13–11(c) shows the equivalent circuit when the D_1 input is at $+5$ V and the others are at ground. This condition represents 0010. Again thevenizing looking from R_8, we get 1.25 V in series with R as shown. This results in a current through R_f of $I = 1.25$ V/$2R$, which gives an output voltage of -1.25 V.

In part (d) of the figure, the equivalent circuit representing the case where D_0 is at $+5$ V and the other inputs are at ground is shown. This condition represents 0001. Thevenizing from R_8 gives an equivalent of 0.625 V in series with R as shown. The resulting current through R_f is $I = 0.625$ V/$2R$, which gives an output voltage of -0.625 V.

Notice that each successively lower weighted input produces an output voltage that is halved, so that the output voltage is proportional to the binary weight of the input bits.

Performance Characteristics of Digital-to-Analog Converters

The performance characteristics of a DAC include resolution, accuracy, linearity, monotonicity, and settling time, each of which is discussed in the following list:

■ *Resolution.* The resolution of a DAC is the reciprocal of the number of discrete steps in the output. This, of course, is dependent on the number of input bits. For example, a 4-bit DAC has a resolution of one part in $2^4 - 1$ (one part in fifteen). Expressed as a percentage, this is $(1/15)100 = 6.67\%$. The total number of discrete steps equals $2^n - 1$, where n is the number of bits. Resolution can also be expressed as the number of bits that are converted.

■ *Accuracy.* Accuracy is a comparison of the actual output of a DAC with the expected output. It is expressed as a percentage of a full-scale, or maximum, output voltage. For example, if a converter has a full-scale output of 10 V and the accuracy is $\pm 0.1\%$, then the maximum error for any output voltage is $(10 \text{ V})(0.001) = 10$ mV. Ideally, the accuracy should be, at most, $\pm\frac{1}{2}$ of an LSB. For an 8-bit converter, 1 LSB is $1/256 = 0.0039$ (0.39% of full scale). The accuracy should be approximately $\pm 0.2\%$.

■ *Linearity.* A linear error is a deviation from the ideal straight-line output of a DAC. A special case is an offset error, which is the amount of output voltage when the input bits are all zeros.

■ *Monotonicity.* A DAC is **monotonic** if it does not take any reverse steps when it is sequenced over its entire range of input bits.

■ *Settling time.* Settling time is normally defined as the time it takes a DAC to settle within $\pm\frac{1}{2}$ LSB of its final value when a change occurs in the input code.

EXAMPLE 13–2

Determine the resolution, expressed as a percentage, of the following:
(a) an 8-bit DAC   (b) a 12-bit DAC

Solution
(a) For the 8-bit converter,

$$\frac{1}{2^8 - 1} \times 100 = \frac{1}{255} \times 100 = 0.392\%$$

(b) For the 12-bit converter,

$$\frac{1}{2^{12} - 1} \times 100 = \frac{1}{4095} \times 100 = 0.0244\%$$

Related Exercise   Calculate the resolution for a 16-bit DAC.

1. What is the disadvantage of the DAC with binary weighted inputs?
2. What is the resolution of a 4-bit DAC?

13–3 ■ ANALOG-TO-DIGITAL (A/D) CONVERSION

As you have seen, analog-to-digital conversion is the process by which an analog quantity is converted to digital form. It is necessary when measured quantities must be in digital form for processing in a computer or for display or storage. Several types of analog-to-digital converters (ADCs) are now examined. After completing this section, you should be able to

□ Explain how a flash ADC works   □ Describe the operation of a digital-ramp ADC
□ Explain the operation of a tracking ADC   □ Discuss single-slope and dual-slope
ADCs   □ Describe the operation of a successive-approximation ADC

Flash (Simultaneous) Analog-to-Digital Converter

The flash method utilizes comparators that compare reference voltages with the analog input voltage. When the analog voltage exceeds the reference voltage for a given comparator, a HIGH is generated. Figure 13–12 shows a 3-bit converter that uses seven comparator circuits; a comparator is not needed for the all-0s condition. A 4-bit converter of this type requires fifteen comparators. In general, $2^n - 1$ comparators are required for conversion to an n-bit binary code. The large number of comparators necessary for a reasonable-sized bi-

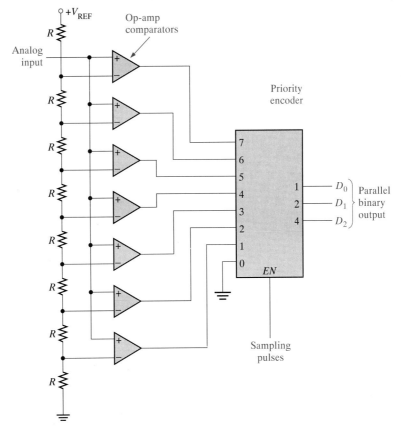

FIGURE 13–12
A 3-bit flash ADC.

nary number is one of the disadvantages of the flash ADC. Its chief advantage is that it provides a fast conversion time.

The reference voltage for each comparator is set by the resistive voltage-divider circuit. The output of each comparator is connected to an input of the priority encoder. The encoder is sampled by a pulse on the enable input, and a 3-bit code representing the value of the analog input appears on the encoder's outputs. The binary code is determined by the highest-order input having a HIGH level.

The sampling rate determines the accuracy with which the sequence of digital codes represents the analog input of the ADC. The more samples taken in a given unit of time, the more accurately the analog signal is represented in digital form.

Example 13–3 illustrates the basic operation of the flash ADC in Figure 13–12.

EXAMPLE 13–3

Determine the binary code output of the 3-bit flash ADC for the analog input signal in Figure 13–13 and the sampling pulses (encoder enable) shown. For this example, $V_{REF} = +8$ V.

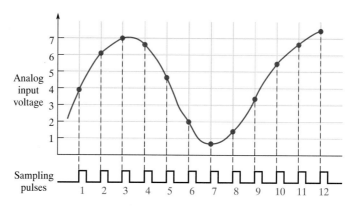

FIGURE 13–13
Sampling of values on an analog waveform for conversion to digital form.

Solution   The resulting digital output sequence is listed as follows and shown in the waveform diagram of Figure 13–14 in relation to the sampling pulses:

100, 110, 111, 110, 100, 010, 000, 001, 011, 101, 110, 111

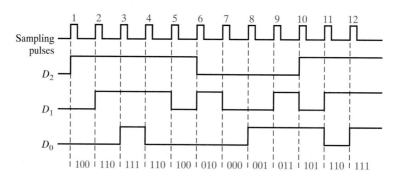

FIGURE 13–14
Resulting digital outputs for sampled values. Output D_0 is the LSB.

Related Exercise   If the sampling pulse frequency in Figure 13–13 is doubled, determine the binary numbers represented by the resulting digital output sequence for 24 pulses.

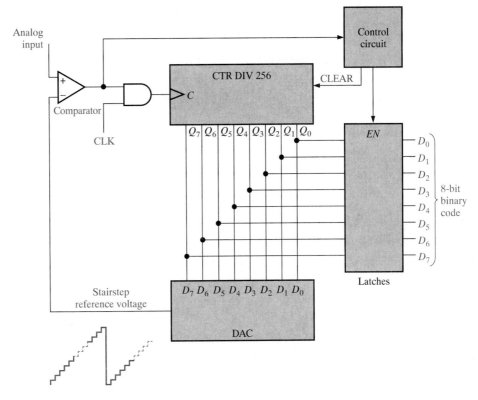

FIGURE 13–15
Digital-ramp ADC (8 bits).

Digital-Ramp Analog-to-Digital Converter

The digital-ramp method of A/D conversion is also known as the *stairstep-ramp* or the *counter* method. It employs a DAC and a binary counter to generate the digital value of an analog input. Figure 13–15 shows a diagram of this type of converter.

Assume that the counter begins RESET and the output of the DAC is zero. Now assume that an analog voltage is applied to the input. When it exceeds the reference voltage (output of DAC), the comparator switches to a HIGH output state and enables the AND gate. The clock pulses begin advancing the counter through its binary states, producing a stairstep reference voltage from the DAC. The counter continues to advance from one binary state to the next, producing successively higher steps in the reference voltage. When the stairstep reference voltage reaches the analog input voltage, the comparator output will go LOW and disable the AND gate, thus cutting off the clock pulses to stop the counter. The binary state of the counter at this point equals the number of steps in the reference voltage required to make the reference equal to or greater than the analog input. This binary number, of course, represents the value of the analog input. The control logic loads the binary count into the latches and resets the counter, thus beginning another count sequence to sample the input value.

The digital-ramp method is slower than the flash method because, in the worst case of maximum input, the counter must sequence through its maximum number of states before a conversion occurs. For an 8-bit conversion, this means a maximum of 256 counter states. Figure 13–16 illustrates a conversion sequence for a 4-bit conversion. Notice that for each sample, the counter must count from zero up to the point at which the stairstep reference voltage reaches the analog input voltage. The conversion time varies, depending on the analog voltage.

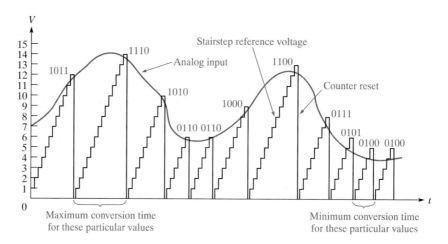

FIGURE 13–16

Example of a 4-bit conversion, showing an analog input and the stairstep reference voltage for a digital-ramp ADC.

Tracking Analog-to-Digital Converter

The tracking method uses an up/down counter and is faster than the digital-ramp method because the counter is not reset after each sample, but rather tends to track the analog input. Figure 13–17 shows a typical 8-bit tracking ADC.

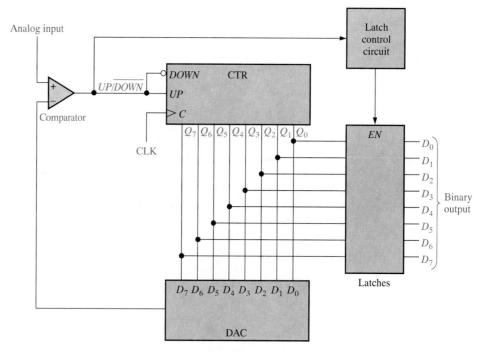

FIGURE 13–17

An 8-bit tracking ADC.

As long as the output reference voltage is less than the analog input, the comparator output is HIGH, putting the counter in the UP mode, which causes it to produce an up sequence of binary counts. This causes an increasing stairstep reference voltage out of the DAC, which continues until the ramp reaches the value of the input voltage.

When the reference voltage equals the analog input, the comparator's output switches LOW and puts the counter in the DOWN mode, causing it to back down one count. If the analog input is decreasing, the counter will continue to back down in its sequence and effectively track the input. If the input is increasing, the counter will back down one count after the comparison occurs and then will begin counting up again. When the input is constant, the counter backs down one count when a comparison occurs. The reference output is now less than the analog input, and the comparator output goes HIGH, causing the counter to count up. As soon as the counter increases one state, the reference voltage becomes greater than the input, switching the comparator to its LOW state. This causes the counter to back down one count. This back-and-forth action continues as long as the analog input is a constant value, thus causing an oscillation between two binary states in the output. This is a disadvantage of this type of converter.

Figure 13–18 illustrates the tracking action of this type of ADC for a 4-bit conversion.

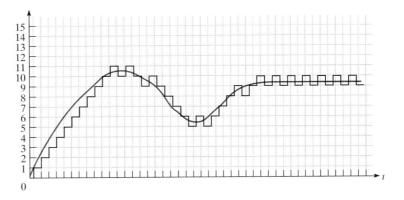

FIGURE 13–18
Tracking action of a tracking ADC.

Single-Slope Analog-to-Digital Converter

Unlike the digital-ramp and tracking methods, the single-slope converter does not require a DAC. It uses a linear ramp generator to produce a constant-slope reference voltage. A diagram is shown in Figure 13–19 (page 668). At the beginning of a conversion cycle, the counter is RESET and the ramp generator output is 0 V. The analog input is greater than the reference voltage at this point and therefore produces a HIGH output from the comparator. This HIGH enables the clock to the counter and starts the ramp generator.

Assume that the slope of the ramp is 1 V/ms. The ramp will increase until it equals the analog input; at this point the ramp is RESET, and the binary or BCD count is stored in the latches by the control logic. Let's assume that the analog input is 2 V at the point of comparison. This means that the ramp is also 2 V and has been running for 2 ms. Since the comparator output has been HIGH for 2 ms, 200 clock pulses have been allowed to pass through the gate to the counter (assuming a clock frequency of 100 kHz). At the point of comparison, the counter is in the binary state that represents decimal 200. With proper scaling and decoding, this binary number can be displayed as 2.00 V. This basic concept is used in some digital voltmeters.

Dual-Slope Analog-to-Digital Converter

The operation of the dual-slope ADC is similar to that of the single-slope type except that a variable-slope ramp and a fixed-slope ramp are both used. This type of converter is common in digital voltmeters and other types of measurement instruments.

A ramp generator (integrator), A_1, is used to produce the dual-slope characteristic. A block diagram of a dual-slope ADC is shown in Figure 13–20 for reference.

Figure 13–21 illustrates dual-slope conversion. Start by assuming that the counter is RESET and the output of the integrator is zero. Now assume that a positive input voltage is

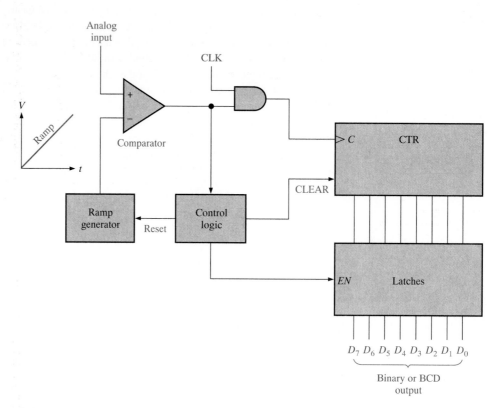

FIGURE 13–19
Single-slope ADC.

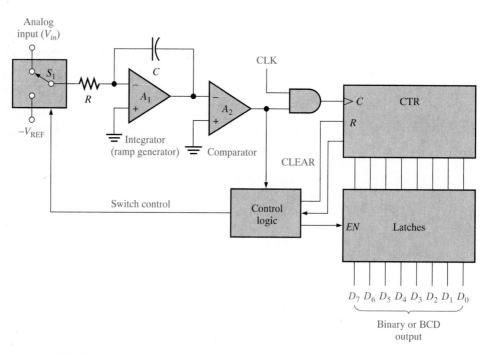

FIGURE 13–20
Dual-slope ADC.

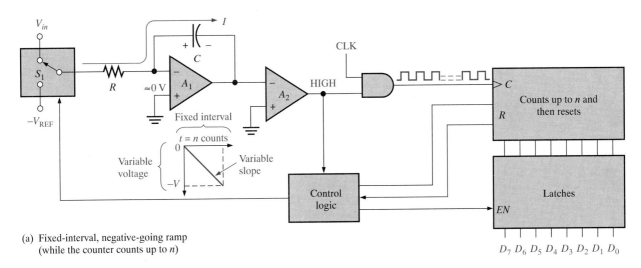

(a) Fixed-interval, negative-going ramp
(while the counter counts up to n)

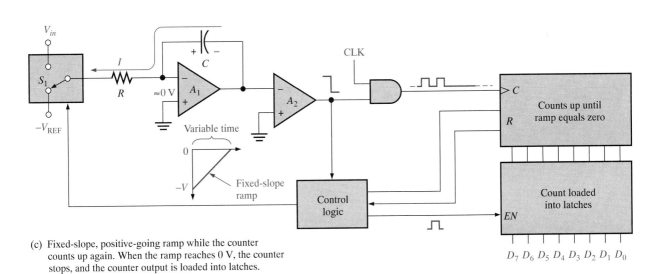

(b) End of fixed-interval when the counter
sends a pulse to control logic to switch S_1

(c) Fixed-slope, positive-going ramp while the counter
counts up again. When the ramp reaches 0 V, the counter
stops, and the counter output is loaded into latches.

FIGURE 13–21
Illustration of dual-slope conversion.

applied to the input through the switch (S_1) as selected by the control logic. Since the inverting input of A_1 is at virtual ground, and assuming that V_{in} is constant for a period of time, there will be constant current through the input resistor R and therefore through the capacitor C. Capacitor C will charge linearly because the current is constant, and as a result, there will be a negative-going linear voltage ramp on the output of A_1, as illustrated in Figure 13–21(a).

When the counter reaches a specified count, it will be RESET, and the control logic will switch the negative reference voltage ($-V_{REF}$) to the input of A_1 as shown in Figure 13–21(b). At this point the capacitor is charged to a negative voltage ($-V$) proportional to the input analog voltage.

Now the capacitor discharges linearly because of the constant current from the $-V_{REF}$ as shown in Figure 13–21(c). This linear discharge produces a positive-going ramp on the A_1 output, starting at $-V$ and having a constant slope that is independent of the charge voltage. As the capacitor discharges, the counter advances from its RESET state. The time it takes the capacitor to discharge to zero depends on the initial voltage $-V$ (proportional to V_{in}) because the discharge rate (slope) is constant. When the integrator (A_1) output voltage reaches zero, the comparator (A_2) switches to the LOW state and disables the clock to the counter. The binary count is latched, thus completing one conversion cycle. The binary count is proportional to V_{in} because the time it takes the capacitor to discharge depends only on $-V$, and the counter records this interval of time.

Successive-Approximation Analog-to-Digital Converter

Successive-approximation is perhaps the most widely used method of A/D conversion. It has a much shorter conversion time than the other methods with the exception of the flash method. It also has a fixed conversion time that is the same for any value of the analog input.

Figure 13–22 shows a basic block diagram of a 4-bit successive-approximation ADC. It consists of a DAC, a successive-approximation register (SAR), and a comparator. The basic operation is as follows: The input bits of the DAC are enabled (made equal to a 1) one at a time, starting with the MSB. As each bit is enabled, the comparator produces an output that indicates whether the analog input voltage is greater or less than the output of the DAC. If the DAC output is greater than the analog input, the comparator's output is LOW, causing the bit in the register to RESET. If the output is less than the analog input, the 1 bit is retained in the register. The system does this with the MSB first, then the next most significant bit, then the next, and so on. After all the bits of the DAC have been tried, the conversion cycle is complete.

FIGURE 13–22
Successive-approximation ADC.

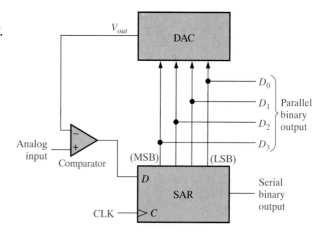

In order to better understand the operation of the successive-approximation ADC, let's take a specific example of a 4-bit conversion. Figure 13–23 illustrates the step-by-step conversion of a constant analog input voltage (5 V in this case). Let's assume that the DAC has the following output characteristic: $V_{out} = 8$ V for the 2^3 bit (MSB), $V_{out} = 4$ V for the 2^2 bit, $V_{out} = 2$ V for the 2^1 bit, and $V_{out} = 1$ V for the 2^0 bit (LSB).

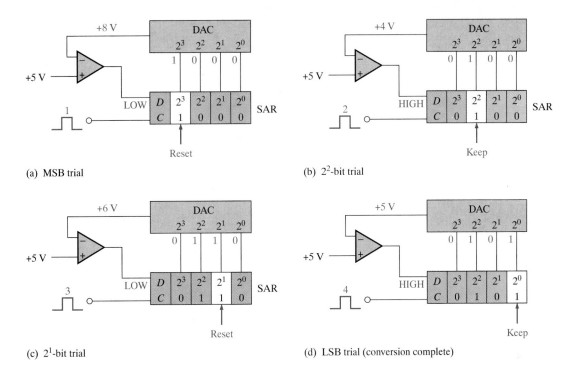

(a) MSB trial

(b) 2^2-bit trial

(c) 2^1-bit trial

(d) LSB trial (conversion complete)

FIGURE 13–23

Illustration of the successive-approximation conversion process.

Figure 13–23(a) shows the first step in the conversion cycle with the MSB = 1. The output of the DAC is 8 V. Since this is greater than the analog input of 5 V, the output of the comparator is LOW, causing the MSB in the SAR to be RESET to a 0.

Figure 13–23(b) shows the second step in the conversion cycle with the 2^2 bit equal to a 1. The output of the DAC is 4 V. Since this is less than the analog input of 5 V, the output of the comparator switches to a HIGH, causing this bit to be retained in the SAR.

Figure 13–23(c) shows the third step in the conversion cycle with the 2^1 bit equal to a 1. The output of the DAC is 6 V because there is a 1 on the 2^2 bit input and on the 2^1 bit input; 4 V + 2 V = 6 V. Since this is greater than the analog input of 5 V, the output of the comparator switches to a LOW, causing this bit to be RESET to a 0.

Figure 13–23(d) shows the fourth and final step in the conversion cycle with the 2^0 bit equal to a 1. The output of the DAC is 5 V because there is a 1 on the 2^2 bit input and on the 2^0 bit input; 4 V + 1 V = 5 V.

The four bits have all been tried, thus completing the conversion cycle. At this point the binary code in the register is 0101, which is the binary value of the analog input of 5 V. Another conversion cycle now begins, and the basic process is repeated. The SAR is cleared at the beginning of each cycle.

A Specific Analog-to-Digital Converter

The ADC0804 is an example of a successive-approximation ADC. A block diagram is shown in Figure 13–24 (next page). This device operates from a +5 V supply and has a resolution of eight bits with a conversion time of 100 μs. Also, it has guaranteed monotonicity and an on-chip clock generator. The data outputs are tristate so that it can be interfaced with a microprocessor bus system.

The basic operation of the device is as follows: The ADC0804 contains the equivalent of a 256 resistor DAC network. The successive-approximation logic sequences the network to match the analog differential input voltage ($V_{in+} - V_{in-}$) with an output from the resistive network. The MSB is tested first. After eight comparisons (sixty-four clock periods), an 8-bit binary code is transferred to output latches, and the interrupt ($\overline{INTR}$) out-

FIGURE 13–24

The ADC0804 analog-to-digital converter.

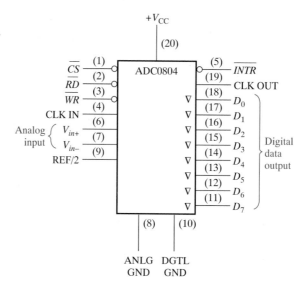

put goes LOW. The device can be operated in a free-running mode by connecting the $\overline{INTR}$ output to the write ($\overline{WR}$) input and holding the conversion start ($\overline{CS}$) LOW. To ensure start-up under all conditions, a LOW $\overline{WR}$ input is required during the power-up cycle. Taking $\overline{CS}$ low anytime after that will interrupt the conversion process.

When the $\overline{WR}$ input goes LOW, the internal successive-approximation register (SAR) and the 8-bit shift register are RESET. As long as both $\overline{CS}$ and $\overline{WR}$ remain LOW, the ADC remains in a RESET state. One to eight clock periods after $\overline{CS}$ or $\overline{WR}$ makes a LOW-to-HIGH transition, conversion starts.

When a LOW is at both the $\overline{CS}$ and $\overline{RD}$ inputs, the tristate output latch is enabled and the output code is applied to the D_0 through D_7 lines. When either the $\overline{CS}$ or the $\overline{RD}$ input returns to a HIGH, the D_0 through D_7 outputs are disabled.

SECTION 13–3 REVIEW

1. What is the fastest method of analog-to-digital conversion?
2. Which A/D conversion method uses an up/down counter?
3. Does the successive-approximation converter have a fixed conversion time?

13–4 ■ TROUBLESHOOTING DACs AND ADCs

Basic testing of DACs and ADCs includes checking their performance characteristics— such as monotonicity, offset, linearity, and gain—and checking for missing or incorrect codes. In this section, the fundamentals of testing these analog interfaces are introduced. After completing this section, you should be able to

☐ Discuss the testing of DACs for nonmonotonicity, differential nonlinearity, low or high gain, and offset error   ☐ Discuss testing ADCs for a missing code, incorrect code, and offset

Testing Digital-to-Analog Converters

The concept of DAC testing is illustrated in Figure 13–25. In this basic method, a sequence of binary codes is applied to the inputs, and the resulting output is observed. The binary code sequence extends over the full range of values from 0 to $2^n - 1$ in ascending order, where n is the number of bits.

FIGURE 13–25
Basic test setup for a DAC.

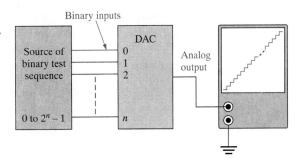

The ideal output is a straight-line stairstep as indicated. As the number of bits in the binary code is increased, the resolution is improved. That is, the number of discrete steps increases, and the output approaches a straight-line linear ramp.

D/A Conversion Errors

Several D/A conversion errors to be checked for are shown in Figure 13–26, which uses a 4-bit conversion for illustration purposes. A 4-bit conversion produces fifteen discrete steps. Each graph in the figure includes an ideal stairstep ramp for comparison with the faulty outputs.

FIGURE 13–26
Illustrations of several D/A conversion errors.

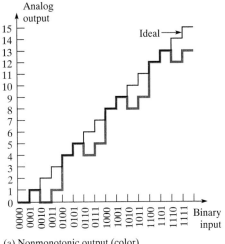

(a) Nonmonotonic output (color)

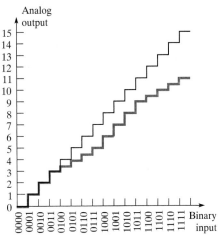

(b) Differential nonlinearity (color)

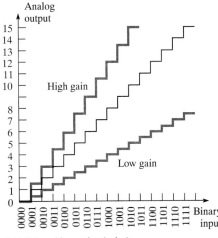

(c) High and low gains (color)

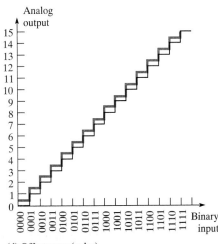

(d) Offset error (color)

Nonmonotonicity The step reversals in Figure 13–26(a) indicate nonmonotonic performance, which is a form of nonlinearity. In this particular case, the error occurs because the 2^1 bit in the binary code is interpreted as a constant 0. That is, a short is causing the bit input line to be stuck LOW.

Differential Nonlinearity Figure 13–26(b) illustrates differential nonlinearity in which the step amplitude is less than it should be for certain input codes. This particular output could be caused by the 2^2 bit having an insufficient weight, perhaps because of a faulty input resistor. We could also see steps with amplitudes greater than normal, if a particular binary weight were greater than it should be.

Low or High Gain Output errors caused by low or high gain are illustrated in Figure 13–26(c). In the case of low gain, all of the step amplitudes are less than ideal. In the case of high gain, all of the step amplitudes are greater than ideal. This situation may be caused by a faulty feedback resistor in the op-amp circuit.

Offset Error An offset error is illustrated in Figure 13–26(d). Notice that when the binary input is 0000, the output voltage is nonzero and that this amount of offset is the same for all steps in the conversion. A faulty op-amp may be the culprit in this situation.

EXAMPLE 13–4

The DAC output in Figure 13–27 is observed when a straight 4-bit binary sequence is applied to the inputs. Identify the type of error, and suggest an approach to isolate the fault.

FIGURE 13–27

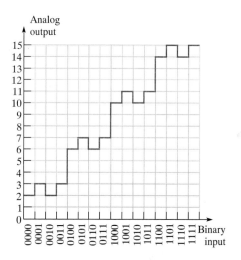

Solution The DAC in this case is nonmonotonic. Analysis of the output reveals that the device is converting the following sequence, rather than the actual binary sequence applied to the inputs.

0010, 0011, 0010, 0011, 0110, 0111, 0110, 0111, 1010, 1011, 1010, 1011, 1110, 1111, 1110, 1111

Apparently, the 2^1 bit is stuck in the HIGH (1) state. To find the problem, first monitor the bit input pin to the device. If it is changing states, the fault is internal, most likely an open. If the external pin is not changing states and is always HIGH, check for an external short to $+V$ that may be caused by a solder bridge somewhere on the circuit board. If no problem is found here, disconnect the source output from the DAC input pin, and see if the output signal is correct. If these checks produce no results, the fault is most likely internal to the DAC, perhaps a short to the supply voltage.

Related Exercise Determine the output of a DAC when a straight 4-bit binary sequence is applied to the inputs and the 2^0 bit is stuck HIGH.

Testing Analog-to-Digital Converters

One method for testing ADCs is shown in Figure 13–28. A DAC is used as part of the test setup to convert the ADC output back to analog form for comparison with the test input.

A test input in the form of a linear ramp is applied to the input of the ADC. The resulting binary output sequence is then applied to the DAC test unit and converted to a stairstep ramp. The input and output ramps are compared for any deviation.

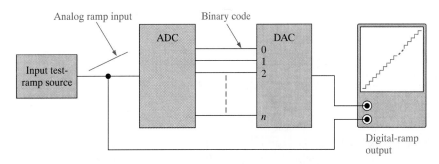

FIGURE 13–28
A method for testing ADCs.

A/D Conversion Errors

Again, a 4-bit conversion is used to illustrate the principles. Let's assume that the test input is an ideal linear ramp.

Missing Code The stairstep output in Figure 13–29(a) indicates that the binary code 1001 does not appear on the output of the ADC. Notice that the 1000 value stays for two intervals and then the output jumps to the 1010 value.

In a flash ADC, for example, a failure of one of the comparators can cause a missing-code error.

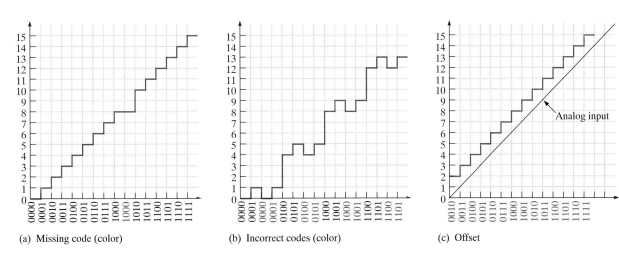

(a) Missing code (color)   (b) Incorrect codes (color)   (c) Offset

FIGURE 13–29
Illustrations of A/D conversion errors.

Incorrect Codes   The stairstep output in Figure 13–29(b) indicates that several of the binary code words coming out of the ADC are incorrect. Analysis indicates that the 2^1-bit line is stuck in the LOW (0) state in this particular case.

Offset   Offset conditions are shown in 13–29(c). In this situation the ADC interprets the analog input voltage as greater than its actual value. This error is probably due to a faulty comparator circuit.

EXAMPLE 13–5   A 4-bit flash ADC is shown in Figure 13–30(a). It is tested with a setup like the one in Figure 13–28. The resulting reconstructed analog output is shown in Figure 13–30(b). Identify the problem and the most probable fault.

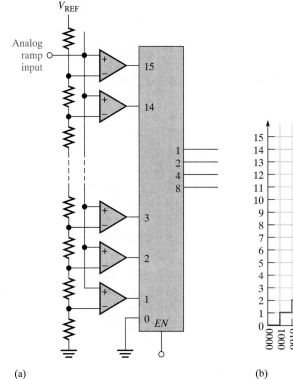

(a)

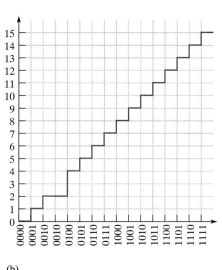

(b)

FIGURE 13–30

Solution   The binary code 0011 is missing from the ADC output, as indicated by the missing step. Most likely, the output of comparator 3 is stuck in its inactive state (LOW).

Related Exercise   Reconstruct the analog output in a test setup like in Figure 13–28 if the ADC in Figure 13–30(a) has comparator 8 stuck in the HIGH output state.

SECTION 13–4 REVIEW

1. How do you detect nonmonotonic behavior in a DAC?
2. What effect does low gain have on a DAC output?
3. Name two types of output errors in an ADC.

13–5 ■ INTERNAL SYSTEM INTERFACING

The concept of the bus was introduced in Chapter 12 in relation to memories. All the components in a given digital system such as a computer are interconnected by buses, which serve as communication paths. Physically, a bus is a set of conductive paths that serves to interconnect two or more functional components of a system or several diverse systems. Electrically, a bus is a collection of specified voltage levels and/or current levels and signals that allow the various devices connected to the bus to work properly together. After completing this section, you should be able to

- ☐ Discuss the concept of a multiplexed bus ☐ Explain the reason for tristate outputs
- ☐ Discuss the Multibus ☐ Discuss the PC bus

Basic Multiplexed Buses

Many digital systems are microprocessor-based. In such systems the microprocessor controls and communicates with other devices, such as memories and input/output (I/O) devices, via the *internal bus structure,* as indicated in Figure 13–31.

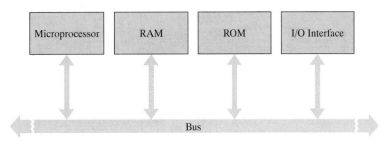

FIGURE 13–31
The interconnection of microprocessor-based system components by a bidirectional, multiplexed bus.

A **bus** is multiplexed so that any of the devices connected to it can either send or receive data to or from one of the other devices. A sending device is often called a **source,** and a receiving device is often called an **acceptor.** At any given time, there is only one source active. For example, the RAM may be sending data to the input/output (I/O) interface under control of the microprocessor.

Bus Signals

With synchronous bus control, the microprocessor usually originates all control and timing signals. The other devices then synchronize their operations to those control and timing signals. With asynchronous bus control, the control and timing signals are generated jointly by a source and an acceptor. The process of jointly establishing communication is called **handshaking.** A simple example of a handshaking sequence is given in Figure 13–32.

An important control function is called **bus arbitration.** Arbitration prevents two sources from trying to use the bus at the same time.

FIGURE 13–32
An example of a handshaking sequence.

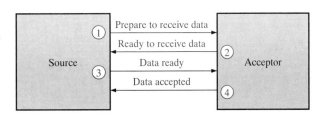

Connecting Devices to the Bus

Tristate or open-collector drivers (see Chapter 15) are normally used to interface the outputs of a source device to the bus. Usually more than one source is connected to a bus, but only one can have access at any given time. All the other sources must be disconnected from the bus to prevent interference.

Tristate circuits are used to connect a source to the bus or disconnect it from the bus, as illustrated in Figure 13–33(a) for the case of two sources. The select input is used to connect either source A or source B but not both at the same time to the bus. When the select input is LOW, source A is connected and source B is disconnected. When the select input is HIGH, source B is connected and source A is disconnected. A switch equivalent of this action is shown in part (b) of the figure.

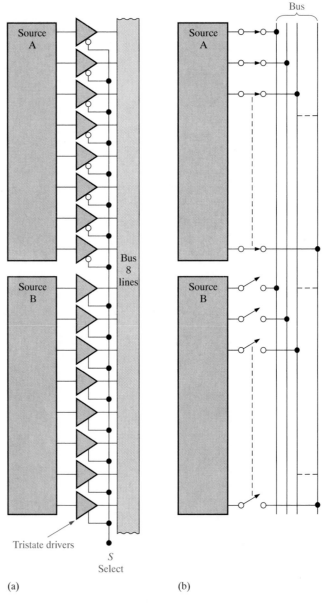

FIGURE 13–33
Tristate driver interface to a bus.

Recall from Chapter 12 that when the enable input of a tristate circuit is not active, the device is in a high-impedance (high-Z) state and acts like an open switch. Many digital ICs provide internal tristate drivers for the output lines. A tristate output is indicated by a ∇ symbol as shown in Figure 13–34.

FIGURE 13–34
Method of indicating tristate outputs on an IC device.

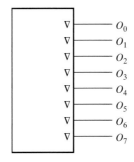

Bus Contention

Bus contention occurs when two or more devices try to output opposite logic levels on the same common bus line. The most common form of bus contention is when one device has not completely turned off before another device connected to the bus line is turned on. This generally occurs in memory systems when switching from the READ mode to the WRITE mode or vice versa and is the result of a timing problem.

Multiplexed I/Os

Some devices that send and receive data have combined input and output lines, called I/O ports, that must be multiplexed onto the data bus. Bidirectional tristate drivers interface this type of device with the bus as illustrated in Figure 13–35(a).

FIGURE 13–35
Multiplexed I/O operation.

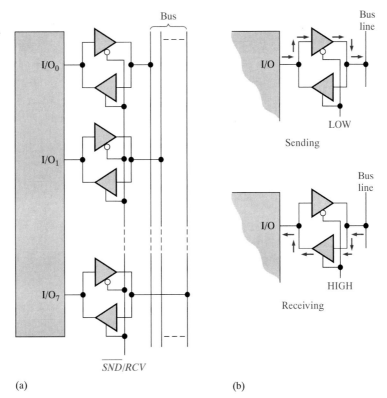

Each I/O port has a pair of tristate drivers. When the $\overline{SND}/RCV$ (Send/Receive) line is LOW, the upper tristate driver in each pair is enabled and the lower one disabled. In this state, the device is acting as a source and sending data to the bus. When the $\overline{SND}/RCV$ line is HIGH, the lower tristate driver in each pair is enabled so that the device is acting as an acceptor and receiving data from the bus. This operation is illustrated in Figure 13–35(b). Some devices provide for multiplexed I/O operation with internal circuitry.

Standard Buses

Several so-called standard buses are used to provide interfacing for the various internal components of digital systems. These standard buses assure compatibility of printed circuit board size, pin numbers, type of signal on each pin, and input and output characteristics. The standard bus allows for system expansion by specifying the conditions under which expansion or replacement units or modules must operate in order to interface properly with existing modules.

To introduce the concept of standard buses, two examples will be discussed briefly. A full, detailed coverage of these buses is beyond the scope of this book.

The Multibus   The Multibus is a general-purpose bus system developed by Intel but used in industry. Some manufacturers offer products that are compatible with this particular bus system. The Multibus provides a flexible interface that can be used to interconnect a wide variety of microcomputer devices or modules. The modules in a Multibus system are designated as *masters* or *slaves.* Masters obtain use of the bus and initiate data transfers. Slaves are devices that cannot transfer data themselves or control the bus. A major feature of the Multibus is that several processors (masters) can be connected to the bus at the same time to implement a multiprocessing operation, in which each processor performs a dedicated task.

The Multibus provides a total of 86 lines. There are 16 bidirectional data lines, 20 address lines, 8 interrupt lines, and various control lines, command lines, and power lines. The Multibus standard defines the physical and electrical parameters for devices using the bus. A generalized Multibus system configuration is shown in Figure 13–36.

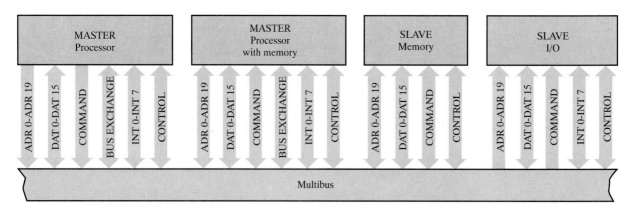

FIGURE 13–36
A Multibus system configuration.

The PC Bus   The IBM PC bus structure actually includes three buses—the address, the data bus, and the control bus—as shown in Figure 13–37 in a greatly simplified representation. The bus runs along the PC system board and connects the computer to expansion slots for adding various compatible boards for memory expansion, video interface, disk interface, and other peripheral interfaces.

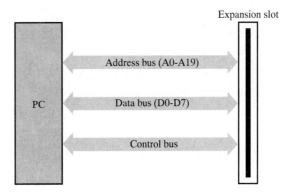

FIGURE 13–37
Basic PC bus.

A diagram of the basic 62-pin PC bus connector is shown in Figure 13–38. There are twenty address lines (A0–A19) and eight data lines (D0–D7). The control bus and various power, ground, and timing signals occupy the remaining line.

FIGURE 13–38
I/O expansion connector for basic PC bus.

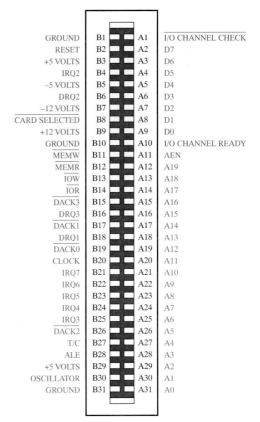

SECTION 13–5 REVEW

1. Why are tristate drivers required to interface digital devices to a bus?
2. What is the purpose of a bus system?

13–6 ■ DIGITAL EQUIPMENT INTERFACING

In the last section, the idea of internal buses was introduced. These buses provide inter-connections for the modules or units within a single digital system. In this section, we introduce buses that are used to interface separate instruments. These buses can be considered external buses. External buses can be used to interface a microcomputer with test instruments or a computer terminal with a printer. After completing this section, you should be able to

☐ Describe the GPIB (IEEE-488)   ☐ Discuss certain aspects of the VXI bus
☐ Describe the RS-232C bus standard   ☐ Discuss SCSI   ☐ Discuss basic bus troubleshooting

The General-Purpose Interface Bus (GPIB)

The **GPIB** is a bus system that provides an orderly and predictable way to transfer parallel data and maintain control between various digital instruments. Other designations, such as **IEEE-488** or **HPIB,** also refer to the GPIB. This bus system is as close to a universal standard as you can get. A common application of the GPIB is the interfacing of test and measurement instruments with a computer to create an automated test system.

GPIB Specifications   One type of GPIB connector is a 24 pin miniature ribbon type as shown in Figure 13–39. The signal name is indicated for each pin.

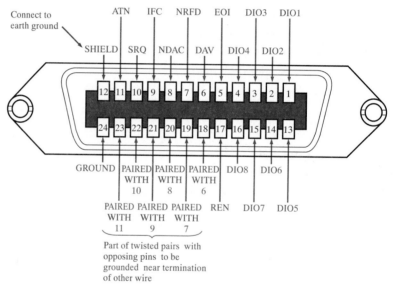

FIGURE 13–39
The GPIB connector.

In a typical GPIB setup, there is a designated controller (usually a computer) and one or more controlled devices (test instruments, for example). When the controller issues a command for a controlled device to perform a specified operation, such as a frequency measurement, it is said that the controller "talks" and the controlled device "listens."

A **listener** is an instrument capable of receiving data over the parallel interface bus when it is addressed by the controller. Examples of listeners are printers, monitors, programmable power supplies, and programmable signal generators. The GPIB allows up to fifteen active listeners on the bus at one time.

A **talker** is an instrument capable of transmitting data over the bus. Examples are DMMs that output data and frequency counters that output data. The GPIB permits only one active talker on the bus at any given time.

Some instruments can transmit and receive and are called *talker/listeners.* An example is the **modem,** a device for interfacing digital systems to the telephone line. The term *modem* is a contraction of *modulator/demodulator.*

The **controller** is an instrument that can specify each of the other instruments on the bus as either a talker or a listener for the purpose of data transfer. A controller generally can be both a talker and a listener itself. A properly configured microcomputer is an example of a controller.

As mentioned, there are a total of twenty-four lines in the GPIB, including both signal lines and ground lines. The signal lines are divided into three functional groups: the data bus, the data transfer control bus, and the interface management bus. Figure 13–40 illustrates the three bus groupings that make up the GPIB in a typical arrangement.

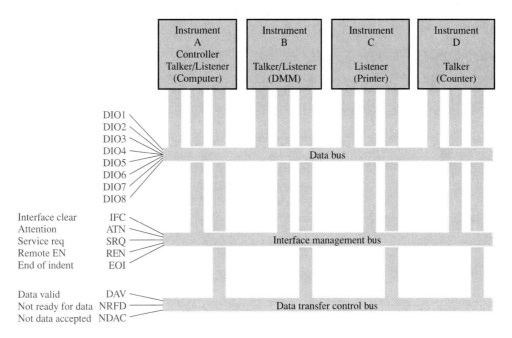

FIGURE 13–40
A typical GPIB interface arrangement showing the three bus groups.

The Data Bus and the Data Transfer Control Bus   Eight data bits are transferred in parallel on the bidirectional data bus (DIO1–DIO8). This type of transfer is a serial-byte–parallel-bit format. Every byte that is transferred undergoes a handshaking operation via the transfer bus. The three active-LOW handshaking lines of the transfer bus indicate if data are valid (DAV), if the addressed instrument is not ready for data (NRFD), or if the data are not accepted (NDAC). More than one instrument can accept data at the same time, and the slowest instrument sets the rate of transfer. Figure 13–41 shows the timing diagram for the GPIB handshaking sequence, and Table 13–1 describes the handshaking signals.

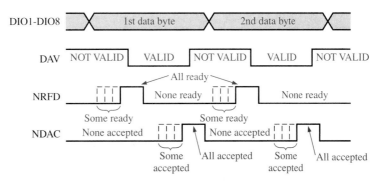

FIGURE 13–41
Timing diagram for the GPIB handshaking sequence.

TABLE 13–1

The GPIB handshaking signals

Name	Description
DAV	**Data Valid:** After the talker detects a HIGH on the NRFD line, a LOW is placed on this line by the talker when the data on its I/O are settled and valid.
NRFD	**Not Ready for Data:** The listener places a LOW on this line to indicate that it is not ready for data. A HIGH indicates that it is ready. The NRFD line will not go HIGH until all addressed listeners are ready to accept data.
NDAC	**Not Data Accepted:** The listener places a LOW on this line to indicate that it has not accepted data. When it accepts data from its I/O, it releases its NDAC line. The NDAC line to the talker does not go HIGH until the last listener has accepted data.

Interface Management Bus These five lines control the orderly flow of data. The ATN (attention) line is monitored by all instruments connected to the bus. When the ATN line is active, the management bus is placed in the command mode. In this mode, the controller selects the specific interface operation, designates the talkers and the listeners, and provides specific addressing for the listeners. Each instrument designed to the GPIB standard has a specific identifying address that is used by the controller in the command mode. Table 13–2 describes the GPIB management lines and their functions.

TABLE 13–2

The GPIB management lines

Name	Description
ATN	**Attention:** Causes all the devices on the bus to interpret data, as a controller command or address and activates the handshaking function.
IFC	**Interface Clear:** Initializes the bus.
SRQ	**Service Request:** Alerts the controller that a device needs to communicate.
REN	**Remote Enable:** Enables devices to respond to remote program control.
EOI	**End or Identify:** Indicates the last byte of data to be transferred.

Limitations of the GPIB There are three basic limitations on the use of the GPIB: (1) the cable length cannot exceed 15 meters, (2) there can be no more than one device per meter, and (3) the capacitive loading cannot exceed 50 pF per device.

The cable-length limitation can be overcome by the use of bus extenders and modems. A bus extender provides for cable-interfacing of instruments that are separated by a distance greater than that allowed by the GPIB cable limitation or for communicating over great distances via modem-interfaced telephone lines. The use of bus extenders is illustrated in Figure 13–42.

The VXI Bus

The **VXI** bus is an extension of a standard bus for modular computer systems called VME. VXI stands for *VME eXtension for Instrumentation*. The concept of the VXI bus is similar to the GPIB (IEEE-488), which is to interface various equipment to form modular instru-

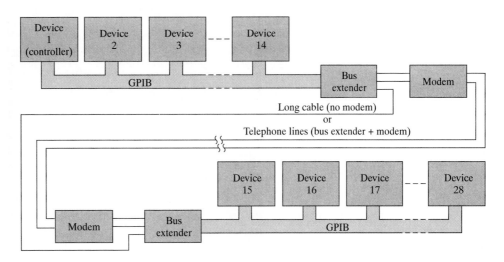

FIGURE 13–42
A bus extender and a modem can be used for interfacing remote GPIB systems.

mentation systems. VXI provides for faster data transfer than does GPIB and also avoids much of the duplication of equipment necessary in GPIB systems. In short, VXI improves on the GPIB.

The VXI bus is made up of several component buses: The VME computer bus, the clock and sync bus, the star bus, the trigger bus, the local bus, the analog sum bus, the module identification bus, and the power distribution bus.

The RS-232C Serial Interface

The RS-232C is the interface standard for serial transfer of data bits. This standard is commonly used for interfacing data terminal equipment (**DTE**) with data communications equipment (**DCE**). An example is a serial computer terminal (DTE) interfaced with a modem (DCE).

This standard specifies twenty-five lines, but in many applications only a few of the lines are used. For example, at its simplest extreme, the RS-232C interface can be implemented with only two wires: a signal and a ground. Figure 13–43 illustrates a more common interface configuration, consisting of eight lines: transmitted data (TD), received data (RD), request to send (RTS), clear to send (CTS), data set ready (DSR), data terminal ready (DTR), protective ground, and signal ground.

DTE		DCE
	Protective ground	
	Transmitted data (TD)	
	Received data (RD)	
	Request to send (RTS)	
	Clear to send (CTS)	
	Data set ready (DSR)	
	Signal ground	
	Data terminal ready (DTR)	

FIGURE 13–43
A typical RS-232C interface.

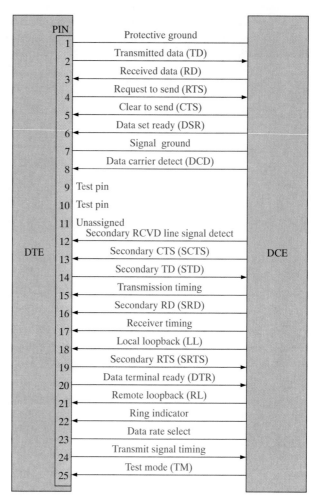

FIGURE 13–44
A full RS-232C interface.

A full 25-line interface showing all specified signals with connector pin numbers is illustrated in Figure 13–44. Electrical and mechanical specifications for the RS-232C interface standard are listed in Table 13–3.

TABLE 13–3
Specifications for the RS-232C

Parameter	Specification
DTE connector	DB-25 male
DCE connector	DB-25 female
Maximum cable length	capacitance limited (2500 pF)
Maximum data rate	20 kbits/s
Number of drivers on line	1 driver
Number of receivers on line	1 receiver
Driver output swing	± 5 V min, ± 15 V max
Driver load	3 kΩ to 7 kΩ
Driver slew rate	30 V/μs max
Receiver input resistance	3 kΩ to 7 kΩ
Receiver input threshold	± 3 V
Receiver input range	± 30 V max

SCSI

SCSI (pronounced *skuh-zee*) stands for Small Computer System Interface and is an industry standard interface for personal computers. The SCSI provides for parallel data communication between a computer and hard disks, tape backup systems, printers, and other devices. The standard SCSI connector format is shown in Figure 13–45 and the signal list is given in Table 13–4.

FIGURE 13–45
SCSI connector.

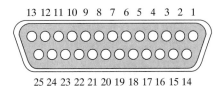

TABLE 13–4
SCSI signals

Pin Number	Signal Name	Signal Description	Pin Number	Signal Name	Signal Description
1	REQ/	Request	14	GND	Signal ground
2	MSG/	Message	15	C/D/	Command/Data
3	I/O/	Input/Output	16	GND	Signal ground
4	RST/	SCSI bus reset	17	ATN/	Attention
5	ACK/	Acknowledge	18	GND	Signal ground
6	BSY/	Busy	19	SEL/	Select
7	GND	Signal ground	20	DBP/	Data parity
8	DB0/	Data bit 0	21	DB1/	Data bit 1
9	GND	Signal ground	22	DB2/	Data bit 2
10	DB3/	Data bit 3	23	DB4/	Data bit 4
11	DB5/	Data bit 5	24	GND	Signal ground
12	DB6/	Data bit 6	25	TPWR	Terminator power
13	DB7/	Data bit 7			

Troubleshooting Buses

Troubleshooting bus systems is very difficult, if not impossible, without specialized equipment, such as a GPIB bus-system analyzer. A GPIB analyzer allows you to see the status of all bus lines as well as the actual codes by completely exercising the talkers, the listeners, or the controller. Normally, the system can be single-stepped for software debugging or run at the normal rate to test for system-related faults. The analyzer can operate as a talker or a listener, and data can be examined and compared at any preselected point in a transfer sequence to check for timing-related problems.

Also, RS-232 bus analyzers are available to ease the problem of troubleshooting these bus systems.

SECTION 13–6 REVIEW

1. True or false: The GPIB provides for parallel data transfer.
2. True or false: The RS-232C provides for serial data transfer.
3. Give an example of a GPIB application.
4. Give an example of an RS-232C application.
5. What does SCSI stand for?

13-7 ■ DIGITAL SYSTEM APPLICATION

In this system application, an ADC is applied in a satellite antenna positioning system. Analog inputs indicating the position of the antenna in terms of elevation and azimuth are fed into an ADC that converts the positions into digital code for display. The antenna can be adjusted remotely until it is in a desired position. After completing this section, you should be able to

☐ Explain the basic operation of an analog multiplexer   ☐ Explain the overall operation of the system   ☐ Discuss the use of the ADC in this particular application
☐ Develop a test procedure

General System Operation

Two angular positions are required to properly aim the antenna at a desired satellite. The azimuth position is along an arc from east to west (or north to south) parallel with the horizon. The elevation position is from horizon to horizon in an overhead arc. These angular movements of the antenna are illustrated in Figure 13–46. Two motors drive the antenna to the proper position, one for azimuth and one for elevation. Also, two position transducers (potentiometers are used as angle sensors in this application) produce voltages proportional to the current directions of the antenna as determined by the angular positions of the motor shafts.

FIGURE 13–46

Antenna dish with position controls and sensors for azimuth and elevation.

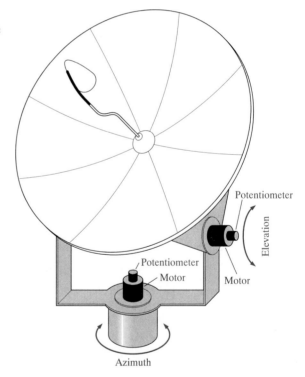

A basic block diagram of the antenna positioning system is shown in Figure 13–47. The analog multiplexer has one input from the azimuth potentiometer and one from the elevation potentiometer. The select input alternately switches the two analog voltages to the ADC for conversion to an 8-bit digital code. During one interval, the analog voltage representing the current azimuth position is converted to digital and displayed. During the next interval, the analog voltage for the current elevation position is converted to digital and displayed. The antenna position is controlled by motor controls that can be located away

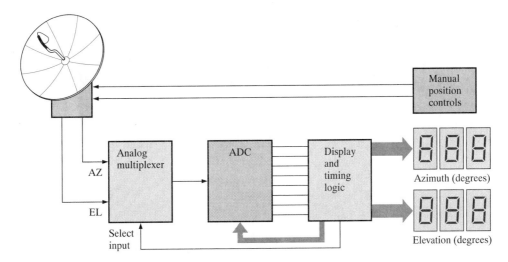

FIGURE 13–47
Simplified block diagram of the antenna positioning system.

from the antenna and connected by cable to the motors. By observing the azimuth and elevation position readouts, the antenna can be brought to the desired position.

The Analog Multiplexer

A general description of analog multiplexers is provided to familiarize you with this type of device before we apply it in the system.

In data acquisition systems where inputs from several different sources must be independently converted to digital form for processing, a technique called *multiplexing* is used. A separate analog switch is used for each analog source as illustrated in Figure 13–48 for a 4-channel system. In this type of application, all of the outputs of the analog switches are connected together to form a common output and only one switch can be closed at a given time. The common switch outputs are connected to the input of an op-amp voltage follower as indicated.

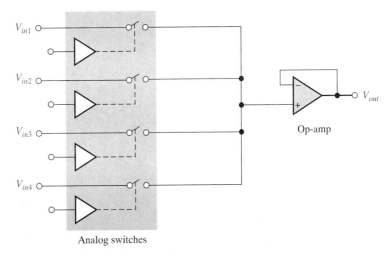

FIGURE 13–48
A basic 4-channel analog multiplexer.

FIGURE 13–49
The AD9300 analog multiplexer.

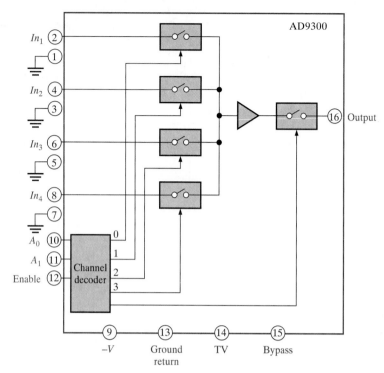

A good example of an IC analog multiplexer is the AD9300 shown in Figure 13–49. This device contains four analog switches that are controlled by a channel decoder. The inputs A_0 and A_1 determine which one of the four switches is on. If A_0 and A_1 are both LOW, input In_1 is selected. If A_0 is HIGH and A_1 is LOW, input In_2 is selected. If A_0 is LOW and A_1 is HIGH, input In_3 is selected. If A_0 and A_1 are both HIGH, input In_4 is selected. The enable input controls the switch that connects or disconnects the output.

The Potentiometer

In this application, the potentiometer is used as a position transducer to convert the angular shaft position of the motor to a proportional dc voltage. The potentiometer is mechanically linked to the associated motor so as the motor shaft turns, the wiper of the potentiometer slides along the resistive element. It is calibrated so that the smallest angle produces the smallest voltage. In Figure 13–50, a simplified diagram shows the basic construction and the schematic indicates minimum and maximum angles corresponding to the wiper position.

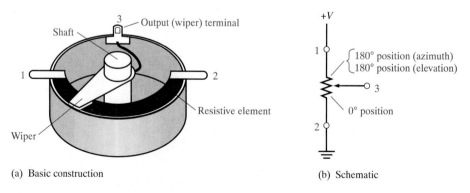

(a) Basic construction

(b) Schematic

FIGURE 13–50
The potentiometer as an angular transducer.

Basic Operation of the Analog-to-Digital Circuits

Figure 13–51 shows the specific analog-to-digital circuits used in the system. The analog voltages from the azimuth and elevation potentiometers on the antenna are fed into two of the inputs to the AD9300 analog multiplexer. The other two analog inputs are not used and are grounded. The digital timing logic produces a square wave that is used as the select input, A_0. When the square wave is LOW, Input 1 is selected and the azimuth voltage is switched to the input of the ADC. When the square wave is HIGH, Input 2 is selected and the elevation voltage is switched to the input of the ADC. Although the AD673 is used here, other ADCs such as the ADC0804 can be used.

As each analog voltage appears on the ADC input, the timing logic issues a convert pulse to initiate a conversion by the ADC. After the conversion is complete, the ADC asserts a LOW on the data ready line to the timing logic, which responds by asserting a LOW on the data enable line. Then the ADC places the 8-bit binary code on its outputs. The timing diagram is illustrated in Figure 13–52.

FIGURE 13–51
Conversion circuits and display/timing logic for the antenna positioning system.

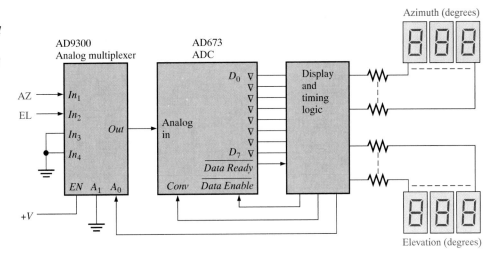

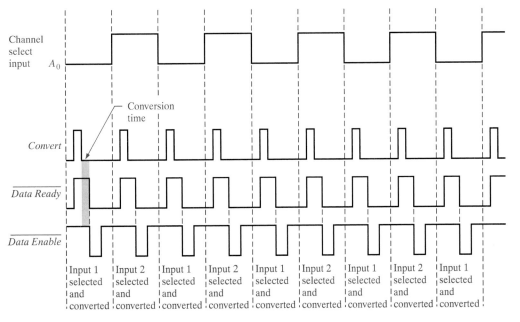

FIGURE 13–52
Basic timing diagram for the system in Figure 13–51.

As the azimuth and elevation analog voltages are alternately converted to a binary code, the display logic stores each code in its associated register. The binary code is then converted to BCD and then to 7-segment format to illuminate the 3-digit display. The displays are updated many times a second by the multiplexing action. The updating rate depends on the frequency of the square wave used as the multiplexer select input.

■ THE DIGITAL WORKBENCH

The basic logic diagram in Figure 13–51 and the timing diagram in Figure 13–52 indicate the general display and timing logic requirements but do not represent the detailed logic necesary to implement the system. In the following workbenches, you will develop, from given specifications, the detailed circuitry required to implement the display and timing logic functions of the system.

■ DIGITAL WORKBENCH 1: Design of Display Logic

This workbench focuses on the design of the display logic for which the basic block diagram is shown below. Reference to earlier chapters and/or device data sheets is necessary.

■ **Activity 1 Binary-to-BCD Code Converter**
The 8-bit binary code from the ADC is to be converted to a 12-bit BCD code (three digits). Specify a circuit or combination of circuits that will accomplish this function.

■ **Activity 2 Azimuth and Elevation Registers**
The BCD for the azimuth position is stored in one register and the BCD for the elevation position is stored in another. The registers are alternately updated. Notice that a parallel-in/parallel-out operation is required. Specify IC registers that will meet the requirements.

■ **Activity 3 BCD-to-7-Segment Decoder**
A BCD-to-7-segment decoder is required for each of the three azimuth position digits and each of the three elevation position digits. Specify an appropriate device. Also specify a 7-segment display and determine the appropriate resistor values for interfacing the decoders to the displays.

■ **Activity 4** Develop a detailed logic diagram using the devices specified in the preceding activities. Show all necessary details including device pin numbers.

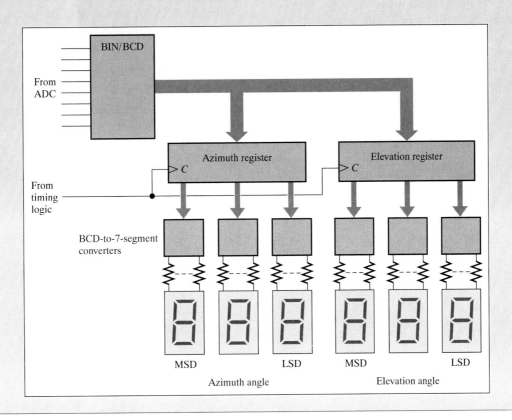

■ DIGITAL WORKBENCH 2: Design of Timing Logic

This workbench focuses on the design of the timing logic for which the basic timing diagram for two conversions is shown below. Reference to earlier chapters and/or device data sheets is necessary.

■ *Activity 1   Channel Select Waveform*
This waveform alternately selects channel 1 (In_1) and channel 2 (In_2) of the ADC for conversion to binary.
Specifications: Square wave with a frequency that will provide a minimum of 100 conversions of each analog input per second. Develop a circuit for generating this square wave.

■ *Activity 2   Convert Waveform*
Each positive-going convert pulse initiates a conversion in the ADC on its trailing edge. Conversion time for the AD673 is a maximum of 30 μs. There must be one pulse for each conversion. Specify a circuit to generate this waveform.
Specifications: Positive-going pulses with a minimum pulse width of 500 ns delayed from each edge of the channel select waveform by a minimum of 100 ns.

■ *Activity 3   Data Enable Waveform*
When the AD673 completes a conversion, it places a LOW on the data ready line which remains LOW until the leading edge of the next convert pulse. The data enable line must remain HIGH during a conversion. At the end of a conversion, the data enable goes LOW to place the binary code on the ADC output lines. Specify a circuit to produce this waveform.
Specifications: Negative-going pulses triggered from the negative-going transition of the data ready line and with a minimum pulse width of 500 ns.

■ *Activity 4   Azimuth and Elevation Register Clock*
Determine the timing of the signal required to properly clock the parallel data into the registers. Specify how it is to be generated.

■ *Activity 5*   Develop a detailed logic diagram of the timing logic and combine with the diagram of the display logic developed in the Workbench 1.

■ *Activity 6*   Write a technical report to provide a detailed description of the operation of the system and a parts list.

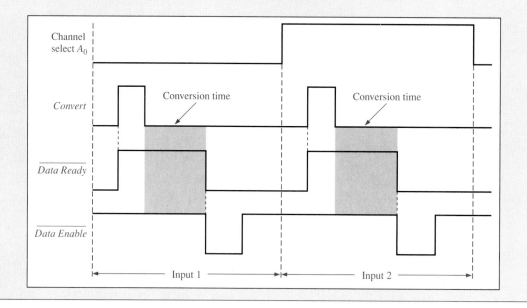

SECTION 13–7
REIEW

1. Why are three digits required for the azimuth and elevation displays?
2. Discuss how the analog voltages representing the azimuth and elevation positions are generated.
3. Why are two registers required in the display logic?

■ **SUMMARY**

- Digital-to-analog converters (DACs): Binary-weighted input, $R/2R$ ladder (most common)
- Analog-to-digital converters (ADCs): Flash (fastest), Digital-ramp, Tracking, Single-slope, Dual-slope, Successive-approximation (most common)
- Internal system buses: Multibus, PC bus
- Equipment buses: GPIB (IEEE-488, HPIB), VXI, RS-232C, SCSI

■ **SELF-TEST**

1. In a binary-weighted digital-to-analog converter (DAC), the resistors on the inputs
   (a) determine the amplitude of the analog signal
   (b) determine the weights of the digital inputs
   (c) limit the power consumption
   (d) prevent loading on the source
2. Each resistor value in a binary-weighted DAC is
   (a) the same as the other values
   (b) half of the feedback resistor
   (c) twice the value of the next lowest value resistor
   (d) half the value of the next lowest value resistor
3. In an $R/2R$ DAC, there are
   (a) four values of resistors   (b) one resistor value
   (c) two resistor values   (d) a number of resistor values equal to the number of inputs
4. An 8-bit DAC has a resolution of
   (a) 0.1%   (b) 0.392%   (c) 1%   (d) 3.92%
5. Ideally, the accuracy of a 4-bit DAC is
   (a) approximately 3.1%   (b) approximately 0.31%
   (c) approximately 6.2%   (d) approximately 0.62%
6. The type of analog-to-digital converter (ADC) with the fastest conversion time is
   (a) flash   (b) digital-ramp   (c) tracking   (d) simultaneous-conversion
7. Most devices are interfaced to a bus with
   (a) totem-pole outputs   (b) tristate drivers   (c) *pnp* transistors   (d) resistors
8. The basic PC bus consists of
   (a) a 20-bit address bus, an 8-bit data bus, and a control bus
   (b) an 8-bit address bus, a 20-bit data bus, and a control bus
   (c) 62 lines
   (d) answers (a) and (c)
9. The devices operating on a GPIB are called
   (a) source and load   (b) talker and listener
   (c) transmitter and receiver (d) donor and acceptor
10. The RS-232C is
    (a) a standard interface for parallel data   (b) a standard interface for serial data
    (c) an enhancement of the IEEE-488 interface   (d) the same as SCSI

■ PROBLEMS

SECTION 13–1   Interfacing the Digital and Analog Worlds

1. The analog curve in Figure 13–53 is sampled at 1 ms intervals. Represent the total curve by a series of 4-bit binary numbers.

FIGURE 13–53

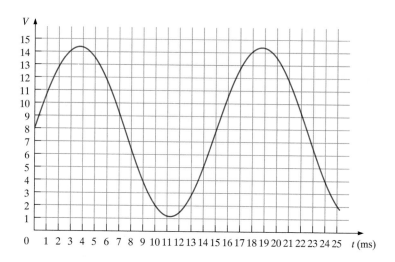

2. Sketch the digital reproduction of the curve in Problem 1.
3. Graph the analog function represented by the following sequence of binary numbers: 1111, 1110, 1101, 1100, 1010, 1001, 1000, 0111, 0110, 0101, 0100, 0101, 0110, 0111, 1000, 1001, 1010, 1011, 1100, 1100, 1100, 1011, 1010, 1001.

SECTION 13–2   Digital-to-Analog (D/A) Conversion

4. The input voltage to a certain op-amp inverting amplifier is 10 mV, and the output is 2 V. What is the closed-loop voltage gain?
5. To achieve a closed-loop voltage gain of 330 with an inverting amplifier, what value of feedback resistor do you use if $R_{in} = 1$ kΩ?
6. In the 4-bit DAC in Figure 13–7 (p. 658), the lowest-weighted resistor has a value of 10 kΩ. What should the values of the other input resistors be?
7. Determine the output of the DAC in Figure 13–54(a) if the sequence of 4-bit numbers in part (b) is applied to the inputs. The data inputs have a low value of 0 V and a high value of +5 V.
8. Repeat Problem 7 for the inputs in Figure 13–55.
9. Determine the resolution expressed as a percentage, for each of the following DACs:
   (a) 3-bit   **(b)** 10-bit   **(c)** 18-bit

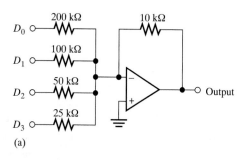

(a)

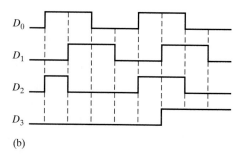

(b)

FIGURE 13–54

FIGURE 13–55

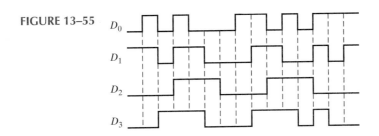

SECTION 13–3 Analog-to-Digital (A/D) Conversion

10. Determine the binary output code of a 3-bit flash ADC for the analog input signal in Figure 13–56. The sampling rate is 100 kHz.

FIGURE 13–56

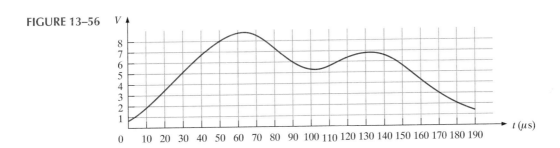

11. Repeat Problem 10 for the analog waveform in Figure 13–57.

FIGURE 13–57

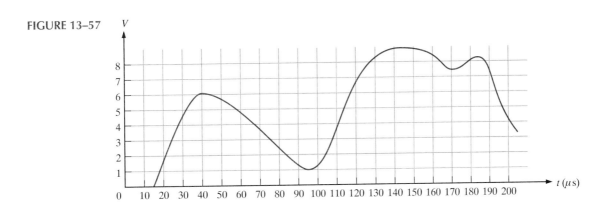

12. For a 3-bit digital-ramp ADC, the reference voltage advances one step every microsecond. Determine the encoded binary sequence for the analog signal in Figure 13–58.

FIGURE 13–58

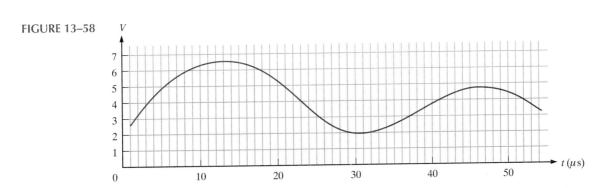

FIGURE 13–59

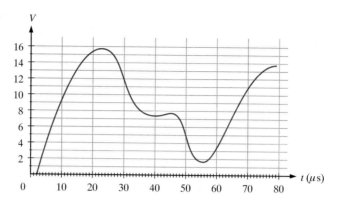

13. For a 4-bit digital-ramp ADC, assume that the clock period is 1 μs. Determine the binary sequence on the output for the analog signal in Figure 13–59.

14. Repeat Problem 12 for a tracking ADC.

15. For a certain 4-bit successive-approximation ADC, the maximum ladder output is +8 V. If a constant +6 V is applied to the analog input, determine the sequence of binary states for the SAR.

 ## SECTION 13–4   Troubleshooting DACs and ADCs

16. Develop a circuit for generating an 8-bit binary test sequence for the test setup in Figure 13–25 (p. 673).

17. A 4-bit DAC has failed in such a way that the MSB is stuck in the 0 state. Draw the analog output when a straight binary sequence is applied to the inputs.

18. A straight binary sequence is applied to a 4-bit DAC, and the output in Figure 13–60 is observed. What is the problem?

FIGURE 13–60   Output

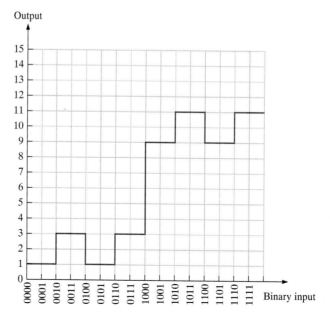

19. An ADC produces the following sequence of binary numbers when an analog signal is applied to its input: 0000, 0001, 0010, 0011, 0100, 0101, 0110, 0111, 0110, 0101, 0100, 0011, 0010, 0001, 0000.

   (a) Reconstruct the input digitally.

   (b) If the ADC failed so that the code 0111 were missing, what would the reconstructed output look like?

SECTION 13–5 Internal System Interfacing

20. In a simple serial transfer of eight data bits from a source device to an acceptor device, the handshaking sequence in Figure 13–61 is observed on the four generic bus lines. By analyzing the time relationships, identify the function of each signal, and indicate if it originates at the source or at the acceptor.

FIGURE 13–61

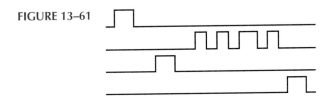

21. Determine the signal on the bus line in Figure 13–62 for the data-input and enable waveforms shown.

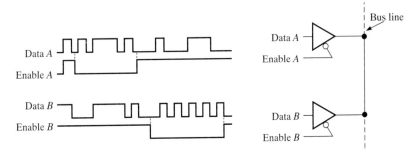

FIGURE 13–62

22. In Figure 13–63(a) data from the two sources are being placed on the data bus under control of the select line. The select waveform is shown in Figure 13–63(b). Determine the data-bus waveforms for the device output codes indicated.

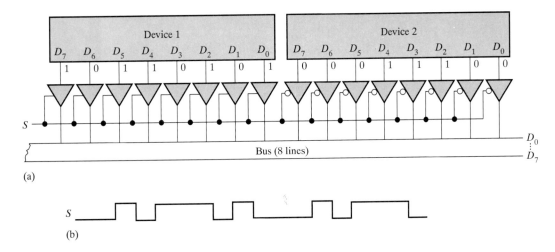

(a)

(b)

FIGURE 13–63

SECTION 13–6   Digital Equipment Interfacing

23. Eight GPIB-compatible instruments are connected to the bus. How many more can be added without exceeding the specifications?

24. Consider the GPIB interface between a talker and a listener as shown in Figure 13–64(a). From the handshaking timing diagram in part (b), determine how many data bytes are actually transferred to the listening device.

FIGURE 13–64

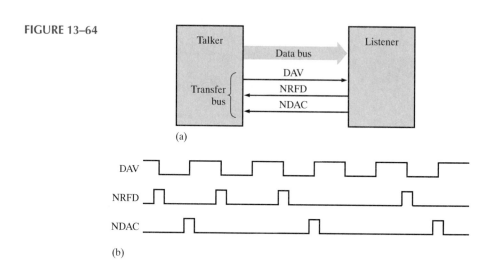

(a)

(b)

25. Describe the operations depicted in the GPIB timing diagram of Figure 13–65. Sketch a basic block diagram of the system involved in this operation.

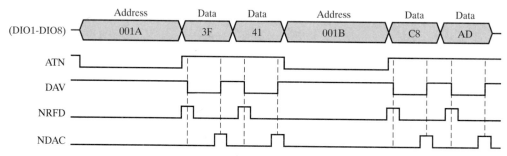

FIGURE 13–65

26. A talker sends a data byte to a listener in a GPIB system. Simultaneously, a DTE sends a data byte to a DCE on an RS-232C interface. Which system will receive the complete data byte first? Why?

■ System Applications

SECTION 13–7   Digital System Application

27. Determine the maximum frequency for the select input waveform with the limitations specified in Workbench 2.

28. Determine the resolution of the ADC used in the antenna positioning system.

29. Determine the smallest incremental change in the azimuth and elevation angles that can be measured assuming ranges from 0° to 180°.

30. The antenna positioning system can be simplified by limiting it to positioning in one dimension at a time; that is, you must first position it in the azimuth and then in the elevation. Develop a basic block diagram for this approach.

▪ ANSWERS TO SECTION REVIEWS

SECTION 13–1

1. In nature, quantities are analog.
2. To convert an analog quantity to digital form is A/D conversion.
3. To convert a digital quantity to analog form is D/A conversion.

SECTION 13–2

1. In a binary-weighted DAC, each resistor has a different value.
2. $[1/(2^4 - 1)]\ 100 = 6.67\%$

SECTION 13–3

1. Simultaneous (flash) method is fastest.
2. Tracking ADC uses an up/down counter.
3. Yes, sucessive approximation has fixed conversion time.

SECTION 13–4

1. A step reversal indicates nonmonotonic behavior in a DAC.
2. Step amplitudes in a DAC are less than ideal with low gain.
3. Missing code and incorrect code are types of ADC output errors.

SECTION 13–5

1. Tristate drivers allow devices to be completely disconnected from the bus when not in use, thus preventing interference with other devices.
2. A bus interconnects all the devices in a system and makes communication between devices possible.

SECTION 13–6

1. True, GPIB provides parallel data transfer.
2. True, RS-232C provides data transfer.
3. Automated test system is a GPIB application.
4. Data communications is an RS-232C application.
5. SCSI is small computer system interface.

SECTION 13–7

1. Three digits are required to represent angles from $0°$ to $180°$.
2. The position of the potentiometer wiper is controlled by the motor shaft position which corresponds to the antenna position. The voltage output is proportional to the wiper position.
3. Two registers are required to store the current azimuth and elevation position codes.

14

INTRODUCTION TO MICRO-PROCESSORS AND MICRO-COMPUTERS

■ CHAPTER OBJECTIVES

☐ Name the basic elements of a microprocessor
☐ Name the basic units of a microcomputer
☐ Discuss the characteristics and development of the Intel microprocessor family
☐ Discuss the characteristics and development of the Motorola microprocessor family
☐ Distinguish between assembly language and machine language
☐ Explain the basic operation of an 8086/8088 CPU
☐ Explain the basic architecture of the 8086/8088 microprocessors
☐ Explain the multiplexed bus operation of the 8086/8088 microprocessor
☐ Describe the function of the bus controller in an 8086/8088 CPU
☐ Analyze a timing diagram for a memory-read cycle and a memory-write cycle
☐ Distinguish between a dedicated I/O port and a memory-mapped I/O port
☐ Compare polled I/O to interrupt-driven I/O
☐ Describe the functions of PIC and PPI devices
☐ Define and explain the advantage of DMA

■ CHAPTER OVERVIEW

This chapter provides a brief introduction to micro-processors and microcomputers. Naturally, a single-chapter coverage must be limited because one or more chapters could easily be devoted to each of the section topics. Keep in mind, however, that the purpose here is to give you a basic acquaintance with the subject in preparation for further study later. Of course, this chapter is optional and some instructors may not wish to cover it at all.

Both the Intel and Motorola microprocessor families are briefly discussed, and the Intel 8088 is used as a "model" to illustrate basic microprocessor concepts. This is a valid approach because the 8086/8088 is the first generation of the Intel 80×86 family; and the newer generations—the 80286, 80386, 80486, and Pentium—although they are more powerful and contain advanced features, are related in architecture and basic functions.

14–1 ■ THE MICROPROCESSOR AND MICROCOMPUTER

The microprocessor is a digital integrated circuit device that can be programmed with a series of instructions to perform specified functions on data. When a microprocessor is connected to a memory device and provided with a means of transferring data to and from the "outside world," you have a microcomputer. After completing this section, you should be able to

□ Define a microprocessor and discuss its basic elements   □ Explain the function of the ALU, the register unit, and the control unit   □ Describe the address bus, the data bus, and the control bus   □ Determine the amount of memory that a microprocessor can access based on the size of its address bus   □ Define *assembly language*   □ Describe the basic elements in a microcomputer   □ Discuss what each part of a microcomputer does   □ Explain what a peripheral device is

Basic Elements of a Microprocessor

In its basic form, a **microprocessor** consists of three elements as shown in Figure 14–1: an arithmetic logic unit (**ALU**), a register unit, and a control unit.

FIGURE 14–1
The basic elements of a microprocessor.

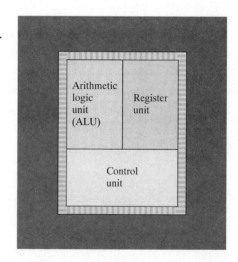

Arithmetic Logic Unit   The ALU performs arithmetic operations such as addition and subtraction and logic operations such as NOT, AND, OR, and exclusive-OR.

Register Unit   During the execution of a program (series of instructions), data are temporarily stored in any of the many registers that make up this unit.

Control Unit   This unit provides the timing and control signals for getting data into and out of the microprocessor, for performing programmed instructions, and for all other operations.

Microprocessor Buses

Typically a microprocessor has three buses for information transfer internally and externally, as shown in Figure 14–2. These buses are the address bus, the data bus, and the control bus.

The Address Bus   The address bus is a "one-way street" over which the microprocessor sends an address code to a memory or other external device. The size or "width" of the address bus is specified by the number of bits that it can handle. Early microprocessors had

FIGURE 14–2
Microprocessor buses.

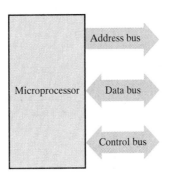

4-bit address buses. This number has increased to 8, 16, 20, 24, and 32 bits as micro-processor technology has advanced.

The more bits there are in the address bus, the more memory locations a given microprocessor can access. With 8 bits, 256 memory locations can be accessed. With 16 bits, 65,536 memory locations can be accessed. With 32 bits, 4,295,000,000 memory locations can be accessed.

The Data Bus   The data bus is a "two-way street" on which data or instruction codes are transferred into the microprocessor or on which the result of an operation or computation is sent out from the microprocessor. Depending on the particular microprocessor, the data bus can handle 8 bits, 16 bits, 32 bits, or 64 bits.

The Control Bus   The control bus is used by the microprocessor to coordinate its operations and to communicate with external devices.

Microprocessor Programming

The **assembly language** that is used to program a microprocessor is classified as a *low-level language* because the English-like instructions contained in a given assembly language represent binary codes that directly control the microprocessor. The binary code instructions are called **machine language** and are all that the microprocessor recognizes. A program called an assembler converts the English-like instructions, called **mnemonics,** in the assembly language into binary patterns called machine language for use by the microprocessor as illustrated in Figure 14–3. Assembly language and the corresponding machine language is specific to the type of microprocessor or microprocessor family.

FIGURE 14–3
Block diagram of microprocessor programming.

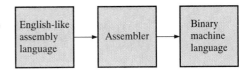

On the other hand, high-level programming languages such as BASIC, Pascal, C, or Fortran are independent of the type of microprocessor in a computer system. A program called a compiler or interpreter translates the high-level program statements into machine language. Programming is discussed further in Section 14–4.

The Microcomputer

When a microprocessor is connected to a memory unit, an input unit, and an output unit, you have a **microcomputer** as shown in Figure 14–4. The CPU (microprocessor) communicates with the memory unit and the input and output units on the buses. Addresses are sent to the memory on the address bus, instructions and/or data are transferred between the

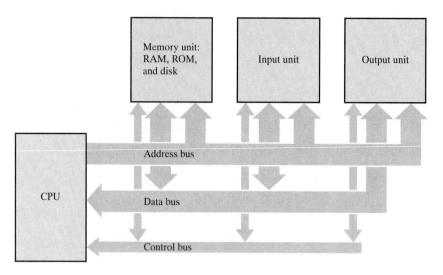

FIGURE 14–4
Microcomputer block diagram.

CPU, memory, and input/output units on the data bus, and signals on the control bus coordinate all operations.

Central Processing Unit (CPU)   The microprocessor and associated support circuits make up the **CPU** which is sometimes called the MPU (microprocessing unit). Basically, the CPU addresses a memory location, obtains **(fetches)** a program instruction that is stored there, and carries out **(executes)** the instruction. After completing one instruction, the CPU moves on to the next one. This fetch and execute process is repeated until all of the instructions in a specific program have been executed. A simple example of an application program is a set of instructions stored in the memory as binary codes that directs the CPU to fetch a series of numbers also stored in the memory, add them, and store the sum back in the memory.

Memory Unit   The memory unit typically consists of RAM, ROM, and disk. The RAM stores data and programs temporarily during processing. Data are numbers or other information in binary form, and programs are lists of instructions that tell the computer what to do. Because the RAM is generally a volatile memory, everything must be backed up in nonvolatile disk storage.

The ROM stores system programs such as BIOS (basic input/output system). These types of programs are used for handling video display graphics, printer communications, power-on self-test, and other routines for servicing peripherals and for general "housekeeping."

Input Unit   The microcomputer receives information from the "outside world" via the input unit which, generally, can handle several external devices called **peripherals.** Examples of peripherals with which a computer communicates through the input unit are the keyboard and the mouse.

Output Unit   The microcomputer sends information to the "outside world" via the output unit which, generally, can handle several peripherals. Examples of peripherals with which a computer communicates through the output unit are the video monitor and the printer. Some peripherals function as both input and output devices; examples are external disk drives, video monitors, and modems.

A Microcomputer System

For a microcomputer to accomplish a given task, it must communicate with the "outside world" by interfacing with people, sensing devices, or devices to be controlled. A typical microcomputer system is shown in Figure 14–5. Usually, there is a keyboard for data entry, a video monitor, and a printer. A mouse is also used on most personal computers. Other types of peripherals such as an external disk drive, modem, scanner, graphics tablet, or voice input are often part of a system.

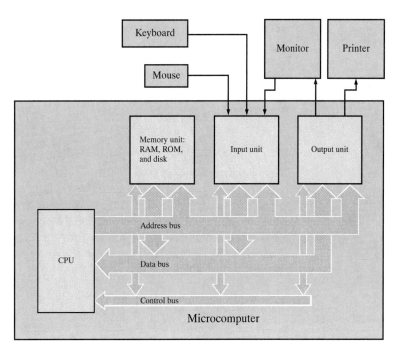

FIGURE 14–5
A typical microcomputer system.

**SECTION 14–1
REVIEW**

1. Define a microprocessor and name the basic elements.
2. How are data transferred from one unit to another in a microcomputer?
3. Name the basic elements of a microcomputer.
4. Define *peripheral*.

14–2 ■ MICROPROCESSOR FAMILIES

The Intel Corporation is a pioneer in the development and marketing of microprocessors; and, today, Intel microprocessor products have a large share of the world market being used in the IBM PC and all PC-compatible computers. This section contains a brief history and description of the evolution of the 80×86 microprocessor family from the 8086/8088 to the Pentium, as well as earlier devices. Motorola also has enjoyed significant success with its line of microprocessors. The first Apple computer was based on the Motorola 6800, and all Apple Macintosh computers today use Motorola microprocessors. As in the 80×86 Intel family, each Motorola microprocessor builds on the basic design established by the first device with added features and improved performance. Our emphasis is on main features and improvement from one generation of devices to the next. Keep in mind that microprocessor technology advances at a fast pace and performance parameters change as current devices are improved or replaced with new versions. After completing this section, you should be able to

☐ Discuss the Intel 80×86 family of microprocessors   ☐ Discuss the Motorola 680×0 family of microprocessors   ☐ Describe a cache memory and state its purpose   ☐ Describe pipelining and state its purpose

The Early Intel Microprocessors

Preceding the 80×86 microprocessors, Intel first introduced in 1971 a microprocessor with a 4-bit data bus designated the 4004. This device was followed by the 8008 which had an 8-bit data bus. Two more 8-bit microprocessors (reference to the number of bits usually refers to the data bus unless stated otherwise), the 8080 and the 8085 were introduced in the mid-1970s. These two devices could address only $2^{16} = 65,536$ memory locations (64 kbytes of memory) with their 16-bit address buses.

The 80×86 Family of Microprocessors

Since its introduction in 1978, the so-called ×86 architecture has undergone five major evolutionary stages. The term *architecture* in relation to microprocessors refers to the internal design and organization of the device. The first generation of the 80×86 family includes the 8086, the 8088, and the 80186. Next came the 80286, followed by the 80386, and then the 80486. The Pentium is the fifth generation Intel microprocessor. Each generation built upon the basic concept of the first with additional features and improved performance.

The 8086/8088 and 80186   Developed in 1978, the 8086 was the first of the 80×86 family and is the basis for all Intel microprocessors that followed. The 8086 was a 16-bit microprocessor (16-bit data bus) and represented a significant departure from the earlier 8-bit devices. The number of address lines in the 8086 increased to twenty, making it capable of addressing $2^{20} = 1,048,576$ memory locations (1 Mbyte). The various versions of the 8086 operated at clock frequencies of 5 MHz, 8 MHz, or 10 MHz.

The 8088 is essentially an 8086 with the 16-bit internal data bus multiplexed down to an 8-bit external data bus. It was intended to meet the demand for applications in simpler 8-bit systems and was used in the original IBM personal computer (PC).

The 80186 is an 8086 with several support functions such as clock generator, system controller, interrupt controller, and direct memory access (DMA) controller integrated on the chip. An increased clock frequency of 12.5 MHz was added and the 5 MHz rate available in the 8086/8088 was dropped, resulting in a selection of 8 MHz, 10 MHz, or 12.5 MHz.

The 80286   In the 80286, which was introduced in 1982, the memory addressing capability was increased to 24 address lines. This provides for addressing $2^{24} = 16,777,216$ memory locations (16 Mbytes). An advanced mode of operation used in all the following microprocessors, called the *protected mode,* was first incorporated in the 80286. This mode allows access to additional memory locations and advanced programming features. The 80286 operates at the same clock frequencies as the 80186.

The 80386   In 1985 a major step was taken when the first Intel 32-bit microprocessor was introduced. The 80386 has both a 32-bit data bus and a 32-bit address bus. With 32 address bits, $2^{32} = 4,294,967,296$ memory locations (4 Gbytes) can be accessed. In addition to some other features, the 80386 was the first Intel microprocessor to enhance processing speed with the use of instruction **pipelining.** This technique allows the microprocessor to start working on a new instruction before it has completed the current one. Versions that operate at clock frequencies of 16 MHz, 20 MHz, 25 MHz, and 33 MHz are available in the 80386.

Other more economical versions of the 80386 include the 80386SX and the 80386SL in which the data bus is multiplexed down to 16 bits and the address bus down to 24 bits. All of the 80386 microprocessors can also operate with a math (numeric) co-processor that can be used to enhance or support the functions of the microprocessor by performing floating-point operations.

The 80486   The next step in the evolutionary process was the introduction of an on-chip 8-kbyte *cache memory* in 1989. Cache memory is located between the processor and the main memory and reduces memory access time by storing recently used instructions or data in a fast SRAM rather than in the relatively slow main memory (DRAM). Also, an internal math coprocessor was incorporated. Certain versions of the 80486 operate at clock frequencies up to 66 MHz.

The Pentium   The Pentium, introduced in 1993, retains the 32-bit address bus of the 80486 but doubles the data bus to 64 bits. The Pentium also has two 8-kbytes cache memories, one for instructions and one for data. A dual pipeline method, known as superscalar architecture, increases the speed at which instructions can be processed above that of single pipelining that are introduced in the 80386. Currently, the Pentium can be operated at clock frequencies up to 166 MHz, which will increase as improved devices are introduced.

The Motorola 6800 and Its Variations

The first microprocessors from Motorola were all 8-bit devices (8-bit data bus). The 6800 appeared in 1975 with a clock frequency of 2 MHz and was capable of addressing 64 kbytes of memory with a 16-bit address bus.

In the 6802, a 128-kbyte RAM was added for use in the place of some registers. The clock frequency was increased in the 6803 to 3.58 MHz and a UART (universal asynchronous receiver/transmitter) was added for serial communications. The last of the 8-bit microprocessors was the 6809 which offered an enhanced instruction set including a multiply instruction. All of the 8-bit devices retained the 16-bit address bus.

The 680×0 Family of Microprocessors

The 68000 was the first of Motorola's 16-bit microprocessors and was introduced in 1978. It had 24 address lines which could access 16 Mbytes of memory compared to the 1-Mbyte capability of the 8086/8088. With a clock frequency of 16 MHz, the 68000 could run faster than the Intel devices at that time. Also, a RAM-based register concept used in some of the 8-bit devices was abandoned in favor of 16 general-purpose registers.

The 68020   Motorola entered the world of 32-bit microprocessors with the 68020. It could address 4 Gbytes of memory and featured a 256-byte cache memory for instructions, an innovative feature at that time. The clock frequency was increased to 33 MHz.

The 68030   For this device, another 256-byte cache was added for data, and the speed of operation was increased to near 50 MHz.

The 68040   The data and instruction caches were increased to 4 kbytes each, and an on-chip math coprocessor was added in the 68040.

The 68060   The 68060 has a superscalar architecture with multiple pipelining of instructions and two 8-kbyte caches. The top clock frequency is 66 MHz.

The PowerPC   The MPC601 or PowerPC is a 64-bit microprocessor with superscalar architecture that can effectively execute up to three instructions per clock cycle. It has 32 kbytes of cache memory and an internal math coprocessor. The PowerPC is a RISC type of microprocessor. **RISC** means *reduced instruction set computing,* an approach that reduces the number of instructions required and results in improved performance.

14–3 ■ THE 8086/8088 MICROPROCESSOR

As you saw in the last section, the 80×86 microprocessor family has undergone a tremendous change from the 8086 to the Pentium. However, the "family resemblance" has been maintained throughout the evolutionary process so that the Pentium, although significantly more complex, still retains many of the basic concepts used in the 8086/8088. The approach in this section and following sections is to use the 8086/8088 as an example to introduce basic concepts of architecture, operation, and programming. Once learned, these concepts can be applied to the study of the more advanced devices such as the Pentium, as well as future devices.

As a reminder, this is a very brief and limited introduction. Entire textbooks are devoted to the 8086/8088 microprocessors alone as well as more recent generations of devices. It takes an extensive amount of material to cover even the simplest microprocessor completely in terms of both its hardware and software. After completing this section, you should be able to

☐ Discuss the basic microprocessor operation   ☐ Discuss the internal organization of the 8086/8088 microprocessors   ☐ Describe the bus interface unit   ☐ State the purpose of the segment registers   ☐ State the purpose of the instruction pointer   ☐ Describe the execution unit   ☐ Describe the general set of registers   ☐ State the purpose of the flag register

Basic Operation

A microprocessor is a VLSI device that executes a **program** (list of instructions) by repeatedly cycling through the following basic steps:

1. Fetch the next **instruction** from memory.
2. Read an **operand** if required by the instruction (an operand is a quantity to be operated on as directed by its associated instruction).
3. Execute the instruction (do what the instruction says).
4. Write the result back into memory (if required by the instruction).

These basic processing steps are performed in the 8086/8088 by two separate internal units, the execution unit (EU), which executes instructions, and the bus interface unit (BIU), which interfaces with the system buses and fetches instructions, reads operands, and writes results. These units are shown in Figure 14–6.

The BIU performs all the bus operations for the EU, such as data transfers from memory or I/O. While the EU is executing instructions, the BIU "looks ahead" and fetches more instructions from memory. This action is called *prefetching.* The prefetched instructions are stored in an internal memory called the instruction **queue** (pronounced "Q"). This queue allows the BIU to keep the EU supplied with instructions so that the EU does not have to wait for the next instruction to be fetched each time it completes the previous one. The queue speeds up the processing by effectively overlapping fetch and execute operations, which, in earlier microprocessors, were handled serially (fetch, then execute, fetch, then execute, etc.). Figure 14–7 compares the serial fetch/execute cycle with the overlapping fetch/execute operation performed by the 8086/8088.

8086/8088 Microprocessor

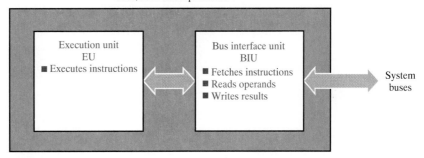

FIGURE 14–6
The 8086/8088 has two separate internal units, the EU and the BIU.

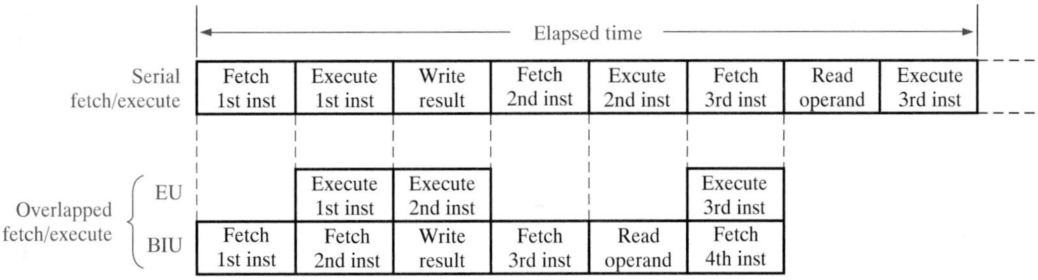

Serial fetch/execute	Fetch 1st inst	Execute 1st inst	Write result	Fetch 2nd inst	Excute 2nd inst	Fetch 3rd inst	Read operand	Execute 3rd inst

Elapsed time

EU		Execute 1st inst	Execute 2nd inst			Execute 3rd inst	
BIU	Fetch 1st inst	Fetch 2nd inst	Write result	Fetch 3rd inst	Read operand	Fetch 4th inst	

Overlapped fetch/execute

FIGURE 14–7
A comparison of serial and overlapped fetch/execute cycles.

Basic 8086/8088 Architecture

Figure 14–8 (next page) is a block diagram of the architecture (internal organization) of an 8088 microprocessor. The 8086 is identical except that it has a 16-bit data bus and a 6-byte instruction queue.

The Bus Interface Unit (BIU)

The major parts of the BIU are the 4-byte instruction queue, the segment registers (ES, CS, and DS), the instruction pointer (IP), and the address summing block (Σ). The internal data buses and the Q bus interconnect the BIU and the EU.

Instruction Queue The instruction queue increases the average speed with which a program is executed (called the **throughput**) by storing up to four prefetched instructions (six in the 8086). As described earlier, this technique allows the 8088 essentially to do two things (fetch and execute) at one time.

Segment Registers The segment registers (CS, DS, and ES) are all 16-bit registers used for addressing the 1 Mbyte (1,048,576 bytes) of memory space.

The memory space is divided into groups of 64-kbyte (65,536-byte) segments. The program assigns every segment a **base address,** which is its starting location in the memory space. The segment registers point to (contain the base addresses of) four currently addressable segments. These registers can be changed by the program to point to other desired segments.

Currently addressable memory segments are those defined by the base addresses contained in the CS (code segment) register, the DS (data segment) register, the SS (stack segment) register, and the ES (extra segment) register.

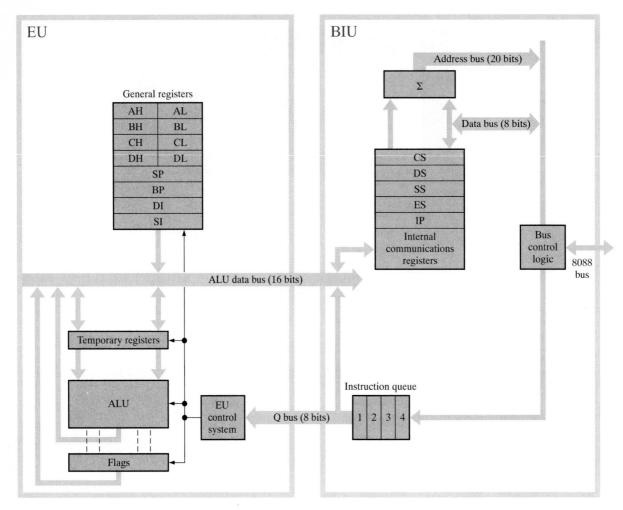

FIGURE 14–8
The internal organization of the 8088 microprocessor.

Instruction Pointer (IP) and Address Summing Block The 16-bit IP is analogous to the program counter in other microprocessors; it points to the next instruction in memory. The IP contains the **offset address** of the next instruction, which is the distance in bytes from the beginning, or base address, of the current code segment (in the CS register).

To achieve the 20-bit **physical address** that goes out on the address bus, the 16-bit offset address in the IP is added to the segment base address, which has been shifted four bits to the left, as indicated in Figure 14–9. This addition is done by the address summing block.

Figure 14–10 illustrates the addressing of a location in memory by the segmented method. In this example, $A000_{16}$ is in the segment register and $A0B0_{16}$ is in the IP. When the segment register is shifted and added to the IP, we get $A0000_{16} + A0B0_{16} = AA0B0_{16}$.

FIGURE 14–9
Formation of the 20-bit address from the segment base address and the offset address.

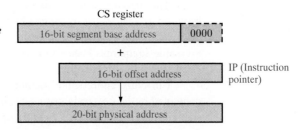

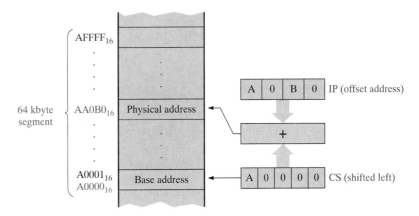

FIGURE 14–10
Illustration of the segmented addressing method.

EXAMPLE 14–1

The hexadecimal contents of the CS register and the IP are shown in Figure 14–11. Determine the physical address in memory of the next instruction.

FIGURE 14–11 $A034_{16}$ CS

$0FF2_{16}$ IP

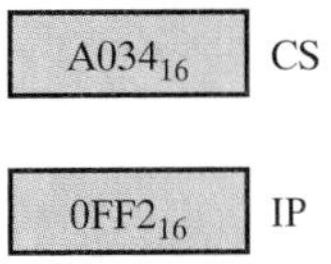

Solution   Shifting the CS base address left four bits (one hex digit), effectively places a 0_{16} in the LSD position, as shown in Figure 14–12. The shifted base address and the offset address are added to produce the 20-bit physical address.

FIGURE 14–12   A 0 3 4 0   +   0 F F 2   =   A 1 3 3 2

Base address   Offset address   Physical address
(shifted left 4 bits)   of next instruction

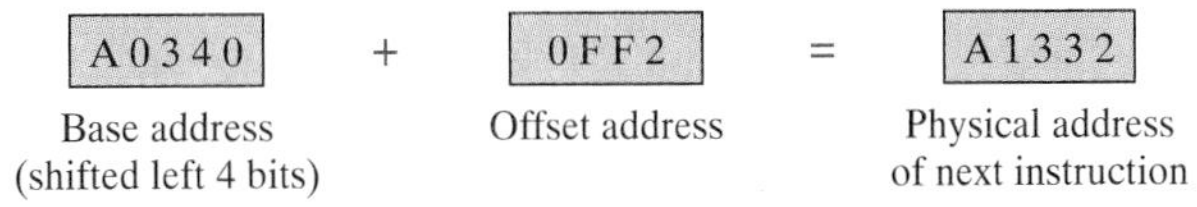

Related Exercise   Determine the physical address if the CS register contains $6B4D_{16}$.

The Execution Unit (EU)

The EU decodes instructions fetched by the BIU, generates appropriate control signals, and executes the instructions. The main parts of the EU are the arithmetic logic unit (ALU), the general registers, and the flags.

The ALU   This unit does all the computational and logic operations, working with either 8-bit or 16-bit operands.

The General Registers   This set of 16-bit registers is divided into two sets of four registers each, as shown in Figure 14–13 (next page). One set consists of the data registers, and the other set consists of the pointer and index registers.

Each of the 16-bit data registers has two separately accessible 8-bit sections and can be used either as a 16-bit register or as two 8-bit registers. The pointer and index registers are used only as 16-bit registers.

The low-order bytes of the data registers are designated as *AL, BL, CL,* and *DL.* The high-order bytes are designated as *AH, BH, CH,* and *DH.* These registers can be used in most arithmetic and logic operations in any manner specified by the programmer for stor-

FIGURE 14–13
The 8088 general register set.

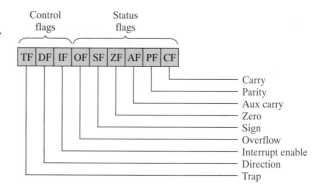

ing data prior to and after processing. Also, some of these registers are used specifically by certain program instructions.

The pointer and index registers are the stack pointer (SP), the base pointer (BP), the source index (SI), and the destination index (DI). These registers are used in various forms of memory addressing under control of the EU.

The Flags The flag register contains nine independent status and control bits **(flags),** as shown in Figure 14–14. A status flag is a one-bit indicator used to reflect a certain condition after an arithmetic or logic operation by the ALU, such as a carry (CF), a zero result (ZF), or the sign of a result (SF), among others. The control flags are used to alter processor operations in certain situations.

FIGURE 14–14
The 8088 status and control flags.

Control flags	Status flags
TF DF IF	OF SF ZF AF PF CF

- Carry
- Parity
- Aux carry
- Zero
- Sign
- Overflow
- Interrupt enable
- Direction
- Trap

SECTION 14–3 REVIEW

1. Name the two operational units within an 8086/8088 microprocessor.
2. What is the purpose of the BIU?
3. Does the EU interface with the system buses?
4. What is the function of the instruction queue?
5. What is the advantage of prefetched instructions?

14–4 ■ MICROPROCESSOR PROGRAMMING

All computers must be programmed to perform even the most elementary tasks. This section focuses on the basic concept of programming in microprocessors. After completing this section, you should be able to

☐ Explain the general purpose of a computer program ☐ Write a simple machine-language program ☐ Discuss how a microprocessor runs a program

The circuitry and physical components (**hardware**) of a microcomputer are useless without programs (**software**). A program is a list of computer instructions arranged to achieve a specific result. To give you a basic idea of how a computer carries out a task under program control, let's look at how a computer runs a very simple program. The steps in our example task are as follows:

1. Input a number from port 2.
2. Add 5 to this number.
3. Output the sum to port 3.

Each step in the task must be implemented with a program statement that instructs the microprocessor what to do. Every microprocessor has an instruction set from which a programmer can choose the most appropriate instructions to accomplish a given task.

For step 1 in our example task, an input instruction, IN, is used to tell the microprocessor to transfer the data byte from the specified input port to the AL register. Let's assume that the IN instruction is represented by the 8-bit code 11100100, or $E4_{16}$. This is the **op code,** or operation code, for this particular instruction, and it is stored in RAM at an address determined by the programmer. The RAM address following the IN op code contains the input port number.

For step 2 an addition instruction, ADD, is used to tell the microprocessor to add an operand (in this case 5) to the contents of the AL register (which now contains the number from the input port) and then place the sum back into the AL register, replacing the previous number. The op code for the ADD instruction is 00000100, or 04_{16}, and is stored at the next RAM address. The address following the ADD op code contains the operand 5.

For step 3 an output instruction, OUT, is used to tell the microprocessor to transfer the contents of the AL register (which is now the sum) to the specified output port. The op code for OUT is 11100110, or $E6_{16}$. The RAM address following the op code address contains the number of the output port.

Figure 14–15 shows how this sample program "looks" in memory. A beginning address of 00000_{16} is assumed for illustration.

FIGURE 14–15

Example of a simple program stored in RAM.

Description of code	RAM	Physical address (hexadecimal)
Op code for IN	11100100	00000
Input port number	00000010	00001
Op code for ADD	00000100	00002
Operand "5"	00000101	00003
Op code for OUT	11100110	00004
Output port number	00000011	00005

When a program is written with the mnemonic designations for the instructions, such as IN, ADD, and OUT, it is an assembly-language program. When a program is written with the binary codes, it is a machine-language program. The sample program is written in both forms as follows:

Assembly-Language Program	*Comments*
IN 02_{16}	Input byte from port 2 and store in AL register.
ADD 05_{16}	Add 5 to AL register and store sum back in register.
OUT 03_{16}	Output sum to port 3.

Machine-Language Program	Comments
11100100 (E4$_{16}$)	IN op code
00000010 (02$_{16}$)	Port number
00000100 (04$_{16}$)	ADD op code
00000101 (05$_{16}$)	Operand
11100110 (E6$_{16}$)	OUT op code
00000011 (03$_{16}$)	Port number

Running the Program

When the microcomputer is commanded to run the program, it goes through the following sequence:

1. Fetch the op code for the IN instruction from memory.
2. Decode the IN op code. This tells the microprocessor to go to the next memory address and get the input port number.
3. Read the input port number from the memory.
4. Access the port with an I/O read operation, and transfer the number that is on the input port lines to the AL register. Let's say the number is binary seven (00000111).
5. Fetch the op code for ADD from the memory.
6. Decode the ADD op code. This tells the microprocessor to go to the next memory address and get the operand (5_{10}) and add it to the contents of the AL register (7_{10}).
7. Read the operand (5_{10}) from the memory, add it to the number 7_{10}, which is in the AL register, and store the sum (12_{10}) in the AL register (replacing the previous number, 7_{10}).
8. Fetch the op code for the OUT instruction from the memory.
9. Decode the OUT op code. This tells the microprocessor to go to the next memory address and get the output port number.
10. Read the output port number from the memory.
11. Access the port with an I/O write operation, and transfer the sum (12_{10}), which is in the AL register, to the output port.

This simple program should give you some feel for microcomputer software operation. Although only three instructions were used, most microprocessors have a large number of instructions in their instruction sets. For example, the 8086/8088 has about 100 instructions. This number allows enormous flexibility and programming power.

Types of Instructions

A coverage of the complete instruction set is obviously beyond the scope of a single section, or even a chapter. Only the basic categories of 8086/8088 instructions and a brief discussion of a few selected instructions are given here. Other categories of instructions are *string, program transfer,* and *processor control.*

Data Transfer   Two data transfer instructions, IN and OUT, were used in the example program. Others include MOV (move), PUSH (push onto stack), POP (take off of stack), and XCHG (exchange). The MOV instruction, for example, can be used in several ways to move a byte or a word (sixteen bits) between various sources and destinations, such as registers, memory, and I/O ports.

Arithmetic   There are a number of instructions for addition, subtraction, multiplication, and division. The ADD instruction was used in the sample program. Others include INC (increment), DEC (decrement), CMP (compare), SUB (subtract), MUL (multiply), and DIV (divide). These instructions allow for specification of operands located in memory, registers, and I/O ports.

Bit Manipulation   This group of instructions includes those used for three classes of operations: logical (Boolean) operations, shifts, and rotations. The logical instructions are NOT, AND, OR, XOR, and TEST. An example of a shift instruction is SAR (shift arithmetic right). An example of a rotate instruction is ROL (rotate left). When bits are shifted out of an operand, they are lost, but when bits are rotated out of an operand, they are looped back into the other end. These logical, shift, and rotate instructions can operate on bytes or words in registers or memory.

SECTION 14–4 REVIEW

1. Define *program*.
2. What is an instruction set?
3. What is an op code?
4. What is an operand?

14–5 ■ THE CENTRAL PROCESSING UNIT (CPU)

A typical CPU (or MPU) consists of a microprocessor and various associated ICs called support circuits. First we will look at the general operation of a CPU, and then we will examine a specific implementation using the 8088 microprocessor as a basic model. After completing this section, you should be able to

☐ Explain the purpose of the CPU in a microcomputer   ☐ Discuss system buses
☐ Describe a basic CPU using an 8088 microprocessor   ☐ Explain the purpose of a bus controller   ☐ Discuss bus interfacing

An 8088-based CPU

A typical CPU (or MPU) is implemented with a microprocessor and some additional support circuits. Figure 14–16 shows an 8088-based CPU.

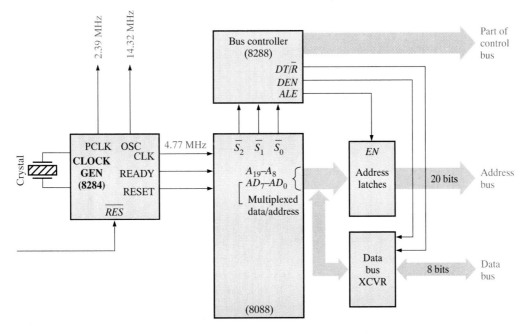

FIGURE 14–16
A simplified 8088-based CPU.

The 8088 Microprocessor As you know, the 8088 is basically the same as the 8086 microprocessor except that the 8088 has an 8-bit external data bus rather than a 16-bit data bus.

As mentioned, the 8088 has an 8-bit data bus, making it compatible with many 8-bit peripheral devices. Internally, however, the 8088 can handle data in 16-bit words, or it can handle data one byte at a time, depending on how it is programmed. Also, the 8088 has a 20-bit address bus, which allows up to 1 Mbyte of memory (actually 1,048,576 bytes) to be addressed.

Actually, in the 8088 the address bus and the data bus are combined. There are a total of twenty lines, AD_0–AD_7 and A_8–A_{19}. The eight lines designated AD_0–AD_7 serve as both data and address lines, using a *multiplexed operation.* The twelve lines designated A_8–A_{19} are used strictly for addressing.

When a 20-bit address is sent out by the microprocessor, all twenty lines are used as the address bus, as illustrated in Figure 14–17(a). When data are sent or received by the microprocessor, the lower eight lines, AD_0–AD_7, are used as a bidirectional data bus, as shown in Figure 14–17(b). The bus multiplexing is controlled by the bus controller. The bus controller is one of the support circuits. Other microprocessor input and output lines will be described in association with the appropriate support circuits.

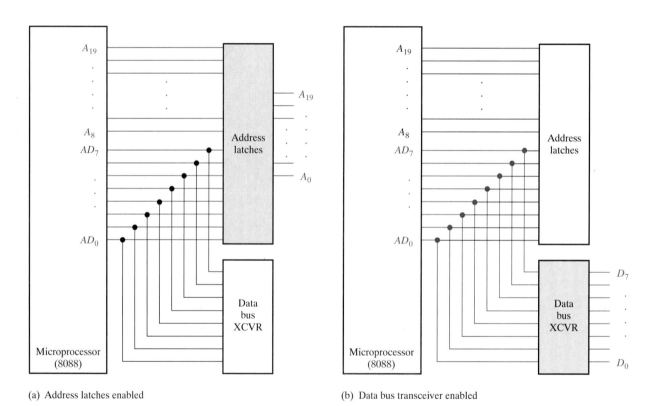

(a) Address latches enabled   (b) Data bus transceiver enabled

FIGURE 14–17
Multiplexed bus operation of the 8088.

The Clock Generator A clock generator circuit, such as the 8284, provides the basic timing signals to the 8088 and to other parts of the system. It also provides a RESET signal to the 8088 to initialize the internal circuitry and a READY signal to synchronize the microprocessor with the rest of the system. An external resonant crystal is used to establish the frequency of oscillation (14.31818 MHz in this case), as shown in Figure 14–18. This frequency is available on the OSC output. A divide-by-three counter inside the clock

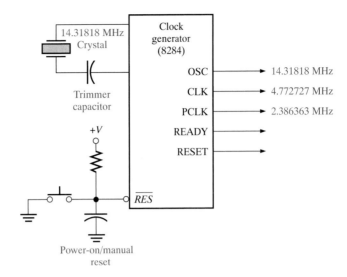

FIGURE 14–18
The 8284 clock generator.

generator produces a 4.772727 MHz clock with a duty cycle of 33% to run the microprocessor. This signal is available on the CLK output. Also, a 2.386363 MHz signal is derived from the CLK frequency and is available on the PCLK output of the clock generator.

The Bus Controller The 8288 bus controller is used to relieve the 8088 microprocessor of most bus control functions in order to maximize the processing efficiency. One function of the bus controller in the CPU is to provide the signals to multiplex the AD_0–AD_7 bus lines coming from the microprocessor. The bus controller generates these control signals based on the codes on the status lines, $\bar{S}_0$, $\bar{S}_1$, and $\bar{S}_2$, from the microprocessor.

When the microprocessor needs to communicate with memory or I/O, it sends a code on the status lines to the bus controller. The bus controller then generates an *ALE* (address latch enable) signal, which latches the address placed on AD_0–AD_7 and A_8–A_{19} by the microprocessor into the address latches. There are twenty latches (A_0–A_{19}), whose tristate outputs form the system address bus. Figure 14–19 (next page) illustrates the steps in this operation. During a memory read/write cycle, a valid address is available on the address bus, and, therefore, the lower eight bits (AD_0–AD_7) of the microprocessor's combined bus are freed up to send or receive data.

After the address has been latched, the microprocessor signals the bus controller via the status lines ($\bar{S}_2$, $\bar{S}_1$, and $\bar{S}_0$) that it wants to read data or write data to the memory location or I/O that is specified by the address that is held on the address bus. The status line codes for the four possible read/write operations are given in Table 14–1.

TABLE 14–1
The 8088 read/write status codes

| Status Code | | | | Active Bus Controller |
$\bar{S}_2$	$\bar{S}_1$	$\bar{S}_0$	Operation	Output Command
0	0	1	Read I/O port	$\overline{IORC}$
0	1	0	Write I/O port	$\overline{IOWC}$
1	0	1	Read memory	$\overline{MRDC}$
1	1	0	Write to memory	$\overline{MWTC}$

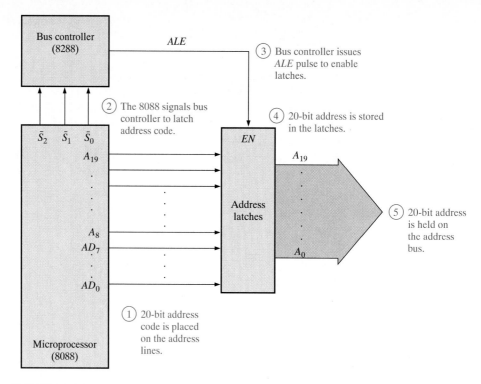

FIGURE 14–19
Steps in placing a valid address on the bus.

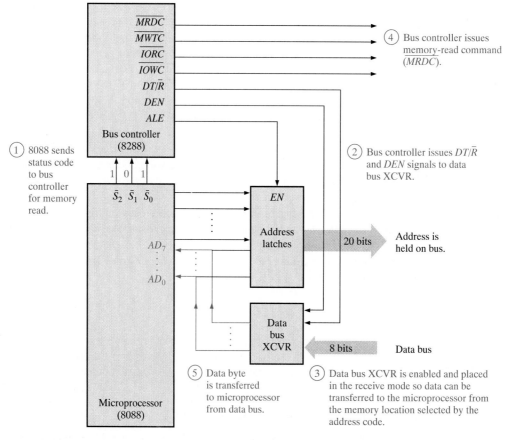

FIGURE 14–20
Example of CPU memory-read operation.

When the bus controller receives one of the status codes, it sends a read command ($\overline{IORC}$ or $\overline{MRDC}$) or a write command ($\overline{IOWC}$ or $\overline{MWTC}$) to the appropriate device, and it also sends two signals to the tristate data bus transceiver (XCVR). These two signals are $DT/\overline{R}$ (data transmit/receive) and DEN (data enable). The DEN signal enables the data bus transceiver, and the $DT/\overline{R}$ signal selects the direction of data on the data bus (D_0–D_7). Figure 14–20 illustrates a memory-read operation, and Figure 14–21 illustrates a memory-write operation. Timing diagrams for the memory read and write cycles are given in Section 14–6.

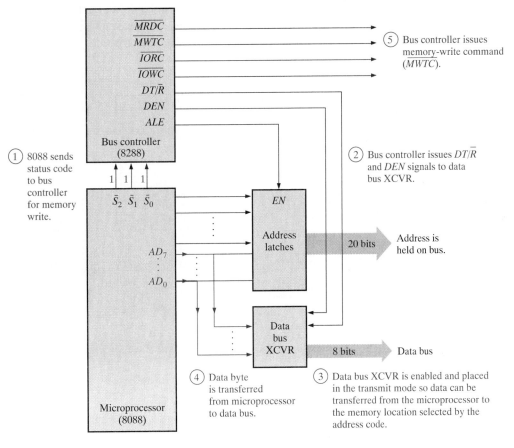

FIGURE 14–21
Example of memory-write operation.

SECTION 14–5 REVIEW	
	1. What is the purpose of the clock generator in the CPU?
	2. List two functions of the bus controller.
	3. Explain why bus multiplexing is necessary in an 8088-based CPU.

14–6 ■ THE MEMORY

This section covers the memory portion of a typical microcomputer and how it is used by the CPU. After completing this section, you should be able to

☐ Explain address allocation in microcomputer memories ☐ Discuss the read and write operations ☐ Describe a memory-read cycle ☐ Describe a memory-write cycle

Address Allocation

As you have seen, the CPU can address up to 1,048,576 locations with a 20-bit address bus (A_0–A_{19}). In a typical microcomputer, a portion of the total address space is allocated to RAM and ROM and a portion to I/O. Input and output ports have unique addresses that are treated like memory locations, discussed in Section 14–7.

Figure 14–22 shows a possible memory **address allocation** in a microcomputer. The RAM may be allocated addresses 00000_{16} through $BFFFF_{16}$ (a total of 786,431 locations), although not all of this RAM memory space is actually used. In the maximum configuration, 640 kbytes of available memory are utilized. The ROM may be allocated address $C0000_{16}$ through $FFFFF_{16}$ (a total of 262,143 locations). Within the RAM and the ROM, the addresses are segmented into areas for various purposes, such as system programs, user programs, data, and various types of system-related information.

FIGURE 14–22

Example of memory address allocation in a microcomputer.

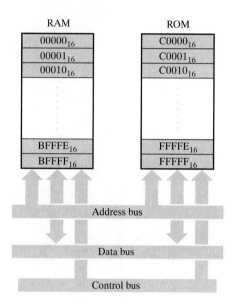

CPU-Memory Operation

The CPU handles two types of memory transfers, *read* and *write*. Basically, during the read operation, the CPU fetches either a program instruction, another address, or data. During the write operation, the CPU sends data resulting from a computation or some other operation back to RAM. A diagram of the CPU and memory portions of a microcomputer is shown in Figure 14–23.

The Memory-Read Cycle The microprocessor initiates a memory-read cycle by sending a status code of $\overline{S_2}\overline{S_1}\overline{S_0} = 101$ to the bus controller and placing a binary address on the AD_0–AD_7 and A_8–A_{19} bus lines. The bus controller then issues an, *ALE* signal, which latches the address onto the address bus (A_0–A_{19}), freeing up the AD_0–AD_7 lines for data transfer. Also, the bus controller issues a low $DT/\overline{R}$ signal to the bus transceiver to set it up for receiving data from the memory and passing it on to the microprocessor.

Next the bus controller issues a memory read command ($\overline{MRDC}$) to the memories. This instructs the memory to place the contents of the selected address on the data bus. The *DEN* signal from the bus controller then enables the bus transceiver, and a byte of data flows into the microprocessor. Figure 14–24 shows the basic timing diagram for a memory-read cycle.

Memory-Write Cycle The microprocessor initiates a memory-write cycle by sending status code $\overline{S_2}\overline{S_1}\overline{S_0} = 110$ to the bus controller and placing a binary address on the bus lines. The bus controller then issues an *ALE* signal to latch the address onto the address bus

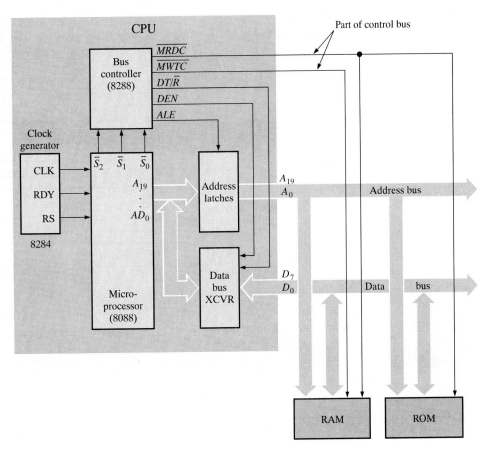

FIGURE 14–23
Basic CPU-memory organization.

FIGURE 14–24
Basic timing diagram for an 8088-based CPU memory-read cycle.

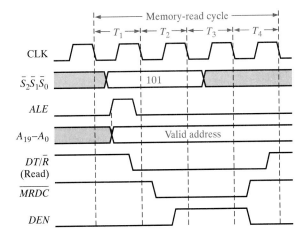

$(A_0–A_{19})$ and a HIGH $DT/\overline{R}$ signal to set up the bus transceiver for sending data to the memory. The microprocessor then places the data byte on the $AD_0–AD_7$ lines. The *DEN* signal from the bus controller then enables the bus transceiver to place the data byte on the data bus $(D_0–D_7)$.

Next the bus controller issues a memory-write command ($\overline{MWTC}$) to the memories. This command instructs the RAM to store the data byte in the selected address. Figure 14–25 shows a basic timing diagram for a memory-write cycle.

FIGURE 14–25
Basic timing diagram for an 8088-based CPU memory-write cycle.

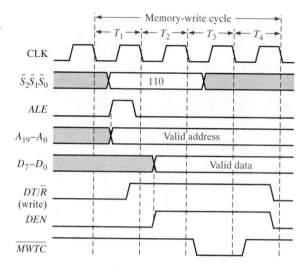

SECTION 14–6 REVIEW

1. Name the two cycles used by the CPU to transfer data to and from the memory.
2. How many clock cycles does it take to read data from the memory in an 8088-based system?
3. When does a valid address appear on the address bus (A_0–A_{19}) during a write cycle?

14–7 ▪ THE INPUT/OUTPUT (I/O) PORT

In this section, you will see generally how the CPU communicates with input and output devices via I/O ports. An I/O port can be thought of as a "window" between the computer and the outside world through which information is passed. After completing this section, you should be able to

☐ Explain what an I/O port is ☐ Discuss the basic types of ports ☐ Explain the difference between dedicated and memory-mapped I/O ports ☐ State the purpose of a programmable peripheral interface

A block diagram of a basic microcomputer system, including I/O ports and typical peripherals, is shown in Figure 14–26. Notice that a **port** can be strictly for input, strictly for output, or bidirectional for both input and output. Although only four I/O ports are shown in the figure, the CPU is capable of handling many more than that.

As mentioned, an I/O port provides an interface between the computer and the "outside world" of peripheral equipment. Depending on the system design, the I/O ports can be either *dedicated* or *memory mapped*. These terms describe the ways in which the ports are accessed by the CPU.

Dedicated I/O Ports

A dedicated I/O port is assigned a unique address within the I/O address space of the computer, and dedicated I/O commands are used for communication. Dedicated I/O ports are accessed by the I/O read and write commands, $\overline{IORC}$ and $\overline{IOWC}$ in an 8086/8088-based system.

Output Port The basic operation of the output port is as follows: The port occupies one location within the I/O address space of the system; that is, it has a unique address different from the addresses of other ports. Typically, only a portion of the 20-bit address bus is

Microcomputer

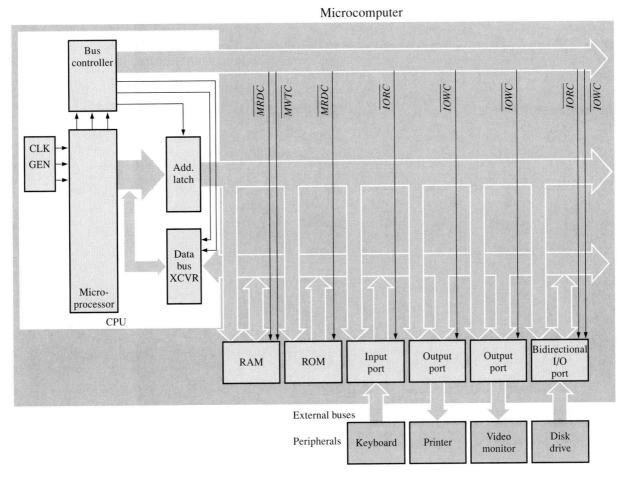

FIGURE 14–26
Basic microcomputer system with I/O ports and peripherals.

used for port addresses because the actual number of I/O ports required is small compared with the number of memory locations.

During a CPU write cycle, the port address is placed on the address bus, and the CPU issues a low $\overline{IOWC}$ (I/O write command) signal to enable the address decoder. The decoder then produces a signal to latch the data byte onto the port output lines. Figure 14–27 shows how a dedicated output port might be implemented.

FIGURE 14–27
A basic dedicated output port.

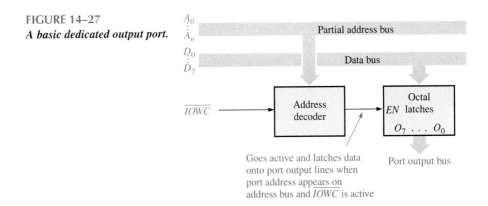

Input Port A dedicated input port is not quite as simple as an output port because the port output lines are connected to the system data bus. This arrangement requires that the port output lines be disabled when the port is not in use to prevent interference with other activity on the data bus. Figure 14–28 shows an example of a basic input port implementation.

FIGURE 14–28
A basic dedicated input port.

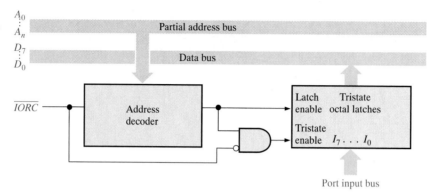

Bidirectional Port The third type of dedicated I/O port is the bidirectional port, which is basically a combination of the input and output ports with additional enabling controls. A bidirectional port allows data to flow to or from a peripheral device such as a disk drive.

Memory-Mapped I/O Ports

A second approach to accessing I/O ports is the memory-mapped port. The port implementation is essentially the same as for the dedicated ports in Figure 14–27 and 14–28 except for two things:

1. The memory-mapped ports are assigned addresses within the computer memory address space and are actually viewed as memory locations by the CPU. All twenty address lines are used for memory-mapped I/O.
2. The CPU accesses a memory-mapped I/O port with the $\overline{MRDC}$ and $\overline{MWTC}$ commands rather than the $\overline{IORC}$ and $\overline{IOWC}$ commands. In other words, the memory-mapped I/O ports are treated exactly as memory locations.

Programmable Peripheral Interface

In many systems, special ICs are used to implement the I/O ports. One such device is the programmable peripheral interface **(PPI),** shown in Figure 14–29. The address decoder is

FIGURE 14–29
A basic programmable peripheral interface (PPI) configuration.

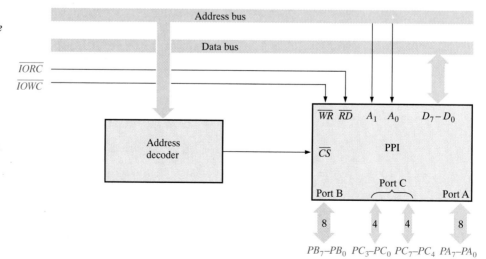

a separate device for enabling the PPI when a port address is decoded. This particular PPI provides three I/O ports, labeled A, B, and C. Ports A and B have 8 lines each, and port C is split into two 4-line groups. The ports can be programmed for various combinations of input ports, output ports, and bidirectional ports and for control and handshaking lines. The PPI can be used for either dedicated or memory-mapped ports.

SECTION 14–7 REVIEW	**1.** What is an I/O port? **2.** Explain the difference between a dedicated I/O port and a memory-mapped I/O port. **3.** What is the purpose of a PPI?

14–8 ■ I/O INTERRUPTS

The concept of an I/O port was presented in Section 14–7, and both dedicated and memory-mapped ports were introduced. In this section, the establishment of communications between a peripheral and the CPU is presented. Two methods are discussed, polled I/O and interrupt-driven I/O. After completing this section, you should be able to

☐ Discuss the need for interrupts in a microcomputer system   ☐ Describe the basic concept of a polled I/O   ☐ Describe the basic concept of an interrupt-driven I/O

In microprocessor-based systems such as the microcomputer, peripheral devices require periodic service from the CPU. The term *service* generally means sending data to or taking data from the device or performing some updating process.

In general, peripheral devices are very slow compared with the CPU. For example, the 8086/8088 CPU can perform about 1,000,000 read or write operations in one second (based on a 4.77 MHz clock frequency). A printer may average only a few characters per second (one character is represented by eight bits), depending on the type of material being printed. A keyboard input may be about one or two characters per second, depending on the operator. So, in between the times that the CPU is required to service a peripheral, it can do a lot of processing, and in most systems this processing time must be maximized by using an efficient method for servicing the peripherals.

Polled I/O

One method of servicing the peripherals is called **polling.** In this method the CPU must test each peripheral device in sequence at certain intervals to see if it needs or is ready for servicing. Figure 14–30 (next page) illustrates the basic polled I/O method.

The CPU sequentially selects each peripheral device via the multiplexer to see if it needs service by checking the state of its ready line. Certain peripherals may need service at irregular and unpredictable intervals, that is, more frequently on some occasions than on others. Nevertheless, the CPU must poll the device at the highest rate. For example, let's say that a certain peripheral occasionally needs service every 1000 μs but most of the time requires service only once every 100 ms. As you can see, precious processing time is wasted if the CPU polls the device, as it must, at its maximum rate (every 1000 μs) because most of the time the device will not need service when it is polled.

Each time the CPU polls a device, it must stop the program that it is currently processing, go through the polling sequence, provide service if needed, and then return to the point where it left off in its current program.

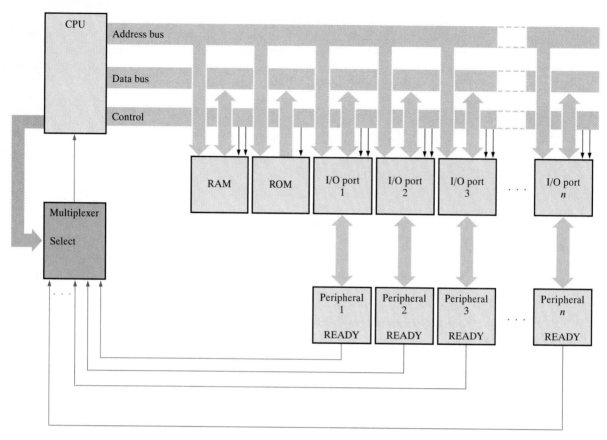

FIGURE 14–30
The basic polled I/O configuration.

Another problem with the sequentially polled I/O approach is that if two or more devices need service at the same time, the first one polled will be serviced first; the other devices will have to wait although they may need servicing much more urgently than the first device polled.

As you can see, polling is suitable only for devices that can be serviced at regular and predictable intervals and only in situations in which there are no priority considerations.

Interrupt-driven I/O

This approach overcomes the disadvantages of the polling method. In the interrupt method the CPU responds to a need for service only when service is requested by a peripheral device. Thus, the CPU can concentrate on running the current program without having to break away unnecessarily to see if a device needs service.

A request for service from a peripheral device is called an **interrupt.** When the CPU receives an I/O interrupt, it temporarily stops its current program and fetches a special program (service routine) from memory for the particular device that has issued the interrupt. When the service routine is complete, the CPU returns where it left off.

A device called a *programmable interrupt controller* (**PIC**) handles the interrupts on a priority basis. It accepts service requests from the peripherals. If two or more devices request service at the same time, the one assigned the highest priority is serviced first, then one with the next highest priority, and so on. After issuing an interrupt (*INTR*) signal to the

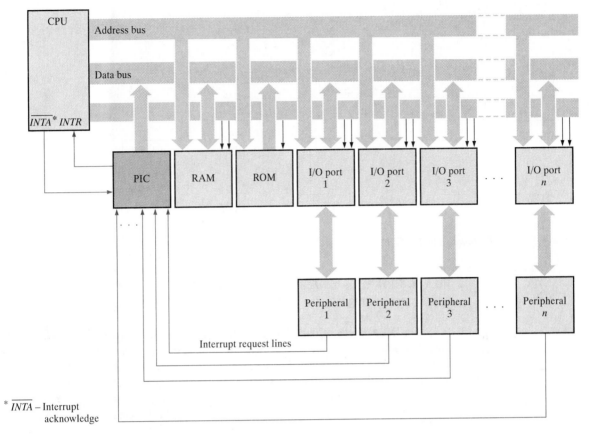

FIGURE 14–31

A basic interrupt-driven I/O configuration.

CPU, the PIC provides the CPU with information that "points" the CPU to the beginning memory address of the appropriate service routine. This process is called *vectoring.* Figure 14–31 shows a basic interrupt-driven I/O configuration.

SECTION 14–8 REVIEW	1. How does an interrupt-driven I/O differ from a polled I/O? 2. What is the main advantage of an interrupt-driven I/O?

14–9 ■ DIRECT MEMORY ACCESS (DMA)

In this short section, we define the technique of data transfer called direct memory access (DMA). A comparison of a CPU-handled transfer and a DMA transfer is presented. After completing this section, you should be able to

☐ Define the term *DMA* ☐ Compare a memory/I/O data transfer handled by the CPU to a DMA transfer

All I/O data transfers discussed so far have passed through the CPU. For example, when data are to be transferred from RAM to a peripheral device, the CPU reads the first data byte from the memory and loads it into an internal register within the microprocessor.

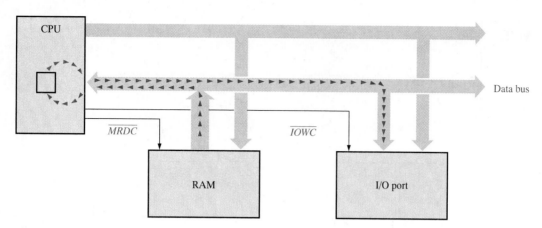

FIGURE 14–32
A memory/I/O transfer handled by the CPU.

Then the CPU writes the data byte to the appropriate I/O port. This read/write operation is repeated for each byte in the group of data to be transferred. Figure 14–32 illustrates this process.

For large blocks of data, intermediate stops in the microprocessor consume a lot of time. For this reason, many systems use a technique called *direct memory access* (**DMA**) to speed up data transfers between RAM and certain peripheral devices. Basically, DMA bypasses the CPU for certain types of data transfers, thus eliminating the time consumed by the normal fetch and execute cycles required for each read or write operation.

For direct memory transfers, a device called the *DMA controller* takes control of the system buses and allows data to flow directly between RAM and the peripheral device, as indicated in Figure 14–33.

Transfers between the disk drive and RAM are particularly suited for DMA because of the large amounts of data involved and the serial nature of the transfers. The DMA controller can handle data transfers several times faster than the CPU.

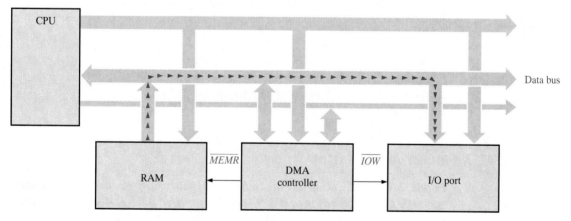

FIGURE 14–33
A DMA transfer.

SECTION 14–9 REVIEW
1. What does DMA stand for?
2. Discuss the advantage of DMA, and give an example of a type of transfer for which it is often used.

■ SUMMARY

■ Basic units of a microcomputer are shown in Figure 14–34.

FIGURE 14–34

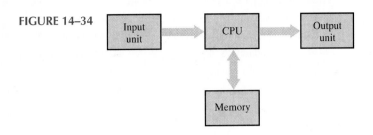

■ Microcomputer buses: address bus, data bus, control bus
■ Typical peripherals: keyboard, video monitor, disk drive, printer, modem
■ 8086/8088 microprocessors
  8086: 16-bit data bus, 20-bit address bus
  8088: 8-bit data bus (16 bits internally), 20-bit address bus
  Execution unit (EU): Executes instructions
  Bus interface unit (BIU): Fetches instructions, reads operands, and writes results
  General registers: Accumulator (AH, AL)   Stack pointer (SP)
                      Base (BH, BL)   Base pointer (BP)
                      Count (CH, CL)   Source index (SI)
                      Data (DH, DL)   Destination index (DI)
  Flags: Trap (TF)   Zero (ZF)
          Direction (DF)   Auxiliary carry (AF)
          Interrupt enable (IF) Parity (PF)
          Overflow (OF)   Carry (CF)
          Sign (SF)
■ Ports
  Physical types: Input, output, input/output (bidirectional)
  Methods of access: Dedicated, memory mapped
  Methods of servicing: Polled, interrupt driven

■ SELF-TEST

1. A basic microcomputer does not include
   (a) an arithmetic logic unit   (b) a control unit
   (c) peripheral units   (d) a memory unit

2. A 20-bit address bus supports
   (a) 100,000 memory addresses   (b) 1,048,576 memory addresses
   (c) 2,097,152 memory addresses   (d) 20,000 memory addresses

3. An example of a peripheral unit is
   (a) the address register   (b) the MPU
   (c) the video monitor   (d) the interface adapter

4. Two types of memory transfers handled by the CPU are
   (a) direct and interrupt   (b) read and write
   (c) bussed and multiplexed   (d) input and output

5. Two types of I/O ports are
   (a) dedicated and memory mapped   (b) DMA and PPI
   (c) tristate and octal   (d) read and write

6. Polling is a method used for
   (a) determining the state of the microprocessor
   (b) establishing communication between the CPU and a peripheral
   (c) establishing a priority for communication with several peripherals
   (d) determining the next instruction

7. DMA stands for

    (a) digital microprocessor address   (b) direct memory access

    (c) data multiplexed access   (d) direct memory addressing

8. A computer program is

    (a) a list of memory addresses that contain data to be used in an operation

    (b) a list of addresses that contain instructions to be used in an operation

    (c) a list of instructions arranged to achieve a specific result

■ ANSWERS TO SECTION REVIEWS

SECTION 14–1

1. A microprocessor is a digital IC that can be programmed with a series of instructions. It consists of an ALU, a register unit, and a control unit.
2. Data are transferred on buses.
3. A microcomputer consists of a CPU, an input unit, an output unit, and a memory.
4. A peripheral is a device external to a computer.

SECTION 14–2

1. An 8-bit microprocessor has an 8-bit data bus.
2. The 8086 was the first 16-bit Intel microprocessor.
3. Cache stores recently used instructions or data to speed up operation.
4. Superscalar refers to a type of architecture that allows more than one instruction to be executed during one clock cycle.

SECTION 14–3

1. An 8086/8088 consists of a bus interface unit (BIU) and an execution unit (EU).
2. The BIU provides addressing and data interface.
3. No, the EU does not interface with the buses.
4. The instruction queue stores prefetched instructions for the EU.
5. Prefetched instructions increase throughput.

SECTION 14–4

1. A program is a list of computer instructions arranged to achieve a specific result.
2. All the instructions available to a particular microprocessor is called the instruction set.
3. An op code is the code for an instruction.
4. A operand is the quantity or element operated on by an instruction.

SECTION 14–5

1. The clock generator provides basic timing signals for the microcomputer.
2. The bus controller issues memory read and write commands, issues I/O read and write commands, and enables the address latches and the data bus transceiver.
3. The twenty bus lines available must be used for both the 20-bit address bus and the 8-bit data bus in the 8088.

SECTION 14–6

1. The two CPU cycles are write and read.
2. Four clock cycles.
3. A valid address appears on the bus after the leading edge of the *ALE* pulse.

SECTION 14–7

1. An I/O port is the interface between the internal data bus and a peripheral.
2. A dedicated I/O port has an address within the I/O address space. A memory-mapped I/O port has an address within the memory address space.
3. A PPI provides I/O ports.

SECTION 14–8

1. For an interrupt-driven I/O, the CPU provides service to a peripheral only when requested to do so by the peripheral. For a polled I/O, the CPU periodically checks a peripheral to see if it needs service.
2. Interrupt-driven I/Os save CPU time.

SECTION 14–9

1. DMA is direct memory access.
2. A DMA transfer of data from memory to I/O or vice versa saves CPU time. Direct memory access is often used in transferring data between RAM and a disk drive.

15

INTEGRATED CIRCUIT TECHNOLOGIES

■ CHAPTER OBJECTIVES

☐ Determine the noise margin of a device from data sheet parameters

☐ Calculate the power dissipation of a device

☐ Explain how propagation delay affects the frequency of operation or speed of a circuit

☐ Interpret the speed-power product as a measure of performance

☐ Use data sheets to obtain information about a specific device

☐ Explain what the fan-out of a gate means

☐ Describe how basic TTL and CMOS gates operate at a component level

☐ Recognize the difference between TTL totem-pole outputs and TTL open-collector outputs and understand the limitations and uses of each

☐ Connect circuits in a wired-AND configuration

☐ Describe the operation of tristate circuits

☐ Properly terminate unused gate inputs

☐ Compare the characteristics of TTL and CMOS

☐ Handle CMOS devices without risk of damage due to electrostatic discharge

☐ Interface TTL and CMOS devices

☐ State the advantages of ECL

☐ Describe PMOS and NMOS circuits

☐ Describe an E^2CMOS cell

■ CHAPTER OVERVIEW

This chapter is intended to be used as a "floating" chapter. It can be covered at any one of several appropriate points throughout the book or completely omitted, depending on course objectives and personal preference. Suggested points that seem most appropriate are following Chapters 3, 6, or 12, but it can be placed after other chapters if desired. Also, selected topics from this chapter can be introduced at various points.

Other approaches are to use this chapter as an optional reading assignment or to use it only for reference. Notice that a tab-edge design on the right-hand pages provides for easy reference to this chapter.

15–1 ■ BASIC OPERATIONAL CHARACTERISTICS AND PARAMETERS

When you work with digital ICs, you should be familiar, not only with their logical operation, but also with such operational properties as voltage levels, noise immunity, power dissipation, fan-out, and propagation delays. In this section, the practical aspects of these properties are discussed. After completing this section, you should be able to

☐ Determine the power and ground connections ☐ Describe the logic levels for TTL and CMOS ☐ Discuss noise immunity ☐ Determine the power dissipation of a logic circuit ☐ Define the propagation delays of a logic gate ☐ Discuss speed-power product and explain its significance ☐ Discuss loading and fan-out of TTL and CMOS

DC Supply Voltage

The nominal value of the dc supply voltage for **TTL** (transistor-transistor logic) and **CMOS** (complementary metal-oxide semiconductor) devices is +5 V. Although omitted from logic diagrams for simplicity, this voltage is connected to the V_{CC} or V_{DD} pin of an IC package, and ground is connected to the GND pin. Both voltage and ground are distributed internally to all elements within the package, as illustrated in Figure 15–1 for a 14-pin package.

FIGURE 15–1

Example of V_{CC} and ground connection and distribution in an IC package. Other pin connections are omitted for simplicity.

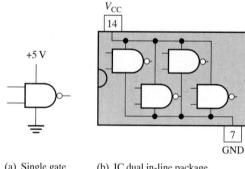

(a) Single gate   (b) IC dual in-line package

TTL Logic Levels

Logic levels were discussed briefly in Chapter 1. For TTL circuits the range of input voltages that can represent a valid LOW (logic 0) is from 0 V to 0.8 V. The range of input voltages that can represent a valid HIGH (logic 1) is from 2 V to V_{CC} (usually 5 V), as indicated in Figure 15–2(a). The range of values between 0.8 V and 2 V is a region of unpredictable performance. When an input voltage is in this range, it can be interpreted as either a HIGH or a LOW by the logic circuit. Therefore, TTL **gates** cannot be operated reliably when input voltages are in this range.

The range of TTL output voltages is shown in Figure 15–2(b). Notice that the minimum HIGH output voltage, $V_{OH(min)}$, is greater than the minimum HIGH input voltage, $V_{IH(min)}$. Also, the maximum LOW output voltage, $V_{OL(max)}$, is less than the maximum LOW input voltage, $V_{IL(min)}$.

CMOS Logic Levels

The input and output logic levels for high-speed CMOS (HCMOS) are given in Figure 15–3 for $V_{DD} = 5$ V.

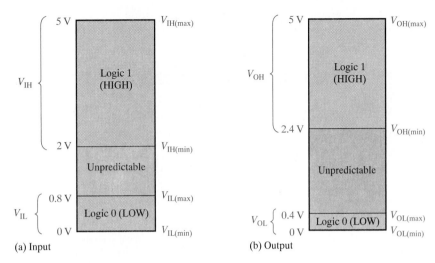

(a) Input   (b) Output

FIGURE 15–2
Input and output logic levels for TTL.

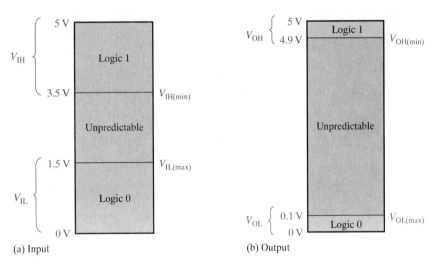

(a) Input   (b) Output

FIGURE 15–3
Input and output logic levels for CMOS.

Noise Immunity

Noise is unwanted voltage that is induced in electrical circuits and can present a threat to the proper operation of the circuit. Wires and other conductors within a system can pick up stray high-frequency electromagnetic radiation from adjacent conductors in which currents are changing rapidly or from many other sources external to the system. Also, power-line voltage fluctuation is a form of low-frequency noise.

In order not to be adversely affected by noise, a logic circuit must have a certain amount of **noise immunity.** This is the ability to tolerate a certain amount of unwanted voltage fluctuation on its inputs without changing its output state. For example, if noise voltage causes the input of a TTL gate to drop below 2 V in the HIGH state, the operation is in the unpredictable region (see Figure 15–2[a]). Thus, the gate may interpret the fluctuation below 2 V as a LOW level, as illustrated in Figure 15–4(a). Similarly, if noise causes a gate input to go above 0.8 V in the LOW state, an uncertain condition is created, as illustrated in part (b).

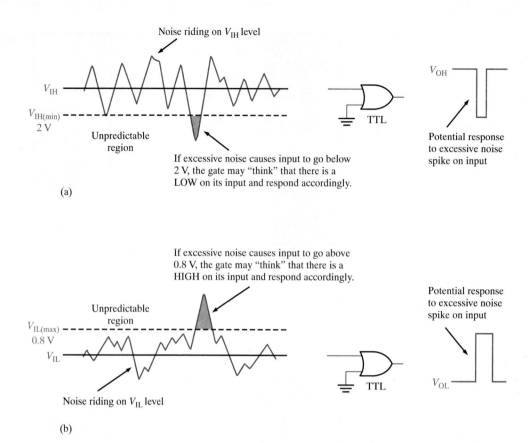

(a)

(b)

FIGURE 15–4
Illustration of the effects of input noise on gate operation.

Noise Margin

A measure of a circuit's noise immunity is called the **noise margin,** which is expressed in volts. There are two values of noise margin specified for a given logic circuit: the HIGH-level noise margin (V_{NH}) and the LOW-level noise margin (V_{NL}). These parameters are defined by the following equations:

$$V_{NH} = V_{OH(min)} - V_{IH(min)} \qquad (15–1)$$

$$V_{NL} = V_{IL(max)} - V_{OL(max)} \qquad (15–2)$$

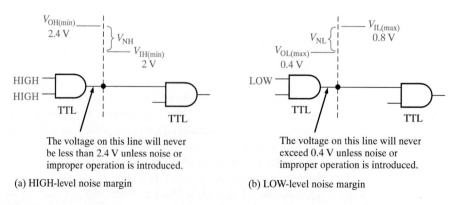

(a) HIGH-level noise margin

(b) LOW-level noise margin

FIGURE 15–5
Noise margins.

From the equations, V_{NH} is the difference between the lowest possible HIGH output from a driving gate ($V_{OH(min)}$) and the lowest possible HIGH input that the load gate can tolerate ($V_{IH(min)}$). Noise margin V_{NL} is the difference between the maximum possible LOW input that a gate can tolerate ($V_{IL(max)}$) and the maximum possible LOW output of the driving gate ($V_{OL(max)}$). Noise margins are illustrated in Figure 15–5.

EXAMPLE 15–1

Determine the HIGH-level and LOW-level noise margins for TTL and for CMOS by using the information in Figures 15–2 and 15–3.

Solution   For TTL,

$$V_{IH(min)} = 2 \text{ V}$$
$$V_{IL(max)} = 0.8 \text{ V}$$
$$V_{OH(min)} = 2.4 \text{ V}$$
$$V_{OL(max)} = 0.4 \text{ V}$$
$$V_{NH} = V_{OH(min)} - V_{IH(min)} = 2.4 \text{ V} - 2 \text{ V} = 0.4 \text{ V}$$
$$V_{NL} = V_{IL(max)} - V_{OL(max)} = 0.8 \text{ V} - 0.4 \text{ V} = 0.4 \text{ V}$$

A TTL gate is immune to up to 0.4 V of noise for both the HIGH and LOW input states.
For CMOS,

$$V_{IH(min)} = 3.5 \text{ V}$$
$$V_{IL(max)} = 1.5 \text{ V}$$
$$V_{OH(min)} = 4.9 \text{ V}$$
$$V_{OL(max)} = 0.1 \text{ V}$$
$$V_{NH} = V_{OH(min)} - V_{IH(min)} = 4.9 \text{ V} - 3.5 \text{ V} = 1.4 \text{ V}$$
$$V_{NL} = V_{IL(max)} - V_{OL(max)} = 1.5 \text{ V} - 0.1 \text{ V} = 1.4 \text{ V}$$

Related Exercise   Based on the preceding noise margin calculations, which family of devices should be used in a high-noise environment?

Power Dissipation

A logic gate draws current from the dc supply voltage source as indicated in Figure 15–6. When the gate is in the HIGH output state, an amount of current designated by I_{CCH} is drawn, and in the LOW output state, a different amount of current, I_{CCL}, is drawn. For a TTL gate, I_{CCL} is greater than I_{CCH}.

FIGURE 15–6
Currents from the dc supply.

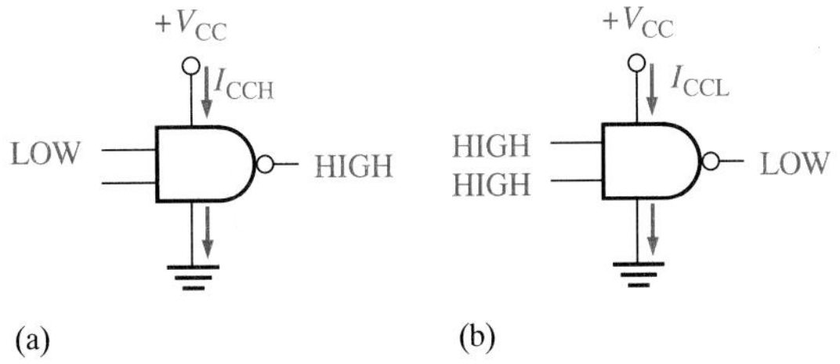

(a)   (b)

As an example, if I_{CCH} is specified as 1.5 mA when V_{CC} is 5 V and if the gate is in a static (nonchanging) HIGH output state, the **power dissipation** (P_D) of the gate is

$$P_D = V_{CC}I_{CCH} = (5 \text{ V})(1.5 \text{ mA}) = 7.5 \text{ mW}$$

When a gate is pulsed, its output switches back and forth between HIGH and LOW, and the amount of supply current varies between I_{CCH} and I_{CCL}. The average power dissipation depends on the duty cycle and is usually specified for a duty cycle of 50%. When the duty cycle is 50%, the output is HIGH half the time and LOW the other half. The average supply current is therefore

$$I_{CC} = \frac{I_{CCH} + I_{CCL}}{2} \tag{15-3}$$

The average power dissipation is

$$P_D = V_{CC}I_{CC} \tag{15-4}$$

EXAMPLE 15–2

A certain gate draws 2 mA when its output is HIGH and 3.6 mA when its output is LOW. What is its average power dissipation if V_{CC} is 5 V and the gate is operated on a 50% duty cycle?

Solution   The average I_{CC} is

$$I_{CC} = \frac{I_{CCH} + I_{CCL}}{2} = \frac{2.0 \text{ mA} + 3.6 \text{ mA}}{2} = 2.8 \text{ mA}$$

The average power dissipation is

$$P_D = V_{CC}I_{CC} = (5 \text{ V})(2.8 \text{ mA}) = 14 \text{ mW}$$

Related Exercise   A certain IC gate has an $I_{CCH} = 1.5$ mA and $I_{CCL} = 2.8$ mA. Determine the average power dissipation for 50% duty cycle operation if V_{CC} is 5 V.

Power dissipation in a TTL circuit is essentially constant over its range of operating frequencies. Power dissipation in CMOS, however, is frequency dependent. It is extremely low under static (dc) conditions and increases as the frequency increases. These characteristics are shown in the general curves of Figure 15–7. For example, the power dissipation of a low-power Schottky (LS) TTL gate is a constant 2 mW. The power dissipation of an HCMOS gate is 0.0000025 mW under static conditions and 0.17 mW at 100 kHz.

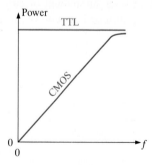

FIGURE 15–7
Power-versus-frequency curves for TTL and CMOS.

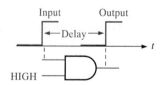

FIGURE 15–8
A basic illustration of propagation delay.

Propagation Delay

When a signal passes (propagates) through a logic circuit, it always experiences a time delay, as illustrated in Figure 15–8. A change in the output level always occurs a short time, called the **propagation delay time,** later than the change in the input level that caused it.

As mentioned in Chapter 3, there are two propagation delay times specified for logic gates:

- t_{PHL}: The time between a designated point on the input pulse and the corresponding point on the output pulse when the output is changing from HIGH to LOW.
- t_{PLH}: The time between a designated point on the input pulse and the corresponding point on the output pulse when the output is changing from LOW to HIGH.

These propagation delay times are illustrated in Figure 15–9, with the 50% points on the pulse edges used as references.

FIGURE 15–9
Propagation delay times.

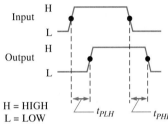

H = HIGH
L = LOW

The propagation delay of a gate limits the frequency at which it can be operated. The greater the propagation delay, the lower the maximum frequency. Thus, a higher-speed circuit is one that has a smaller propagation delay. For example, a gate with a delay of 3 ns is faster than one with a 10 ns delay.

Speed-Power Product

The speed-power product provides a basis for the comparison of logic circuits when *both* propagation delay and power dissipation are important considerations in the selection of the type of logic to be used in a certain application. The unit of speed-power product is the picojoule (pJ).

The lower the speed-power product, the better. The CMOS circuits typically have much lower values for this parameter than do TTL circuits because of much lower power dissipation. For example, HCMOS has a speed-power product of 1.4 pJ at 100 kHz while LS TTL has a value of 20 pJ.

Loading and Fan-Out

When the output of a logic gate is connected to one or more inputs of other gates, a load on the driving gate is created, as shown in Figure 15–10. There is a limit to the number of load gate inputs that a given gate can drive. This limit is called the **fan-out** of the gate.

FIGURE 15–10
Loading a gate output with gate inputs.

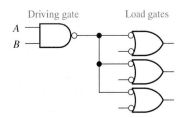

TTL Loading   A TTL driving gate sources current to a load gate input in the HIGH state (I_{IH}) and sinks current from the load gate in the LOW state (I_{IL}). **Current sourcing** and **current sinking** are illustrated in simplified form in Figure 15–11, where the resistors represent the internal input and output resistance of the gate.

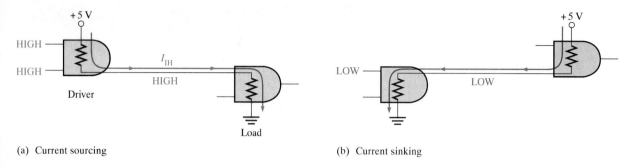

(a) Current sourcing   (b) Current sinking

FIGURE 15–11
Basic illustration of current sourcing and current sinking in logic gates.

As more load gates are connected to the driving gate, the loading on the driving gate increases. The total source current increases with each load gate input that is added, as illustrated in Figure 15–12. As this current increases, the internal voltage drop of the driving gate increases, causing the output, V_{OH}, to decrease. If an excessive number of load gate inputs are connected, V_{OH} drops below $V_{OH(min)}$, and the HIGH-level noise margin is reduced, thus compromising the circuit operation. Also, as the total source current increases, the power dissipation of the driving gate increases.

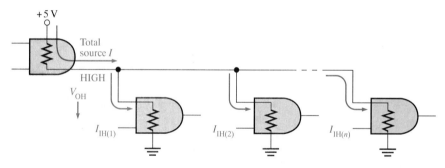

FIGURE 15–12
HIGH-state TTL loading.

The fan-out is the maximum number of load gate inputs that can be connected without adversely affecting the specified operational characteristics of the gate. For example, low-power Schottky (LS) TTL has a fan-out of 20 unit loads. One input of the same logic family as the driving gate is called a **unit load.**

The total sink current also increases with each load gate input that is added, as shown in Figure 15–13. As this current increases, the internal voltage drop of the driving gate increases, causing V_{OL} to increase. If an excessive number of loads are added, V_{OL} exceeds $V_{OL(max)}$, and the LOW-level noise margin is reduced.

In TTL, the current-sinking capability (LOW state) is the limiting factor in determining the fan-out.

CMOS Loading   Loading in CMOS differs from that in TTL because the field-effect transistors (**FETs**) used in CMOS logic present a predominantly capacitive load to the driving gate, as illustrated in Figure 15–14. In this case, the limitations are the charging and discharging times associated with the output resistance of the driving gate and the input capacitance of the load gates. When the output of the driving gate is HIGH, the input capac-

FIGURE 15–13
LOW-state TTL loading.

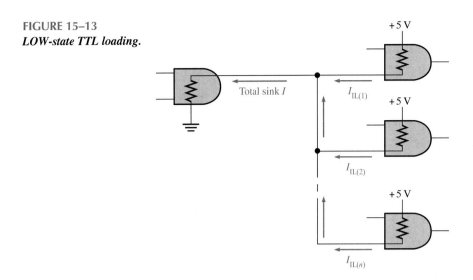

itance of the load gate is charging through the output resistance of the driving gate. When the output of the driving gate is LOW, the capacitance is discharging, as indicated in Figure 15–14.

When more load gate inputs are added to the driving gate output, the total capacitance increases because the input capacitances effectively appear in parallel. This increase in capacitance increases the charging and discharging times, thus reducing the maximum frequency at which the gate can be operated. Therefore, the fan-out of a CMOS gate depends on the frequency of operation. The fewer the load gate inputs, the greater the maximum frequency.

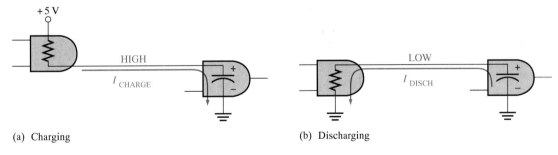

(a) Charging

(b) Discharging

FIGURE 15–14
Capacitive loading of a CMOS gate.

SECTION 15–1 REVIEW

1. Define V_{IH}, V_{IL}, V_{OH}, and V_{OL}.
2. Is it better to have a lower value of noise margin or a higher value?
3. Gate A has a greater propagation delay time than gate B. Which gate can operate at a higher frequency?
4. How does excessive loading affect the noise margin of a gate?

15–2 ■ TTL CIRCUITS

The internal circuit operation of standard TTL logic gates with totem-pole outputs is covered in this section. Other TTL series, including 74L, 74S, 74LS, and 74ALS, are also presented and are compared in power consumption and speed. Also, the operation of TTL gates with open-collector outputs and the operation of tristate gates are cov-ered. After completing this section, you should be able to

□ Identify a bipolar junction transistor (BJT) by its symbol   □ Describe the switching action of a BJT   □ Describe the basic operation of a TTL inverter circuit   □ Describe the basic operation of TTL AND, NAND, OR, and NOR gate circuits   □ Explain what a totem-pole output is   □ Explain the operation and use a TTL gate with an open-collector output   □ Explain the operation of a gate with a tristate output   □ Discuss the LS, S, AS, and ALS TTL circuits

The Bipolar Junction Transistor

The bipolar junction transistor (**BJT**) is the active switching element used in all TTL circuits. Figure 15–15 shows the symbol for an *npn* BJT with its three terminals: **base, emitter,** and **collector.** A BJT has two **junctions,** the base-emitter junction and the base-collector junction.

The basic switching operation is as follows: When the base is approximately 0.7 V more positive than the emitter and when sufficient current is provided into the base, the **transistor** turns on and goes into saturation. In saturation, the transistor ideally acts like a closed switch between the collector and the emitter, as illustrated in Figure 15–16(a). When the base is less than 0.7 V more positive than the emitter, the transistor turns off and becomes an open switch between the collector and the emitter, as shown in part (b). To summarize in general terms, a HIGH on the base turns the transistor on and makes it a closed switch. A LOW on the base turns the transistor off and makes it an open switch. In TTL, some BJTs have multiple emitters.

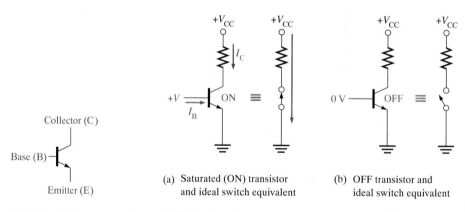

(a) Saturated (ON) transistor and ideal switch equivalent

(b) OFF transistor and ideal switch equivalent

FIGURE 15–15
The symbol for a BJT.

FIGURE 15–16
The ideal switching action of the BJT.

TTL Inverter

Transistor-transistor logic (TTL or T²L) is one of the most widely used integrated circuit technologies. The logic function of an inverter or any type of gate is always the same, regardless of the type of circuit technology that is used. Using TTL technology is only one way to build an inverter or any other logic circuit. Figure 15–17 shows a standard TTL circuit for an inverter. In this figure Q_1 is the input coupling transistor, and D_1 is the input clamp **diode.** Transistor Q_2 is called a *phase splitter,* and the combination of Q_3 and Q_4 forms the output circuit often referred to as a **totem-pole** arrangement.

When the input is a HIGH, the base-emitter junction of Q_1 is **reverse biased,** and the base-collector junction is **forward biased.** This condition permits current through R_1 and the base-collector junction of Q_1 into the base of Q_2, thus driving Q_2 into saturation. As a

FIGURE 15–17
A standard TTL inverter circuit.

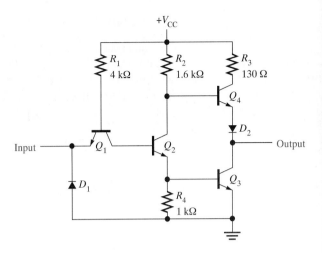

result, Q_3 is turned on by Q_2, and its collector voltage, which is the output, is near ground potential. We therefore have a LOW output for a HIGH input. At the same time, the collector of Q_2 is at a sufficiently low voltage level to keep Q_4 off.

When the input is LOW, the base-emitter junction of Q_1 is forward biased, and the base-collector junction is reverse biased. There is current through R_1 and the base-emitter junction of Q_1 to the LOW input. A LOW provides a path to ground for the current. There is no current into the base of Q_2, so it is off. The collector of Q_2 is HIGH, thus turning Q_4 on. A saturated Q_4 provides a low-resistance path from V_{CC} to the output; we therefore have a HIGH on the output for a LOW on the input. At the same time, the emitter of Q_2 is at ground potential, keeping Q_3 off.

Diode D_1 in the TTL circuit prevents negative spikes of voltage on the input from damaging Q_1. Diode D_2 ensures that Q_4 will turn off when Q_2 is on (HIGH input). In this condition the collector voltage of Q_2 is equal to the base-to-emitter voltage, V_{BE}, of Q_3 plus the collector-to-emitter voltage V_{CE}, of Q_2. Diode D_2 provides an additional V_{BE} equivalent drop in series with the base-emitter junction of Q_4 to ensure its turn-off when Q_2 is on.

The operation of the TTL inverter for the two input states is illustrated in Figure 15–18. In the circuit in part (a), the base of Q_1 is 2.1 V above ground, so Q_2 and Q_3 are on. In the circuit in part (b), the base of Q_1 is about 0.7 V above ground—not enough to turn Q_2 and Q_3 on.

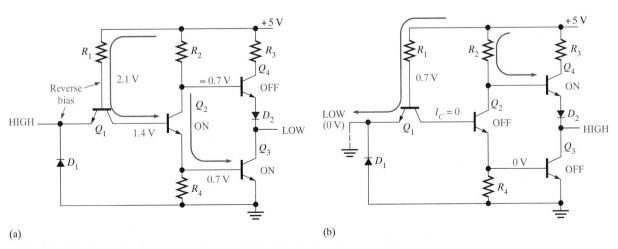

(a)   (b)

FIGURE 15–18
Operation of a TTL inverter.

TTL NAND Gate

A 2-input standard TTL NAND gate is shown in Figure 15–19. Basically, it is the same as the inverter circuit except for the additional input emitter of Q_1. In TTL technology multiple-emitter transistors are used for the input devices. These multiple-emitter transistors can be compared to the diode arrangement, as shown in Figure 15–20.

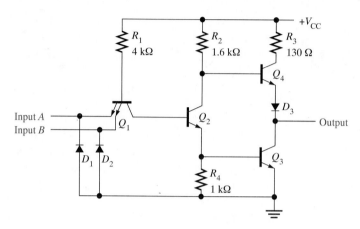

FIGURE 15–19
A standard TTL NAND gate circuit.

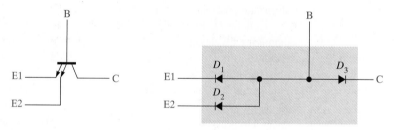

FIGURE 15–20
Diode equivalent of a TTL multiple-emitter transistor.

Perhaps you can understand the operation of this circuit better by visualizing Q_1 in Figure 15–19 replaced by the diode arrangement. A LOW on either input A or input B forward-biases the respective diode and reverse-biases D_3 (Q_1 base-collector junction). This action keeps Q_2 off and results in a HIGH output in the same way as described for the TTL inverter. Of course, a LOW on both inputs will do the same thing.

A HIGH on both inputs reverse-biases both input diodes and forward-biases D_3 (Q_1 base-collector junction). This action turns Q_2 on and results in a LOW output in the same way as described for the TTL inverter.

You should recognize this operation as that of the NAND function: The output is LOW if and only if all inputs are HIGH.

TTL NOR Gate

A 2-input standard TTL NOR gate is shown in Figure 15–21. Compare this with the NAND gate circuit, and notice the additional transistors. Both Q_1 and Q_2 are input transistors. Transistors Q_3 and Q_4 are in parallel and act as a phase splitter. You should recognize Q_5 and Q_6 as the ordinary TTL totem-pole output circuit.

FIGURE 15–21
A standard TTL NOR gate circuit.

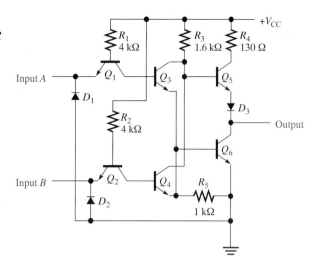

The operation of the TTL NOR gate is as follows:

- When both inputs are LOW, the forward-biased base-emitter junctions of the input transistors pull current away from the phase-splitter transistors, Q_3 and Q_4, keeping them off. As a result, Q_5 is on and Q_6 is off, producing a HIGH output.
- When input A is LOW and input B is HIGH, Q_3 is off and Q_4 is on. Transistor Q_4 turns on Q_6 and turns off Q_5, producing a LOW output.
- When input A is HIGH and input B is LOW, Q_3 is on and Q_4 is off. Transistor Q_3 turns on Q_6 and turns off Q_5, producing a LOW output.
- When both inputs are HIGH, both Q_3 and Q_4 are on. Having both on has the same effect as having either one on, turning Q_6 on and Q_5 off. The result is still a LOW output.

You should recognize this operation as that of the NOR function: The output is LOW when any of the inputs is HIGH.

TTL AND Gates and OR Gates

Figure 15–22 (next page) shows a standard TTL 2-input AND gate and a standard 2-input OR gate. Compare these to the NAND and NOR gate circuits, and notice the additional circuitry that is shaded. In each case, this transistor arrangement provides an inversion so that the NAND becomes AND and the NOR becomes OR.

Open-Collector Gates

The TTL gates described in the previous sections all had the totem-pole output circuit. Another type of output available in TTL integrated circuits is the **open-collector** output. A standard TTL inverter with an open-collector is shown in Figure 15–23(a). The other types of gates are also available with open-collector outputs.

Notice that the output is the collector of transistor Q_3 with nothing connected to it, hence the name *open collector*. In order to get the proper HIGH and LOW logic levels out of the circuit, an external pull-up resistor must be connected to V_{CC} from the collector of Q_3, as shown in Figure 15–23(b). When Q_3 is off, the output is pulled up to V_{CC} through the external resistor. When Q_3 is on, the output is connected to near-ground through the saturated transistor.

The ANSI/IEEE standard symbol that designates an open-collector output is shown in Figure 15–24 for an inverter.

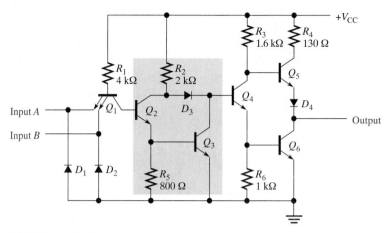

(a) AND gate circuit

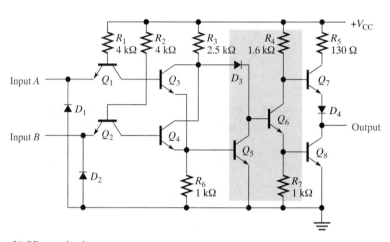

(b) OR gate circuit

FIGURE 15–22
Standard TTL AND gate and OR gate circuits.

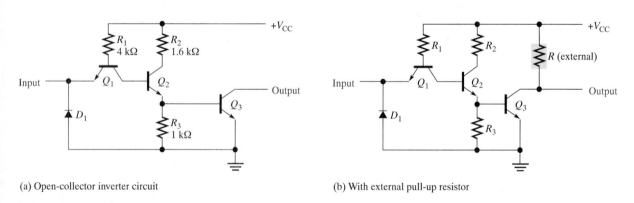

(a) Open-collector inverter circuit

(b) With external pull-up resistor

FIGURE 15–23
Standard TTL inverter with open-collector output.

FIGURE 15–24
Open-collector symbol in an inverter.

Gates with Tristate Outputs

The **tristate** output combines the advantages of the totem-pole and open-collector circuits. The three output states are HIGH, LOW, and high-impedance **(high-Z).** When selected for normal logic-level operation, as determined by the state of the enable input, a tristate circuit operates in the same way as a regular gate. When a tristate circuit is selected for high-Z operation, the output is effectively disconnected from the rest of the circuit. Figure 15–25 illustrates the operation of a tristate circuit. The inverted triangle (∇) designates a tristate output.

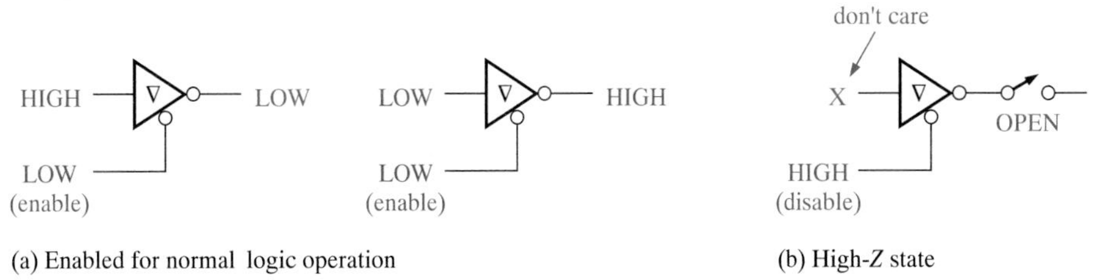

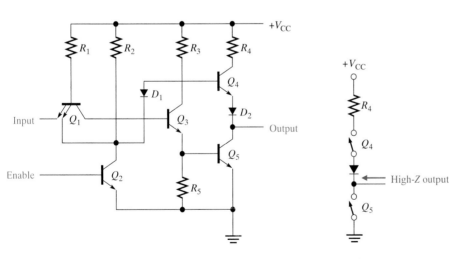

(a) Enabled for normal logic operation

(b) High-Z state

FIGURE 15–25
The three states of a tristate circuit.

Figure 15–26 shows the basic circuit for a TTL tristate inverter. When the enable input is LOW, Q_2 is off, and the output circuit operates as a normal totem-pole configuration, in which the output state depends on the input state. When the enable input is HIGH, Q_2 is on. There is thus a LOW on the second emitter of Q_1, causing Q_3 and Q_5 to turn off, and diode D_1 is forward biased, causing Q_4 also to turn off. When both totem-pole transistors are off, they are effectively open, and the output is completely disconnected from the internal circuitry, as illustrated in Figure 15–27.

FIGURE 15–26
Basic tristate inverter circuit.

FIGURE 15–27
An equivalent circuit for the tristate totem-pole in the high-Z state.

Other TTL Series

The basic TTL NAND gate circuit was discussed earlier. It is a current-sinking type of logic that draws current from the load when in the LOW output state and sources negligible current to the load when in the HIGH output state. Figure 15–28 is a standard 54/74 TTL 2-input NAND gate, and it will be used for comparison with the other types of TTL circuits.

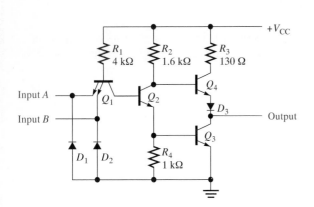

FIGURE 15–28
A 54/74 series standard TTL NAND gate.

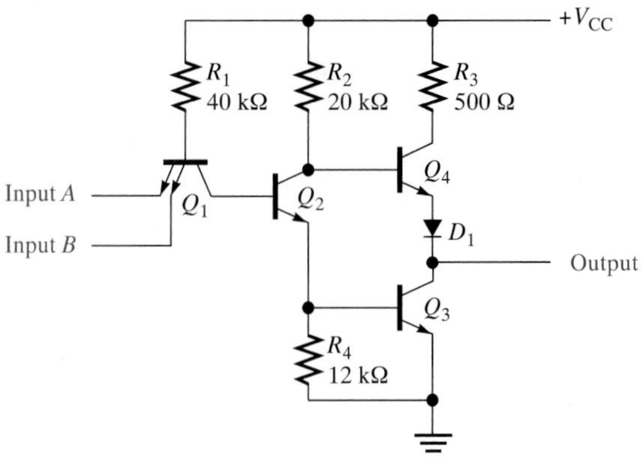

FIGURE 15–29
A 54L/74L series TTL NAND gate.

Low-Power TTL (54L/74L) The 54L/74L series of TTL circuits is designed for low power consumption. A typical gate circuit is shown in Figure 15–29. Notice that the circuit's resistor values are considerably higher than those of the standard gate in Figure 15–28. The higher resistance, of course, results in less current and, therefore, less power but increases the switching time of the gate. The typical power dissipation of a standard 54/74 gate is 10 mW, and that of a 54L/74L gate is 1 mW. The saving of power, however, is paid for in loss of speed. A typical standard 54/74 gate has a propagation delay time of 10 ns, compared with 33 ns for a 54L/74L gate.

Schottky TTL (54S/74S) The 54S/74S series of TTL circuits provides a faster switching time by incorporating Schottky diodes to prevent the transistors from going into saturation, thereby decreasing the time for a transistor to turn on or off. Figure 15–30 shows a Schottky gate circuit. Notice the symbols for the Schottky transistor and Schottky diodes. Typical propagation delay for a 54S/74S gate is 3 ns, and the power dissipation is 19 mW.

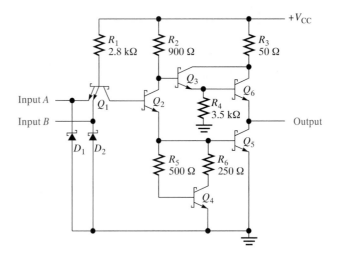

FIGURE 15–30
A 54S/74S series TTL NAND gate.

Low-Power Schottky TTL (54LS/74LS) This TTL series compromises speed to obtain a lower power dissipation. A NAND gate circuit is shown in Figure 15–31. Notice that the input uses Schottky diodes rather than the conventional transistor input. The typical power dissipation for a gate is 2 mW, and the propagation delay is 10 ns.

FIGURE 15–31

A 54LS/74LS series TTL NAND gate.

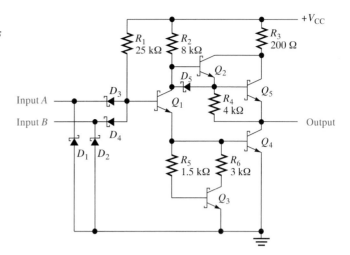

Advanced Schottky and Advanced Low-Power Schottky (AS/ALS) These technologies are advanced versions of the S and LS series. The typical static power dissipation is 8.5 mW for the AS series and 1 mW for the ALS series. The typical propagation delay time is 1.5 ns for the AS and 4 ns for the ALS. One version of the AS series is called the F or FAST series.

SECTION 15–2 REVIEW

1. True or false: An *npn* BJT is on when the base is more negative than the emitter.
2. In terms of switching action, what do the on and off states of a BJT represent?
3. What are the two major types of output circuits in TTL?
4. Explain how tristate logic differs from normal, two-state logic.

15–3 ■ PRACTICAL CONSIDERATIONS IN THE USE OF TTL

We will now examine several practical considerations in the use and application of TTL circuits. After completing this section, you should be able to

☐ Describe current sinking and current sourcing ☐ Use an open-collector circuit for wired-AND operation ☐ Describe the effects of connecting two or more totem-pole outputs ☐ Use open-collector gates to drive LEDs and lamps ☐ Explain what to do with unused TTL inputs

Current Sinking and Current Sourcing

The concepts of current sinking and current sourcing were introduced in Section 15–1. Now that you are familiar with the totem-pole-output circuit configuration used in TTL, let's look closer look at the sinking and sourcing action.

Figure 15–32 shows a standard TTL inverter with a totem-pole output connected to the input of another TTL inverter. When the driving gate is in the HIGH output state, the driver is sourcing current to the load, as shown in Figure 15–32(a). The input to the load gate is like a reverse-biased diode, so there is practically no current required by the load. Actually, since the input is nonideal, there is a maximum of 40 μA from the totem-pole output of the driver into the load gate input.

When the driving gate is in the LOW output state, the driver is sinking current from the load, as shown in Figure 15–32(b). This current is 1.6 mA maximum for standard TTL and is indicated on the **data sheet** with a negative value because it is *out* of the input.

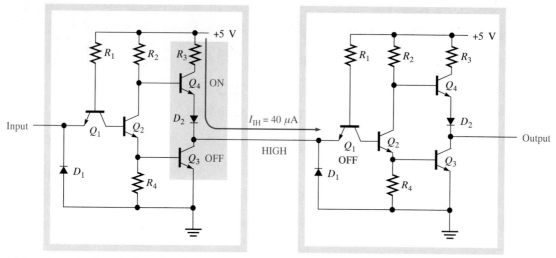

(a) Current sourcing (I_{IH} value is maximum)

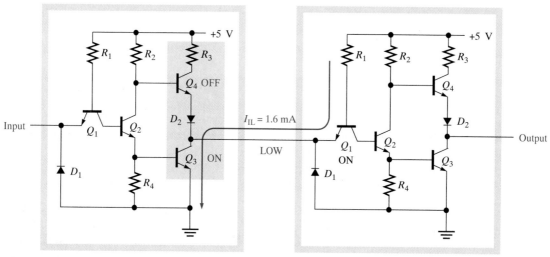

(b) Current sinking (I_{IL} value is maximum)

FIGURE 15–32
Current sinking and sourcing action in TTL.

EXAMPLE 15–3

When a standard TTL NAND gate drives five standard TTL inputs, how much current does the driver output source, and how much does it sink?

Solution   Total source current (in HIGH output state):

$$I_{IH(max)} = 40 \ \mu A \text{ per input}$$

$$I_{T(source)} = (5 \text{ inputs})(40 \ \mu A/\text{input}) = 5(40 \ \mu A) = 200 \ \mu A$$

Total sink current (in LOW output state):

$$I_{IL(max)} = -1.6 \text{ mA per input}$$

$$I_{T(sink)} = (5 \text{ inputs})(-1.6 \text{ mA/input}) = 5(-1.6 \text{ mA}) = -8.0 \text{ mA}$$

Related Exercise   Repeat the calculations for an LS TTL gate. Refer to a data sheet.

EXAMPLE 15–4

Refer to the data sheet in Appendix A, and determine the fan-out of the 7400 NAND gate.

Solution   According to the data sheet, the current parameters are as follows:

$$I_{IH(max)} = 40\ \mu A$$
$$I_{IL(max)} = 1.6\ mA$$
$$I_{OH(max)} = -400\ \mu A$$
$$I_{OL(max)} = 16\ mA$$

Fan-out for the HIGH output state is figured as follows: Current $I_{OH(max)}$ is the maximum current that the gate can source to a load. Each load input requires an $I_{IH(max)}$ of 40 μA. The HIGH-state fan-out is

$$\left| \frac{I_{OH(max)}}{I_{IH(max)}} \right| = \frac{400\ \mu A}{40\ \mu A} = 10$$

For the LOW output state, fan-out is calculated as follows: $I_{OL(max)}$ is the maximum current that the gate can sink. Each load input produces an $I_{IL(max)}$ of 1.6 mA. The LOW-state fan-out is

$$\left| \frac{I_{OL\,(max)}}{I_{IL\,(max)}} \right| = \frac{16\ mA}{1.6\ mA} = 10$$

In this case both the HIGH-state fan-out and the LOW-state fan-out are the same.

Related Exercise   Determine the fan-out for a 74LS00 NAND gate.

Using Open-Collector Gates for Wired-AND Operation

The outputs of open-collector gates can be wired together to form what is called a *wired-AND* configuration. Figure 15–33 illustrates how four inverters are connected to produce a 4-input negative-AND gate. A single external pull-up resistor, R_p, is required in all wired-AND circuits.

FIGURE 15–33
A wired-AND (actually negative-AND) configuration of four inverters.

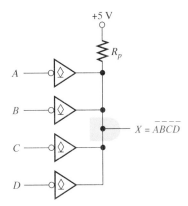

When one (or more) of the inverter inputs is HIGH, the output X is pulled LOW because an output transistor is on and acts as a closed switch to ground, as illustrated in Figure 15–34(a). In this case only one inverter has a HIGH input, but this is sufficient to pull the output LOW through the saturated output transistor Q_1 as indicated.

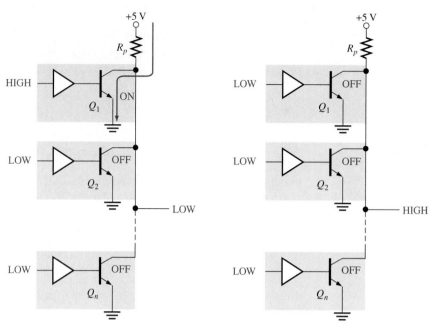

(a) When one or more output transistors are on, the output is LOW.

(b) When all output transistors are off, the output is HIGH.

FIGURE 15–34
Open-collector wired negative-AND operation with inverters.

For the output X to be HIGH, *all* inverter inputs must be LOW so that all the open-collector output transistors are off as indicated in Figure 15–34(b). When this condition exists, the output X is pulled HIGH through the pull-up resistor. Thus, the output X is HIGH only when *all* the inputs are LOW. Therefore, we have a negative-AND function, as expressed in the following equation:

$$X = \overline{A}\,\overline{B}\,\overline{C}\,\overline{D}$$

EXAMPLE 15–5

Write the output expression for the wired-AND configuration of open-collector AND gates in Figure 15–35.

FIGURE 15–35

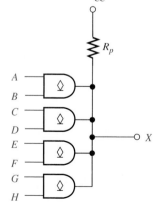

Solution The output expression is

$$X = ABCDEFGH$$

The wired-AND connection of the four 2-input AND gates creates an 8-input AND gate.

Related Exercise   Determine the output expression if NAND gates are used in Figure 15–35.

EXAMPLE 15–6

Three open-collector AND gates are connected in a wired-AND configuration as shown in Figure 15–36. Assume that the wired-AND circuit is driving four standard TTL inputs (-1.6 mA each).
(a) Write the logic expression for X.
(b) Determine the minimum value of R_p if $I_{OL(max)}$ for each gate is 30 mA and $V_{OL(max)}$ is 0.4 V.

FIGURE 15–36

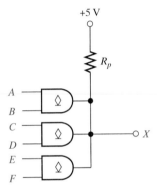

Solution
(a) $X = ABCDEF$
(b) $4(1.6 \text{ mA}) = 6.4 \text{ mA}$

$$I_{R_p} = I_{OL(max)} - 6.4 \text{ mA} = 30 \text{ mA} - 6.4 \text{ mA} = 23.6 \text{ mA}$$

$$R_p = \frac{V_{CC} - V_{OL(max)}}{I_{R_p}} = \frac{5 \text{ V} - 0.4 \text{ V}}{23.6 \text{ mA}} = 195 \text{ }\Omega$$

Related Exercise   Show the wired-AND circuit for a 10-input AND function using 74LS09 quad 2-input AND gates.

Connection of Totem-Pole Outputs

Totem-pole outputs cannot be connected together because such a connection might produce excessive current and result in damage to the devices. For example, in Figure 15–37 (next page), when Q_1 in device A and Q_2 in device B are both on, the output of device A is effectively shorted to ground through Q_2 of device B.

Open-Collector Buffer/Drivers

A TTL circuit with a totem-pole output is limited in the amount of current that it can sink in the LOW state ($I_{OL(max)}$) to 16 mA for standard TTL and 20 mA for AS TTL. In many special applications, a gate must drive external devices, such as LEDs, lamps, or relays, that may require more current than that.

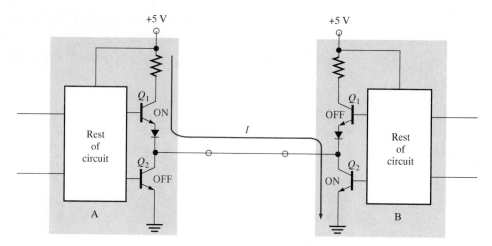

FIGURE 15–37

Totem-pole outputs wired together. Such a connection may cause excessive current through Q_1 of device A and Q_2 of device B.

Because of their higher voltage and current-handling capability, circuits with open-collector outputs are generally used for driving LEDs, lamps, or relays. However, totem-pole outputs can be used, as long as the output current required by the external device does not exceed the amount that the TTL driver can sink.

With an open-collector TTL gate, the collector of the output transistor is connected to an LED or incandescent lamp as illustrated in Figure 15–38. In part (a) the limiting resistor is used to keep the current below maximum LED current. When the output of the gate is LOW, the output transistor is sinking current, and the **LED** is on. The LED is off when the output transistor is off and the output is HIGH. A typical open-collector buffer gate can sink up to 40 mA. In part (b) of the figure, the lamp requires no limiting resistor because the filament is resistive. Typically, up to +30 V can be used on the open collector, depending on the particular logic family.

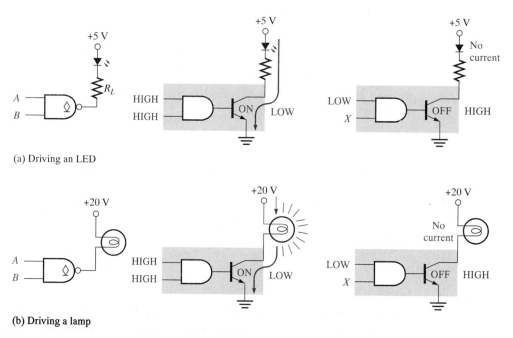

(a) Driving an LED

(b) Driving a lamp

FIGURE 15–38

Some applications of open-collector drivers.

EXAMPLE 15–7

Determine the value of the limiting resistor, R_L, in the open-collector circuit of Figure 15–39 if the LED current is to be 20 mA. Assume a 0.7 V drop across the LED when it is forward biased and a LOW-state output voltage of 0.1 V at the output of the gate.

FIGURE 15–39

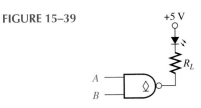

Solution

$$V_{R_L} = 5\ \text{V} - 0.7\ \text{V} - 0.1\ \text{V} = 4.2\ \text{V}$$

$$R_L = \frac{V_{R_L}}{I} = \frac{4.2\ \text{V}}{20\ \text{mA}} = 210\ \Omega$$

Related Exercise Determine the value of the limiting resistor, R_L, if the LED requires 35 mA.

Unused TTL Inputs

An unconnected input on a TTL gate acts as a HIGH because an open input results in a reverse-biased emitter junction on the input transistor, just as a HIGH level does. This effect is illustrated in Figure 15–40. However, because of noise sensitivity, it is best not to leave unused TTL inputs unconnected (open). There are several alternative ways to handle unused inputs.

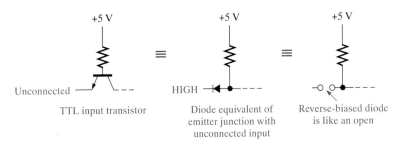

FIGURE 15–40
Comparison of an open TTL input and a HIGH-level input.

Tied-Together Inputs   The most common method for handling unused gate inputs is to connect them to a used input of the same gate. For AND gates and NAND gates, all tied-together inputs count as one unit load in the LOW state; but for OR gates and NOR gates, each input tied to another input counts as a separate unit load in the LOW state. In the HIGH state, each tied-together input counts as a separate load for all types of TTL gates. In Figure 15–41(a) are two examples of the connection of two unused inputs to a used input.

The reason that AND and NAND gates present only a single unit load no matter how many inputs are tied together whereas OR and NOR gates present a unit load for each tied-together input can be seen by looking back at Figures 15–19 (p. 746) and 15–21 (p. 747). The NAND gate uses a multiple-emitter input transistor. No matter how many inputs are LOW, the total LOW-state current is limited by R_1 to a fixed value. The NOR gate uses a separate transistor for each input, so the LOW-state current is the sum of the currents from all the tied-together inputs.

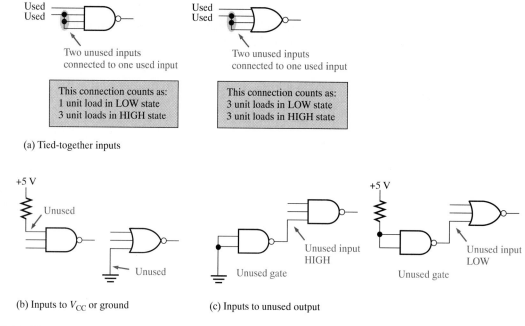

(a) Tied-together inputs

(b) Inputs to V_{CC} or ground

(c) Inputs to unused output

FIGURE 15–41
Methods for handling unused TTL inputs.

Inputs to V_{CC} or Ground Unused inputs of AND and NAND gates can be connected to V_{CC} through a 1 kΩ resistor. This connection pulls the unused inputs to a HIGH level. Unused inputs of OR and NOR gates can be connected to ground. These methods are illustrated in Figure 15–41(b).

Inputs to Unused Output A third method of terminating unused inputs may be appropriate in some cases when an unused gate or inverter is available. The unused gate output must be a constant HIGH for unused AND and NAND inputs and a constant LOW for unused OR and NOR inputs, as illustrated in Figure 15–41(c).

SECTION 15–3 REVIEW

1. In what output state does a TTL circuit sink current from a load?
2. Why does a TTL circuit source less current into a TTL load than it sinks?
3. Why can TTL circuits with totem-pole outputs not be connected together?
4. What type of TTL circuit must be used for a wired-AND configuration?
5. What type of TTL circuit would you use to drive a lamp?
6. True or false: An unconnected TTL input acts as a LOW.

15–4 ■ CMOS CIRCUITS

Basic internal CMOS circuitry and its operation are presented in this section. The abbreviation CMOS stands for complementary metal-oxide semiconductor. The term complementary refers to the use of two types of transistors in the output circuit in a configuration similar to the totem-pole in TTL. An n-channel MOSFET (MOS field-effect transistor) and a p-channel MOSFET are used. After completing this section, you should be able to

□ Identify a MOSFET by its symbol   □ Discuss the switching action of a MOSFET
□ Describe the basic operation of a CMOS inverter circuit   □ Describe the basic operation of CMOS NAND and NOR gates   □ Explain the operation of a CMOS

gate with an open-drain output ☐ Discuss the operation of tristate CMOS gates ☐ List the precautions required when handling CMOS devices

The MOSFET

Metal-oxide semiconductor field-effect transistors (**MOSFETs**) are the active switching elements in CMOS circuits. These devices differ greatly in construction and internal operation from BJTs but the switching action is basically the same: they function ideally as open or closed switches, depending on the input.

Figure 15–42(a) shows the symbols for both n-channel and p-channel MOSFETs. As indicated, the three terminals of a MOSFET are **gate, drain,** and **source.** When the gate voltage of an n-channel MOSFET is more positive than the source, the MOSFET is on, and there is a low value of resistance between the drain and the source (R_{ON}). When the gate-to-source voltage is zero, the MOSFET is off, and there is an extremely high resistance between the drain and the source (R_{OFF}). This operation is illustrated in Figure 15–42(b). The p-channel MOSFET operates with opposite voltage polarities, as shown in part (c). Ideally, R_{ON} and R_{OFF} can be neglected, and the MOSFET can be considered as a closed switch when it is on and an open switch when it is off.

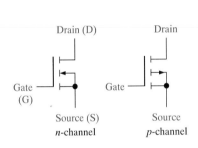

(a) MOSFET symbols

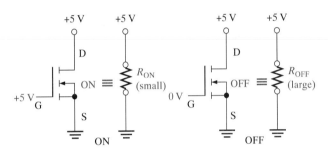

(b) n-channel switch

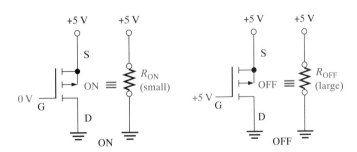

(c) p-channel switch

FIGURE 15–42
Basic symbols and switching action of MOSFETs.

CMOS Inverter

Complementary MOS (CMOS) logic uses the MOSFET as its basic element. It uses both p-channel and n-channel enhancement MOSFETs, as shown in the inverter circuit in Figure 15–43.

When a HIGH is applied to the input, the p-channel MOSFET Q_1 is off, and the n-channel MOSFET Q_2 is on. This condition connects the output to ground through the *on* resistance of Q_2, resulting in a LOW output. When a LOW is applied to the input, Q_1 is on

FIGURE 15–43
A CMOS inverter circuit.

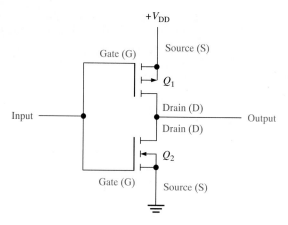

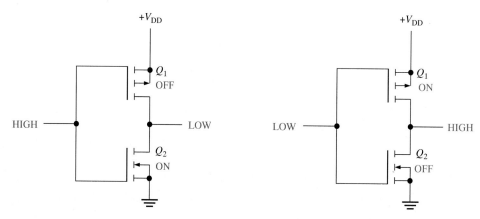

FIGURE 15–44
Operation of a CMOS inverter.

and Q_2 is off. This condition connects the output to $+V_{DD}$ through the *on* resistance of Q_1, resulting in a HIGH output. The operation is illustrated in Figure 15–44.

CMOS NAND Gate

Figure 15–45 shows a CMOS NAND gate with two inputs. Notice the arrangement of the complementary pairs (*n*-channel and *p*-channel MOSFETs).

The operation of a CMOS NAND gate is as follows:

- When both inputs are LOW, Q_1 and Q_2 are on, and Q_3 and Q_4 are off. The output is pulled HIGH through the *on* resistance of Q_1 and Q_2 in parallel.
- When input A is LOW and input B is HIGH, Q_1 and Q_4 are on, and Q_2 and Q_3 are off. The output is pulled HIGH through the low *on* resistance of Q_1.
- When input A is HIGH and input B is LOW, Q_1 and Q_4 are off, and Q_2 and Q_3 are on. The output is pulled HIGH through the low *on* resistance of Q_2.
- Finally, when both inputs are HIGH, Q_1 and Q_2 are off, and Q_3 and Q_4 are on. In this case, the output is pulled LOW through the *on* resistance of Q_3 and Q_4 in series to ground.

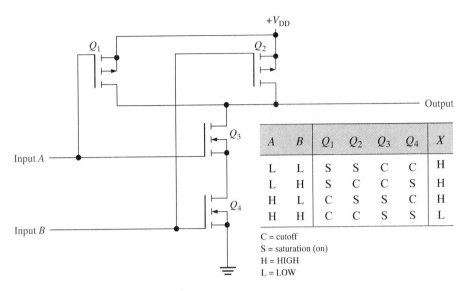

A	B	Q_1	Q_2	Q_3	Q_4	X
L	L	S	S	C	C	H
L	H	S	C	C	S	H
H	L	C	S	S	C	H
H	H	C	C	S	S	L

C = cutoff
S = saturation (on)
H = HIGH
L = LOW

FIGURE 15–45
A CMOS NAND gate circuit.

CMOS NOR Gate

Figure 15–46 shows a CMOS NOR gate with two inputs. Notice the arrangement of the complementary pairs.

The operation of a CMOS NOR gate is as follows:

- When both inputs are LOW, Q_1 and Q_2 are on, and Q_3 and Q_4 are off. As a result, the output is pulled HIGH through the *on* resistance of Q_1 and Q_2 in series.
- When input A is LOW and input B is HIGH, Q_1 and Q_4 are on, and Q_2 and Q_3 are off. The output is pulled LOW through the low *on* resistance of Q_4 to ground.
- When input A is HIGH and input B is LOW, Q_1 and Q_4 are off, and Q_2 and Q_3 are on. The output is pulled LOW through the *on* resistance of Q_3 to ground.
- When both inputs are HIGH, Q_1 and Q_2 are off, and Q_3 and Q_4 are on. The output is pulled LOW through the *on* resistance of Q_3 and Q_4 in parallel to ground.

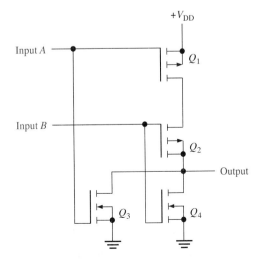

A	B	Q_1	Q_2	Q_3	Q_4	X
L	L	S	S	C	C	H
L	H	S	C	C	S	L
H	L	C	S	S	C	L
H	H	C	C	S	S	L

C = cutoff
S = saturation (on)
H = HIGH
L = LOW

FIGURE 15–46
A CMOS NOR gate circuit.

Open-Drain Gates

An open-drain gate is the CMOS counterpart of an open-collector TTL gate. An open-drain output circuit is a single n-channel MOSFET as shown in Figure 15–47(a). As with open-collector TTL, a pull-up resistor must be used, as shown in part (b), to produce a HIGH output state. Also, like open-collector TTL, open-drain outputs can be connected in a wired-AND configuration.

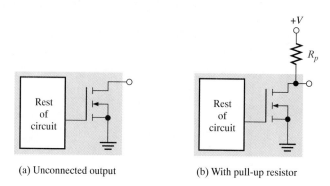

(a) Unconnected output   (b) With pull-up resistor

FIGURE 15–47
Open-drain gates.

Tristate Gates

The circuitry in a tristate CMOS gate, as shown in Figure 15–48, allows each of the output transistors Q_1 and Q_2 to be turned off at the same time, thus disconnecting the output from the rest of the circuit.

When the enable input is LOW, the device is enabled for normal logic operation. When the enable is HIGH, both Q_1 and Q_2 are off, and the circuit is in the high-Z state.

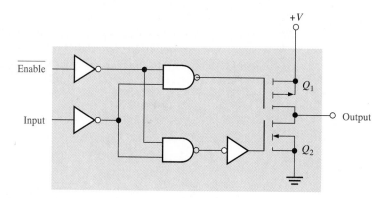

FIGURE 15–48
A tristate CMOS inverter.

Precautions for Handling CMOS

All CMOS devices are subject to damage from electrostatic discharge (ESD). Therefore, they must be handled with special care. Note the following precautions:

1. All CMOS devices are shipped in conductive foam to prevent electrostatic charge buildup. When they are removed from the foam, the pins should not be touched.
2. The devices should be placed with pins down on a grounded surface, such as a metal plate, when removed from protective material. Do not place CMOS devices in polystyrene foam or plastic trays.

3. All tools, test equipment, and metal workbenches should be earth-grounded. A person working with CMOS devices should, in certain environments, have his or her wrist grounded with a length of cable and a large-value resistor. The resistor prevents severe shock should the person come in contact with a voltage source.

4. Do not insert CMOS devices (or any other ICs) into sockets or PC boards with the power on.

5. All unused inputs should be connected to the supply voltage or ground as indicated in Figure 15–49. If left open, an input can acquire electrostatic charge and "float" to unpredicted levels.

6. After assembly on PC boards, protection should be provided by storing or shipping boards with their connectors in conductive foam. The CMOS input and output pins may also be protected with large-value resistors connected to ground.

FIGURE 15–49
Handling unused CMOS inputs.

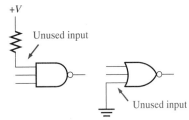

SECTION 15–4
REVIEW

1. What type of transistor is used in CMOS logic?
2. What is meant by the term *complementary MOS?*
3. Why must CMOS devices be handled with great care?

15–5 ■ COMPARISON OF CMOS AND TTL CHARACTERISTICS

In this section, the main operational and performance characteristics of HCMOS (high-speed CMOS), which is the 74HC series, are compared with those of the various TTL series, particularly low-power Schottky (74LS). After completing this section, you should be able to

☐ Compare TTL and CMOS devices in terms of power dissipation, propagation delay, maximum clock frequency, speed/power product, and fan-out

Table 15–1 (next page) gives a broad performance comparison of HCMOS and several TTL series. Also, the CMOS 4000 series is listed for comparison. Keep in mind that parameter values for specific devices within a series may vary, so you should always consult the appropriate data sheet. The maximum clock frequency, f_s, is the maximum rate at which flip-flops can be clocked reliably. All values are typical and are specified for a supply voltage of $+5$ V. Fan-outs are specified for same-series loads (standard TTL to standard TTL, LS TTL to LS TTL, etc.) and also for LS loads as a basis for comparison among the various devices.

The HCMOS devices can operate at speeds comparable to LS TTL. At higher frequencies the power consumption of HCMOS approaches that of LS TTL. The extremely low static (no-switching) power consumption of CMOS makes it ideal for battery-powered and battery-backup systems in which the current drain on the battery is a major consideration.

An HCMOS circuit has a better noise immunity than any of the TTL series, as indicated in Table 15–2.

TABLE 15–1

Performance comparison of CMOS and TTL logic families

Technology	CMOS		TTL				
	Silicon-Gate	Metal-Gate	Std.	Low-Power Schottky	Schottky	Advanced Low-Power Schottky	Advanced Schottky
Device series	74HC	4000	74	74LS	74S	74ALS	74AS
Power dissipation (mW/gate):							
Static	0.0000025	0.001	10	2	19	1	8.5
At 100 kHz	0.17	0.1	10	2	19	1	8.5
Propagation delay time (ns) ($C_L = 15$ pF)	8	50	10	10	3	4	1.5
Maximum clock frequency (MHz) ($C_L = 15$ pF)	40	12	35	40	125	70	200
Speed/Power product (pJ) (at 100 kHz)	1.4	11	100	20	57	4	13
Minimum output drive I_{OL} (mA) ($V_O = 0.4$ V)	4	1.6	16	8	20	8	20
Fan-out:							
LS loads	10	4	40	20	50	20	50
Same-series	*	*	10	20	20	20	40
Maximum input current, I_{IL} (mA) ($V_1 = 0.4$ V)	±0.001	−0.001	−1.6	−0.4	−2.0	−0.1	−0.5

*Fan-out is frequency dependent.

TABLE 15–2

Comparison of typical noise margins

Noise Margin	HCMOS	STD. TTL	LS TTL	S TTL	AS TTL
V_{NH}	1.4 V	0.4 V	0.7 V	0.7 V	0.7 V
V_{NL}	0.9 V	0.4 V	0.4 V	0.4 V	0.4 V

SECTION 15–5 REVIEW

1. True or false: A CMOS circuit consumes less power than a TTL circuit.
2. Does the power consumption of CMOS increase or decrease with frequency?
3. True or false: A TTL circuit has better noise immunity than an HCMOS circuit.

15–6 ■ INTERFACING LOGIC FAMILIES

In this section, the interfacing of TTL and CMOS logic devices is considered. When two different types of technologies are interfaced, the input and output voltages and currents of each are important parameters. In fact, the difference in these parameters creates the interfacing problem. After completing this section, you should be able to

☐ Interface CMOS to TTL   ☐ Interface TTL to CMOS

Table 15–3 shows typical worst-case values of the input and output parameters for a CMOS family and several TTL families.

TABLE 15–3

Worst-case values of interfacing parameters

Parameter	74HC CMOS	74 TTL	74LS TTL	74S TTL	74AS TTL
$V_{IH(min)}$	3.15 V	2 V	2 V	2 V	2 V
$V_{IL(max)}$	1 V	0.8 V	0.8 V	0.8 V	0.8 V
$V_{OH(min)}$	4.9 V	2.4 V	2.7 V	2.7 V	2.7 V
$V_{OL(max)}$	0.1 V	0.4 V	0.4 V	0.5 V	0.5 V
$I_{IH(max)}$	1 μA	40 μA	20 μA	50 μA	200 μA
$I_{IL(max)}$	−1 μA	−1.6 mA	−400 μA	−2 mA	−2 mA
$I_{OH(max)}$	−4 mA	−400 μA	−400 μA	−1 mA	−2 mA
$I_{OL(max)}$	4 mA	16 mA	8 mA	20 mA	20 mA

CMOS-to-TTL Interfacing

This discussion is for the case in which CMOS is driving TTL. Table 15–3 shows that the minimum HIGH output voltage $V_{OH(min)}$ for CMOS is 4.9 V. Since this value exceeds the minimum HIGH input voltage $V_{IH(min)}$ of 2 V that is required by TTL, the CMOS is compatible with TTL in the HIGH state.

A CMOS has a maximum LOW output voltage $V_{OL(max)}$ of 0.1 V. Since this value is less than the maximum LOW input voltage $V_{IL(max)}$ of 0.8 V required by TTL, the CMOS is also compatible with TTL in the LOW state.

In terms of currents, CMOS can sink 4 mA, $I_{OL(max)}$, in the LOW output state with a guaranteed output voltage. When driving standard TTL, the CMOS gate must be able to sink 1.6 mA ($I_{IL(max)}$) from each TTL input. This limits the fan-out of the CMOS gate to two TTL inputs (2 × 1.6 mA = 3.2 mA), as shown in Figure 15–50(a).

When driving low-power Schottky TTL (74LS), the CMOS gate must be able to sink 400 μA from each TTL input. This limits the fan-out of the CMOS gate to ten LS TTL inputs (10 × 400 μA = 4 mA), as shown in Figure 15–50(b).

When driving Schottky TTL (74S), the CMOS gate must be able to sink 2 mA from each TTL input. This limits the fan-out of the CMOS gate to two S TTL inputs (2 × 2 mA = 4 mA), as shown in Figure 15–50(c).

Finally, when driving advanced Schottky TTL (74AS), the CMOS gate must be able to sink 2 mA from each TTL input. This limits the fan-out of the CMOS gate to two AS TTL inputs (2 × 2 mA = 4 mA), as shown in Figure 15–50(d).

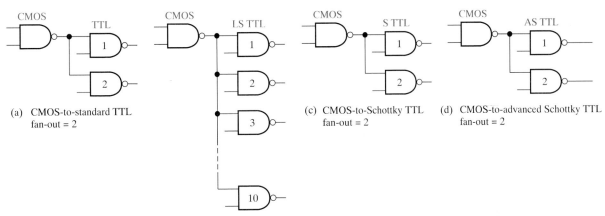

(a) CMOS-to-standard TTL
fan-out = 2

(b) CMOS-to-low-power Schottky TTL
fan-out = 10

(c) CMOS-to-Schottky TTL
fan-out = 2

(d) CMOS-to-advanced Schottky TTL
fan-out = 2

FIGURE 15–50

Interfacing of CMOS to TTL.

TTL-to-CMOS Interfacing

When TTL is driving CMOS, the interface is not as simple as CMOS-to-TTL. As you can see in Table 15–3, the TTL families have minimum HIGH output voltages $V_{OH(min)}$ of 2.4 V to 2.7 V. The minimum HIGH input voltage required by CMOS is 3.15 V. Thus, the output voltage of TTL is not sufficient to drive CMOS in the HIGH state. In the LOW state, the voltages are compatible, as you verify in Table 15–3.

For TTL to be properly interfaced to CMOS, a pull-up resistor (R_p) to V_{CC} must be added as shown in Figure 15–51.

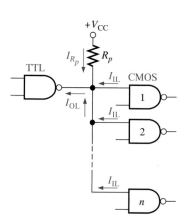

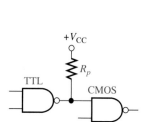

FIGURE 15–51
Proper TTL-to-CMOS interfacing.

FIGURE 15–52
LOW-state current sinking.

In the LOW state, the TTL driving gate must sink current from the resistor as well as from the CMOS inputs to which it is connected, as indicated in Figure 15–52. This requirement sets the value of R_p according to the following equation:

$$R_p = \frac{V_{CC} - V_{OL(min)}}{I_{OL(TTL)} + nI_{IL(CMOS)}} \qquad (15\text{–}5)$$

where n is the number of CMOS inputs being driven and $I_{OL(TTL)} + nI_{IL(CMOS)} = I_{R_p}$.

EXAMPLE 15–8

A 74LS TTL gate drives four 74HC CMOS gates. The minimum V_{CC} is 4.75 V. Determine the minimum value of pull-up resistor for interfacing these devices.

Solution From Table 15–3, $I_{OL} = 8$ mA for the 74LS gate, $I_{IL} = -1$ μA, and $V_{OL(min)}$ is unspecified for a 74HC gate. Therefore, a value of 0 for $V_{OL(min)}$ is assumed, to ensure that $I_{OL(max)}$ is never exceeded. Thus,

$$R_p = \frac{V_{CC} - V_{OL(min)}}{I_{OL} + 4I_{IL}} = \frac{4.75 \text{ V} - 0 \text{ V}}{8 \text{ mA} - 4 \text{ }\mu\text{A}} = \frac{4.75 \text{ V}}{7.996 \text{ mA}} = 594 \text{ }\Omega$$

Related Exercise Determine R_p for interfacing a 74LS gate with ten 74HC gates.

SECTION 15–6
REVIEW

1. When is a pull-up resistor required in interfacing TTL and CMOS?
2. True or false: A higher fan-out requires a lower value of R_p.

15–7 ■ ECL CIRCUITS

The abbreviation ECL stands for emitter-coupled logic, which, like TTL, is a bipolar technology. The typical ECL circuit consists of a differential amplifier input circuit, a bias circuit, and emitter-follower outputs. ECL is much faster than TTL because the transistors do not operate in saturation. After completing this section, you should be able to

☐ Describe how ECL differs from TTL and CMOS   ☐ Discuss the basic operation of an ECL circuit   ☐ Explain the advantages and disadvantages of ECL

An **ECL** OR/NOR gate is shown in Figure 15–53(a). The emitter-follower outputs provide the OR logic function and its NOR complement, as indicated by Figure 15–53(b).

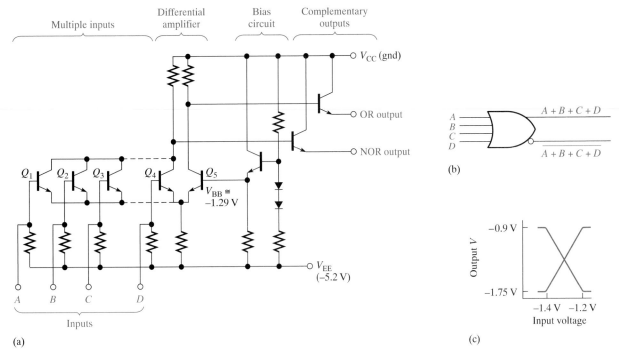

FIGURE 15–53
An ECL OR/NOR gate circuit.

Because of the low output impedance of the emitter-follower and the high input impedance of the differential amplifier input, high fan-out operation is possible. In this type of circuit, saturation is not possible. The lack of saturation results in higher power consumption and limited voltage swing (less than 1 V), but it permits high-frequency switching.

The V_{CC} pin is normally connected to ground, and the V_{EE} pin is connected to -5.2 V from the power supply for best operation. Notice that in Figure 15–53(c) the output varies from a LOW level of -1.75 V to a HIGH level of -0.9 V with respect to ground. In positive logic a 1 is the HIGH level (less negative), and a 0 is the LOW level (more negative).

ECL Circuit Operation

The basic ECL gate operation is shown in Figure 15–54. Beginning with LOWs on all inputs (typically -1.75 V), assume that Q_1 through Q_4 are off because the base-emitter junctions are reverse biased, and that transistor Q_5 is conducting (but not saturated). The bias

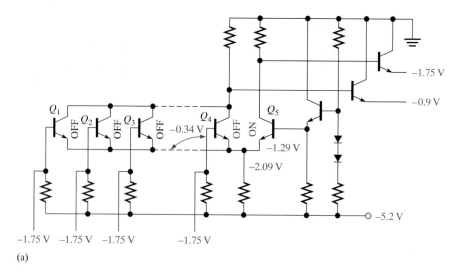

(a)

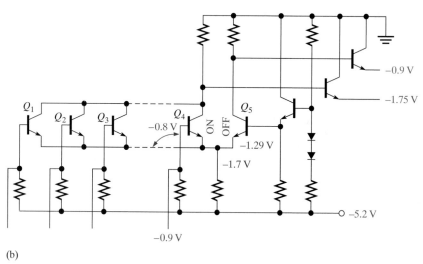

(b)

FIGURE 15–54
Basic ECL gate operation.

circuit holds the base of Q_5 at -1.29 V, and, therefore, the emitter of Q_5 is approximately 0.8 V below the base at -2.09 V. The voltage differential from base to emitter of the input transistors Q_1 through Q_4 is -2.09 V $- (-1.75$ V$) = -0.34$ V. This is less than the forward-biased voltage of these transistors, and they are therefore off. This condition is shown in Figure 15–54(a).

When any one or all of the inputs are raised to the HIGH level $(-0.9$ V), that transistor (or transistors) will conduct. When this happens, the voltage at the emitters of Q_1 through Q_5 increases from -2.09 V to -1.7 V (one base-emitter drops below the -0.9 V base). Since the base of Q_5 is held at a constant -1.29 V by the bias circuit, Q_5 turns off. The resulting collector voltages are coupled through the emitter-followers to the output terminals. Because of the differential action of Q_1 through Q_4 with Q_5, Q_5 is off when one or more of the Q_1 through Q_4 transistors conduct, thus providing simultaneous complementary outputs. This operation is shown in Figure 15–54(b).

When all the inputs are returned to the LOW state, Q_1 through Q_4 are again cut off, and Q_5 conducts.

Noise Margin

As you know, the noise margin of a gate is the measure of its immunity to undesired voltage fluctuations (noise). Typical ECL circuits have noise margins from about 0.2 V to 0.25 V. These are less than for TTL and make ECL unreliable in high-noise environments.

Comparison of ECL with Advanced Schottky TTL

Table 15–4 shows a comparison of typical values of key parameters for advanced Schottky (AS) TTL and ECL.

TABLE 15–4

	Propagation Delay Time	Power Dissipation Per Gate	Voltage Swing	DC Supply Voltage	Flip-Flop Clock Freq.
74AS	1.5 ns	8.5 mW	3.0 V	+5 V	200 MHz
ECL	1 ns	60 mW	0.85 V	−5.2 V	500 MHz

SECTION 15–7 REVIEW	1. What is the primary advantage of ECL over TTL? 2. Name two disadvantages of ECL compared with TTL.

15–8 ■ PMOS, NMOS, AND E²CMOS

The PMOS and NMOS circuits are used largely in LSI functions, such as long shift registers, large memories, and microprocessor products. Such use is a result of the low power consumption and very small chip area required for MOS transistors. E²CMOS is used in reprogrammable PLDs. After completing this section, you should be able to

☐ Describe a basic PMOS gate   ☐ Describe a basic NMOS gate   ☐ Describe a basic E²CMOS cell

PMOS

One of the first high-density **MOS** circuit technologies to be produced was **PMOS.** It utilizes enhancement-mode *p*-channel MOS transistors to form the basic gate building blocks. Figure 15–55 (next page) shows a basic PMOS gate that produces the NOR function in positive logic.

The operation of the PMOS gate is as follows: The supply voltage V_{GG} is a negative voltage, and V_{CC} is a positive voltage or ground (0 V). Transistor Q_3 is permanently biased to create a constant drain-to-source resistance. Its sole purpose is to function as a current-limiting resistor. If a HIGH (V_{CC}) is applied to input A or B, then Q_1 or Q_2 is off, and the output is pulled down to a voltage near V_{GG}, which represents a LOW. When a LOW voltage (V_{GG}) is applied to both input A and input B, both Q_1 and Q_2 are turned on. This causes the output to go to a HIGH level (near V_{CC}). Since a LOW output occurs when either or both inputs are HIGH, and a HIGH output occurs only when all inputs are LOW, we have a NOR gate.

FIGURE 15–55
Basic PMOS gate.

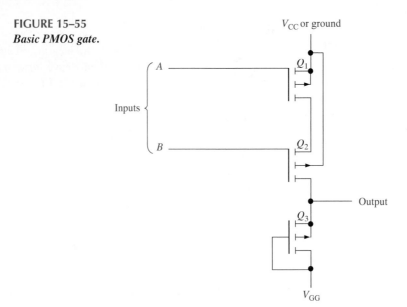

NMOS

The **NMOS** devices were developed as processing technology improved, and now most memories and microprocessors use NMOS. The *n*-channel MOS transistor is used in NMOS circuits, as shown in Figure 15–56 for a NAND gate and a NOR gate.

In Figure 15–56(a), Q_3 acts as a resistor to limit current. When a LOW (V_{GG} or ground) is applied to one or both inputs, then at least one of the transistors (Q_1 or Q_2) is off, and the output is pulled up to a HIGH level near V_{CC}. When HIGHs (V_{CC}) are applied to both A and B, both Q_1 and Q_2 conduct, and the output is LOW. This action, of course, identifies this circuit as a NAND gate.

In Figure 15–56(b), Q_3 again acts as a resistor. A HIGH on either input turns Q_1 or Q_2 on, pulling the output LOW. When both inputs are LOW, both transistors are off, and the output is pulled up to a HIGH level.

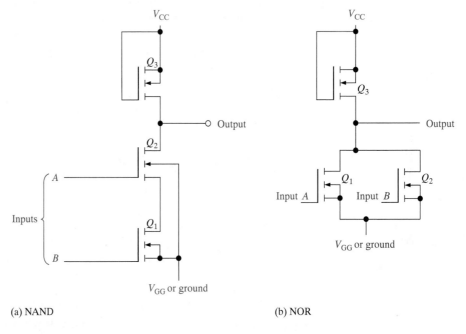

(a) NAND   (b) NOR

FIGURE 15–56
Two NMOS gates.

E²CMOS

E²CMOS technology is based on a combination of CMOS and NMOS technologies and is used in generic array logic devices (GALs) such as those discussed in Chapters 7 and 11. An E²CMOS cell is built around an MOS transistor with a floating gate which is externally charged or discharged by a small programming current. A schematic of this type of cell is shown in Figure 15–57.

FIGURE 15–57
An E²CMOS cell.

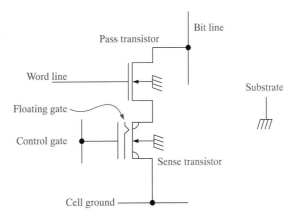

When the floating gate is charged to a positive potential by removing electrons, the sense transistor is turned on, storing a binary zero. When the floating gate is charged to a negative potential by placing electrons on it, the sense transistor is turned off, storing a binary 1. The control gate controls the potential of the floating gate. The pass transistor isolates the sense transistor from the array during read and write operations which use the word and bit lines.

The cell is programmed by applying a programming pulse to either the control gate or the bit line of a cell that has been selected by a voltage on the word line. During the programming cycle, the cell is first erased by applying a voltage to the control gate to make the floating gate negative. This leaves the sense transistor in the *off* state (storing a 1). A write pulse is applied to the bit line of a cell in which a 0 is to be stored. This will charge the floating gate to a point where the sense transistor is on (storing a 0). The bit stored in the cell is read by sensing presence or absence of a small cell current in the bit line. When a 1 is stored, there is no cell current because the sense transistor is off. When a 0 is stored, there is a small cell current because the sense transistor is on. Once a bit is stored in a cell, it will remain indefinitely unless the cell is erased or a new bit is written into the cell.

SECTION 15–8 REVIEW

1. What is the main feature of NMOS and PMOS technology in integrated circuits?
2. What is the mechanism for charge storage in a E²CMOS cell?

■ SUMMARY

- *Formulas*

    (15–1): $V_{\text{NH}} = V_{\text{OH(min)}} - V_{\text{IH(min)}}$ High-level noise margin

    (15–2): $V_{\text{NL}} = V_{\text{IL(max)}} - V_{\text{OL(max)}}$ Low-level noise margin

    (15–3): $I_{\text{CC}} = \dfrac{I_{\text{CCH}} + I_{\text{CCL}}}{2}$ Average dc supply current

    (15–4): $P_{\text{D}} = V_{\text{CC}} I_{\text{CC}}$ Power dissipation

    (15–5): $R_p = \dfrac{V_{\text{CC}} - V_{\text{OL(min)}}}{I_{\text{OL(TTL)}} + n I_{\text{IL(CMOS)}}}$ Pull-up resistor for TTL-to-CMOS interface

- Totem-pole outputs of TTL cannot be connected together.
- Open-collector and open-drain outputs can be connected for wired-AND.
- An HCMOS device offers lower power dissipation than any of the TTL series.
- The HCMOS operating speed is comparable to that of LS TTL. Some TTL series (S TTL, AS TTL, and ALS TTL) offer greater operating speed.
- The $V_{IH(min)}$ of HCMOS is not compatible with the $V_{OH(min)}$ of TTL. When TTL is driving CMOS, use HCT MOS (HCMOS that is TTL compatible) or use pull-up resistors.
- The output voltages of HCMOS are TTL compatible.
- The output current capability of HCMOS is not as large as for TTL.
- An HCMOS device has a smaller fan-out to LS devices than the TTL series.
- An HCMOS device has a wide operating supply-voltage range (2 V to 6 V).
- A TTL device is not as vulnerable to electrostatic discharge (ESD) as is a CMOS device.
- Because of ESD, CMOS devices must be handled with great care.
- E^2CMOS is u. 'd in GAL devices.

■ SELF-TEST

1. Which of the following is not a TTL circuit?
   (a) 7400   (b) 74S00   (c) 74HC00   (d) 74AS00
2. An open TTL NOR gate input
   (a) acts as a LOW   (b) acts as a HIGH
   (c) should be grounded   (d) should be connected to V_{CC} through a resistor
   (e) answers (b) and (c)   (f) answers (a) and (c)
3. A standard TTL gate can drive a maximum of
   (a) ten LS unit loads   (b) forty standard unit loads
   (c) forty LS unit loads   (d) unlimited TTL unit loads
4. If two unused inputs of a standard TTL gate are connected to an input being driven by another standard TTL gate, the total number of remaining unit loads that can be driven by this gate is
   (a) ten   (b) eight   (c) seven   (d) unlimited
5. When a standard TTL gate is driving seven standard unit loads, it is sinking a maximum of
   (a) 11.2 mA   (b) 1.6 mA   (c) 400 μA   (d) 2.8 μA
6. When the frequency of the input signal to a CMOS gate is increased, the average power dissipation
   (a) decreases   (b) increases
   (c) does not change   (d) decreases exponentially
7. CMOS operates more reliably than TTL in a high-noise environment because of its
   (a) lower noise margin   (b) input capacitance
   (c) higher noise margin   (d) smaller power dissipation
8. Proper handling of a CMOS device is necessary because of its
   (a) fragile construction   (b) high-noise immunity
   (c) susceptibility to electrostatic discharge   (d) low power dissipation
9. The main advantage of ECL over TTL or CMOS is
   (a) ECL is less expensive   (b) ECL consumes less power
   (c) ECL is available in a greater variety of circuit types   (d) ECL is faster
10. ECL cannot be used in
    (a) high-noise environments   (b) damp environments   (c) high-frequency applications
11. The basic mechanism for storing a data bit in an E^2CMOS cell is
    (a) control gate   (b) floating drain   (c) floating gate   (d) cell current

■ PROBLEMS

SECTION 15–1   Basic Operational Characteristics and Parameters

1. A certain logic gate has a $V_{OH(min)} = 2.2$ V, and it is driving a gate with a $V_{IH(min)} = 2.5$ V. Are these gates compatible for HIGH-state operation? Why?
2. A certain logic gate has a $V_{OL(max)} = 0.45$ V, and it is driving a gate with a $V_{IL(max)} = 0.75$ V. Are these gates compatible for LOW-state operation? Why?

3. A TTL gate has the following actual voltage level values: $V_{IH(min)} = 2.25$ V, $V_{IL(max)} = 0.65$ V. Assuming it is being driven by a gate with $V_{OH(min)} = 2.4$ V and $V_{OL(max)} = 0.4$ V, what are the HIGH- and LOW-level noise margins?

4. What is the maximum amplitude of noise spikes that can be tolerated on the inputs in both the HIGH state and the LOW state for the gate in Problem 3?

5. Voltage specifications for three types of logic gates are given in Table 15–5. Select the gate that you would use in a high-noise industrial environment.

TABLE 15–5

	$V_{OH(min)}$	$V_{OL(max)}$	$V_{IH(min)}$	$V_{IL(max)}$
Gate A	2.4 V	0.4 V	2 V	0.8 V
Gate B	3.5 V	0.2 V	2.5 V	0.6 V
Gate C	4.2 V	0.2 V	3.2 V	0.8 V

6. A certain gate draws a dc supply current from a +5 V source of 2 mA in the LOW state and 3.5 mA in the HIGH state. What is the power dissipation in the LOW state? What is the power dissipation in the HIGH state? Assuming a 50% duty cycle, what is the average power dissipation?

7. Each gate in the circuit of Figure 15–58 has a t_{PLH} and a t_{PHL} of 4 ns. If a positive-going pulse is applied to the input as indicated, how long will it take the output pulse to appear?

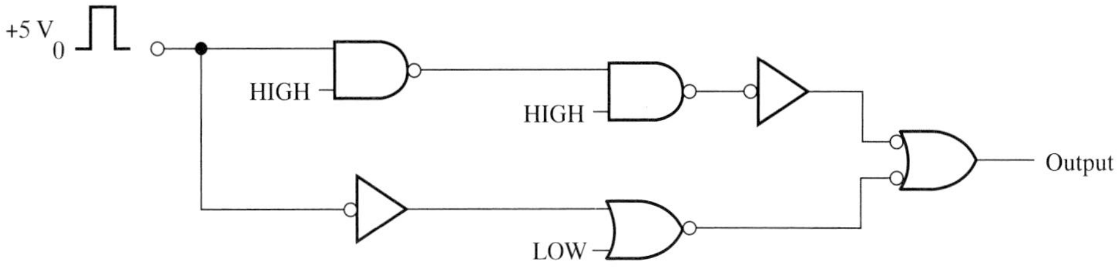

FIGURE 15–58

8. For a certain gate, $t_{PLH} = 3$ ns and $t_{PHL} = 2$ ns. What is the average propagation delay time?

9. Table 15–6 lists parameters for three types of gates. Basing your decision on the speed-power product, which one would you select for best performance?

TABLE 15–6

	t_{PLH}	t_{PHL}	P_D
Gate A	1 ns	1.2 ns	15 mW
Gate B	5 ns	4 ns	8 mW
Gate C	10 ns	10 ns	0.5 mW

10. Which gate in Table 15–6 would you select if you wanted the gate to operate at the highest possible frequency?

FIGURE 15–59

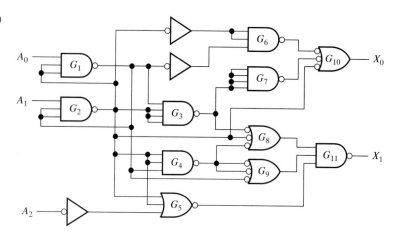

11. A standard TTL gate has a fan-out of 10. Are any of the gates in Figure 15–59 over-loaded? If so, which ones?

12. Which CMOS gate ᴉ ᴉtwork in Figure 15–60 can operate at the highest frequency?

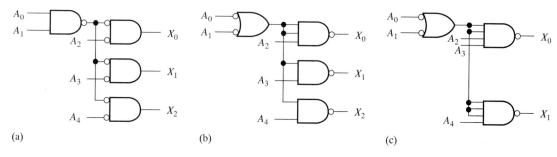

(a) (b) (c)

FIGURE 15–60

SECTION 15–2 TTL Circuits

13. Determine which BJTs in Figure 15–61 are off and which are on.

14. Determine the output state of each TTL gate in Figure 15–62.

FIGURE 15–61

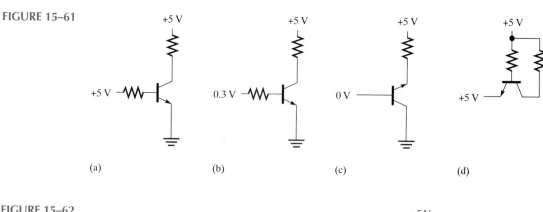

(a) (b) (c) (d)

FIGURE 15–62

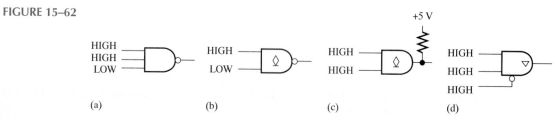

(a) (b) (c) (d)

SECTION 15–3 Practical Considerations in the Use of TTL

15. Determine the output level of each TTL gate in Figure 15–63.

16. For each part of Figure 15–64, tell whether each driving gate is sourcing or sinking current. Specify the maximum current out of or into the output of the driving gate or gates in each case. All gates are standard TTL.

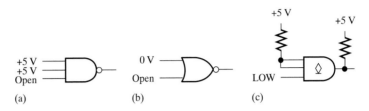

FIGURE 15–63

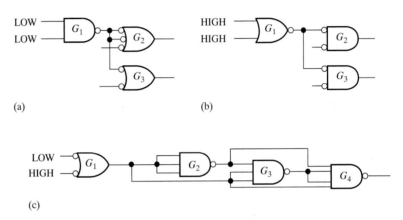

FIGURE 15–64

17. Use open-collector inverters to implement the following logic expressions:
   (a) $X = \overline{A}\,\overline{B}\,\overline{C}$ **(b)** $X = A\,\overline{B}\,C\,\overline{D}$ **(c)** $X = A\,B\,C\,\overline{D}\,\overline{E}\,\overline{F}$

18. Write the logic expression for each of the circuits in Figure 15–65.

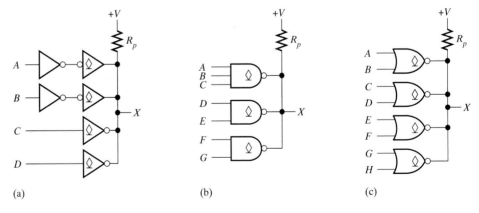

FIGURE 15–65

19. Determine the minimum value for the pull-up resistor in each circuit in Figure 15–65 if $I_{OL(max)} = 40$ mA and $V_{OL(max)} = 0.25$ V for each gate. Assume that 10 standard TTL unit loads are being driven from output X and the supply voltage is 5 V.

20. A certain relay requires 60 mA. Devise a way to use open-collector NAND gates with $I_{OL(max)} = 40$ mA to drive the relay.

SECTION 15–4 CMOS Circuits

21. Determine the state (on or off) of each MOSFET in Figure 15–66.

22. The CMOS gate network in Figure 15–67 is incomplete. Indicate the changes that should be made.

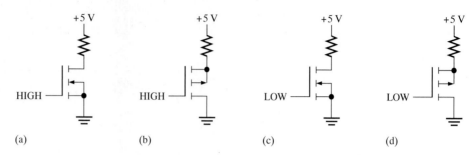

(a)   (b)   (c)   (d)

FIGURE 15–66

FIGURE 15–67

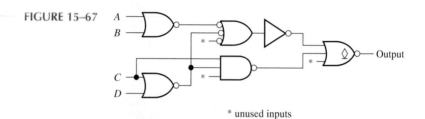

* unused inputs

23. Devise a circuit, using appropriate CMOS logic gates and/or inverters, with which signals from four different sources can be connected to a common line at different times without interfering with each other.

SECTION 15–5 Comparison of CMOS and TTL Characteristics

24. Determine from Table 15–1 (p. 764) the logic family that is most appropriate for each of the following requirements:

  (a) highest speed
  (b) lowest static power

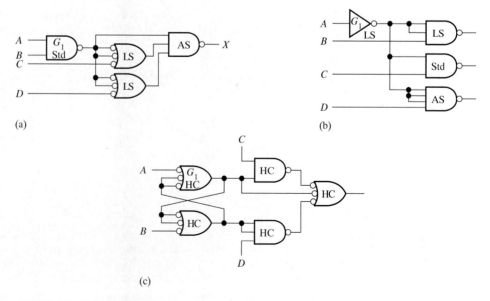

(a)   (b)

(c)

FIGURE 15–68

(c) lowest power at 100 kHz

(d) optimum compromise between speed and power

(e) highest fan-out into LS loads

(f) highest fan-out into same-series loads

25. Determine the amount of current that gate G_1 is sinking when its output is LOW for each circuit in Figure 15–68.

26. One of the flip-flops in Figure 15–69 has an erratic output. Which is it and why?

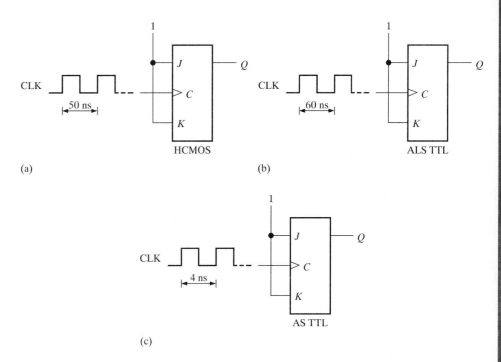

(a) (b)

(c)

FIGURE 15–69

SECTION 15–6 Interfacing Logic Families

27. The circuit in Figure 15–70 is a mixture of IC technologies interfaced together. Determine whether the circuit is properly designed from an interface point of view, and if not, make the necessary design corrections.

28. Repeat Problem 27 for the gate network in Figure 15–71.

29. A standard TTL gate drives five HCMOS gate inputs. Determine the minimum value of the pull-up resistor for $V_{CC} = 5$ V. Sketch the circuit.

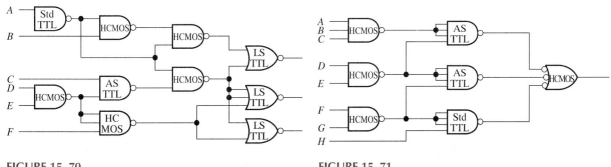

FIGURE 15–70 **FIGURE 15–71**

SECTION 15–7 ECL Circuits

30. What is the basic difference between ECL circuitry and TTL circuitry?

31. Select ECL, HCMOS, or the appropriate TTL series for each of the following requirements:

(a) highest speed

(b) lowest power

(c) best compromise between high speed and low power

(d) high-noise environment applications

■ **ANSWERS TO SECTION REVIEWS**

SECTION 15–1

1. V_{IH}: HIGH level input voltage; V_{IL}: LOW level input voltage; V_{OH}: HIGH level output voltage; V_{OL}: LOW level output voltage
2. A higher value of noise margin is better.
3. Gate B can operate at a higher frequency.
4. Excessive loading reduces the noise margin of a gate.

SECTION 15–2

1. False, the *npn* BJT is off.
2. The *on* state of a BJT is a closed switch; the *off* state is an open switch.
3. Totem-pole and open-collector are types of TTL outputs.
4. Tristate logic provides a high-impedance state, in which the output is disconnected from the rest of the circuit.

SECTION 15–3

1. Sink current occurs in a LOW state.
2. Source current is less than sink current because a TTL load looks like a reverse-biased diode in the HIGH state.
3. The totem-pole transistors cannot handle the current when one output tries to go HIGH and the other is LOW.
4. Wired-AND must use open-collector.   **5.** Lamp driver must be open-collector.
6. False, an unconnected TTL input acts as a HIGH.

SECTION 15–4

1. MOSFETs are used in CMOS logic.
2. A complementary output circuit consists of an *n*-channel and a *p*-channel MOSFET.
3. Because electrostatic discharge can damage CMOS devices

SECTION 15–5

1. True CMOS uses less power than TTL.   **2.** Power increases with frequency in CMOS.
3. False, HCMOS has better noise immunity.

SECTION 15–6

1. A pull-up resistor is used when interfacing TTL to CMOS.   **2.** False

SECTION 15–7

1. ECL is faster than TTL.   **2.** ECL has more power and less noise margin than TTL.

SECTION 15–8

1. NMOS and PMOS are high density.
2. The floating gate is the mechanism for storing charge in an E^2CMOS cell.

A

DATA SHEETS

SN5400, SN54LS00, SN54S00, SN7400, SN74LS00, SN74S00
QUADRUPLE 2-INPUT POSITIVE-NAND GATES

DECEMBER 1983 – REVISED MARCH 1988

- Package Options Include Plastic "Small Outline" Packages, Ceramic Chip Carriers and Flat Packages, and Plastic and Ceramic DIPs
- Dependable Texas Instruments Quality and Reliability

description

These devices contain four independent 2-input-NAND gates.

The SN5400, SN54LS00, and SN54S00 are characterized for operation over the full military temperature range of −55°C to 125°C. The SN7400, SN74LS00, and SN74S00 are characterized for operation from 0°C to 70°C.

FUNCTION TABLE (each gate)

INPUTS		OUTPUT
A	B	Y
H	H	L
L	X	H
X	L	H

logic symbol[†]

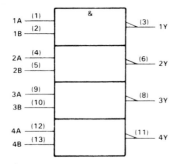

[†]This symbol is in accordance with ANSI/IEEE Std. 91-1984 and IEC Publication 617-12.
Pin numbers shown are for D, J, and N packages.

SN5400 . . . J PACKAGE
SN54LS00, SN54S00 . . . J OR W PACKAGE
SN7400 . . N PACKAGE
SN74LS00, SN74S00 . . . D OR N PACKAGE
(TOP VIEW)

```
1A  [1   U  14] VCC
1B  [2      13] 4B
1Y  [3      12] 4A
2A  [4      11] 4Y
2B  [5      10] 3B
2Y  [6       9] 3A
GND [7       8] 3Y
```

SN5400 . . . W PACKAGE
(TOP VIEW)

```
1A  [1   U  14] 4Y
1B  [2      13] 4B
1Y  [3      12] 4A
VCC [4      11] GND
2Y  [5      10] 3B
2A  [6       9] 3A
2B  [7       8] 3Y
```

SN54LS00, SN54S00 . . . FK PACKAGE
(TOP VIEW)

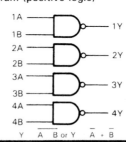

NC   No internal connection

logic diagram (positive logic)

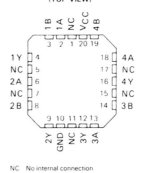

SN5400, SN54LS00, SN54S00,
SN7400, SN74LS00, SN74S00
QUADRUPLE 2-INPUT POSITIVE-NAND GATES

schematics (each gate)

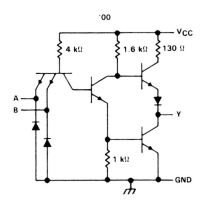

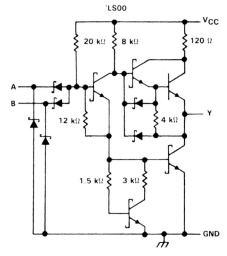

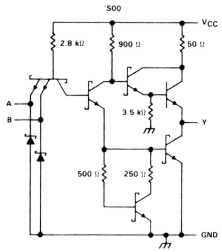

Resistor values shown are nominal.

absolute maximum ratings over operating free-air temperature range (unless otherwise noted)

Supply voltage, V_CC (see Note 1) . . . . 7 V
Input voltage: '00, 'S00 . . . . 5.5 V
              'LS00 . . . . 7 V
Operating free-air temperature range: SN54' . . . . −55 °C to 125 °C
                                   SN74' . . . . 0 °C to 70 °C
Storage temperature range . . . . −65 °C to 150 °C

NOTE 1: Voltage values are with respect to network ground terminal.

SN5400, SN7400
QUADRUPLE 2-INPUT POSITIVE-NAND GATES

recommended operating conditions

		SN5400 MIN	NOM	MAX	SN7400 MIN	NOM	MAX	UNIT
V_{CC}	Supply voltage	4.5	5	5.5	4.75	5	5.25	V
V_{IH}	High-level input voltage	2			2			V
V_{IL}	Low-level input voltage			0.8			0.8	V
I_{OH}	High-level output current			− 0.4			− 0.4	mA
I_{OL}	Low-level output current			16			16	mA
T_A	Operating free-air temperature	− 55		125	0		70	°C

electrical characteristics over recommended operating free-air temperature range (unless otherwise noted)

PARAMETER	TEST CONDITIONS †	SN5400 MIN	TYP‡	MAX	SN7400 MIN	TYP‡	MAX	UNIT
V_{IK}	V_{CC} = MIN, I_I = − 12 mA			− 1.5			− 1.5	V
V_{OH}	V_{CC} = MIN, V_{IL} = 0.8 V, I_{OH} = − 0.4 mA	2.4	3.4		2.4	3.4		V
V_{OL}	V_{CC} = MIN, V_{IH} = 2 V, I_{OL} = 16 mA		0.2	0.4		0.2	0.4	V
I_I	V_{CC} = MAX, V_I = 5.5 V			1			1	mA
I_{IH}	V_{CC} = MAX, V_I = 2.4 V			40			40	µA
I_{IL}	V_{CC} = MAX, V_I = 0.4 V			− 1.6			− 1.6	mA
I_{OS}§	V_{CC} = MAX	− 20		− 55	− 18		− 55	mA
I_{CCH}	V_{CC} = MAX, V_I = 0 V		4	8		4	8	mA
I_{CCL}	V_{CC} = MAX, V_I = 4.5 V		12	22		12	22	mA

† For conditions shown as MIN or MAX, use the appropriate value specified under recommended operating conditions.
‡ All typical values are at V_{CC} = 5 V, T_A = 25°C.
§ Not more than one output should be shorted at a time.

switching characteristics, V_{CC} = 5 V, T_A = 25°C

PARAMETER	FROM (INPUT)	TO (OUTPUT)	TEST CONDITIONS	MIN	TYP	MAX	UNIT
t_{PLH}	A or B	Y	R_L = 400 Ω, C_L = 15 pF		11	22	ns
t_{PHL}	A or B	Y	R_L = 400 Ω, C_L = 15 pF		7	15	ns

SN54LS00, SN74LS00
QUADRUPLE 2-INPUT POSITIVE-NAND GATES

recommended operating conditions

		SN54LS00			SN74LS00			UNIT
		MIN	NOM	MAX	MIN	NOM	MAX	
V_{CC}	Supply voltage	4.5	5	5.5	4.75	5	5.25	V
V_{IH}	High-level input voltage	2			2			V
V_{IL}	Low-level input voltage			0.7			0.8	V
I_{OH}	High-level output current			−0.4			−0.4	mA
I_{OL}	Low-level output current			4			8	mA
T_A	Operating free-air temperature	−55		125	0		70	°C

electrical characteristics over recommended operating free-air temperature range (unless otherwise noted)

PARAMETER	TEST CONDITIONS †			SN54LS00			SN74LS00			UNIT
				MIN	TYP‡	MAX	MIN	TYP‡	MAX	
V_{IK}	V_{CC} = MIN,	I_I = −18 mA				−1.5			−1.5	V
V_{OH}	V_{CC} = MIN,	V_{IL} = MAX,	I_{OH} = −0.4 mA	2.5	3.4		2.7	3.4		V
V_{OL}	V_{CC} = MIN,	V_{IH} = 2 V,	I_{OL} = 4 mA		0.25	0.4		0.25	0.4	V
	V_{CC} = MIN,	V_{IH} = 2 V,	I_{OL} = 8 mA					0.35	0.5	
I_I	V_{CC} = MAX,	V_I = 7 V				0.1			0.1	mA
I_{IH}	V_{CC} = MAX,	V_I = 2.7 V				20			20	µA
I_{IL}	V_{CC} = MAX,	V_I = 0.4 V				−0.4			−0.4	mA
I_{OS}§	V_{CC} = MAX			−20		−100	−20		−100	mA
I_{CCH}	V_{CC} = MAX,	V_I = 0 V			0.8	1.6		0.8	1.6	mA
I_{CCL}	V_{CC} = MAX,	V_I = 4.5 V			2.4	4.4		2.4	4.4	mA

† For conditions shown as MIN or MAX, use the appropriate value specified under recommended operating conditions.
‡ All typical values are at V_{CC} = 5 V, T_A = 25°C
§ Not more than one output should be shorted at a time, and the duration of the short-circuit should not exceed one second.

switching characteristics, V_{CC} = 5 V, T_A = 25°C

PARAMETER	FROM (INPUT)	TO (OUTPUT)	TEST CONDITIONS	MIN	TYP	MAX	UNIT
t_{PLH}	A or B	Y	R_L = 2 kΩ, C_L = 15 pF		9	15	ns
t_{PHL}					10	15	ns

**HIGH-SPEED
CMOS LOGIC**

**TYPES SN54HC00, SN74HC00
QUADRUPLE 2-INPUT POSITIVE-NAND GATES**

D2684, DECEMBER 1982 REVISED MARCH 1984

- **Package Options Include Both Plastic and Ceramic Chip Carriers in Addition to Plastic and Ceramic DIPs**

- **Dependable Texas Instruments Quality and Reliability**

description

These devices contain four independent 2-input NAND gates. They perform the Boolean functions $Y = \overline{A \cdot B}$ or $Y = \overline{A} + \overline{B}$ in positive logic.

The SN54HC00 is characterized for operation over the full military temperature range of $-55\,°C$ to $125\,°C$. The SN74HC00 is characterized for operation from $-40\,°C$ to $85\,°C$.

FUNCTION TABLE (each gate)

INPUTS		OUTPUT
A	B	Y
H	H	L
L	X	H
X	L	H

logic symbol

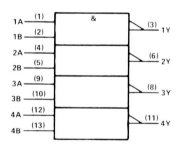

Pin numbers shown are for J and N packages.

SN54HC00 . . . J PACKAGE
SN74HC00 . . . J OR N PACKAGE
(TOP VIEW)

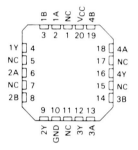

SN54HC00 . . . FH OR FK PACKAGE
SN74HC00 . . . FH OR FN PACKAGE
(TOP VIEW)

NC No internal connection

TYPES SN54HC00, SN74HC00
QUADRUPLE 2-INPUT POSITIVE-NAND GATES

switching characteristics over recommended operating free-air temperature range (unless otherwise noted), C_L = 50 pF

PARAMETER	FROM (INPUT)	TO (OUTPUT)	V_{CC}	T_A = 25°C			SN54HC00		SN74HC00		UNIT
				MIN	TYP	MAX	MIN	MAX	MIN	MAX	
t_{pd}	A or B	Y	2 V		45	90		135		115	ns
			4.5 V		9	18		27		23	
			6 V		8	15		23		20	
t_t		Y	2 V		38	75		110		95	ns
			4.5 V		8	15		22		19	
			6 V		6	13		19		16	

C_{pd}	Power dissipation capacitance per gate	No load, T_A = 25°C	20 pF typ

SN5474, SN54LS74A, SN54S74, SN7474, SN74LS74A, SN74S74
DUAL D-TYPE POSITIVE-EDGE-TRIGGERED FLIP-FLOPS WITH PRESET AND CLEAR
DECEMBER 1983 – REVISED MARCH 1988

- Package Options Include Plastic "Small Outline" Packages, Ceramic Chip Carriers and Flat Packages, and Plastic and Ceramic DIPs

- Dependable Texas Instruments Quality and Reliability

description

These devices contain two independent D-type positive-edge-triggered flip-flops. A low level at the preset or clear inputs sets or resets the outputs regardless of the levels of the other inputs. When preset and clear are inactive (high), data at the D input meeting the setup time requirements are transferred to the outputs on the positive-going edge of the clock pulse. Clock triggering occurs at a voltage level and is not directly related to the rise time of the clock pulse. Following the hold time interval, data at the D input may be changed without affecting the levels at the outputs.

The SN54' family is characterized for operation over the full military temperature range of −55°C to 125°C. The SN74' family is characterized for operation from 0°C to 70°C.

SN5474 . . . J PACKAGE
SN54LS74A, SN54S74 . . . J OR W PACKAGE
SN7474 . . . N PACKAGE
SN74LS74A, SN74S74 . . . D OR N PACKAGE
(TOP VIEW)

```
1CLR  [1   U  14] VCC
1D    [2      13] 2CLR
1CLK  [3      12] 2D
1PRE  [4      11] 2CLK
1Q    [5      10] 2PRE
1Q    [6       9] 2Q
GND   [7       8] 2Q
```

SN5474 . . . W PACKAGE
(TOP VIEW)

```
1CLK  [1   U  14] 1PRE
1D    [2      13] 1Q
1CLR  [3      12] 1Q
VCC   [4      11] GND
2CLR  [5      10] 2Q
2D    [6       9] 2Q
2CLK  [7       8] 2PRE
```

SN54LS74A, SN54S74 . . . FK PACKAGE
(TOP VIEW)

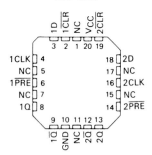

NC - No internal connection

FUNCTION TABLE

INPUTS				OUTPUTS	
PRE	CLR	CLK	D	Q	Q
L	H	X	X	H	L
H	L	X	X	L	H
L	L	X	X	H[†]	H[†]
H	H	↑	H	H	L
H	H	↑	L	L	H
H	H	L	X	Q0	Q0

[†] The output levels in this configuration are not guaranteed to meet the minimum levels in V_{OH} if the lows at preset and clear are near V_{IL} maximum. Furthermore, this configuration is nonstable, that is, it will not persist when either preset or clear returns to its inactive (high) level.

logic symbol[‡]

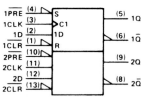

```
1PRE  (4)  S
1CLK  (3)  C1      (5)  1Q
1D    (2)  1D
1CLR  (1)  R       (6)  1Q
2PRE  (10)
2CLK  (11)         (9)  2Q
2D    (12)
2CLR  (13)         (8)  2Q
```

[‡] This symbol is in accordance with ANSI/IEEE Std 91-1984 and IEC Publication 617-12.
Pin numbers shown are for D, J, N, and W packages.

logic diagram (positive logic)

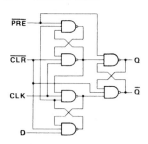

SN5474, SN54LS74A, SN54S74,
SN7474, SN74LS74A, SN74S74
DUAL D-TYPE POSITIVE-EDGE-TRIGGERED FLIP-FLOPS WITH PRESET AND CLEAR

schematic

'LS74A

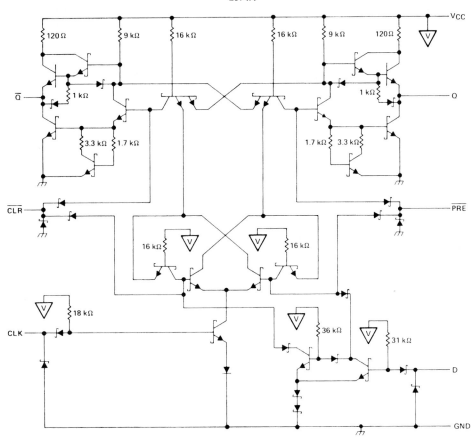

absolute maximum ratings over operating free-air temperature range (unless otherwise noted)

Supply voltage, V$_{CC}$ (see Note 1) . . . . 7 V
Input voltage: '74, 'S74 . . . . 5.5 V
            'LS74A . . . . 7 V
Operating free-air temperature range: SN54' . . . . −55 °C to 125 °C
                                  SN74' . . . . 0 °C to 70 °C
Storage temperature range . . . . −65 °C to 150 °C

NOTE 1: Voltage values are with respect to network ground terminal.

SN5474, SN7474
DUAL D-TYPE POSITIVE-EDGE-TRIGGERED FLIP-FLOPS WITH PRESET AND CLEAR

recommended operating conditions

			SN5474			SN7474			UNIT
			MIN	NOM	MAX	MIN	NOM	MAX	
V_{CC}	Supply voltage		4.5	5	5.5	4.75	5	5.25	V
V_{IH}	High-level input voltage		2			2			V
V_{IL}	Low-level input voltage				0.8			0.8	V
I_{OH}	High-level output current				−0.4			−0.4	mA
I_{OL}	Low-level output current				16			16	mA
t_w	Pulse duration	CLK high	30			30			ns
		CLK low	37			37			
		$\overline{PRE}$ or $\overline{CLR}$ low	30			30			
t_{su}	Input setup time before CLK ↑		20			20			ns
t_h	Input hold time-data after CLK ↑		5			5			ns
T_A	Operating free-air temperature		−55		125	0		70	°C

electrical characteristics over recommended operating free-air temperature range (unless otherwise noted)

PARAMETER		TEST CONDITIONS[†]		SN5474			SN7474			UNIT
				MIN	TYP[‡]	MAX	MIN	TYP[‡]	MAX	
V_{IK}		V_{CC} = MIN,	I_I = −12 mA			−1.5			−1.5	V
V_{OH}		V_{CC} = MIN, V_{IH} = 2 V, V_{IL} = 0.8 V, I_{OH} = −0.4 mA		2.4	3.4		2.4	3.4		V
V_{OL}		V_{CC} = MIN, V_{IH} = 2 V, V_{IL} = 0.8 V, I_{OL} = 16 mA			0.2	0.4		0.2	0.4	V
I_I		V_{CC} = MAX,	V_I = 5.5 V			1			1	mA
I_{IH}	D	V_{CC} = MAX,	V_I = 2.4 V			40			40	μA
	$\overline{CLR}$					120			120	
	All Other					80			80	
I_{IL}	D	V_{CC} = MAX,	V_I = 0.4 V			−1.6			−1.6	mA
	$\overline{PRE}$[§]					−1.6			−1.6	
	$\overline{CLR}$[§]					−3.2			−3.2	
	CLK					−3.2			−3.2	
I_{OS}[¶]		V_{CC} = MAX		−20		−57	−18		−57	mA
I_{CC}[#]		V_{CC} = MAX,	See Note 2		8.5	15		8.5	15	mA

[†]For conditions shown as MIN or MAX, use the appropriate value specified under recommended operating conditions.
[‡]All typical values are at V_{CC} = 5 V, T_A = 25°C.
[§]Clear is tested with preset high and preset is tested with clear high.
[¶]Not more than one output should be shown at a time.
[#]Average per flip-flop.
NOTE 2: With all outputs open, I_{CC} is measured with the Q and $\overline{Q}$ outputs high in turn. At the time of measurement, the clock input is grounded.

switching characteristics, V_{CC} = 5 V, T_A = 25°C

PARAMETER	FROM (INPUT)	TO (OUTPUT)	TEST CONDITIONS		MIN	TYP	MAX	UNIT
f_{max}			R_L = 400 Ω,	C_L = 15 pF	15	25		MHz
t_{PLH}	$\overline{PRE}$ or $\overline{CLR}$	Q or $\overline{Q}$					25	ns
t_{PHL}							40	ns
t_{PLH}	CLK	Q or $\overline{Q}$				14	25	ns
t_{PHL}							20	40	ns

SN54LS74A, SN74LS74A
DUAL D-TYPE POSITIVE-EDGE-TRIGGERED FLIP-FLOPS WITH PRESET AND CLEAR

recommended operating conditions

			SN54LS74A			SN74LS74A			UNIT
			MIN	NOM	MAX	MIN	NOM	MAX	
V_{CC}	Supply voltage		4.5	5	5.5	4.75	5	5.25	V
V_{IH}	High-level input voltage		2			2			V
V_{IL}	Low-level input voltage				0.7			0.8	V
I_{OH}	High-level output current				−0.4			−0.4	mA
I_{OL}	Low-level output current				4			8	mA
f_{clock}	Clock frequency		0		25	0		25	MHz
t_w	Pulse duration	CLK high	25			25			ns
		PRE or CLR low	25			25			
t_{su}	Setup time-before CLK ↑	High-level data	20			20			ns
		Low-level data	20			20			
t_h	Hold time-data after CLK ↑		5			5			ns
T_A	Operating free-air temperature		−55		125	0		70	°C

electrical characteristics over recommended operating free-air temperature range (unless otherwise noted)

PARAMETER		TEST CONDITIONS†			SN54LS74A			SN74LS74A			UNIT
					MIN	TYP‡	MAX	MIN	TYP‡	MAX	
V_{IK}		V_{CC} = MIN,	I_I = −18 mA				−1.5			−1.5	V
V_{OH}		V_{CC} = MIN, I_{OH} = −0.4 mA	V_{IH} = 2 V,	V_{IL} = MAX,	2.5	3.4		2.7	3.4		V
V_{OL}		V_{CC} = MIN, I_{OL} = 4 mA	V_{IL} = MAX,	V_{IH} = 2 V,		0.25	0.4		0.25	0.4	V
		V_{CC} = MIN, I_{OL} = 8 mA	V_{IL} = MAX,	V_{IH} = 2 V,					0.35	0.5	
I_I	D or CLK	V_{CC} = MAX,	V_I = 7 V				0.1			0.1	mA
	CLR or PRE						0.2			0.2	
I_{IH}	D or CLK	V_{CC} = MAX,	V_I = 2.7 V				20			20	µA
	CLR or PRE						40			40	
I_{IL}	D or CLK	V_{CC} = MAX,	V_I = 0.4 V				−0.4			−0.4	mA
	CLR or PRE						−0.8			−0.8	
I_{OS}§		V_{CC} = MAX,	See Note 4		−20		−100	−20		−100	mA
I_{CC} (Total)		V_{CC} = MAX,	See Note 2			4	8		4	8	mA

† For conditions shown as MIN or MAX, use the appropriate value specified under recommended operating conditions.
‡ All typical values are at V_{CC} = 5 V, T_A = 25°C.
§ Not more than one output should be shorted at a time, and the duration of the short circuit should not exceed one second.
NOTE 2: With all outputs open, I_{CC} is measured with the Q and Q̄ outputs high in turn. At the time of measurement, the clock input is grounded.
NOTE 4: For certain devices where state commutation can be caused by shorting an output to ground, an equivalent test may be performed with V_O = 2.25 V and 2.125 V for the 54 family and the 74 family, respectively, with the minimum and maximum limits reduced to one half of their stated values.

switching characteristics, V_{CC} = 5 V, T_A = 25°C

PARAMETER	FROM (INPUT)	TO (OUTPUT)	TEST CONDITIONS		MIN	TYP	MAX	UNIT
f_{max}			R_L = 2 kΩ,	C_L = 15 pF	25	33		MHz
t_{PLH}	CLR, PRE or CLK	Q or Q̄				13	25	ns
t_{PHL}						25	40	ns

SN54LS138, SN54S138, SN74LS138, SN74S138A
3-LINE TO 8-LINE DECODERS/DEMULTIPLEXERS

DECEMBER 1972 – REVISED MARCH 1988

- **Designed Specifically for High-Speed:**
  Memory Decoders
  Data Transmission Systems

- **3 Enable Inputs to Simplify Cascading**
  and/or Data Reception

- **Schottky-Clamped for High Performance**

description

These Schottky-clamped TTL MSI circuits are designed to be used in high-performance memory decoding or data-routing applications requiring very short propagation delay times. In high-performance memory systems, these docoders can be used to minimize the effects of system decoding. When employed with high-speed memories utilizing a fast enable circuit, the delay times of these decoders and the enable time of the memory are usually less than the typical access time of the memory. This means that the effective system delay introduced by the Schottky-clamped system decoder is negligible.

The 'LS138, SN54S138, and SN74S138A decode one of eight lines dependent on the conditions at the three binary select inputs and the three enable inputs. Two active-low and one active-high enable inputs reduce the need for external gates or inverters when expanding. A 24-line decoder can be implemented without external inverters and a 32-line decoder requires only one inverter. An enable input can be used as a data input for demultiplexing applications.

All of these decoder/demultiplexers feature fully buffered inputs, each of which represents only one normalized load to its driving circuit. All inputs are clamped with high-performance Schottky diodes to suppress line-ringing and to simplify system design.

The SN54LS138 and SN54S138 are characterized for operation over the full military temperature range of −55°C to 125°C. The SN74LS138 and SN74S138A are characterized for operation from 0°C to 70°C.

SN54LS138, SN54S138 . . . J OR W PACKAGE
SN74LS138, SN74S138A . . . D OR N PACKAGE
(TOP VIEW)

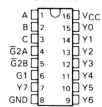

SN54LS138, SN54S138 . . . FK PACKAGE
(TOP VIEW)

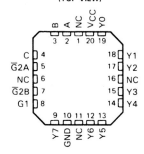

NC – No internal connection

logic symbols†

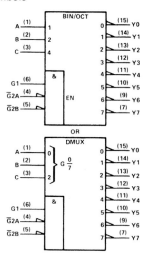

†These symbols are in accordance with ANSI/IEEE Std 91-1984
and IEC Publication 617-12.
Pin numbers shown are for D, J, N, and W packages.

SN54LS138, SN54S138, SN74LS138, SN74S138A
3-LINE-TO 8-LINE DECODERS/DEMULTIPLEXERS

logic diagram and function table

'LS138, SN54S138, SN74S138A

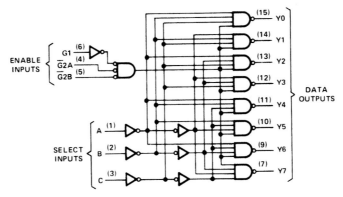

Pin numbers shown are for D, J, N, and W packages.

'LS138, SN54138, SN74S138A
FUNCTION TABLE

INPUTS					OUTPUTS							
ENABLE		SELECT										
G1	$\overline{G2}$*	C	B	A	Y0	Y1	Y2	Y3	Y4	Y5	Y6	Y7
X	H	X	X	X	H	H	H	H	H	H	H	H
L	X	X	X	X	H	H	H	H	H	H	H	H
H	L	L	L	L	L	H	H	H	H	H	H	H
H	L	L	L	H	H	L	H	H	H	H	H	H
H	L	L	H	L	H	H	L	H	H	H	H	H
H	L	L	H	H	H	H	H	L	H	H	H	H
H	L	H	L	L	H	H	H	H	L	H	H	H
H	L	H	L	H	H	H	H	H	H	L	H	H
H	L	H	H	L	H	H	H	H	H	H	L	H
H	L	H	H	H	H	H	H	H	H	H	H	L

* $\overline{G2}$ = $\overline{G2A}$ + $\overline{G2B}$
H = high level, L = low level, X = irrelevant

SN54LS138, SN54S138, SN74LS138, SN74S138A
3-LINE TO 8-LINE DECODERS/DEMULTIPLEXERS

schematics of inputs and outputs

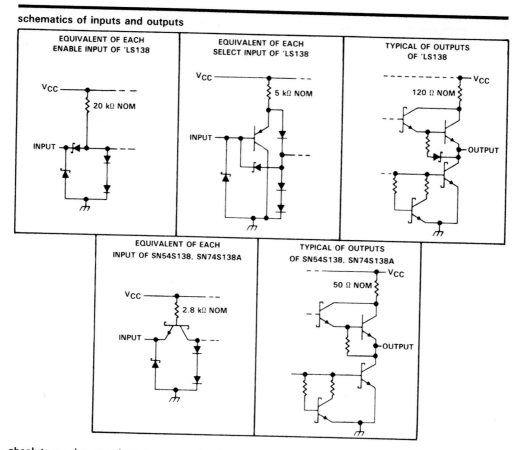

absolute maximum ratings over operating free-air temperature range (unless otherwise noted)

Supply voltage, V_{CC} (see Note 1) . . . . 7 V
Input voltage . . . . 7 V
Operating free-air temperature range: SN54LS138, SN54S138 . . . . −55°C to 125°C
                                   SN74LS138, SN74S138A . . . . 0°C to 70°C
Storage temperature range . . . . −65°C to 150°C

NOTE 1: Voltage values are with respect to network ground terminal.

SN54LS138, SN74LS138
3-LINE TO 8-LINE DECODERS/DEMULTIPLEXERS

recommended operating conditions

		SN54LS138			SN74LS138			UNIT
		MIN	NOM	MAX	MIN	NOM	MAX	
V_{CC}	Supply voltage	4.5	5	5.5	4.75	5	5.25	V
V_{IH}	High-level input voltage	2			2			V
V_{IL}	Low-level input voltage			0.7			0.8	V
I_{OH}	High-level output current			-0.4			-0.4	mA
I_{OL}	Low-level output current			4			8	mA
T_A	Operating free-air temperature	-55		125	0		70	°C

electrical characteristics over recommended operating free-air temperature range (unless otherwise noted)

PARAMETER	TEST CONDITIONS[†]		SN54LS138			SN74LS138			UNIT
			MIN	TYP[‡]	MAX	MIN	TYP[‡]	MAX	
V_{IK}	V_{CC} = MIN, I_I = -18 mA				-1.5			-1.5	V
V_{OH}	V_{CC} = MIN, V_{IH} = 2 V, V_{IL} = MAX, I_{OH} = -0.4 mA		2.5	3.4		2.7	3.4		V
V_{OL}	V_{CC} = MIN, V_{IH} = 2 V, V_{IL} = MAX	I_{OL} = 4 mA		0.25	0.4		0.25	0.4	V
		I_{OL} = 8 mA					0.35	0.5	
I_I	V_{CC} = MAX, V_I = 7 V				0.1			0.1	mA
I_{IH}	V_{CC} = MAX, V_I = 2.7 V				20			20	μA
I_{IL}	V_{CC} = MAX, V_I = 0.4 V	Enable			-0.4			-0.4	mA
		A, B, C			-0.2			-0.2	
I_{OS}[§]	V_{CC} = MAX		-20		-100	-20		-100	mA
I_{CC}	V_{CC} = MAX, Outputs enabled and open			6.3	10		6.3	10	mA

[†] For conditions shown as MIN or MAX, use the appropriate value specified under recommended operating conditions.
[‡] All typical values are at V_{CC} = 5 V, T_A = 25 °C.
[§] Not more than one output should be shorted at a time, and duration of the short-circuit test should not exceed one second.

switching characteristics, V_{CC} = 5 V, T_A = 25 °C

PARAMETER[¶]	FROM (INPUT)	TO (OUTPUT)	LEVELS OF DELAY	TEST CONDITIONS	SN54LS138 SN74LS138			UNIT
					MIN	TYP	MAX	
t_{PLH}	Binary Select	Any	2	R_L = 2 kΩ, C_L = 15 pF,		11	20	ns
t_{PHL}						18	41	ns
t_{PLH}			3			21	27	ns
t_{PHL}						20	39	ns
t_{PLH}	Enable	Any	2			12	18	ns
t_{PHL}						20	32	ns
t_{PLH}			3			14	26	ns
t_{PHL}						13	38	ns

[¶] t_{PLH} = propagation delay time, low-to-high-level ouput
t_{PHL} = propagation delay time, high-to-low-level output

SN54150, SN54151A, SN54LS151, SN54S151, SN74150, SN74151A, SN74LS151, SN74S151
DATA SELECTORS/MULTIPLEXERS

DECEMBER 1972 – REVISED MARCH 1988

- '150 Selects One-of-Sixteen Data Sources
- Others Select One-of-Eight Data Sources
- All Perform Parallel-to-Serial Conversion
- All Permit Multiplexing from N Lines to One Line
- Also For Use as Boolean Function Generator
- Input-Clamping Diodes Simplify System Design
- Fully Compatible with Most TTL Circuits

TYPE	TYPICAL AVERAGE PROPAGATION DELAY TIME DATA INPUT TO W OUTPUT	TYPICAL POWER DISSIPATION
'150	13 ns	200 mW
'151A	8 ns	145 mW
'LS151	13 ns	30 mW
'S151	4.5 ns	225 mW

description

These monolithic data selectors/multiplexers contain full on-chip binary decoding to select the desired data source. The '150 selects one-of-sixteen data sources; the '151A, 'LS151, and 'S151 select one-of-eight data sources. The '150, '151A, 'LS151, and 'S151 have a strobe input which must be at a low logic level to enable these devices. A high level at the strobe forces the W output high, and the Y output (as applicable) low.

The '150 has only an inverted W output; the '151A, 'LS151, and 'S151 feature complementary W and Y outputs.

The '151A and '152A incorporate address buffers that have symmetrical propagation delay times through the complementary paths. This reduces the possibility of transients occurring at the output(s) due to changes made at the select inputs, even when the '151A outputs are enabled (i.e., strobe low).

SN54150 . . . J OR W PACKAGE
SN74150 . . . N PACKAGE
(TOP VIEW)

```
       ___  ___
E7  [1        24]  Vcc
E6  [2        23]  E8
E5  [3        22]  E9
E4  [4        21]  E10
E3  [5        20]  E11
E2  [6        19]  E12
E1  [7        18]  E13
E0  [8        17]  E14
G̅   [9        16]  E15
W   [10       15]  A
D   [11       14]  B
GND [12       13]  C
```

SN54151A, SN54LS151, SN54S151 . . . J OR W PACKAGE
SN74151A . . . N PACKAGE
SN74LS151, SN74S151 . . . D OR N PACKAGE
(TOP VIEW)

```
       ___  ___
D3  [1        16]  Vcc
D2  [2        15]  D4
D1  [3        14]  D5
D0  [4        13]  D6
Y   [5        12]  D7
W   [6        11]  A
G̅   [7        10]  B
GND [8         9]  C
```

SN54LS151, SN54S151 . . . FK PACKAGE
(TOP VIEW)

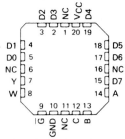

NC - No internal connection

SN54150, SN54151A, SN54LS151, SN54S151, SN74150, SN74151A, SN74LS151, SN74S151
DATA SELECTORS/MULTIPLEXERS

logic symbols†

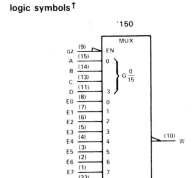

'150

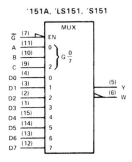

'151A, 'LS151, 'S151

†These symbols are in accordance with ANSI/IEEE Std. 91-1984 and IEC Publication 617-12.
Pin numbers shown are D, J, N, and W packages.

'150
FUNCTION TABLE

INPUTS				OUTPUT
SELECT			STROBE	W
D	C	B	A	$\overline{G}$	

D	C	B	A	$\overline{G}$	W
X	X	X	X	H	H
L	L	L	L	L	$\overline{E0}$
L	L	L	H	L	$\overline{E1}$
L	L	H	L	L	$\overline{E2}$
L	L	H	H	L	$\overline{E3}$
L	H	L	L	L	$\overline{E4}$
L	H	L	H	L	$\overline{E5}$
L	H	H	L	L	$\overline{E6}$
L	H	H	H	L	$\overline{E7}$
H	L	L	L	L	$\overline{E8}$
H	L	L	H	L	$\overline{E9}$
H	L	H	L	L	$\overline{E10}$
H	L	H	H	L	$\overline{E11}$
H	H	L	L	L	$\overline{E12}$
H	H	L	H	L	$\overline{E13}$
H	H	H	L	L	$\overline{E14}$
H	H	H	H	L	$\overline{E15}$

'151A, 'LS151, 'S151
FUNCTION TABLE

INPUTS				OUTPUTS	
SELECT			STROBE	Y	W
C	B	A	$\overline{G}$		

C	B	A	$\overline{G}$	Y	W
X	X	X	H	L	H
L	L	L	L	D0	$\overline{D0}$
L	L	H	L	D1	$\overline{D1}$
L	H	L	L	D2	$\overline{D2}$
L	H	H	L	D3	$\overline{D3}$
H	L	L	L	D4	$\overline{D4}$
H	L	H	L	D5	$\overline{D5}$
H	H	L	L	D6	$\overline{D6}$
H	H	H	L	D7	$\overline{D7}$

H = high level, L = low level, X = irrelevant
$\overline{E0}$, $\overline{E1}$ . . . $\overline{E15}$ = the complement of the level of the respective E input
D0, D1 . . . D7 = the level of the D respective input

SN54150, SN54151A, SN54LS151, SN54S151,
SN74150, SN74151A, SN74LS151, SN74S151
DATA SELECTORS/MULTIPLEXERS

logic diagrams (positive logic)

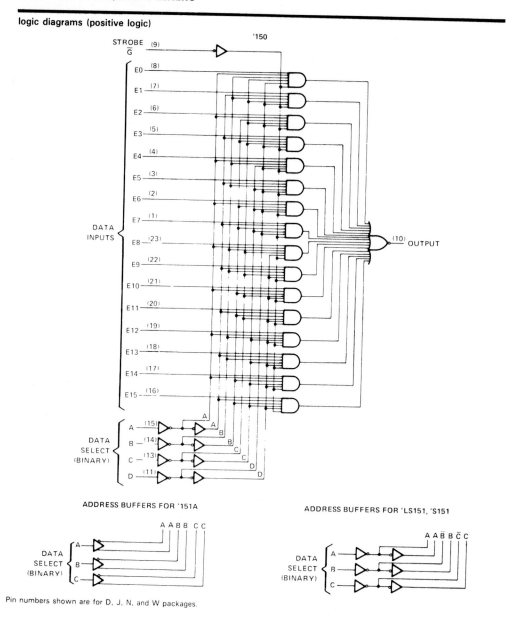

ADDRESS BUFFERS FOR '151A

ADDRESS BUFFERS FOR 'LS151, 'S151

Pin numbers shown are for D, J, N, and W packages.

SN54150, SN54151A, SN54LS151, SN54S151, SN74150, SN74151A, SN74LS151, SN74S151
DATA SELECTORS/MULTIPLEXERS

'151A, 'LS151, 'S151

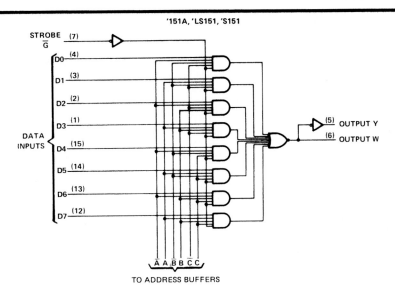

TO ADDRESS BUFFERS

absolute maximum ratings over operating free-air temperature range (unless otherwise noted)

Supply voltage, V_{CC} (see Note 1) . . . . 7 V
Input voltage (see Note 2): '150, '151A, 'S151 . . . . 5.5 V
'LS151 . . . . 7 V
Operating free-air temperature range: SN54' . . . . −55°C to 125°C
SN74' . . . . 0°C to 70°C
Storage temperature range . . . . −65°C to 150°C

NOTES: 1: Voltage values are with respect to network ground terminal.
2. For the '150, input voltages must be zero or positive with respect to network ground terminal.

SN54150, SN54151A, SN74150, SN74151A
DATA SELECTORS/MULTIPLEXERS

recommended operating conditions

	SN54'			SN74'			UNIT
	MIN	NOM	MAX	MIN	NOM	MAX	
Supply voltage, V_{CC}	4.5	5	5.5	4.75	5	5.25	V
High-level output current, I_{OH}			−800			−800	μA
Low-level output current, I_{OL}			16			16	mA
Operating free-air temperature, T_A	−55		125	0		70	C

electrical characteristics over recommended operating free-air temperature range (unless otherwise noted)

PARAMETER		TEST CONDITIONS†	'150			'151A			UNIT
			MIN	TYP‡	MAX	MIN	TYP‡	MAX	
V_{IH}	High-level input voltage		2			2			V
V_{IL}	Low-level input voltage				0.8			0.8	V
V_{IK}	Input clamp voltage	V_{CC} = MIN, I_I = −8 mA			−1.5			−1.5	V
V_{OH}	High-level output voltage	V_{CC} = MIN, V_{IH} = 2 V, V_{IL} = 0.8 V, I_{OH} = −800 μA	2.4	3.4		2.4	3.4		V
V_{OL}	Low-level output voltage	V_{CC} = MIN, V_{IH} = 2 V, V_{IL} = 0.8 V, I_{OL} = 16 mA		0.2	0.4		0.2	0.4	V
I_I	Input current at maximum input voltage	V_{CC} = MAX, V_I = 5.5 V			1			1	mA
I_{IH}	High-level input current	V_{CC} = MAX, V_I = 2.4 V			40			40	μA
I_{IL}	Low-level input current	V_{CC} = MAX, V_I = 0.4 V			−1.6			−1.6	mA
I_{OS}	Short-circuit output current§	V_{CC} = MAX SN54'	−20		−55	−20		−55	mA
		SN74'	−18		−55	−18		−55	
I_{CC}	Supply current	V_{CC} = MAX, See Note 3		40	68		29	48	mA

† For conditions shown as MIN or MAX, use the appropriate value specified under recommended operating conditions for the applicable device type.
‡ All typical values at V_{CC} = 5 V, T_A = 25°C.
§ Not more than one output of the '151A should be shorted at a time.
NOTE 3: I_{CC} is measured with the strobe and data select inputs at 4.5 V, all other inputs and outputs open.

switching characteristics, V_{CC} = 5 V, T_A = 25°C

PARAMETER¶	FROM (INPUT)	TO (OUTPUT)	TEST CONDITIONS	'150			'151A			UNIT
				MIN	TYP	MAX	MIN	TYP	MAX	
t_{PLH}	A, B, or C	Y						25	38	ns
t_{PHL}	(4 levels)							25	38	
t_{PLH}	A, B, C, or D	W			23	35		17	26	ns
t_{PHL}	(3 levels)				22	33		19	30	
t_{PLH}	Strobe $\overline{G}$	Y	C_L = 15 pF, R_L = 400 Ω,					21	33	ns
t_{PHL}								22	33	
t_{PLH}	Strobe $\overline{G}$	W			15.5	24		14	21	ns
t_{PHL}					21	30		15	23	
t_{PLH}	D0 thru D7	Y						13	20	ns
t_{PHL}								18	27	
t_{PLH}	E0 thru E15, or	W			8.5	14		8	14	ns
t_{PHL}	D0 thru D7				13	20		8	14	

¶ t_{PLH} = propagation delay time, low-to-high-level output
t_{PHL} = propagation delay time, high-to-low-level output

SN54LS151, SN74LS151
DATA SELECTORS/MULTIPLEXERS

recommended operating conditions

	SN54LS151			SN74LS151			UNIT
	MIN	NOM	MAX	MIN	NOM	MAX	
Supply voltage, V_{CC}	4.5	5	5.5	4.75	5	5.25	V
High-level output current, I_{OH}			−400			−400	μA
Low-level output current, I_{OL}			4			8	mA
Operating free-air temperature, T_A	−55		125	0		70	°C

electrical characteristics over recommended operating free-air temperature range (unless otherwise noted)

PARAMETER		TEST CONDITIONS[†]	SN54LS151			SN74LS151			UNIT
			MIN	TYP[‡]	MAX	MIN	TYP[‡]	MAX	
V_{IH}	High-level input voltage		2			2			V
V_{IL}	Low-level input voltage				0.7			0.8	V
V_{IK}	Input clamp voltage	V_{CC} = MIN, I_I = −18 mA			−1.5			−1.5	V
V_{OH}	High-level output voltage	V_{CC} = MIN, V_{IH} = 2 V, V_{IL} = V_{IL}max, I_{OH} = −400 μA	2.5	3.4		2.7	3.4		V
V_{OL}	Low-level output voltage	V_{CC} = MIN, V_{IH} = 2 V, I_{OL} = 4 mA, V_{IL} = V_{IL}max		0.25	0.4		0.25	0.4	V
		I_{OL} = 8 mA					0.35	0.5	
I_I	Input current at maximum input voltage	V_{CC} = MAX, V_I = 7 V			0.1			0.1	mA
I_{IH}	High-level input current	V_{CC} = MAX, V_I = 2.7 V			20			20	μA
I_{IL}	Low-level input current	V_{CC} = MAX, V_I = 0.4 V			−0.4			−0.4	mA
I_{OS}	Short-circuit output current[§]	V_{CC} = MAX	−20		−100	−20		−100	mA
I_{CC}	Supply current	V_{CC} = MAX, Outputs open, All inputs at 4.5 V		6.0	10		6.0	10	mA

[†] For conditions shown as MIN or MAX, use the appropriate value specified under recommended operating conditions for the applicable device type.
[‡] All typical values are at V_{CC} = 5 V, T_A = 25°C.
[§] Not more than one output should be shorted at a time and duration of short-circuit should not exceed one second.

switching characteristics, V_{CC} = 5 V, T_A 25°C

PARAMETER[¶]	FROM (INPUT)	TO (OUTPUT)	TEST CONDITIONS	MIN	TYP	MAX	UNIT
t_{PLH}	A, B, or C (4 levels)	Y			27	43	ns
t_{PHL}					18	30	
t_{PLH}	A, B, or C (3 levels)	W			14	23	ns
t_{PHL}					20	32	
t_{PLH}	Strobe $\overline{G}$	Y	C_L = 15 pF, R_L = 2 kΩ,		26	42	ns
t_{PHL}					20	32	
t_{PLH}	Strobe $\overline{G}$	W			15	24	ns
t_{PHL}					18	30	
t_{PLH}	Any D	Y			20	32	ns
t_{PHL}					16	26	
t_{PLH}	Any D	W			13	21	ns
t_{PHL}					12	20	

[¶] t_{PLH} = propagation delay time, low-to-high-level output
t_{PHL} = propagation delay time, high-to-low-level output

SN54190, SN54191, SN54LS190, SN54LS191, SN74190, SN74191, SN74LS190, SN74LS191
SYNCHRONOUS UP/DOWN COUNTERS WITH DOWN/UP MODE CONTROL
DECEMBER 1972–REVISED MARCH 1988

- Counts 8-4-2-1 BCD or Binary
- Single Down/Up Count Control Line
- Count Enable Control Input
- Ripple Clock Output for Cascading
- Asynchronously Presettable with Load Control
- Parallel Outputs
- Cascadable for n-Bit Applications

SN54190, SN54191, SN54LS190,
SN54LS191 . . . J PACKAGE
SN74190, SN74191 . . . N PACKAGE
SN74LS190, SN74LS191 . . . D OR N PACKAGE
(TOP VIEW)

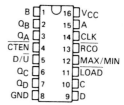

SN54LS190, SN54LS191 . . . FK PACKAGE
(TOP VIEW)

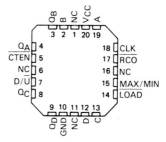

NC - No internal connection

TYPE	AVERAGE PROPAGATION DELAY	TYPICAL MAXIMUM CLOCK FREQUENCY	TYPICAL POWER DISSIPATION
'190,'191	20ns	25MHz	325mW
'LS190,'LS191	20ns	25MHz	100mW

description

The '190, 'LS190, '191, and 'LS191 are synchronous, reversible up/down counters having a complexity of 58 equivalent gates. The '191 and 'LS191 are 4-bit binary counters and the '190 and 'LS190 are BCD counters. Synchronous operation is provided by having all flip-flops clocked simultaneously so that the outputs change coincident with each other when so instructed by the steering logic. This mode of operation eliminates the output counting spikes normally associated with asynchronous (ripple clock) counters.

The outputs of the four master-slave flip-flops are triggered on a low-to-high transition of the clock input if the enable input is low. A high at the enable input inhibits counting. Level changes at the enable input should be made only when the clock input is high. The direction of the count is determined by the level of the down/up input. When low, the counter count up and when high, it counts down. A false clock may occur if the down/up input changes while the clock is low. A false ripple carry may occur if both the clock and enable are low and the down/up input is high during a load pulse.

These counters are fully programmable; that is, the outputs may be preset to either level by placing a low on the load input and entering the desired data at the data inputs. The output will change to agree with the data inputs independently of the level of the clock input. This feature allows the counters to be used as modulo-N dividers by simply modifying the count length with the preset inputs.

The clock, down/up, and load inputs are buffered to lower the drive requirement which significantly reduces the number of clock drivers, etc., required for long parallel words.

Two outputs have been made available to perform the cascading function: ripple clock and maximum/minimum count. The latter output produces a high-level output pulse with a duration approximately equal to one complete cycle of the clock when the counter overflows or underflows. The ripple clock output produces a low-level output pulse equal in width to the low-level portion of the clock input when an overflow or underflow condition exists. The counters can be easily cascaded by feeding the ripple clock output to the enable input of the succeeding counter if parallel clocking is used, or to the clock input if parallel enabling is used. The maximum/minimum count output can be used to accomplish look-ahead for high-speed operation.

Series 54' and 54LS' are characterized for operation over the full military temperature range of −55°C to 125°C; Series 74' and 74LS' are characterized for operation from 0°C to 70°C.

SN54190, SN54191, SN54LS190, SN54LS191,
SN74190, SN74191, SN74LS190, SN74LS191
SYNCHRONOUS UP/DOWN COUNTERS WITH DOWN/UP MODE CONTROL

logic symbols†

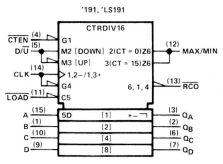

† These symbols are accordance with ANSI/IEEE Std 91-1984 and IEC Publication 617-12.
Pin numbers shown are for D, J, and N packages.

SN54190, SN54LS190, SN74190, SN74LS190
SYNCHRONOUS UP/DOWN COUNTERS WITH DOWN/UP MODE CONTROL

'190, 'LS190 DECADE COUNTERS

typical load, count, and inhibit sequences

Illustrated below is the following sequence:

1. Load (preset) to BCD seven.
2. Count up to eight, nine (maximum), zero, one, and two.
3. Inhibit.
4. Count down to one, zero (minimum), nine, eight, and seven.

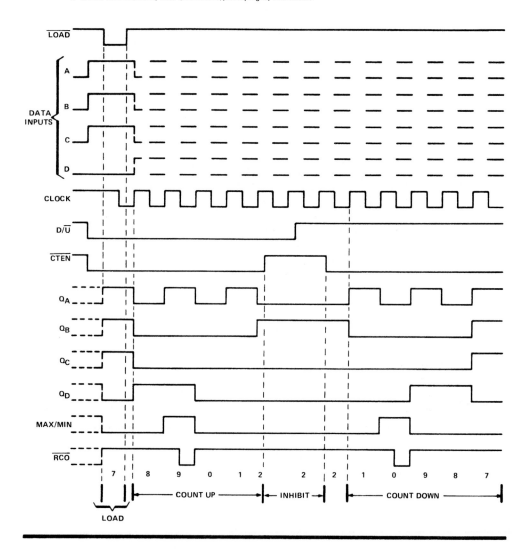

SN54191, SN54LS191, SN74191, SN74LS191
SYNCHRONOUS UP/DOWN COUNTERS WITH DOWN/UP MODE CONTROL

'191, 'LS191 BINARY COUNTERS

typical load, count, and inhibit sequences

Illustrated below is the following sequence:

1. Load (preset) to binary thirteen.
2. Count up to fourteen, fifteen (maximum), zero, one, and two.
3. Inhibit.
4. Count down to one, zero (minimum), fifteen, fourteen, and thirteen.

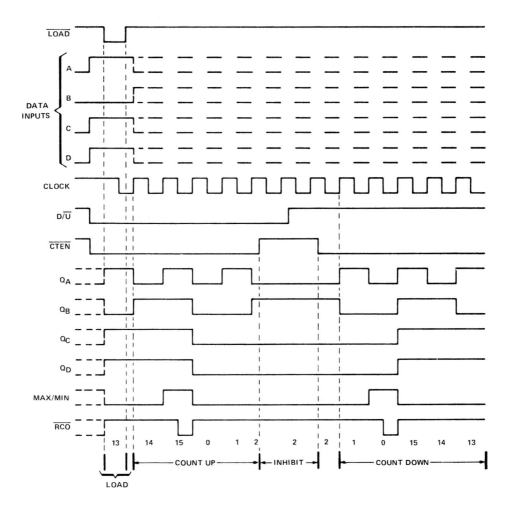

SN54190, SN54191, SN54LS190, SN54LS191, SN74190, SN74191, SN74LS190, SN74LS191
SYNCHRONOUS UP/DOWN COUNTERS WITH DOWN/UP MODE CONTROL

absolute maximum ratings over operating free-air temperature range (unless otherwise noted)

Supply voltage, V_{CC} (see Note 1) . . . . 7 V
Input voltage: SN54', SN74' Circuits . . . . 5.5 V
SN54LS', SN74LS' Circuits . . . . 7 V
Operating free-air temperature range: SN54', SN54LS' Circuits . . . . $-55°C$ to $125°C$
SN74', SN74LS' Circuits . . . . $0°C$ to $70°C$
Storage temperature range . . . . $-65°C$ to $150°C$

NOTE 1: Voltage values are with respect to network ground terminal.

recommended operating conditions

		SN54190, SN54191			SN74190, SN74191			UNIT
		MIN	NOM	MAX	MIN	NOM	MAX	
V_{CC}	Supply voltage	4.5	5	5.5	4.75	5	5.25	V
I_{OH}	High-level output current			-0.8			-0.8	mA
I_{OL}	Low-level output current			16			16	mA
f_{clock}	Input clock frequency	0		20	0		20	MHz
$t_{w(clock)}$	Width of clock input pulse	25			25			ns
$t_{w(load)}$	Width of load input pulse	35			35			ns
t_{su}	Setup time	Data, high or low (See Figure 1 and 2)	20			20			ns
		Load inactive state	20			20			
t_{hold}	Data hold time	0			0			ns
T_A	Operating free-air temperature	-55		125	0		70	°C

electrical characteristics over recommended operating free-air temperature range (unless otherwise noted)

PARAMETER		TEST CONDITIONS[†]	SN54190, SN54191			SN74190, SN74191			UNIT
			MIN	TYP[‡]	MAX	MIN	TYP[‡]	MAX	
V_{IH}	High-level input voltage	$V_{CC} = MIN$	2			2			V
V_{IL}	Low-level input voltage	$V_{CC} = MIN$			0.8			0.8	V
V_{IK}	Input clamp voltage	$V_{CC} = MIN$, $I_I = -12$ mA			-1.5			-1.5	V
V_{OH}	High-level output voltage	$V_{CC} = MIN$, $V_{IH} = 2$ V, $V_{IL} = 0.8$ V, $I_{OH} = -0.8$ mA	2.4	3.4		2.4	3.4		V
V_{OL}	Low-level output voltage	$V_{CC} = MIN$, $V_{IH} = 2$ V, $V_{IL} = 0.8$ V, $I_{OL} = 16$ mA		0.2	0.4		0.2	0.4	V
I_I	High-level input current at maximum input voltage	$V_{CC} = MAX$, $V_I = 5.5$ V			1			1	mA
I_{IH}	High-level input current at any input except enable	$V_{CC} = MAX$, $V_I = 2.4$ V			40			40	μA
I_{IH}	High-level input current at enable input				120			120	μA
I_{IL}	Low-level input current at any input except enable	$V_{CC} = MAX$, $V_I = 0.4$ V			-1.6			-1.6	mA
I_{IL}	Low-level input current at enable input				-4.8			-4.8	mA
I_{OS}	Short-circuit output current[§]	$V_{CC} = MAX$	-20		-65	-18		-65	mA
I_{CC}	Supply current	$V_{CC} = MAX$, See Note 2		65	99		65	105	mA

[†] For conditions shown as MAX or MIN, use appropriate value specified under recommended operating conditions.
[‡] All typical values are at $V_{CC} = 5$ V, $T_A = 25°C$.
[§] Not more than one output should be shorted at a time.
NOTE 2: I_{CC} is measured with all inputs grounded and all outputs open.

SN54190, SN54191, SN74190, SN74191
SYNCHRONOUS UP/DOWN COUNTERS WITH DOWN/UP MODE CONTROL

switching characteristics, V_{CC} = 5 V, T_A = 25°C

PARAMETER[†]	FROM (INPUT)	TO (OUTPUT)	TEST CONDITIONS	'190, '191			UNIT
				MIN	TYP	MAX	
f_{max}				20	25		MHz
t_{PLH}	$\overline{Load}$	Q_A, Q_B, Q_C, Q_D			22	33	ns
t_{PHL}					33	50	
t_{PLH}	Data A, B, C, D	Q_A, Q_B, Q_C, Q_D			14	22	ns
t_{PHL}					35	50	
t_{PLH}	CLK	$\overline{RCO}$	C_L = 15 pF, R_L = 400 Ω,		13	20	ns
t_{PHL}			See Figures 1 and 3 thru 7		16	24	
t_{PLH}	CLK	Q_A, Q_B, Q_C, Q_D			16	24	ns
t_{PHL}					24	36	
t_{PLH}	CLK	Max/Min			28	42	ns
t_{PHL}					37	52	
t_{PLH}	$D/\overline{U}$	$\overline{RCO}$			30	45	ns
t_{PHL}					30	45	
t_{PLH}	$D/\overline{U}$	Max/Min			21	33	ns
t_{PHL}					22	33	

[†] f_{max} ≡ maximum clock frequency
t_{PLH} ≡ propagation delay time, low-to-high-level output
t_{PHL} ≡ propagation delay time, high-to-low-level output

schematics of inputs and outputs

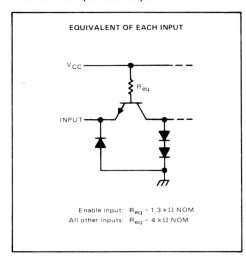

EQUIVALENT OF EACH INPUT

Enable input: R_{eq} = 1.3 kΩ NOM
All other inputs: R_{eq} = 4 kΩ NOM

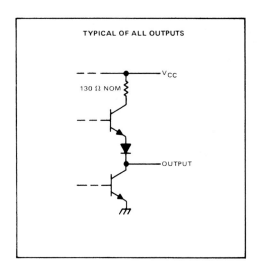

TYPICAL OF ALL OUTPUTS

130 Ω NOM

SN54LS190, SN54LS191, SN74LS190, SN74LS191
SYNCHRONOUS UP/DOWN COUNTERS WITH DOWN/UP MODE CONTROL

recommended operating conditions

		SN54LS190 SN54LS191			SN74LS190 SN74LS191			UNIT
		MIN	NOM	MAX	MIN	NOM	MAX	
V_{CC}	Supply voltage	4.5	5	5.5	4.75	5	5.25	V
I_{OH}	High-level output current			−0.4			−0.4	mA
I_{OL}	Low-level output current			4			8	mA
f_{clock}	Clock frequency	0		20	0		20	MHz
$t_{w(clock)}$	Width of clock input pulse	25			25			ns
$t_{w(load)}$	Width of load input pulse	35			35			ns
t_{su}	Data setup time (See Figures 1 and 2)	20			20			ns
t_{su}	Load inactive state setup time	30			30			ns
t_h	Data hold time	5			5			ns
t_h	Enable hold time	0			0			ns
t_{enable}	Count enable time (see Note 3)	40			40			ns
T_A	Operating free-air temperature	−55		125	0		70	°C

electrical characteristics over recommended operating free-air temperature range (unless otherwise noted)

PARAMETER		TEST CONDITIONS†		SN54LS190 SN54LS191			SN74LS190 SN74LS191			UNIT
				MIN	TYP‡	MAX	MIN	TYP‡	MAX	
V_{IH} High-level input voltage				2			2			V
V_{IL} Low-level input voltage						0.7			0.8	V
V_{IK} Input clamp voltage		V_{CC} = MIN,	I_I = −18 mA			−1.5			−1.5	V
V_{OH} High-level output voltage		V_{CC} = MIN, V_{IH} = 2 V, V_{IL} = V_{IL} max, I_{OH} = −400 µA		2.5	3.4		2.7	3.4		V
V_{OL} Low-level output voltage		V_{CC} = MIN, V_{IH} = 2 V, V_{IL} = V_{IL} max	I_{OL} = 4 mA		0.25	0.4		0.25	0.4	V
			I_{OL} = 8 mA					0.35	0.5	
I_I High-level input current at maximum input voltage	Enable	V_{CC} = MAX, V_I = 7 V				0.3			0.3	mA
	Others					0.1			0.1	
I_{IH} High-level input current	Enable	V_{CC} = MAX, V_I = 2.7 V				60			60	µA
	Others					20			20	
I_{IL} Low-level input current	Enable	V_{CC} = MAX, V_I = 0.4 V				−1.2			−1.2	mA
	Others					−0.4			−0.4	
I_{OS} Short-circuit output current§		V_{CC} = MAX,		−20		−100	−20		−100	mA
I_{CC} Supply current		V_{CC} = MAX,	See Note 2		20	35		20	35	mA

†For conditions shown as MAX or MIN, use appropriate value specified under recommended operating conditions for the applicable device type.

‡All typical values are at V_{CC} = 5 V, T_A = 25°C.

§Not more than one output should be shorted at a time, and duration of the short-circuit should not exceed one second.

NOTES: 2. I_{CC} is measured with all inputs grounded and all outputs open.

3. Minimum count enable time is the interval immediately preceeding the rising edge of the clock pulse during which interval the count enable input must be low to ensure counting.

SN54LS190, SN54LS191, SN74LS190, SN74LS191
SYNCHRONOUS UP/DOWN COUNTERS WITH DOWN/UP MODE CONTROL

switching characteristics, V_{CC} = 5 V, T_A = 25°C

PARAMETER[†]	FROM (INPUT)	TO (OUTPUT)	TEST CONDITIONS	'LS190, 'LS191– MIN	TYP	MAX	UNIT
f_{max}				20	25		MHz
t_{PLH}	$\overline{Load}$	Q_A, Q_B, Q_C, Q_D			22	33	ns
t_{PHL}					33	50	
t_{PLH}	Data A, B, C, D	Q_A, Q_B, Q_C, Q_D			20	32	ns
t_{PHL}					27	40	
t_{PLH}	CLK	$\overline{RCO}$	C_L = 15 pF, R_L = 2 kΩ, See Figures 1 and 3 thru 7		13	20	ns
t_{PHL}					16	24	
t_{PLH}	CLK	Q_A, Q_B, Q_C, Q_D			16	24	ns
t_{PHL}					24	36	
t_{PLH}	CLK	Max/Min			28	42	ns
t_{PHL}					37	52	
t_{PLH}	$D/\overline{U}$	$\overline{RCO}$			30	45	ns
t_{PHL}					30	45	
t_{PLH}	$D/\overline{U}$	Max/Min			21	33	ns
t_{PHL}					22	33	
t_{PLH}	$\overline{CTEN}$	$\overline{RCO}$			21	33	ns
t_{PHL}					22	33	

[†] f_{max} ≡ maximum clock frequency
t_{PLH} ≡ propagation delay time, low-to-high-level output
t_{PHL} ≡ propagation delay time, high-to-low-level output

schematics of inputs and outputs

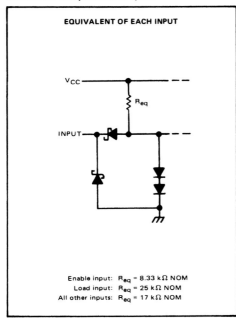

EQUIVALENT OF EACH INPUT

Enable input: R_{eq} = 8.33 kΩ NOM
Load input: R_{eq} = 25 kΩ NOM
All other inputs: R_{eq} = 17 kΩ NOM

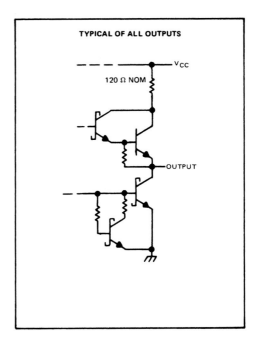

TYPICAL OF ALL OUTPUTS

120 Ω NOM

B

ERROR DETECTION AND CORRECTION CODES

B–1 ■ ERROR-DETECTION CODES

The *2-out-of-5 code* is sometimes used in communications work. It utilizes five bits to represent the ten decimal digits, so it is a form of BCD code. Each code word has exactly two 1s, a convention that facilitates decoding and provides for better error detection than the single-parity-bit method. If more or less than two 1s appear, an error is indicated.

The *63210 BCD code* is also characterized by having exactly two 1s in each of the 5-bit groups. Like the 2-out-of-5 code, it provides reliable error detection and is used in some applications.

The *biquinary* (two-five) *code* is used in certain counters and is composed of a 2-bit group and a 5-bit group, each with a single 1. Its weights are 50 43210. The 2-bit group, having weights of five and zero, indicates whether the number represented is less than, equal to, or greater than 5. The 5-bit group indicates the count above or below 5.

The *ring counter code* has ten bits, one for each decimal digit, and a single 1 makes error detection possible. It is easy to decode but wastes bits and requires more circuitry to implement than the 4-bit or 5-bit codes. The name is derived from the fact that the code is generated by a certain type of shift register, a ring counter. Its weights are 9876543210.

Each of these codes is listed in Table B–1. You should realize that this is not an exhaustive coverage of all codes but simply an introduction to some of them.

TABLE B–1
Some codes with error-detection properties

Decimal	2-out-of-5	63210	5043210	9876543210
0	00011	00110	01 00001	0000000001
1	00101	00011	01 00010	0000000010
2	00110	00101	01 00100	0000000100
3	01001	01001	01 01000	0000001000
4	01010	01010	01 10000	0000010000
5	01100	01100	10 00001	0000100000
6	10001	10001	10 00010	0001000000
7	10010	10010	10 00100	0010000000
8	10100	10100	10 01000	0100000000
9	11000	11000	10 10000	1000000000

B–2 ▪ HAMMING ERROR-CORRECTION CODE

This section discusses a method, generally known as the *Hamming code,* that not only provides for the detection of a bit error, but also identifies the bit that is in error so that it can be corrected. The code uses a number of parity bits (dependent on the number of information bits), located at certain positions in the code group.

The Hamming code construction that follows is for single-error correction.

Number of Parity Bits

If the number of information bits is designated *m,* then the number of parity bits, *p,* is determined by the following relationship:

$$2^p \geq m + p + 1 \qquad\qquad \textbf{(B–1)}$$

For example, if we have four information bits, then *p* is found by trial and error with Equation (B–1). Let $p = 2$. Then

$$2^p = 2^2 = 4$$

and

$$m + p + 1 = 4 + 2 + 1 = 7$$

Since 2^p must be equal to or greater than $m + p + 1$, the relationship in Equation (B–1) is *not* satisfied. We have to try again. Let $p = 3$. Then

$$2^p = 2^3 = 8$$

and

$$m + p + 1 = 4 + 3 + 1 = 8$$

This value of *p* satisfies the relationship of Equation (B–1), so three parity bits are required to provide single-error correction for four information bits. It should be noted here that error detection and correction are provided for *all* bits, both parity and information, in a code group.

Placement of the Parity Bits in the Code

Now that we have found the number of parity bits required in our particular example, we must arrange the bits properly in the code. At this point you should realize that in this example the code is composed of the four information bits and the three parity bits. The leftmost bit is designated *bit 1,* the next bit is *bit 2,* and so on as follows:

bit 1,   bit 2,   bit 3,   bit 4,   bit 5,   bit 6,   bit 7

The parity bits are located in the positions that are numbered corresponding to ascending powers of two (1, 2, 4, 8, . . .), as indicated:

$$P_1, \quad P_2, \quad M_1, \quad P_3, \quad M_2, \quad M_3, \quad M_4$$

The symbol P_n designates a particular parity bit, and M_n designates a particular information bit.

Assignment of Parity Bit Values

Finally, we must properly assign a 1 or 0 value to each parity bit. Since each parity bit provides a check on certain other bits in the total code, we must know the value of these others in order to assign the parity bit value. To find the bit values, first number each bit position in binary, that is, write the binary number for each decimal position number (as shown in the second two rows of Table B–2). Next, indicate the parity and information bit loca-

TABLE B–2

Bit position table for a 7-bit error-correcting code

Bit Designation Bit Position Binary Position Number	P_1 1 001	P_2 2 010	M_1 3 011	P_3 4 100	M_2 5 101	M_3 6 110	M_4 7 111
Information bits (M_n)							
Parity bits (P_n)							

tions, as shown in the first row of Table B–2. Notice that the binary position number of parity bit P_1 has a 1 for its right-most digit. *This parity bit checks all bit positions, including itself, that have 1s in the same location in the binary position numbers.* Therefore, parity bit P_1 checks bit positions 1, 3, 5, and 7.

The binary position number for parity bit P_2 has a 1 for its middle bit. It checks all bit positions, including itself, that have 1s in this same position. Therefore, parity bit P_2 checks bit positions 2, 3, 6, and 7.

The binary position number for parity bit P_3 has a 1 for its left-most bit. It checks all bit positions, including itself, that have 1s in this same position. Therefore, parity bit P_3 checks bit positions 4, 5, 6, and 7.

In each case, the parity bit is assigned a value to make the quantity of 1s in the set of bits that it checks odd or even, depending on which is specified. The following examples should make this procedure clear.

EXAMPLE B–1

Determine the single-error-correcting code for the BCD number 1001 (information bits), using even parity.

Solution

Step 1. Find the number of parity bits required. Let $p = 3$. Then

$$2^p = 2^3 = 8$$

$$m + p + 1 = 4 + 3 + 1 = 8$$

Three parity bits are sufficient.

$$\text{Total code bits} = 4 + 3 = 7$$

Step 2. Construct a bit position table, and enter the information bits. Parity bits are determined in the following steps.

Bit Designation Bit Position Binary Position Number	P_1 1 001	P_2 2 010	M_1 3 011	P_3 4 100	M_2 5 101	M_3 6 110	M_4 7 111
Information bits			1		0	0	1
Parity bits	0	0		1			

Step 3. Determine the parity bits as follows:

Bit P_1 checks bit positions 1, 3, 5, and 7 and must be a 0 for there to be an even number of 1s (2) in this group.

Bit P_2 checks bit positions 2, 3, 6, and 7 and must be a 0 for there to be an even number of 1s (2) in this group.

Bit P_3 checks bit positions 4, 5, 6, and 7 and must be a 1 for there to be an even number of 1s (2) in this group.

Step 4. These parity bits are entered in the table, and the resulting combined code is 0011001.

EXAMPLE B–2

Determine the single-error-correcting code for the information code 10110 for odd parity.

Solution

Step 1. Determine the number of parity bits required. In this case the number of information bits, m, is five. From the previous example we know that $p = 3$ will not work. Try $p = 4$:

$$2^p = 2^4 = 16$$
$$m + p + 1 = 5 + 4 + 1 = 10$$

Four parity bits are sufficient.

$$\text{Total code bits} = 5 + 4 = 9$$

Step 2. Construct a bit position table, and enter the information bits. Parity bits are determined in the following steps.

Bit Designation Bit Position Binary Position Number	P_1 1 0001	P_2 2 0010	M_1 3 0011	P_3 4 0100	M_2 5 0101	M_3 6 0110	M_4 7 0111	P_4 8 1000	M_5 9 1001
Information bits			1		0	1	1		0
Parity bits	1	0		1				1	

Step 3. Determine the parity bits as follows:

Bit P_1 checks bit positions 1, 3, 5, 7, and 9 and must be a 1 for there to be an odd number of 1s (3) in this group.

Bit P_2 checks bit positions 2, 3, 6, and 7 and must be a 0 for there to be an odd number of 1s (3) in this group.

Bit P_3 checks bit positions 4, 5, 6, and 7 and must be a 1 for there to be an odd number of 1s (3) in this group.

Bit P_4 checks bit positions 8 and 9 and must be a 1 for there to be an odd number of 1s (1) in this group.

Step 4. These parity bits are entered in the table, and the resulting combined code is 101101110.

B–3 ■ DETECTING AND CORRECTING AN ERROR

Now that a method for constructing an error-correcting code has been covered, how do we use it to locate and correct an error? Each parity bit, along with its corresponding group of bits, must be checked for the proper parity. If there are three parity bits in a code word, then three parity checks are made. If there are four parity bits, four checks must be made, and so on. Each parity check will yield a good or a bad result. The total result of all the parity checks indicates the bit, if any, that is in error, as follows:

Step 1. Start with the group checked by P_1.
Step 2. Check the group for proper parity. A 0 represents a good parity check, and 1 represents a bad check.
Step 3. Repeat step 2 for each parity group.
Step 4. The binary number formed by the results of all the parity checks designates the position of the code bit that is in error. This is the *error position code.* The first parity check generates the least significant bit (LSB). If all checks are good, there is no error.

EXAMPLE B–3

Assume that the code word in Example B–1 (0011001) is transmitted and that 0010001 is received. The receiver does not "know" what was transmitted and must look for proper parities to determine if the code is correct. Designate any error that has occurred in transmission if even parity is used.

Solution   First, make a bit position table:

Bit Designation Bit Position Binary Position Number	P_1 1 001	P_2 2 010	M_1 3 011	P_3 4 100	M_2 5 101	M_3 6 110	M_4 7 111
Received code	0	0	1	0	0	0	1

First parity check:
   Bit P_1 checks positions 1, 3, 5, and 7.
   There are two 1s in this group.
   Parity check is good. ———————————————————→ 0 (LSB)
Second parity check:
   Bit P_2 checks positions 2, 3, 6, and 7.
   There are two 1s in this group.
   Parity check is good. ———————————————————→ 0
Third parity check:
   Bit P_3 checks positions 4, 5, 6, and 7.
   There is one 1 in this group.
   Parity check is bad. ———————————————————→ 1 (MSB)
Result:
   The error position code is 100 (binary four). This says that the bit in position 4 is in error. It is a 0 and should be a 1. The corrected code is 0011001, which agrees with the transmitted code.

EXAMPLE B–4

The code 101101010 is received. Correct any errors. There are four parity bits, and odd parity is used.

Solution   First, make a bit position table:

Bit Designation Bit Position Binary Position Number	P_1 1 0001	P_2 2 0010	M_1 3 0011	P_3 4 0100	M_2 5 0101	M_3 6 0110	M_4 7 0111	P_4 8 1000	M_5 9 1001
Received code	1	0	1	1	0	1	0	1	0

First parity check:

    Bit P_1 checks positions 1, 3, 5, 7, and 9.

    There are two 1s in this group.

    Parity check is bad. ⟶ 1 (LSB)

Second parity check:

    Bit P_2 checks positions 2, 3, 6, and 7.

    There are two 1s in this group.

    Parity check is bad. ⟶ 1

Third parity check:

    Bit P_3 checks positions 4, 5, 6, and 7.

    There are two 1s in this group.

    Parity check is bad. ⟶ 1

Fourth parity check:

    Bit P_4 checks positions 8 and 9.

    There is one 1 in this group.

    Parity check is good. ⟶ 0 (MSB)

Result:

    The error position code is 0111 (binary seven). This says that the bit in position 7 is in error. The corrected code is therefore 101101110.

C CONVERSIONS

Decimal	BCD(8421)	Octal	Binary	Decimal	BCD(8421)	Octal	Binary	Decimal	BCD(8421)	Octal	Binary
0	0000	0	0	34	00110100	42	100010	68	01101000	104	1000100
1	0001	1	1	35	00110101	43	100011	69	01101001	105	1000101
2	0010	2	10	36	00110110	44	100100	70	01110000	106	1000110
3	0011	3	11	37	00110111	45	100101	71	01110001	107	1000111
4	0100	4	100	38	00111000	46	100110	72	01110010	110	1001000
5	0101	5	101	39	00111001	47	100111	73	01110011	111	1001001
6	0110	6	110	40	01000000	50	101000	74	01110100	112	1001010
7	0111	7	111	41	01000001	51	101001	75	01110101	113	1001011
8	1000	10	1000	42	01000010	52	101010	76	01110110	114	1001100
9	1001	11	1001	43	01000011	53	101011	77	01110111	115	1001101
10	00010000	12	1010	44	01000100	54	101100	78	01111000	116	1001110
11	00010001	13	1011	45	01000101	55	101101	79	01111001	117	1001111
12	00010010	14	1100	46	01000110	56	101110	80	10000000	120	1010000
13	00010011	15	1101	47	01000111	57	101111	81	10000001	121	1010001
14	00010100	16	1110	48	01001000	60	110000	82	10000010	122	1010010
15	00010101	17	1111	49	01001001	61	110001	83	10000011	123	1010011
16	00010110	20	10000	50	01010000	62	110010	84	10000100	124	1010100
17	00010111	21	10001	51	01010001	63	110011	85	10000101	125	1010101
18	00011000	22	10010	52	01010010	64	110100	86	10000110	126	1010110
19	00011001	23	10011	53	01010011	65	110101	87	10000111	127	1010111
20	00100000	24	10100	54	01010100	66	110110	88	10001000	130	1011000
21	00100001	25	10101	55	01010101	67	110111	89	10001001	131	1011001
22	00100010	26	10110	56	01010110	70	111000	90	10010000	132	1011010
23	00100011	27	10111	57	01010111	71	111001	91	10010001	133	1011011
24	00100100	30	11000	58	01011000	72	111010	92	10010010	134	1011100
25	00100101	31	11001	59	01011001	73	111011	93	10010011	135	1011101
26	00100110	32	11010	60	01100000	74	111100	94	10010100	136	1011110
27	00100111	33	11011	61	01100001	75	111101	95	10010101	137	1011111
28	00101000	34	11100	62	01100010	76	111110	96	10010110	140	1100000
29	00101001	35	11101	63	01100011	77	111111	97	10010111	141	1100001
30	00110000	36	11110	64	01100100	100	1000000	98	10011000	142	1100010
31	00110001	37	11111	65	01100101	101	1000001	99	10011001	143	1100011
32	00110010	40	100000	66	01100110	102	1000010				
33	00110011	41	100001	67	01100111	103	1000011				

Power of Two

2^n	n	2^{-n}
1	0	1.0
2	1	0.5
4	2	0.25
8	3	0.125
16	4	0.062 5
32	5	0.031 25
64	6	0.015 625
128	7	0.007 812 5
256	8	0.003 906 25
512	9	0.001 953 125
1 024	10	0.000 976 562 5
2 048	11	0.000 488 281 25
4 096	12	0.000 244 140 625
8 192	13	0.000 122 070 312 5
16 384	14	0.000 061 035 156 25
32 768	15	0.000 030 517 578 125
65 536	16	0.000 015 258 789 062 5
131 072	17	0.000 007 629 394 531 25
262 144	18	0.000 003 814 697 265 625
524 288	19	0.000 001 907 348 632 812 5
1 048 576	20	0.000 000 953 674 316 406 25
2 097 152	21	0.000 000 476 837 158 203 125
4 194 304	22	0.000 000 238 418 579 101 562 5
8 388 608	23	0.000 000 119 209 289 550 781 25
16 777 216	24	0.000 000 059 604 644 775 390 625
33 554 432	25	0.000 000 029 802 322 387 695 312 5
67 108 864	26	0.000 000 014 901 161 193 847 656 25
134 217 728	27	0.000 000 007 450 580 596 923 828 125
268 435 456	28	0.000 000 003 725 290 298 461 914 062 5
536 870 912	29	0.000 000 001 862 645 149 230 957 031 25
1 073 741 824	30	0.000 000 000 931 322 574 615 478 515 625
2 147 483 648	31	0.000 000 000 465 661 287 307 739 257 812 5
4 294 967 296	32	0.000 000 000 232 830 643 653 869 628 906 25
8 589 934 592	33	0.000 000 000 116 415 321 826 934 814 453 125
17 179 869 184	34	0.000 000 000 058 207 660 913 467 407 226 562 5
34 359 738 368	35	0.000 000 000 029 103 830 456 733 703 613 281 25
68 719 476 736	36	0.000 000 000 014 551 915 228 366 851 806 640 625
137 438 953 472	37	0.000 000 000 007 275 957 614 183 425 903 320 312 5
274 877 906 944	38	0.000 000 000 003 637 978 807 091 712 951 660 156 25
549 755 813 888	39	0.000 000 000 001 818 989 403 545 856 475 830 078 125
1 099 511 627 776	40	0.000 000 000 000 909 494 701 772 928 237 915 039 062 5
2 199 023 255 552	41	0.000 000 000 000 454 747 350 886 464 118 957 519 531 25
4 398 046 511 104	42	0.000 000 000 000 227 373 675 443 232 059 478 759 765 625
8 796 093 022 208	43	0.000 000 000 000 113 686 837 721 616 029 739 379 882 812 5
17 592 186 044 416	44	0.000 000 000 000 056 843 418 860 808 014 869 689 941 406 25
35 184 372 088 832	45	0.000 000 000 000 028 421 709 430 404 007 434 844 970 703 125
70 368 744 177 664	46	0.000 000 000 000 014 210 854 715 202 003 717 422 485 351 562 5
140 737 488 355 328	47	0.000 000 000 000 007 105 427 357 601 001 858 711 242 675 781 25
281 474 976 710 656	48	0.000 000 000 000 003 552 713 678 800 500 929 355 621 337 890 625
562 949 953 421 312	49	0.000 000 000 000 001 776 356 839 400 250 464 677 810 668 945 312 5
1 125 899 906 842 624	50	0.000 000 000 000 000 888 178 419 700 125 232 338 905 334 472 656 25
2 251 799 813 685 248	51	0.000 000 000 000 000 444 089 209 850 062 616 169 452 667 236 328 125
4 503 599 627 370 496	52	0.000 000 000 000 000 222 044 604 925 031 308 084 726 333 618 164 062 5
9 007 199 254 740 992	53	0.000 000 000 000 000 111 022 302 462 515 654 042 363 166 809 082 031 25
18 014 398 509 481 984	54	0.000 000 000 000 000 055 511 151 231 257 827 021 181 583 404 541 015 625
36 028 797 018 963 968	55	0.000 000 000 000 000 027 755 575 615 628 913 510 590 791 702 270 507 812 5
72 057 594 037 927 936	56	0.000 000 000 000 000 013 877 787 807 814 456 755 295 395 851 135 253 906 25
144 115 188 075 855 872	57	0.000 000 000 000 000 006 938 893 903 907 228 377 647 697 925 567 626 953 125
288 230 376 151 711 744	58	0.000 000 000 000 000 003 469 446 951 953 614 188 823 848 962 783 813 476 562 5
576 460 752 303 423 488	59	0.000 000 000 000 000 001 734 723 475 976 807 094 411 924 481 391 906 738 281 25
1 152 921 504 606 846 976	60	0.000 000 000 000 000 000 867 361 737 988 403 547 205 962 240 695 953 369 140 625
2 305 843 009 213 693 952	61	0.000 000 000 000 000 000 433 680 868 994 201 773 602 981 120 347 976 684 570 312 5
4 611 686 018 427 387 904	62	0.000 000 000 000 000 000 216 840 434 497 100 886 801 490 560 173 988 342 285 156 25
9 223 372 036 854 775 808	63	0.000 000 000 000 000 000 108 420 217 248 550 443 400 745 280 086 994 171 142 578 125
18 446 744 073 709 551 616	64	0.000 000 000 000 000 000 054 210 108 624 275 221 700 372 640 043 497 085 571 289 062 5
36 893 488 147 419 103 232	65	0.000 000 000 000 000 000 027 105 054 312 137 610 850 186 320 021 748 542 785 644 531 25
73 786 976 294 838 206 464	66	0.000 000 000 000 000 000 013 552 527 156 088 805 425 093 160 010 874 271 392 822 265 625
147 573 952 589 676 412 928	67	0.000 000 000 000 000 000 006 776 263 578 034 402 712 546 580 005 437 135 696 411 132 812 5
295 147 905 179 352 825 856	68	0.000 000 000 000 000 000 003 388 131 789 017 201 356 273 290 002 718 567 848 205 566 406 25
590 295 810 358 705 651 712	69	0.000 000 000 000 000 000 001 694 065 894 508 600 678 136 645 001 359 283 924 102 783 203 125
1 180 591 620 717 411 303 424	70	0.000 000 000 000 000 000 000 847 032 947 254 300 339 068 322 500 679 641 962 051 391 601 562 5
2 361 183 241 434 822 606 848	71	0.000 000 000 000 000 000 000 423 516 473 627 150 169 534 161 250 339 820 981 025 695 800 781 25
4 722 366 482 869 645 213 696	72	0.000 000 000 000 000 000 000 211 758 236 813 575 084 767 080 625 169 910 490 512 847 900 390 625

ANSWERS TO SELF-TESTS

CHAPTER 1

1. (b)	**2.** (d)	**3.** (a)	**4.** (c)	**5.** (c)	**6.** (d)	**7.** (b)	**8.** (d)
9. (a)	**10.** (e)	**11.** (d)	**12.** (d)				

CHAPTER 2

1. (d)	**2.** (a)	**3.** (b)	**4.** (c)	**5.** (c)	**6.** (a)	**7.** (d)	**8.** (b)
9. (d)	**10.** (a)	**11.** (c)	**12.** (d)	**13.** (b)	**14.** (c)	**15.** (a)	**16.** (c)
17. (a)							

CHAPTER 3

1. (d)	**2.** (d)	**3.** (a)	**4.** (e)	**5.** (c)	**6.** (a)	**7.** (d)	**8.** (b)
9. (d)	**10.** (d)	**11.** (c)	**12.** (c)				

CHAPTER 4

1. (d)	**2.** (a)	**3.** (b)	**4.** (c)	**5.** (d)	**6.** (b)	**7.** (a)	**8.** (b)
9. (d)	**10.** (d)	**11.** (c)	**12.** (b)	**13.** (b)	**14.** (c)	**15.** (a)	**16.** (c)
17. (a)	**18.** (b)						

CHAPTER 5

1. (d)	**2.** (b)	**3.** (c)	**4.** (a)	**5.** (d)	**6.** (b)	**7.** (a)	**8.** (d)
9. (d)	**10.** (e)						

CHAPTER 6

1. (a)	**2.** (b)	**3.** (c)	**4.** (a)	**5.** (d)	**6.** (b)	**7.** (c)	**8.** (b)
9. (a)	**10.** (d)	**11.** (c)	**12.** (f)				

CHAPTER 7

1. (c)	**2.** (b)	**3.** (f)	**4.** (c)	**5.** (c)	**6.** (d)	**7.** (b)	**8.** (b)
9. (c)	**10.** (f)	**11.** (b)	**12.** (c)	**13.** (d)	**14.** (b)	**15.** (d)	**16.** (a)
17. (c)	**18.** (c)	**19.** (c)	**20.** (b)				

CHAPTER 8

1. (a)	**2.** (c)	**3.** (d)	**4.** (b)	**5.** (d)	**6.** (d)	**7.** (a)	**8.** (b)
9. (d)	**10.** (d)	**11.** (c)	**12.** (f)				

CHAPTER 9

1. (a)	**2.** (b)	**2.** (b)	**4.** (c)	**5.** (a)	**6.** (c)	**7.** (b)	**8.** (c)
9. (d)	**10.** (a)	**11.** (c)	**12.** (b)	**13.** (b)	**14.** (d)		

CHAPTER 10

1. (b)	**2.** (c)	**3.** (a)	**4.** (c)	**5.** (a)	**6.** (d)	**7.** (c)	**8.** (a)
9. (b)	**10.** (c)						

CHAPTER 11

1. (b)   **2.** (b)   **3.** (d)   **4.** (a)   **5.** (c)   **6.** (d)   **7.** (e)   **8.** (b)
9. (d)   **10.** (c)

CHAPTER 12

1. (b)   **2.** (c)   **3.** (c)   **4.** (d)   **5.** (a)   **6.** (d)   **7.** (c)   **8.** (a)
9. (b)   **10.** (f)

CHAPTER 13

1. (b)   **2.** (c)   **3.** (c)   **4.** (b)   **5.** (a)   **6.** (a)   **7.** (b)   **8.** (d)
9. (b)   **10.** (b)

CHAPTER 14

1. (c)   **2.** (b)   **3.** (c)   **4.** (b)   **5.** (a)   **6.** (b)   **7.** (b)   **8.** (c)

CHAPTER 15

1. (c)   **2.** (e)   **3.** (c)   **4.** (c)   **5.** (a)   **6.** (b)   **7.** (c)   **8.** (c)
9. (d)   **10.** (a)   **11.** (c)

ANSWERS TO RELATED EXERCISES FOR EXAMPLES

CHAPTER 1

1–1 $f = 6.67$ kHz; Duty cycle $= 16.7\%$

1–2 Parallel transfer: 100 ns; Serial transfer: $1.6\,\mu s$

CHAPTER 2

2–1 9 has a value of 900, 3 has a value of 30, 9 has a value of 9.

2–2 6 has a value of 60, 7 has a value of 7, 9 has a value of 9/10 (0.9), 2 has a value of 2/100 (0.02), 4 has a value of 4/1000 (0.004).

2–3 $10010001 = 128 + 16 + 1 = 145$ **2–4** $10.111 = 2 + 0.5 + 0.25 + 0.125 = 2.875$

2–5 $125 = 64 + 32 + 16 + 8 + 4 + 1 = 1111101$ **2–6** $39 = 32 + 4 + 2 + 1 = 100111$

2–7 $1111 + 1100 = 11011$ **2–8** $111 - 100 = 011$ **2–9** $110 - 101 = 001$

2–10 $1101 \times 1010 = 10000010$ **2–11** $1100 \div 100 = 11$ **2–12** 00110101

2–13 01000000   **2–14** See Table RE–1.

TABLE RE–1

	Sign-Magnitude	1's comp	2's comp
$+19$	00010011	00010011	00010011
-19	10010011	11101100	11101101

2–15 $01110111 = +119_{10}$ **2–16** $11101011 = -20_{10}$ **2–17** $111010111 = -41_{10}$

2–18 01010101   **2–19** 00010001   **2–20** 1001000110

2–21 $(83)(-59) = -4897$ (10110011011111 in 2's comp)   **2–22** $100 \div 25 = 4$ (0100)

2–23 $4F79C_{16}$ **2–24** 01101011110100011_2

2–25 $6BD_{16} = 011010111101 = 2^{10} + 2^9 + 2^7 + 2^5 + 2^4 + 2^3 + 2^2 + 2^0$
$= 1024 + 512 + 128 + 32 + 16 + 8 + 4 + 1 = 1725_{10}$

2–26 $60A_{16} = (6 \times 256) + (0 \times 16) + (10 \times 1) = 1546_{10}$

2–27 $2591_{10} = A1F_{16}$ **2–28** $4C_{16} + 3A_{16} = 86_{16}$ **2–29** $BCD_{16} - 173_{16} = A5A_{16}$

2–30 (a) $001011_2 = 11_{10} = 13_8$ (b) $010101_2 = 21_{10} = 25_8$
(c) $001100000_2 = 96_{10} = 140_8$ (d) $111101010110_2 = 3926_{10} = 7526_8$

2–31 1250762_8 **2–32** 1001011001110011

2–33 $82,276_{10}$ **2–34** 1001100101101000   **2–35** 10000010

2–36 (a) 111011 (Gray)   (b) 111010_2 **2–37** 110001011011 (excess-3)

2–38 The sequence of codes for 80 INPUT Y is $38_{16}30_{16}20_{16}49_{16}4E_{16}50_{16}55_{16}54_{16}20_{16}59_{16}$

2–39 01001011   **2–40** Yes

CHAPTER 3

3–1 The timing diagram is not affected.   **3–2** See Table RE–2.

3–3 See Figure RE–1.   **3–4** The output waveform is the same as input A.

3–5 See Figure RE–2.   **3–6** See Figure RE–3.   **3–7** See Figure RE–4.

3–8 See Figure RE–5.   **3–9** See Figure RE–6.   **3–10** See Figure RE–7.

TABLE RE–2

Imputs ABCD	Output X	Inputs ABCD	Output X
0000	0	1000	0
0001	0	1001	0
0010	0	1010	0
0011	0	1011	0
0100	0	1100	0
0101	0	1101	0
0110	0	1110	0
0111	0	1111	1

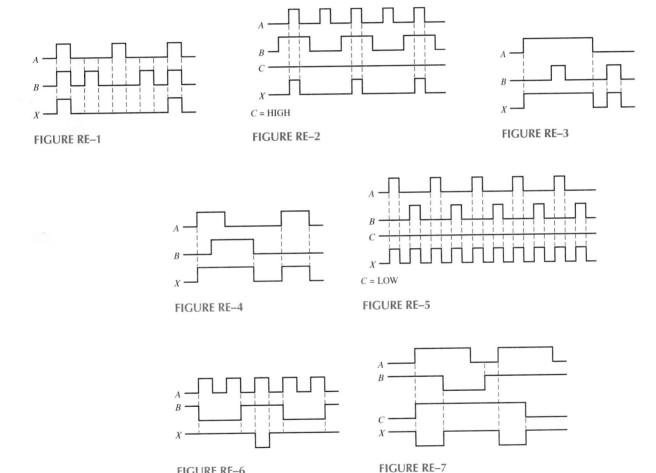

FIGURE RE–1

FIGURE RE–2

C = HIGH

FIGURE RE–3

FIGURE RE–4

FIGURE RE–5

C = LOW

FIGURE RE–6

FIGURE RE–7

3–11 Use a 3-input NAND gate.

3–12 Use a 4-input NAND gate operating as a negative-OR gate.　　**3–13** See Figure RE–8.

3–14 See Figure RE–9.　　**3–15** See Figure RE–10.

3–16 Use a 2-input NOR gate.　　**3–17** A 3-input NAND gate.

3–18 The output is always LOW. The timing diagram is a straight line.

3–19 The exclusive-OR gate will not detect simultaneous failures if both circuits produce the same outputs.

3–20 The outputs are unaffected.

3–21 The gate with 4 ns t_{PLH} and t_{PHL} can operate at the highest frequency.

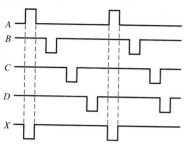

FIGURE RE–8

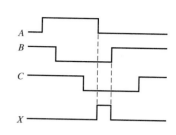

FIGURE RE–9

FIGURE RE–10

3–22 The gate output or pin 13 input is internally open.

3–23 The display will show an erratic readout because the counter continues until reset.

3–24 The enable pulse is too short or the counter is reset too soon.

CHAPTER 4

4–1 $\overline{A} + B = 0$ when $A = 1$ and $B = 0$.   **4–2** $\overline{A}\,\overline{B} = 1$ when $A = 0$ and $B = 0$.   **4–3** XYZ

4–4 $W + X + Y + Z$ **4–5** $ABC\overline{D}\,\overline{E}$ **4–6** $(A + \overline{B} + \overline{C}D)\overline{E}$

4–7 $\overline{ABCD} = \overline{A} + \overline{B} + \overline{C} + \overline{D}$ **4–8** $A\overline{B}$ **4–9** $BCD + \overline{A}CD$ **4–10** $AB\overline{C} + \overline{A}C + \overline{A}\,\overline{B}$

4–11 $\overline{A} + \overline{B} + \overline{C}$ **4–12** $\overline{A}B\overline{C} + AB + A\overline{C} + A\overline{B} + \overline{B}\,\overline{C}$

4–13 $W\overline{X}YZ + W\overline{X}Y\overline{Z} + WX\overline{Y}Z + \overline{W}\,\overline{X}YZ + WXYZ + WX\overline{Y}\,\overline{Z}$

4–14 011, 101, 110, 010, 111. Yes   **4–15** $(A + \overline{B} + C)(A + \overline{B} + \overline{C})(A + B + C)(\overline{A} + B + C)$

4–16 010, 100, 001, 111, 011. Yes   **4–17** SOP and POS expressions are equivalent.

4–18 See Table RE–3.   **4–19** See Table RE–4.

TABLE RE–3

A	B	C	X
0	0	0	0
0	0	1	0
0	1	0	1
0	1	1	0
1	0	0	0
1	0	1	1
1	1	0	0
1	1	1	0

TABLE RE–4

A	B	C	X
0	0	0	1
0	0	1	0
0	1	0	0
0	1	1	1
1	0	0	1
1	0	1	1
1	1	0	1
1	1	1	0

4–20 The SOP and POS expressions are equivalent.   **4–21** See Figure RE–11.

4–22 See Figure RE–12.   **4–23** See Figure RE–13.   **4–24** See Figure RE–14.

4–25 No other ways   **4–26** $X = B + \overline{A}C + A\overline{C}D + C\overline{D}$ **4–27** $X = \overline{D} + AB\overline{C} + B\overline{C} + \overline{A}B$

4–28 $Q = X + Y$ **4–29** $Q = \overline{X}\,\overline{Y}\,\overline{Z} + W\overline{X}Z + \overline{W}YZ$ **4–30** See Figure RE–15.

FIGURE RE–11

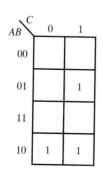

AB \ C	0	1
00		
01		1
11		
10	1	1

FIGURE RE–12

AB \ CD	00	01	11	10
00				
01			1	
11	1		1	1
10				

FIGURE RE–13

AB \ C	0	1
00	1	
01	1	1
11		1
10		

FIGURE RE–14

AB\CD	00	01	11	10
00		1		
01		1		1
11	1	1	1	1
10	1	1	1	1

FIGURE RE–15

AB\CD	00	01	11	10
00	0	0		
01				0
11				
10				0

4–31 $Q = (X + \overline{Y})(X + \overline{Z})(\overline{X} + Y + Z)$

4–32 $Q = (\overline{X} + \overline{Y} + Z)(\overline{W} + \overline{X} + Z)(W + X + Y + Z)(W + \overline{X} + Y + \overline{Z})$

4–33 $Q = \overline{Y}\,\overline{Z} + \overline{X}\,\overline{Z} + \overline{W}Y + \overline{X}\,\overline{Y}Z$

4–34 $Y = \overline{D}\,\overline{E} + \overline{A}\,\overline{E} + \overline{B}\,\overline{C}\,\overline{E}$

CHAPTER 5

5–1 $Y = AB + AC + BC$

5–2 $Y = \overline{AB + AC + BC}$

        If $A = 0$ and $B = 0$, $Y = \overline{0 \cdot 0 + 0 \cdot 1 + 0 \cdot 1} = \overline{0} = 1$

        If $A = 0$ and $C = 0$, $Y = \overline{0 \cdot 1 + 0 \cdot 0 + 1 \cdot 0} = \overline{0} = 1$

        If $B = 0$ and $C = 0$, $Y = \overline{1 \cdot 0 + 1 \cdot 0 + 0 \cdot 0} = \overline{0} = 1$

5–3 Cannot be simplified   **5–4** Cannot be simplified   **5–5** $X = A + B + C + D$ is valid.

5–6 See Figure RE–16.   **5–7** $X = \overline{(\overline{ABC})(\overline{DEF})} = (\overline{AB})C + (\overline{DE})F = (\overline{A} + \overline{B})C + (\overline{D} + \overline{E})F$

5–8 See Figure RE–17.

5–9 $X = \overline{\overline{(\overline{A + B} + C)} + \overline{(\overline{D + E} + F)}} = (\overline{A + B} + C)(\overline{D + E} + F) = (\overline{A}\,\overline{B} + C)(\overline{D}\,\overline{E} + F)$

5–10 See Figure RE–18.   **5–11** See Figure RE–19.

5–12 See Figure RE–20.   **5–13** See Figure RE–21.   **5–14** See Figure RE–22.

FIGURE RE–16

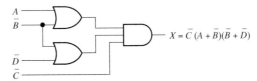

$X = \overline{C}\,(A + \overline{B})(\overline{B} + \overline{D})$

FIGURE RE–17

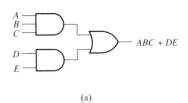

$ABC + DE$

(a)

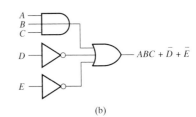

$ABC + \overline{D} + \overline{E}$

(b)

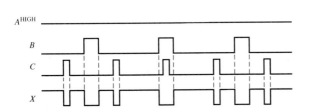

FIGURE RE–18

FIGURE RE–19

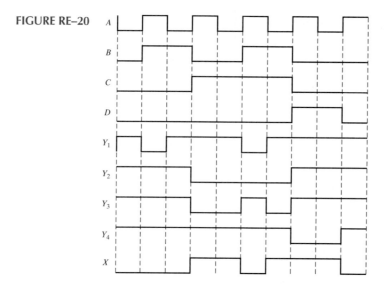

FIGURE RE–20

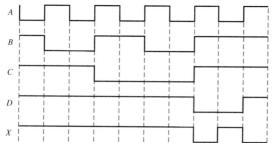

FIGURE RE–21

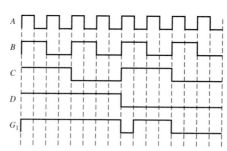

FIGURE RE–22

CHAPTER 6

6–1 Replace the AND gates with NAND gates. Replace the OR gates with NAND gates used as negative-OR gates. Replace the inverters with NAND gates with inputs connected together.

6–2 $\Sigma_1 = (1 \oplus 0) \oplus 0 = 1 \oplus 0 = 1$
$\Sigma_2 = [(1 \oplus 0) \cdot 0 + (1 \cdot 0)] \oplus (1 \oplus 1) = (0 + 0) \oplus 0 = 0 \oplus 0 = 0$
$\Sigma_3 = (1 \oplus 1) \oplus [(1 \oplus 0) \cdot 0 + (1 \cdot 0)] = 0 \oplus (0 + 1) = 0 \oplus 1 = 1$

6–3 $1011 + 1010 = 10101$ **6–4** See Figure RE–23.   **6–5** See Figure RE–24.

FIGURE RE–23

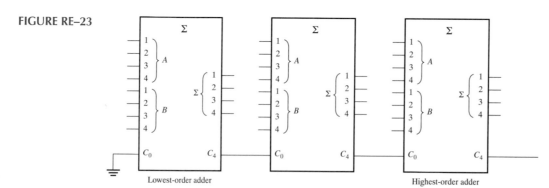

Lowest-order adder   Highest-order adder

FIGURE RE–24

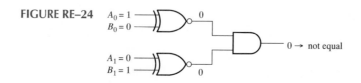

$A_0 = 1$
$B_0 = 0$ 0

$A_1 = 0$
$B_1 = 1$ 0

$0 \rightarrow$ not equal

6–6 See Table RE–5.

6–7 See Figure RE–25. **6–8** See Figure RE–26.

6–9 Output 22 **6–10** See Figure RE–27.

6–11 All inputs LOW: $\overline{A}_0 = 0, \overline{A}_1 = 1, \overline{A}_2 = 1, \overline{A}_3 = 0$
All inputs HIGH: All outputs HIGH.

6–12 BCD 01000001

	00000001	1
	00101000	40
Binary	00101001	41

6–13 Seven exclusive-OR gates

TABLE RE–5

$A_3A_2A_1A_0$	$B_3B_2B_1B_0$	G_1	G_2	G_3	G_4	G_5	G_6	G_7	G_8	G_9	G_{10}	G_{11}	G_{12}	G_{13}	G_{14}	G_{15}
0010	1000	0	1	0	1	0	0	0	0	0	0	1	0	0	0	1

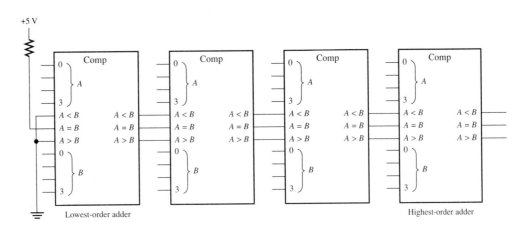

FIGURE RE–25

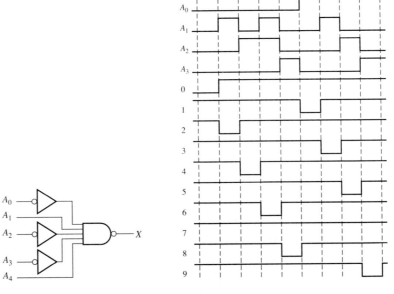

FIGURE RE–26   **FIGURE RE–27**

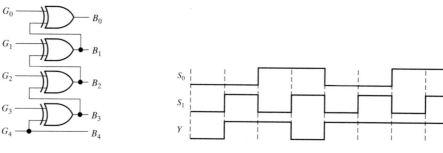

FIGURE RE–28　　　　　　　　　　　　　　**FIGURE RE–29**

6–14 See Figure RE–28.　　**6–15** See Figure RE–29.

6–16 D_0: $S_4 = 0, S_3 = 0, S_2 = 0, S_1 = 0, S_0 = 0$
D_8: $S_4 = 0, S_3 = 1, S_2 = 0, S_1 = 0, S_0 = 0$
D_{21}: $S_4 = 1, S_3 = 0, S_2 = 1, S_1 = 0, S_0 = 1$
D_{29}: $S_4 = 1, S_3 = 1, S_2 = 1, S_1 = 0, S_0 = 1$

6–17 See Figure RE–30.　　**6–18** See Figure RE–31.　　**6–19** See Figure RE–32.

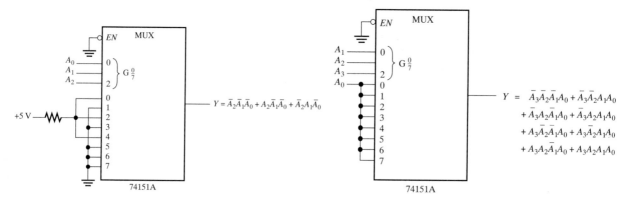

FIGURE RE–30　　　　　　　　　　　　　　**FIGURE RE–31**

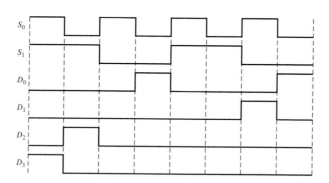

FIGURE RE–32

CHAPTER 7

7–1 $X = \overline{B}C + \overline{A}B\overline{C} + \overline{A}\,\overline{B} + C$

7–2 A PAL14H4 has 14 inputs and 4 active-HIGH outputs.

7–3 $X = B\overline{C} + BC + BC + A\overline{B} = \overline{B}C + BC + A\overline{B}$

7–4 A GAL18V10 has up to 18 inputs and 10 outputs.

7–5 $X = \overline{\overline{A}BCD + \overline{A}B\overline{C}\overline{D} + A\overline{B}C\overline{D} + \overline{A}B\overline{C}D + A\overline{B}CD + AB\overline{C}D}$

7–6 One more product term can be added.

7–7 W = !A & B & !C # A # !B # C;

7–8 Y = (S == 3) & A # (S == 2) & B # (S == 1) & C;

7–9
```
TRUTH_TABLE (BCD -> [a, b, c, d, e, f, g])
              0  -> [0, 0, 0, 0, 0, 0, 1];
              1  -> [1, 0, 0, 1, 1, 1, 1];
              2  -> [0, 0, 1, 0, 0, 1, 0];
              3  -> [0, 0, 0, 0, 1, 1, 0];
              4  -> [1, 0, 0, 1, 1, 0, 0];
              5  -> [0, 1, 0, 0, 1, 0, 0];
              6  -> [0, 1, 0, 0, 0, 0, 0];
              7  -> [0, 0, 0, 1, 1, 1, 1];
              8  -> [0, 0, 0, 0, 0, 0, 0];
              9  -> [0, 0, 0, 0, 1, 0, 0];
```

7–10
```
BIT0 = [D0, C0, B0, A0];
BIT1 = [D1, C1, B1, A1];
BIT2 = [D2, C2, B2, A2];
BIT3 = [D3, C3, B3, A3];
   S = [S1, S0];
 OUT = [Dout, Cout, Bout, Aout];
 OUT = (S == 0) & BIT0 # (S == 1) & BIT1 # (S == 2) & BIT2 # (S == 3) & BIT3;
```

7–11 18 pins are used as inputs. Four pins are used as outputs.

CHAPTER 8

8–1 The Q output is the same as shown in Figure 8–5(b).

8–2 See Figure RE–33. **8–3** See Figure RE–34.

8–4 See Figure RE–35. **8–5** See Figure RE–36.

8–6 See Figure RE–37.

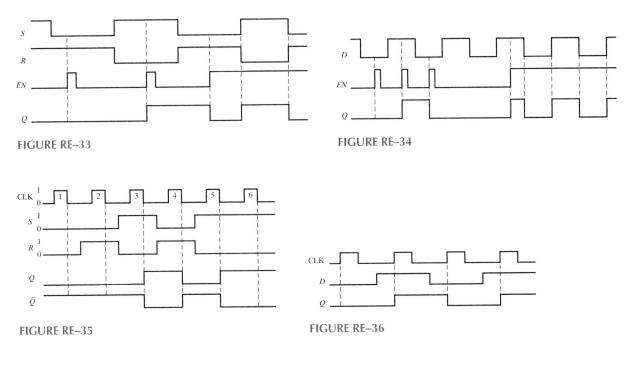

FIGURE RE–33

FIGURE RE–34

FIGURE RE–35

FIGURE RE–36

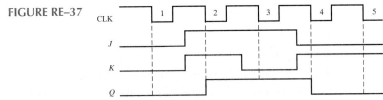

FIGURE RE–37

8–7 See Figure RE–38.   **8–8** See Figure RE–39.

8–9 See Figure RE–40.   **8–10** See Figure RE–41.

8–11 $2^5 = 32$. Five flip-flops are required.

8–12 Sixteen states require four flip-flops ($2^4 = 16$).

8–13 $C_{EXT} = 7143$ pF (6800 pF ∥ 330 pF) connected from CX to RX/CX of the 74121.

8–14 $C_{EXT} = 560$ pF, $R_{EXT} = 27$ kΩ. See Figure RE–42.

8–15 $R_1 = 91$ kΩ   **8–16** Duty cycle ≅ 32%

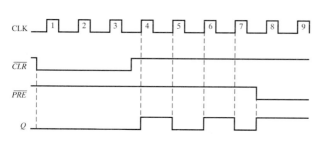

FIGURE RE–38

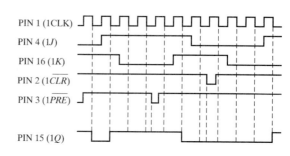

FIGURE RE–39

FIGURE RE–40

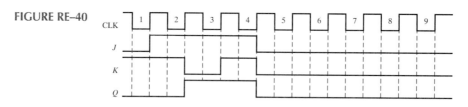

FIGURE RE–41

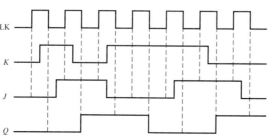

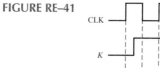

FIGURE RE–42

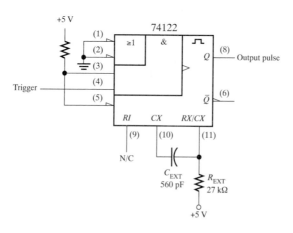

CHAPTER 9

9–1 See Figure RE–43.

9–2 Connect Q_0 to the NAND gate as a third input (Q_2 and Q_3 are two of the inputs). Connect the $\overline{CLR}$ line to the $\overline{CLR}$ input of FF0 as well as FF2 and FF3.

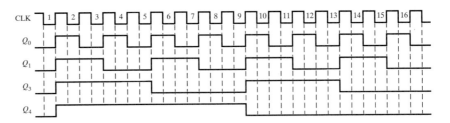

FIGURE RE–43

9–3 See Figure RE–44. **9–4** See Figure RE–45. **9–5** See Table RE–6.

9–6 Application of Boolean algebra to the logic in Figure 9–37 shows that the output of each OR gate agrees with the expression in step 5.

9–7 Five decade counters are required. $10^5 = 100,000$

9–8 $f_{Q0} = 1\ \text{MHz}/[(10)(2)] = 50\ \text{kHz}$

FIGURE RE–44

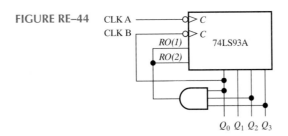

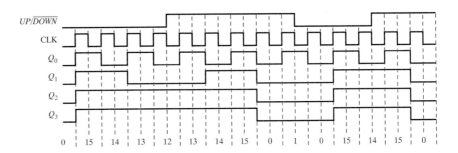

FIGURE RE–45

TABLE RE–6

Present Invalid State			J-K Inputs						Next State			
Q_2	Q_1	Q_0	J_2	K_2	J_1	K_1	J_0	K_0	Q_2	Q_1	Q_0	
0	0	0	0	0	1	1	1	1	0	1	1	
0	1	1	1	1	1	1	1	1	1	0	0	
1	0	0	0	0	1	1	1	0	1	1	1	valid state
1	1	0	1	1	1	1	1	0	0	0	1	valid state

9–9 See Figure RE–46.

9–10 $8AC0_{16}$ would be loaded. $16^4 - 8AC0_{16} = 65,536 - 32,520 = 30,016$
$f_{TC4} = 10\ \text{MHz}/30,016 = 333.2\ \text{Hz}$

9–11 See Figure RE–47.

FIGURE RE–46

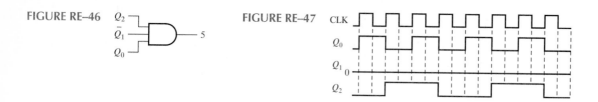

FIGURE RE–47

CHAPTER 10

10–1 See Figure RE–48.

10–2 The state of the register after three additional clock pulses is 0000.

10–3 See Figure RE–49. **10–4** See Figure RE–50.

10–5 See Figure RE–51. **10–6** $f = 1/3\,\mu s = 333$ kHz

FIGURE RE–48

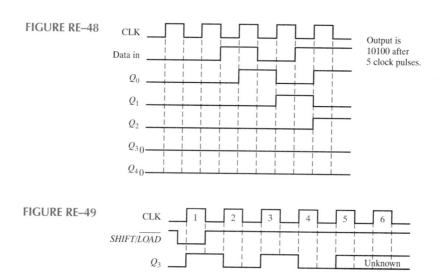

Output is 10100 after 5 clock pulses.

FIGURE RE–49

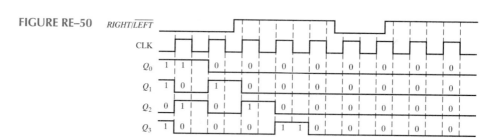

FIGURE RE–50

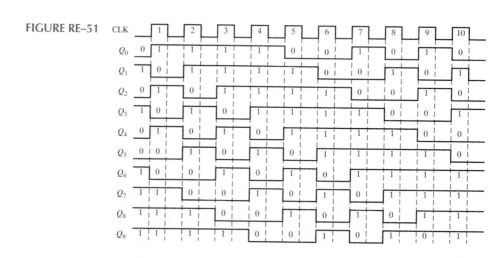

FIGURE RE–51

CHAPTER 11

11–1 `Q0,Q1,Q2,Q3,Q4,Q5,Q6,Q7  PIN 22,21,20,19,18,17,16,15 ISTYPE 'reg,invert';`

11–2
```
'Pin Declaration
Clock, Clear        pin 1,2;
SHLD                pin 3;
D0, D1, D2, D3, D4  pin 4,5,6,7,8 ISTYPE 'reg,buffer';
Q0, Q1, Q2, Q3, Q4  pin 14,15,16,17,18 ISTYPE 'reg,buffer';


Equations
Q0 := D0;
Q1 := Q0 & SHLD # D1 & !SHLD;
Q2 := Q1 & SHLD # D2 & !SHLD;
Q3 := Q2 & SHLD # D3 & !SHLD;
Q4 := Q3 & SHLD # D4 & !SHLD;


[Q0, Q1, Q2, Q3, Q4].CLK = Clock;
[Q0, Q1, Q2, Q3, Q4].AR = !Clear;
```

11–3 Line 1 clears the counter to 000.
Line 2 decrements the counter to 100.
Line 3 decrements the counter to 101.
Line 4 decrements the counter to 111.
Line 5 decrements the counter to 110.
Line 6 decrements the counter to 010.
Line 7 decrements the counter to 011.
Line 8 decrements the counter to 001.
Line 9 decrements the counter to 000.
Line 10 increments the counter to 001.
Line 11 increments the counter to 011.
Line 12 increments the counter to 010.
Line 13 increments the counter to 110.
Line 14 increments the counter to 111.
Line 15 increments the counter to 101.
Line 16 increments the counter to 100.

11–4 The .x. don't-care constant in the first line allows the truth table to show that the outputs are reset when clear is LOW regardless of the states of Q_0, Q_1, and Q_2.

11–5 State A: if !Y then H else B

11–6 The test vector lines trace the paths as shown in Figure RE–52.

FIGURE RE–52

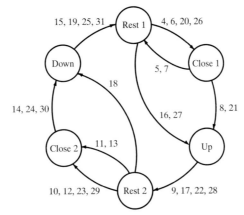

11–7
```
Module      Elevator_display_logic
Title       'Display logic for the two-story elevator system'
"Device declaration
    Elevator_disp    DEVICE    'P16V8'
```

```
"Pin declarations
    MOTION, DIR          PIN 2,3;
    SEGADEG, SEGB, SEGC  PIN 12,13,14 ISTYPE 'com,invert';
    UPARR, DWNARR        PIN 18,19 ISTYPE 'com,invert';

Equations
    UPARR    =  MOTION & !DIR;
    DWNARR   =  MOTION & DIR;
    SEGADEG  =  !MOTION & DIR;
    SEGB     =  !MOTION;
    SEGC     =  !MOTION & !DIR;

Test_vectors
    ([MOTION,DIR] -> [SEGADEG, SEGB, SEGC, UPARR, DWNARR])
    [  0 , 0 ] -> [  1 ,  0 ,  0 ,  1 ,  1  ];
    [  0 , 1 ] -> [  0 ,  1 ,  1 ,  1 ,  1  ];
    [  1 , 0 ] -> [  1 ,  1 ,  1 ,  0 ,  1  ];
    [  1 , 1 ] -> [  1 ,  0 ,  0 ,  1 ,  0  ];

END
```

CHAPTER 12

12–1 $G_3G_2G_1G_0 = 1110$

12–2 Connect eight 256×1 ROMs in parallel to form a 256×8 ROM.

12–3 Sixteen 256×1 ROMs

12–4 See Figure RE–53.

12–5 ROM 1: 0 to 255;   ROM 2: 256 to 511

FIGURE RE–53

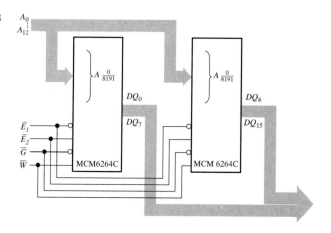

FIGURE RE–54

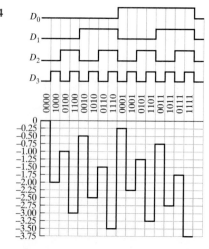

CHAPTER 13

13–1 See Figure RE–54.

13–2 $[1/(2^{16} - 1)]\,100\% = 0.00153\%$

13–3 010, 100, 101, 110, 110, 111, 110, 110, 101, 100, 011, 010, 000, 000, 001, 001, 010, 011, 100, 101, 110, 110, 111, 111

13–4 See Figure RE–55.   **13–5** See Figure RE–56.

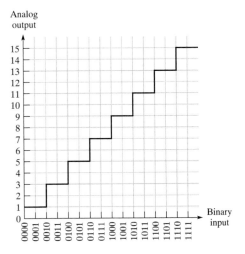

FIGURE RE–55

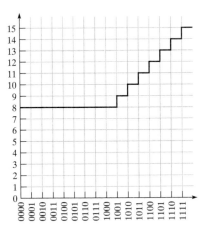

FIGURE RE–56

CHAPTER 14

14–1 $6C4C2_{16}$

CHAPTER 15

15–1 CMOS   **15–2** 10.75 mW

15–3 $I_{T(\text{source})} = 5(20\,\mu A) = 100\,\mu A$
$I_{T(\text{sink})} = 5(-0.4\,\text{mA}) = -2.0\,\text{mA}$

15–4 Fan-out = 20   **15–5** $X = (\overline{AB})(\overline{CD})(\overline{EF})(\overline{GH}) = (\overline{A} + \overline{B})(\overline{C} + \overline{D})(\overline{E} + \overline{F})(\overline{G} + \overline{H})$

15–6 See Figure RE–57.   **15–7** $R_L = 120\,\Omega$ **15–8** $R_p = 594\,\Omega$

FIGURE RE–57

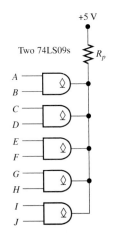

ANSWERS TO SELECTED ODD-NUMBERED PROBLEMS

CHAPTER 1

1. Digital can be transmitted and stored more efficiently and reliably.

3. (a) 11010001   **(b)** 000101010

5. (a) 550 ns   **(b)** 600 ns   **(c)** 2.7 μs   **(d)** 10 V

7. 250 Hz   **9.** 50%   **11.** 8 μs; 1 μs   **13.** AND gate

15. (a) adder   **(b)** multiplier   **(c)** multiplexer   **(d)** comparator

17. 01010000

19. DIP pins go through holes in circuit board. SMT pins connect to surface pads.

21. 7 V   **23.** A collection of circuits interconnected to perform a specified function

25. Enter a new value on the keypad.

CHAPTER 2

1. (a) 1   **(b)** 100   **(c)** 100,000

3. (a) 400; 70; 1   **(b)** 9000; 300; 50; 6   **(c)** 100,000; 20,000; 5000; 0; 0; 0

5. (a) 3   **(b)** 4   **(c)** 7   **(d)** 8   **(e)** 9   **(f)** 12   **(g)** 11   **(h)** 15

7. (a) 51.75   **(b)** 42.25   **(c)** 65.875   **(d)** 120.625
(e) 92.65625   **(f)** 113.0625   **(g)** 90.625   **(h)** 127.96875

9. (a) 5 bits   **(b)** 6 bits   **(c)** 6 bits   **(d)** 7 bits
(e) 7 bits   **(f)** 7 bits   **(g)** 8 bits   **(h)** 8 bits

11. (a) 1010   **(b)** 10001   **(c)** 11000   **(d)** 110000
(e) 111101   **(f)** 1011101   **(g)** 1111101   **(h)** 10111010

13. (a) 1111   **(b)** 10101   **(c)** 11100   **(d)** 100010
(e) 101000   **(f)** 111011   **(g)** 1000001   **(h)** 1001001

15. (a) 100   **(b)** 100   **(c)** 1000   **(d)** 1101   **(e)** 1110   **(f)** 11000

17. (a) 1001   **(b)** 1000   **(c)** 100011
(d) 110110   **(e)** 10101001   **(f)** 10110110

19. (a) 010   **(b)** 001   **(c)** 0101   **(d)** 00101000   **(e)** 0001010   **(f)** 11110

21. (a) 00011101   **(b)** 11010101   **(c)** 01100100   **(d)** 11111011

23. (a) 00001100   **(b)** 10111100   **(c)** 01100101   **(d)** 10000011

25. (a) -102 **(b)** $+116$ **(c)** -64

27. (a) 00110000   **(b)** 00011101   **(c)** 11101011   **(d)** 100111110

29. (a) 11000101   **(b)** 11000000

31. 100111001010

33. (a) 00111000   **(b)** 01011001   **(c)** 101000010100
(d) 010111001000   **(e)** 0100000100000000   **(f)** 1111101100010111
(g) 1000101010011101

35. (a) 35   **(b)** 146   **(c)** 26   **(d)** 141
(e) 243   **(f)** 235   **(g)** 1474   **(h)** 1792

37. (a) 60_{16} **(b)** $10B_{16}$ **(d)** $1BA_{16}$

39. (a) 10   **(b)** 23   **(c)** 46   **(d)** 52   **(e)** 67
(f) 367   **(g)** 115   **(h)** 532   **(i)** 4085

41. (a) 001011 **(b)** 101111 **(c)** 001000001
(d) 011010001 **(e)** 101100000 **(f)** 100110101011
(g) 001011010111001 **(h)** 100101110000000 **(i)** 001000000010001011

43. (a) 00010000 **(b)** 00010011 **(c)** 00011000 **(d)** 00100001
(e) 00100101 **(f)** 00110110 **(g)** 01000100 **(h)** 01010111
(i) 01101001 **(j)** 10011000 **(k)** 000100100101 **(l)** 000101010110

45. (a) 000100000100 **(b)** 000100101000 **(c)** 000100110010
(d) 000101010000 **(e)** 000110000110 **(f)** 001000010000
(g) 001101011001 **(h)** 010101000111 **(i)** 0001000001010001

47. (a) 80 **(b)** 237 **(c)** 346 **(d)** 421 **(e)** 754
(f) 800 **(g)** 978 **(h)** 1683 **(i)** 9018 **(j)** 6667

49. (a) 00010100 **(b)** 00010010 **(c)** 00010111 **(d)** 00010110
(e) 01010010 **(f)** 000100001001 **(g)** 000110010101 **(h)** 0001001001101001

51. The Gray code makes only one bit change at a time when going from one number in the sequence to the next.

53. (a) 1100 **(b)** 00011 **(c)** 10000011110

55. (a) 0 **(b)** 6 **(c)** 4 **(d)** 13 **(e)** 49 **(f)** 52

57. 48 65 6C 6C 6F 2E 20 48 6F 77 20 61 72 65 20 79 6F 75 3F

59. (b)

61. (a) 110100100 **(b)** 000001001 **(c)** 111111110

63. (a) 00100101 (25 BCD) **(b)** 00011001 (25 binary)
(c) 00001111 (15) **(d)** 0110000110010010 (24,975)
(e) 0110000110100001 (24,990)

CHAPTER 3

1. See Figure P–1. **3.** See Figure P–2. **5.** See Figure P–3.
7. See Figure P–4. **9.** See Figure P–5. **11.** See Figure P–6.

FIGURE P–1
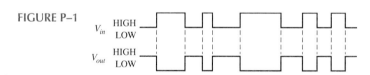

FIGURE P–2
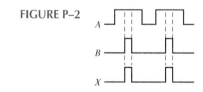

FIGURE P–3
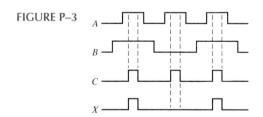

FIGURE P–4
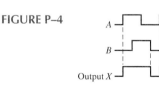

FIGURE P–5
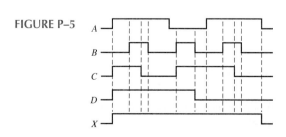

FIGURE P–6
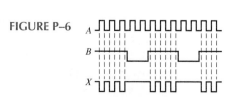

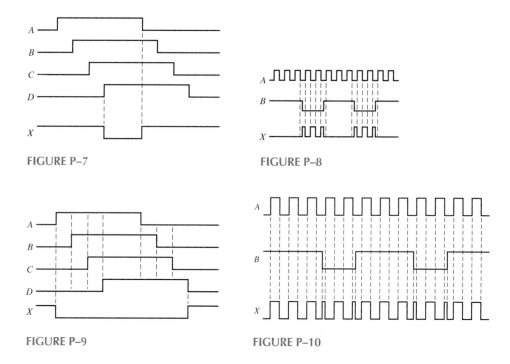

FIGURE P–7

FIGURE P–8

FIGURE P–9

FIGURE P–10

13. See Figure P–7.   **15.** See Figure P–8.   **17.** See Figure P–9.

19. $XOR = A\overline{B} + \overline{A}B$; $OR = A + B$ **21.** See Figure P–10.

23. CMOS   **25.** $t_{PLH} = 4.3$ ns; $t_{PHL} = 10.5$ ns   **27.** 20 mW

29. The gates in parts (b), (c), (e) are faulty.

31. **(a)** defective output (stuck LOW or open)   **(b)** Pin 4 input or pin 6 output internally open.

33. The seatbelt input to the AND gate is open.

35. See Figure P–11.

FIGURE P–11

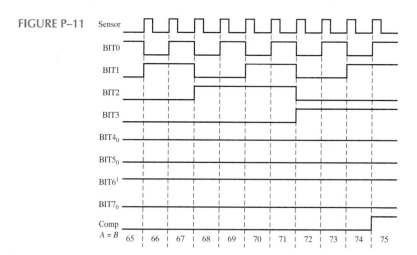

37. BIT5 input stuck at a 1 or open which could be caused by an open input to the corresponding XOR gate or BIT6 input stuck at a 0 due to short.

39. See Figure P–12.

41. Add an inverter to the enable input line of the AND gate.

43. See Figure P–13.

45. The valve and conveyor line must be taken from the NAND gate output. The register line still comes from inverter.

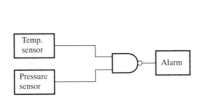

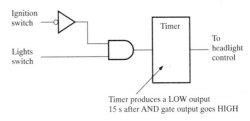

Timer produces a LOW output
15 s after AND gate output goes HIGH

FIGURE P–12   **FIGURE P–13**

CHAPTER 4

1. $X = A + B + C + D$ **3.** $X = \overline{A} + \overline{B} + \overline{C}$

5. **(a)** $AB = 1$ when $A = 1, B = 1$ **(b)** $A\overline{B}C = 1$ when $A = 1, B = 0, C = 1$

  (c) $A + B = 0$ when $A = 0, B = 0$ **(d)** $\overline{A} + B + \overline{C} = 0$ when $A = 1, B = 0, C = 1$

  (e) $\overline{A} + \overline{B} + C = 0$ when $A = 1, B = 1, C = 0$ **(f)** $\overline{A} + B = 0$ when $A = 1, B = 0$

  (g) $A\overline{B}\,\overline{C} = 1$ when $A = 1, B = 0, C = 0$

7. **(a)** Commutative   **(b)** Commutative   **(c)** Distributive

9. **(a)** $\overline{AB}$ **(b)** $A + \overline{B}$ **(c)** $\overline{A}\,\overline{B}\,\overline{C}$ **(d)** $\overline{A} + \overline{B} + \overline{C}$

  (e) $\overline{A} + \overline{B}\,\overline{C}$ **(f)** $\overline{A} + \overline{B} + \overline{C} + \overline{D}$ **(g)** $(\overline{A} + \overline{B})(\overline{C} + \overline{D})$ **(h)** $\overline{AB} + \overline{CD}$

11. **(a)** $(\overline{A} + \overline{B} + \overline{C})(\overline{E} + \overline{F} + \overline{G})(\overline{H} + \overline{I} + \overline{J})(\overline{K} + \overline{L} + \overline{M})$

  (b) $\overline{A}\overline{B}\overline{C}(\overline{C} + \overline{D}) + BC = \overline{A}\overline{B}\overline{C} + BC$ **(c)** $\overline{A}\,\overline{B}\,\overline{C}\,\overline{D}\,\overline{E}\,\overline{F}\,\overline{G}\,\overline{H}$

13. **(a)** $X = ABCD$ **(b)** $X = AB + C$ **(c)** $X = \overline{\overline{A}\overline{B}}$ **(d)** $X = (A + B)C$

15. See Figure P–14.

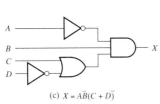

(a) $X = AB\overline{} + A\overline{B}$

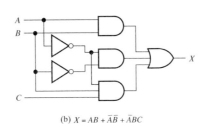

(b) $X = AB + \overline{A}\overline{B} + \overline{A}BC$

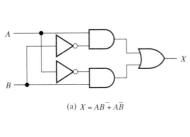

(c) $X = A\overline{B}(C + \overline{D})$

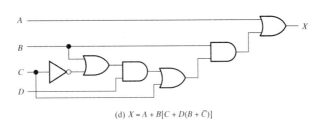

(d) $X = A + B[C + D(B + \overline{C})]$

FIGURE P–14

17. **(a)** A **(b)** AB **(c)** C **(d)** A **(e)** $\overline{A}C + \overline{B}C$

19. **(a)** $BD + BE + \overline{D}F$ **(b)** $\overline{A}\,\overline{B}C + \overline{A}\,\overline{B}D$ **(c)** B **(d)** $AB + CD$ **(e)** ABC

21. **(a)** $A\overline{B} + AC + BC$ **(b)** $AC + \overline{B}C$ **(c)** $AB + AC$

23. **(a)** Domain: A, B, C

  Standard SOP: $A\overline{B}C + A\overline{B}\,\overline{C} + ABC + \overline{A}BC$

  (b) Domain: A, B, C

  Standard SOP: $ABC + A\overline{B}C + \overline{A}\,\overline{B}C$

  (c) Domain: A, B, C

  Standard SOP: $ABC + AB\overline{C} + A\overline{B}C$

25. (a) $101 + 100 + 111 + 011$ **(b)** $111 + 101 + 001$ **(c)** $111 + 110 + 101$

27. (a) $(A + B + C)(A + B + \overline{C})(A + \overline{B} + C)(\overline{A} + \overline{B} + C)$

    (b) $(A + B + C)(A + \overline{B} + C)(A + \overline{B} + \overline{C})(\overline{A} + B + C)(\overline{A} + \overline{B} + C)$

    (c) $(A + B + C)(A + B + \overline{C})(A + \overline{B} + C)(A + \overline{B} + \overline{C})(\overline{A} + B + C)$

29. (a) See Table P–1.   **(b)** See Table P–2.

31. (a) See Table P–3.   **(b)** See Table P–4.

33. (a) See Table P–5.   **(b)** See Table P–6.

35. See Figure P–15.   **37.** See Figure P–16.

TABLE P–1

A	B	C	X
0	0	0	0
0	0	1	0
0	1	0	1
0	1	1	0
1	0	0	0
1	0	1	1
1	1	0	0
1	1	1	1

TABLE P–2

X	Y	Z	Q
0	0	0	1
0	0	1	1
0	1	0	0
0	1	1	1
1	0	0	0
1	0	1	1
1	1	0	1
1	1	1	0

TABLE P–3

A	B	C	X
0	0	0	1
0	0	1	0
0	1	0	1
0	1	1	1
1	0	0	0
1	0	1	1
1	1	0	1
1	1	1	0

TABLE P–4

W	X	Y	Z	Q
0	0	0	0	1
0	0	0	1	1
0	0	1	0	1
0	0	1	1	1
0	1	0	0	0
0	1	0	1	1
0	1	1	0	1
0	1	1	1	0
1	0	0	0	1
1	0	0	1	1
1	0	1	0	1
1	0	1	1	1
1	1	0	0	0
1	1	0	1	1
1	1	1	0	1
1	1	1	1	1

TABLE P–5

A	B	C	X
0	0	0	0
0	0	1	0
0	1	0	0
0	1	1	1
1	0	0	1
1	0	1	1
1	1	0	1
1	1	1	1

TABLE P–6

A	B	C	D	X
0	0	0	0	1
0	0	0	1	0
0	0	1	0	1
0	0	1	1	1
0	1	0	0	0
0	1	0	1	0
0	1	1	0	0
0	1	1	1	0
1	0	0	0	1
1	0	0	1	0
1	0	1	0	0
1	0	1	1	1
1	1	0	0	1
1	1	0	1	1
1	1	1	0	1
1	1	1	1	1

FIGURE P–15

AB＼C	0	1
00	000	001
01	010	011
11	110	111
10	100	101

FIGURE P–16

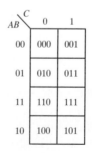

AB＼C	0	1
00	$\overline{A}\,\overline{B}\,\overline{C}$	$\overline{A}\,\overline{B}C$
01	$\overline{A}B\overline{C}$	$\overline{A}BC$
11	$AB\overline{C}$	ABC
10	$A\overline{B}\,\overline{C}$	$A\overline{B}C$

39. (a) No simplification   **(b)** AC **(c)** $\overline{D}\,\overline{F} + E\overline{F}$

41. (a) $AB + AC$

(b) $A + BC$

(c) $B\overline{C}D + A\overline{C}D + BC\overline{D} + AC\overline{D}$

(d) $A\overline{B} + CD$

43. $\overline{B} + C$ **45.** $\overline{A}\,\overline{B}\,\overline{C}D + C\overline{D} + BC + A\overline{D}$

47. (a) $(A + \overline{B} + C + \overline{D})(\overline{A} + B + \overline{C} + D)(\overline{A} + \overline{B} + \overline{C} + \overline{D})$

(b) $(W + \overline{Z})(W + X)(\overline{Y} + \overline{Z})(X + \overline{Y})$

49. $(A + C + D)(A + \overline{B} + C)(\overline{A} + B + \overline{D})(B + \overline{C} + \overline{D})(\overline{A} + \overline{B} + \overline{C} + D)$

51. $X = \overline{A}\,\overline{B}\,\overline{C}\,\overline{D}\,\overline{E} + \overline{A}BCD\,\overline{E} + ABC\,\overline{D}\,\overline{E} + \overline{A}\,BDE + \overline{A}BD\overline{E} + \overline{A}\,\overline{C}DE + AB\overline{C}D$

53. LED. LEDs emit light, LCDs do not.   **55.** One less inverter and six fewer gates.

57. Add an inverter to the output of the OR gate in each of the segment logic circuits.

59. See Figure P–17.

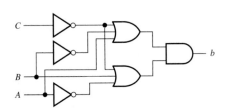

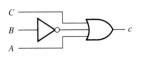

segment $b = (\overline{C} + B + \overline{A})(\overline{C} + \overline{B} + A)$ segment $c = C + \overline{B} + A$

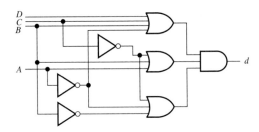

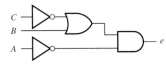

segment $d = (D + C + B + \overline{A})(\overline{C} + B + A)(\overline{C} + \overline{B} + \overline{A})$ segment $e = \overline{A}(\overline{C} + B)$

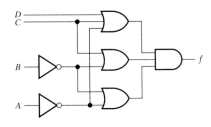

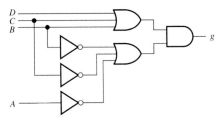

segment $f = (D + C + \overline{A})(C + \overline{B})(\overline{B} + \overline{A})$ segment $g = (D + C + B)(\overline{C} + \overline{B} + \overline{A})$

FIGURE P–17

CHAPTER 5

1. See Figure P–18.

3. (a) $X = ABB$ **(b)** $X = AB + B$ **(c)** $X = \overline{A} + B$

(d) $X = (A + B) + AB$ **(e)** $X = \overline{\overline{\overline{A}\,\overline{B}}\,\overline{B}C}$ **(f)** $X = (A + B)(\overline{B} + C)$

FIGURE P–18

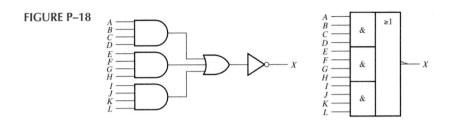

5. (a)

A	B	X
0	0	0
0	1	0
1	0	0
1	1	1

(b)

A	B	X
0	0	0
0	1	1
1	0	0
1	1	1

(c)

A	B	X
0	0	1
0	1	1
1	0	0
1	1	1

(d)

A	B	X
0	0	0
0	1	1
1	0	1
1	1	1

(e)

A	B	C	X
0	0	0	0
0	0	1	0
0	1	0	0
0	1	1	0
1	0	0	0
1	0	1	0
1	1	0	0
1	1	1	0

(f)

A	B	C	X
0	0	0	0
0	0	1	0
0	1	0	0
0	1	1	1
1	0	0	1
1	0	1	1
1	1	0	0
1	1	1	1

7. $X = \overline{A\bar{B} + \bar{A}B} = (\bar{A} + B)(A + \bar{B})$ **9.** See Figure P–19.

11. See Figure P–20.   **13.** $X = AB$

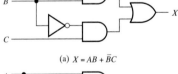

(a) $X = AB + \bar{B}C$

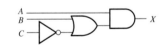

(b) $X = A(B + \bar{C})$

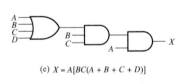

(c) $X = A\bar{B} + AB$

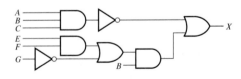

(d) $X = \overline{ABC} + B(EF + \bar{G})$

(e) $X = A[BC(A + B + C + D)]$

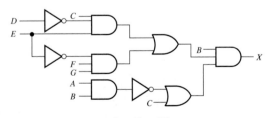

(f) $X = B(C\bar{D}E + \bar{E}FG)(\overline{AB} + C)$

FIGURE P–19

FIGURE P–20

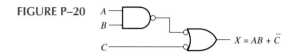

$X = AB + \bar{C}$

15. (a) No simplification (b) No simplification
    (c) $X = A$ (d) $X = \bar{A} + \bar{B} + \bar{C} + EF + \bar{G}$
    (e) $X = ABC$ (f) $X = BC\bar{D}E + \bar{A}B\bar{E}FG + BC\bar{E}FG + \bar{A}B\bar{C}DE$

17. (a) $X = AC + AD + BC + BD$ (b) $X = \bar{A}CD + \bar{B}CD$
    (c) $X = ABD + CD + E$ (d) $X = \bar{A} + B + D$ (e) $X = ABD + \bar{C}D + \bar{E}$
    (f) $X = \bar{A}\,\bar{C} + \bar{A}\,\bar{D} + \bar{B}\,\bar{C} + \bar{B}\,\bar{D} + \bar{E}\,\bar{G} + \bar{E}\,\bar{H} + \bar{F}\,\bar{G} + \bar{F}\,\bar{H}$

19. See Figure P–21. 21. See Figure P–22. 23. See Figure P–23.

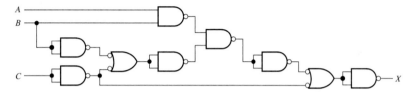

FIGURE P–21

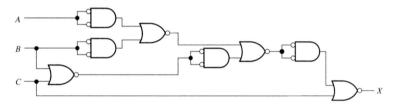

FIGURE P–22

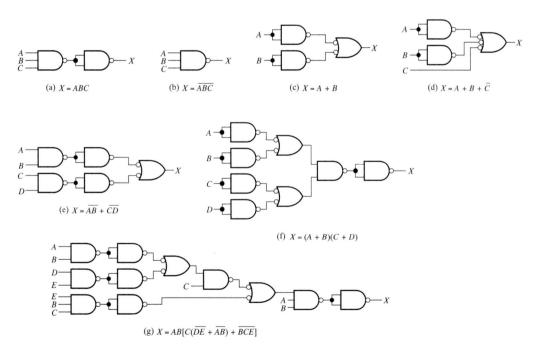

(a) $X = ABC$

(b) $X = \overline{ABC}$

(c) $X = A + B$

(d) $X = A + B + \bar{C}$

(e) $X = \overline{AB} + \overline{CD}$

(f) $X = (A + B)(C + D)$

(g) $X = AB[C(\overline{DE} + \overline{AB}) + \overline{BCE}]$

FIGURE P–23

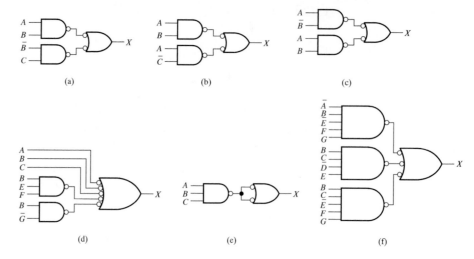

FIGURE P–24

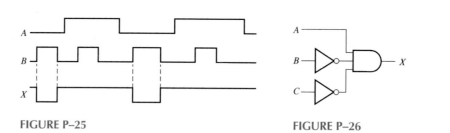

FIGURE P–25

FIGURE P–26

25. See Figure P–24.   **27.** $X = A + \bar{B}$; see Figure P–25.   **29.** $X = AB\,\bar{C}$; see Figure P–26.

31. The output pulse width is greater than the specified minimum.

33. $X = ABC + D\bar{E}$. Since X is the same as the G_3 output, either G_1 or G_2 has failed, with its output stuck LOW.

35. Pin 10 of the right 7400 is shorted to ground.

37. **(a)** $X = E$. See Figure P–27.   **(b)** $X = E$ **(c)** $X = E$ **39.** See Figure P–28.

41. See Figure P–29 for modified M_4 logic. Other motor logic remains the same.

43. $X =$ lamp on, $A =$ front door switch on, $B =$ back door switch on. See Figure P–30.

45. See Figure P–31. Inverters (not shown) are used to convert each HIGH key closure to LOW.

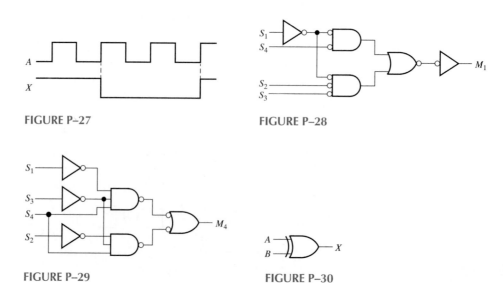

FIGURE P–27

FIGURE P–28

FIGURE P–29

FIGURE P–30

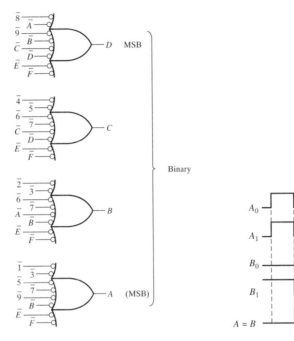

FIGURE P–31

FIGURE P–32

CHAPTER 6

1. (a) $A \oplus B = 0, \Sigma = 1, (A \oplus B)C_{in} = 0, AB = 1, C_{out} = 1$
   (b) $A \oplus B = 1, \Sigma = 0, (A \oplus B)C_{in} = 1, AB = 0, C_{out} = 1$
   (c) $A \oplus B = 1, \Sigma = 1, (A \oplus B)C_{in} = 0, AB = 0, C_{out} = 0$

3. no simplification for sum logic: $C_{out} = AB + BC_{in} + AC_{in}$

5. 11100

7. $\Sigma_1 = 01101110$; $\Sigma_2 = 10110100$; $\Sigma_3 = 01101000$;
   $\Sigma_4 = 00010100$; $\Sigma_5 = 10101010$

9. 225 ns

11. $A = B$ is HIGH when $A_0 = B_0$ and $A_1 = B_1$; see Figure P–32.

13. (a) G_1 output = 1   (b) G_1 output = 1   (c) G_1 output = 1
       G_2 output = 0 G_2 output = 1 G_2 output = 1
       G_3 output = 1 G_3 output = 0 G_3 output = 1
       G_4 output = 0 G_4 output = 0 G_4 output = 1
       G_5 output = 0 G_5 output = 0 G_5 output = 1
       G_6 output = 0 G_6 output = 0 G_6 output = 0
       G_7 output = 1 G_7 output = 0 G_7 output = 0
       G_8 output = 0 G_8 output = 0 G_8 output = 0
       G_9 output = 0 G_9 output = 0 G_9 output = 0
       G_{10} output = 1 G_{10} output = 0 G_{10} output = 0
       G_{11} output = 0 G_{11} output = 0 G_{11} output = 0
       G_{12} output = 0 G_{12} output = 0 G_{12} output = 0
       G_{13} output = 0 G_{13} output = 1 G_{13} output = 0
       G_{14} output = 0 G_{14} output = 0 G_{14} output = 0
       G_{15} output = 0 G_{15} output = 1 G_{15} output = 0
           $A > B$ $A < B$ $A = B$

15. See Figure P–33.

17. $X = A_3A_2\overline{A}_1\overline{A}_0 + \overline{A}_3\overline{A}_2\overline{A}_1A_0 + A_3\overline{A}_2A_1$

19. See Figure P–34.

21. $A_3A_2A_1A_0 = 1011$, invalid BCD

23. **(a)** $0010 \rightarrow 0010_2$ **(b)** $1000 \rightarrow 1000_2$

     (c) $00010011 \rightarrow 1101_2$ **(d)** $00100110 \rightarrow 11010_2$

     (e) $00110011 \rightarrow 100001_2$

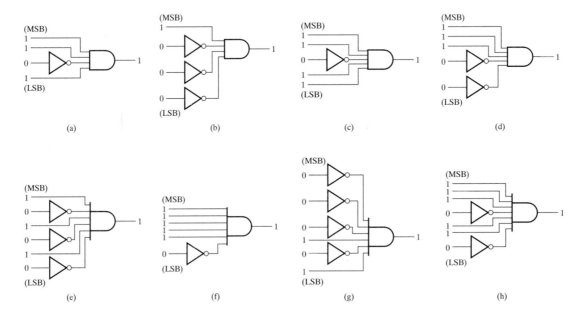

FIGURE P–33

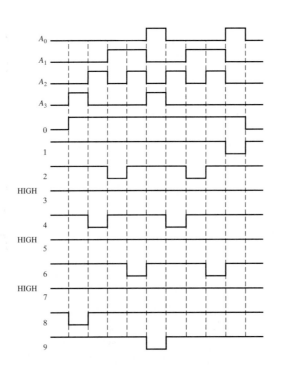

FIGURE P–34

25. (a) 1010000000 Gray
 1100000000 binary
(b) 0011001100 Gray
 0010001000 binary
(c) 1111000111 Gray
 1010000101 binary
(d) 0000000001 Gray
 0000000001 binary
See Figure P–35.

27. See Figure P–36. **29.** See Figure P–37. **31.** See Figure P–38.

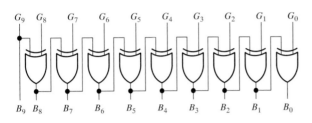

FIGURE P–35

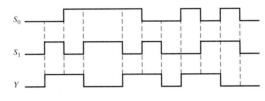

FIGURE P–36

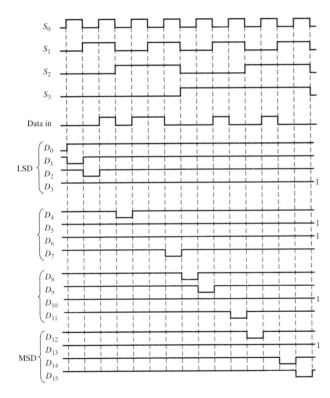

FIGURE P–37

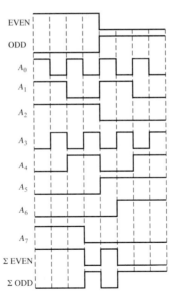

FIGURE P–38

33. (a) OK **(b)** segment g burned out **(c)** Segment b output stuck LOW

35. (a) The A_1 input of the top adder is open:
 All binary values corresponding to a BCD number having a value of 0, 1, 4, 5, 8, or 9 will be off by 2. This will first be seen for a BCD value of 0000 0000.

 (b) The carry out of the top adder is open:
 All values not normally involving an output carry will be off by 32. This will first be seen for a BCD value of 0000 0000.

 (c) The Σ_4 output of the top adder is shorted to ground:
 Same binary values above 1s will be short by 16. The first BCD value to indicate this will be 0001 1000.

 (d) The Σ_3 output of the bottom adder is shorted to ground:
 Every other set of 16 values starting with 16 will be short 16. The first BCD value to indicate this will be 0001 0110.

TABLE P–7

S_2	S_1	S_0	Y	$\bar{Y}$
0	0	0	1	0
0	0	1	0	1
0	1	0	0	1
0	1	1	0	1
1	0	0	0	1
1	0	1	0	1
1	1	0	0	1
1	1	1	0	1

37. 1. Place a LOW on pin 7 (Enable).
2. Apply a HIGH to D_0 and a LOW to D_1 through D_7.
3. Go through the binary sequence on the select inputs and check Y and $\bar{Y}$ according to Table P–7.
4. Repeat the binary sequence of select inputs for each set of data inputs listed in Table P–8. A HIGH on the Y output should occur only for the corresponding combinations of select inputs shown.

TABLE P–8

D_0	D_1	D_2	D_3	D_4	D_5	D_6	D_7	Y	$\bar{Y}$	S_2	S_1	S_0
L	H	L	L	L	L	L	L	1	0	0	0	1
L	L	H	L	L	L	L	L	1	0	0	1	0
L	L	L	H	L	L	L	L	1	0	0	1	1
L	L	L	L	H	L	L	L	1	0	1	0	0
L	L	L	L	L	H	L	L	1	0	1	0	1
L	L	L	L	L	L	H	L	1	0	1	1	0
L	L	L	L	L	L	L	H	1	0	1	1	1

39. Apply a HIGH in turn to each Data input, D_0 through D_7 with LOWs on all the other inputs. For each HIGH applied to a data input, sequence through all eight binary combinations of select inputs ($S_2 S_1 S_0$) and check for a HIGH on the corresponding data output and LOWs on all the other data outputs.

41. Replace each 74LS00 with a 74LS08 AND gate.   **43.** See Figure P–39.

45. See the block diagram in Figure P–40.   **47.** See Figure P–41.   **49.** See Figure P–42.

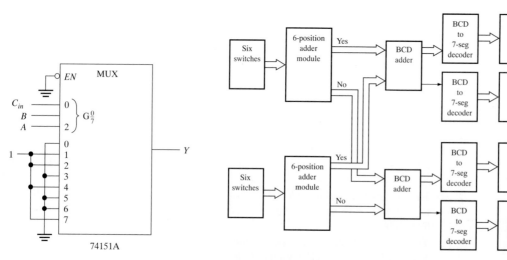

FIGURE P–39   FIGURE P–40

FIGURE P–41

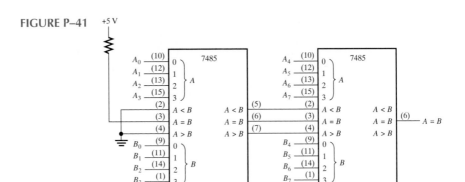

FIGURE P–42

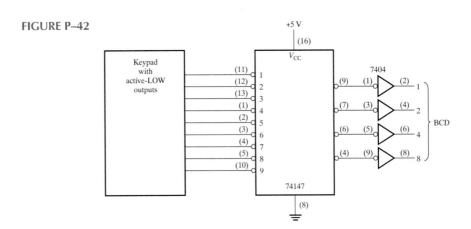

CHAPTER 7

1. 1. Programmable read-only memory (PROM): fixed AND array and programmable OR array
  2. Programmable logic array (PLA): programmable AND and OR arrays
  3. Programmable array logic (PAL): programmable AND and fixed OR arrays
  4. Generic array logic (GAL): reprogrammable AND and fixed OR arrays

3. For $X_1 = \overline{A}BC$, blow: Row 1, Columns 1, 3, 4, 5, 6
  Row 2, Columns 1, 2, 4, 5, 6
  Row 3, Columns 1, 2, 3, 4, 6

  For $X_2 = AB\overline{C}$, blow: Row 4, Columns 2, 3, 4, 5, 6
  Row 5, Columns 1, 2, 4, 5, 6
  Row 6, Columns 1, 2, 3, 4, 5

  For $X_3 = \overline{A}B\overline{C}$, blow: Row 7, Columns 1, 3, 4, 5, 6
  Row 8, Columns 1, 2, 4, 5, 6
  Row 9, Columns 1, 2, 3, 4, 5

5. See Figure P–43.

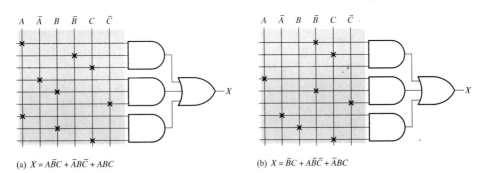

(a) $X = AB\overline{C} + \overline{A}B\overline{C} + ABC$

(b) $X = \overline{B}C + AB\overline{C} + \overline{A}BC$

FIGURE P–43

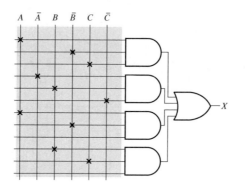 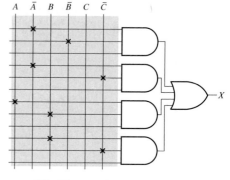

(a) $X = A\bar{B}C + \bar{A}B\bar{C} + A\bar{B} + BC$

(b) $X = \bar{A}\bar{B} + \bar{A}C + AB + B\bar{C}$

FIGURE P–44

7. The programmed polarity output in a PAL works by selectively opening or leaving intact a cell on one input of an XOR gate, causing the gate to invert its other input or not.

9. See Figure P–44.   **11.** 7 available outputs

13. The select inputs determine combinational or registered mode, input or output, inverted or non-inverted output.

15. **(a)** 8   **(b)** 8   **(c)** 16

17. 8 product terms, 15 variables

19. Computer, programmer, compiler

21. The ispGAL22V10 can be programmed in the circuit, whereas the GAL22V10 must be inserted in a programmer.

23. **(a)** !A & !B & !C;   **(b)** A # B # !C;   **(c)** A & !(B # C);

25. **(a)** X = A & !B & C & !D & E # A & !B & C # !A & !B;

(b) Y = A & B & !C # D & E & F # !G & H & !I # J & !K & L;

(c) Z = (!A # !B # C # D) & (E # !F) # G;

27. TRUTH_TABLE ([A, B] –> [X])
                [0, 0] –> [1];
                [0, 1] –> [0];
                [1, 0] –> [0];
                [1, 1] –> [1];

29. Test_Vectors (BCD –> [a, b, c, d, e, f, g, h, i, j])
                0 –> [1, 0, 0, 0, 0, 0, 0, 0, 0, 0];
                1 –> [0, 1, 0, 0, 0, 0, 0, 0, 0, 0];
                2 –> [0, 0, 1, 0, 0, 0, 0, 0, 0, 0];
                3 –> [0, 0, 0, 1, 0, 0, 0, 0, 0, 0];
                4 –> [0, 0, 0, 0, 1, 0, 0, 0, 0, 0];
                5 –> [0, 0, 0, 0, 0, 1, 0, 0, 0, 0];
                6 –> [0, 0, 0, 0, 0, 0, 1, 0, 0, 0];
                7 –> [0, 0, 0, 0, 0, 0, 0, 1, 0, 0];
                8 –> [0, 0, 0, 0, 0, 0, 0, 0, 1, 0];
                9 –> [0, 0, 0, 0, 0, 0, 0, 0, 0, 1];

31. Yes, the design requires only 2 inputs and 8 outputs.

33. Module   Decimal_to_BCD_Priority_Encoder
    Title   'Decimal-to-BCD Priority Encoder in a GA16V8'

    Encoder_1   device   'P16V8';

    I0, I1, I2, I3, I4   pin 1,2,3,4,5;
    I5, I6, I7, I8, I9   pin 6,7,8,9,12;
    Q0, Q1, Q2, Q3   pin 16,17,18,19;

    Equations
        Q0 = I1 & !I2 & !I4 & !I6 & !I8 # I3 & !I4 & !I6 & !I8 # I5 & !I6 & !I8 #
            I7 & !I8 # I9;
        Q1 = I2 & !I4 & !I5 & !I8 & !I9 # I3 & !I4 & !I5 & !I8 & !I9 #
            I6 & !I8 & !I9 # I7 & !I8 & !I9;
        Q2 = I4 & !I8 & !I9 # I5 & !I8 & !I9 # I6 & !I8 & !I9 # I7 & !I8 & !I9;
        Q3 = I8 # I9;

CHAPTER 8

1. See Figure P–45. **3.** See Figure P–46. **5.** See Figure P–47.
7. See Figure P–48. **9.** See Figure P–49. **11.** See Figure P–50.
13. See Figure P–51. **15.** See Figure P–52.

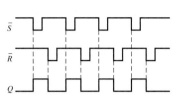

FIGURE P–45

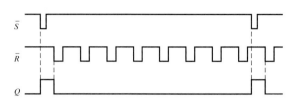

FIGURE P–46

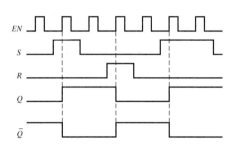

FIGURE P–47

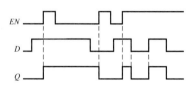

FIGURE P–48

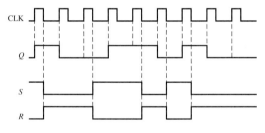

FIGURE P–49

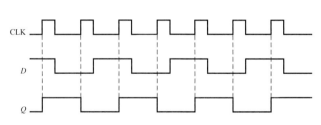

FIGURE P–50

FIGURE P–51
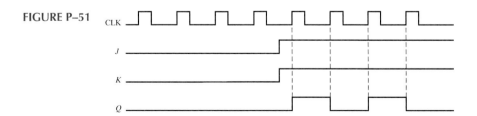

FIGURE P–52
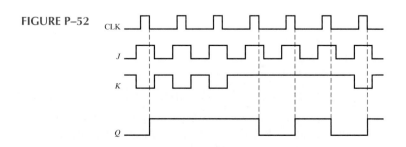

17. See Figure P–53.   **19.** See Figure P–54.   **21.** See Figure P–55.
23. See Figure P–56.   **25.** See Figure P–57.   **27.** 14.9 MHz   **29.** 150 mA, 750 mW
31. divide-by-2; see Figure P–58.   **33.** 4.62 μs
35. $C_1 = 1\,\mu$F, $R_1 = 227$ kΩ (use 220 kΩ). See Figure P–59.

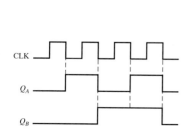

FIGURE P–53

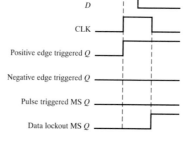

FIGURE P–54

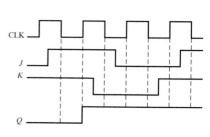

FIGURE P–55

FIGURE P–56

FIGURE P–57

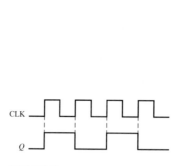

FIGURE P–58

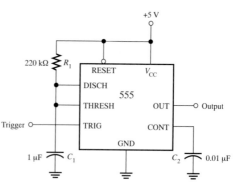

FIGURE P–59

37. $R_1 = 18$ kΩ, $R_2 = 9.1$ kΩ.
39. The wire from pin 6 to pin 10 and ground wire are reversed on the protoboard.
41. $\overline{CLR}$ shorted to ground.   **43.** See Figure P–60. Delays not shown.
45. See Figure P–61.
    4 s: $C_1 = 1\,\mu$F, $R_1 = 3.63$ MΩ (use 3.9 MΩ)
    25 s: $C_1 = 2.2\,\mu$F, $R_1 = 10.3$ MΩ (use 10 MΩ)
47. See Figure P–62.
49. Changes include (1) a 15 s timer (2) decoding to trigger the 15 s timer (3) a decoder to incorporate the turn signal state.

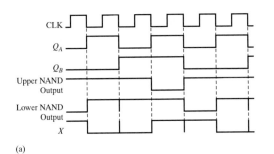

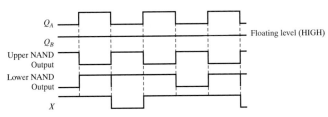

(a)

(b) Same as (a)

Floating level (HIGH)

(c)

(d) X = LOW if Q_B = 1; $X = \bar{Q}_A$ if Q_B = 0

FIGURE P–60

FIGURE P–61

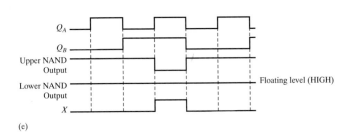

FIGURE P–62

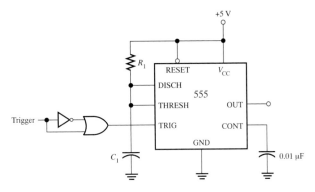

CHAPTER 9

1. See Figure P–63.

3. Worst-case delay is 24 ns; it occurs when all flip-flops change state from 011 to 100 or from 111 to 000.

5. 8 ns

FIGURE P–63

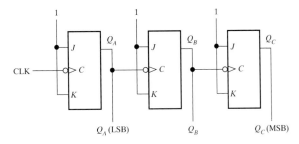

7. Initially, each flip-flop is reset.

At CLK1: $J_0 = K_0 = 1$ Therefore Q_0 goes to a 1.
$J_1 = K_1 = 0$ Therefore Q_1 remains a 0.
$J_2 = K_2 = 0$ Therefore Q_2 remains a 0.
$J_3 = K_3 = 0$ Therefore Q_3 remains a 0.

At CLK2: $J_0 = K_0 = 1$ Therefore Q_0 goes to a 0.
$J_1 = K_1 = 1$ Therefore Q_1 goes to a 1.
$J_2 = K_2 = 0$ Therefore Q_2 remains a 0.
$J_3 = K_3 = 0$ Therefore Q_3 remains a 0.

At CLK3: $J_0 = K_0 = 1$ Therefore Q_0 goes to a 1.
$J_1 = K_1 = 0$ Therefore Q_1 remains a 1.
$J_2 = K_2 = 0$ Therefore Q_2 remains a 0.
$J_3 = K_3 = 0$ Therefore Q_3 remains a 0.

A continuation of this procedure for the next seven clock pulses will show that the counter progresses through the BCD sequence.

9. See Figure P–64.

11. See Figure P–65.

13. See Figure P–66.

15. The sequence is 0000, 1111, 1110, 1101, 1010, 0101. The counter "locks up" in the 1010 and 0101 states and alternates between them.

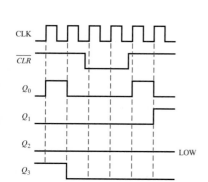

FIGURE P–64

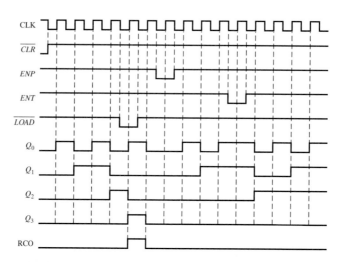

FIGURE P–65

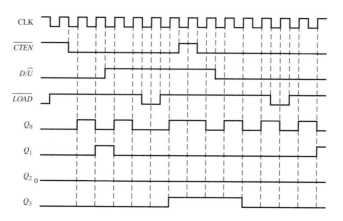

FIGURE P–66

17. See Figure P–67.   **19.** See Figure P–68.

21. See Figure P–69.   **23.** See Figure P–70.

25. CLK2, output 0; CLK4, outputs 2, 0; CLK6, output 4; CLK8, outputs 6, 4, 0; CLK10, output 8; CLK12, outputs 10, 8; CLK14, output 12; CLK16, outputs 14, 12, 8

27. A glitch of the AND gate output occurs on the 111 to 000 transition. Eliminate by ANDing $\overline{CLK}$ with counter outputs (strobe) or use Gray code.

29. Hours tens: 0001
Hours units: 0010
Minutes tens: 0000
Minutes units: 0001
Seconds tens: 0000
Seconds units: 0010

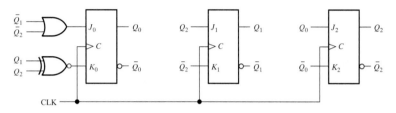

FIGURE P–67

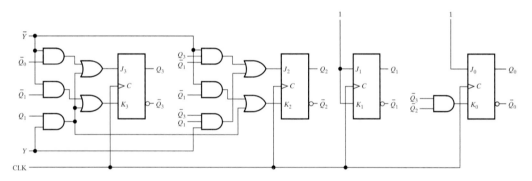

FIGURE P–68

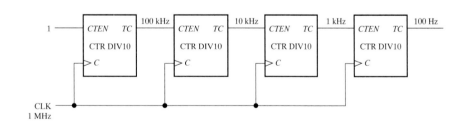

FIGURE P–69

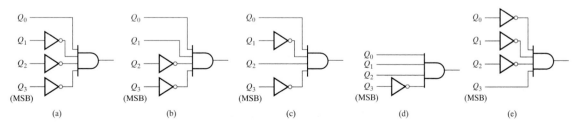

FIGURE P–70

31. 64

33. (a) Q_0 and Q_1 will not change from their initial state.

    (b) normal operation except Q_0 floating

    (c) Q_0 waveform is normal; Q_1 remains in initial state.

    (d) normal operation

    (e) The counter will not change from its initial state.

35. The K input of FF1 must be connected to ground rather than to the J input. Check for a wiring error.

37. Q_0 input to AND gate open and acting as a HIGH

39. See Table P–9.

TABLE P–9

Stage	Open	Loaded Count	f_{out}
1	0	63C1	250.006 Hz
1	1	63C2	250.012 Hz
1	2	63C4	250.025 Hz
1	3	63C8	250.050 Hz
2	0	63D0	250.100 Hz
2	1	63E0	250.200 Hz
2	2	63C0	250 Hz
2	3	63C0	250 Hz
3	0	63C0	250 Hz
3	1	63C0	250 Hz
3	2	67C0	256.568 Hz
3	3	6BC0	263.491 Hz
4	0	73C0	278.520 Hz
4	1	63C0	250 Hz
4	2	63C0	250 Hz
4	3	E3C0	1.383 kHz

41. The decode 6 gate interprets count 4 as a 6 (0110) and clears the counter back to 0 (actually 0010 since Q_1 is open). The apparent sequence of the tens portion of the counter is 0010, 0011, 0010, 0011, 0110.

43. See Figure P–71.

45. Increase the $R_{EXT}C_{EXT}$ time constant of the 25 s one-shot by 2.4 times.

47. See Figure P–72.

49. See Figure P–73.

51. See Figure P–74.

53. See Figure P–75.

FIGURE P–71

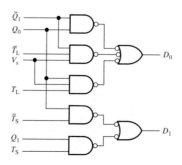

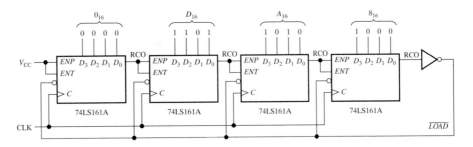

FIGURE P–72

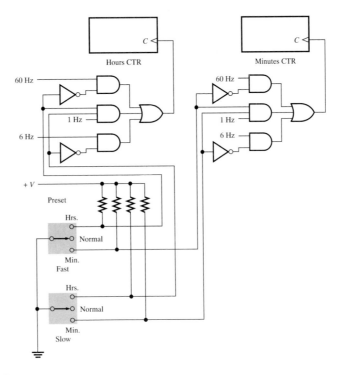

FIGURE P–73

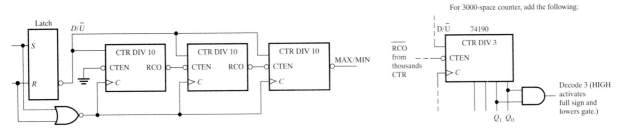

FIGURE P–74

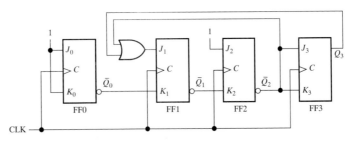

FIGURE P–75

FIGURE P–76

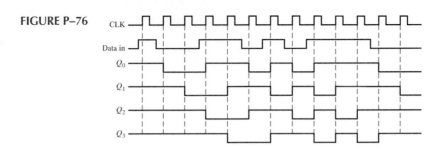

CHAPTER 10

1. Shift registers store binary data.   **3.** See Figure P–76.

5. Initially: 101001111000
   CLK1:   010100111100
   CLK2:   001010011110
   CLK3:   000101001111
   CLK4:   000010100111
   CLK5:   100001010011
   CLK6:   110000101001
   CLK7:   111000010100
   CLK8:   011100001010
   CLK9:   001110000101
   CLK10: 000111000010
   CLK11: 100011100001
   CLK12: 110001110000

7. See Figure P–77.   **9.** See Figure P–78.   **11.** See Figure P–79.

FIGURE P–77

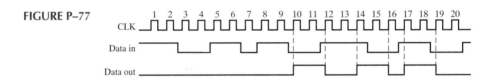

FIGURE P–78

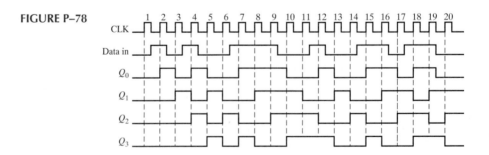

FIGURE P–79

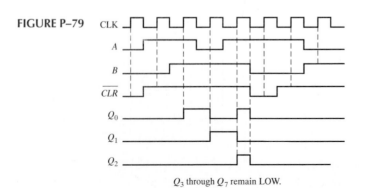

Q_3 through Q_7 remain LOW.

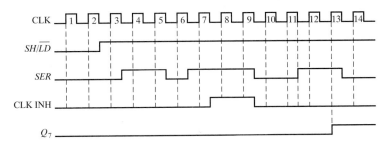

FIGURE P–80

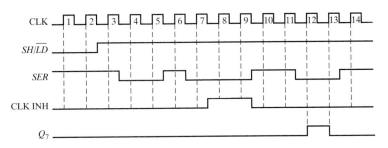

FIGURE P–81

FIGURE P–82

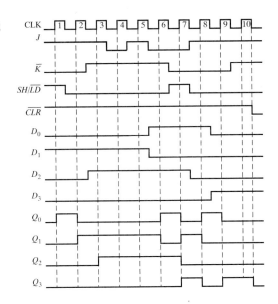

13. See Figure P–80.   **15.** See Figure P–81.   **17.** See Figure P–82.

19. Initially (76):   01001100

CLK1:	10011000	left
CLK2:	01001100	right
CLK3:	00100110	right
CLK4:	00010011	right
CLK5:	00100110	left
CLK6:	01001100	left
CLK7:	00100110	right
CLK8:	01001100	left
CLK9:	00100110	right
CLK10:	01001100	left
CLK11:	10011000	left

21. See Figure P–83.

23. **(a)** 3   **(b)** 5   **(c)** 7   **(d)** 8   **(e)** 10   **(f)** 12   **(g)** 18

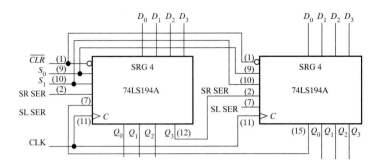

FIGURE P–83

FIGURE P–84

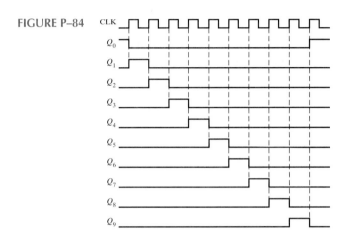

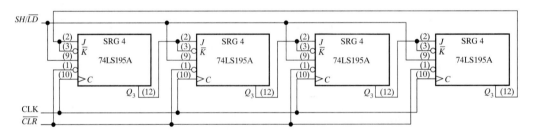

FIGURE P–85

25. See Figure P–84.

27. See Figure P–85.　　**29.** An incorrect code may be produced.　　**31.** D_3 input open

33. **(a)** No clock at switch closure because of faulty NAND (negative-OR) gate or one-shot; open clock (C) input to key code register; open $SH/\overline{LD}$ input to key code register

　　(b) Diode in third row open; Q_2 output of ring counter open

　　(c) The NAND (negative-OR) gate input connected to the first column is open or shorted.

　　(d) The "2" input to the column encoder is open.

35. **(a)** Contents of data output register remain constant.

　　(b) Contents of both registers do not change.

　　(c) Third stage output of data output register remains HIGH.

　　(d) Clock generator is disabled after each pulse by the flip-flop being continuously SET and then RESET.

37. shift register A: 1001
　　shift register B: 11100

39. Gate G_1 has an open or shorted output or an open input. One-shot A is faulty.

41. Control flip-flop: 7476
Clock generator: 555
Counter: 74LS163
Data input register: 74LS164
Data output register: 74LS199
One-shot: 74121

43. See Figure P–86. **45.** See Figure P–87.

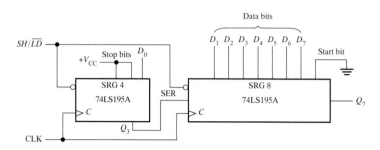

FIGURE P–86

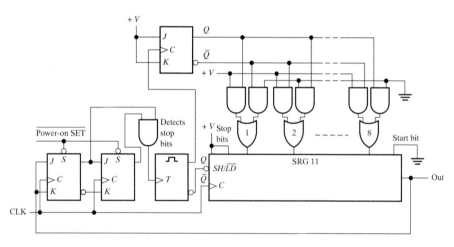

FIGURE P–87

CHAPTER 11

1. Combinational with active-LOW outputs, combinational with active-HIGH outputs, registered with active-LOW outputs, and registered with active-HIGH outputs

3. Multiplexer

5. (a) Pin 15 is a combinational output associated with the variable X.

(b) Pin 20 is a registered output associated with the variable Q_2.

(c) Pin 18 is an active-LOW combinational output associated with the variable A.

(d) Pin 21 is an active-HIGH registered output associated with the variable Q_1.

7. Q1.CLK = Clock;

9.
```
Test_vectors ([Clock,Clear,Din]->[Q0,Q1,Q2,Q3,Q4,Q5,Q6,Qout])
              [ .x. ,  0  ,.x.]->[ 0, 0, 0, 0, 0, 0, 0,  0 ];
              [ .c. ,  1  , 1 ]->[ 1, 0, 0, 0, 0, 0, 0,  0 ];
              [ .c. ,  1  , 0 ]->[ 0, 1, 0, 0, 0, 0, 0,  0 ];
              [ .c. ,  1  , 1 ]->[ 1, 0, 1, 0, 0, 0, 0,  0 ];
              [ .c. ,  1  , 0.]->[ 0, 1, 0, 1, 0, 0, 0,  0 ];
              [ .c. ,  1  , 1 ]->[ 1, 0, 1, 0, 1, 0, 0,  0 ];
              [ .c. ,  1  , 0 ]->[ 0, 1, 0, 1, 0, 1, 0,  0 ];
              [ .c. ,  1  , 1 ]->[ 1, 0, 1, 0, 1, 0, 1,  0 ];
              [ .c. ,  1  , 0 ]->[ 0, 1, 0, 1, 0, 1, 0,  1 ];
```

11. The test vectors are identical to those for the GAL22V10 in Problem 9.

13. The 8-bit register in Problem 12 cannot be implemented in a GAL16V8 because of insufficient input pins (9 are required; the GAL16V8 has 8).

15. 8-bit counter

17. For an up-only Gray code counter the logic equations are (using ABEL notation)

Q0 =   !Q2 & !Q1 & !Q0 # !Q2 & !Q1 & Q0 # Q2 & Q1 & !Q0 # Q2 & Q1 & Q0
  = > !Q2 & !Q1 # Q2 & Q1

Q1 =   !Q2 & !Q1 & Q0 # !Q2 & Q1 & Q0 # !Q2 & Q1 & !Q0 # Q2 & Q1 & !Q0
  => !Q2 & Q0 # Q1 & !Q0

Q2 =   !Q2 & Q1 & !Q0 # Q2 & Q1 & !Q0 # Q2 & Q1 & Q0 # Q2 & !Q1 & Q0
  => Q1 & !Q0 # Q2 & Q0

An ABEL input file for a GAL22V10 circuit is then

```
Module      3-bit_Gray_code_up_counter

Title       '3-bit Gray code up counter in a GAL22V10'

"Device declaration
      Counter    DEVICE    'P22V10'

"Pin Declarations
      Clock,Clear    PIN 1,2;
      Q0,Q1,Q2       PIN 21,22,23 ISTYPE 'reg,buffer';

Equations
      [Q0,Q1,Q2].CLK = Clock;
      [Q0,Q1,Q2].AR = !Clear;
      Q0 := !Q2.FB&!Q1.FB # Q2.FB&Q1.FB;
      Q1 := !Q2.FB&Q0.FB # Q1.FB&!Q0.FB;
      Q2 := Q1.FB&Q0.FB # Q2.FB&Q0.FB;

Test_vectors ([Clock,Clear]->[Q2,Q1,Q0])
                [ .x. ,  0  ]->[ 0 , 0 , 0 ];
                [ .c. ,  1  ]->[ 0 , 0 , 1 ];
                [ .c. ,  1  ]->[ 0 , 1 , 1 ];
                [ .c. ,  1  ]->[ 0 , 1 , 0 ];
                [ .c. ,  1  ]->[ 1 , 1 , 0 ];
                [ .c. ,  1  ]->[ 1 , 1 , 1 ];
                [ .c. ,  1  ]->[ 1 , 0 , 1 ];
                [ .c. ,  1  ]->[ 1 , 0 , 0 ];
                [ .c. ,  1  ]->[ 0 , 0 , 0 ];

END
```

19.
```
Module      3-bit_Gray_code_up_counter

Title       '3-bit Gray code up counter in a GAL22V10'

"Device declaration
      Counter    DEVICE    'P22V10'

"Pin Declarations
      Clock,Clear    PIN 1,2;
      Q0,Q1,Q2       PIN 21,22,23 ISTYPE 'reg,buffer';
      COUNT = [Q2,Q1,Q0];
      A      = ^B000;
      B      = ^B001;
      C      = ^B011;
      D      = ^B010;
      E      = ^B110;
      F      = ^B111;
      G      = ^B101;
      H      = ^B100;

Equations
      COUNT.CLK = Clock;
      COUNT.AR = !Clear;
```

```
State_diagram COUNT
    State A : GOTO B;
    State B : GOTO C;
    State C : GOTO D;
    State D : GOTO E;
    State E : GOTO F;
    State F : GOTO G;
    State G : GOTO H;
    State H : GOTO A;

Test_vectors ([Clock,Clear]->[Q2,Q1,Q0])
            [ .x. ,  0  ]->[ 0 , 0 , 0 ];
            [ .c. ,  1  ]->[ 0 , 0 , 1 ];
            [ .c. ,  1  ]->[ 0 , 1 , 1 ];
            [ .c. ,  1  ]->[ 0 , 1 , 0 ];
            [ .c. ,  1  ]->[ 1 , 1 , 0 ];
            [ .c. ,  1  ]->[ 1 , 1 , 1 ];
            [ .c. ,  1  ]->[ 1 , 0 , 1 ];
            [ .c. ,  1  ]->[ 1 , 0 , 0 ];
            [ .c. ,  1  ]->[ 0 , 0 , 0 ];

END
```

21.
```
Module      3-bit_Gray_code_up_counter

Title       '3-bit Gray code up counter in a GAL16V8'

"Device declaration
    Counter   DEVICE    'P16V8'

"Pin Declarations
    Clock,Clear   PIN 1,11;
    Q0,Q1,Q2      PIN 17,18,19 ISTYPE 'reg,buffer';

Equations
    [Q0,Q1,Q2].CLK = Clock;
    [Q0,Q1,Q2].AR  = !Clear;

Truth_table ([Clock,Q2,Q1,Q0]:>[Q2,Q1,Q0])
            [ .c. , 0, 0, 0]:>[ 0, 0, 1];
            [ .c. , 0, 0, 1]:>[ 0, 1, 1];
            [ .c. , 0, 1, 1]:>[ 0, 1, 0];
            [ .c. , 0, 1, 0]:>[ 1, 1, 0];
            [ .c. , 1, 1, 0]:>[ 1, 1, 1];
            [ .c. , 1, 1, 1]:>[ 1, 0, 1];
            [ .c. , 1, 0, 1]:>[ 1, 0, 0];
            [ .c. , 1, 0, 0]:>[ 0, 0, 0];

Test_vectors ([Clock,Clear]->[Q2,Q1,Q0])
            [ .x. ,  0  ]->[ 0 , 0 , 0 ];
            [ .c. ,  1  ]->[ 0 , 0 , 1 ];
            [ .c. ,  1  ]->[ 0 , 1 , 1 ];
            [ .c. ,  1  ]->[ 0 , 1 , 0 ];
            [ .c. ,  1  ]->[ 1 , 1 , 0 ];
            [ .c. ,  1  ]->[ 1 , 1 , 1 ];
            [ .c. ,  1  ]->[ 1 , 0 , 1 ];
            [ .c. ,  1  ]->[ 1 , 0 , 0 ];
            [ .c. ,  1  ]->[ 0 , 0 , 0 ];

END
```

23. **(a)** CLOSE1 state: System goes to REST1 state.

(b) REST1 state: System stays in REST1 state.

(c) UP state: System stays in UP state until "ARRIVE" signal.

(d) REST2 state: System stays in REST2 state.

(e) CLOSE2 state: System goes to REST2 state.

(f) DOWN state: System stays in DOWN state until "ARRIVE" signal.

25.
```
Module      Elevator_state_control

Title       'Control logic for a two-story elevator system'

"Device declaration
    Elevator_cont   DEVICE    'P22V10'
```

```
"Pin declarations
    CLK, !OE                PIN 1,13;
    REQ1,REQ2               PIN 2,3;
    UP,DOWN,OPEN,ARRIVE     PIN 4,5,6,7;
    L1PB,L1PB_              PIN 20,21 ISTYPE 'com,buffer';
    L2PB,L2PB_              PIN 22,23 ISTYPE 'com,buffer';
    DOOR,MOTION,DIR         PIN 14,15,16 ISTYPE 'reg,buffer';

"State definitions
    CONSTATE  =  [DOOR,MOTION,DIR];
    REST1     =  ^B000;
    OPEN1     =  ^B100;
    UP        =  ^B110;
    REST2     =  ^B001;
    OPEN2     =  ^B101;
    DOWN      =  ^B111;

Equations
    CONSTATE.CLK  =  CLK;
    CONSTATE.AR   =  OE;
    L1PB          =  REQ1 # !DIR.FB&OPEN # DIR.FB&DOWN # !L1PB_;
    L1PB_         =  !DOOR.FB&!MOTION.FB&!DIR.FB # !L1PB;
    L2PB          =  REQ2 # DIR.FB&OPEN # !DIR.FB&UP # !L2PB_;
    L2PB_         =  !DOOR.FB&!MOTION.FB&DIR.FB # !L2PB;

State_diagram CONSTATE
    State REST1  : if (L2PB) then UP else CLOSE1;
    State CLOSE1 : if (L2PB) then UP else if (L1PB) then REST1 else CLOSE1;
    State UP     : if (ARRIVE) then REST2 else UP;
    State REST2  : if (L1PB) then DOWN else CLOSE2;
    State CLOSE2 : if (L1PB) then DOWN else if (L2PB) then REST2 else CLOSE2;
    State DOWN   : if (ARRIVE) then REST1 else DOWN;

Test_vectors
    ([CLK,REQ1,REQ2,UP,DOWN,OPEN,ARRIVE]->[DOOR,MOTION,DIR])
    [.c., 0 , 0 ,0, 0 , 0 , 1 ]->[ .x. ,  .x. , .x.];
    [.c., 1 , 0 ,0, 0 , 0 , 1 ]->[ .x. ,  .x. , .x.];
    [.c., 1 , 0 ,0, 0 , 0 , 1 ]->[ 0 ,   0 , 0 ];
    [.c., 0 , 0 ,0, 0 , 0 , 0 ]->[ 1 ,   0 , 0 ];
    [.c., 1 , 0 ,0, 0 , 0 , 0 ]->[ 0 ,   0 , 0 ];
    [.c., 0 , 0 ,0, 0 , 0 , 1 ]->[ 1 ,   0 , 0 ];
    [.c., 0 , 0 ,0, 0 , 1 , 0 ]->[ 0 ,   0 , 0 ];
    [.c., 0 , 1 ,0, 0 , 0 , 0 ]->[ 1 ,   1 , 0 ];
    [.c., 0 , 0 ,0, 0 , 0 , 1 ]->[ 0 ,   0 , 1 ];
    [.c., 0 , 0 ,0, 0 , 0 , 0 ]->[ 1 ,   0 , 1 ];
    [.c., 0 , 1 ,0, 0 , 0 , 0 ]->[ 0 ,   0 , 1 ];
    [.c., 0 , 0 ,0, 0 , 0 , 0 ]->[ 1 ,   0 , 1 ];
    [.c., 0 , 0 ,0, 0 , 1 , 0 ]->[ 0 ,   0 , 1 ];
    [.c., 1 , 0 ,0, 0 , 0 , 0 ]->[ 1 ,   1 , 1 ];
    [.c., 0 , 0 ,0, 0 , 0 , 1 ]->[ 0 ,   0 , 0 ];
    [.c., 0 , 0 ,1, 0 , 0 , 0 ]->[ 1 ,   1 , 0 ];
    [.c., 0 , 0 ,0, 0 , 0 , 1 ]->[ 0 ,   0 , 1 ];
    [.c., 0 , 0 ,0, 1 , 0 , 0 ]->[ 1 ,   1 , 1 ];
    [.c., 0 , 0 ,0, 0 , 0 , 1 ]->[ 0 ,   0 , 0 ];
    [.c., 0 , 0 ,0, 0 , 0 , 0 ]->[ 1 ,   0 , 0 ];
    [.c., 0 , 1 ,0, 0 , 0 , 0 ]->[ 1 ,   1 , 0 ];
    [.c., 0 , 0 ,0, 0 , 0 , 1 ]->[ 0 ,   0 , 1 ];
    [.c., 0 , 0 ,0, 0 , 0 , 0 ]->[ 1 ,   0 , 1 ];
    [.c., 1 , 0 ,0, 0 , 0 , 0 ]->[ 1 ,   1 , 1 ];
    [.c., 0 , 0 ,0, 0 , 0 , 1 ]->[ 0 ,   0 , 0 ];
    [.c., 0 , 0 ,0, 0 , 0 , 0 ]->[ 1 ,   0 , 0 ];
    [.c., 0 , 0 ,1, 0 , 0 , 0 ]->[ 1 ,   1 , 0 ];
    [.c., 0 , 0 ,0, 0 , 0 , 1 ]->[ 0 ,   0 , 1 ];
    [.c., 0 , 0 ,0, 0 , 0 , 0 ]->[ 1 ,   0 , 1 ];
    [.c., 0 , 0 ,0, 1 , 0 , 0 ]->[ 1 ,   1 , 1 ];
    [.c., 0 , 0 ,0, 0 , 0 , 1 ]->[ 0 ,   0 , 0 ];

END
```

27. No, Ten inputs are required and the GAL16V8 has only eight.

29. The first state will be skipped if there is a vehicle on the side street. The third state will always be skipped.

31. Module   8-bit_SISO_shift_register

    Title   '8-bit SISO shift register in a GAL16V8'

    "Device declaration
        Register   DEVICE   'P16V8'

    "Pin declarations
        Clock,Clear   PIN 1,11;
        Din   PIN 2;
        Q0,Q1,Q2,Q3   PIN 12,13,14,15 ISTYPE 'reg,buffer';
        Q4,Q5,Q6,Qout   PIN 16,17,18,19 ISTYPE 'reg,buffer';

    Equations
        [Q0,Q1,Q2,Q3,Q4,Q5,Q6,Qout].CLK = Clock;
        [Q0,Q1,Q2,Q3,Q4,Q5,Q6,Qout].AR = !Clear;
        Q0 := Din;
        Q1 := Q0.FB;
        Q2 := Q1.FB;
        Q3 := Q2.FB;
        Q4 := Q3.FB;
        Q5 := Q4.FB;
        Q6 := Q5.FB;
        Qout := Q6.FB;

```
    Test_vectors ([Clock,Clear,Din]->[Q0,Q1,Q2,Q3,Q4,Q5,Q6,Qout])
                 [ .x. ,  0  ,.x.]->[ 0 , 0 , 0 , 0 , 0 , 0 , 0 ,  0 ];
                 [ .c. ,  1  , 1 ]->[ 1 , 0 , 0 , 0 , 0 , 0 , 0 ,  0 ];
                 [ .c. ,  1  , 0 ]->[ 0 , 1 , 0 , 0 , 0 , 0 , 0 ,  0 ];
                 [ .c. ,  1  , 1 ]->[ 1 , 0 , 1 , 0 , 0 , 0 , 0 ,  0 ];
                 [ .c. ,  1  , 0 ]->[ 0 , 1 , 0 , 1 , 0 , 0 , 0 ,  0 ];
                 [ .c. ,  1  , 1 ]->[ 1 , 0 , 1 , 0 , 1 , 0 , 0 ,  0 ];
                 [ .c. ,  1  , 0 ]->[ 0 , 1 , 0 , 1 , 0 , 1 , 0 ,  0 ];
                 [ .c. ,  1  , 1 ]->[ 1 , 0 , 1 , 0 , 1 , 0 , 1 ,  0 ];
                 [ .c. ,  1  , 0 ]->[ 0 , 1 , 0 , 1 , 0 , 1 , 0 ,  1 ];
```

    END

33. Module   4-bit_Excess-3_code_counter

    Title   '4-bit Excess-3 code counter in a GAL22V10'

    "Device declaration
        Counter   DEVICE   'P22V10'

    "Pin declarations
        Clock,Reset   PIN 1,2;
        Q0,Q1,Q2,Q3   PIN 12,13,14,15 ISTYPE 'reg,buffer';
        COUNT = [Q3,Q2,Q1,Q0];

    Equations
        COUNT.CLK = Clock;

```
    Truth_table([Clock,Reset, COUNT ]:>[ COUNT ])
                [ .C. ,  0  ,  .X.  ]:>[ ^B0011 ];
                [ .C. ,  1  , ^B0000 ]:>[ ^B0011 ];
                [ .C. ,  1  , ^B0001 ]:>[ ^B0011 ];
                [ .C. ,  1  , ^B0010 ]:>[ ^B0011 ];
                [ .C. ,  1  , ^B0011 ]:>[ ^B0100 ];
                [ .C. ,  1  , ^B0100 ]:>[ ^B0101 ];
                [ .C. ,  1  , ^B0101 ]:>[ ^B0110 ];
                [ .C. ,  1  , ^B0110 ]:>[ ^B0111 ];
                [ .C. ,  1  , ^B0111 ]:>[ ^B1000 ];
                [ .C. ,  1  , ^B1000 ]:>[ ^B1001 ];
                [ .C. ,  1  , ^B1001 ]:>[ ^B1010 ];
                [ .C. ,  1  , ^B1010 ]:>[ ^B1011 ];
                [ .C. ,  1  , ^B1011 ]:>[ ^B1100 ];
                [ .C. ,  1  , ^B1100 ]:>[ ^B0000 ];
                [ .C. ,  1  , ^B1101 ]:>[ ^B0011 ];
                [ .C. ,  1  , ^B1110 ]:>[ ^B0011 ];
                [ .C. ,  1  , ^B1111 ]:>[ ^B0011 ];
```

```
Test_vectors([Clock,Reset ]->[ COUNT])
            [ .C. ,  0   ]->[^B0011];
            [ .C. ,  1   ]->[^B0100];
            [ .C. ,  1   ]->[^B0101];
            [ .C. ,  1   ]->[^B0110];
            [ .C. ,  1   ]->[^B0111];
            [ .C. ,  1   ]->[^B1000];
            [ .C. ,  1   ]->[^B1001];
            [ .C. ,  1   ]->[^B1010];
            [ .C. ,  1   ]->[^B1011];
            [ .C. ,  1   ]->[^B1100];
            [ .C. ,  1   ]->[^B0011];

        END
```

CHAPTER 12

1. (a) ROM   (b) RAM

3. *Address bus* provides for transfer of address code to memory for accessing any memory location in any order for a read or write operation. *Data bus* provides for transfer of data between the microprocessor and the memory or I/O. *Control bus* provides for transfer of control signals between microprocessor, memory, and I/O.

5. See Table P–10.   **7.** See Figure P–88.

TABLE P–10

Inputs		Outputs			
A_1	A_0	O_3	O_2	O_1	O_0
0	0	0	1	0	1
0	1	1	0	0	1
1	0	1	1	1	0
1	1	0	0	1	0

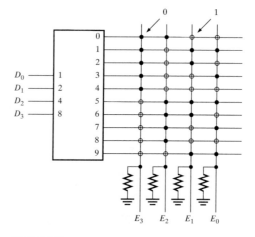

FIGURE P–88

9. Blown links: 1–17, 19–23, 25–31, 34, 37, 38, 40–47, 53, 55, 58, 61, 62, 63, 65, 67, 69

11. (a) RESET (0)   (b) RESET (0)   (c) SET (1)

13. 512 rows × 128 8-bit columns

15. A SRAM stores bits in latches indefinitely as long as power is applied. A DRAM stores bits in capacitors which must be refreshed periodically to retain the data.

17. Use eight 16k × 4 RAMs with sixteen address lines. Three of the address lines are decoded to enable the selected memory chips. Four data lines go to each chip.

19. 8 bits, 64k words; 4 bits, 256k words

21. lowest address: $FC0_{16}$
highest address: FFF_{16}

23. See Figure P–89.

25. 9

27. Checksum content is in error.

29. (a) ROM2   (b) ROM1   (c) All ROMs

31. 010

33. PROM will retain code when power is off. The code in PROM cannot be readily changed.

35. See Figure P–90.

GLOSSARY

ABEL Advanced Boolean Expression Language. A software compiler language for PLD programming; a type of hardware description language.

Acceptor A receiving device on a bus.

Access time The time from the application of a valid address code on the inputs until the appearance of valid output data.

Addend In addition, the number that is added to another number called the augend.

Adder A logic circuit used to add two binary numbers.

Address The location of a given storage cell or group of cells in a memory.

Address allocation The arrangement of memory space (addresses) in a computer.

Adjacency Characteristic of cells in a Karnaugh map in which there is a single-variable change from one cell to another cell next to it on any of its four sides.

Alphanumeric Consisting of numerals, letters, and other characters.

ALU Arithmetic logic unit; the part of a microprocessor that processes information by using arithmetic and logic operations.

Amplitude In a pulse waveform, the height or maximum value of the pulse as measured from its low level.

Analog Being continuous or having continuous values, as opposed to having a set of discrete values.

Analog-to-digital (A/D) conversion The process of converting an analog signal to digital form.

Analog-to-digital converter (ADC) A device used to convert an analog signal to a sequence of digital codes.

AND A basic logic operation in which a true (HIGH) output occurs only if all the input conditions are true (HIGH).

AND gate A logic circuit that produces a HIGH output only when all of the inputs are HIGH.

ANSI American National Standards Institute.

Architecture The internal functional arrangement of the elements that give a device its unique characteristics.

Array In a PLD, a matrix formed by rows of product-term lines and columns of input lines with a programmable cell at each junction.

ASCII American Standard Code for Information Interchange; the most widely used alphanumeric code.

Assembly language A form of computer program language in which statements expressed in mnemonics represent the instructions.

Associative law In addition (ORing) and multiplication (ANDing) of three or more variables, the order in which the variables are grouped makes no difference.

Astable Having no stable state. An astable multivibrator oscillates between two quasi-stable states.

Asynchronous Having no fixed time relationship.

Asynchronous counter A type of counter in which each stage is clocked from the output of the preceding stage.

Augend In addition, the number to which the addend is added.

Base One of the three regions in a bipolar junction transistor.

Base address The beginning address of a segment of memory.

BCD Binary coded decimal; a digital code in which each of the decimal digits, 0 through 9, is represented by a group of four bits.

Bidirectional Having two directions. In a bidirectional shift register, the stored data can be shifted right or left.

Binary Having two values or states; describes a number system that has a base of two and utilizes 1 and 0 as its digits.

Bipolar Having two opposite charge carriers within the transistor structure.

Bistable Having two stable states. Flip-flops and latches are bistable multivibrators.

Bit A binary digit, which can be either a 1 or 0.

Bit time The interval of time occupied by a single bit in a sequence of bits; the period of the clock.

BJT Bipolar junction transistor; a semiconductor device used for switching or amplification. A BJT has two junctions, the base-emitter junction and the base-collector junction.

Boolean addition In Boolean algebra, the OR operation.

Boolean algebra The mathematics of logic circuits.

Boolean expression An algebraic expression used to analyze the operation of logic circuits.

Boolean multiplication In Boolean algebra, the AND operation.

Buffer A circuit that prevents loading of an input or output.

Bus A set of interconnections that interface components of a digital system or systems based on standardized specifications.

Bus arbitration The process that prevents two sources from using a bus at the same time.

Byte A group of eight bits.

Capacity   The total number of data units that a memory can store.

Carry   The digit generated when the sum of two binary digits exceeds 1.

Carry generation   The process of producing an output carry in a full-adder when both input bits are 1s.

Carry propagation   The process of rippling an input carry to become the output carry in a full-adder when either or both of the input bits are 1s and the input carry is a 1.

Cascade   To connect "end-to-end" as when several counters are connected from the terminal count output of one counter to the enable input of the next counter.

Cascading   Connecting the output of one device to the input of a similar device, allowing one device to drive another in order to expand the operational capability.

CCD   Charge-coupled device; a type of semiconductor memory that stores data in the form of charge packets.

CD-ROM   A disk storage device on which data are permanently stored and can only be read.

Cell   An area on a Karnaugh map that represents a unique combination of variables in product form; a single storage element in a memory; the programmable element in a PLD array.

Character   A symbol, letter, or numeral.

Circuit   An arrangement of electrical and/or electronic components interconnected in such a way as to perform a specified function.

Clear   An asynchronous input used to reset a flip-flop; to place a register or counter in the state in which it contains all 0s.

Clock   The basic timing signal in a digital system.

CMOS   Complementary metal oxide semiconductor; a type of transistor circuit which uses field-effect transistors.

Code   A set of bits arranged in a unique pattern and used to represent such information as numbers, letters, and other symbols.

Code converter   A digital device that converts one type of coded information into another coded form.

Collector   One of the three regions in a bipolar transistor.

Combinational logic   A combination of logic gates interconnected to produce a specified Boolean function with no storage or memory capability; sometimes called *combinatorial logic.*

Commutative law   In addition (ORing) and multiplication (ANDing) of two variables, the order the variables are ORed or ANDed makes no difference.

Comparator   A digital circuit that compares the magnitudes of two quantities and produces an output indicating the relationship of the quantities.

Compiler   Software that translates from high-level language that uses words or symbols, such as ABEL or CUPL, into low-level machine language (1s and 0s).

Complement   The inverse or opposite of a number; in Boolean algebra, the inverse function, expressed with a bar over the variable. The complement of a 1 is a 0, and vice versa.

Complementation   Inversion. LOW is the complement of HIGH, and 0 is the complement of 1.

Controller   An instrument that can specify each of the other instruments on the bus as either a talker or a listener for the purpose of data transfer.

Counter   A digital circuit capable of counting electronic events, such as pulses, by progressing through a sequence of binary states.

CPU   Central processing unit; a major component of all computers that controls the internal operations and processes data.

CUPL   Compiler for Universal Programmable Logic. A software compiler language for PLD programming; a type of hardware description language.

Current sinking   The action of a circuit in which it accepts current into its output from a load.

Current sourcing   The action of a circuit in which it sends current out of its output and into a load.

DAT   Digital audiotape.

Data   Information in numeric, alphabetic, or other form.

Data selector   A circuit that selects data from several inputs in a time sequence and places them on the output; also called a multiplexer.

Data sheet   A document that specifies parameter values and operating conditions for an integrated circuit or other device.

DCE   Data communications equipment.

Decade   Characterized by ten states or values.

Decade counter   A digital counter having ten states.

Decimal   Describes a number system with a base of ten.

Decoder   A digital circuit (device) that converts coded information into another (familiar) form.

Decrement   To decrease the binary state of a counter by one.

D flip-flop   A type of bistable multivibrator in which the output follows the state of the D input.

Demultiplexer (DEMUX)   A circuit (digital device) that switches digital data from one input line to several output lines in a specified time sequence.

Dependency notation   A notational system for logic symbols that specifies input and output relationships, thus fully defining a given function; an integral part of ANSI/IEEE Std. 91–1984.

Difference   The result of a subtraction.

Digit   A symbol used to express a quantity.

Digital   Related to digits or discrete quantities; having a set of discrete values as opposed to continuous values.

Digital-to-analog (D/A) conversion   The process of converting a sequence of digital codes to an analog form.

Digital-to-analog converter (DAC)   A device in which information in digital form is converted to analog form.

Diode   A semiconductor device that conducts current in only one direction.

Distributive law   The law that states that ORing several variables and then ANDing the result with a single variable is equivalent to ANDing the single variable with each of the several variables and the ORing the product.

FIGURE P–89

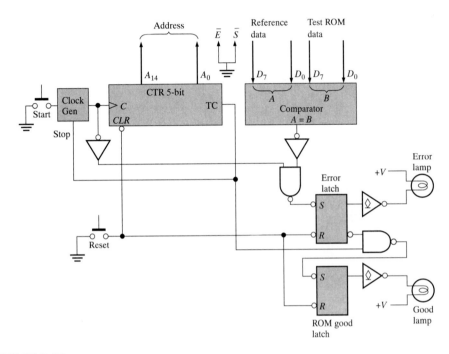

FIGURE P–90

37. To accommodate a 5-bit entry code, shift register B must be loaded with five 0s instead of four. Connect pin 14 of the 74LS165 to ground rather than to the HIGH.

CHAPTER 13

1. 1000, 1010, 1101, 1110, 1110, 1110, 1100, 1001, 0110, 0100, 0010, 0001, 0001, 0011, 0101, 1000, 1011, 1101, 1110, 1110, 1110, 1100, 1001, 0110, 0100, 0010.

3. See Figure P–91.   **5.** 330 kΩ   **7.** See Figure P–92.

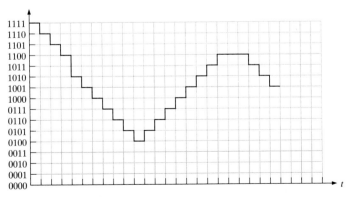

FIGURE P–91

FIGURE P–92

9. **(a)** 14.29%   **(b)** 0.098%   **(c)** 0.00038%

11. 000, 001, 100, 101, 101, 100, 011, 010, 001, 001, 011, 110, 111, 111, 111, 111, 111, 111, 111, 100

13. 0000, 0000, 1110, 1100, 0111, 0110, 0011, 0010, 1100

15. See Table P–11.   **17.** See Figure P–93.

19. See Figure P–94.   **21.** See Figure P–95.   **23.** seven

25. A controller is sending data to two listeners. The first two bytes of data (3F and 41) go to the listener with address 001A. The second two bytes go to the listener with address 001B. The handshaking signals (DAV, NRFD, and NDAC) indicate the data transfer is successful. See Figure P–96.

27. 32.2 kHz   **29.** 0.703°

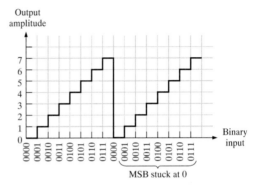

FIGURE P–93

TABLE P–11

SAR	Comment
1000	Greater than V_{in}, reset MSB
0100	Less than V_{in}, keep the 1
0110	Equal to V_{in}, keep the (final state)

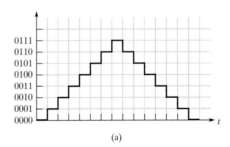

(a)   (b)

FIGURE P–94

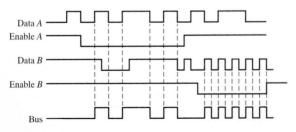

FIGURE P–95

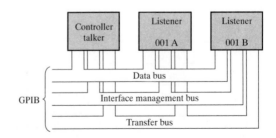

FIGURE P–96

CHAPTER 15

1. No; $V_{OH(min)} < V_{IH(min)}$

3. 0.15 V in HIGH state; 0.25 V in LOW state.

5. Gate C **7.** 16 ns   **9.** Gate C **11.** Yes, G_2

13. (a) on   **(b)** off   **(c)** off   **(d)** off

15. (a) LOW   **(b)** LOW   **(c)** LOW   **17.** See Figure P–97.

19. (a) $R_p = 198 \ \Omega$ **(b)** $R_p = 198 \ \Omega$ **(c)** $R_p = 198 \ \Omega$

21. (a) on   **(b)** off   **(c)** off   **(d)** on

23. See Figure P–98 for one possible circuit.

25. (a) -1.3 mA   **(b)** -2.5 mA   **(c)** $-4 \ \mu A$

27. See Figure P–99.   **29.** 313 Ω

31. (a) ECL   **(b)** HCMOS   **(c)** HCMOS   **(d)** HCMOS

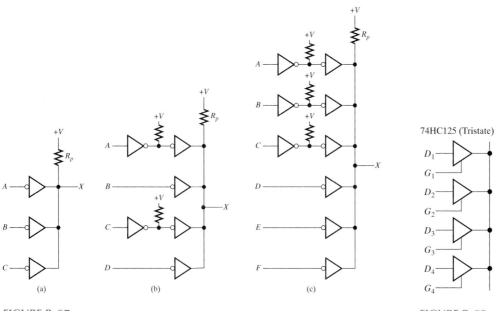

FIGURE P–97

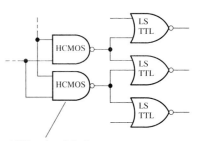

FIGURE P–98

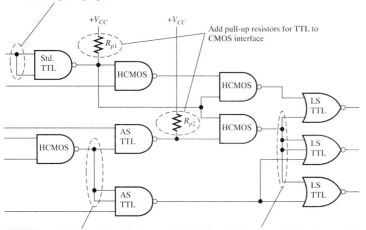

All unused inputs should be tied together. We are assuming a 2-input gate.

Add pull-up resistors for TTL to CMOS interface

HCMOS fan-out of two is not exceeded here because the tied-together TTL NAND inputs act as only one unit load in the LOW state.

HCMOS fan-out of two is exceeded here because it is driving four unit loads. NOR gate inputs that are tied together each act as a unit load. A second CMOS gate in parallel can be used to split the loading as shown below.

Addition of parallel gates. Now each CMOS gate drives only two unit loads.

FIGURE P–99

Dividend In a division operation, the quantity that is being divided.

Divisor In a division operation, the quantity that is divided into the dividend.

DMA Direct memory assess; a technique in which the CPU is bypassed for certain types of data transfers.

Documentation file The information from a computer that documents the final design after the input file has been processed.

Domain All of the variables in a Boolean expression.

"Don't care" A combination of input literals that cannot occur and are used as 1s or 0s on a Karnaugh map.

Drain One of the terminals of a field-effect transistor.

DRAM Dynamic random-access semiconductor memory; a volatile read/write memory.

DTE Data terminal equipment.

Dual-in-line package (DIP) A type of IC package whose leads must pass through holes to the other side of a circuit board.

Dual symbols The negative-AND is the dual symbol for the NOR gate, and the negative-OR is the dual symbol for the NAND gate.

Duty cycle The ratio of pulse width to period expressed as a percentage.

Dynamic memory A memory having cells that tend to lose stored data over a period of time and, therefore, must be refreshed. Typically the storage elements are capacitors.

ECL Emitter-coupled logic; a type of integrated circuit that uses nonsaturating bipolar junction transistors.

E²CMOS Electrically-erasable CMOS (EECMOS). The circuit technology used for the reprogrammable cells in a GAL.

Edge-triggered flip-flop A type of flip-flop in which the data are entered and appear on the output on the same clock edge.

EEPROM Electrically erasable programmable read-only memory.

Emitter One of the three regions in a bipolar junction transistor.

Enable To activate or put into an operational mode.

Encoder A digital circuit (device) that converts information to a coded form.

EPROM Erasable programmable read-only memory.

Error detection The process of detecting bit errors in a digital code.

Even parity The condition of having an even number of 1s in every group of bits.

Excess-3 An unweighted digital code in which each of the decimal digits is represented by a 4-bit code by adding 3 to the decimal digit and converting to binary.

Exclusive-NOR (XNOR) gate A logic circuit in which a LOW output is produced only when the two inputs are at opposite levels.

Exclusive-OR (XOR) A basic logic operation in which a HIGH occurs when the two inputs are at opposite levels.

Exclusive-OR (XOR) gate A logic circuit in which a HIGH output occurs only when the two inputs are at opposite levels.

Execute A CPU process in which an instruction is carried out.

Fall time The time interval between the 90% point and the 10% point on the negative-going edge of a pulse.

Fan-out The number of equivalent gate inputs of the same family series that a logic gate can drive.

Feedback The output voltage or a portion of it that is connected back to the input of a circuit.

FET Field-effect transistor.

Fetch A CPU process in which an instruction is obtained from the memory.

FIFO First in—first out memory.

Flag A bit used to control or indicate the status of an operation in the CPU.

Flash ADC A simultaneous analog-to-digital converter.

Flash memory A nonvolatile read/write random-access semiconductor memory.

Flat pack An SMT package whose leads extend straight out from the body.

Flip-flop A basic storage circuit that can store only one bit at a time; a synchronous bistable device.

Floppy disk A magnetic storage device. Typically, a 3.5 in. flexible disk enclosed in a rigid case or a 5.25 in. flexible Mylar® disk enclosed in a flexible jacket.

Forward bias A voltage polarity condition that allows a semiconductor *pn* junction in a transistor or diode to conduct current.

Frequency The number of pulses in one second for a periodic waveform. The unit of frequency is the hertz.

Full-adder A digital device that adds two bits and an input carry to produce a sum and an output carry.

Fuse The programmable element in certain types of PLDs; also called a fusible link.

GAL Generic Array Logic. A device with a reprogrammable AND array, a fixed OR array, and programmable output logic macrocells.

Gate A logic circuit that performs a specified logic operation, such as AND or OR; one of the three terminals of a field-effect transistor.

GPIB General-purpose interface bus.

Glitch A voltage or current spike of short duration, usually unintentionally produced and unwanted.

Global cell A programmable cell in a PLD array that affects all of the OLMCs when programmed.

Gray code An unweighted digital code characterized by a single bit change between adjacent code numbers in a sequence.

Half-adder A digital device that adds two bits and produces a sum and an output carry. It cannot handle input carries.

Handshaking The method or process of signal interchange by which two digital devices or systems jointly establish communication.

Hard disk A magnetic storage device mounted in a sealed housing.

Hardware The circuitry and mechanical components of an electronic system.

HDL Hardware Description Language. ABEL and CUPL are examples.

Hexadecimal Describes a number system with a base of 16.

High-Z The high-impedance state of a tristate circuit in which the output is effectively disconnected from the rest of the circuit.

Hold time The time interval required for the control levels to remain on the inputs to a flip-flop after the triggering edge of the clock in order to reliably activate the device.

HPIB Hewlett-Packard interface bus; same as GPIB (general-purpose interface bus).

Hysteresis A characteristic of a threshold-triggered circuit, such as the Schmitt trigger, where the device turns on and off at different input levels.

IEEE Institute of Elecrical and Electronic Engineers.

IEEE-488 bus Same as GPIB (general-purpose interface bus).

I^2L Integrated injection logic; an IC technology.

Increment To increase the binary state of a counter by one.

Input The signal or line going into a circuit; a signal that controls the operation of a circuit.

Input file The information entered in a computer that describes a logic design using a PLD programming language such as ABEL.

Input/Output (I/O) A terminal of a device that can be used as either an input or as an output.

Instruction One step in a computer program; a unit of information that tells the CPU what to do.

Integrated circuit (IC) A type of circuit in which all of the components are integrated on a single semiconductor chip of very small size.

Interfacing The process of making two or more electronic devices or systems operationally compatible with each other so that they function properly together.

Interrupt A request for service from a peripheral device.

Inversion Conversion of a HIGH level to a LOW level or vice versa.

Inverter A NOT circuit; a circuit that changes a HIGH to a LOW or vice versa.

JEDEC Joint Electronic Device Engineering Council.

JEDEC file A standard software file generated from the compiler software that is used by a programming device to implement a logic design in a PLD; also called a fuse map or cell map.

J-K flip-flop A type of flip-flop that can operate in the SET, RESET, no-change, and toggle modes.

Johnson counter A type of register in which a specific pattern of 1s and 0s is shifted through the stages, creating a unique sequence of bit patterns.

Junction The boundary between an n region and a p region in a BJT.

Karnaugh map An arrangement of cells representing the combinations of literals in a Boolean expression and used for a systematic simplification of the expression.

Latch A bistable digital device used for storing a bit.

LCCC Leadless ceramic chip carrier; an SMT package that has metallic contacts molded into its body.

LCD Liquid crystal display.

Leading edge The first transition of a pulse.

Least significant bit (LSB) Generally, the right-most bit in a binary whole number or code.

LED Light-emitting diode.

LIFO Last in—first out memory, a memory stack.

Listener An instrument capable of receiving data on a GPIB (general-purpose interface bus).

Literal A variable or the complement of a variable.

Load To enter data into a shift register.

Local cell A programmable cell in a PLD array that affects individual OLMCs when programmed.

Logic In digital electronics, the decision-making capability of gate circuits, in which a HIGH represents a true statement and a LOW represents a false one.

Look-ahead carry A method of binary addition whereby carries from preceding adder stages are anticipated, thus eliminating carry propagation delays.

LSI Large-scale integration; a level of IC complexity in which there are 100 to 9999 equivalent gates per chip.

Machine language The representation of computer program instructions by a binary code.

Magneto-optic disk A storage device that uses electromagnetism and a laser beam to read and write data.

Magnitude The size or value of a quantity.

Master-slave flip-flop A type of flip-flop in which the input data are entered into the device on the leading edge of the clock pulse and appear at the output on the trailing edge.

Memory array An array of memory cells.

Microcomputer A computer in which the CPU is built around a microprocessor.

Microprocessor A VLSI or ULSI device that can be programmed to perform arithmetic, logic and other operations and to process data in a specified manner.

Minimization The process that results in an SOP or POS Boolean expression containing the fewest possible terms with the fewest possible literals per term.

Minuend The number from which another number is subtracted.

Mnemonic In a computer programming language, the "shorthand" representation of an instruction.

Modem A modulator/demodulator for interfacing digital devices to analog transmission systems such as telephone lines.

Modulus The number of states in a counter sequence.

Monostable Having only one stable state. A monostable multivibrator, commonly called a one-shot, produces a single pulse in response to a triggering input.

Monotonicity The characteristic of a DAC defined by the absence of any incorrect step reversals; one type of digital-to-analog linearity.

MOS Metal-oxide semiconductor; a type of transistor technology.

MOSFET Metal-oxide semiconductor field-effect transistor.

Most significant bit (MSB) The left-most bit in a binary whole number or code.

MSI Medium-scale integration; a level of IC complexity in which there are 12 to 99 equivalent gates per chip.

Multiplexer (MUX) A circuit (digital device) that switches digital data from several input lines onto a single output line in a specified time sequence.

Multiplicand The number that is being multiplied by another number.

Multiplier The number that multiplies the multiplicand.

Multivibrator A class of digital circuits in which the output is connected back to the input (an arrangement called feedback) to produce either two stable states, one stable state, or no stable states, depending on the configuration.

NAND gate A logic circuit in which a LOW output occurs if and only if all the inputs are HIGH.

Negative-AND A NOR gate equivalent operation in which there is a HIGH output when all inputs are LOW.

Negative-OR A NAND gate equivalent operation in which there is a HIGH output when any input is LOW.

Nibble Four bits.

NMOS An n-channel metal-oxide semiconductor.

Node A common connection point in a circuit in which a gate output is connected to one or more gate inputs.

Noise immunity The ability of a circuit to reject unwanted signals.

Noise margin The difference between the maximum LOW output of a gate and the maximum acceptable LOW input of an equivalent gate; also, the difference between the minimum HIGH output of a gate and the minimum HIGH input of an equivalent gate.

Nonvolatile A term that describes a memory that can retain stored data when the power is removed.

NOR gate A logic circuit in which the output is LOW when any or all of the inputs are HIGH.

NOT A basic logic operation that performs inversions.

Numeric Related to numbers.

Octal Describes a number system with a base of eight.

Odd parity The condition of having an odd number of 1s in every group of bits.

Offset address The distance in number of bytes of a physical address from the base address.

OLMC Output Logic Macrocell. The programmable output logic in a GAL.

One-shot A monostable multivibrator.

Op code Operation code; the code representing a particular microprocessor instruction.

Open-collector A type of output in a logic circuit in which the collector of the output transistor is left disconnected from any internal circuitry and is available for external connection; normally used for driving higher-current or higher-voltage loads.

Operand A quantity or element operated on during the execution of an instruction.

Operational amplifier (op-amp) A device with two differential inputs that has very high gain, very high input impedance, and very low output impedance.

OR A basic logic operation in which a true (HIGH) output occurs when one or more of the input conditions are true (HIGH).

OR gate A logic circuit that produces a HIGH output when any of the inputs is HIGH.

Oscillator An electronic circuit that is based on the principle of regenerative feedback and produces a repetitive output waveform; a signal source.

Output The signal or line coming out of a circuit.

Overflow The condition that occurs when the number of bits in a sum exceeds the number of bits in each of the numbers added.

PAL Programmable Array Logic. A device with a programmable AND array and a fixed OR array.

Parallel In digital systems, data occurring simultaneously on several lines; transferring or processing several bits simultaneously.

Parity In relation to binary codes, the condition of evenness or oddness of the number of 1s in a code group.

Parity bit A bit attached to each group of information bits to make the total number of 1s odd or even for every group of bits.

Period The time required for a periodic waveform to repeat itself.

Periodic Describes a waveform that repeats itself at a fixed interval.

Peripheral A device or instrument that provides communication with a computer or provides auxiliary services or functions for the computer.

Physical address The actual location of a data unit in memory.

PIC Programmable interrupt controller.

Pipelining A technique that allows a microprocessor to work on more than one instruction at a time.

PLA Programmable logic array. A device with programmable AND and OR arrays.

PLCC Plastic leaded chip carrier; an SMT package whose leads are turned up under its body in a J-type shape.

PLD Programmable logic device.

PMOS A p-channel metal-oxide semiconductor.

Polling The process in which the CPU periodically checks peripheral devices to see if they need service.

Port The physical interface between a computer and a peripheral.

Positive logic The system of representing a binary 1 with a HIGH and a binary 0 with a LOW.

Power dissipation The product of the dc supply voltage and the dc supply current in an electronic circuit; the amount of power required by a circuit.

PPI Programmable peripheral interface.

Preset An asynchronous input used to set a flip-flop.

Priority encoder An encoder in which only the highest value input digit is encoded and any other active input is ignored.

Product The result of a multiplication.

Product-of-sums (POS) A form of Boolean expression that is basically the ANDing of ORed terms.

Product term The Boolean product of two or more literals equivalent to an AND operation.

Program A list of computer instructions arranged to achieve a specific result; software.

Programmer An instrument that programs a PLD using a JEDEC file downloaded from a computer running hardware descriptive language (HDL) software.

PROM Programmable read-only semiconductor memory. A device with a fixed AND array and programmable OR array; used as a memory device and normally not as a logic circuit device.

Propagation delay time The time interval between the occurrence of an input transition and the occurrence of the corresponding output transition in a logic circuit.

Pull-up resistor A resistor used to keep a given point in a circuit HIGH when in the inactive state.

Pulse A sudden change from one level to another, followed after a time by a sudden change back to the original level.

Pulse width The time interval between the 50% points of the leading and trailing edges of the pulse; the duration of the pulse.

Queue A file or lineup of elements, for example, prefetched instruction.

Quotient The result of a division.

Race A condition in a logic network in which the difference in propagation times through two or more signal paths in the network can produce an erroneous output.

RAM Random-access memory; read/write types of memory.

Read The process of retrieving data from a memory.

Recycle To undergo transition (as a counter) from the final or terminal state back to the initial state.

Refresh To renew the contents of a dynamic memory.

Register A digital circuit capable of storing and shifting binary information; typically used as a temporary storage device.

Registered A PLD output configuration where the output comes from a flip-flop.

Remainder The amount left over after a division.

RESET The state of a flip-flop or latch when the output is 0; the action of producing a RESET state.

Reverse bias A voltage polarity condition that prevents a *pn* junction of a transistor or diode from conducting current.

Ring counter A register in which a certain pattern of 1s and 0s is continuously recirculated.

Ripple carry A method of binary addition in which the output carry from each adder becomes the input carry of the next higher-order adder.

Ripple counter An asynchronous counter.

RISC Reduced instruction set computing.

Rise time The time required for the positive-going edge of a pulse to go from 10% of its full value to 90% of its full value.

ROM Read-only semiconductor memory, accessed randomly; also referred to as mask-ROM.

Schottky A specific type of transistor-transistor logic circuit technology.

SCSI Small computer system interface.

Sequence The order in which several things occur in a specified time relationship.

Sequential circuit A digital circuit whose logic states depend on a specified time sequence.

Serial Having one element following another, as in a serial transfer of bits; occuring, as pulses, in sequence rather than simultaneously.

SET The state of a flip-flop or latch when the output is 1; the action of producing a SET state.

Set-up time The time interval required for the control levels to be on the inputs to a digital circuit, such as a flip-flop, prior to the triggering edge of the clock pulse.

Shift To move binary data from stage to stage within a shift register or other storage device or to move binary data into or out of the device.

Signal tracing A troubleshooting technique in which waveforms are observed in a step-by-step manner beginning at the input and working toward the output or vice versa. At each point the observed waveform is compared with the correct signal for that point.

Sign bit The left-most bit of a binary number that designates whether the number is positive (0) or negative (1).

SIMM Single-in-line memory module.

Software Computer programs; programs that instruct a computer what to do in order to carry out a given set of tasks.

SOIC Small-outline integrated circuit; an SMT package that resembles a small DIP but has its leads bent out in a "gull-wing" shape.

Source A sending device of a bus; one of the terminals of a field-effect transistor.

Speed-power product A performance parameter that is the product of the propagation delay time and the power dissipation in a digital circuit.

SRAM Static random access semiconductor memory; a volatile read/write memory.

S-R flip-flop A SET-RESET flip-flop.

SSI Small-scale integration; a level of IC complexity in which there are twelve fewer equivalent gates per chip.

Stage One storage element (flip-flop) in a register.

State diagram A graphic depiction of a sequence of states or values.

State machine A logic system exhibiting a sequence of states conditioned by internal logic and external inputs; any sequential circuit exhibiting a specified sequence of states.

Static memory A memory composed of storage elements such as latches or magnetic elements and capable of retaining data indefinitely without refreshing.

Storage The capability of a digital device to retain bits; the process of retaining digital data for later use.

Strobing A process of using a pulse to sample the occurrence of an event at a specified time in relation to the event.

Subtracter A logic circuit used to subtract two binary numbers.

Subtrahend The number that is being subtracted from the minuend.

Sum The result when two or more numbers are added together.

Sum-of-products (SOP) A form of Boolean expression that is basically the ORing of ANDed terms.

Sum term The Boolean sum of two or more literals equivalent to an OR operation.

Surface-mount technology (SMT) An IC package technique in which the packages are smaller than DIPs and are mounted on the printed surface of the circuit board.

Synchronous Having a fixed time relationship.

Synchronous counter A type of counter in which each stage is clocked by the same pulse.

Syntax The prescribed format and symbology used in describing a category of functions.

Synthesis The software process of converting a circuit description to a standard JEDEC file.

Talker An instrument capable of transmitting data on a GPIB (general-purpose interface bus).

Terminal count The final state in a counter's sequence.

Test vectors A section of a PLD program that consists of a set of input levels and the resulting output levels and used to test a given logic design prior to committing it to hardware.

Throughput The average speed with which a program is executed.

Timer A circuit that can be used as a one-shot or as an oscillator.

Timing diagram A graph of digital waveforms showing the proper time relationship of all the waveforms and how each waveform changes in relation to the others.

Toggle The action of a flip-flop when it changes state on each clock pulse.

Totem-pole A type of output in TTL circuits.

Trailing edge The second transition of a pulse.

Transistor A semiconductor device exhibiting current and/or voltage gain. When used as a switching device, it approximates an open or closed switch.

Trigger A pulse used to initiate a change in the state of a logic circuit.

Tristate A type of output in logic circuits that exhibits three states: HIGH, LOW, and open (disconnected). The open state is called the *high-Z state.*

Tristate buffer A logic circuit having three output states: HIGH, LOW, and high-impedance (open).

Troubleshooting The technique of systematically identifying, isolating, and correcting a fault in a circuit or system.

Truncated Shortened.

Truncated sequence A sequence that does not include all of the possible states of a counter.

Truth table A table showing the input and corresponding output levels of a logic device.

TTL Transistor-transistor logic; a type of integrated circuit that uses bipolar junction transistors.

ULSI Ultra large-scale integration; a level of IC complexity in which there are more than 100,000 equivalent gates per chip.

Unit load A measure of fan-out. One gate input represents a unit load to the output of a gate within the same IC family.

Universal gate Either a NAND gate or a NOR gate. The term *universal* refers to the property of a gate that permits any logic operation to be implemented by that gate or by a combination of gates of that kind.

Universal shift register A register that has both serial and parallel input and output capability.

Up/Down counter A counter that can progress in either direction through a certain sequence.

UV EPROM Ultraviolet erasable programmable ROM.

Variable A symbol used to represent a logical quantity that can have a value of 1 or 0, usually designated by an italic letter.

VLSI Very large-scale integration; a level of IC complexity in which there are 10,000 to 99,999 equivalent gates per chip.

Volatile A term that describes a memory that loses its stored data when the power is removed.

VXI A modular instrumentation interface.

Weight The value of a digit in a number based on its position in the number.

Word A complete unit of binary data.

Word capacity The number of words in a memory.

Word length The number of bits in a word.

WORM Write once–read many disk memory.

Write The process of storing data in a memory.

Zero suppression The process of blanking out leading or trailing zeros in a digital readout.

ZIF socket Zero Insertion Force socket. A type of socket used in most programmers that accepts a PLD package.

INDEX